2

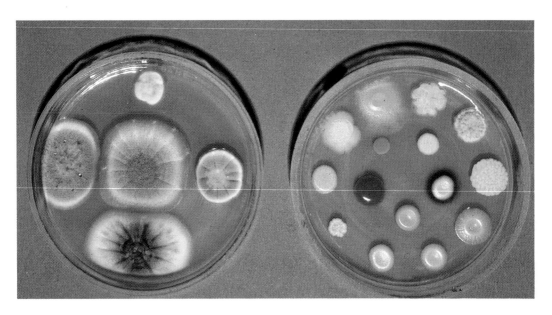

Cultures of molds and yeasts on nutrient agar in glass Petri dishes. From H.
Phaff, Industrial microorganisms, *Scientific American,* September 1981.

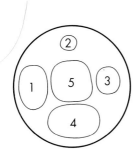

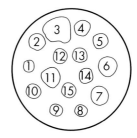

MOLDS
1 *Penicillium chrysogenum*
2 *Monascus purpurea*
3 *Penicillium notatum*
4 *Aspergillus niger*
5 *Aspergillus oryzae*

YEASTS
1 *Saccharomyces cerevisiae*
2 *Candida utilis*
3 *Aureobasidium pullulans*
4 *Trichosporon cutaneum*
5 *Saccharomycopsis capsularis*
6 *Saccharomycopsis lipolytica*
7 *Hanseniaspora guilliermondii*
8 *Hansenula capsulata*
9 *Saccharomyces carlsbergensis*
10 *Saccharomyces rouxii*
11 *Rhodotorula rubra*
12 *Phaffia rhodozyma*
13 *Cryptococcus laurentii*
14 *Metschnikowia pulcherrima*
15 *Rhodotorula pallida*

MICROBIAL BIOTECHNOLOGY

FUNDAMENTALS OF APPLIED MICROBIOLOGY

ALEXANDER N. GLAZER

HIROSHI NIKAIDO

University of California, Berkeley

W.H. FREEMAN AND COMPANY
New York

Cover photograph credits

Background: Copyrighted material used with permission of
Hoefer Scientific Instruments

Inset: ©USDA/Scientific Source/Photo Researchers

Library of Congress Cataloging in Publication Data

Glazer, Alexander N.
    Microbial biotechnology : fundamentals of applied microbiology / Alexander N. Glazer and
Hiroshi Nikaido.
        p.   cm.
        Includes index.
        ISBN 0-7167-2608-4
        1. Microbial biotechnology. I. Nikaido, Hiroshi. II. Title.
    TP248.27.M53G57 1994
        660'.62 — dc20   94-11367
                                                                                                                  CIP

© 1995 by W.H. Freeman and Company

Printed in the United States of America

Second printing, 1998

WE DEDICATE THIS BOOK TO

the memory of Roger Yate Stanier (1916–1982),
valued friend and teacher,
who brought order into our vision of the microbial world.

Contents in Brief

Contents

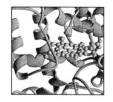

Preface

This book is unique because we make no artificial attempt to confine our discussion to a single scientific discipline. Instead, we treat real-world questions, and the treatment ranges freely from microbiology to genetics, molecular biology, biochemistry, and chemistry in order to define the problems and to explain how these diverse disciplines can come together in solutions. We hope that undergraduate and graduate students in all these fields may find the book useful and that our approach will encourage them to synthesize what they have learned in traditional courses on their way to making creative contributions in science. We also hope that scientists already working in related fields will gain an appreciation for the vast sweep of microbial biotechnology.

A glance at the table of contents illustrates the importance and breadth of societal issues affected by this aspect of modern biology. Some understanding of the promise and the limits of microbial biotechnology is thus essential for many policymakers, regulators, and corporate decision makers throughout the world. As a result, while our treatment encompasses the state of the art, we have attempted to ensure that much of the book will also prove accessible to those whose primary training lies outside the physical and biological sciences.

For the past eight years, we have offered a course in applied microbiology and biochemistry to Berkeley undergraduates, and we thank our students for the inspiration to develop much of the material herein. Professor Lubert Stryer gave us essential early encouragement and advice. In the initial phase of writing, Professor Jeremy Thorner and Dr. Jonathan Raymond made major editorial and substantive contributions.

Dr. Judy Glazer and John Glazer provided invaluable critical input from their viewpoint as experienced modern textbook consumers. Near the end of the process, Professors Julian Davies and Jane Gibson read the entire manuscript and made many insightful suggestions. We also thank our colleagues throughout the world who graciously answered our sometimes unschooled inquiries, promptly granted us permission to use published material, and generously provided original illustrations.

Finally, we are grateful to many at W. H. Freeman and Company without whom the book would have been a far lesser work, in particular Moira Lerner, who crafted elegant prose from our raw writing, and Penelope D. Hull, who assembled this complex volume with remarkable patience and tact.

Alexander N. Glazer
Hiroshi Nikaido
University of California, Berkeley
August 1994

Preamble

Il n'y a pas des sciences appliquées . . . mais il y a des applications de la science. (There are no applied sciences . . . but there are the applications of science.)
                                                              —Louis Pasteur

Biotechnology is defined generally as the use of living organisms to produce products beneficial to mankind. It is the application of biological organisms to technical and industrial processes. It involves the use of "novel" microbes, which have been altered or manipulated by humans through techniques of genetic engineering.
                                    —U.S. Environmental Protection Agency

Microorganisms are the most versatile and adaptable forms of life on Earth, and they have existed here for some 3.5 billion years. Indeed for the first 2 billion years of their existence, bacteria alone ruled the biosphere, colonizing every accessible ecological niche from glacial ice to the hydrothermal vents of the deep sea bottoms. As these early bacteria evolved, they developed the major metabolic pathways characteristic of all living organisms today—as well as various other metabolic processes, such as nitrogen fixation, which is still restricted to bacteria alone. Over their long period of global dominance, bacteria also changed Earth, transforming its anaerobic atmosphere to one rich in oxygen and generating massive amounts of organic compounds. Eventually they created an environment able to sustain more complex forms of life.

Today the biochemistry and physiology of bacteria and other microorganisms reflect several billion years' worth of genetic responses to an ever-changing world. At the same time, because of their physiologic and metabolic versatility and their ability to survive in small niches, microorganisms are much less affected by the depredations of human technology than are larger, more complex forms of life. Thus it is likely that most of the microbial species that existed before humans are still here to be explored.

Such exploration is by no means a purely academic pursuit. The many thousands of microorganisms already available in pure culture and the thousands of others yet to be discovered represent a large fraction of the total gene pool of the living world, and this tremendous genetic diversity is the raw material of genetic engineering, the direct manipulation of the heritable characteristics of living organisms. Biologists are now able to accelerate greatly the acquisition of desired traits in an organism by directly modifying its genetic make-up through manipulations of its DNA, rather than through the traditional methods of breeding and selection at the level of the whole organism. The various techniques of manipulation summarized under the rubric of "recombinant DNA technology" can take the form of removing genes, adding genes from a different organism, modifying genetic control mechanisms, and introducing synthetic DNA, sometimes enabling a cell to perform functions that are totally new to the living world. In these ways, new stable, heritable traits have by now been introduced into viruses, bacteria, plants, and animals. One result has been a significant enhancement of the already considerable practical value of applied microbiology. Applied microbiology covers a broad spectrum of activities—the production of fermented foods and beverages, the production of antibiotics, wastewater treatment, and bioremediation, to name but a few. The ability to manipulate the genetic make-up of organisms has led to explosive progress in all areas of this field.

The purpose of this book is to provide a rigorous, unified treatment of all facets of microbial biotechnology. We shall freely cross the boundaries of formal disciplines in order to do so: Microbiology supplies the raw materials; genetic engineering provides the blueprints; biochemistry, chemistry, and process engineering furnish the tools; and many other scientific fields serve as important reservoirs of information. Moreover, unlike a textbook in biochemistry, microbiology, molecular biology, organic chemistry, or some other vast basic field, which must concentrate solely on teaching general principles and patterns in order to provide an overview, this one will continually emphasize the importance of diversity and uniqueness. In applied microbiology, one is more likely to seek the unusual: a producer of a novel antibiotic, a parasitic organism that specifically infects a particularly widespread and noxious pest, a thermophilic bacterium that might serve as a source of enzymes active at 100°C. David Perlman's *Laws of Applied Microbiology* are both amusing and accurate:

The microorganism is $\begin{cases} \textit{always right} \\ \textit{your friend} \\ \textit{a sensitive partner} \end{cases}$

There are no *stupid microorganisms*

Microorganisms $\begin{cases} \textit{can} \\ \\ \textit{will} \end{cases}$ *do anything*

Microorganisms are $\begin{cases} \textit{smarter} \\ \textit{wiser} \\ \textit{more energetic} \end{cases} \begin{matrix} \textit{than} \\ \textit{chemists,} \\ \textit{engineers, etc.} \end{matrix}$

If you take care of your microbial friends, they will take care of your future (and you will live happily ever after). Perlman, D. (1980), Some problems on the new horizons of applied microbiology, *Developments in Industrial Microbiology*, vol. 21, pp. xv–xxiii (Society for Industrial Microbiology).

In short, this book examines the fundamental principles and facts that underlie current practical applications of bacteria, fungi, and other microorganisms, describes those applications, and examines future prospects for related technologies. It consists of 15 chapters grouped into 5 parts.

Part One, *Microbial Diversity and Its Applications*, introduces microbial biotechnology in the context of the vast diversity of microorganisms. Chapter 1, "Microbial Biotechnology: Scope, Techniques, Examples," gives an account of the current status of microbial biotechnology and discusses some of its potential contributions to medicine, agriculture, wastewater treatment, hazardous waste management, and the production of feedstock chemicals. It also addresses the debate surrounding the release of genetically engineered microorganisms into the environment. The chapter ends with a brief overview of the economics of biotechnology—a reminder that new ideas and products are subject to the demanding test of competition in the marketplace. Chapter 1 should be read first, because it puts the material presented in subsequent chapters into the perspective of the entire field, present and future. It can be read with profit again at the end, as a retrospective summary of many of the book's important themes. Chapter 2, "Microbial Diversity," introduces the extraordinary genetic and biochemical variety of bacteria and fungi, their classification, and key aspects of their metabolism. This chapter presents the fundamentals of microbiology that will be necessary for understanding later chapters.

Part Two, *Living Factories for Macromolecules*, addresses the most widely exploited attribute of engineered microorganisms, their ability to synthesize large amounts of complex biopolymers such as proteins, polysaccharides, and polyesters. Chapters 3 and 4 introduce the general techniques of genetic engineering and the associated biochemical techniques whereby bacteria and yeasts are made to produce desired macromolecules by either overexpression of endogenous genes or expression of foreign genes. The technical approaches presented in this section are widely applicable and will be encountered repeatedly throughout the book. These general chapters are followed by chapters treating individual classes of biopolymers: synthetic vaccines (Chapter 5), microbial insecticides (Chapter 6), microbial enzymes (Chapter 7), and polysaccharides and polyesters (Chapter 8).

Part Three, *Microorganisms in Plant Biotechnology*, illustrates how the techniques of genetic manipulation discussed earlier are exploited for the stable introduction into crop plants of genes such as those of herbicide and pest resistance (genes encoding microbial insecticides).

The conversion of biomass and starch into fuels and feedstock chemicals is frequently proposed as a future alternative to fossil fuels. In Part Four, *From Biomass to Fuels*, we describe the biochemistry of lignin, cellulose, and hemicellulose, the major components of land-plant biomass (Chapter 10). We also examine the enzymatic conversion of plant polysaccharides to simple sugars, the raw materials in the manufacture of organic solvents. Chapter 11 describes microbial fermentations leading to ethanol and the issues surrounding the use of ethanol produced by fermentation as a fuel.

The exploitation of microorganisms as "bioreactors" in the manufacture of valuable bioorganic compounds is described in Part Five, *Metabolites from Microorganisms*. Here we see how genetic engineering can be used to tailor the metabolic pathways of microorganisms to overproduce specific metabolites; our focus is on amino acids (Chapter 12). Chapter 13, "Antibiotics," discusses an area of microbial biotechnology that has long been recognized as of great importance to the welfare of humanity. The cornucopia of different natural products that exhibit antibiotic activity illustrates the enormous value of microbial diversity.

Part Six, *Organic Synthesis and Degradation*, encompasses two seemingly unrelated fields: microorganisms in organic chemistry and the environmental applications of microbiology. In these fields, the wealth of enzymes present in microorganisms is exploited either to synthesize a variety of organic compounds, most of which differ from those in the living world, or to degrade a wide range of such xenobiotic ("foreign") compounds. Chapter 14 describes the wide use of microbial enzymes in catalyzing steps in the synthesis of nonbiological organic molecules, particularly where requirements of stereospecificity and/or mild reaction conditions preclude the application of conventional chemical methods. Chapter 15, "Environmental Applications," gives an account of the role of microorganisms in destroying toxic compounds in the environment and chronicles the attempts to construct microorganisms able to degrade recalcitrant pollutants. Examination of the role of microorganisms in wastewater treatment and in mineral recovery brings the chapter and the book to a close.

The application of biotechnology to medicine, agriculture, the chemical industry, and the environment is changing all aspects of everyday life, and the pace of that change is increasing. Basic understanding of the many facets of microbial biotechnology is important to scientists and nonscientists alike. We hope that both will find this book a useful source of information. Although a strong technical background may be necessary to assimilate the fine points described herein, we have tried to make the fundamental concepts and issues accessible to readers whose background in the life sciences is quite modest. The attempt is vital, for only an *informed* public can distinguish desirable biotechnological options from undesirable choices and those that are likely to succeed from those likely to result in costly failure.

PART

MICROBIAL DIVERSITY AND ITS APPLICATIONS

one

Crystals of human insulin produced in *Escherichia coli*. Courtesy of Dr. Ronald E. Chance, Eli Lilly and Co., Indianapolis

Microbial Biotechnology:
Scope, Techniques, Examples

The umbrella of microbial biotechnology covers many scientific activities, ranging from production of recombinant human hormones to that of microbial insecticides, from mineral leaching to bioremediation of toxic wastes. This book describes many of these endeavors in some detail. Before embarking bravely on fourteen chapters of fairly technical and widely varied information, however, even the most intrepid traveler will wish to consult a map. This chapter provides such a map by addressing the following questions.

- What current and anticipated products and activities depend on microbial biotechnology?
- What are some of the concerns raised by the use of genetic engineering in applied microbiology and how can they be addressed?
- In what ways do economic considerations influence the technologies discussed in this book?

This chapter also treats in general terms the major techniques used in microbial biotechnology, offers examples, and discusses probable future developments. Areas discussed extensively elsewhere in the book are touched on briefly, whereas those treated here alone are presented in greater detail.

Horizons of Microbial Biotechnology

In 1973 Herbert Boyer, Stanley Cohen, and their collaborators carried out the experiments that led to the development of recombinant DNA technology. The first major commercial fruit of that technology, human insulin produced in bacteria, was approved for clinical use in 1982. Microbial biotechnology is a mosaic of concepts and techniques from many different fields of science. However, virtually all the activities it encompasses have been revolutionized by genetic engineering. The history of microbial biotechnology can thus be divided into two periods: the pre-recombinant DNA era, through 1981, and the post-recombinant DNA era, from 1982 onward.

Table 1.1 and Figure 1.1 give information on the main products of microbial biotechnology in 1981, just before the first commercial use of recombinant DNA. At that time, food products represented some 80% of the market value of all biotechnological applications. The metabolic activities of certain microorganisms have long been exploited in the production of alcoholic beverages and of foods derived by fermentations

TABLE 1.1

Major Products Dependent on Microbial Biotechnology and Their Primary End Uses Before the Advent of Genetic Engineering

Product	Major Uses
Fermented juices and distilled liquors	Beverages
Cheese	Food
Antibiotics	Drugs
Industrial alcohol	Fuel additive (gasohol)
High-fructose syrups	Sweeteners
Amino acids	Feed additives, food enrichment and flavoring agents, artificial sweetener (aspartame), feed preservatives
Baker's yeast	Food additive, enrichment agent
Steroids	Therapeutic agents, animal growth promotion
Vitamins	Feed and food enrichment additives
Citric acid	Food additive
Enzymes	Food processing, laundry detergents
Vaccines	Disease prevention
Polysaccharide gums	Food emulsifiers, thickeners and stabilizers, enhanced oil recovery

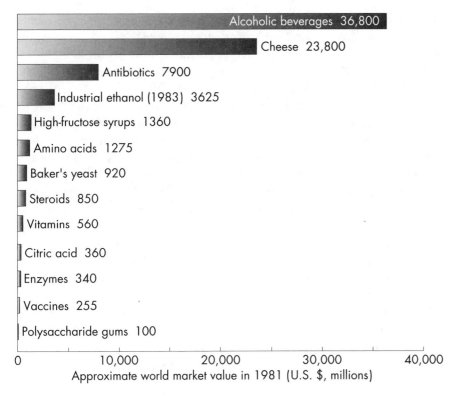

Figure 1.1
Major products dependent on microbial biotechnology immediately before the commercial application of genetic engineering.

or other microbial transformations. The food industry continues to dominate the biotechnology scene today.

All the products that appear in Table 1.1 and Figure 1.1 continue to be important, although the methods of manufacture have changed in some instances to incorporate advances made possible by genetic engineering. The post-1982 era, however, is remarkable for the appearance in rapid succession of widely used protein therapeutic agents whose production depends entirely on recombinant DNA technology.

Human Therapeutics

One of the most dramatic and immediate impacts of genetic engineering was the production in large amounts of proteins encoded by human genes. In 1982 insulin, expressed from human insulin genes on plasmids inserted into *Escherichia coli*, became the first genetically engineered therapeutic agent to be approved for clinical use in humans. This bacterially produced insulin, now used widely in the treatment of diabetes, is indistinguishable in its structure and clinical effects from natural insulin.

Human growth hormone (hGH), a protein made naturally by the pituitary gland, was the second such product. Inadequate secretion of hGH in children results in dwarfism. Before the advent of recombinant DNA technology, hGH was prepared from pituitaries removed from human cadavers. The supply of such preparations was limited and the cost prohibitive. Furthermore, dangers inherent in administering them led to their withdrawal from the market. Some patients treated with injections of pituitary hGH developed a disease caused by a contaminating slow virus, Jakob–Creutzfeldt syndrome, which leads to dementia and death. But hGH free from such contaminants can be produced in genetically engineered *E. coli* in large amounts and at relatively little cost. Moreover, the resulting abundant supply of this hormone has made it possible to explore new applications in clinical trials. One result is the finding that hGH may have therapeutic value in wound healing.

Human tissue plasminogen activator (tPA), a proteolytic enzyme (a "serine" protease) with an affinity for fibrin clots, is another therapeutic agent made available in large amounts as a consequence of recombinant DNA technology. At the surface of fibrin clots, tPA cleaves a single peptide bond in plasminogen to form another serine protease, plasmin, which then degrades the clots. This clot-degrading property of tPA makes it a life-saving drug in the treatment of patients with acute myocardial infarction (damage to heart muscle due to arterial blockage).

Recombinant human insulin, hGH, and tPA offer impressive proof of the clinical efficacy and safety of human proteins made by engineered microorganisms. The roster of recombinant human gene products expressed in bacteria or fungi is growing exponentially (see Table 1.2).

Microbial biotechnology may also supply safe vaccines against many diseases. For example, vaccines are under development for protection of humans against hepatitis A and B, "non-A" and "non-B," herpes simplex, influenza, poliomyelitis, rabies, and HIV, the viral agent in acquired immune deficiency syndrome (AIDS). In the veterinary field, a genetically engineered vaccine against diarrhea in pigs was introduced in 1982. Work is in progress on vaccines against the causative agents of other diseases, such as foot-and-mouth disease virus.

Agriculture

Methods that depend on microbial biotechnology greatly increase the diversity of genes that can be incorporated into crop plants and dramatically shorten the time required for the production of new varieties of plants. It is now possible to transfer foreign genes into plant cells. Transgenic plants that are viable and fertile can be regenerated from these transformed cells, and the genes that have been introduced into these transgenic plants are as stable as other genes in the plant nuclei and show a normal pattern of inheritance. Transgenic plants are most

Examples of Human Proteins Cloned in *E. coli:* Their Biological Functions and Current or Envisaged Therapeutic Use

TABLE 1.2

Protein	Function(s)	Therapeutic Use(s)
α-1 antitrypsin	Protease inhibitor	Treatment of emphysema
Calcitonin	Influences Ca^{2+} and phosphate metabolism	Treatment of osteomalacia
Colony-stimulating factors	Stimulate hematopoiesis	Antitumor
Epidermal growth factor	Epithelial cell growth, tooth eruption	Wound healing
Erythropoietin	Stimulates hematopoiesis	Treatment of anemia
Factor VIII	Blood-clotting factor	Prevention of bleeding in hemophiliacs
Factor IX	Blood-clotting factor	Prevention of bleeding in hemophiliacs
Growth hormone-releasing factor	Stimulates secretion of growth hormone	Growth promotion
Interferons (α, β, γ)	A family of 20 to 25 low-molecular-weight proteins that cause cells to become resistant to the growth of a wide variety of viruses	Antiviral, antitumor (?), antiinflammatory (?)
Interleukins 1, 2, and 3	Stimulators of cells in the immune system	Antitumor; treatment of immune disorders
Lymphotoxin	A bone-resorbing factor produced by leukocytes	Antitumor
Somatomedin C (IGF-I)	Sulfate uptake by cartilage	Growth promotion
Serum albumin	Major protein constituent of plasma	Plasma supplement
Superoxide dismutase	Decomposes superoxide free radicals in the blood	Prevention of damage when O_2-rich blood enters O_2-deprived tissues; has applications in cardiac treatment and organ transplantation
Tumor necrosis factor	A product of mononuclear phagocytes cytotoxic to certain tumor cell lines	Antitumor
Urogastrone	Control of gastrointestinal secretion	Antiulcerative
Urokinase	Plasminogen activator	Anticoagulant (dissolution of blood clots)

commonly generated by exploiting a plasmid vector carried by *Agrobacterium tumefaciens*, a bacterium that we will discuss in detail in Chapter 9 (plasmids are described in Chapter 2). Foreign DNA carrying from 1 to 50 genes can be introduced into plants in this manner, with the donor DNA originating from different plant species, animal cells, or microorganisms.

Higher plants have genes whose expression shows precise temporal and spatial regulation in various parts of plants—for example, the leaves, floral organs, and seeds that appear at specific times during plant development and/or at specific locations or whose expression is regulated by light. Other plant genes respond to different stimuli, such as plant hormones, nutrients, lack of oxygen (anaerobiosis), heat shock, and wounding. It should therefore be possible to insert the control sequence(s) from such genes into transgenic plants to confine the expression of foreign genes to specific tissues and to determine when such expression begins and how long it lasts. Microorganisms that live on or within plants can be manipulated to control insect pests and fungal disease or to establish new symbioses, such as those between nitrogen-fixing bacteria and plants.

What are some of the objectives of plant microbial biotechnology? We provide an overview here and follow with a detailed discussion in Chapter 9.

Herbicide Tolerance. Many otherwise effective broad-spectrum herbicides do not distinguish between weeds and crops, but crop plants can be modified to become resistant to particular herbicides. When an appropriate herbicide is applied to a weed-infested field of such genetically modified plants, it then acts as a selective weed killer.

Resistance to Insect Pests. Certain strains of the bacterium *Bacillus thuringiensis* produce a protein endotoxin that drastically increases the permeability of the epithelial cells in the gut of the larvae of lepidopteran insects, moths, and butterflies (Chapter 6). The gene encoding the *B. thuringiensis* endotoxin has been transferred into and expressed in tobacco, cotton, and tomato plants. In field tests, the transgenic tomato and tobacco plants were only slightly damaged by caterpillar larvae under conditions that led to total defoliation of control plants. A different approach to achieving the same end is to transfer the *B. thuringiensis* endotoxin gene into bacteria such as *Clavibacter xyli* subsp. *cynodontis*, which colonizes the interior of plants. This organism is generally found inside Bermuda grass plants, but it can reach population sizes in excess of 10^8 per gram of stem tissue when purposefully inoculated into other monocotyledonous species such as corn. Recombinant *C. xyli* strains expressing the endotoxin show promise in controlling leaf- and stem-feeding lepidopteran larvae.

Resistance to Viral Diseases. The coat protein (capsid) gene of tobacco mosaic virus has been introduced into tobacco, tomato, and potato plants. Its expression provides increased resistance not only against tobacco mosaic virus but also against several other closely related plant RNA viruses. The coat protein appears to exert its protective mechanism by interfering with the uncoating of virus particles within the cells of the transgenic plants. Such uncoating is a prerequisite to the productive translation and replication of the viral nucleic acid.

Control of Pathogenic Bacteria, Fungi, and Parasitic Nematodes. The cell walls of many plant pests, such as insects and fungi, contain chitin (poly-*N*-acetylglucosamine) as a major structural component. Many bacteria (for example, species of *Serratia*, *Streptomyces*, and *Vibrio*) produce chitin-degrading enzymes (chitinases). The control of some fungal diseases by such bacteria has been correlated with the production of chitinases. Genes encoding chitinases from several different soil bacteria have been cloned into *Pseudomonas fluorescens*, an efficient colonizer of plant roots. The effectiveness of these recombinant strains in controlling fungal disease is not yet known.

Decreasing the Expression of Selected Normal Gene(s) in a Transgenic Plant. The expression of specific genes has been eliminated or reduced by introducing genes that encode antisense RNA. An antisense gene specifies an RNA whose sequence is complementary to that of the messenger RNA (mRNA) encoded by the target gene. Hybridization of the antisense RNA with the mRNA prevents translation of the latter or leads to its degradation.

Chalcone synthase is a key enzyme in the synthesis of naringenin chalcone, which forms the skeleton of the anthocyanin and flavonoid flower pigments (Box 1.1). Introduction of a gene that encodes antisense RNA against chalcone synthase in petunia led to flowers with altered colors. This example illustrates the power of an approach that allows alteration of a single trait in the transgenic organism.

The antisense RNA approach is also applicable to selective modulation of the flow of metabolic pathways in plants. Such modulation could improve food quality by producing, for example, seeds with altered protein or oil composition.

Nitrogen Fixation. Leguminous plants, including important crops such as soybeans, form symbiotic associations with species of *Rhizobium*, *Bradyrhizobium*, and *Frankia* bacteria that fix atmospheric molecular nitrogen. Free-living rhizobia are found in the soil. Natural infection of host plants by the bacteria leads to formation of root nodules within which the rhizobia proliferate. For almost a hundred years, commercially produced rhizobia have been added to soil as legume inoculants to reduce the need for nitrogenous fertilizer. No adverse effects of such applications have been observed, so no adverse consequences should attend large-scale applications of genetically engineered strains of rhizobia.

Strains of *Bradyrhizobium japonicum* and *Rhizobium meliloti*, engineered to increase the expression of certain genes that are important to nitrogen fixation, have been shown to give greater increases in the biomass of their respective host plants under greenhouse conditions than the wild-type bacterial strains. Because of the very high population of free-living rhizobia in the soil, new strains have to be introduced at very high concentrations to overcome competition from the resident bacte-

BOX 1.1 Chalcone Synthase

Chalcone synthase is the first enzyme in the flavonoid biosynthetic branch of the phenylpropanoid pathway in plants that leads to the production of flower pigments.

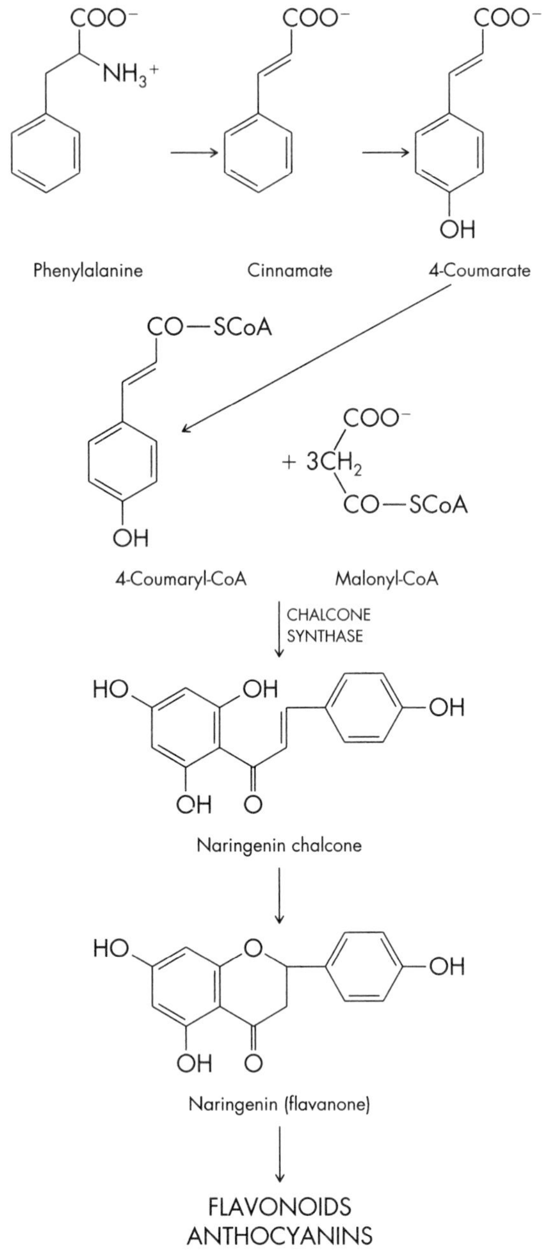

COO⁻
NH3+
COO⁻
COO⁻
OH
Phenylalanine
Cinnamate
4-Coumarate
CO—SCoA
OH
COO⁻
+ 3CH2
CO—SCoA
4-Coumaryl-CoA
Malonyl-CoA
CHALCONE
SYNTHASE
HO
OH
OH
OH
O
Naringenin chalcone
HO
O
OH
OH
O
Naringenin (flavanone)
FLAVONOIDS
ANTHOCYANINS

The expression of the structural gene for chalcone synthase is induced by ultraviolet light. It acts as a defense gene against UV-induced damage by initiating the production of pigments that absorb UV radiation.

ria. This leads to high inoculant cost. Studies of the mechanism of infection and of the biochemical determinants of *Rhizobium* competitiveness may reveal ways of resolving this difficulty.

Transfering the genes for nodule formation to *Agrobacterium* enables the recombinant organism to initiate nodulation on nonlegumes, which suggests that it may be possible to extend nitrogen fixation to nonleguminous plants. Achieving this goal will require manipulation of the host plant as well as of the bacterial genes.

Ability to Grow in Harsh Environments. It may be possible to extend the habitat range for plants by imparting traits such as tolerance to cold, to heat, and to drought, ability to withstand high moisture or high salt concentrations, and resistance to iron deficiency in very alkaline soils. However, tolerances to environmental stresses are likely to be polygenic traits and so may be difficult to transfer from one kind of plant to another.

The exploitation of microbial biotechnology in agriculture is driven by the realization that agricultural practices that rely heavily on expensive nitrogenous fertilizers and widespread use of pesticides are no longer sustainable.

Wastewater Treatment

Living organisms are about 70% water. A human being, for instance, has to consume an average of 1.5 liters of water per day to survive. The volume of water being contaminated and the need to reclaim wastewater are both increasing with growth in population and in the industrial use of water. Treatment of wastewater is essential to prevent the contamination of drinking water and the entry of contaminants into the food chain. Wastewater originates from four primary sources: sewage, industrial effluents, agricultural runoff, and storm water and urban runoff. Given the number and variety of contaminants listed in Table 1.3, the success of some of the current treatments for the reclamation of pure water is little short of amazing.

Many different microorganisms already participate in the breakdown of organic matter at various stages in the wastewater treatment process. However, much of the basic biochemical and genetic information on ways in which genetic engineering might be employed to improve wastewater treatment is not yet available. Current methods of wastewater treatment are discussed in Chapter 15; some of the efforts to develop new approaches are outlined here.

Primary treatment of sewage consists of removing suspended solids. The secondary treatment of sewage reduces the biochemical oxygen demand (Box 1.2). This is accomplished by lowering, through microbial oxidation, the organic compound content of the effluent from the primary treatment. Bacteria of *Zoogloea* species play an important role in

Types of Pollutants Found in Wastewater

TABLE 1.3

Organic residues that can be used as nutrients by microorganisms

Domestic sewage
Waste water from food-processing industries
Effluent from manure heaps and cattle yards
Slaughterhouse wastes
Wastewater from laundries
Wastewater and residues from paper mills

Inorganic residues some of which can be used as nutrients by microorganisms

Ammonia, nitrates, other nitrogen compounds (primarily from domestic sewage and fertilizers)
Orthophosphate, other phosphate compounds (primarily from domestic sewage and fertilizers)
Sulfates
Sulfides, sulfites and ferrous salts (from certain industries)
Silica

Toxic substances (largely from industrial wastewater)

Strong acids and alkalis
Heavy metals
Oil
Toxic organic compounds (from certain chemical industries, use of pesticides, etc.)

Inert suspensions of particles

Finely divided particles from mining, quarrying, and washing processes

Bacterial contamination

Domestic sewage (primary source)
Food-processing industry
Manure heaps and cattle yards

the aerobic secondary stage of sewage treatment. These organisms produce abundant extracellular polysaccharide and, as a result, form aggregates called flocs. Such aggregates efficiently adsorb organic matter, part of which is then metabolized by the bacteria. The flocs settle out and are transferred to an anaerobic digestor, where other bacteria complete the degradation of the adsorbed organic matter.

Several genes responsible for the production of extracellular polysaccharides have been cloned from *Z. ramigera* strains. The ability to link the genes for the extracellular polysaccharide synthesis with those

BOX 1.2 **Biological Oxygen Demand**

Maintenance of a high oxygen concentration in aquatic ecosystems is essential for the survival of fish and other aquatic organisms. Decomposition of organic matter may rapidly deplete the oxygen. When organic matter such as untreated sewage is added to an aquatic ecosystem, it is rapidly attacked by bacteria that degrade it, using up oxygen in the process. The biological oxygen demand (BOD) is related to the concentration of organic matter in the water. Usually, the oxygen consumption is measured over a period of 5 days and is abbreviated BOD_5. BOD_5 for municipal wastewater generally ranges from 80 to 250 mg O_2 per liter. Appropriate secondary treatment decreases the BOD_5 to less than 20 mg O_2 per liter.

for the degradation of organic compounds may both improve flocculation and increase the rate of substrate use. The degradation of organic compounds by the floc-forming bacteria is frequently limited by the availability of oxygen. The gene for a bacterial hemoglobin present in the bacterium *Vitreoscilla* has been cloned and expressed in *E. coli*. Transfer of this gene to the floc-forming bacteria may enhance their ability to scavenge oxygen from their environment.

Removal of heavy metals from wastewater by bacteria represents a particularly attractive target for genetic engineering. Numerous organisms synthesize small cysteine-rich proteins called metallothioneins in response to exposure to heavy metals such as Cu^{2+}, Zn^{2+}, and Cd^{2+}. Metallothioneins have both a high affinity and a high capacity for binding heavy metal ions. Various metallothioneins have been cloned and expressed in *E. coli*. A direct correlation has been observed between the level of the expressed metallothionein and accumulation of Cd^{2+} and Cu^{2+} within the cells. Cells of such engineered bacteria could be immobilized and used to remove heavy metal ions from wastewater.

Hazardous Waste Management

A great many industrial chemicals are released into the environment deliberately—to function as pesticides, for example, or to preserve wood or insulate electric transformers. Others are released accidentally or emitted as wastes. Many of these chemicals, produced on a large scale as part of the normal activities of industrialized societies, are toxic to plants and animals. The Environmental Protection Agency estimated that in 1985, industry in the United States alone generated at least 569 million metric tons of hazardous waste. Because the definition of hazardous waste used by the Environmental Protection Agency is more restrictive than that adopted by other government agencies, this is a minimum estimate (Box 1.3).

The persistence of industrial chemicals in soil and water varies widely. Those that can be degraded by microorganisms or decompose spontaneously may disappear in weeks; others may persist for years. Many pesticides once used in massive amounts have long half-lives in the soil. For example, the environmental persistence of the insecticide DDT (*p,p'*-dichlorodiphenyltrichloroethane) ranges from 3 to 10 years, that of chlordane from 2 to 4 years.

Over 1200 hazardous waste sites in the United States have been placed on the National Priorities List for clean-up, and the total number of such sites identified by the U.S. Environmental Protection Agency exceeds 33,000. Most hazardous waste dumps are unlined, with no barrier between the waste and ground water. A dump site may contain solid or liquid waste or both; a single compound may be present, or there may be a mixture of closely related compounds or an unknown

> ### BOX 1.3 What is Hazardous Waste?
>
> The Office of Technology Assessment defines hazardous waste as "all non-product hazardous outputs from an industrial operation into all environmental media, even though they may be within permitted or licensed limits" (Congress of the United States, Office of Technology Assessment. "Serious Reduction of Hazardous Waste: Summary," OTA-ITE-318, U.S. Government Printing Office, September 1986).
>
> This definition is much broader than that accorded the term under the Resource Conservation and Recovery Act enacted by the U.S. Congress in 1976. It is also more inclusive than that of the term *hazardous substances,* as used in the Federal Superfund program. For the latter, see Wagner, T.P. (1992), *The Hazardous Waste Q & A* (Van Nostrand Reinhold).
>
> The following definition of a "hazardous material" was adopted by the California Department of Health Services: a substance or a combination of substances which, because of its quantity, concentration, or physical, chemical, or infectious characteristics, may either (1) cause or significantly contribute to an increase in mortality or an increase in serious irreversible, or incapacitating reversible, illness or (2) pose a substantial present or potential hazard to human health or environment when improperly treated, stored, transported or disposed of or otherwise managed.
>
> Illustrative examples of hazardous wastes are waste oils, halogenated solvents, polychlorobiphenyls, hospital wastes, unused pesticides, heavy metals, and spent catalysts from oil refining.

combination of unrelated substances. Cleaning up such sites poses a variety of technical challenges, and the projected costs are daunting. The Office of Technology Assessment has estimated that it will cost $300 billion over the next 50 years to clean up waste already generated in the United States. It is evident that successful new approaches to toxic waste management will represent a signal service to society—and will be profitable as well.

Considerable regulatory pressure encourages the adoption of waste management alternatives to burial, the traditional means of disposing of solid and liquid wastes. Approaches such as air stripping (to remove volatile compounds) and incineration have been used. However, where one or more compounds contaminate a large area in low but significant concentrations, such methods are either very costly or simply not feasible. In such cases, microorganisms may provide an effective alternative. A massive oil spill, such as that from the *Exxon Valdez* off the coast of Alaska in 1989, is an obvious example. There are numerous microorganisms capable of degrading oil. Consequently, gradual clean-up of such pollution can result from temporary amplification of the relevant microbial population.

Such "bioremediation" is already being used to degrade or transform certain common pollutants in waste dumps (Table 1.4). In general, the approaches have been empirical. Nutrients are added to the waste, and the environmental conditions are modified to encourage the microbial population indigenous to the polluted site to degrade the pollutants.

**T
A
B
L
E**

1.4

On-Site Bacterial Biodegradation of Some Common Organic Pollutants

Chemical	Major Source(s)
Anthracene	Chemical industry
Benzene	Chemical industry (petrochemical used as an intermediate in the manufacture of styrene, phenol, and cyclohexane)
Benzo(a)pyrene	Coal tar hydrocarbon
Naphthalene	Chemical industry (coal tar hydrocarbon used for the manufacture of resins, other chemicals, and mothballs)
Pentachlorophenol	Wood preservative
Phenanthrene	Coal tar hydrocarbon
Phenol	Chemical industry (starting material for synthesis of numerous important industrial chemicals)
Polychlorinated biphenyls (PCBs)	Transformer oil, softener in paints and plastics
Toluene	Chemical industry (starting material for synthesis of numerous important industrial chemicals)
Trichloroethylene (TCE)	Dry cleaning solvent, metal degreasing
Xylenes	Chemical industry (starting material for synthesis of numerous important industrial chemicals)

SOURCE: New Developments in Biotechnology: U.S. Investment in Biotechnology—Special Report, p. 226. OTA-BA-360 (U.S. Government Printing Office, July 1988).

Or microorganisms are taken from another location where the pollutant is present, and then the appropriate species are enriched in laboratory cultures grown with the pollutant as a major nutrient. In some cases, the microbial population is exposed to a mutagen in an effort to produce strains with enhanced ability to utilize the pollutant. The enriched microbial population can then be grown on a large scale, harvested, and introduced into appropriate waste sites with nutrients added as required.

There is considerable promise for biotechnological treatments that remove undesirable compounds from waste streams at the source. For example, 4000–6000 lb of phenol is released every day into the coke oven waste stream of the Bethlehem Steel Company's plant at Sparrows

Point, Maryland. The stream passes through a treatment facility where it is seeded with sludge from local sewage that contains a wide variety of microorganisms. Microbial degradation reduces the phenol by about 99.9%, resulting in an outflow from the treatment facility that contains about 2 lb of phenol per day.

Many toxic compounds produced by human activities are resistant to microbial attack, but genetic engineering is being applied to generate microorganisms capable of degrading them. Among such compounds are polychlorinated biphenyls, chlorinated benzenes, and chlorinated phenols. Two strategies have been applied to the development of novel microbial degradative pathways. The first approach, termed *pathway restructuring*, is used when the resistant compound is similar in structure to a chemical compound that is degradable. It consists of determining which enzyme-catalyzed steps of the known degradation pathway are unable to handle these related compounds and attempting to modify the enzymes. The second approach is called *pathway assembly*. Soil and water microorganisms collectively exhibit a wide range of degradative activities. By considering the sum of such activities, it is sometimes possible to plan a pathway that would utilize enzymes from more than one organism and then, by means of genetic engineering, to assemble those enzymes within a single host or in a consortium and place them under the appropriate metabolic regulation. This important area of microbial biotechnology is examined in detail in Chapter 15.

Feedstock Chemicals

Feedstock chemicals are the basic building blocks that serve as the raw materials used to synthesize other chemicals, ranging from small molecules to plastics and rubber, or that are used as solvents in a variety of industrial processes. Natural gas and the primary products of petroleum refining, such as ethylene, propylene, benzene, toluene, and xylenes, are the dominant feedstocks for the chemical industry. These compounds and their derivatives account for over 97% of synthetic organic chemicals; their production in the United States exceeds 200 billion pounds. Alternative renewable sources of feedstock chemicals are needed to conserve world oil reserves and, because of concern about global warming, to minimize the increase in atmospheric carbon dioxide. What are the prospects of our finding alternative sources for bulk organic chemicals?

An immense supply of carbohydrate-rich materials is produced by plants. These include lignocellulose, the main structural component of wood; the starch in corn, wheat, potatoes, cassava, and so on; and the sugars in corn syrup and molasses. In principle, plant matter represents an abundant, inexpensive source of organic matter that could be converted to primary feedstock chemicals by a combination of microbial

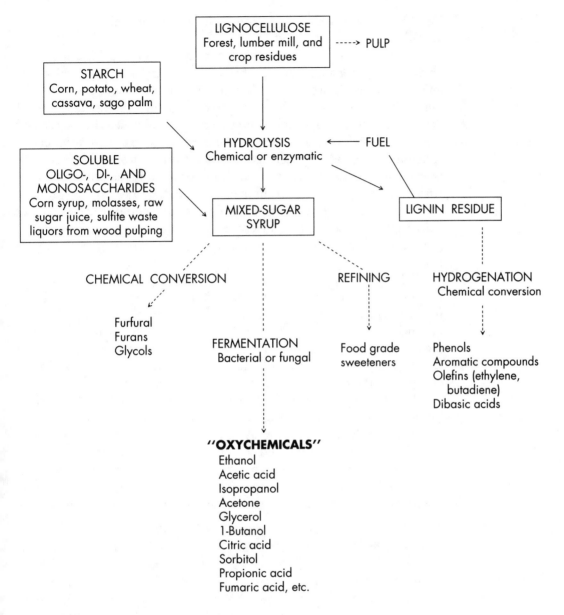

Figure 1.2
Possible renewable sources of key feedstock chemicals. Based on Busche,
R.M. (1985), The business of biomass, *Biotechnology Progress* 1:165–180.

fermentation and chemical processes (Figure 1.2). In several countries,
under abnormal conditions of supply and demand during World Wars I
and II, certain organic chemicals were produced on a large scale by
microbial fermentation. In 1975 in Brazil, a combination of high oil
prices, the need to conserve foreign currency, and a greatly depressed
world market price for sugar triggered the creation of a large-scale gov-
ernment-sponsored industry to produce ethanol by fermentation.

Worldwide, however, only a very small fraction of available biomass is actually utilized to such ends.

The major primary feedstocks (ethylene, propylene, benzene, toluene, xylenes, propane, ethane, ethylene, butadiene, *n*-butane, cyclohexane, isobutane, isoprene) are all hydrocarbons. The end products obtained by direct bacterial or fungal degradation of biomass are all "oxychemicals" (compounds containing oxygen as well as hydrogen and carbon). Therefore, a dehydration step is required in the conversion of a fermentation product (such as ethanol) to a hydrocarbon feedstock (such as ethylene). In order for such use of organic matter to compete with petrochemicals, the combined cost of the fermentation process, including recovery of the fermentation product, and of the subsequent dehydration process must not exceed that of using petrochemicals. At prevailing oil and biomass feedstock prices, the petrochemical process is cheaper.

When the oxychemicals themselves are desirable as feedstocks, the economic picture is more favorable. Some useful or potentially useful ones are listed in Table 1.5. Although a number of these compounds are produced from petrochemicals at lower cost, all are obtainable by fermentation. For example, fumaric acid, once manufactured by large-scale fermentation with a strain of the fungus *Rhizopus*, is produced more cheaply via the catalytic oxidation of benzene or butane.

The compounds listed in Table 1.5 and their derivatives represent, by weight, close to one-half of the amount of the 100 industrial organic chemicals made in the largest quantity. Any future shift toward greater production by a biomass-based chemical industry will depend strictly on economics rather than on feasibility. As illustrated by the following examples, microbial biotechnology has the potential to change the economic picture by decreasing the cost of fermentation products.

Incorporation of Genes for Degradative Enzymes. Many industrial microorganisms are able to utilize only a limited number of simple sugars. A relatively simple substrate, such as starch, has to be subjected to enzymatic hydrolysis before its monosaccharide constituents can be fermented by such organisms. Starch subjected to hydrolysis is not completely degraded to monosaccharides. About 1–2% remains as "limit dextrins," short-chain polysaccharides containing the branch points of the molecule. Similarly, utilization of cellulose by the majority of microorganisms requires both a physical pretreatment to decrease the crystallinity of this polysaccharide and enzymatic degradation with cellulases. If the ability to degrade starch and cellulose completely could be engineered into industrial microorganisms, a greater variety of carbon sources could be utilized and higher yields of products could be obtained at lower cost. This subject is discussed further in Chapters 8, 10, and 11.

Genetic Modification of Metabolic Pathways. The goal of the applied microbiologist is to obtain the most efficient conversion of the feed-

Industrial "Oxychemicals" Obtainable by Microbial Fermentation

TABLE 1.5

Chemical	Primary Bacterial or Fungal Source	Major Actual or Potential Uses	U.S. Market Value ($ millions)*
Ethanol	Saccharomyces	Octane enhancer, beverages, industrial solvent, intermediate for production of vinegar, ethylene, butadiene, esters, and ethers	9000
Acetic acid	Acetobacter	Industrial solvent and intermediate in the production of many organic chemicals, food acidulant	620
Isopropanol	Clostridium	Industrial solvent, cosmetics, antifreeze inks	500
Acetone	Clostridium	Industrial solvent, intermediate for production of many organic compounds	460
Acrylic acid	Bacillus	Industrial intermediate for plastics	360
Glycerol	Saccharomyces	Solvent, plasticizer, sweetener, explosives manufacture, printing, cosmetics, soaps, antifreeze	250
Propylene glycol	Bacillus	Antifreeze, solvent, synthetic resin manufacture, mold inhibitor	220
1-Butanol	Clostridium	Industrial solvent, intermediate for production of many organic chemicals	200
Citric acid	Aspergillus	Food acidulant	190
Sorbitol	Acetobacter	Confectionery industry	90
Propionic acid	Propionibacteria	Food preservative	35
Fumaric acid	Rhizopus	Manufacture of lacquers, dyes, and esters for perfumes	25

* Figures for 1979.

SOURCES: Busche, R.M. (1985), *Biotechnology Progress* 1:165–180; Hatch, R.T., and Hardy, R. (1989), in Marx, J.L. (ed.), *A Revolution in Biotechnology*, pp. 28–41 (Cambridge University Press).

stock to the desired end product. The objective of the microorganism is to multiply. Unfortunately, the accumulation of microbial cell mass represents competition for end product and frequently poses problems of disposal as well. Modern genetic engineering techniques make it possible to manipulate cells to maximize the formation of product while permitting only a low level of cell growth.

Recombinant Organisms With Favorable Growth Characteristics. Microorganisms used in industrial fermentations often grow relatively slowly. It is important to reduce the length of the fermentation cycle. A shorter cycle reduces the chances of contamination by phage, other bacteria, or mutants and increases productivity. The workhorse of molecular biology, *Escherichia coli*, is a fast-growing organism, and its genetics, metabolism, and physiology are exceptionally well understood. Consequently, there are advantages in transferring the capability of producing the desired product to *E. coli*. The organism can be engineered in such a way that the fermentation pathway of interest requires special activation conditions—for example, an elevation of temperature. Cells engineered in this manner are allowed to grow rapidly to a predetermined cell density. Growth is then inhibited, and synthesis of the desired product is initiated by a temperature shift. This strategy is particularly valuable where the desired end product inhibits cell growth.

Visions of the Future

In principle, each of the vast diversity of small molecules and polymers made by countless living organisms can be produced on a large scale, investigated, and modified. The limitations in the future are likely to be imposed more by the imagination of the scientist and the constraints of economics than by the capabilities of technology. The following studies of two classes of small molecules and two types of polymers should convey a sense of the discoveries to come.

Screening Cultures for New Drugs

A traditional approach to the discovery of new naturally occurring bioactive molecules utilizes "screens." A screen is an assay procedure by which numerous compounds are tested for a particular activity. The most extensive use of screens is in the search for compounds with selective toxicity for bacteria, fungi, or protozoa. The recent discoveries of avermectins and zaragozic acids testify to the value of this approach.

Avermectins. Many microorganisms indigenous to the soil—especially actinomycete bacteria and many fungi—produce biologically active secondary metabolites. Secondary metabolites are compounds an organism produces, whose absence does not adversely affect that organism under laboratory culture conditions. Intensive screening of culture supernatants (usually called fermentation broths) rich in secondary metabolites has led to the discovery of numerous clinically valuable antibiotics, penicillin being the most famous example. The structures of newly characterized compounds that are derived from soil microorganisms and have herbicidal, insecticidal, and nematocidal properties are described in the scientific literature at a rate of several hundred each year.

The avermectins were discovered in the early 1980s as a result of a deliberate search for antihelminthic compounds produced by soil microorganisms. Helminths are parasitic worms that infect the intestines of any animal unfortunate enough to ingest their eggs. There were two particularly notable features of the screening program. First, the microbial fermentation broths were tested by being administered in the diet to mice infested with the nematode *Nematospiroides dubius*. Nematodes are a subclass of helminths that includes roundworms, or threadworms. Although such an *in vivo* assay was expensive, it simultaneously tested for efficacy of the preparation against the nematode and toxicity to the host. Second, to increase the chance of discovering new types of compounds, the selection of microorganisms for testing was biased toward those with unusual morphological traits and nutritional requirements. The morphological characteristics of *Streptomyces avermitilis*, the producer of avermectins, were unlike those of other known *Streptomyces* species. *S. avermitilis* produces a family of closely related macrocyclic lactones (Figure 1.3), compounds that are active against certain nematodes and

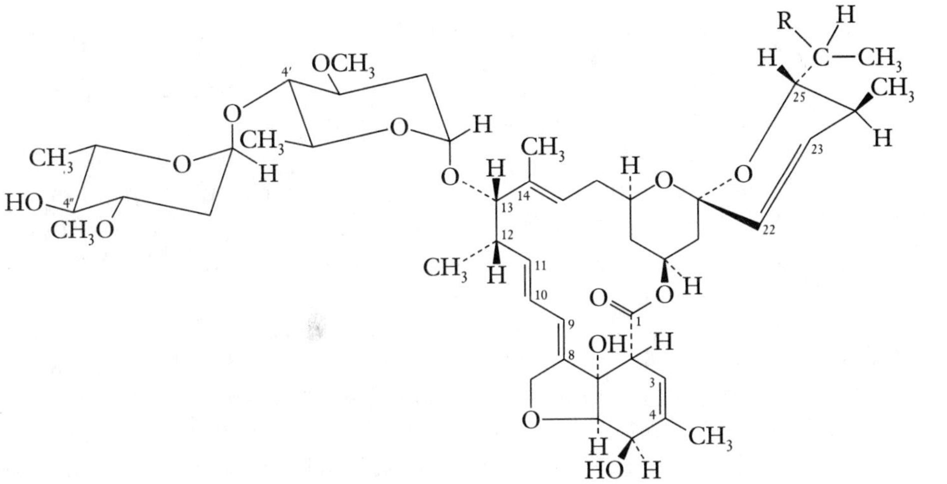

Figure 1.3

Avermectin B_1. This component is the major macrocyclic lactone produced by *Streptomyces avermitilis*. It is a mixture of two homologous avermectins with $R = C_2H_5 \geq 80\%$ and $R = CH_3 \leq 20\%$. In nematodes and insects, avermectins stimulate the release of γ-aminobutyric acid from nerve endings and then potentiate the binding of γ-aminobutyric acid to receptors on the postsynaptic membrane of an inhibitory motor neuron in the case of nematodes and on the postjunction membrane of a muscle cell in the case of insects and other arthropods. The enhanced γ-aminobutyric acid binding results in an increased flow of chloride ions into the cell, which causes hyperpolarization and the end of signal transmission. References: Babu, J.R. (1988), Avermectins: Biological and pesticidal activities, in Cutler, H.G. (ed.), *Biologically Active Natural Products: Potential Use in Agriculture*, pp. 91-108 (American Chemical Society); Dybas, R.A., Hilton, N.J., Babu, J.R., Preiser, F.A., and Dolce, G.J. (1989), Novel second-generation avermectin insecticides and miticides for crop protection, in Demain, A.L., Somkuti, G.A., Hunter-Cevera, J.C., and Rossmoore, H.W. (eds.), *Novel Microbial Products for Medicine and Agriculture*, pp. 203-212 (Elsevier).

arthropods at extremely low doses but have relatively low toxicity to mammals. These avermectins, as the compounds came to be called, are highly effective in veterinary use and show promise for treating infestations in humans.

The avermectins act on invertebrates by potentiating the activity of the inhibitory neurotransmitter γ-aminobutyric acid at its receptor. This results in irreversible paralysis and death. Their selective toxicity (they do not harm vertebrates) led to the conclusion that their toxicity involves a specific cellular target either absent or inaccessible in the resistant organisms. The avermectins do not migrate in soils from the site of application and are subject to both rapid photodegradation and microbial decomposition. Consequently, avermectins do not persist even on repeated application. The biological activity and selective toxicity of the avermectins could not have been anticipated even if the structures of these compounds had been known.

Zaragozic Acids. Over 93% of the cholesterol in the human body is located in cells, where it performs indispensable structural and metabolic roles. The remaining 7% circulates in the plasma, where it contributes to atherosclerosis (formation of plaques on the walls of the arteries supplying the heart, the brain, and other vital organs). For delivery to tissues, plasma cholesterol is packaged in lipoprotein particles; two-thirds is associated with low-density lipoproteins (LDL) and the balance with high-density lipoproteins.

The disorder *familial hypercholesterolemia* occurs in 1 individual per 500 of the population and results in elevated plasma levels of cholesterol-bearing LDL. Male heterozygotes with dominant alleles of familial hypercholesterolemia have an 85% chance of suffering heart attacks (myocardial infarction) before the age of 60. (Homozygotes of either sex die of heart disease at an early age.) A much larger number of people, who do not have familial hypercholesterolemia, have plasma levels of LDL at the upper limit of the normal range and are also at high risk for atherosclerosis. The goal of therapy in these subjects is to reduce the level of LDL without impairing cholesterol delivery to cells. This is achieved by partial inhibition of cholesterol biosynthesis.

Cholesterol is a product of the isoprenoid pathway in mammals. In addition to cholesterol and other steroids, this pathway produces several key metabolic intermediates essential to cells: dolichol, ubiquinone, the farnesyl and geranyl moieties of prenylated proteins, and the isopentenyl side chain of isopentenyl adenine. The pathways for the synthesis of these compounds diverge from the synthesis of cholesterol either at or before the farnesyl diphosphate branch point (Figure 1.4). The first committed step in cholesterol biosynthesis is the squalene synthase-catalyzed conversion of 2 moles of farnesyl pyrophosphate to 1 mole of squalene. Therefore, squalene synthase is an attractive target for selective inhibition of cholesterol biosynthesis.

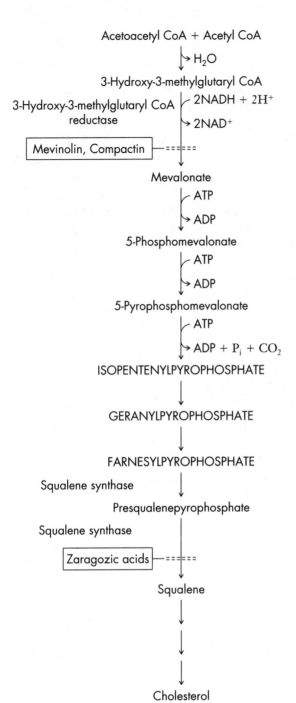

Acetoacetyl CoA + Acetyl CoA

$\searrow H_2O$

3-Hydroxy-3-methylglutaryl CoA

3-Hydroxy-3-methylglutaryl CoA
reductase

$\quad\searrow 2NADH + 2H^+$

$\quad\searrow 2NAD^+$

| Mevinolin, Compactin |─======

Mevalonate

$\nearrow$ ATP

$\searrow$ ADP

5-Phosphomevalonate

$\nearrow$ ATP

$\searrow$ ADP

5-Pyrophosphomevalonate

$\nearrow$ ATP

$\searrow ADP + P_i + CO_2$

ISOPENTENYLPYROPHOSPHATE

$\downarrow$

GERANYLPYROPHOSPHATE

$\downarrow$

FARNESYLPYROPHOSPHATE

Squalene synthase

$\downarrow$

Presqualenepyrophosphate

Squalene synthase

| Zaragozic acids |─======

$\downarrow$

Squalene

$\downarrow$

$\downarrow$

$\downarrow$

Cholesterol

Figure 1.4
Biosynthetic pathway leading to cholesterol in humans. Isopentenyl-, geranyl-, and farnesylpyrophosphate are precursors not only of sterols but also of several other important isoprenoid derivatives. The fungal fermentation products mevinolin (from *Aspergillus terreus*) and compactin (from *Penicillium* spp.) are highly effective drugs used to reduce serum cholesterol in humans. These compounds are potent inhibitors of 3-hydroxy-3-methylglutaryl CoA reductase and block formation of all the products of the mammalian polyisoprenoid pathway. In contrast, the zaragozic acids inhibit squalene synthase, which catalyzes the first committed step in sterol synthesis, and do not affect the formation of other isoprenoids.

Three structurally related and very potent inhibitors of squalene synthase were discovered by screening fungal cultures. Zaragozic acid A (Figure 1.5) was obtained from an unidentified fungus found in a water sample taken from the Jalon river in Zaragoza, Spain—hence the name. Soon after, zaragozic acids B and C were obtained from fungi isolated elsewhere: *Sporomiella intermedia* and *Leptodontium elatius*, respectively.

Squalene synthase catalyzes a two-step reaction. Farnesyl pyrophosphate is converted to presqualene diphosphate and thence to squalene. The zaragozic acids are potent inhibitors of squalene synthase competitive with farnesyl pyrophosphate. Their K_i values are extraordinarily low (about 10^{-11} M), and they are at least 10^3 times more potent inhibitors of the catalytic activity of squalene synthase than any previously described compound. Structural comparisons suggest that the zaragozic acids bind to squalene synthase in a manner similar to that of presqualene pyrophosphate (see Figure 1.5). Experiments in laboratory animals indicate that zaragozic acids are promising therapeutic agents for hypercholesterolemia.

The zaragozic acids add another entry to the long list of chemotherapeutic agents isolated from bacteria or fungi. Moreover, they differ greatly in chemical structure from every other known compound in the medical arsenal.

The discoveries of avermectins and zaragozic acids suggest that, in spite of our increasingly sophisticated ability to design drugs in the laboratory, the screening of microbial fermentation cultures for possible new drugs will retain its importance for many years to come.

Creating New Polymers

"Amber Waves of Plastics." Solid-waste disposal poses a costly problem in technologically advanced countries. In 1984 the United States alone generated 126.5 million tons of municipal solid waste, of which 7.2%

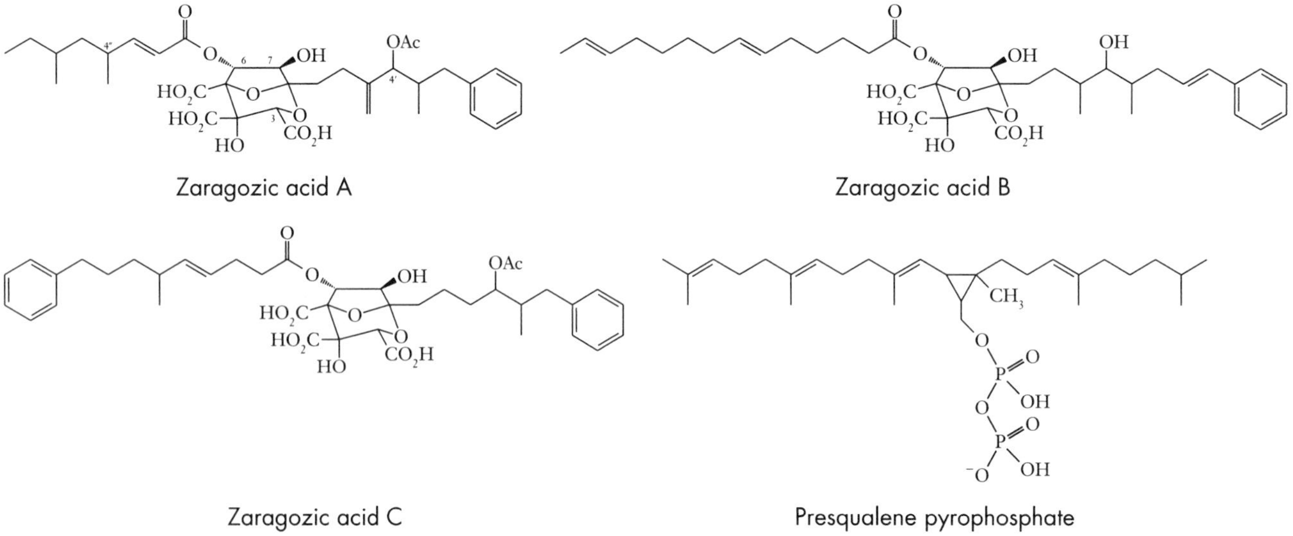

Zaragozic acid A

Zaragozic acid B

Zaragozic acid C

Presqualene pyrophosphate

Figure 1.5
Structure of zaragozic acids and of presqualene pyrophosphate. From Wilson, K.E., Burk, R.M., Biftu, T., Ball, R.G., and Hoogsteen, K. (1992), Zaragozic acid A, a potent inhibitor of squalene synthase: Initial chemistry and absolute stereochemistry, *J. Org. Chem.* 57:7151–7158.

(over 9 million tons) was plastic. This fraction is growing rapidly, almost doubling from 1975 to 1984. Commonplace plastic objects—the bottles, wrappers, and food containers we see everywhere—are highly resistant to degradation in the environment, persist as unsightly litter, and are often dangerous to wildlife.

There is hope that new families of biodegradable plastics made by microorganisms will provide a partial solution to the problem of plastic waste. For example, numerous microorganisms form polyhydroxybutyrate (PHB), a storage material for organic carbon that plays the same role in these bacteria that starch plays in plants and fat plays in humans. Bacteria cultured in a medium that is very low in nitrogen compounds but contains a high concentration of an energy-rich carbon source, such as glucose, stop growing and utilize the glucose to make PHB.

$$\underset{\text{poly-}\beta\text{-hydroxybutyrate}}{\left(\text{O}-\underset{\overset{|}{\text{CH}_3}}{\text{CH}}-\text{CH}_2-\overset{\overset{\text{O}}{\|}}{\text{C}}\right)_x}$$

In fact, this polyester represents 80% of the dry weight of the common bacterium *Alcaligenes eutrophus* cultured on a pilot plant scale. It is a stiff, brittle plastic with a high melting temperature (170°C) and is well suited to the manufacture of soft-drink bottles. As we shall see in Chapter 8, a variety of other polyesters with very different physical properties can be made by changing the composition of the nutrient medium.

Certain of these polymers have properties similar to those of polypropylene and can be used for plastic sheeting and bags. All are readily biodegradable, being broken down by microorganisms in soil or sewage sludge to carbon dioxide and water (Figure 1.6). Some may find uses as medical sutures or drug delivery devices that will gradually break down in the body by chemical or enzymic hydrolysis.

The anticipation of a giant future market for these types of microbial polymers has given rise to a vision of "plastic corn" and "plastic potatoes" whereby the polymers can be made in massive amounts. Transgenic plants might be engineered to contain the bacterial genes

Figure 1.6
Microbial degradation of polyesters of 3-hydroxybutyrate and 4-hydroxybutyrate in soil. Disks of polyester films, initially 8 cm in diameter and 0.07 mm thick, were exposed to unsterilized soil for 6 weeks at 20–25°C. Samples 1–3 were films of poly-3-hydroxybutyrate, random copolyester of 3-hydroxybutyrate (91%)-4-hydroxybutyrate (9%), and random copolyester of 3-hydroxybutyrate (50%) and 3-hydroxyvalerate (50%), respectively. Reproduced with permission from Kunioka, M., Kawaguchi, Y., and Doi, Y. (1989), Production of biodegradable copolyesters of 3-hydroxybutyrate by *Alcaligenes eutrophus*, *Appl. Microbiol. Biotechnol.* 30:569–573.

Figure 1.7
Space-filling representation of tussah silk structure with pleated sheets normal to the plane of the page and the polypeptide chain direction vertical. Note the tight side-chain packing. This silk is produced by the oriental silkworm *Antheraea paphia* syn. *mylitta*. After Marsh, R.E., Corey, R.B., and Pauling, L. (1955), The structure of tussah silk fibroin, *Acta Cryst.* 8:710.

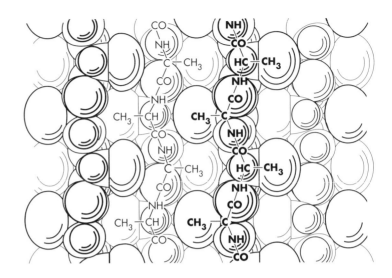

for polyester production. In such plants starch production would be turned off. Thus agriculturally produced polymers, which might cost no more than potato or corn starch, might replace petrochemical feedstocks for plastics.

Spider's Silk. Spiders secrete different kinds of silks, varying in adhesiveness, flexibility, and strength, that are designed for entrapping prey, forming egg cases, and spinning webs. Drag line silk is the strongest of these fibers. Large spiders float suspended in space from a single thin strand of drag line silk. The various silks are made up of polypeptides arranged as antiparallel β sheets, with short amino acid side chains that allow tight packing and strong hydrophobic interactions (Figure 1.7).

Reports of large-scale production of spider silk proteins using genetically engineered *E. coli* have opened the way to making new fiber materials of exceptional strength. The possibility of introducing selected amino acids with appropriate reactive side chains into these microbially produced silks by recombinant DNA methodology suggest that the silks might be attached to other polymers, possibly generating new classes of composite materials.

PUBLIC CONCERNS ABOUT THE USES OF MICROBIAL BIOTECHNOLOGY

Profound moral and political questions surround a technology that has the power to modify germ cells, assign detailed genetic profiles to individuals, and create new biological weapons. Advances in agriculture and animal husbandry made possible by the new biology promise to alter the nature of farming in ways that may have fundamental ecological, societal, and economic consequences.

The quest to understand and control the natural world has a long history. Likewise, concern over misuse of technological capabilities has ancient origins. The alchemists desired to keep the powerful mysteries of their material and spiritual research from the greedy and unworthy. The following lines are from a fifteenth-century poem by Thomas Norton of Bristol, England, entitled *Ordinal of Alchemy*.

> *This art must ever secret be,*
> *The cause whereof is this, as ye may see;*
> *If one evil man had thereof his will,*
> *All Christian peace he might easily spill,*
> *And with his pride he might pull down*
> *Rightful kings and princes of renown.*

For over a thousand years, until about A.D. 1700, alchemists searched for the philosopher's stone, a magical substance with the power to transform other metals to gold, and the elixir of life that would cure the ills of the flesh: disease and old age. By manipulating DNA, the blueprint of life, molecular geneticists may succeed in the latter quest where the alchemists failed. Modern molecular genetics offers horizons beyond some of the alchemists' wildest imaginings, but at the risk of changing the living world forever. At present, the ultimate scope of genetic engineering cannot be fully envisioned.

The social and ethical dilemmas posed by the new biology have led to deep controversy about its future. Public responses to the realization that the evolution of life may have passed from the realm of chance to control by human design range from euphoria to alarm. In a fictional scenario published in 1986, microbial biotechnology sustains humanity through a long, dark winter in the aftermath of a nuclear war.

> *The biotechnologists also cultured the necessary microorganisms and explained how to keep them going and avoid contamination. Yeasts for growing food on many kinds of substances, methane-makers for biogas tanks, nitrogen-fixers, and pesticidal bacteria to replace agrochemicals, single-cell algae for growing in water—these were basic items in every microbiological package. As the winter farmers grew more expert with their cultures and vats and immobilized cells they were clamoring for microbes and enzymes that opened the way to a do-it-yourself chemical industry based on lignocellulose, the stuff of wood and straw. Many microbiologists liked to end their homilies by saying: "If in doubt, make*

alcohol. It's a very efficient way of producing calories and if need be you can pass the nuclear winter in drunken stupor."

Nigel Calder, *The Green Machines*, p. 196 (G.P. Putnam's Sons, 1986).

Quite a different supposition prevails in the Founding Appeal of the international Gen-Ethic Network, a not-for-profit organization incorporated in West Berlin:

> *"Today it is possible to set into motion global processes which can neither be reversed nor reliably restrained, and which may end in the annihilation of our planet. It is the ruling paradigm of scientific and technical 'progress' which has brought us to this level of potential destruction—a paradigm which dictates that whatever can be done must be done, and that it is only in the application of new knowledge that the consequences become evident. The notion of unrestrained 'progress' based on trial-and-error is clearly no longer adequate to the discussion of the problems we face today. We must develop a new ethic for dealing with our knowledge and cannot entrust this task only to scientists, politicians and so-called experts, nor can we leave it to the mechanisms of the free market and international competition."*
>
> —Cited by K. Heusler (1988) in *Biotechnology and the Changing Role of Government*, p. 110 (Organization for Economic Cooperation and Development).

Both of these perspectives are valid. Any significant application of a new technology should be preceded by systematic evaluation of its consequences: practical, environmental, societal, and ethical. A probability of risk is then assigned, and all must recognize that such risk assessment is based on current knowledge that may well be inadequate. This is not a novel process in our world: Despite the omnipresent dangers of fire and explosion, we manufacture and use gasoline and natural gas; we do not abandon them merely because using them entails risk. Evaluation of the benefits and risks of microbial biotechnology has been the subject of intense debate. So far, the argument has centered on the development of risk assessment and regulatory policies for the planned introduction of genetically engineered organisms.

Risk Assessment

The optimist sees opportunity in every danger, the pessimist sees danger in every opportunity.

—Winston Churchill

He who hesitates is sometimes saved.

—James Thurber

It is difficult to make predictions, especially about the future.
—H.L. Mencken

Many of the applications of microbial biotechnology require the release and maintenance of microorganisms in specific ecological niches. Here are some examples:

- Potentially valuable vaccines have been produced by using avirulent genetically engineered mutants of certain pathogenic bacteria. These engineered strains must be administered live to animals and humans to immunize them against the virulent bacteria.
- Protection of plants against some insect pests involves the introduction of engineered bacteria that live either on or within the plants.
- The establishment of new symbioses between nitrogen-fixing bacteria and plants requires massive application to the soil of either appropriate nonindigenous wild-type strains or engineered microorganisms.
- The decomposition of certain toxic pollutants in the environment requires large-scale seeding of the affected area with natural or suitably constructed microorganisms.
- The use of engineered bacteria to leach ores at certain locations requires massive inoculation into mine tailings or low-grade deposits.

The potential benefits to be gained from deliberate large-scale releases of genetically engineered organisms are enormous. However, every such project requires an initial conceptual assessment of potential risks, followed by careful testing under controlled conditions that simulate as closely as possible the intended ecological niche for the transgenic organism. Recombinant organisms, of course, were produced long before the advent of genetic engineering. Microorganisms for specific purposes, such as increased antibiotic production, were generated by mutation and selection. Crop plants were hybridized with their wild relatives to increase resistance to disease, drought, or frost. Domesticated animals were improved through selective breeding. It has been argued that the recombinant organisms made by genetic engineering are in fact better defined than those generated by the older approaches, because the nature of the genetic alteration is known precisely. Even so, combining genetic information from very distantly related organisms produces phenotypes that were not accessible by traditional methods. This condition may have consequences beyond those that genetic engineers have foreseen. For example, the habitat available to the recombinant organism may be broader than desired (Table 1.6). The consensus

Attributes of Organisms and Environments for Possible Consideration in Risk Evaluation
Part A. Features of the Genetic Alteration and of the Recipient Wild-type Organism

	Level of Possible Scientific Consideration		
	Less → → → → → → → → *More*		

Attributes of genetic alteration			
Characterization	Fully characterized		Poorly characterized or unknown
Genetic stability of alteration	High (e.g., chromosomal)		Low (e.g., extra chromosomal)
Nature of alteration	Gene deletions (unless host range altered)	Single gene added	Multiple gene added
Function	None (no expression or regulation)	Regulation of existing gene product	Synthesis of gene product new to parent organism
Source of inserted DNA	Same species	Closely related species	Unrelated species
Vector	None	Non-self-transmissible	Self-transmissible
Source of vector	Same species; nonpathogen	Closely related species	Unrelated species or pathogen
Vector DNA or RNA	Absent	Present, but nonfunctional	Functional

Attributes of parent (wild-type) organism			
Level of domestication	Unable to reproduce without human aid	Semi-domesticated; wild populations known	Self-propagating; wild
Ease of subsequent control	Control agents known		No known control agents
Origin		Indigenous	Exotic
Habit	Free-living		Pathogenic, parasitic, or symbiotic
Pest status	Relative not pests	Relative pests	Pest itself
Survival under adverse conditions	Short-term		Long-term (e.g., spores, cysts, seeds, dormancy)
Geographic range, range of habitats	Narrow		Broad or unknown
Prevalence of gene exchange in natural populations	None		Frequent

(continued)

Part B. Phenotypic Attributes of the Engineered Organism in Comparison with Parent (Wild-type) Organism and Attributes of the Environment of the Engineered Organism

Level of Possible Scientific Consideration

Less → → → → → → → → More

Phenotypic attributes of engineered organism in comparison with parent organism			
Fitness	Reduced irreversibly	Reduced reversibly	Increased
Infectivity, virulence, pathogenicity, or toxicity	Reduced irreversibly	Reduced reversibly	Increased
Host range	Unchanged		Shifted or broadened
Substrate resource	Unchanged	Altered	Expanded
Environmental limits to growth or reproduction (habitat, microhabitat)	Narrowed but not shifted		Broadened or shifted
Resistance to disease, parasitism, herbivory, or predation	Decreased	Unchanged	Increased
Susceptibility to control by antibiotics or biocides, by absence of substrate, or by mechanical means	Increased	Unchanged	Decreased
Expression of trait	Independent of environmental context		Dependent on environmental context

Attributes of the environment			
Selection pressure for the engineered trait	Absent		Present
Wild, weedy, or feral relatives within dispersal capability of organism or its genes	Absent		Present
Vectors or agents of dissemination or dispersal (mites, insects, rodents, birds, humans, machines, wind, water, etc.)	Absent or controllable		Present, uncontrollable
Direct involvement in basic ecosystem processes (e.g., nutrient cycling)	Not involved	Marginally involved	Key species
Alternative hosts (partners), if organism is involved in symbiosis (mutualism)	Absent		Present
Range of environments for testing or use; potential geographic range	Very restricted		Broad, widespread
Simulation of test conditions	Not difficult to simulate realistically		Very difficult to simulate realistically
Public access to test site	Tightly controlled	Limited	Uncontrolled
Effectiveness of monitoring and mitigation plans	Proven effective		Untested or unlikely to be effective

Reproduced with permission from Tiedje, J.M., Colwell, R.K., Grossman, Y.L., Hodson, R.E., Lenski, R.E., Mack, R.N., and Regal, P.J. (1989), The planned introduction of genetically engineered organisms: Ecological considerations and recommendations, *Ecology* 70:298–315.

"Although there is a lot of controversy on the question of whether ecology can be a predictive science, ecologists can offer one general prediction with certainty about the biological fate of genetically engineered organisms. The universal law or principle is this: an engineered species that is successful economically will be successful biologically as well. This is because humans—often with the aid of their ecological expertise—will manage or eradicate any wild species that offers resistance to a profit-making domesticated one. The principle can be stated in this way as well: nothing in nature can resist a species that is produced and sold at a profit, even in the short run . . . Ecologists may look for economic opportunities themselves in testing the safety of engineered organisms—but the main chance lies elsewhere. The future for ecologists lies in environmental engineering, not environmental protection; it lies in making nature safe for biotechnology, not the other way around. The invisible hand of the market has superseded the invisible hand of evolution. The genetic code, for all intents and purposes, might as well be written on ticker tape." Sagoff, M. (1991), On making nature safe for biotechnology, in Ginzburg, L. R. (ed.), *Assessing Ecological Risks of Biotechnology*, pp. 341–365 (Butterworth-Heinemann).

is that the emphasis in safety assessment should be on the nature of the recombinant organism and of the environment into which it is to be introduced, not on the method by which it was modified.

The strongest plea for caution in releasing genetically engineered organisms has come from ecologists. Although they concur in the main that introduction of most engineered microorganisms would pose minimal risk of adverse outcomes, they stress that some deliberate releases may have undesirable consequences. A few potentially unfavorable scenarios follow.

Gene Transfer from a Released Recombinant Microorganism to Microorganisms in the Natural Environment. The transfer of genetic material between bacteria occurs by three known mechanisms: transformation, the uptake of naked DNA; conjugation, which occurs when two bacteria connect by a bridge and DNA is moved from one to the other (a process dependent on plasmids that encode the information needed for their transmission from one cell to another); and transduction, the transfer of a bacterial gene from one bacterium to another by a bacteriophage vector (a bacterial virus; see Chapter 3). Intra- and intergeneric exchange of genetic information by one or another of these three mechanisms has been observed at low frequency among bacteria in soil and water and on or within plants and animals. Such transfer results in the production of novel recombinants. The novel organisms and their newly acquired DNA sequences are then subject to evolutionary change with results that are difficult to predict.

How likely is it that gene transfer in the environment will lead to undesirable consequences? Genetic transfer between distantly related microorganisms is far less frequent than that between closely related microorganisms. Moreover, the likelihood that successful transfer will result in a thriving new species is low in nature, because possession of the new gene(s) rarely confers a strong advantage on the recipient. If a novel gene had been deliberately linked in the original engineered organism to a selectable marker, such as a resistance gene for an agent commonly used in the environment or, in human populations, for a frequently prescribed antibiotic, the rare interspecies transfers would be provided with a selective advantage. Such constructions can be avoided, however, in which case the persistence of organisms that acquire novel genes by transfer is considered unlikely.

Creation of New Pests. A new transgenic rice endowed with enhanced salt tolerance might become an undesirable invader of new habitats like saline estuaries. However, modifications that have been introduced into plants by classical breeding techniques, such as frost or drought tolerance, would seem just as likely to increase a plant's invasiveness, and this outcome has apparently not occurred.

Enhancement of Competitiveness of Existing Pests. Practically all crop plants have wild relatives at some taxonomic level with which they are interfertile. Insect tolerance created in a plant by introducing a bacterial toxin gene might be transferred from the transgenic crop plant to a related weed. Examples of nondeliberate gene transfer between closely related plant species are known. Forty years ago, plant breeders in India selected for increased production of the purple pigment anthocyanin in cultivated rice so that paddy workers could readily distinguish, by their purple leaves, seedlings of cultivated rice from those of wild rice. Within a few plant generations, the trait had been transferred to the wild rice.

Harm to Nontarget Species. Insect viruses with a broadened host range might infect beneficial insects as well as the targeted pest. This type of risk can be minimized by appropriate choice of virus and strategic genetic manipulation. Baculoviruses, which specifically infect arthropods, have been used since the nineteenth century to control certain insect pests. A particular merit of these viruses is that they affect only a few species of insects, the "permissive hosts." Genetic engineering of the baculoviruses could yield genetically crippled recombinants that act quickly on their caterpillar hosts and then rapidly become inactive in the soil. This approach could lead to viruses that are even safer insecticides than the unmodified viruses being used for that purpose today.

Disruptive Effects on Biotic Communities. The enhanced fitness of particular individuals might shift the population balance within a community of competing organisms. For example, introducing a gene for an insecticidal toxin into a forest tree could alter the composition of a plant community if the toxin were expressed in the seeds of the transgenic trees or of wild relatives that acquired the toxin genes by hybridization. Such seeds would be less subject to insect damage and would have an increased chance for survival.

These are only a sampling of the many considerations that must enter into an assessment of risk when any deliberate release of genetically engineered organisms is being considered. A conceptual framework for identifying and assessing potential risk was formulated in 1989 by a committee of The Ecological Society of America (see Table 1.6A and B). The degree of detailed scrutiny required increases as the organism's attributes fall farther toward the right-hand end of the scale in one or more categories in these tables. *Potential risks can be minimized if they are recognized in advance and taken into account in the choice of goals to be achieved and in the design of the recombinant organisms required to meet those goals.* Given this premise, there is reason to believe that applications such as those described at the beginning of this discussion can be employed safely without unforeseen negative effects.

Microcosms—Model Ecosystems

Many of the factors that affect the potential success of deliberate release of a genetically engineered microorganism, and the risks associated with doing so, can be examined with microcosms. Microcosms are fully contained laboratory ecosystems designed to correspond to the particular environment(s) in which the engineered microorganism is to survive in competition with that environment's complete repertoire of interacting microbial species. Microcosms help to define the parameters that influence the survival and propagation of the genetically engineered microorganism and the long-term fate of the novel DNA it carries.

Survival in a Given Environment. The population level at which the genetically engineered organism persists in an appropriate microcosm is of central importance to the function it is meant to perform. In general, microorganisms that have been grown under laboratory culture conditions for many generations do not adjust well when returned to a natural environment. The natural environment may differ significantly from the artificial conditions in temperature, moisture content, salinity, pH, nature and concentration of nutrients, and oxygen level. A microcosm environment is made to resemble nature in all these particulars, as well as in the characteristics of competing organisms that may be present. The performance of the engineered microorganism in the microcosm is then assessed systematically with respect to each of these factors. The results may indicate the need to modify the organism further in the laboratory to make it more ecologically fit to survive and function in a natural environment. The microcosm is also likely to reveal the influence of any predatory organisms on the survival of the engineered microorganism. Such direct antagonists include phagocytic organisms (such as protozoa), bacteria that are obligately parasitic on other bacteria (such as *Bdellovibrio* spp.), and bacteriophages.

Some microcosms are samples of a natural environment brought into containment within the laboratory. Others are ingenious simulations of the natural environment. For example, a microcosm has been designed for examining the ability of engineered organisms to establish themselves in the digestive tract of humans. For this purpose, the complex fecal flora of humans was transferred to germ-free mice. Keeping the mice in an isolation chamber prevented contamination by other microbes so that the gross composition of the transferred flora in the mouse gut remained similar to that of the human donor's microflora for at least several weeks. Thus, using the digestive tract microcosm described above, investigators were able to assess the survival of the recombinant bacteria they had introduced and to study the consequences of lateral gene transfer between the recombinant organisms and the hundreds of different species of microorganisms already present in the human gut.

Stability of the "Novel" Gene(s) and Consequences of Lateral Transfer.
A microcosm experiment can be used to assess the long-term stability of
the novel gene in an engineered microorganism over many generations
of growth in a natural environment that may lack strong selection pressure for the trait conferred by the novel gene. The microcosm also
facilitates the assessment of lateral gene transfer. Probes derived from
the DNA sequence of the "novel" gene and methods for amplifying that
DNA make it possible to follow the transfer of a gene from one organism to another, even if the gene is not always expressed.

Ability to Function in a Natural Environment. Let's consider a microorganism engineered to degrade a pollutant. It is possible that the microorganism will not function effectively in nature because the pollutant
is strongly bound to soil particles and, as a result, is not readily available
to be acted on. If the genes that were engineered into the microorganism need to be induced by the pollutant in order to be expressed, the
levels of pollutant in the natural environment (frequently in parts per
billion) may be too low to activate the expression and consequent translation of these genes into pollutant-degrading enzymes, or the particular environment in which the pollutant is found may have a low oxygen
level that severely limits the metabolism and growth of the engineered
microorganism. Well-designed microcosms should reveal whether
these environmental factors do indeed impede the function of a released
engineered microorganism.

Negative Effects on the Ecosystem. The large-scale introduction of a
new organism into an ecosystem has the potential to affect the nutrient
cycling, the pH, and the oxygen level of the environment. Such changes
may lead to the displacement of microorganisms indigenous to the ecosystem. The microcosm affords investigators an opportunity to monitor
such changes and their impact.

ECONOMICS OF MICROBIAL BIOTECHNOLOGY

In the preface to his excellent monograph "Economic Aspects of Biotechnology," Andrew Hacking (1986) notes, "The comment made by
J.B.S. Haldane in 1926: 'Why trouble to make compounds yourself
when a bug will do it for you?' may be answered in part 'because it's
cheaper that way.' Many biotechnological processes are not attractive
when compared to the alternatives under prevailing conditions."

Economics and government policies play a decisive role in determining the areas in which microbial biotechnology is likely to advance.
A comparison of pharmaceutical products with those of the food in-

dustry is revealing. Considerable effort, time, and expense are required to discover a new drug and to demonstrate its effectiveness and safety. However, once these conditions have been met, people are prepared to pay high prices for medicines. Moreover, in many countries, the cost of medicines to the patient is reduced by government-subsidized health care (that is, part of the cost is distributed among the population as a whole). A high *value added* in the production of health care products also provides an incentive to conduct the expensive research and development necessary in the pharmaceutical industry. The value added by a manufacturer is the difference between its sales revenue and the sum of the costs of the materials, utilities, and services required to generate that revenue. For the pharmaceutical industry, value added is approximately 45%.

In the food industry, the technology required to provide nutritious foods already exists. Unlike a new drug or a new diagnostic test for a disease, a new food product does not address a fundamental unmet need. Instead, a new technology-intensive food product has to compete in price with foods that are already well established in the market place. At the same time, the costs of research and development and of demonstrating the safety of foods made by novel processes are comparable to those for pharmaceutical products. No preferential government subsidy supports the consumption of novelty processed foods over those produced by conventional methods, and the value added in the food industry is only about 25%. Furthermore, the processes and products of food biotechnology are more difficult to protect via patents than those of the pharmaceutical industry. The innovator of a new food product has much less protection from invasion of the market by a cheaper product from another company that has simply copied the novel process without bearing the costs of development and running the risks of market resistance.

Prospects for Novel Food Technologies

As a consequence, ventures directed at producing *novel* microbial foods for human consumption have generally failed. Microbial biotechnology has contributed instead to the *improvement* of processes through the bulk use of bacterial and fungal enzymes and to the flavoring of foods through the production of compounds such as citric acid and monosodium glutamate. The economic importance of such contributions should not be minimized, however. Over 4 million tons (dry weight) of high-fructose corn syrup are produced annually by processes that depend entirely on the consecutive treatment of starch with three microbial enzymes: α-amylase, amyloglucosidase, and glucose isomerase.

Figure 1.8
Quorn, the commercial *Fusarium graminearum* mycoprotein food ingredient, a dish prepared with Quorn, and a recipe leaflet. Reproduced with permission courtesy of Wearne Public Relations Ltd., London.

This syrup represents 30% of the total sweetener in food in the United States.

At the same time, there are some signs that microbes may play very important new roles in future food technologies. In 1985 an unrestricted clearance for human consumption was granted by the British Ministry of Agriculture, Fisheries and Food for the sale of "mycoprotein," a preparation made from the filamentous fungus *Fusarium graminearum* (Figure 1.8). The successful introduction of this fungal food product on the market was an important landmark. It demonstrated that microorganisms might solve the problem of protein shortage should future agricultural capacities be insufficient to meet the needs of an ever-increasing population. A chicken gains 49 g of protein for each kilogram of protein-containing feed, and its weight doubles every 3 weeks. The fungus grows on a medium containing only glucose and inorganic salts. *Fusarium graminearum* converts each kilogram of "feed" to 136 g of cell protein and doubles in mass every 6 h! Mycoprotein and chicken protein are of comparable nutritional quality. *Fusarium graminearum* grown on a large scale would probably be competitive in price with conventional sources of high-quality protein.

Disposal of Whey—A Case Study

The disposal of whey, the major waste product in the manufacture of cheese, offers an instructive example of the interplay between the theoretically possible and the economically viable. Hacking (1986) notes that whey presents a classic problem in waste management. It is dilute, it has high collection costs, and its production fluctuates seasonally. These fluctuations, resulting from the much higher level of milk production in the summer than in the winter, pose problems for industries that require a stable supply of raw materials. Some 20 million metric tons of whey accumulate annually in the United States alone. Approximately half of the whey is discarded in industrial and municipal waste treatment facilities. The discharge into municipal sewage facilities of 1000 gal of raw whey, the daily output of a small creamery, imposes a load equal to the organic waste generated by 1800 people. The obstacles to the complete "recycling" of whey are economic.

It takes about 10 kg of milk to produce 1 kg of cheese. In the manufacture of cheese, the major family of proteins in the milk (casein) has to be separated from the other components. A liter of milk contains approximately 25 g of casein. To coagulate the casein, milk is first subjected to low-temperature pasteurization (10 s at 74°C), cooled to 30°C, and placed in a vat. A starter culture of lactic acid bacteria is added to the vat, and the mixture is stirred automatically to distribute the bacteria throughout the milk. Rennet extract, obtained from the lining of a calf's fourth stomach, is then added, together with some calcium chloride. Rennet acts as a curdling agent because it contains the proteolytic enzyme chymosin, which catalyzes a site-specific cleavage of one of the casein proteins, κ-casein. This cleavage leads to the rapid coagulation of casein into a soft, semisolid mass—the curd. The liquid that can be drained away from the curd some 90 min after injection of the starter culture is the whey. The volume of whey is over 90% of the starting volume of milk in the vat.

Whey is 93% water, 0.7% protein, 0.3% fat, 4–5% lactose, and 0.5–0.6% salts, with slight variations depending on the animals that produced the milk. This high-quality medium for microbial growth offers an abundant energy supply (lactose) and fixed nitrogen (protein). More than half of the nutrients of the milk used in cheese production remain in the whey. In principle, these nutrients could serve as feedstocks in various industrial microbial fermentations.

The total global consumption of feedstocks, such as starch and sucrose, by fermentation processes in 1983 was approximately 30 million tons. The available lactose waste in whey was 6–7% of that amount, about 2 million tons. This amount of lactose could be converted to a million tons of ethanol or could meet the total raw material needs for citric acid production per annum. Whey is already used to make some of these products. A fermentation plant at a large cheese factory in Ballineen, Ireland, produces 1 million gallons of potable alcohol yearly from whey by a yeast fermentation. Protein prepared by ultrafiltration from whey is used in foods. After hydrolysis of lactose to glucose and galactose with β-galactosidase, syrups similar to molasses or glucose syrups can be produced and can compete with the latter in some applications in sugar confectionary and ice cream.

Before whey can be used on a much larger scale, however, there are some important disadvantages to overcome.

- Most of the organisms commonly used in commercial fermentations either do not metabolize lactose or metabolize it poorly.

- Unlike glucose or molasses (the residual syrup that remains after crystallization of raw sucrose), lactose is poorly soluble and crystallizes out of solution above a concentration of 20%. Thus it cannot be concentrated, transported, and sold as a 70–80% syrup with minimal microbial decomposition during storage. Microorganisms grow readily at the relatively low osmotic pressure of the mother liquor of lactose crystallization but cannot tolerate the very high osmotic pressure of syrups that are more than 70% glucose or sucrose.

- The output of whey is high during the summer months, when adequate pasture yields a surplus of milk that can be converted to cheese, but is generally lower during the winter. If there is a constant demand for whey year-round, an unsaleable surplus is generated in the summer. Because whey does not store well, this surplus cannot be used to compensate for shortfalls in supply during the winter. These factors contribute to keeping the demand for whey low.

- Creameries tend to be relatively small operations: an annual production of 5000–10,000 tons of whey solids is large, and the median in the United States is approximately 2500 tons a year. Creameries are not centrally located but rather are scattered throughout dairying regions to reduce milk transport costs. Therefore, unless it is processed on site, the collection, concentration, and transport of whey generate additional costs.

- Antipollution legislation has had relatively little impact on creameries because of their small size and rural location. Also, in rural locations, whey can often be discharged directly onto farmland to avoid polluting water courses. Consequently, there is less pressure to find uses for whey to avoid the costs of disposal.

The methods of microbial biotechnology are powerful, and they give us the means to design new processes, generate new products, and solve environmental problems. In many cases, however, the sheer technical virtuosity of the new technology is so appealing that a substantial investment is made in novel applications without adequate economic analysis. Although feasible applications of microbial biotechnology are endless, resources are limited. An economic analysis forces the critical examination of all facets of a proposed project, as well as of existing and potential alternative approaches.

SUMMARY

The role of microorganisms in biotechnology has broadened greatly since the beginning of the recombinant DNA era in 1982, when human insulin produced in bacteria was approved for the treatment of diabetes. In addition to insulin, engineered microorganisms produce many human proteins used as therapeutic agents, such as growth hormone for the treatment of pituitary dwarfism and tissue plasminogen activator for the rapid dissolution of blood clots. Food products (including fermented beverages) continue to represent some 80% of the market value of microbial biotechnology. Microorganisms play an indispensable role in the wastewater treatment necessary to ensure the safety of drinking water. Microbial conversion of plant materials (biomass) to simple organic molecules may provide a renewable source of building blocks for the chemical industry. Novel microbial polymers show promise as biodegradable plastics. "Screening" of bacterial and fungal culture broths for antibiotics and other bioactive compounds continues to lead to the discovery of valuable drugs with unusual structures, exemplified recently by zaragozic acids (selective inhibitors of cholesterol biosynthesis) and avermectins (antihelminthic agents).

The release of genetically engineered microorganisms is a focus of public concern and of skepticism about objectivity in risk assessment and the adequacy of oversight procedures. Microcosms (model ecosystems) make it possible to explore many of the concerns in detail in the laboratory. The consensus is that the emphasis in safety assessment should fall on the particular recombinant organism and the environment into which it is to be introduced, not on the method by which it was modified. Release of engineered organisms into the environment requires case-by-case risk assessment. In cases of recombinant microorganisms such as the Ice$^-$ strain of *Pseudomonas syringae*, the potential risks appear to be minimal.

Cost considerations loom large in determining whether particular biotechnological applications are practical, especially in terms of novel food products. We examined the disposal of whey as a case study in the interplay between technical capacity and economic feasibility.

"... carefully conducted studies with released GEMS (genetically engineered microorganisms) suggest that their release will have little deleterious effect. By summarizing the history of releases of nonengineered microbes and the current status of the releases of GEMS, we hope to have shown that GEMS can be released without deleterious perturbation of a given habitat, and that their release, in combination with continued cautious and thorough study, can contribute to superior strategies for their effective application to solve field-based problems." Wilson, M., and Lindow, S.E. (1993), Release of recombinant microorganisms, *Ann. Rev. Microbiol.* 47:913–944.

SELECTED REFERENCES

Overview of Biotechnology

Marx, J.L. (ed.), 1989. *A Revolution in Biotechnology*. Cambridge University Press.

Moses, V., and Cape, R.E. (eds.), 1991. *Biotechnology: The Science and the Business*. Harwood Academic Publishers.

Treatises in Biotechnology

Rehm, H.-J., and Reed, G. (eds.), 1982–1989. *Biotechnology: A Comprehensive Treatise in 8 Volumes*. Verlag Chemie.

Moo-Young, M. (ed.), 1985. *Comprehensive Biotechnology: The Principles, Applications and Regulations of Biotechnology in Industry, Agriculture and Medicine*. Vols. 1–4. Pergamon Press.

Terminology of Biotechnology

Coombs, J., 1986. *Dictionary of Biotechnology*. Elsevier.

Walker, J.M., and Cox, M., 1988. *The Language of Biotechnology: A Dictionary of Terms*. American Chemical Society.

Ethical and Social Concerns About Biotechnology

Suzuki, D., and Knudtson, P., 1989. *Genethics: The Clash Between the New Genetics and Human Values*. Harvard University Press.

Sarink, H., 1989. Biotechnology and agriculture: Shared views and collective actions. *Tr. Biotechnol.* 7:S8–S13.

Prospects of Microbial Biotechnology

Busche, R.M., 1985. The business of biomass. *Biotechnol. Prog.* 1:165–180.

Congress of the United States, Office of Technology Assessment, 1988. *New Developments in Biotechnology: U.S. Investment in Biotechnology—Special Report*, OTA-BA-360. U.S. Government Printing Office.

Lindow, S.E., Panopoulos, N.J., and McFarland, B.L., 1989. Genetic engineering of bacteria from managed and natural habitats. *Science* 244:1300–1307.

Gasser, C.S., and Fraley, R.T., 1989. Genetically engineering plants for crop improvement. *Science* 244:1293–1299.

Kung, S., Bills, D.D., and Quatrano, R. (eds.), 1989. *Biotechnology and Food Quality*. Butterworth.

Tabor, J.M. (ed.), 1989. *Genetic Engineering Technology in Industrial Pharmacy: Principles and Applications*. Marcel Dekker.

Demain, A.L., Somkuti, G.A., Hunter-Cevera, J.C., and Rossmoore, H.W. (eds.), 1989. *Novel Microbial Products for Medicine and Agriculture*. Elsevier.

Murray, D.R. (ed.), 1991. *Advanced Methods in Plant Breeding and Biotechnology*. C.A.B. International.

Gray, J., Picton, S., Shabbeer, J., Schuch, W., and Grierson, D., 1992. Molecular biology of fruit ripening and its manipulation with antisense genes. *Plant Mol. Biol.* 19:69–87.

Deliberate Release of Genetically Engineered Microorganisms

Davis, B.D., 1987. Bacterial domestication: Underlying assumptions. *Science* 235:1329–1335.

Sussman, M., Collins, C.H., Skinner, F.A., and Stewart-Tull, D.E., 1988. *The Release of Genetically-engineered Micro-organisms*. Academic Press.

U.S. Congress, Office of Technology Assessment, 1988. *New Developments in Biotechnology: Field-Testing Engineered Organisms: Genetic and Ecological Issues*. OTA-BA-350. U.S. Government Printing Office.

National Research Council (Board on Biology), 1989. *Field Testing Genetically Modified Organisms: Framework for Decisions*. National Academy Press.

Tiedje, J.M., Colwell, R.K., Grossman, Y.L., Hodson, R.E., Lenski, R.E., Mack, R.N., and Regal, P.J., 1989. The planned introduction of genetically engineered organisms: Ecological considerations and recommendations. *Ecology* 70:298–315.

Veal, D.A., Stokes, H.W., and Daggard, G., 1992. Genetic exchange in natural microbial communities. *Adv. Microb. Ecol.* 12:383–430.

Wilson, M., and Lindow, S.E., 1993. Release of recombinant microorganisms. *Ann. Rev. Microbiol.* 47:913–944.

Economics of Microbial Biotechnology

Hacking, A.J., 1986. *Economic Aspects of Biotechnology*. Cambridge University Press.

Angold, R., Beech, G., and Taggart, J., 1989. *Food Biotechnology*. Cambridge University Press.

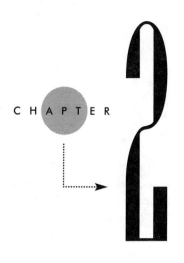

Microbial Diversity

A practical definition of microorganisms is that they are unicellular or, if multicellular, are not composed of differentiated tissues. Such organisms fall into five major groups: bacteria, viruses, fungi, algae, and protozoa. In this book we focus primarily on the bacteria and the fungi, whose contributions to microbial biotechnology have far exceeded those of the algae and the protozoa. We will also devote some attention to the uses of viruses, as well as to the problems they pose in certain technological contexts.

The term *bacteria* encompasses a huge number of organisms that differ in their sources of energy, their sources of cell carbon or nitrogen, their metabolic pathways, the end products of their metabolism, and their ability to attack various naturally occurring organic compounds. Different bacteria have adapted to every available climate and microenvironment on Earth. Halophilic bacteria grow in brine ponds encrusted with salt, thermophilic bacteria survive on smoldering coal piles or in volcanic hot springs, and barophilic bacteria live under enormous pressure in the depths of the seas. Some bacteria are symbionts of plants, others live as intracellular parasites inside mammalian cells, and some form stable consortia with other microorganisms. The seemingly limitless diversity of bacteria provides an immense pool of raw material for applied microbiology.

The variety of organisms classified as fungi rivals even that of the bacteria. Fungi are particularly effective in colonizing dry wood, and by secreting powerful extracellular enzymes to degrade biopolymers (proteins, polysaccharides, and lignin), they bring about most

> Viruses differ from all other organisms in three major respects: They contain only one kind of nucleic acid, either deoxyribonucleic acid (DNA) or ribonucleic acid (RNA); only the nucleic acid is necessary for their reproduction; and they are unable to reproduce outside of a host's living cell. Viruses are not described further in this chapter, but we will encounter them later, when we discuss bacteriophages (Chapter 3), vaccines (Chapter 6), and microbial insecticides (Chapter 7).

decomposition of plant materials. Fungi produce a huge number of small organic molecules of unusual structure, including many important antibiotics. On the other hand, fungi as a group lack some of the metabolic capabilities of the bacteria. In particular, fungi do not carry out photosynthesis or nitrogen fixation, and they are unable to exploit the oxidation of inorganic compounds as a source of energy. Fungi are unable to use inorganic compounds other than oxygen as terminal electron acceptors in respiration. Fungi as a group are also less versatile than bacteria in the range of organic compounds they can use as sole sources of cell carbon. Frequently, fungi and bacteria complement each other's abilities in degrading complex organic materials.

A consortium is a system of organisms (frequently two) in which each organism contributes something the others need. Many fundamental processes in nature are the outcome of such interactions among microorganisms, which thus affect the biosphere on a worldwide scale. For example, consortia of bacteria and fungi play an indispensable role in the cycling of organic matter. By decomposing the organic byproducts and the remains of plants and animals, they release nutrients that sustain the growth of all living things. The top 6 inches of fertile soil may contain over 2 tons of fungi and bacteria per acre. In fact, the respiration of bacteria and fungi has been estimated to account for over 90% of the carbon dioxide production in the biosphere. Technology, too, takes advantage of the special abilities of mixed cultures of microorganisms, employing them in beverage, food, and dairy fermentations, for example, and in biotreatment processes for wastewater.

Lately, the challenges of cleaning up massive oil spills and decontaminating toxic waste sites to minimize permanent damage to the environment have focused attention on the powerful degradative capabilities of consortia of microorganisms. Experience suggests that encouraging the growth of natural mixed microbial populations at the site of contamination can enhance the degradation of undesirable organic compounds in diverse ecological settings more effectively than can the introduction of a single ingeniously engineered recombinant microorganism with new metabolic capabilities. However, as Cornelius B. van Niel, one of the great teachers of comparative microbiology in this century, wrote in 1955, "It must . . . be recognized that in nature the conditions are seldom simple. Hence we must learn to study more carefully the effects of complicating circumstances. . . . This will require much imaginative work, and correlation of many kinds of observations." We are still far from an adequate understanding of microbial interactions in natural environments.

This chapter has a dual purpose: to explore the importance of the diversity of microorganisms to biotechnology and to provide a guide to the relative positions of important microorganisms on the taxonomic map of the microbial world. First we describe the major categories into which cellular organisms are divided for taxonomic purposes and explain the characteristics on the basis of which they are further subdivided within those major categories, especially the principal classification schemes that are applied to bacteria. Here we pause to emphasize the importance of consistent classification and to glance into its past—and its future. Then we list and describe groups of bacteria that are currently most important in applied microbiology. Next, more briefly, we explain how fungi are classified and cite applications of some particularly useful species. Finally, we note the importance of microbial diversity as a source of enzymes and comment on the ways in which bacterial and fungal cultures are preserved and made available.

PROKARYOTES AND EUKARYOTES

Cellular organisms fall first of all into two classes that differ from each other in the fundamental internal organization of their cells. The cells of eukaryotes contain a true, membrane-bounded nucleus (karyon), which in turn contains a set of chromosomes that serve as the major repositories of genetic information in the cell. Eukaryotic cells also contain other membrane-bounded organelles that carry genetic information, namely mitochondria and chloroplasts. In the prokaryotes, the chromosome (nucleoid) is a closed circular DNA molecule that lies in the cytoplasm, is not surrounded by a nuclear membrane, and contains all the information necessary for the reproduction of the cell. Prokaryotes have no other membrane-bounded organelles whatsoever. Bacteria are prokaryotes, whereas fungi are eukaryotes. The choice of whether to use a fungus (such as the yeast *Saccharomyces cerevisiae*) or a bacterium (such as *Escherichia coli*) for a particular application often depends on the basic genetic, biochemical, and physiological differences between prokaryotes and eukaryotes.

The Two Groups of Prokaryotes

Among prokaryotes, a general distinction is made between two groups of bacteria: the eubacteria and the archaebacteria (or Archaea). The evolutionary distance that separates the eubacteria, the archaebacteria, and the eukaryotes, estimated from the divergence in their ribosomal RNA sequences, is so great that many believe these three groups diverged from an ancient progenitor rather than evolving from one another. With respect to many molecular features, the archaebacteria are almost as different from the eubacteria as the latter are from eukaryotes (Table 2.1). For example, the cell wall structure of eubacteria is based on a cross-linked polymer called peptidoglycan with an [*N*-acetylglucosa-

A Comparison of Archaebacterial, Eubacterial, and Eukaryotic Cells

	Eubacteria	Archaebacteria	Eukaryotes
Structural features			
Chromosome number	One	One	More than one
Nuclear membrane	Absent	Absent	Present
Nucleolus	Absent	Absent	Present
Mitotic apparatus	Absent	Absent	Present
Microtubules	Absent	Absent	Present
Membrane lipids	Glycerol diesters	Glycerol diethers or glycerol tetraethers	Glycerol diesters
Membrane sterols	Rare	Rare	Nearly universal
Peptidoglycan	Present	Absent	Absent
Gene structure, transcription, and translation			
Introns in genes	Absent	Rare	Common
Transcription coupled with translation	Yes	?	No
Polygenic mRNA	Yes	?	No
Terminal polyadenylylation of mRNA	Absent	Present	Present
Ribosome subunit sizes (sedimentation coefficient)	30S, 50S	30S, 50S	40S, 60S (cytoplasmic)
Amino acid carried by initiator tRNA	Formylmethionine	Methionine	Methionine
Metabolic processes			
Oxidative phosphorylation	Membrane-dependent	Membrane-dependent	In mitochondria
Photosynthesis	Membrane-dependent	Membrane-dependent	In chloroplasts
Reduced inorganic compounds as energy source	May be used	May be used	Not used
Nonglycolytic pathways for anaerobic energy generation	May occur	May occur	Do not occur
Poly-β-hydroxybutyrate as organic reserve material	May occur	May occur	Does not occur
Nitrogen fixation	May occur	May occur	Does not occur
Other processes			
Exo-and endocytosis	Does not occur	Does not occur	May occur
Amoeboid movement	Does not occur	Does not occur	May occur

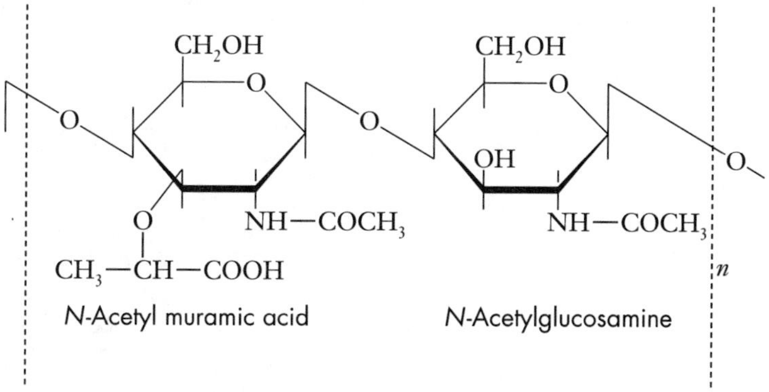

Figure 2.1
Repeating unit of the polysaccharide backbone of the peptidoglycan layer in
the cell wall of eubacteria.

mine-*N*-acetylmuramic acid] repeating unit (Figure 2.1). Because of the
virtually universal presence of peptidoglycan in eubacteria and its absence in eukaryotes, the presence of muramic acid is considered a eubacterial "signature." The different archaebacteria have a variety of cell
wall polymers, but none of them incorporates muramic acid. The most
dramatic difference between these organisms is in the nature of the
glycerol lipids that make up the cytoplasmic membrane. The hydrophobic moieties in the archaebacteria are ether-linked and branched
aliphatic chains, whereas those of eubacteria and eukaryotes are ester-
linked straight aliphatic chains (Figure 2.2).

Classifying Archaebacteria

The archaebacteria include three distinct kinds of bacteria, all found in
extreme environments: the methanogens, the extreme halophiles, and
the extreme thermophiles. The methanogens live only in oxygen-free
environments and generate methane by the reduction of carbon dioxide.
The halophiles require very high concentrations of salt to survive and
are found in such natural habitats as the Great Salt Lake and the Dead
Sea, as well as in salt evaporation ponds that people create. The extreme
thermophiles are found in such environments as hot springs and marine
thermal vents with temperatures between 80 and 100°C, and they utilize
sulfur for energy metabolism; some live in strongly acidic environments
(pH < 2). Because of the properties they have evolved in becoming
adapted to unusual and extreme conditions, the archaebacteria are a
unique future resource of organisms and macromolecules for biotechnological processes. For example, the extreme thermophile *Sulfolobus*

EUBACTERIAL LIPID

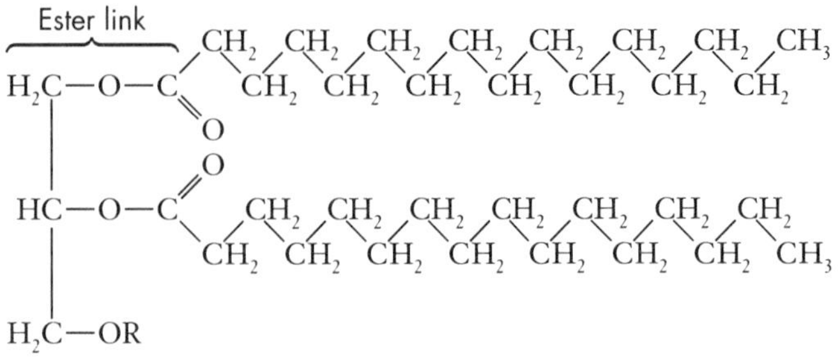

ARCHAEBACTERIAL LIPIDS

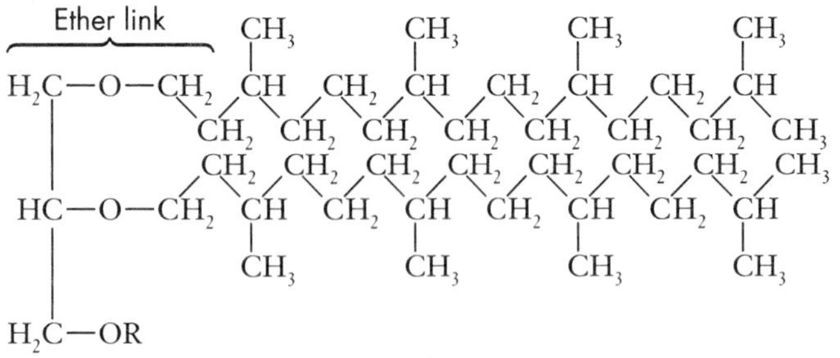

Diether

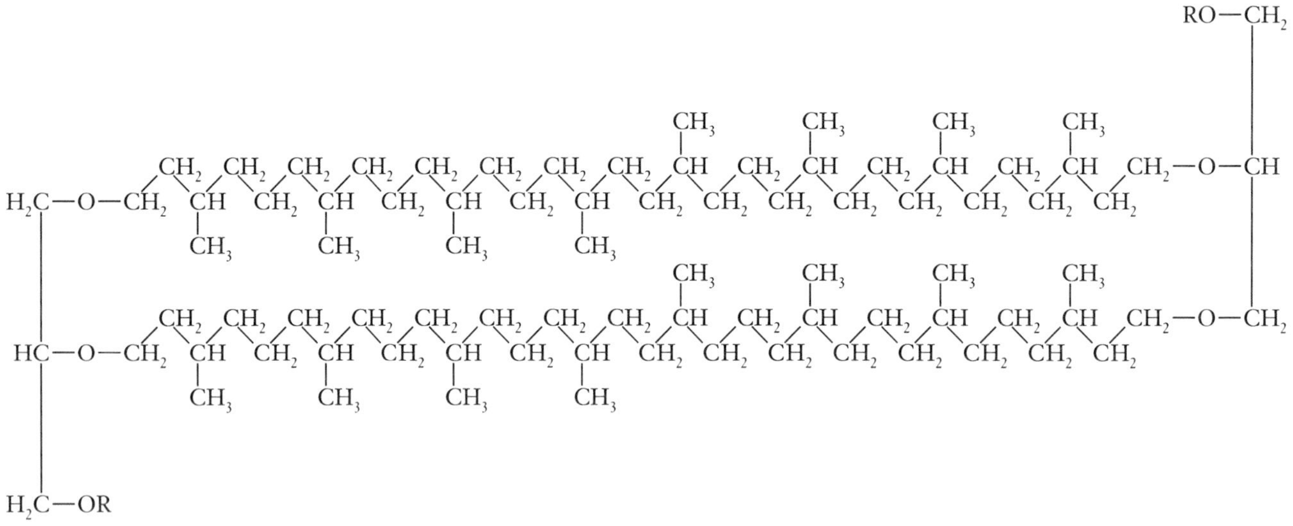

Tetraether

Figure 2.2

Membrane lipids of eubacteria and eukaryotes are glycerol esters of straight-chain fatty acids such as palmitate. Archaebacterial membrane lipids are diethers or tetraethers in which the glycerol unit is linked by an ether link to phytanols, branched-chain hydrocarbons. Moreover, the configuration about the central carbon atom of the glycerol unit is D in the ester-linked lipids but L in the ether-linked lipids. R is phosphate or phosphate esters in phospholipids and sugars in glycolipids.

spp. are excellent candidates for use in the microbial leaching of metals from ores.

Classifying Eubacteria

Gram Stain Method

The Gram stain procedure was described by the Danish physician Hans Christian Gram in 1884 and has survived in virtually unmodified form. Gram worked at the morgue of the City Hospital of Berlin, where he developed a method to detect bacteria in tissues by differential staining. In a widely used version of his empirical procedure, a heat-fixed tissue sample or smear of bacteria on a glass slide is stained first with a solution of the dye crystal violet and then with a dilute solution of iodine to form an insoluble crystal violet–iodine complex. The preparation is then washed with either alcohol or acetone. Bacteria that are rapidly decolorized by this means are said to be Gram-negative; those that remain violet are said to be Gram-positive. The ease of dye elution, and consequently the Gram staining behavior of eubacteria, correlates with the structure of the cell walls. Gram-positive bacteria have a thick cell wall of highly crosslinked peptidoglycan, whereas Gram-negative bacteria usually have a thin peptidoglycan layer covered by an outer membrane. The outer membrane is an asymmetric lipid bilayer membrane: A lipopolysaccharide forms the exterior layer, and phospholipid forms the inner layer (Figure 2.3).

The presence of the outer membrane on Gram-negative bacteria confers a higher resistance to antibiotics, such as penicillin, and to degradative enzymes, such as lysozyme. Eubacteria are almost equally divided between Gram-positive and Gram-negative types, and the result of the Gram stain remains a valuable character in bacterial classification.

Principal Mode of Metabolism

Organisms that use organic compounds as their major source of cell carbon are called *heterotrophs;* those that use carbon dioxide as the major source are called *autotrophs.* Organisms that use chemical bond energy for the generation of ATP are called *chemotrophs,* whereas those that use light energy for this purpose are called *phototrophs.* These descriptions lead to the division of microorganisms into the four types listed in Table 2.2. Those chemoautotrophs that obtain energy from the oxidation of inorganic compounds are also called *chemolithotrophs.*

All organisms need energy and reducing power in order to conduct the biosynthetic reactions required for growth. In all cases, the energy-generating processes produce adenosine triphosphate (ATP, a molecule with high phosphate group donor potential); reducing power is stored in

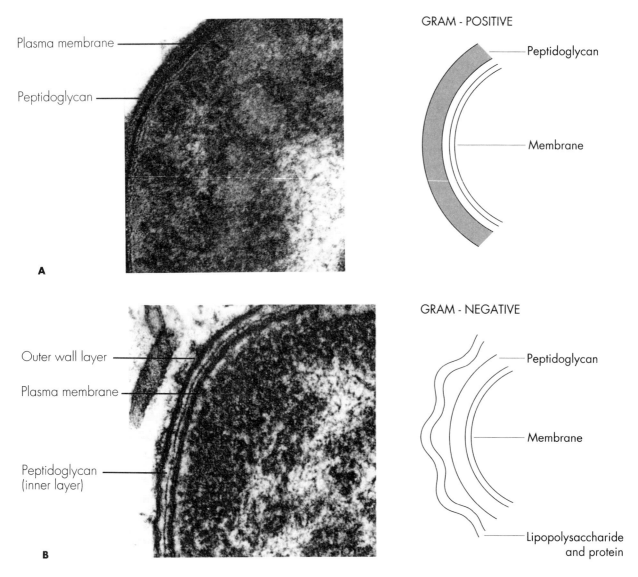

Figure 2.3
Electron micrographs of illustrative bacterial cell walls. **A.** Gram-positive, *Arthrobacter cystallopoietes*. Magnification 126,000×. **B.** Gram-negative, *Leucothrix mucor*. Magnification 165,000×. Courtesy of Dr. Thomas D. Brock and Dr. Jack L. Pate. Based on Brock, T.D., and Madigan, M.T. (1988), *Biology of Microorganisms*, 5th ed., Fig. 3.22 (Prentice Hall).

Principal Modes of Metabolism

TABLE 2.2

Type	Prokaryotes	Eukaryotes
Chemoautotrophs	yes	none
Chemoheterotrophs	yes	yes ("animals," fungi)
Photoautotrophs	yes	yes ("plants")
Photoheterotrophs	yes	none

nicotinamide adenine dinucleotides (NADH and NADPH, molecules with high electron donor potential). Bacteria exhibit a wider range of energy-generating schemes than eukaryotes. The three types of processes that lead to the formation of ATP in bacteria are reviewed very briefly below and summarized in Table 2.3 (pages 58–59).

Extraction of Chemical Bond Energy From Preformed Organic Compounds (Chemoheterotrophy). Catabolic pathways are sequences of chemical reactions in which carbon compounds are degraded. The molecules are altered or broken into small fragments, usually by reactions involving the removal of electrons (that is, by oxidations). The enzymes that catalyze catabolic reactions are usually located in the cytoplasm. There are two classes of energy-producing catabolic pathways: fermentations and respirations.

Fermentations are catabolic pathways that operate when no exogenous electron acceptor is present and in which the structures of carbon compounds are rearranged, thereby releasing free energy that is used to make ATP. It is essential to distinguish between the biological meaning of fermentation as presented here and its meaning in the common parlance of applied microbiology. To the biotechnologist, a fermentation is any process mediated by microorganisms that involves a transformation of organic substances. The rigorous, chemical definition of a fermentation is that it is a process in which no net oxidation-reduction occurs; the electrons of the substrate are distributed among the products. For example, in a lactic acid fermentation, 1 mole of glucose is converted to 2 moles of lactic acid (Figure 2.4). The process whereby some of the released free energy is conserved in activated compounds formed in the course of catabolism and then used to generate ATP is called *substrate-level phosphorylation.*

Respirations are catabolic pathways by which organic compounds can be completely oxidized to carbon dioxide (mainly via the tricarboxylic acid cycle) because an exogenous terminal electron acceptor is present. Released free energy is conserved in the form of a protonic potential, or a proton motive force, generated by the vectorial (unidirectional) translocation of protons across a membrane within which components of an

$$C_6H_{12}O_6 \longrightarrow 2\ CH_3CHOHCOOH$$
Glucose   Lactic acid

Figure 2.4
Overall equation for the fermentation reaction sequence in which glucose is converted to lactic acid (homolactic fermentation).

Summary of the Principal Modes of Microbial Metabolism

Source of Energy Utilized	Major Source of Carbon Assimilated	Generation of ATP and NADH (NADPH)			Physiological Group of Microorganisms
		Process	Electron Donor $\downarrow -e^-$ Oxidized Donor	Electron Acceptor $\downarrow +e^-$ Reduced Acceptor	
Chemical bond energy ("chemotrophs")	Organic compounds ("chemoorganotrophs")	Fermentation	Organic compound $\downarrow$ Oxidized organic compound (and, in some cases, CO_2)	Organic compound $\downarrow$ Reduced organic compound (and, in some cases, H_2)	Many obligately anaerobic and many facultative chemoorganotrophic bacteria; some fungi, such as yeasts
		Respiration	Organic compound $\downarrow$ CO_2	O_2 $\downarrow$ H_2O	Many obligately aerobic and many facultative chemoorganotrophic bacteria; many fungi and protozoa
		Anaerobic respiration		NO_3^- $\downarrow$ NO_2^-	Nitrate reducers*
				NO_2^- $\downarrow$ N_2	Denitrifiers*
				SO_4^{2-} $\downarrow$ H_2S	Sulfate reducers

(continued)

electron transport chain are contained. The vectorial translocation of protons is driven by the passage of electrons along the electron transport chain to the molecule that serves as the terminal electron acceptor. ATP is generated at the expense of the proton gradient upon return of the protons through a transmembrane enzyme complex, an F_0-F_1-type adenosine triphosphatase (ATPase). This process is called *oxidative phosphorylation*.

In *aerobic* respiration, molecular oxygen (O_2) is utilized as the terminal electron acceptor. In *anaerobic* respiration, other oxidized substances are used as terminal electron acceptors for electron transport chains. Such molecules include nitrate (NO_3^-), sulfur (S), sulfate (SO_4^{2-}), carbonate (CO_3^{2-}), ferric ion (Fe^{3+}), and even such organic compounds as fumarate ion and trimethylamine N-oxide.

Table 2.3 *(continued)*

Source of Energy Utilized	Major Source of Carbon Assimilated	Generation of ATP and NADH (NADPH)			Physiological Group of Microorganisms
		Process	*Electron Donor* $\downarrow -e^-$ *Oxidized Donor*	*Electron Acceptor* $\downarrow +e^-$ *Reduced Acceptor*	
	CO_2 ("chemolitho-trophs")	Respiration	H_2 $\downarrow$ H_2O	O_2 $\downarrow$ H_2O	Hydrogen bacteria
			NH_3 $\downarrow$ NO_2^-		Ammonia oxidizers (e.g., *Nitrosomonas*)
			NO_2^- $\downarrow$ NO_3^-		Nitrite oxidizers (e.g., *Nitrobacter*)
			H_2S S $\downarrow$ or $\downarrow$ S SO_4^{2-}		Sulfur oxidizers (e.g., *Thiobacillus*)
		Anaerobic respiration	H_2 $\downarrow$ H_2O	CO_2 $\downarrow$ CH_4	Methanogenic bacteria
Radiant light energy ("photo-trophs')	Organic compound ("photoorgano-trophs")	Phototrans-duction	Organic compound $\downarrow$ Oxidized organic compound	Bacteriorhodopsin	*Halobacterium**
					Purple nonsulfur* and gliding green* bacteria
	CO_2 ("photolitho-trophs")	Photosynthesis	H_2S S $\downarrow$ or $\downarrow$ S SO_4^{2-}	NADP $\downarrow$ NADPH	Green sulfur and purple sulfur bacteria
			H_2O $\downarrow$ O_2	NADP $\downarrow$ NADPH	Cyanobacteria (blue-green algae; eukaryotic algae, some protozoa)

* These bacteria utilize the alternative pathways of metabolism indicated in the table when they are in the absence of oxygen (O_2).

Extraction of Chemical Bond Energy from Inorganic Compounds (Chemolithotrophy). Certain bacteria use reduced inorganic compounds, such as hydrogen (H_2), ammonia (NH_3), nitrite (NO_2^-), sulfur, or hydrogen sulfide (H_2S), as electron donors to specific electron transfer chains, commonly with O_2 as terminal electron acceptor but in some instances with CO_2 or sulfate, to generate ATP by oxidative phosphorylation.

Conversion of Light Energy to Chemical Energy (Phototrophy). *Photosynthesis* is performed within membrane-bounded macromolecular complexes containing pigments (bacteriochlorophylls, chlorophylls, carotenoids, bilins) that absorb light energy. The absorbed energy is conveyed to reaction centers, where it produces a charge separation in a special pair of chlorophyll (or bacteriochlorophyll) molecules. Reaction centers are specialized electron transport chains. The charge separation initiates electron flow within reaction centers, and the light-energy-driven electron flow generates a vectorial proton gradient in a manner analogous to that described for respiratory electron flow.

Some bacteria perform photosynthesis only under anaerobic conditions. This is termed *anoxygenic photosynthesis.* In other bacteria, photosynthesis is accompanied by the light-driven evolution of oxygen (similar to the photosynthesis in chloroplasts). Such photosynthesis is termed *oxygenic photosynthesis.*

Halobacteria perform a unique type of photosynthesis when the oxygen partial pressure is low. In these organisms, absorption of light drives the isomerization of a retinal molecule covalently bound to an abundant cytoplasmic membrane protein called *bacteriorhodopsin,* after which the retinal molecule rapidly returns to its original conformation. The retinal photocycle results in a vectorial pumping of protons by bacteriorhodopsin to the exterior of the cell with the generation of a proton motive force. ATP is generated at the expense of the proton gradient.

Different bacteria use one or another of the foregoing processes as their preferred mode of energy generation. However, almost all bacteria are able to switch from one form of energy production to another, depending on the nature of the available substrates and on the environmental conditions. For example, purple nonsulfur bacteria grow on a variety of organic acids as substrates and obtain energy from respiration when oxygen is present. However, under anaerobic conditions and in the presence of light, these organisms synthesize intracellular membranes that possess the complexes needed for photosynthesis, and then they use light energy to generate ATP. Under aerobic conditions, the enteric bacterium *Escherichia coli* oxidizes substrates such as succinate and lactate and utilizes an electron transport system with ubiquinone, cytochrome *b*, and cytochrome *o* as components and with O_2 as a terminal electron acceptor. Under anaerobic conditions, with formate as a substrate, *E. coli* utilizes an electron transport system with ubiquinone and cytochrome *b* as components and with nitrate as a terminal electron acceptor. When *E. coli* is growing on oxaloacetate as a substrate under anaerobic conditions, the sequence of carriers is NADH, flavoprotein, menaquinone, and cytochrome *b*, and fumarate is the terminal electron acceptor. There are hundreds of other examples of such metabolic versatility among bacteria. This flexibility in mode of energy generation is limited to the prokaryotes and gives these organisms a virtual monopoly on the colonization of certain ecological niches.

IMPORTANCE OF IDENTIFICATION AND CLASSIFICATION OF MICROORGANISMS

One does not have to be an expert taxonomist to distinguish a horse from a zebra. However, microorganisms are very small and are difficult to tell apart by simple inspection under the light microscope. Partly as a consequence, different names have often been assigned to the same microorganism. It is disconcerting to read current papers that, taken together, seem to describe similar processes involving several apparently different yeast strains and then to discover, by referring to a register of yeast names that all are probably one and the same. For example, strains previously designated as *Saccharomyces diastaticus* clearly belong within *S. cerevisiae*, but *S. diastaticus* still crops up in the literature. *Zygosaccharomyces rouxii* was previously known as *Saccharomyces rouxii*, and *Kluyveromyces marxianus* was also referred to as *K. fragilis*. The profusion of names for the same organism inspired Barnett, Payne, and Yarrow to preface *A Guide to Identifying and Classifying Yeasts* (first published in 1979) with the opening lines of T.S. Eliot's *Old Possum's Book of Practical Cats:*

> *The Naming of Cats is a difficult matter,*
> *It isn't just one of your holiday games;*
> *You may think at first I'm as mad as a hatter*
> *When I tell you a cat must have THREE DIFFERENT NAMES.*

In the search for organisms to assist in a technical process or to produce unusual metabolites, the unintentional rediscovery and renaming of previously described organisms represents an unnecessary duplication of effort. Conversely, each time a new organism can be placed within a well-studied genus, strong and readily testable predictions can be made about many of its genetic, biochemical, and physiological characteristics.

To classify a microorganism, one must first obtain a large, uniform population of individuals, a *pure culture*. In the traditional methods of taxonomy, one then examines the organism's phenotypic characters— that is, the properties that result from the expression of its genotype, which is defined as the complete set of genes it possesses. Phenotype includes morphological characteristics such as the size and shape of individual cells and their arrangement in multicellular clusters, the occurrence and arrangement of flagella, and the nature of membrane and cell wall layers; behavioral characteristics such as motility, chemotactic responses, and phototactic responses; cultural characteristics such as colony shape and size, optimal growth temperature and pH range, tolerance of the presence of oxygen and of high concentrations of salts; and the ability to resist adverse conditions by forming spores. The range

▶ "Taxonomy (the science of classification) is often undervalued as a glorified form of filing—with each species in its folder, like a stamp in its prescribed place in an album; but taxonomy is a fundamental and dynamic science, dedicated to exploring the causes of relationships and similarities among organisms. Classifications are theories about the basis of natural order, not dull catalogues compiled only to avoid chaos." Gould, S.J. (1989), *Wonderful Life: The Burgess Shale and the Nature of History* (Norton).

of compounds that support the growth of a given organism, the way these compounds are degraded, and the nature of the end products (including the involvement of oxygen in the process) represent an important set of phenotypic characters.

It is customary to examine dozens of characters; in the computer-based method of numerical taxonomy, hundreds of characters may be examined. Armed with such information, one can then consult the ninth edition of *Bergey's Manual of Systematic Bacteriology* (Williams & Wilkins, 1984–1988), the recognized authority on bacterial taxonomy. The identification of a bacterium is thus relatively straightforward. However, some difficulty arises when one tries to deduce phylogenetic relationships between organisms on the basis of the classification scheme presented in *Bergey's Manual*. The same is true of the classification of fungi.

Classification and Phylogeny

In principle, any group of organisms can be classified according to any set of criteria, as long as the scheme results in reproducible identification of new strains. However, a classification scheme based on totally arbitrary criteria is likely to be of very limited practical use. Thus taxonomists group together apparently similar, presumably related species into a genus, and presumably related genera into a family, in the hope that this classification accurately reflects the evolutionary, or phylogenetic, relationships between various organisms. A hierarchical classification of this type is also adopted by *Bergey's Manual*. But how does one build such a classification scheme? In basing a classification scheme on phenotypic characters, a taxonomist must decide which characters are more fundamental and thus useful for dividing organisms into major groups, such as families, and which characters are more variable and thus suitable for dividing the major groups into smaller ones, such as species. In traditional taxonomy, the shape of the bacterial cell, for example, has been used for dividing bacteria into large groups. Thus of the lactic acid bacteria (which, as we will see later, characteristically obtain energy by fermenting hexoses into lactic acid plus sometimes ethanol and carbon dioxide), those with round cells and those with rod-shaped cells are placed in two completely different groups in *Bergey's Manual*.

More recently, quantitative information on the phylogenetic relationships between organisms has become available through comparison of their DNA sequences. Because the bacterial world is so diverse, however, this method is useful only for comparing species of bacteria that are very closely related. Otherwise, the DNA sequences are so dissimilar that no data of significance are obtained. Thus it was the use of ribosomal RNA sequences for comparison, pioneered by Carl Woese in the early 1970s, that revolutionized the field. Ribosomal RNA is present and performs an identical function in every cellular organism, and more important, its sequence has changed extremely slowly during the course of evolution. It is therefore an ideal marker for comparing distantly related organisms. Characteristic sequences of nucleotides, or "signature" sequences, may be conserved for a long time in a given branch of the phylogenetic tree and enable scientists to assign organisms to different branches with great confidence.

Undoubtedly, the future classification of the bacteria will be based on their phylogenetic relationships. However, when the ninth edition of *Bergey's Manual* was written, few of these relationships were well understood, so classification in the *Manual* was largely based on the traditional model. This unhappy compromise has created problems such as the situation described above: Although the round lactic acid bacteria were placed far from the rod-shaped bacteria in *Bergey's Manual*, their ribosomal RNA sequences show that many of the former are actually very closely related to the latter.

At present, the classification in *Bergey's Manual*, imperfect as it is, is the only universally accepted system. Is there any value in using the phylogenetic scheme of Woese and his coworkers as well? Certainly, but we must always keep in mind the vast time scale we are dealing with when we consider the evolution of bacteria. Even bacteria that are thought to be closely related phylogenetically can be quite distant, on the evolutionary time scale, relative to the changes that have taken place among higher organisms. Thus if we are looking at characteristics that change rapidly during the course of evolution, then the phylogenetic relationship may not offer much help. However, it will certainly help us in the study of slowly changing characters. An example is the organization and regulation of biosynthetic pathways. Because the prokaryotic world is so diverse, different pathways are seen in the biosynthesis of even such common compounds as amino acids. The distribution and the mechanism of control of these pathways, which we need to know in order to use bacteria to produce amino acids (see Chapter 12), clearly follow the phylogenetic lines.

Plasmids and the Classification of Bacteria

The genetic information of a bacterial cell is contained not only in the main chromosome but also in extrachromosomal DNA elements called plasmids. Plasmids are self-replicating within a cell, and many plasmids have a block of genes that enable them to move from one bacterial cell to another. Loss of its plasmids has no effect on the essential functions of a bacterial cell. Consequently, the cell is seen to act as *host* to the plasmids. Similar to bacterial chromosomes but much smaller, plasmids are circular double-stranded DNA molecules. Plasmid DNA often replicates at a different rate and sometimes on a different schedule from the chromosomal DNA, and cells may contain multiple copies of specific plasmids. Some plasmids encode resistance to certain antibiotics or heavy metal ions, or to ultraviolet radiation. Others, surprisingly, carry genes coding for functions that have been thought to be distinguishing characteristics of the host species. For example, the most characteristic trait of the fluorescent *Pseudomonas*, which we will discuss soon, is thought to be its ability to degrade a wide range of organic compounds; however, many of the genes that make these degradations possible are located on plasmids. The same is true of the genes for nitrogen fixation in *Rhizobium*, the species that carries out much of the biological nitrogen fixation on Earth. And it is true of the genes for disease-causing factors (toxins, proteases, and hemolysins—the proteins that lyse red blood cells and other animal cells) in many pathogenic bacteria. Because plasmids sometimes confer highly noticeable phenotypic traits on their hosts, they may influence the classification of the host organism. For example, certain strains of *Streptococcus lactis*, classified as *S. lactis* subsp. *diacetylactis*, carry a plasmid that enables them to utilize citrate. These are the strains responsible for the characteristic aroma of cultured butter, which results from the diacetyl they produce when fermenting the citrate in milk.

Some plasmids have the ability to transfer themselves from one bacterial host cell into another. Sometimes the host can be of a different species or genus. On the other hand, the plasmid genes can become integrated into the host's chromosome and are thereafter a part of the permanently inherited genetic makeup of the cell. This "lateral" transfer of genetic information into different groups of bacteria, if it were to occur frequently, would make every bacterium an extremely complex hodgepodge of genes from many different sources. Experimental studies, however, have shown that lateral exchange certainly has not occurred to the extent of obliterating the phylogenetic lines of descent of various organisms.

The ability of plasmids to replicate themselves has been utilized in the construction of cloning vectors, many of which contain a replication function derived from plasmids and can therefore be maintained indefinitely in the cytoplasm of the host bacteria. However, in many cases, the replication of plasmids requires the participation of host functions too. This is one of the reasons why plasmids can survive in only a limited range of hosts. One way to construct a vector that can replicate in a wide range of hosts is to use the replication genes from a plasmid that has a broad host range. Here again, knowledge of phylogenetic relationships will help us predict the range of host bacteria that would support the replication of such vectors. For example, many broad-host-range plasmids isolated from the Gram-negative bacteria of the "purple bacteria" group are likely to replicate in most of the members of this group, or at least in the members of the same subgroup.

TAXONOMIC DIVERSITY OF USEFUL BACTERIA

As we noted at the beginning of this chapter, bacteria as a group show remarkable metabolic diversity. The eubacteria, in particular, are characterized by their tremendous metabolic versatility. Archaebacteria, although they are adapted to extreme environments and often obtain energy in rather unexpected ways, are not particularly diverse in metabolism. Eubacteria include organisms that degrade a number of unusual compounds, that fix atmospheric nitrogen, that obtain energy by oxidizing nitrogen-containing inorganic compounds, that produce a number of useful organic compounds as end products of fermentation, and so on and on. The list is almost endless. Consequently, most of the bacteria that have so far been utilized for biotechnological applications are eubacteria. Below we list some of the groups of eubacteria that are particularly important in biotechnology, in order to familiarize readers with their names and properties and to demonstrate the importance of bacterial diversity to biotechnology. Our list follows the phylogenetic divisions of Woese and his coworkers. One striking observation is that all of the organisms listed, except *Thermus*, come from only two out of the eleven major branches of the eubacteria, the purple bacteria and the Gram-positive bacteria (Figure 2.5). This is partly because many branches contain very few representatives, but it is also true that many of the less familiar organisms have received less attention than they deserve. In looking for unusual products, or metabolic capabilities, it may make sense to cast a wide net, drawing organisms from hitherto neglected branches of the phylogenetic scheme.

Purple Bacteria

Photosynthesis is a complex process, yet photosynthetic bacteria are found in many branches of the phylogenetic scheme. It therefore seems likely that photosynthesis was "invented" very early during the evolution of eubacteria and that the present-day nonphotosynthetic bacteria evolved simply by losing the photosynthetic machinery. The purple bacteria comprise not only the purple photosynthetic bacteria but also many (even most) of the well-studied Gram-negative bacteria of the nonphotosynthetic type. The purple bacteria (also called proteobacteria) branch is divided into four subdivisions: α, β, γ, and δ.

Gamma (γ) Division

The γ division contains both the family of enteric bacteria (which includes the well-known *E. coli*) and some of the best-known *Pseudomonas* species.

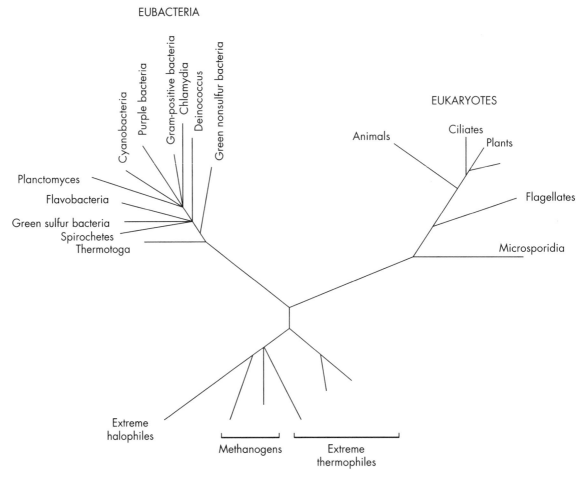

Figure 2.5
The universal phylogenetic tree. The lengths of the lines are proportional to distances calculated by comparing rRNA sequences. Based on Woese, C.R. (1987), Bacterial evolution. *Microbiol. Rev.* 51:221–271, Figs. 4 and 11.

E. coli. This inhabitant of the intestinal tract of higher animals is the most extensively studied living organism. It is one of a group of closely related organisms called enterics or *Enterobacteriaceae,* most of which are rod-shaped Gram-negative bacteria with peritrichous flagella (that is, the flagella are more or less uniformly distributed over the surface of the cell). The hallmark of the enterics' metabolism is that they can generate ATP either by oxidative degradation of organic compounds in the presence of air or by fermentation of simple sugars under anaerobic conditions. The range of compounds they can metabolize is limited, as expected for intestinal bacteria that encounter only food of the type ingested by the host animal. However, their metabolism is extremely well regulated. This too is a necessity for intestinal bacteria; they live a "feast-or-famine" existence and have to compete with many other organisms in a confined environment. These are important points: Much of our knowledge of metabolic regulation was derived from *E. coli*, and we tend to think, often incorrectly, that the *E. coli* model applies to other organisms living in totally different ways (see the discussion of amino acid production in Chapter 12).

Because it grows rapidly on well-defined simple media, and because we know a great deal about its genetics, biochemistry, and physiology, *E. coli* has been a favorite choice for producing foreign proteins via recombinant DNA technology. Human insulin and growth hormone are prime examples (see Chapter 1).

Fluorescent Pseudomonads. The Gram-negative bacteria included in the genus *Pseudomonas* in *Bergey's Manual* differ from *E. coli* in many ways. Although they too are rod-shaped, their flagella are polarly located. They cannot make ATP by fermentation; they are obligate aerobes. In marked contrast to *E. coli*, some of the species are famous for using a wide range of organic compounds as energy sources—in some cases more than a hundred. These properties are ideally suited to the pseudomonads' existence as "generalists" in soil and water.

The group classically recognized as the genus *Pseudomonas*, however, is extremely heterogeneous. In addition to the fluorescent members (which include *P. aeruginosa*, *P. putida*, *P. fluorescens*, and *P. syringae*) belonging to Woese's γ division, it contains species from both the β and the α divisions. The fluorescent pseudomonads excrete yellowish-green fluorescent compounds into the culture medium. The evolutionary distance between the fluorescent group and these other species is as wide as that separating the most disparate members of the animal kingdom. It is clear that lumping such remotely related strains into a single "genus" will eventually create more confusion than understanding. It is very misleading to discuss the metabolism, physiology, and genetics of the group designated *Pseudomonas* as though it were made up of closely related organisms, so the practice must be avoided.

Many members of the fluorescent group degrade compounds such as camphor, toluene, and octane, as well as certain substances that humans have created, such as halogenated aromatic compounds. Thus both naturally occurring and laboratory-engineered fluorescent strains are the subjects of active study as possible candidates for the reclamation of sites that have been contaminated with high levels of toxic organic compounds. Interestingly, the genes for the enzymes that degrade camphor or octane are usually found on plasmids, although the enzymes that degrade aromatic compounds through the "classical" or "ortho cleavage" pathway, which involves opening a catechol ring between the two hydroxyl groups of catechol, are apparently coded for by chromosomal genes.

The fluorescent group includes one plant pathogen, *P. syringae*. The outer membrane of this pseudomonad contains a protein complex that nucleates the formation of ice crystals. These organisms cause frost to form on leaves at temperatures only slightly below the freezing point

($-3°C$ rather than $-10°C$), damaging the outer tissue so that the bacteria can invade the plant. The resulting damage to crops is estimated to exceed $1 billion annually in the United States alone. To decrease this damage, "ice minus" *P. syringae* strains were made by using recombinant DNA methods to inactivate one of the genes necessary for the formation of the ice nucleation complex. When the mutant organisms are sprayed on plants, they compete with the wild type and decrease the chances that ice will form. On the other hand, using wild-type *P. syringae* strains in manufacturing snow at ski resorts has yielded appreciable savings in the cost of cooling. Genetic engineering has produced strains able to form ice crystals at even higher temperatures; thus science has found ways both to increase this microbe's powers and to disarm it.

Xanthomonas. This plant pathogen is related to the fluorescent pseudomonads. It produces characteristic yellow pigments, which gave the genus its name (the Greek *xanthos* means "yellow"). Like many other animal and plant pathogens, this organism secretes polysaccharide into the medium. The *X. campestris* polysaccharide, xanthan, has been put to many uses in the food industry as well as in enhanced oil recovery (see Chapter 8).

Beta (β) Division

Thiobacillus. This "genus" consists of small, rod-shaped, chemoautotrophic bacteria that live by oxidizing reduced sulfur compounds such as H_2S and thiosulfate and do not accumulate intracellular sulfur deposits. That it is an extremely heterogeneous group is evident from the GC content of the members' DNA, which spans the extremely wide range of 34–70%. In other bacterial genera the span is typically 5% or so. Moreover, ribosomal RNA sequences show that although many species (such as *T. ferrooxidans*, *T. intermedius*, *T. thiooxidans*, and *T. denitrificans*) belong to the β division, other *Thiobacillus* species are found in the α and γ divisions. As in the case of the "genus" *Pseudomonas*, calling this heterogeneous assembly a genus is a source of confusion.

The *Thiobacillus* metabolism generates sulfuric acid, and many strains grow well under strongly acidic conditions (pH 1.5–2.5). Some show remarkable metabolic versatility. For example, *T. ferrooxidans* is also able to derive energy from the oxidation of ferrous to ferric iron. In the absence of oxygen, this organism is able to oxidize reduced forms of sulfur, using ferric iron as an alternative electron acceptor.

Thiobacilli are important organisms in the industrial extraction of copper and uranium ores (Chapter 15). Efforts are under way to improve such strains through recombinant DNA technology. An understanding of the phylogenetic relationships within this very heterogeneous group will be essential to that effort.

Alpha (α) Division

The α division contains organisms with some unusual properties. One of its subgroups consists of three members, *Agrobacterium*, *Rhizobium*, and *Rickettsia*, that all interact closely with eukaryotic hosts, the first two with plants and the last with animal cells. Interestingly, ribosomal RNA sequence data show that the ancestor of the mitochondria in animal cells was an organism belonging to the α division of the purple bacteria branch.

Rhizobium. These bacteria are flagellated Gram-negative rods. *Rhizobium* strains are aerobic chemoheterotrophs that live in the soil and invade the root hairs of leguminous plants, where they form root nodules within which they fix nitrogen largely for the plant's benefit. The recognition between a *Rhizobium* species and its plant host is very specific (see Chapter 9). The practical importance of this genus is evident from the fact that, while a total of about 30 million metric tons of synthetic nitrogen fertilizers are produced per year, nitrogen-fixing microorganisms yearly convert about 200 million tons of nitrogen to ammonia, and the major portion of this biological nitrogen fixation is carried out by the symbiotic nitrogen fixers, such as *Rhizobium*.

Agrobacterium. These flagellated Gram-negative rods are also aerobic chemoheterotrophs abundant in soil. They carry a large plasmid, the Ti plasmid, which encodes various functions for the transfer of a small portion of the plasmid DNA, called T-DNA, into plant cells. The T-DNA becomes integrated into the plant chromosomal DNA and stimulates the synthesis of a plant growth hormone, thereby causing the growth of galls or tumors in the host plant.

The ability of these strains to transfer genes into plant cells is, to date, the only known example of natural gene transfer between a prokaryote and a eukaryote. It is a phenomenon of immense potential importance in biotechnology, because it opens the door to the stable transfer of foreign genes into crop plants. One can imagine the Ti plasmid being used to introduce into cereal grains genes for engineered storage proteins, enriched in certain essential amino acids, or to transfer genes for fixing nitrogen or to introduce into plants genes for resistance to specific diseases or herbicides. The Ti plasmid system is described in detail in Chapter 9.

Zymomonas. These Gram-negative rods with polar flagella are found in sugar-rich fermenting plant extracts such as palm wine (which is made from palm sap), sugar cane extract, and apple ciders. They can grow either by fermentation or by respiration. However, sugars are fermented not by the Embden-Meyerhof pathway used by the enteric species, but by the Entner-Doudoroff pathway, with ethanol as virtually the only end product rather than a mixture of lactate, ethanol, formate,

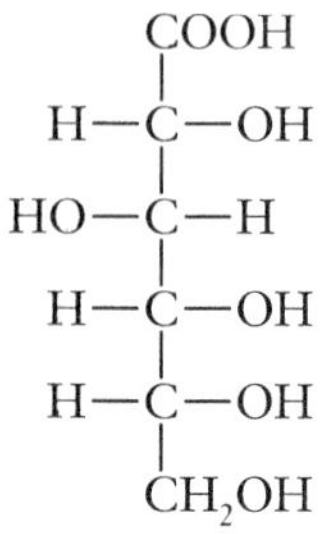

```
        COOH
         |
   H—C—OH
         |
  HO—C—H
         |
   H—C—OH
         |
   H—C—OH
         |
        CH₂OH
```

Figure 2.6
Gluconic acid.

acetate, and other end products. In certain respects, discussed in Chapter 11, *Zymomonas* offers advantages over yeasts in large-scale ethanol production.

Gluconobacter. The cells of this genus are ellipsoidal to rod-shaped, and many strains are motile with polar flagella. They are obligately aerobic chemoheterotrophs that characteristically obtain energy by oxidizing ethanol to acetic acid, but the acetic acid is not oxidized further. This is remarkable, for in almost every oxidative degradation pathway in other organisms, the substrate is always oxidized completely to CO_2. Thanks to this property, *Gluconobacter* is very useful in the manufacture of vinegar. It can also oxidize glucose to gluconic acid (Figure 2.6), a product of considerable commercial importance.

Gram-Positive Bacteria

Many branches of this cluster contain endospore-forming, obligate anaerobes traditionally classified in the Gram-positive genus *Clostridium*. Endospores are thick-walled spores formed within the bacterial cell (the Greek *endon* means "within"). The production of the endospore is an extremely complex process that appears to have been "invented" only once during biological evolution; it is found only in the Gram-positive branch. Thus it is reasonable to hypothesize that the ancestor of this branch was an obligately anaerobic chemoheterotroph capable of endospore production and that some members later became adapted for an aerobic mode of life and some lost the capacity for sporulation. Another observation of interest is that when defined on the basis of ribosomal RNA sequence, this branch turns out to contain two sub-branches apparently with Gram-negative cell walls. Gram-negative cell walls are also present in all other branches of eubacteria, so it may again be argued that the ancestral eubacteria had a cell wall of Gram-negative type and that the Gram-positive cell wall arose by the loss of the outer membrane structure.

Several lines within the traditional Gram-positive branch have low-GC DNA, and most of them contain clostridia. In the paragraphs that follow, we discuss these clostridia lines as the "genus" *Clostridium*, but remember that the group is phylogenetically very "deep," or ancient, and thus is extremely heterogeneous. The evolutionary distance between one *Clostridium* species and another may be as great as the distance between animals and plants.

One other line of low-GC organisms is made up largely of facultatively aerobic organisms such as *Bacillus*, lactic acid bacteria, and *Staphylococcus*. Finally, we describe the rather tight cluster of essentially aerobic organisms with high-GC DNA called the actinomycete line.

Clostridial Branches: "Genus" Clostridium

These rod-shaped, usually flagellated, Gram-positive bacteria are strictly anaerobic and form endospores under unfavorable conditions. Phylogenetic data suggest that they are ancient groups, and some of them preserve the fermentation pathways that were prevalent when the Earth's atmosphere was largely devoid of oxygen and that have since disappeared in other branches of life. These pathways are obviously of interest to comparative biochemists. Some are useful in biotechnological applications because they terminate in useful products such as ethanol, acetylmethylcarbinol, butanol, and acetone.

Circumstances brought about by the outbreak of World War I led to a practical interest in clostridial fermentations. At that time, acetone was an important ingredient in the manufacture of smokeless powder (cordite), and before 1914, acetone was prepared starting from wood. Dry distillation (pyrolysis) of wood yielded a liquid distillate containing 10% acetic acid as well as other volatile products. Acetic acid was separated by distillation into a calcium hydroxide solution to form calcium acetate, and dried calcium acetate was then decomposed by heating to produce acetone and calcium carbonate. The wartime demand for acetone far outstripped the supply available from this process. Chaim Weizmann, a chemist for the firm of Strange and Graham Ltd. in Manchester, England, happened to be working on the microbial production of acetone and butanol by bacterial fermentation of starch, in order to obtain butanol for the manufacture of rubber. Among the organisms he screened, Weizmann discovered a bacterium, later named *Clostridium acetobutylicum*, that produced 12 tons of acetone from 100 tons of molasses. *C. acetobutylicum* fermentation became a major source of acetone by 1916. Later the production of organic solvents by the petroleum industry slowly eroded the market for the fermentation product, and in 1982 the last operating clostridial fermentation plant, in South Africa, was closed down. However, the advent of genetic engineering has raised the possibility that clostridial fermentation will once again become an important source of acetone.

Lactic Acid Bacteria–Staphylococcus–Bacillus Cluster

Unlike clostridia, most of the organisms in this cluster can be classified as facultative anaerobes, growing in the absence as well as the presence of oxygen. However, their relationship with oxygen varies from that of the lactic acid bacteria, which tolerate its presence but carry out the same fermentative metabolism of sugar regardless of the presence or absence of air, to that of *Staphylococcus* and *Bacillus*, which switch from fermentative to respiratory metabolism in response to oxygen level. Among them, only *Bacillus* still retains the presumedly ancestral capability of forming endospores. Four major groups are found, all of interest to biotechnologists.

Lactobacillus, Pediococcus, and Leuconostoc. These genera form part of the group commonly called lactic acid bacteria. They obtain energy by fermentation of simple sugars such as glucose, producing lactic acid in some *Lactobacillus* species and *Pediococcus*, and lactic acid, ethanol, and carbon dioxide in *Leuconostoc* and the so-called heterofermentative species of *Lactobacillus*. All lack flagella. *Lactobacillus* cells are rod-shaped, whereas *Leuconostoc* and *Pediococcus* cells are round. Because cell shape is one of the basic criteria by which organisms are categorized in traditional taxonomy schemes, *Bergey's Manual* places these genera in widely separated groups, yet ribosomal RNA studies reveal that they are very closely related. They grow best at low oxygen tension, in habitats rich in soluble sugars, peptides, purines, pyrimidines, and vitamins. These bacteria tolerate acidic conditions well and are not inhibited by the drop in pH that accompanies the conversion of glucose to lactic acid. The growth of many other bacteria slows down when pH is low, so they present minimal competition to the lactobacilli under acidic conditions.

Various strains of these genera are used in starter cultures, together with appropriate strains of *Streptococcus*, to produce cheeses and fermented milk products such as butter, buttermilk, and yogurt.

Streptococcus. This is another genus belonging to the lactic acid bacteria. The cells of streptococci are spherical, and they generate ATP by converting glucose into two molecules of lactic acid. Unlike *Leuconostoc* and *Pediococcus*, this genus appears to be only distantly related to the genus *Lactobacillus*. Some streptococci are associated with higher animals, and some are pathogens. Others occur in association with plants. *S. cremoris* is the main organism used for the manufacture of hard-pressed cheeses such as Gouda and Cheddar, as well as soft-ripened cheeses such as Camembert. Streptococci are also important in the production of other fermented milk products. Together with *Lactobacillus* and its relatives, the streptococci account for a world output of dairy products in excess of 20 million metric tons per year, valued at about $50 billion. The strains of streptococci and lactobacilli are supplied to the dairy and meat industries by commercial enterprises that specialize in the production of starter cultures.

Bacillus. These rod-shaped, motile organisms form endospores when conditions are unfavorable for growth. It was the latter property that first called attention to this genus: Robert Koch showed, in classic studies culminating in 1876, that *B. anthracis* was the causative organism of anthrax, the killer of cattle and sheep, and that the long persistence of anthrax infections in certain pastures was due to the resistance of the spores of *B. anthracis* to drying and to prolonged residence in soil. The majority of *Bacillus* strains are harmless saprophytes (organisms that feed on decaying organic matter).

Bacillus strains are all chemoheterotrophic, and unlike the other group of endospore-forming bacteria, *Clostridium*, they can grow in the presence of air. *Bacillus* is not a phylogenetically deep group, again in contrast to *Clostridium*. This is probably to be expected for a group that is likely to have evolved after oxygen became abundant on Earth—a rather recent occurrence in relation to the history of the planet. Many *Bacillus* strains can switch between fermentative and respiratory modes of metabolism; others employ respiration alone. Many are inhabitants of soil, have rather simple nutritional requirements, and grow rapidly in synthetic media. Some strains are thermophilic and grow well at 65–75°C. A number of *Bacillus* species produce extracellular hydrolytic enzymes that break down proteins, nucleic acids, polysaccharides, and lipids. Some of these enzymes are produced commercially in large amounts: The proteolytic enzymes are used in laundry detergents, and the polysaccharide-hydrolyzing enzymes are used in the degradation of starch (Chapter 7). Some species are insect pathogens, and one of these, *B. thuringiensis*, is the only bacterium exploited on a large scale as a biological insecticide (Chapter 6). Antibiotics synthesized by some *Bacillus* strains are produced on a commercial scale—for example, bacitracin from *B. subtilis* and polymyxin from *B. polymyxa*.

Staphylococcus. These bacteria are spherical, nonmotile cells that grow in irregular clusters. They are related to *Bacillus* but do not form endospores. They can switch between fermentative and respiratory modes of metabolism, and they use sugars as the chief source of energy. The main habitat of *Staphylococcus* is the skin of humans and animals; their remarkable tolerance of high salt concentration makes this possible (the drying of sweat is likely to concentrate salt on the skin). Because they are relatively resistant to drying, they are also found in secondary locations such as meat, poultry, animal feeds, and dust and air inside homes. *S. aureus* is the species that causes skin infections, as well as other, more serious diseases, including endocarditis and osteomyelitis.

One of the virulence factors produced by *S. aureus* is called protein A. As is well known, much of an animal's defense against bacterial infection depends on its production of antibodies, proteins with an antigen-binding domain that recognizes and binds to specific structures found on the surface of invading bacteria. Many of the beneficial protective consequences of the binding of antibody to antigen, however, are evoked through conformational changes at the other, nonspecific end of the antibody molecule, which is called the Fc region. By binding tightly to its Fc domain, protein A prevents one class of antibody, the immunoglobulin G class, from causing these effects. Protein A is used extensively for protein purification and analytical procedures, because if an antibody to a molecule one wishes to isolate is available, either protein A or whole-cell preparations of *S. aureus* can be used to selectively bind to the complex formed by the antibody and the target molecule.

Organisms of the Actinomycete Line

One group of Gram-positive bacteria exhibit a rather idiosyncratic set of properties. Their DNA has a high-GC content; most are essentially aerobic soil bacteria with respiratory metabolisms; most of them lack flagella; most are rod-shaped, often slender and long, and have a tendency to divide irregularly and form branched filaments. That these organisms are quite closely related to each other was apparent even to practitioners of the traditional taxonomic methods.

The actinomycetes are nevertheless divided into three major subgroups based on 16S RNA homologies. The first includes organisms with slender, rod-shaped cells, such as *Arthrobacter* and *Cellulomonas*, as well as a genus with spherical cells, *Micrococcus*. The second contains the *Corynebacterium–Mycobacterium–Nocardia* group, which are basically aerobic soil organisms with a very characteristic cell wall: Its polysaccharide, arabinogalactan, is substituted with fatty acids of exceptionally long chain lengths called mycolic acids. Together with *Pseudomonas* species, these organisms are suspected to be very important in the degradation of unusual organic compounds in the soil. Unfortunately, our knowledge of this group is quite limited, except for those atypical species that cause human diseases. The third group consists of *Streptomyces* and its relatives, organisms that grow as clusters of highly branched filaments. We will briefly examine important members from each of the groups.

Cellulomonas. Like some other members of their subgroup, the members of this genus are irregular rods with a respiratory metabolism. The main distinguishing feature of *Cellulomonas* strains is their ability to decompose cellulose. The cellulose-degrading enzymes of *Cellulomonas* have been closely studied in recent years because of interest in the use of cellulose-rich plant matter as a source of feedstock for the production of alcohol and proteins (Chapter 10).

Corynebacterium. One species, *C. glutamicum*, became famous when it was discovered to have the ability to convert a very large fraction of its feedstock into glutamic acid and excrete it into the medium. This process, which involves some remarkable features in its regulation of amino acid biosynthetic pathways, combined with an accidental undersupply of a vitamin (biotin) and of oxygen, is described in detail in Chapter 12. *C. glutamicum* and its relatives appear to have a relatively simple regulatory mechanism for amino acid biosynthesis under ordinary conditions, a mechanism that has been exploited for the production of other amino acids.

Streptomyces. *Streptomyces* strains grow as branching filaments called hyphae, which form convoluted networks called mycelia. As the mycelium ages, filaments called sporophores, or aerial hyphae, form and project above the surface of the colony. The aerial hyphae divide by forming internal crosswalls, and the individual cells mature into spores (conidia). These spores are quite different from the endospores formed within the cells of clostridia and bacilli. Although the streptomyces look very much like fungi both macroscopically and microscopically, they are totally different organisms, *Streptomyces* being prokaryotic.

Streptomyces, like most members of the actinomycete line, is an inhabitant of soil. Several traits have made it successful in this habitat. *Streptomyces* strains degrade polymeric substrates such as polysaccharides (starch, pectin, and chitin), as well as proteins. They have simple growth requirements, and their alternating spore-mycelium-spore life cycle enables them to survive rapid changes in moisture, temperature, and aeration and allows them to be dispersed by wind.

In 1944 Selman Waksman and his collaborators discovered a potent antibacterial substance, streptomycin, released into the growth medium by *S. griseus*. This was the second antibiotic of very high utility to be characterized, soon after the characterization of penicillin. Since then, many other antibiotics have been isolated from streptomycetes. They include tetracycline, erythromycin, neomycin, and gentamicin. The subject is treated further in Chapter 13.

Deinococcus Branch

Until recently, this branch contained only *Deinococcus*, an unusual organism with an extremely high resistance to ionizing radiation. *Deinococcus*, a Gram-positive chemoheterotroph, has been isolated from soil, ground meat, and dust. The cells are bright red or pink because of their high carotenoid content and are surrounded by an outer membrane layer, which is normally absent from Gram-positive bacteria. However, this outer membrane is chemically distinctive in that it does not contain the lipopolysaccharide that is characteristic of the outer membranes of Gram-negative bacteria. The thermophile *Thermus* has now been grouped with *Deinococcus* on the basis of homology in ribosomal RNA sequences.

Thermus. Cells of *Thermus* strains are variable in shape and are nonmotile. The organism was first found in hot springs and is one of the most thermophilic eubacteria known. Most species have an optimal temperature for growth of 70–72°C and can grow at significantly higher temperatures. Their habitat is not limited to hot springs, however. One investigator found that the best source for isolation is the hot water tanks in homes and institutions.

Thermus aquaticus is currently used as the source of a thermostable DNA polymerase (Taq polymerase) that is valuable in the amplification of genes by the polymerase chain reaction. This enzyme has been cloned and expressed in *E. coli* and is manufactured on a large scale.

THE FUNGI

The classification of the Kingdom Fungi recognizes two divisions, the Myxomycota, or slime molds (wall-less fungi), and the Eumycota (true

BOX 2.1 Fungi: Glossary of Pertinent Terms

ascocarp (a type of **sporocarp**) Ascus-bearing structure (see **ascus**), or "fruiting body."

ascus In ascomycetes, the ascospores are produced within the cell wall and membrane of a diploid cell that has undergone meiosis and sporulation; the resulting sac-like structure that surrounds the ascospores is called the ascus.

asexual reproduction Production of progeny identical to the parent by mitotic cell division.

basidium Fungal cell that bears spores terminally and singly in extensions of its wall after karyogamy (see Fig. 2.7) and meiosis.

conidium Type of asexual spore that represents a separate portion of a hypha (see **thallus**).

diploid Having two sets of chromosomes (2x), as opposed to one set (haploid) or more (polyploid).

gametangium (plural gametangia) Cell that produces gametes (a gamete is a sex cell capable of fusing with another gamete, generally of opposite mating type, to form a zygote).

mycotoxins In general, low-molecular-weight fungal metabolites capable of eliciting a toxic response in humans and animals.

phagotroph Organism that ingests solid food particles.

saprophyte Organism that lives on decaying organic matter.

septa Transverse walls dividing hyphae into compartments.

sporocarps Certain fungi reproduce sexually by a process of conjugation that results in the formation of **zygospores.** Structures in which the zygospores occur in clusters surrounded by sterile hyphae are called sporocarps.

thallus The part of a fungus that grows and absorbs nutrients and eventually produces the reproductive part, the "fruiting body." The thallus is composed of microscopic tubular vegetative filaments that branch and rebranch. These vegetative filaments are called hyphae, and a thallus made up of hyphae is termed the mycelium.

yeast Fungus that is mainly unicellular.

zygospore Thick-walled resting spore resulting from conjugation of two cells of opposite sex (mating type).

---

fungi). The evolutionary relationship between the Myxomycota and the Eumycota is distant. Slime molds appear to be more closely related to protozoa and will not be considered further here. We concentrate instead on the Eumycota. This division encompasses an extraordinary diversity of organisms: bread molds, yeasts, powdery mildews, cup and sponge fungi, smuts, rusts, puffballs, and mushrooms. Some are invisible to the naked eye. Others grow to over 2 feet in diameter. Whatever their differences, however, all fungi have certain important properties in common (Box 2.1).

- They are eukaryotic.
- They produce spores by means of sexual and asexual reproduction.
- They grow as hyphae or as yeasts, the hyphae exhibiting apical growth.

- They are heterotrophic and do not perform photosynthesis. Most fungi are saprophytes or symbionts, but some are parasites of humans, other animals, or plants.
- They absorb nutrients through their cell membranes. Phagotrophy (ingestion of solid food particles) is a very rare property among the fungi. To utilize particulate or high-molecular-weight substrates, the fungi secrete various degradative extracellular enzymes.
- They generally have rigid, polysaccharide-rich cell walls.

Classification of the Fungi

Estimates of the number of species of fungi range from 100,000 to 250,000. If the second figure is more nearly correct, the number of fungal species approaches that of all flowering plants. However, fungal taxonomy is not as well developed as bacterial classification, and some strains reported as "new" may have been described earlier under a different name.

The true fungi (Eumycota) are grouped into five subdivisions largely on the basis of differences in the morphology of their reproductive structures, the nature of their reproductive stages, and the composition of their cell walls. The cell walls typically contain 80–90% polysaccharide polymers, and most of the remainder consists of protein and lipid (Figure 2.7). The features typical of the five subdivisions of the Eumycota are described briefly in the following paragraphs.

Mastigomycotina. These fungi produce flagellate asexual spores and can exist in unicellular or mycelial forms. The cell wall is composed either of cellulose (a polymer of D-glucose) and other glucans (as in the class *Oomycetes*) or of chitin, a β-$(1 \rightarrow 4)$ linked polymer of *N*-acetylglucosamine, and glucans (as in the class *Chytridiomycetes*). A notorious representative of this subdivision is *Phytophora infestans*, cause of late blight of potatoes, which invaded Europe from Peru. Between 1845 and 1847, this fungus brought famine to the working-class population of Ireland, who depended on potatoes as their major source of food. Another member of this subdivision, *Rhizophlyctis rosea*, is a commonly encountered decomposer of cellulose in soils.

Zygomycotina. The Zygomycotina produce nonmotile asexual spores (zygospores) formed in a sporocarp. The thallus is usually mycelial and typically aseptate (lacking crosswalls). The cell wall is composed of chitosan (a poorly acetylated or nonacetylated polymer of glucosamine) and chitin. Representative organisms are the soil saprophytes *Mucor* and *Rhizopus*. *Rhizopus nigricans* has long been used in the production of citric acid. *Entomophthora* is an important common parasite of insects such as house flies and aphids.

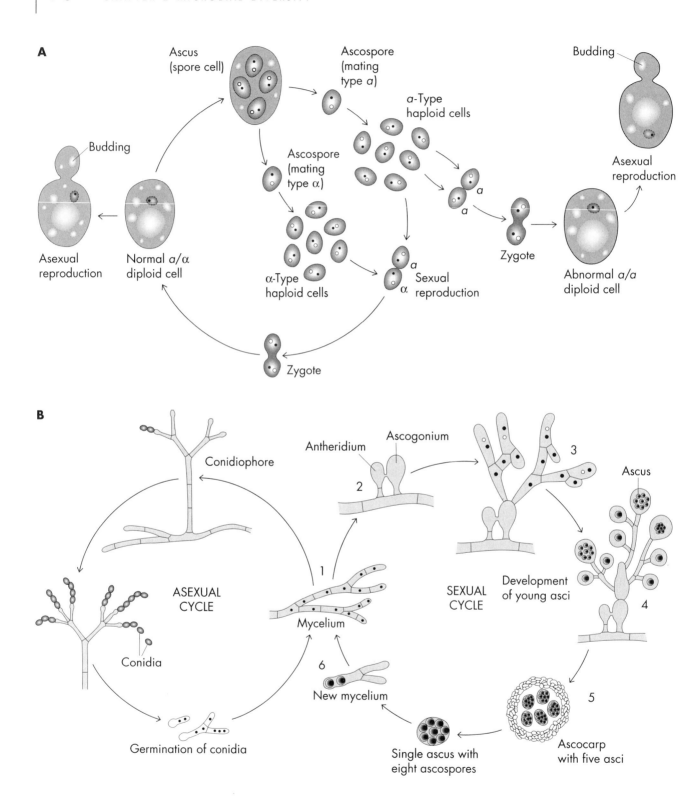

A

Ascus (spore cell)

Ascospore (mating type *a*)

a-Type haploid cells

Budding

Asexual reproduction

Abnormal *a/a* diploid cell

Zygote

a
a

Sexual reproduction

a
α

Ascospore (mating type α)

α-Type haploid cells

Budding

Asexual reproduction

Normal *a/α* diploid cell

Zygote

B

Conidiophore

Antheridium

Ascogonium

3

Ascus

ASEXUAL CYCLE

SEXUAL CYCLE

Development of young asci

4

Conidia

1

Mycelium

2

6

New mycelium

5

Germination of conidia

Single ascus with eight ascospores

Ascocarp with five asci

Ascomycotina. This largest subdivision of fungi contains some 15,000 species. The vegetative structure consists either of single cells (as in yeasts) or of septate (segmented) filaments, most segments containing several nuclei. The cell walls are composed of chitin and glucans (and mannan, in many yeasts). Sexual reproduction leads to the formation of spores in an ascus. Two organisms from this subdivision, *Neurospora* (bread mold) and *Saccharomyces* (baker's and brewer's yeast), are especially familiar to geneticists; others include *Schizosaccharomyces*, *Ceratocystis ulmi* (the cause of Dutch elm disease), and *Erisyphe graminis*, a powdery mildew fungus that parasitizes cereals.

Over 40 species of *Saccharomyces* are recognized. *S. cerevisiae* strains grow on the surface of grapes and other sugar-rich plants. These unicellular organisms multiply by budding. *S. cerevisiae* stocks are used extensively in the fermentation of certain beers and wines, in the production of baker's yeast, and in many biotechnological applications. An

---

Figure 2.7

A. The reproduction of yeast is normally asexual, proceeding by the formation of buds on the cell surface, but sexual reproduction can be induced under special conditions. In the sexual cycle, a normal diploid cell divides by meiosis, and sporulation gives rise to asci, or spore cells, that contain four haploid ascospores. The ascospores are of two mating types: *a* and *α*. Each type can develop by budding into other haploid cells. The mating of an *a* haploid cell and an *α* haploid cell yields a normal *α* diploid cell. Haploid cells of the same sex also unite occasionally to form abnormal diploid cells (*a/a* or *α/α*) that can reproduce only asexually, by budding in the usual way. The majority of industrial yeasts reproduce by budding. **B.** Reproduction of a multicellular fungus, such as one of the higher ascomycetes, can be asexual or sexual. The details vary with genus and species. The branched vegetative structure common to both reproductive cycles is the mycelium, composed of hyphae (**1**). In the asexual cycle, the mycelium gives rise to conidiophores that bear the spores called conidia, which are dispersed by the wind. In the sexual cycle, the mycelium develops gametangial structures (**2**), each consisting of an antheridium (containing "+" nuclei) and an ascogonium (containing "−" nuclei). The nuclei pair in the ascogonium but do not fuse. Ascogenous binucleate hyphae develop from the fertilized ascogonium (**3**), and the pairs of nuclei undergo mitosis, which replicates the newly paired chromosomes. Finally, some pairs of nuclei fuse, a process called karyogamy (**4**), at the tips of the ascogenous hyphae. This is the only diploid stage in the life cycle. Soon afterward the diploid nuclei (large dots) undergo meiosis, or reduction division. The result is eight haploid nuclei (small dots), each of which develops into an ascospore. At the same time, the developing asci are enclosed by mycelial hyphae in an ascocarp (**5**). In the example shown here, the ascocarp is a cleistothecium, a closed structure. Ascospores germinate to yield binucleate or multinucleate mycelia (**6**). After Phaff, H. (1981), Industrial microorganisms, *Scientific American* 245 (Sept.):76–89. Copyright © 1981 by Scientific American, Inc. All rights reserved.

African beer called pombe, made from millet, and an oriental one called arrack, made from molasses, rice, or cocoa palm sap, are both products of fermentation by *Schizosaccharomyces pombe*. *Schizosaccharomyces* yeast divide by binary fission and are termed fission yeasts.

Basidiomycotina. These fungi form sexual spores on a special cell known as the basidium. As in the Ascomycotina, the vegetative structure is either unicellular (yeasts) or a septate mycelium. The cell walls are composed of glucans and chitin. Representatives of this subdivision include *Serpula lacrymans*, a dry-rot fungus causing wood decay; rust and smut fungi such as *Puccinia graminis*, the cause of black stem rust in grasses and cereals; and *Ustilago maydis*, which afflicts corn plants. The *Agaricus* species, the mushrooms most commonly cultivated for human consumption in the Western world, are included among the Basidiomycotina. Mushrooms are grown commercially on organic composts.

Deuteromycotina. Members of this subdivision have a vegetative structure that is either unicellular (yeasts) or a septate mycelium similar to those of Ascomycotina and Basidiomycotina. These fungi are also known as fungi imperfecti, because in them the sexual state is absent, unknown, or lost, and only vegetative reproduction is present, carried out by asexual reproductive structures known as conidia. The polysaccharides of the cell wall are glucans and chitin. This subdivision includes genera of considerable economic importance, such as *Aspergillus* and *Penicillium*.

A. niger is used in the production of citric and gluconic acids. *A. oryzae* is used in the food industry in fermentations of rice and soya products and in the industrial production of proteolytic and amylolytic enzymes. However, some strains of *Aspergillus* are pathogens of plants —for example, crown rot of groundnuts and boll rot of cotton. Infestation of dried fruits, groundnut meal, or peanuts by *A. flavus* may result in the production of aflatoxin B_1 (Figure 2.8), a mycotoxin known to induce liver cancer in humans and poultry. *Penicillium* species grow on all kinds of decaying materials and are cosmopolitan in their distribution. Their spores are almost universally present in the air and frequently contaminate cultures of other microorganisms. In 1928, Alexander Fleming found that a petri dish in which he was culturing staphylococci had become contaminated by a growth of *Penicillium*. He noticed that the growth of the staphylococci was inhibited in the region of the plate close to the fungal colony. Studies stimulated by this phenomenon led to isolation and purification of penicillin and laid the foundation of antibiotic therapy (see Chapter 13; Figure 13.4). The *Penicillium notatum* strain purified by Fleming was the first used for penicillin production, although as a result of intensive screening it has since been replaced by other strains. *P. griseo-fulvum* is a source of griseofulvin, which is given orally to treat fungal infections of the skin or nails. (In fungi sensitive to griseofulvin, the antibiotic binds to proteins

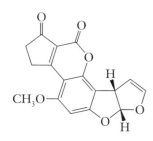

Figure 2.8
Aflatoxin B_1.

involved in the assembly of tubulin into microtubules. This prevents the separation of the chromosomes in mitosis, and hyphal growth ceases.)

Several species of *Penicillium* are important in the food industry. For example, *P. camemberti* and *P. roqueforti* are used in the manufacture of the cheeses that bear their names. Not all *Penicillium* species are sources of benefit, however: *P. italicum* and *P. digitatum* cause rotting of citrus fruit, and *P. expansum* causes a brown rot of apples. *Penicillium* species are also major producers of mycotoxins, common contaminants of grains and other foods. Mycotoxins are secondary metabolites produced by species of several genera of fungi, including *Fusarium* and *Aspergillus*, as well as *Penicillium*.

Yeasts, the Most Exploited of Fungi

Hundreds of thousands of tons of yeast are grown yearly. Many of these unicellular fungi are put to practical use in wine-making, brewing, and baking and as sources of enzymes. Yeasts recovered as by-products of alcohol fermentations are sold for animal feed. *Torulopsis* and *Candida* strains are grown specifically for feed on molasses or on the spent sulfite liquor that is a by-product of paper pulp manufacture. Yeasts that utilize hydrocarbons and methanol are grown for the production of protein. Baker's yeast, *Saccharomyces cerevisiae*, is produced in large amounts.

Yeasts are classified on the basis of (1) the microscopic appearance of the cells, (2) the mode of sexual reproduction, (3) certain physiological features (especially metabolic capabilities and nutritional requirements), and (4) biochemical features (cell wall chemistry and type of ubiquinone present in the mitochondrial respiratory electron transport chain).

The physiological features that distinguish different yeasts include the range of carbohydrates (mono-, di-, tri-, and polysaccharides) that a given organism can use as a source of carbon and energy under semian-aerobic and aerobic conditions, the relative ability to grow in the presence of 50–60% (wt/vol) D-glucose or 10% (wt/vol) sodium chloride plus 5% (wt/vol) glucose (a measure of osmotolerance), and the relative ability to hydrolyze and utilize lipids. These properties help investigators determine which yeast strains merit investigation for a particular application. Thus, as with the bacteria, detailed taxonomic studies of yeasts and other fungi are of considerable importance.

Yeasts grow well at lower pH values than those that are optimal for most bacteria, and they are insensitive to antibiotics that inhibit bacterial growth. Consequently, large-scale cultures of yeasts can be kept free from contamination by fast-growing bacteria. Because of their larger size, yeasts are more easily and cheaply harvested than bacteria. Industrial yeasts in current use do not present public health problems. Given these advantages and with the advent of genetic engineering, the range of applications of yeasts is expanding rapidly.

MICROBIAL DIVERSITY AS A RICH RESERVOIR OF SPECIFIC ENZYMES

We have seen that at all environmental extremes, microorganisms exist whose enzymes and metabolic pathways are optimized to make use of the available nutrients and cope with the physical conditions (temperature, salt concentration) prevailing in that particular habitat. This helps explain why even closely related microorganisms differ in ways that can have important technological implications. Microorganisms meet other environmental challenges as well. Bacteria, for example, have evolved an enzymic defense mechanism against the encroachment of genetic information from foreign DNA molecules that might enter their cells. The following two examples illustrate the incalculable practical value of the diversity in the properties of enzymes that catalyze similar reactions in different microorganisms.

Restriction Enzymes

The number of commercially available restriction enzymes from different bacterial strains approaches 200. Nearly all bacteria that have been examined possess so-called restriction-modification systems, sets of enzymes that enable them to distinguish their own DNA from invading foreign DNA. These systems consist, first, of specific methyltransferases ("modification enzymes") that produce a distinctive pattern of methylation at particular sites along the DNA sequence of the bacterial chromosome and, second, of sequence-specific endonucleases ("restriction enzymes") that can cleave DNA at exactly those same sites. When either DNA strand is methylated at the sequence recognized by a given restriction enzyme, the sequence is not cleaved. Foreign DNA has a different methylation pattern, so it is not protected from cleavage by the restriction enzyme. The majority of restriction enzymes of different bacterial species recognize different palindromic sequences and palindromic sequences of different lengths (Box 2.2).

Restriction enzymes make the manipulations of molecular genetics possible, enabling scientists to cut and splice large DNA molecules at a few precisely known sites. There are bacterial strains whose major claim to fame is that they are a source of a particularly valuable restriction enzyme. For example, restriction endonucleases that recognize a palindromic sequence longer than 6 bases are very rare, but *Streptomyces fimbriatus* strain ATCC15051 produces a restriction endonuclease, *Sfi*I, that cleaves the 13-base sequence GGCCNNNN ↓ NGGCC where N

BOX 2.2 Restriction-Modification Systems and Target Sequences of Restriction Enzymes

The restriction site for the *Hemophilus influenzae* (strain RD) restriction endonuclease *Hind*III, which recognizes the palindromic sequence AAGCTT, is shown below (where N is either a purine or a pyrimidine base).

cleavage site
↓   *Hind*III
5′ N—A—A—G—C—T—T—N—N—3′ → 5′ N—A   A—G—C—T—T—N—N—3′
3′ N—T—T—C—G—A—A—N—N—5′   3′ N—T—T—C—G—A   A—N—N—5′

↑
cleavage site

A shorthand notation for this reaction is A ↓ AGCTT. By convention, only one strand is given, reading from 5′ → 3′. The arrow indicates where the cut on that strand is made. Any DNA molecule with this sequence would be cleaved by *Hind*III. In *H. influenzae* (strain RD), an adenine at the cleavage site on each strand of this sequence is methylated, preventing the site from being cleaved by *Hind*III. Because DNA replication is semiconservative, one strand is always methylated, even during replication.

"old" strand   5′ N—A^{CH_3}—A—G—C—T—T—N—N—3′
"new" strand 3′ N—T—— T—C—G—A—A—N—N—5′

After replication, the specific DNA-methylation activity of the restriction-modification system of *H. influenzae* strain RD brings the DNA to its fully methylated form.

5′ N—A^{CH_3}—A—G—C—T—T——N—N—3′
3′ N—T—— T—C—G—A—A^{CH_3}—N—N—5′

EXAMPLES OF OTHER RESTRICTION ENZYMES

Enzyme	Sequence	Organism
*Bam*HI	G ↓ GATCC	*Bacillus amyloliquefaciens* H
*Eco*RI	G ↓ AATTC	*Escherichia coli* RY13
*Sal*I	GG ↓ TCGAC	*Streptomyces albus* G
*Xba*I	T ↓ CTAGA	*Xanthomonas badrii*

---

is any base. *Nocardia otitidis-caviarum* strain ATCC14630 endonuclease *Not*I cleaves the 8-base sequence GC ↓ GGCCGC. Sequences of such lengths do not occur with great frequency within a molecule of DNA, so enzymes like *Sfi*I and *Not*I are invaluable for mapping large DNA molecules, such as human chromosomes. They make relatively few scissions, keeping the number of fragments manageable. Specificity of restriction enzymes is not predictable from phylogenetic relationships between strains of microorganisms. For example, *N. otitidis-caviarum* strain ATCC14629 produces a restriction endonuclease, *Noc*I, that

Thermostable *Bacillus* α-Amylases

Source	Optimal Temperature (°C)	Optimal pH	pH stability
B. acidocaldarius	70	3.5	4–5.5
B. stearothermophilus	65–73	5–6	6–11
B. licheniformis	90	7–9	7–10
B. subtilis	95–98	6–8	5–11

recognizes the shorter sequence CTGCA ↓ G, which is different from the sequence recognized by *Not*I but the same as that recognized by an enzyme, *Pst*I, from the unrelated bacterium *Providencia stuartii* strain 164.

Amylolytic Enzymes

Sometimes different strains of the same genus produce variants of a given enzyme with corresponding differences in properties that can be of major practical importance. The properties of four thermostable α-amylases from different *Bacillus* species are given in Table 2.4. α-Amylase hydrolyzes α-1,4-glucosidic linkages in starch but does not attack those α-1,4-glucosidic bonds that are adjacent to α-1,6-glucosidic linkages. α-Amylases are used in very large amounts in the initial stages of starch degradation for the production of glucose and maltose and in removing from the final product the starch used in cloth manufacture. In these applications, the enzyme is exposed to elevated temperatures for very short periods of time. The high thermostability and near-neutral pH optimum of the *B. subtilis* α-amylase makes it the enzyme of choice for such applications. For example, a 1000-kg batch of 35% by weight cornstarch slurry is liquefied by a 3- to 5-minute exposure to *B. subtilis* α-amylase at an enzyme/substrate ratio of 0.1% by weight at 105–107°C. The enzyme remains active at the high temperature because of stabilization by the substrate.

AVAILABILITY AND PRESERVATION OF MICROORGANISMS

Users of microorganisms require reliable sources of pure, authenticated cultures. Worldwide, there are over 500 culture collections that make strains of bacteria and fungi available, generally for a modest fee. These collections obtain most of their strains from microbiologists working in universities or research institutes; other strains come from industries that no longer have use for them. Moreover, law now requires that if a process that uses a microorganism is to be patented, a culture of the microorganism must be deposited with a recognized culture collection. In the United Kingdom alone, the national culture collections hold over 27,000 strains of bacteria and fungi. The American Type Culture Collections include over 35,000 strains of bacteria, fungi, yeasts, viruses, and plasmids.

No single preservation procedure is appropriate for all organisms. Microbial cells can be maintained for short periods on slants and stabs of appropriate nutrient-containing agar or stored for longer periods in freeze-dried or other frozen form. For organisms that produce spores, the latter can be preserved in dry form on solid supports. Thus, there are four basic methods that differ in cost and convenience.

- The simplest procedure is to transfer cultures periodically to fresh solid slants of agar in the appropriate medium and to incubate them at a suitable growth temperature. Once the slant cultures are well established, they are kept in a refrigerator at 5°C, enclosed in a container to avoid desiccation. This is the least expensive procedure and it keeps cells viable for many months, but there is danger that mutants or contaminants may accumulate in such cultures.

- Lyophilization (freeze-drying) is a particularly convenient preservation method. Microbial cells are mixed with a medium containing skim milk powder (at 20% wt/vol) or sucrose (at 12% wt/vol) and frozen, after which the water is removed from them by sublimation under partial vacuum. Lyophilized samples remain viable for many years and can be shipped without refrigeration.

- Cells can be stored for prolonged periods of time at liquid nitrogen temperature. In this procedure, the cells are placed in ampules with media containing (by volume) either 10% glycerol or 5% dimethylsulfoxide and are slowly frozen; their temperature is decreased by 1°C to 2°C per minute until it reaches about −50°C. The ampules are then stored at −156°C to −196°C in a liquid nitrogen refrigerator. Additives such as skim milk, sucrose, glycerol, and dimethylsulfoxide minimize damage to the cells by preventing ice crystals from forming during the freezing process.

- Many spore-forming bacteria and fungi can be preserved by slowly air-drying the spores at ambient temperature on the surface of sterilized soil, silica gel, or glass beads.

SUMMARY

The terms *bacteria* and *fungi* describe a huge number of organisms that differ in their sources of energy, cell carbon, and nitrogen; in their metabolic pathways; in the end products of their metabolism, and in their ability to attack various naturally occurring compounds. Only a small fraction of known bacteria and fungi have been studied extensively, and a still smaller fraction put to practical use. Cellular organisms fall into two classes that differ from each other in the fundamental internal organization of their cells. Prokaryotes have no membrane-bounded organelles, whereas eukaryotes contain membrane-bounded nuclei as well as other organelles (mitochondria, chloroplasts) that possess genetic information. Bacteria are prokaryotes, fungi eukaryotes. Prokaryotes fall into two kingdoms: eubacteria and archaebacteria. With respect to many molecular features, the archaebacteria are almost as different from the eubacteria as the latter are from eukaryotes. The archaebacteria include three distinct kinds of bacteria, all found in extreme environments: the methanogens, the extreme halophiles, and the extreme thermophiles.

Living organisms can be subdivided into four classes on the basis of their principal modes of metabolism. Those that use organic compounds as their major source of cell carbon are called *heterotrophs*; those that use carbon dioxide as the major source are called *autotrophs*. Organisms that use the energy stored in chemical bonds to generate ATP are called *chemotrophs*, whereas those that use light energy for this purpose are called *phototrophs*. Various bacteria exhibit one or more of these four modes of metabolism. In contrast, plants utilize light energy for the generation of ATP, use carbon dioxide as the major source of cell carbon, and are exclusively *photoautotrophs*. Fungi and animals use organic compounds as the major source of cell carbon, exploit the chemical bond energy of such compounds for the generation of ATP, and are exclusively *chemoheterotrophs*. Correct identification and classification of bacteria and fungi is important, because the unintentional rediscovery and renaming of previously described organisms represents an unnecessary duplication of effort. And each time a new organism can be placed within a well-studied genus, strong and readily testable predictions can be made about many of its genetic, biochemical, and physiological characteristics. Taxonomy of microorganisms now relies largely on the comparison of genomic DNA sequences. Sequences of slowly evolving macromolecules (ribosomal RNAs) make possible the classification of distantly related microorganisms.

Bacteria that are useful in biotechnology come from many different branches of the phylogenetic tree based on 16S RNA sequences. Plasmids, self-replicating extrachromosomal DNA elements within bacterial cells, sometimes confer highly noticeable phenotypic traits on their hosts and may complicate the classification of the host organism. The classification of fungi is not so far advanced as that of the bacteria. Yeasts are the most widely exploited of the fungi; they are grown in the hundreds of tons in wine making, brewing, and baking and as sources of enzymes. Microbial diversity is a source of an immense variety of enzymes. Such variety is illustrated by the restriction endonucleases, the "molecular scissors" of the genetic engineer. Culture collections make available pure cultures of tens of thousands of different bacterial and fungal strains. Methods have been developed for the long-term preservation of bacteria and fungal spores.

SELECTED REFERENCES

Taxonomy

Balows, A., et al. (eds.), 1992. *The Prokaryotes. A Handbook on the Biology of Bacteria: Ecophysiology, Isolation, Identification, Applications*. Vols. 1–4. Springer-Verlag.

Barnett, J.A., Payne, R.W., and Yarrow, D., 1990. *Yeasts: Characteristics and Identification*. 2d ed. Cambridge University Press.

Holt, J.G. (ed.), 1984–1988. *Bergey's Manual of Systematic Bacteriology*. Vols. 1–4. Williams & Wilkins.

Woese, C.R., 1987. Bacterial evolution. *Microbiol. Rev.* 51:221–271.

Olsen, G.J., Woese, C.R., and Overbeek, R., 1994. The winds of (evolutionary) change: Breathing new life into microbiology. *J. Bacteriol.* 176:1–6.

Hawksworth, D.L., Sutton, B.C., and Ainsworth, G.C., 1983. *Ainsworth & Bisby's Dictionary of the Fungi*. 7th ed. C.A.B. International Mycological Institute.

Microbiology and Microbial Metabolism

Brock, T.D., and Madigan, M.T., 1991. *Biology of Microorganisms*. 6th ed. Prentice-Hall.

Stanier, R.Y., Ingraham, J.L., Wheelis, M.L., and Painter, P.R., 1986. *The Microbial World*. 5th ed. Prentice-Hall.

Gottschalk, G., 1986. *Bacterial Metabolism*. 2d ed. Springer-Verlag.

Demain, A.L., and Solomon, N.A. (eds.), 1986. *Manual of Industrial Microbiology and Biotechnology*. American Society for Microbiology.

Demain, A.L., and Solomon, N.A. (eds.), 1985. *Biology of Industrial Microorganisms*. Benjamin/Cummings.

Mycology

Berry, D.R. (ed.), 1988. *Physiology of Industrial Fungi*. Blackwell Scientific Publications.

Webster, J., 1980. *Introduction to the Fungi*. 2d ed. Cambridge University Press.

Microbial Communities

Bull, A.T., and Slater, J.H. (eds.), 1982. *Microbial Interactions and Communities*. Vol. I. Academic Press.

Culture Collections

Staines, J.E., McGowan, V.F., and Skerman, V.B.D. (eds.), 1986. *World Directory of Collections of Cultures of Microorganisms*. 3d ed. World Data Center, University of Queensland.

MICROBES: LIVING FACTORIES FOR MACROMOLECULES

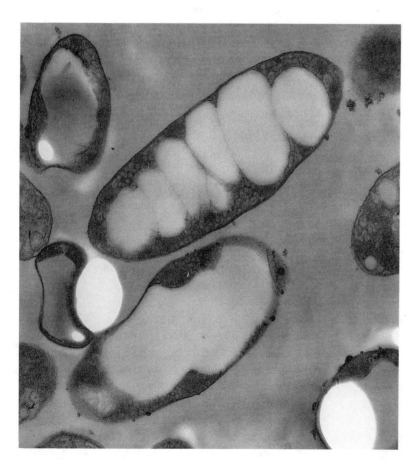

Accumulation of large granular inclusions of polyhydroxyalkanoate in cells of *Rhodobacter sphaeroides* grown on acetate. Magnification: × 35,000. Courtesy of Dr. R. Clinton Fuller.

Production of Proteins in Bacteria

T he human body functions properly only when thousands of bioactive peptides and proteins—hormones, lymphokines, interferons, various enzymes—are produced in precisely regulated amounts, and serious diseases result whenever any of these macromolecules are in short supply. Until 1982, however, the only available pharmaceutical preparations of these peptides and proteins for the treatment of such diseases were obtained from animal sources, and they were sometimes prohibitively expensive. Bioactive proteins and peptides typically occur at low concentrations in animal tissues, so it was difficult to purify significant amounts for medical use. Some important proteins, such as pituitary growth hormone, differ in animals and humans to the extent that a preparation of animal extraction is useless for treating humans. Finally, it was extremely difficult to isolate labile macromolecules from human and animal tissues without running some risk that the products might be contaminated by viral particles and viral nucleic acids.

The introduction of recombinant DNA techniques brought about a revolution in the production of these compounds (Chapter 1). It is now possible to clone a DNA segment coding for a protein and introduce the cloned fragment into a suitable microorganism, such as *E. coli* or the yeast *Saccharomyces cerevisiae*. The "engineered" microorganism then works as a living factory, producing very large amounts of rare peptides and proteins from the inexpensive ingredients of the culture medium. And with such products obtained in this way from pure cultures of microorganisms, there is no chance of contamination by viruses harmful to humans.

For several reasons, bacteria were the first microorganisms to be chosen for use as living factories. To begin with, a great deal was known about their genetics, physiology, and biochemistry. After *Homo sapiens*, the bacterium *E. coli* is the most thoroughly studied and best-understood organism in the living world. Furthermore, it is easy to culture bacteria in large amounts in inexpensive media, and bacteria can multiply very rapidly. For example, *E. coli* doubles its mass every 20 minutes or so in a rich medium. Finally, bacteria are so small that up to a billion cells can fit on a single petri dish only 10 cm in diameter. This permits us to test very large populations in order to find extremely rare mutants or recombinants—an enormous help at many stages of genetic and recombinant DNA manipulations.

INTRODUCTION OF DNA INTO BACTERIA

The field of bacterial genetics grew explosively in the mid-twentieth century, laying much of the groundwork for the development of procedures that efficiently introduce foreign DNA into bacteria. The three basic approaches take advantage of the three modes by which bacteria are known to exchange genetic information. There are two aspects of a genetic exchange: DNA (1) leaves a donor cell and (2) enters a recipient cell. It is the latter process, the uptake of DNA by a cell, that is all-important to biotechnologists.

Direct Introduction by Transformation

Transformation was the first process of genetic exchange to be discovered in bacteria. In 1928, Frederick Griffith injected living cells of *noncapsulated* pneumococcus (*Streptococcus pneumoniae*) together with heat-killed cells of a *capsulated* pneumococcus strain into mice and found that the noncapsulated strain then acquired, presumably from the capsulated strain, the ability to produce a capsule (Figure 3.1). These experiments thus showed that genetic information can be transferred into living bacterial cells from a preparation containing no living donor cells. In 1944, the substance that carried the genetic information in the transformation process was identified as DNA in the famous work of Oswald T. Avery, Colin M. MacLeod, and Maclyn McCarty. This discovery led to the development of modern molecular biology.

We now know of several species of bacteria that, like pneumococcus, have a natural ability to undergo transformation (Table 3.1). The mechanism by which they acquire an exogenous piece of DNA has been studied in great detail in two organisms, pneumococcus and *Haemophilus*

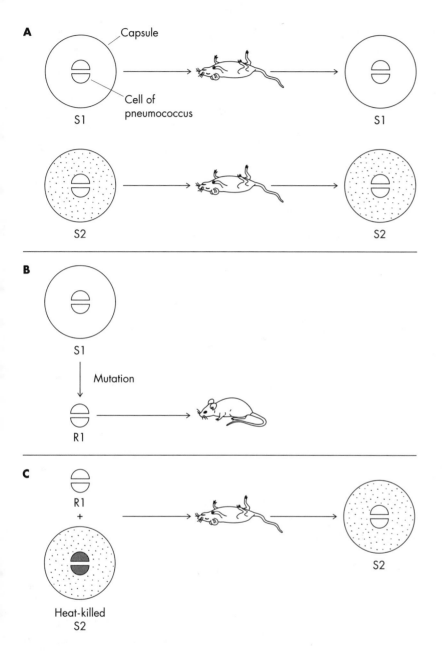

Figure 3.1

Discovery of bacterial transformation by Griffith. **A.** When an encapsulated ("smooth") pneumococcal strain producing either type 1 (S1) or type 2 (S2) capsule is injected into mice, an infection occurs that results in the death of the animals, and more cells of the same strain are recovered from the bodies of the animals. **B.** In contrast, nonencapsulated ("rough") mutants (such as R1, derived from S1) are avirulent and cannot cause successful infection in mice. **C.** However, when a nonencapsu-lated pneumococcal strain (here R1) is injected together with heat-killed cells of an encapsulated strain (here S2), lethal infection ensues, and the bacteria that are recovered from the animals are encapsulated with poly-saccharide of the type present in the killed encapsulated strain (here type 2). These data indicate that the rough mutant strain of type 1 somehow acquired, from the heat-killed cells, genetic information for synthesizing capsule of a different specific type, type 2.

TABLE 3.1

Diverse Groups of Bacteria That Undergo Natural Transformation

Gram-positive bacteria

Streptococcus pneumoniae, S. sanguis
Bacillus subtilis, B. cereus, B. licheniformis, B. stearothermophilus
Streptomyces spp.
Mycobacterium spp.

Gram-negative bacteria

Neisseria gonorrhoeae
Acinetobacter calcoaceticus
Moraxella osloensis, M. urethralis
Achromobacter spp.
Azotobacter spp.
Synechococcus spp.
Haemophilus influenzae, H. parainfluenzae
Pseudomonas stutzeri, P. alcaligenes, P. pseudoalcaligenes,
    P. mendocina

SOURCE: Stewart, G.J., and Carlson, C.A. (1986), The biology of natural transformation, *Ann. Rev. Microbiol.* 40:211–235.

influenzae. In both organisms, DNA is taken up via elaborate machinery produced by the recipient cell, suggesting that the uptake is an active process. The ability to take up DNA, which is called *competence,* is typically developed only under special conditions. In *H. influenzae* the specificity is very strict, and the cells usually take up DNA from the same or related species only. This specificity is achieved by the recognition, on the part of a cell surface protein, of an 11-base repeat sequence found several hundred times in the chromosome of *H. influenzae* but not in chromosomes of other organisms. In contrast, the DNA uptake mechanism operating in pneumococcus is relatively nonspecific as to the sources of the DNA it will accept. Pneumococcus takes up only one strand of DNA, whereas the *H. influenzae* cells take up double-stranded DNA.

The genetics and physiology of naturally transformable species are not well known, however, with the exception perhaps of *Bacillus subtilis.* Thus it was fortunate for biotechnological applications that the best-studied bacterium, *Escherichia coli,* was found to accept exogenous DNA in an artificial transformation process. Here *E. coli* cells are first converted into a competent state by resuspension in buffer solutions containing very high concentrations (typically 30 mM) of $CaCl_2$ at 0°C. The effect of Ca^{2+} on a membrane bilayer with a high content of acidic lipids is to "freeze" the hydrocarbon interior, presumably by binding tightly to the negatively charged head groups of the lipids. Because the outer membrane of Gram-negative bacteria such as *E. coli* (see Figure 2.3) contains a large number of acidic groups at a very high density, this membrane becomes frozen and brittle, with cracks through which macromolecules, including DNA, can pass. After DNA is added to the suspension, the cells are heated to 42°C and then chilled. Under these conditions, cells have been found to take up pieces of DNA through the cytoplasmic membrane, but the molecular mechanisms of the process still remain obscure.

Transformation can be achieved by similar means in certain other bacteria, but there are many species for which this method does not work. Several other methods can be tried on these organisms:

- Enzymes can be used to hydrolyze the rigid cell wall (that is, the peptidoglycan layer) of the bacteria and to convert the cells into protoplasts bounded only by the cytoplasmic membrane. These forms are then able to take up exogenous DNA, although at low efficiency.

- Applying short electrical pulses of very high voltage is believed to reorient asymmetric membrane components that carry charged groups, thus creating transient holes in the membrane. DNA fragments can then enter through these openings, presumably by spontaneous diffusion. This procedure, called *electroporation*, is often quite effective in producing transformation in organisms that are resistant to the classical method and in increasing the efficiency of transformation by many orders of magnitude, even in organisms such as *E. coli* that can be transformed by the classical method. Electroporation has been successfully used to introduce DNA into many bacterial species, including *Mycobacterium smegmatis*, which is covered by a thick, impermeable cell wall, as well as into cells of many fungi, plants, and animals.

- Miniscule metal "bullets" can be coated with DNA and shot into bacterial cells with a gunlike device. This method is described in Chapter 9 (pages 322–323).

Introduction by Conjugation

We have said that it is difficult to introduce DNA directly into certain species of bacteria. In such cases, taking an indirect route sometimes achieves the desired result. First, a piece of DNA is introduced into an organism (such as *E. coli*) that *can* receive DNA by transformation. This piece of DNA is then transferred from the *E. coli* into the species of interest by another form of genetic exchange in bacteria, conjugation.

The *conjugational transfer* of genes in bacteria was discovered by Joshua Lederberg and Edward L. Tatum in 1946. Subsequent work has

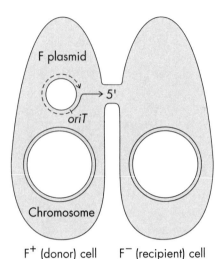

Figure 3.2
Conjugational transfer of the F-plasmid. One of the strands of the F-plasmid is cut at a specific position (*oriT*, for origin of transfer). This strand becomes elongated by rolling-circle replication (broken line), gradually displacing the old part of this strand, which enters into the F⁻ cell 5′-end first. A complementary strand is synthesized in the cytoplasm of the recipient cell, and the plasmid is then circularized, converting the recipient cell from F⁻ to F⁺.

shown it to be a unidirectional transfer from a cell containing a sex plasmid, or F-plasmid (for "fertility"), into a cell lacking that plasmid. The transfer of chromosomal genes by conjugation occurs only in rare donor cells, in which the sex plasmid has become integrated into the chromosome. A more frequent process, which occurs with nearly a 100% efficiency, is the transfer of just the F-plasmid from a donor to a recipient (Figure 3.2). Conjugation requires that the donor and recipient cells join to form a stable pair connected, at least in the beginning, by a filamentous apparatus (sex pilus). The mating pairs are held together weakly, so the culture must not be shaken vigorously. Using high concentrations of donor and recipient cells promotes the formation of mating pairs. For maximal efficiency of mating, the mixed cell suspension may be filtered through a membrane filter to produce a thin layer of immobilized cells on the filter.

As we shall see, the first step in the cloning of a fragment of DNA is to insert it into a suitable *vector DNA*, and plasmids are the most frequently used vectors. However, sex plasmids are *not* used as vectors. If they were, the job of transferring the recombinant plasmids to other strains and species would be easy, because all the proteins needed for such a transfer are encoded on the plasmid itself. But the procedure would also be dangerous, because if a plasmid-containing strain were to escape into the environment, the recombinant plasmid with the foreign DNA could conceivably start to spread into other, naturally occurring bacteria. The current practice, therefore, is to use as vectors only *non-conjugative* or *non-self-transferring* plasmids (plasmids that lack the information for the cell-to-cell transfer). For these plasmids to be transferred by conjugation, the missing information must be supplied from another plasmid. This procedure is called *plasmid mobilization*. It is useful when DNA must be transferred into strains that cannot be made to receive it by transformation.

Injection of Bacteriophage DNA and Transduction

The major problem with the transformation process is its low efficiency. With *E. coli* as the recipient, the usual frequency of transformation suggests that only one out of hundreds of thousands of the exogenous DNA molecules enters the cell. In contrast, when bacteriophage (bacterial virus) infects bacterial cells, *every* virus particle adsorbs to a susceptible host cell and injects it with the DNA contained in the virus head at a very high efficiency, often close to 100%. (The general features of the bacteriophage replication cycle are described in Figure 3.3.) Scientists have been able to take advantage of this natural process to inject foreign DNA into bacterial cells, thanks to a third type of genetic exchange in bacteria, transduction.

In *generalized transduction*, a piece of bacterial chromosome is transferred into a recipient cell by means of a bacteriophage. The chromo-

Figure 3.3

Multiplication of a virulent bacteriophage (bacterial virus) within a bacterial cell. The bacteriophage first adsorbs to a specific structure on the cell surface (step 1). The phage DNA is then injected into the cytoplasm, in some cases driven by contraction of the tail sheath (step 2). Within the cytoplasm, phage DNA and phage capsid (head as well as tail) proteins are synthesized separately (step 3). With most phages, DNA is synthesized as a concatemer containing many repeats of the genomic sequence. Finally, the DNA is cut to the length that corresponds to one phage genome (arrows in step 3) and becomes packaged into phage heads (step 4). The cell is then lysed (step 5). Thus when a mixture of phages and a larger number of host bacterial cells is spread as a lawn on the surface of a solid medium, phages released by the lysis of one cell infect neighboring cells, causing cycles of lysis and infection and finally producing a small area of clearing (a plaque) where most of the host cells have lysed. This course of events occurs with *virulent* phages, which always cause lytic infections. With *temperate* phages, such as lambda or P1, the infection may result in the *lysogenic response*, in which the phage DNA is replicated in step with the host genome without exhibiting the runaway replication of the lytic response. Temperate phages usually produce turbid plaques, because some host cells within the plaques survive as lysogenic bacteria.

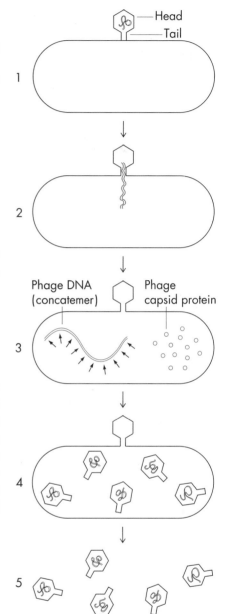

somal DNA gets into the phage head by the mechanism illustrated in Figure 3.4. Once there, the fragment is injected into the cytoplasm of a new host cell in exactly the same way as the phage DNA. The phage head simply injects any DNA it happens to be carrying, regardless of the nature or the source of that DNA. Recombinant DNA technologies utilize this feature of the virus infection process by packaging recombinant DNA into phage heads *in vitro*. The specific vectors used for this type of delivery, phage lambda and cosmids, are described in more detail below.

USE OF VECTORS

Let us assume that we have isolated a fragment of DNA coding for a commercially valuable protein and we want to convert *E. coli* into a factory that produces large amounts of this protein. Our first inclination might be to inject this piece of foreign DNA directly into *E. coli* cells by using one of the methods just described. Unfortunately, that approach would not work. A random piece of DNA floating in the cytoplasm would not be replicated. Only DNA that contains a special *replication origin* sequence is recognized and replicated by *E. coli*, and there is almost no chance that a fragment of foreign DNA will contain such a sequence. It is true that the foreign DNA fragment would be replicated if it got inserted into the bacterial chromosome and became a part of it—that is, if it became successfully "integrated" into the chromosome.

A

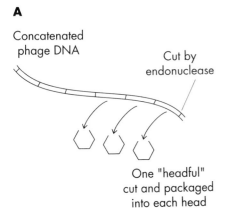

Concatenated phage DNA

Cut by endonuclease

One "headful" cut and packaged into each head

B

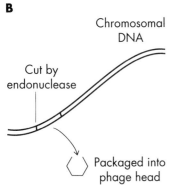

Chromosomal DNA

Cut by endonuclease

Packaged into phage head

Figure 3.4

The generation of transducing phage particles. **A.** In the normal infection cycle, the DNA of such phages as P22 is synthesized as a long concatemer, which is then cleaved by a phage-coded endonuclease at a specific site, the *pac* site. Then each newly assembled phage head becomes packed with a "headful" length of DNA. **B.** If the chromosomal DNA is cut by an endonuclease—perhaps because it carries a sequence similar to that of the *pac* site, perhaps for other reasons—the headful packaging mechanism incorporates fragments of chromosomal DNA into newly assembled phage heads. In *generalized transduction*, these transducing particles subsequently inject such fragments of host DNA into other bacteria, and the DNA recombines with the chromosomes of the recipients to generate *transductants*.

(We rely on a similar process of integration when we introduce fragments of foreign DNA into higher plants and higher animals to create *transgenic* plants and animals.) In bacteria, however, the chromosomal integration of unrelated pieces of DNA is a rare event. Even if our fragment did become integrated into some part of the bacterial chromosome, the genes in the fragment would exist in the cell as single copies only, so they would not be expressed very strongly. Furthermore, the large size of the chromosome would prevent us from manipulating the fragment further—for example, by cutting it out for *subcloning*.

For these reasons, it is usually necessary to insert a cloned foreign gene into a vector—typically a plasmid or phage DNA that is much smaller in size than the bacterial chromosomes—that replicates autonomously in host microorganisms and acts as a carrier of the inserted foreign DNA sequence. There are hundreds of cloning vectors now available, each with its advantages and disadvantages. However, before we discuss the properties of each type of cloning vector, we must start by drawing a general picture of the cloning process itself.

Global Strategy for Cloning

Say that we are going to clone, in *E. coli*, a gene *X* coding for a protein X from a "foreign" organism (that is, an organism other than *E. coli*). The coding region of an average prokaryotic gene is only 1 or 2 kilobases (kb) long. In contrast, the genome of a bacterium has a length of thousands of kilobases, and that of a higher eukaryote a length of millions of kilobases. Thus gene *X* makes up only a small part (1 in thousands to 1 in millions) of the genome. The usual first step in a cloning effort is therefore to clone random segments of the genome (this is often called *shotgun cloning*) so that the subsequent isolation and identification of a clone containing the gene *X*, but not much else, will become possible (Figure 3.5). At this stage, it is advantageous to use vectors that can accommo-

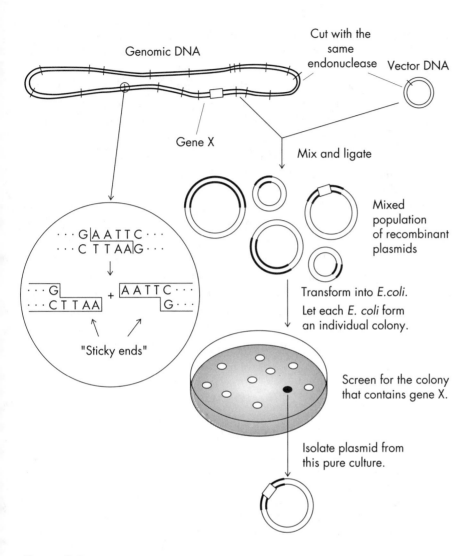

Figure 3.5

Cloning of genomic DNA in *E. coli*. In the first step, one restriction endonuclease is used both to cut and open the vector plasmid DNA and to create fragments of genomic DNA. With most endonucleases, this procedure creates complementary "sticky ends" (see enlargement, here illustrating the ends created by restriction endonuclease *Eco*RI), which facilitate the end-to-end attachment of fragments by the complementary annealing of hanging protrusions. In the second step, the opened vector DNA is mixed with the fragments of the donor DNA. Many of the ends of the donor fragments will then anneal to the open ends of the vector DNA because of the complementary overhanging sequences. Addition of DNA ligase results in the covalent connection between the ends of DNA strands, producing a library of recombinant DNA. In the next step, the recombinant DNA pieces are introduced into *E. coli*, and the bacteria are spread on an agar plate containing a suitable growth medium so that each bacterium will produce a colony—a pure clone—well separated from other colonies. When the vector contains an antibiotic resistance gene, the antibiotic is added to the medium so that only those *E. coli* cells that have received the recombinant plasmid (or the resealed vector plasmid) will grow to produce colonies. Because transformation is a rare event, each clone will contain only one plasmid species. The colony containing the desired gene can then be identified by one of the methods discussed in the text. A pure preparation of the recombinant plasmid, amplified to billions of copies, can now be isolated from this *E. coli* strain, and the fragment can be "subcloned" further in different vectors for the purpose of expression, sequencing, or mutagenesis.

date large DNA fragments, because that dramatically decreases the number of recombinant DNA clones that must be examined in order to find the one containing the gene X (Box 3.1).

The large fragment cloned in this first step—the primary cloning— contains many genes in addition to gene X. Such complex pieces of DNA are not suitable for use in expression, sequencing, or site-directed mutagenesis. This is why it is necessary to pull out a small portion of the DNA, corresponding to only a little more than gene X. This essential step is called *subcloning*, and several different types of vectors are available for the purpose.

Genes from higher eukaryotes usually contain one or more intervening sequences, or *introns*, that do not code for the amino acid sequence of the protein product (Figure 3.6). As a rather extreme example, the gene for thyroglobulin has a size of 300 kb, but that includes 36 introns; the actual coding regions represent only 3% of the total gene length. When RNA transcripts are made from the DNA sequence, they still contain the sequences corresponding to introns. These sequences are then removed from the transcripts by *splicing*, and the mature mRNA molecules that leave the nucleus and enter the cytoplasm do not

---

BOX 3.1 **Fragment Size and the Probability of Finding a Desired Gene in a Set of DNA Fragments**

Let us assume that we use a vector that can accommodate up to 40-kb DNA to clone fragments of a 4000-kb bacterial genome. We use a restriction endonuclease with rare recognition sites so that the genomic DNA is cut into about 100 distinct fragments with an average size of 40-kb. Among these, only one fragment (say, fragment 29) contains the gene X. So when we randomly examine clones to find the one containing gene X, how certain can we be of success? If we had the 100 fragments from the single chromosome of one bacterium in a box, then that set of 100 would certainly contain fragment 29. In actual practice, however, we will be using fragments generated from a mixture of many DNA molecules obtained from billions (or even more) of bacteria. Thus when we pick just 100 fragments of these molecules (or 100 clones), we are likely to have gathered multiple copies of some fragments and no copies of others (possibly including fragment 29). Statistical calculation shows that in order to have a probability P of finding the fragment containing X, one has to examine N clones, which is expressed by

$$N = \frac{\ln (1\text{-}P)}{\ln (1\text{-}R)}$$

where R is the ratio of the fragment size (here 40 kb) to the genome size (4000 kb). If one wants a 99% probability ($P = 0.99$) of fragment 29 being included in the collection, one has to examine 465 clones. This equation shows that if the size of the fragments cloned into vectors is 10 times smaller (4 kb), then the number of clones that must be examined increases to 4500. It is thus advantageous, in a *primary cloning* (that is, in the production of a "genomic library"), to use a vector with a large insert size.

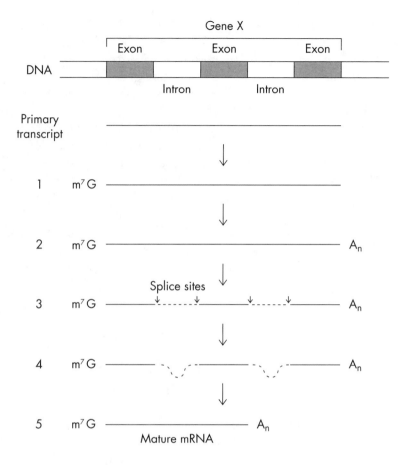

Figure 3.6
The processing of RNA transcripts in eukaryotes. A eukaryotic gene, especially one from a higher animal, is likely to contain many intervening sequences, or *introns*. Primary RNA transcripts of eukaryotic genes are processed first by "capping"—that is, by the addition of 7-methyl-GMP units at the 5′-end through 5′-5′ linkage and by the shortening of the 3′-end (stage 1). A polyA tail is then added to the 3′-end (stage 2). Finally, the RNA sequences that correspond to the introns in the DNA are spliced out (stage 3), producing the mature mRNA.

contain the intervening sequences. The mRNAs are also modified usually at the 3′ terminus by the addition of polyadenylate "tails" (see Figure 3.6).

To determine the nucleotide sequence of a particular gene (say for the purpose of identifying genetic defects in an inherited disease), it is necessary to clone the gene from the genomic DNA so that the intron sequences are included as well. This cloning of intron-containing genes is difficult; luckily, for most biotechnological applications, it is also undesirable. Bacterial DNAs do not contain introns, and bacteria cannot carry out the splicing reactions. In Chapter 4 we will see that even a eukaryotic microorganism such as yeast cannot be relied on to recognize all the splicing signals to be found in the RNA transcripts of genes of higher animals and plants. These eukaryotic genes, therefore, may not be expressed properly in microorganisms. Furthermore, it is often difficult to clone these very large sequences in a single vector molecule. Consequently, in these cases a better template for cloning is usually the mature mRNA, which does not contain the intervening sequences. In such a procedure, the mRNA is first converted to a double-stranded DNA through use of the enzyme reverse transcriptase, which was originally found as a product of RNA viruses (Figure 3.7). Because each eukaryotic mRNA usually contains coding information for only one protein, each of these DNAs, called cDNAs (for "complementary DNA"), also codes for one protein. For this reason, cDNA molecules can then be inserted directly into specialized vectors, such as expression vectors, often circumventing the need for subcloning.

Cloning Vectors

Some cloning vectors are used only for general-purpose cloning, such as the primary cloning and identification of the coding segments. Plasmids are used most commonly for such purposes, although phage lambda-derived vectors and cosmids are advantageous in situations that require the cloning of large segments of DNA. Vectors derived from single-stranded DNA phages are used for sequencing of the cloned DNA and for site-directed mutagenesis. We discuss below some features of these vectors. Expression vectors, which are used for the high-level expression of cloned genes, are addressed later in this chapter (pages 123–127).

Plasmids

One of the first generation of plasmid vectors is pBR322 (Figure 3.8). It is still very frequently used, and many other plasmid vectors have been derived from it by the introduction of additional desirable properties. In the following description, we shall use pBR322 as an example and shall examine the various features that make it a good, general-purpose cloning vector.

The first important feature of this and any cloning vector is the presence of an origin of replication (*ori* in Figure 3.8), obtained for pBR322 from a naturally occurring colicin plasmid (Box 3.2). This origin of replication is recognized by the *E. coli* DNA replication machinery, which then initiates replication of the vector (and its foreign DNA inserts). A second feature of pBR322, and indeed of practically all the plasmid vectors, is the presence of an antibiotic resistance gene. In

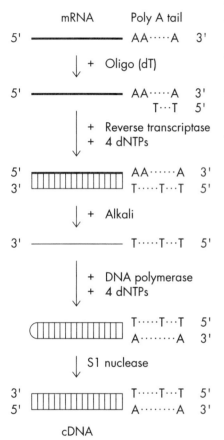

Figure 3.7
Production of cDNA from mRNA. With oligo(dT) as the primer, reverse transcriptase is used to synthesize a single strand of DNA. The template mRNA is then digested with alkali, and DNA polymerase is used to synthesize a complementary DNA sequence on the first strand. Finally, treatment with S1 nuclease cuts the looped end of the DNA, generating a double-stranded cDNA.

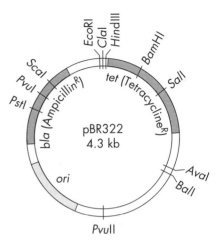

Figure 3.8
Structure of a plasmid vector, pBR322. Note that the vector has only a single susceptible site for each of the commonly used restriction endonucleases, such as *Eco*RI, *Bam*HI, and so on.

fact, pBR322 contains two such genes: *bla*, coding for β-lactamase, which degrades penicillins (including ampicillin) and cephalosporins and thereby produces resistance against these compounds, and *tet*, which codes for a membrane protein that acts as an exit pump for tetracycline, thus producing resistance to tetracycline and its relatives. These resistance markers are needed because, when plasmid DNA is introduced into *E. coli* cells by transformation, only one out of tens of thousands of cells receives a plasmid. Isolating this extremely rare cell would be practically impossible if there were no genetic markers to facilitate its selection (Box 3.3) out of the large excess of cells that failed to acquire the plasmid. Antibiotic resistance is an ideal positive selection marker, because all one has to do after transformation is to spread a large population of cells onto plates containing adequate concentrations of the antibiotics (Figure 3.9). The only cells to survive will be those that have acquired the plasmid, with its resistance genes.

The antibiotic resistance genes also serve a second purpose in pBR322. During the attempt to insert a piece of foreign DNA into a vector DNA that has been opened up by a restriction enzyme (see Figure 3.9), the vector DNA very often recircularizes (closes up again) without incorporating the foreign DNA. This is because unimolecular reactions, which are required for recircularization, occur much more frequently than the bimolecular reactions that are needed for the insertion of another piece of DNA. Reclosure of the vector DNA can be

Figure 3.9
Cloning of foreign DNA segments in pBR322. The vector DNA is cut open by a restriction endonuclease and then treated with phosphatase (see Figure 3.10) in order to prevent its religation. The addition of foreign DNA cut with the same restriction endonuclease results in the annealing of the foreign DNA to the complementary ends of the cut vector. After ligation and transformation into *E. coli*, the cells are plated on a suitable selective medium. In the example shown, the insert was cloned into the *Bam*HI site, thus destroying the *tet* gene. The plasmid-containing cells were therefore selected on ampicillin-containing plates (by using their ampicillin-resistant—Amp^r— phenotype), and the presence of inserts in the plasmids was detected by the inability of certain colonies to grow on tetracycline-containing plates (by using their tetracycline-susceptible—Tet^s— phenotype). This screening can be conveniently accomplished by the replica plating technique (see Box 3.4). When sites within *bla* genes (such as *Pst*I or *Pvu*I) are used in cloning, the tetracycline-resistant (Tet^r), ampicillin-susceptible (Amp^s) cells that contain the recombinant plasmids are selected on tetracycline-containing plates, and the presence of inserts is scored on ampicillin-containing plates.

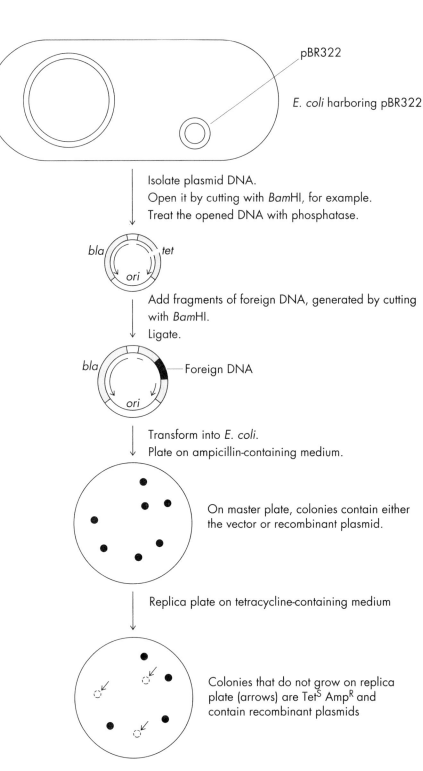

minimized by treating the opened vector with phosphatase (Figure 3.10), but it is difficult to prevent recircularization entirely. Thus it is important to have a quick way of telling, from the phenotype of the transformed strains (transformants), whether the plasmids contain any inserted foreign DNA. Again, the resistance markers in pBR322 provide the needed information. For example, if one opens up the vector DNA by using *Bam*HI or *Sal*I restriction endonuclease (the cleavage sites for which lie within the *tet* tetracycline resistance gene), then the successful insertion of the cloned DNA will interrupt that gene and create transformants that are susceptible to tetracycline (see Figure 3.9). Screening for such transformants can be achieved conveniently by replica plating (Box 3.4). (By selecting for ampicillin resistance, we can still select for transformants that have successfully acquired plasmids.)

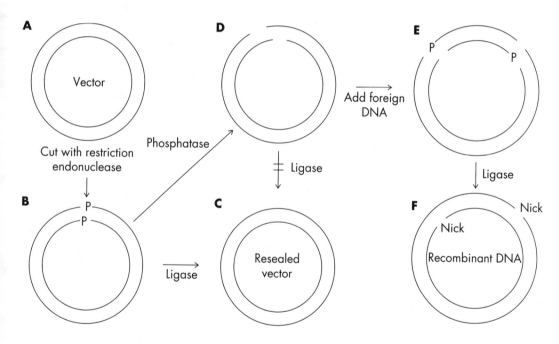

Figure 3.10
Preventing the religation of opened vector DNA. When the vector DNA (**A**) is opened by cutting with a restriction endonuclease, the open ends are usually staggered, with the 5′-phosphate groups still in place (**B**). These 5′-phosphate groups can react with 3′-OH ends of other DNA strands in the presence of DNA ligase, producing closed strands linked with phosphodiester bonds. Without further treatments, it is difficult to use this DNA in the construction of recombinant DNA, because ligation will cause much of the vector DNA simply to reseal (**C**). To prevent this, the opened vector DNA is treated with phosphatase. The treated vector DNA (**D**) cannot reseal on itself, because it lacks the 5′-phosphate groups needed for the formation of phosphodiester bonds. If foreign DNA cut with the same endonuclease is added, its staggered ends, with phosphate groups attached, become annealed with the staggered ends of the vector DNA (**E**). Finally, ligase connects the foreign DNA strands at the end containing the 5′-phosphate (**F**). Although the recombinant DNA created still contains nicks, these are readily repaired once it is transformed into the host cell.

> **BOX** 3.4 **Replica Plating**
>
> This method, developed by Joshua and Esther Lederberg, permits the screening of many colonies in one operation. For example, if we want to screen a population of *E. coli* for their susceptibility to tetracycline, we first spread the population on an agar medium without tetracycline so that a few hundred colonies arise, after incubation, on a single plate (the master plate). The surface of the master plate is lightly "stamped" with a flat, sterile piece of velvet, and then the velvet is momentarily placed on the surface of a new plate that contains tetracycline (the replica plate). From each colony on the master plate, a few cells are transferred onto the replica plate by this operation. After incubation of the replica plate, colonies that exist on the master plate but do not develop at corresponding locations on the replica plate are noted: They correspond to tetracycline-susceptible clones.

The third characteristic of pBR322 that makes it so useful as a cloning vector is that it contains only one site of cleavage for many commonly used restriction enzymes. (The precursor plasmid to pBR322 did contain multiple restriction sites for some of these enzymes, and the extra sites were eliminated by complex operations.) This feature is found in most of the widely used cloning vectors and is very important. If the vector contained, say, three sites for the restriction enzyme *Eco*RI, religation of a mixture containing the three fragments produced from the vector and one fragment of foreign DNA will create many species of recombinant products (Figure 3.11). In contrast, with pBR322 containing a single *Eco*RI site, a large proportion of the product will be the desired recombinant plasmid, containing complete sequences of the vector and the foreign DNA (see Figure 3.11). Commonly the foreign DNA is cut using the same restriction enzyme that is used in cutting the vector. Then all the ends of DNA will have the same hanging protrusions ("sticky ends"), which base-pair exactly with each other, increasing the chance of insertion of the foreign DNA (see Figure 3.10).

Lambda Phage Vectors

As we have seen already, plasmids are convenient vectors. However, they are not ideal for every application. For example, when very large (>20 kb) pieces of DNA are inserted into plasmid vectors, it becomes difficult to introduce the large, recombinant plasmid into a host by transformation. This is a problem when one wants to clone random fragments of genomic DNA in search of a particular gene, because the odds that the gene of interest will appear in any given fragment plummet when the average size of the cloned fragment decreases (see Box 3.1). The need to clone large fragments becomes especially acute when one is working with the genomic DNA of higher animals and plants, because such eukaryotic genes are interrupted frequently by introns, and so only very large pieces of DNA can contain a complete gene.

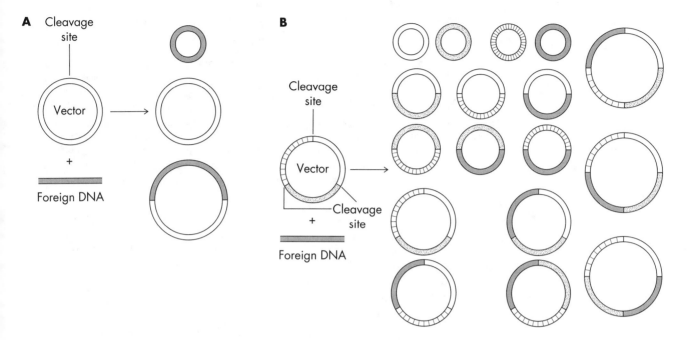

Figure 3.11
Vectors containing single or multiple cleavage sites for a restriction endonuclease. If a vector contains a single cleavage site for an endonuclease (**A**), then annealing and ligation with a segment of foreign DNA produce only three species of circular DNA, one of which is the desired recombinant containing the foreign DNA and the vector sequence. In contrast, if a vector is cut at three places by an endonuclease, annealing and ligation with foreign DNA produce many species of circular DNA (**B**), only a small fraction of which is the desired recombinant species. The situation is far worse in reality, because for simplicity the figure does not show the species in which multiple copies of one fragment are present within a single molecule. Clearly, it is a major disadvantage for a vector to have more than one cleavage site for each of the commonly used endonucleases.

Some of the lambda-derived vectors are more useful than plasmids for this type of situation.

Phage lambda is a well-known temperate bacteriophage (Box 3.5) containing linear, double-stranded DNA. It was originally discovered in some strains of *E. coli* K-12, the standard bacteria used in bacterial genetics. The entire lambda phage genome is 50 kb long, but an 8-kb region of this DNA (the *b*-region, Figure 3.12) has no known function. Another, adjoining region about 7 kb long and containing *att*, *int*, and *xis* (see Figure 3.12) is not needed for the lytic growth (see Box 3.5) of the phage. These two segments can be removed and replaced with a segment of foreign DNA without affecting the phage multiplication. In a lambda-derived vector such as EMBL3 (see Figure 3.12), the insert can be significantly longer (up to 20 kb or even slightly more) than the length of these two deleted lambda fragments (15 kb), for two reasons. (1) Two additional short segments (KH54 and *nin5*), totaling about 5 kb

> **BOX** 3.5   Lytic and Lysogenic Responses in Phage Infection
>
> Bacteriophages are classified as either virulent (such as T4 and T5) or temperate (such as P1, P22, and λ). When a bacterial host is infected by a *virulent* phage, a lytic response is inevitable: The phage multiplies extensively within the cell, which ultimately bursts (lyses) and dies. Infection by a *temperate* phage brings either a lytic or a lysogenic response. In the latter, replication of the phage genome is limited, and the phage genome continues to coexist within the host as a ''prophage,'' either a separate, plasmid-like piece of DNA (as in the case of P1) or a part of the host chromosome (as in the case of phage λ). Many prophages can be ''induced'' to initiate a lytic cycle by inactivation of repressor proteins.

and representing regions not needed for lytic growth, were deleted to create EMBL3. (2) The head of the lambda phage can package a piece of DNA that is slightly longer (by about 2 kb) than the length of the normal lambda DNA.

As with most bacteriophages, lambda phage particles are produced during the last stage of infection: The phage DNA is packaged into proteinaceous phage capsids that have been assembled in the cytoplasm of the infected cell (see Figure 3.3). We take advantage of this packaging reaction in using lambda-derived cloning vectors. In practice, the fragment of foreign DNA is inserted into the vector DNA by cutting with restriction endonuclease, annealing the ends, and then ligating with DNA ligase. When one mixes the recombinant DNA thus produced with a mixture of the proteins that form the phage capsids, the capsid is assembled and the DNA is packaged spontaneously into lambda particles *in vitro*. After packaging, the new phages containing recombinant DNA are used to infect the host bacteria, a process in which DNA enters the bacteria with an efficiency of nearly 100% (rather than 0.001% or less, which is typical of the transformation process). With some vectors, the recombinant DNA may become integrated into the host chromosome as a prophage and can be stably maintained as such until the prophage is induced to initiate the lytic cycle. With others, however, the part of the phage genome that is required for integration has been deleted (for an example in EMBL3, see Figure 3.12), and all infection events result in extensive multiplication of the phage, followed by cell lysis.

Some products of foreign genes are very toxic to the host, and it is difficult to clone such genes by using a plasmid vector, even when the plasmid exists in small numbers of copies per cell (''has a low copy number''). This is because we can isolate and identify plasmid-containing bacterial strains only when the plasmids coexist with the host bacteria for many generations. For the cloning of such deleterious genes, the lambda phage vectors of the nonintegrating type are ideal; with such vectors the infected host cells are soon killed anyway, and the toxicity of

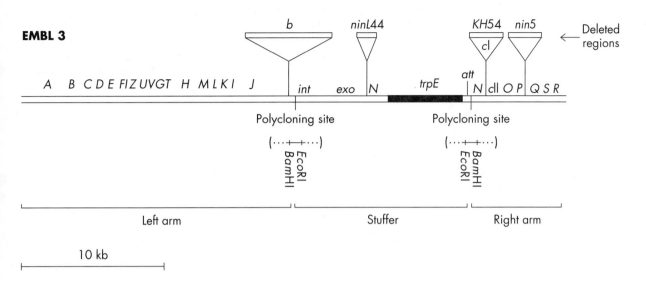

λ **(wild type)**

EMBL 3

Left arm

Stuffer

Right arm

10 kb

Figure 3.12
Phage lambda and the EMBL3 vector. The lambda-based vectors require that both the vector DNA itself and the recombinant DNA be packed efficiently into the lambda phage heads. Packaging demands that the DNA have a length between 78% and 105% of the length of the normal lambda phage DNA, so *replacement vectors* such as EMBL3 contain a *stuffer* segment, which is replaced by a foreign DNA segment of the same or somewhat larger size in the recombinant DNA constructs. More specifically, to create EMBL3, several deletions and one insertion (the *trpE* gene) were made in the lambda genome. The stuffer sequence between the two polycloning sites increases the size of the vector DNA itself so that it is packaged efficiently into phage heads, allowing workers to prepare sufficient quantities of vector DNA by propagating the vector as a phage. The cloning is performed by cutting the vector DNA at the two polycloning sites preferably with *Bam*HI, removing the stuffer fragment, and then ligating the insert DNA (cut partially with *Sau*3A, which generates the same overhanging ends as *Bam*HI) in between the two polycloning sites. The inserted fragment thus replaces the stuffer fragment in the vector, producing recombinant DNA large enough to be packaged into phage heads. Instead of physically removing the stuffer fragment by electrophoresis, one can cut the mixture of the three fragments of vector DNA further with *Eco*RI (there is no other *Eco*RI site in the vector), thereby preventing the stuffer, now with *Eco*RI ends, from becoming religated in the middle of the vector. In the recombinant DNA, the sequences necessary for lysogenic integration into chromosomal DNA (*att* and *int*) are deleted with the stuffer segment. Thus the "phage" particle containing the recombinant DNA can cause only lytic infection of the host. Modified from Sambrook, J., Fritsch, F.F., and Maniatis, T. (1989), *Molecular Cloning: A Laboratory Manual*, 2d ed. (Cold Spring Harbor Laboratory Press).

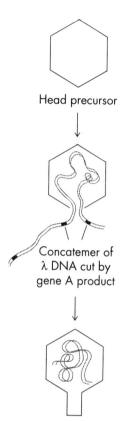

Head precursor

Concatemer of
λ DNA cut by
gene A product

Phage particle completed
by the addition of other
capsid proteins and the tail

Figure 3.13
Packaging of DNA into phage head. Normally lambda DNA is produced as concatemers. An enzyme associated with the phage head (gene *A* product) cuts the DNA at each *cos* site, and the linear DNA is then packaged into the phage particle, together with the tail that has been assembled separately.

> **BOX 3.6   Termination of mRNA Transcription in *E. coli***
>
> The elongation of mRNA is terminated by two major mechanisms. In *rho-dependent termination*, a protein factor, rho, recognizes a complex nucleotide sequence and releases the RNA polymerase from the DNA helix. In *rho-independent termination*, a simpler nucleotide sequence, forming a short loop followed by a succession of U in the mRNA, is recognized by the RNA polymerase itself as the termination signal.

the cloned protein does not make much difference. Lambda-based vectors are also very effective at expressing foreign genes (translating them into protein products), because the promoters found in the lambda genome are very powerful. Moreover, lambda produces an antiterminator protein, N, so that rho-dependent termination of transcription (Box 3.6) can be suppressed. Phage λgt11 is an example of a vector that is useful when the screening of the clones is dependent on the expression of foreign genes.

Cosmids

Lambda DNA is synthesized in the cytoplasm of the infected cells as a polysequence, or concatemer, containing several repeats of the lambda genome. A lambda-coded enzyme recognizes the *cos* (or *co*hesive *s*ite) sequences that correspond to the proper ends of the genome and cuts the DNA at these points, preparing it to be packaged into the head (Figure 3.13). *Cosmid vectors* are vectors that contain lambda *cos* sites but little other material derived from lambda genome. Foreign DNA inserts are cloned between the two *cos* sequences, which then initiate the *in vitro* packaging of the recombinant DNA, composed of the cosmid and its insert, into lambda phage heads. Cosmids also contain a plasmid origin of replication, so that they can be replicated as plasmids, and an antibiotic resistance marker, so that cosmid-containing cells can be selected for (Figure 3.14). Because the cosmid vector is so small (typically only several kilobases), it is possible to clone up to 40 kb of foreign DNA into cosmids and deliver the recombinant DNA very efficiently via lambda phage particles assembled *in vitro*. Because of their ability to incorporate larger pieces of foreign DNA, cosmids are significantly better than lambda vectors for cloning genomic DNA of higher eukaryotes. However, because cosmids have to be propagated as plasmids, it is difficult to use them for cloning genes (or cDNAs) that code for proteins that are very toxic for the *E. coli* hosts.

Derivatives of Single-Stranded DNA Phages

One closely related family of phages (f1, fd, M13) infects only those *E. coli* cells that contain the F sex factor. A remarkable feature of these

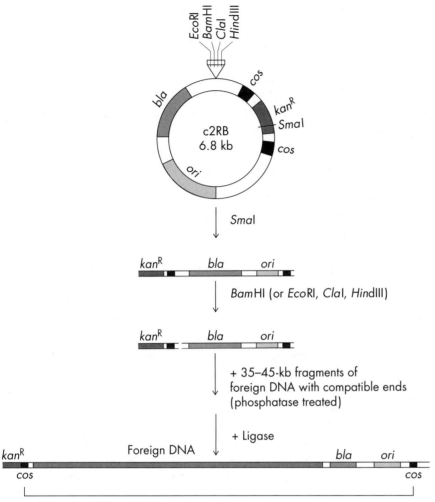

Packaged into λ heads

Figure 3.14
Cloning with a cosmid vector. The cosmid vector c2RB, shown as an example here, contains two *cos* sites, a plasmid origin of replication, a polycloning site (a short stretch of DNA containing cleavage sites for several restriction endonucleases), and two antibiotic resistance markers (Amp[r] and Kan[r]). Cutting the cosmid with, say, *Sma*I and *Bam*HI produces two cosmid halves. Ligation with 40-kb fragments of foreign DNA partially digested with *Mbo*I or *Sau*3A (which produce the ends complementary to those produced by *Bam*HI) creates the construct shown. This is then packaged *in vitro* and introduced into *E. coli*. The strains containing recombinant DNA should be both ampicillin-resistant (because of the *bla* gene) and kanamycin-sensitive, because packaging eliminates the kanamycin resistance gene. The latter feature is useful for eliminating plasmids made of multiple copies of the fragments of the vector. Modified from Sambrook, J., Fritsch, F.F., and Maniatis, T. (1989), *Molecular Cloning: A Laboratory Manual*, 2d ed. (Cold Spring Harbor Laboratory Press).

phages is that they continuously produce progeny phages within the growing cells without causing the lysis and death of the host. These phages contain a circular, single-stranded DNA about 6.4 kb long that is replicated as a double-stranded, plasmid-like entity in the *E. coli* cell. The phage particle itself is filamentous, so insertion of foreign DNA at an intergenic site within the phage DNA simply results in an elongation of the phage particle. In other words, there is no stringent limit on the length of the DNA insert, although larger inserts tend to suffer spontaneous deletions at an alarming frequency.

The merits of these vectors all stem from the single-stranded nature of the recombinant DNA recovered from the phage particles. For example, single-stranded DNA is effective for sequencing by the dideoxy chain termination method (Box 3.7), although methods using double-stranded DNA have recently become much improved. The single-stranded DNA phages are the vectors of choice for site-directed mutagenesis, because one can anneal a chemically synthesized oligonucleotide with a mismatched sequence to a single-stranded recombinant DNA containing the wild-type gene, complete the synthesis of double-stranded DNA, and infect the recipient cells with this material (Figure 3.15).

Two convenient features that were first introduced into M13 vectors (Figure 3.16) are now present in many vectors of other types. The first is a system for distinguishing between recombinant clones and the original vectors. It consists of a fragment of the *lacZ* gene that contains the portion coding for the N-terminal fifth of the LacZ protein. When this truncated LacZ fragment is expressed in a host cell that contains a *lacZ* gene lacking the 5'-terminal part of the gene, both fragments can assemble together to produce a functioning enzyme (α-complementation). Thus, when a cell harboring this vector phage is placed on a plate

BOX 3.7   Sequencing of DNA by the Dideoxy Chain Termination Method

This ingenious method, developed by Frederick Sanger, makes possible the sequencing of fairly long stretches of DNA. The first step is to anneal an oligonucleotide primer to the single-stranded DNA one wishes to sequence. DNA polymerase then synthesizes the complementary strand as a 3' extension of the primer. To each of four such reaction mixtures, one adds low concentrations of an unnatural nucleoside triphosphate containing 2,3-dideoxyribose rather than 2-deoxyribose. This causes chain elongation to stop occasionally, when the unnatural nucleotide is incorporated into the DNA strand. If the template strand contains, for example, C at positions 50, 55, and 60, then the newly made complementary strand becomes truncated when dideoxyguanosine phosphate is incorporated at the corresponding positions. Thus DNA of 50, 55, and 60 nucleotides in length will be made only in the reaction mixture to which dideoxyguanosine triphosphate was added. Analysis of the products by gel electrophoresis, on the basis of their length, thus permits unequivocal sequencing of the DNA.

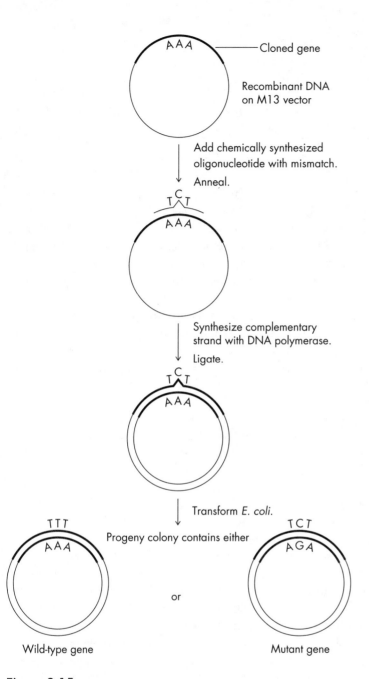

Figure 3.15

Site-directed mutagenesis using an M13 vector. An oligonucleotide with an intentional mismatch (here,—TCT—) is added to a preparation of single-stranded recombinant DNA containing the target gene sequence (here,—AAA—). After the annealing of the oligonucleotide, the rest of the complementary strand is synthesized *in vitro*, and the double-stranded DNA is introduced into *E. coli* by transformation. The phage DNA is replicated in a double-stranded form (replicative form), with each strand of the original DNA serving as a template. Thus one would expect that half of the progeny colonies of *E. coli* would contain the mutated gene (with AGA). In practice, the mutated gene is recovered much less frequently. Various methods have been used to improve the frequency of recovery of the strand containing the mutation. One calls for making the original single-stranded DNA (containing the wild-type gene) so that it contains U instead of T. Transforming the duplex DNA into wild-type *E. coli* cells then results in the preferential degradation of the strand containing the wild-type gene, because it contains uracil.

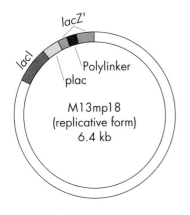

Figure 3.16
An example of M13-derived vectors. The M13mp18 vector contains a polylinker sequence near the 5'-terminus of the sequence coding for a fragment of the *lacZ* gene called *lacZ'*. Precise sequences of the promoter region and one version of the polylinker of this type are given in Figure 3.21.

containing 5-bromo-4-chloro-3-indolyl-β-D-galactopyranoside (X-gal), hydrolysis of X-gal by β-galactosidase (LacZ protein) produces indoxyl, which is oxidized to indigo that stains the colony blue. When a segment of foreign DNA becomes inserted into the cloning site, the coding sequence of the truncated *lacZ* gene is interrupted, the functional N-terminal LacZ fragment is not produced, and the colony stays white. (In principle, the same effect can be achieved by inserting the whole *lacZ* gene in the vector. However, *lacZ* is a large gene, and introducing large DNA fragments makes the recombinant M13 construction rather unstable.) The second feature is the insertion, close to the beginning of the *lacZ* gene, of a short sequence called *polylinker*, or *multiple cloning site*, designed to contain cleavage sites for many popular restriction enzymes. This sequence serves as a convenient site of insertion of foreign DNA. Its proximity to the efficient *lac* promoter ensures good expression of the cloned gene, as long as the gene is in the correct orientation and there are no transcription termination sites between the *lac* promoter and the foreign gene. (The advantage of this construction for gene expression is further discussed in the section dealing with expression vectors).

Recently, *phagemids* have become popular cloning vectors. These chimeric vectors contain two origins of replication, one from a plasmid and the other from f1 or some other phage. These vectors multiply as plasmids, not as phages, in the host cell, because they lack genes needed for replication as phage DNA and for the assembly of phage particles. However, once the missing phage functions are supplied by superinfecting the host with helper phages, they are replicated as phage DNA, packaged, and released into the medium as phage-like particles. Before the invention of phagemids, the usual method for sequencing involved first cloning in plasmid vectors and then recloning the fragment in single-stranded DNA vectors. With phagemids, the tedious recloning step is unnecessary. One example of these vectors is pBluescript, which is cut out from λZAP as shown in Figure 3.17.

More complex are the *diphagemids*, with two phage origins of replication. λZAP (see Figure 3.17) is an example of this class and contains both the lambda origin and the f1 origin of replication, as well as a plasmid origin. Most of its DNA sequence (about 40 kb) comes from phage lambda. Up to 10 kb of foreign DNA can be inserted into this vector, and the recombinant DNA can be packaged *in vitro* into phage particles and injected into *E. coli* host cells, as described earlier. This vector contains sequences needed for integration of lambda DNA as a prophage into the *E. coli* chromosome, so the cloned sequence is maintained as a part of the prophage. Thus for primary cloning, we utilize the lambda properties of this vector. But because the cloning sites of this vector are in the middle of the phagemid pBluescript sequence, which is contained within the larger λZAP vector, simply superinfecting the cell with helper f1 phage excises the phagemid sequence containing the insert, replicates the sequence as single-stranded phage DNA, and pack-

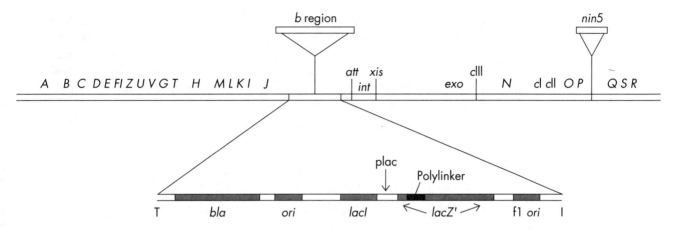

Figure 3.17

Map of λ-ZAP. In this vector, much of the *b* region as well as the *nin5* region is deleted, as in the other λ-based vectors (see Figure 3.12). The important feature in this vector is the insertion, between the J gene and the *att* site, of a sequence of the phagemid (a hybrid of f1 phage DNA and a plasmid, described earlier) vector pBluescript M13-. This region begins with the initiator (I) and ends with the terminator (T) signals for replication by the polymerase coded for by a single-stranded DNA phage f1 (or M13). Because of this, when *E. coli* is infected with λ-ZAP-derived recombinant phage together with an fl (M13) helper phage, this region becomes excised and replicates to produce single-stranded DNA molecules that are packaged as f1-like phage particles and that serve as a good source of single-stranded DNA for use in sequencing. Bacteria exposed to these f1-like particles also perpetuate the recombinant DNA as double-stranded, circular plasmids, because the pBluescript sequence contains a plasmid origin of replication (*ori*). These recombinant plasmids are suitable for the overexpression of the cloned genes. Other convenience features of λ-ZAP include the presence of multiple cloning sites within the polylinker, which is inserted near the 5'-terminus of the coding region for a part of the *lacZ* gene. This type of construction was originally developed in M13 vectors (and later in pUC plasmid vectors) by Joachim Messing and associates. Its use with single-stranded phage vectors is shown in Figure 3.15, and a detailed example is given in Figure 3.21. Based on Sambrook, J., Fritsch, F.F., Maniatis, T. (1989), *Molecular Cloning: A Laboratory Manual*, 2d ed. (Cold Spring Harbor Laboratory Press).

ages the DNA into f1-like particles. In this manner, one can directly obtain the single-stranded DNA containing the insert, suitable for sequencing or site-directed mutagenesis, without going through any recloning steps. Finally, if *E. coli* host strains are infected with these f1-like particles, the recombinant DNA is injected into the cells, is made into a double-stranded, circular DNA, and is replicated as plasmids by utilizing the plasmid origin of replication (just like the recombinant plasmids made with pBluescript SK vector). This situation is beneficial if high-level expression of the cloned genes is desired, because plasmids are usually present in high numbers of copies—in contrast to λZAP, which exists as a single copy on the chromosome. Without the complex vectors of this type, separate recloning steps by restriction digestion, purification of the fragments, ligation, and transformation are required for obtaining high-level expression of the cloned gene and for DNA sequencing.

DETECTION OF THE CLONE CONTAINING THE DESIRED FRAGMENT

Cloning fragments of genomic DNA or cDNA is not a difficult task. Many restriction endonucleases are commercially available, as are numerous vectors of sophisticated design, such as those we have described. Usually the most challenging step in cloning is the detection, among many clones, of the ones that contain the fragment of interest. The magnitude of this task becomes clear when we realize that even with vectors that can accept a 20-kb piece of DNA, and even when the source is a bacterial genome (5000 kb), to have a 99% probability of finding 1 clone with the desired gene, we have to examine about 1000 clones (see Box 3.1). The thought of attempting the same task with the genome of a higher eukaryote, which could be almost three orders of magnitude larger than that of *E. coli*, is daunting to say the least. One would have to examine almost a million recombinant clones in order to be 99% certain of recovering the gene of interest. Thus careful strategic planning becomes necessary for the identification of desired clones.

Importance of Using a Better Template

If one wants to express eukaryotic proteins in bacteria, which cannot carry out the splicing reaction, it is best to use mRNA as the template, because the sequences corresponding to the intron sequences are already spliced out, as described earlier. In such cases, the usual procedure is to obtain the specific types of cells in which the target gene is expressed strongly and then to use the mRNA from those cells as the source of genetic information. This approach exploits a source in which the sequence of interest has been very strongly amplified. Large amounts of stable RNA (ribosomal RNAs, transfer RNAs, and so on) are also present in any cell, but they can be removed easily by taking advantage of the fact that eukaryotic mRNA molecules have a polyA "tail" at the 3'-terminus. Once the mRNA fraction is isolated, it can be purified according to size in order to obtain a fraction enriched in the sequence of interest. The mRNA is then converted into double-stranded cDNA as described above (see Figure 3.7) for insertion into a cloning vector. Most of the sequences coding for the production of animal and human peptides and proteins have been cloned by using mRNA preparations.

Clone Identification Based on Protein Products

When a cloned gene is expected to be transcribed and translated in the host bacterium (see the discussion of expression vectors that follows), the task of identification, and perhaps even selection (see Box 3.3), of the cells containing the right clone is fairly straightforward. In the simplest case, one can test for the function of the protein coded by the cloned gene. Let us assume that we want to clone from some organism (call it Organism A) the gene for anthranilate synthase, an enzyme involved in the synthesis of tryptophan, for the purpose of improving the commercial production of this important amino acid (see Chapter 12). *E. coli*, like most bacteria, can synthesize all the usual amino acids from simple carbon sources and ammonia, and it contains a gene, *trpE*, that codes for anthranilate synthase. We first introduce a mutation into the *E. coli trpE* gene. The mutant strain cannot synthesize tryptophan and thus cannot grow unless we add tryptophan to the growth medium. We now introduce into this strain, by transformation, recombinant plasmids containing fragments of the DNA of Organism A and spread a large number of transformant cells on a solid medium that does not contain tryptophan. Most of the cells contain either no plasmid or plasmids with irrelevant pieces of DNA and are unable to grow. The only cells that grow and form visible colonies are those that contain the rare recombinant plasmid with the *trpE* homolog from Organism A. In this manner we achieve an efficient selection of these rare plasmids.

In the foregoing example, the desired gene had a function required in many microorganisms. In some cases, however, the desired gene would have a significant function in the source organism, Organism A, but not in *E. coli*. An example is an attempt to clone a gene coding for one of the enzymes of the xylene degradation pathway from *Pseudomonas putida*. Many strains of this organism contain a series of enzymes that lead to the complete oxidation of an aromatic hydrocarbon, xylene, but one of these enzymes can perform no useful function in *E. coli*, which does not contain any other enzymes of this series. *Shuttle vectors*, which contain origins of replication of both *E. coli* and some other microorganism, are useful in such situations. We can then screen for the clone that expresses the desired function in Organism A, because the recombinant plasmids will be replicated in this organism. At the same time, we can propagate the plasmids in *E. coli*, where subcloning and other procedures can be carried out more easily.

In many cases, though, a *complementation assay* such as the one described would be difficult or even impossible to perform. For example, if we are trying to clone a eukaryotic gene coding for a hormone that has no homologs in unicellular bacteria, complementation cannot be used as a method of detection. A frequently used approach in these cases is to detect production of the desired protein by its reactivity with specific antibodies. Unfortunately, this usually involves screening, rather than selection, of the recombinant clones. However, if the screening can be carried out on plates with hundreds of colonies on each, it is not so difficult to test tens of thousands of recombinant clones in a single experiment. Lambda phage vectors are especially convenient for this method, because within each plaque (see the legend of Figure 3.3) generated by lytic infection by a recombinant phage or by induced lysis of an *E. coli* strain lysogenic for a recombinant phage, the cells will have been lysed already, releasing into the medium the proteins expressed from the recombinant fragment. Furthermore, because a single lysing cell contains hundreds of copies of the phage genome, each including the cloned piece, the expression of the cloned genes is strongly enhanced. The λgt11 expression vector was especially constructed for screening of this type.

Clone Identification Based on DNA Sequence

The methods we have examined depend on successful expression of the cloned genes. But this is not always assured, especially when the cloned DNA comes from a source phylogenetically distant from the bacterial host. The RNA polymerase of the host bacteria does not recognize the promoter and other regulatory elements of eukaryotic genes, nor even those of remotely related bacteria. Pieces of cDNA lack such regulatory "upstream" sequences altogether, and genomic DNA from eukaryotes will not result in the production of whole proteins because of the presence of introns.

Because of these problems, it often becomes necessary to identify the clone containing the desired fragment by its DNA sequence. Scoring for such clones can be done by hybridization with suitable DNA probes, labeled either with a radioactive isotope or with chemical substituents that can be detected by nonradioactive methods, such as fluorescence. The major hurdle in this procedure is finding the requisite DNA probe, especially when the exact sequence of the clone is not yet known. This is not an insurmountable problem, however. If the sought-after gene has homologs in related organisms, and if their sequences are known, it is possible to design probes that correspond to the most conserved regions of the aligned sequences and use conditions of low stringency for hybridization. In fact, this is probably the most frequently used method for cloning genes and cDNAs from eukaryotes, because the evolutionary divergence between higher eukaryotes tends to be quite small in comparison with that between prokaryotic groups. Alternatively, if at least a partial sequence of the protein is known, one can deduce the DNA sequences that would code for such an amino acid sequence and use a "degenerate" probe that contains a mixture of these possible DNA sequences. Using this approach is particularly advantageous when the peptide sequence does not contain amino acids that, like leucine or arginine, are coded by many codons.

In practice, the cells containing recombinant plasmids are spread on plates so that there will be a few hundred colonies per plate. These colonies are replica-plated onto a filter and placed on a fresh plate. After the cells have grown, the filter is lifted out and treated with a NaOH solution to lyse the cells and denature the DNA. The proteins are digested by a protease, and the DNA is fixed onto the filter by "baking" at 80°C. The filter is then incubated with the labeled probe DNA, and any probe that anneals to the DNA on the filter is detected after suitable washing (Figure 3.18). Although this is only a screening method, in this way one can test a fairly large number of colonies in a short time.

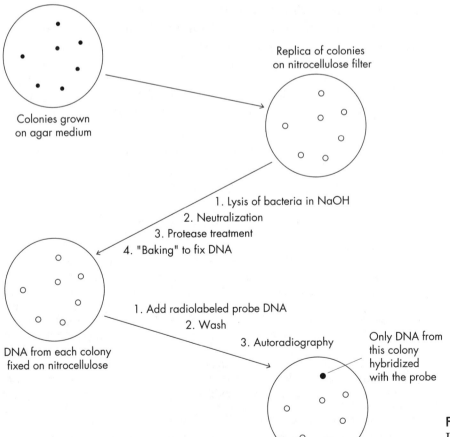

Colonies grown
on agar medium

Replica of colonies
on nitrocellulose filter

1. Lysis of bacteria in NaOH
2. Neutralization
3. Protease treatment
4. "Baking" to fix DNA

DNA from each colony
fixed on nitrocellulose

1. Add radiolabeled probe DNA
2. Wash
3. Autoradiography

Only DNA from
this colony
hybridized
with the probe

Figure 3.18
Use of a DNA probe to detect the desired recombinant clone.

Combined Detection of the DNA Sequence and the Protein Product

In some cases, it is possible to combine the two methods we have outlined. For example, to clone gene X from a bacterium very distantly related to *E. coli*, one might begin by making random transposon insertions (Box 3.8) in the chromosomes of the bacterium containing gene X. If a transposon inserts into gene X, it will disrupt this gene, generally resulting in a recognizable phenotype. Genomic DNA from this mutant organism is then cloned into a plasmid vector, and the recombinant plasmids are introduced into an *E. coli* host strain by transformation. Most transposons contain a resistance marker that codes for antibiotic resistance or resistance to other toxic compounds, such as mercury (Figure 3.19). Moreover, because a transposon is a piece of "selfish" DNA that propagates itself in diverse species of bacteria (see Box 3.8), its resistance genes are designed to be expressed efficiently in many bacterial species. Thus the resistance gene, located within the transposon in the recombinant plasmid, will be expressed in the *E. coli* host, and it should therefore be possible to select for this plasmid.

A clone of gene X still exists in two pieces within the plasmid, flanking the transposon. The cloned DNA can be cut out of the plasmid, and the fragments of gene X DNA can be used as probes in the next phase. One then clones random fragments from the wild-type genome (which does not contain the transposon) into a plasmid or other vector. Screening of these recombinant clones with the DNA probes of gene X described above will lead to identification of the clone that contains the wild-type version of the gene, uninterrupted by the transposon sequence.

BOX 3.8 Transposons

A transposon is a segment of DNA that has the ability to insert its *copy* at random sites in the genome. Thus a transposon has a natural tendency to increase the number of its copies under favorable conditions and can be considered an example of a piece of "selfish DNA." It is bounded by inverted repeat sequences, and it usually contains a drug resistance gene. It also contains gene(s) for enzymes that catalyze the transpositional insertion. Because of these properties, transposons played a major part in the natural formation of "R plasmids," which code for resistance to many antibacterial agents. Transposons are also very useful as genetic tools. For example, introduction of a transposon into a new cell results in the insertion of copies of the transposon at various places on the chromosome. Insertion of such large fragments in the middle of a gene totally inactivates the gene, so *transposon mutagenesis* is a convenient method for generating null mutants (mutants in which the gene function is utterly obliterated). Furthermore, the mutations are easy to analyze genetically, and the alleles can be easily cloned, because of the presence of antibiotic resistance markers within the transposon sequences.

Tn 3 (4.96 kb)

Figure 3.19
An example of a transposon, Tn3. The genes *tnpA* and *tnpR* code for trans-
posase (cointegrase) and resolvase, two enzymes needed for the insertion of a
copy of the transposon into a new site on the DNA duplex. [For the mechanism, see Grindley, N.D.F., and Reed, R.R. (1985), Transpositional recombination in prokaryotes, *Ann. Rev. Biochem.* 54:863–896.] The gene *bla* codes
for a β-lactamase, which produces resistance to β-lactam antibiotics such
as ampicillin and cephaloridine. The ends of the transposon contain inverted
repeats (arrows).

POLYMERASE CHAIN REACTION (PCR)

In cases where we know at least short stretches of nucleotide sequence
either within the gene of interest or in an area flanking that gene, it is
possible to isolate the desired clone without going through the painstaking cloning procedures we have described. This is done with a technique known as polymerase chain reaction (PCR), and in 1993 its in-
ventor, Kary Mullis, received a Nobel Prize in chemistry for devising it.
In this method, we first synthesize oligonucleotide primers that are
complementary to opposite strands of these short stretches of DNA
(Figure 3.20). We then add a large excess of these primers to a denatured preparation of genomic DNA or cDNA and let the primers anneal
to the complementary sequences. Adding a heat-resistant DNA polymerase and a mixture of deoxyribonucleoside triphosphates results in the
elongation of primers into complementary strands of DNA, as shown in
Figure 3.20, step 1. The mixture is next heated to denature the DNA
again, and then it is rapidly cooled. Under these conditions, annealing
occurs predominantly between primers and DNA strands (some of
which are newly synthesized ones), rather than between long DNA
strands, because the latter takes place more slowly. Because the polymerase is heat-resistant, DNA synthesis begins again by utilizing the
primers (Figure 3.20, step 2). In the first round of DNA synthesis, the
newly made DNA strands have random ends. In the second round, the
mixture becomes enriched for strands that begin and end at sequences
corresponding to the two primers, because some primers have annealed
to the strands synthesized in the first round (Figure 3.20, step 2). After
the DNA synthesis and DNA denaturation/annealing steps are continued for many cycles, most of the newly made strands will have a finite
length and will correspond only to the limited region of the DNA

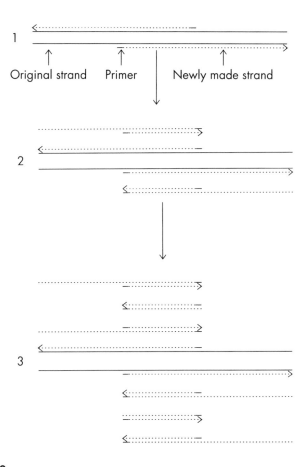

Figure 3.20
Amplification of a defined segment of genome by PCR. In step 1, primers
(short lines) are annealed to complementary sequences of genomic DNA
(continuous lines). Addition of DNA polymerase and deoxyribonucleoside
triphosphates results in elongation of the primer (dotted lines). In step 2,
the reaction mixture is heat-denatured and then renatured, causing some of
the primers to anneal to newly made strands (dotted lines). Elongation produces new strands, two of which now have a limited length, terminating at
positions corresponding to the primer sequences. In the third cycle, after
denaturation, renaturation, and elongation again, 8 out of the total of 16
strands have this limited length. After further cycles, practically all of the
newly made DNA will have the finite, short length.

between the two primers—that is, to the gene of interest if the primers
corresponding to the regions flanking the gene were used.

The crucial factor for the success of the polymerase chain reaction
was the discovery of a thermostable DNA polymerase that can withstand many cycles of heating and cooling. In theory, the usual heat-

labile enzyme should suffice if it is added fresh at the beginning of every cycle; however, a large number of cycles are required to achieve a high degree of amplification, and impurities brought in each time the enzyme is added eventually inhibit the reaction. The heat-stable enzyme commonly used is derived from a thermophilic Gram-negative eubacterium, *Thermus aquaticus*, which grows optimally at around 70–80°C. Lately, even more thermostable DNA polymerases, isolated from archaebacteria living in marine thermal vents at temperatures of 98–104°C, are being used in PCR procedures.

The PCR procedure offers important advantages. As we have seen, one can totally circumvent the complicated cloning steps as well as the steps involved in identifying the clone that contains the desired gene. Moreover, because the degree of amplification is very large, the process is extremely sensitive. In theory, this procedure could amplify a single copy of a gene, and in many experiments the results have approached this limit. Of course, PCR has potential disadvantages as well. First, errors occur during DNA synthesis, and if they occur in the early cycles, the amplified DNA will differ in sequence from that of the original template. The frequency of error, however, is reported to be significantly lower with the more thermostable archaebacterial enzymes. Second, PCR depends on knowing the sequence either within or around the gene of interest. In a procedure used commonly to circumvent this problem, the protein product is isolated and the amino acid sequence of the N-terminus of the whole protein is determined, as are the sequences of internal peptides generated by the enzymatic or chemical cleavage of the protein. Primers are then made on the basis of these amino acid sequences. These primers are mixtures containing various degenerate codons. PCR using these primers will amplify a DNA sequence that corresponds to a portion of the gene coding for the protein. This amplified sequence can then be used as a probe to identify the clone containing the entire gene.

EXPRESSION OF CLONED GENES

The usual reason for cloning a gene is to obtain the protein product in substantial quantities. Even when that is not the case, identification of the correct clone often requires expression of the cloned gene. However, many of the general-purpose cloning vectors are not designed for strong expression of cloned genes. There are many reasons why genes in the fragments cloned into pBR322, for instance, are often expressed only at a low level. Frequently, the foreign promoter in a fragment from another organism is not efficiently recognized by *E. coli* RNA polymerase. In such a case, successful transcription must start from promoters recognized well by *E. coli*—those for the *tet* and *bla* genes—and continue onto the cloned segment of the recombinant plasmid. The problem is that there may be sequences in between that act as transcription terminators. Even if the mRNA is successfully produced, it may not contain the proper ribosome-binding sequence (see below) at a proper place. These difficulties indicate that a different arrangement is needed to ensure a high level of expression of foreign genes in a reproducible manner.

Vectors of a special class, called *expression vectors,* are designed for this purpose. Most expression vectors are plasmids, because multiple copies of plasmids can exist stably in the cell. More plasmids carrying a given gene means a higher production of specific mRNA, because each copy of the gene is transcribed independently. This principle is sometimes called the *gene dosage effect.*

Expression vectors must have a strong promoter. *E. coli* promoters contain two "consensus" sequences: TTGACA, about 35 nucleotides upstream from the transcription start site, and TATAAT, about 10 nucleotides upstream. These two sequences are therefore separated by a 16- to 18-bp intervening region. When the vector's promoter has sequences that closely resemble the host bacteria's, the genes downstream of it tend to be expressed strongly. Strong promoters are also found in phage genomes, because phage life cycle depends on its proteins being produced in very large amounts during the short period of phage infection. Lambda promoters pL and pR are often used, and there are reports that these promoters express the downstream foreign gene at least several times more strongly than cellular promoters such as pLac.

Strong promoters introduce a problem, however. When *E. coli* cells produce a very large amount of a protein that not only does not contribute to cell growth but also tends to be deleterious, cells that have lost the plasmid, and cells whose plasmids have been altered and have ceased to produce the protein, have a competitive advantage and will eventually become the predominant members of the population. This instability can be a severe problem in industrial-scale production, because extensive scale-up means that *E. coli* must go through a proportionately larger number of generations, significantly increasing the likelihood that non-producing cells will appear. For this reason, it is usually preferable to use promoters whose expression can be regulated, so that the production of the foreign protein can be delayed until the culture has reached a high density. Common regulatable promoters used in *E. coli* include pLac (from the lactose operon), pTrp (from the tryptophan operon), and λpL and λpR (from the lambda phage). The lactose promoter is easy to induce, but its uninduced (basal) level of transcription is often significant and can create problems when the foreign gene products being expressed are strongly toxic to *E. coli*. Accordingly, it is a common practice to use this promoter in the presence of a highly efficient, mutant version of the *lac* repressor, coded for by the *lacI^q* allele, to suppress the uninduced level of transcription. Lambda promoters have an exceptionally low basal level of transcription; this makes sense, because successful transcription would induce the multiplication of phages during the lysogenic phase, killing the host cell prematurely. Thus they are well suited for expressing toxic proteins, but their induction requires the heat-inactivation of a mutant repressor protein, cI857, and rapidly raising the temperature of a large fermentation tank may not be easy.

Good expression vectors need to have a Shine–Dalgarno sequence, or ribosome-binding sequence (RBS), a sequence complementary to a part of the 3′-terminal segment of 16S rRNA. This complementarity allows the mRNA to associate with the 30S ribosomal subunit of *E. coli*. If RBS occurs close to the first codon, AUG, of the gene, this association leads to an efficient initiation of translation. The Shine–Dalgarno sequence is absent in eukaryotic mRNA. If the cloned fragment comes from such an organism, it is necessary to insert the *E. coli* RBS into the vector, and to place the 5′-terminus of the cloned gene close to this RBS, to ensure the efficient translation of the mRNA.

Some of these features are illustrated by the vectors of the pUC series (Figure 3.21). The segment that contains the regulatable promoter, pLac, and the 5′-terminal portion of the *lacZ* gene comes intact from the lactose operon of *E. coli*. Thus the promoter is at a proper (natural) distance from the transcription initiation site. The Shine-Dalgarno sequence is located at its natural distance from the initiation codon of the *lacZ* gene. At the very beginning of the *lacZ* gene, the polylinker developed earlier for the M13 series of vectors (see page 114) provides many cloning sites within a very short stretch of DNA. Because of this arrangement, it is possible to express the cloned protein very efficiently as a fusion protein containing only a few of the N-terminal amino acids of LacZ. For this purpose, however, the reading frame of the cloned gene has to be the same as that of the *lacZ* fragment: Some derivatives of pUC vectors are designed in such a manner that the *lacZ* fragment is present in all three reading frames relative to the inserted gene. A final convenient feature is that the disruption, by the cloned fragment, of the 5′-terminal fragment of the *lacZ* gene makes it possible to distinguish cells containing recombinant clones from those containing resealed vectors only, as explained in connection with the M13 vectors.

Somewhat different strategies are used in the expression vector pAS1 (Figure 3.22), with its phage-derived λpL promoter. As we have noted, transcription terminators, or sequences that produce similar polarity effects, may intervene between the promoter provided by the vector and the cloned gene, hampering the expression of the cloned gene. The pAS1 system utilizes the antitermination function of the lambda N protein produced from the prophage genome in the host. This antiterminator function depends on the presence of Nut ("N utilization") sites, and the pAS1 vector contains two such sequences between λpL and the cloning site. The RBS is present at the requisite distance from

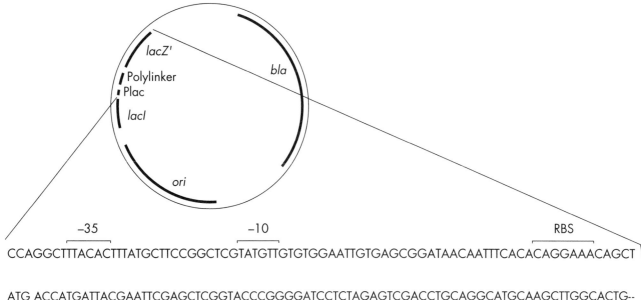

CCAGGCTTTACACTTTATGCTTCCGGCTCGTATGTTGTGTGGAATTGTGAGCGGATAACAATTTCACACAGGAAACAGCT

ATG ACCATGATTACGAATTCGAGCTCGGTACCCGGGGATCCTCTAGAGTCGACCTGCAGGCATGCAAGCTTGGCACTG--
Met Thr Met Ile Thr Asn Ser Ser Ser Val Pro Gly Asp Pro Leu Glu Ser Thr Cys Arg His Ala Ser Leu Ala **Leu**

Polylinker region

Figure 3.21
Structure of the *E. coli* expression vector pUC18. In addition to the origin of replication (*ori*) and an antibiotic resistance marker (*bla*), this vector contains a portion of the *E. coli lac* operon. The latter includes the repressor gene (*lacI*), the promoter region (Plac), and the 5′-terminal portion of the *lacZ* gene, coding for about 60 amino acid residues. As shown at the bottom, a polylinker region (containing restriction sites for more than 10 endonucleases) is inserted inside the *lacZ* gene. The amino acids present in the LacZ protein are shown in boldface type; those coded by the polylinker sequence in standard type. The polylinker does not contain any nonsense codons and is inserted in phase, so the vector codes for a complete N-terminal fragment of LacZ with an 18-amino-acid insert. The −35 and −10 regions of the promoter, as well as the RBS sequence, are indicated. The RBS sequence here (CAGGAAA) deviates somewhat from the consensus sequence in *E. coli* (AAGGAGG).

the ATG initiation codon; cutting the plasmid with *Bam*HI and removing the overhanging end produce a flush end terminating with the ATG sequence. If the foreign DNA is cut somewhere upstream from the 5′-end of the gene of interest and then is shortened so that the gene's 5′-terminus is in phase, insertion of the fragment into the vector cloning site produces a recombinant plasmid that produces high levels of foreign proteins, even eukaryotic proteins (10% to 30% of total *E. coli* protein in host cells lysogenic for a lambda prophage coding for a temperature-sensitive repressor protein, such as cI857).

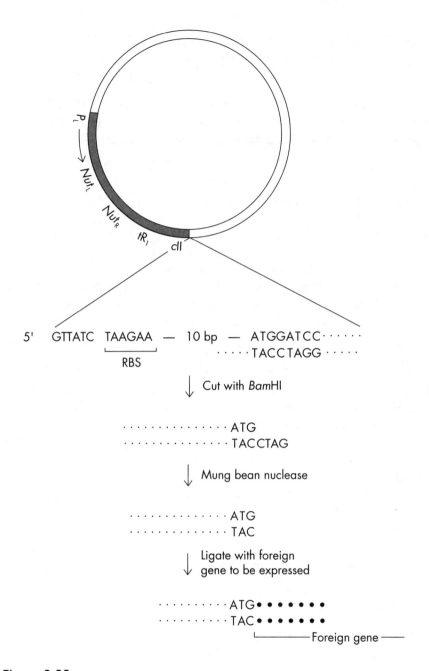

5' GTTATC TAAGAA — 10 bp — ATGGATCC······
                               ······TACCTAGG·····

          RBS

↓ Cut with *Bam*HI

··············ATG
··············TACCTAG

↓ Mung bean nuclease

··············ATG
··············TAC

↓ Ligate with foreign
↓ gene to be expressed

············ATG●●●●●●●
············TAC●●●●●●●
            └———Foreign gene———

Figure 3.22
Structure of the *E. coli* expression vector pAS1. Transcription begins with the
λpL promoter. The presence of the two Nut sites makes the transcription
resistant to the termination sequence tR1. The cloning site is opened with
*Bam*HI, and the protruding end is removed with mung bean exonuclease.
If a foreign gene in a flush-ended fragment is joined in phase, then the ATG
codon just upstream of the cloning site acts as the first codon, together with
the properly positioned RBS. Adapted from Rosenberg, M., Ho, Y.-S., and
Shatzman, A. (1983), The use of pKC30 and its derivatives for controlled
expression of genes, *Meth. Enzymol.* 101:123–138.

RECOVERY AND PURIFICATION OF EXPRESSED PROTEINS

Even when the cloned gene is successfully expressed in a bacterial host, product recovery is not always a simple matter. Potential problems and some approaches to solving them are discussed below.

Expression of Fusion Proteins

When short peptides are expressed in *E. coli*, they are likely to be rapidly degraded by the various and plentiful peptidases in the bacterial cytoplasm. To protect these products, the DNA sequences coding for them are usually fused to genes that code for proteins endogenous to *E. coli*. On expression of the resulting fusion protein, the small foreign peptide is folded as a portion of the large endogenous protein and generally escapes proteolytic degradation.

Selective site-specific cleavage of the fusion protein is required to separate these peptides from the "carrier" proteins. Some of the conditions and reagents that cleave proteins at specific sites are listed in Table 3.2. When the peptides do not contain internal bonds that would be cleaved by trypsin, CNBr, or acid, it is safe to generate a cleavage site for one of these agents at the peptide–carrier junction by altering DNA sequence. Peptides that are fairly large are likely to contain sites susceptible to such simple agents; in these cases a protease, such as Factor Xa, with its very stringent amino acid sequence specificity, is used to cleave at the desired site.

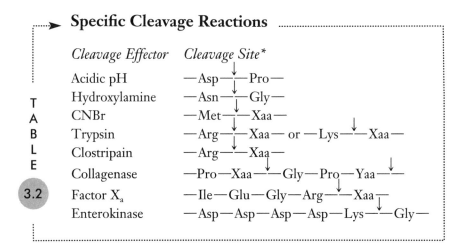

Specific Cleavage Reactions

TABLE 3.2

Cleavage Effector	Cleavage Site*
Acidic pH	—Asp—↓—Pro—
Hydroxylamine	—Asn—↓—Gly—
CNBr	—Met—↓—Xaa—
Trypsin	—Arg—↓—Xaa— or —Lys—↓—Xaa—
Clostripain	—Arg—↓—Xaa—
Collagenase	—Pro—Xaa—↓—Gly—Pro—Yaa—↓—
Factor X$_a$	—Ile—Glu—Gly—Arg—↓—Xaa—
Enterokinase	—Asp—Asp—Asp—Asp—Lys—↓—Gly—

* Xaa and Yaa indicate any amino acid residue. —↓— indicates position of cleaved peptide bond.

> **BOX 3.9   Affinity Columns**
>
> Traditional methods of protein purification rely on gross, physicochemical properties of the proteins, such as electrical charge (in ion exchange chromatography), size (gel filtration chromatography), and hydrophobicity (hydrophobic chromatography). A single fraction obtained after such purification procedures tends to contain more than one protein, if the starting material is a complex mixture. In contrast, affinity chromatography relies on *specific* interaction of the proteins with specific ligand molecules. For example, to purify an enzyme, either the substrate of the enzyme or a substrate analog is covalently linked to a granular matrix material, and the granules are packed into a column. When a crude mixture of hundreds of proteins is passed through this affinity column, only the enzyme is bound specifically to its substrate immobilized in the column; all the other proteins pass through the column unretarded. Elution of the column by procedures that decrease the affinity of the enzyme to the substrate, perhaps by altering the enzyme conformation, results in the one-step purification of the desired enzyme.

Another advantage of expressing peptides and proteins as fusion products is that it facilitates product purification. For example, if the foreign gene is fused with a sequence coding for an IgG antibody-binding domain of protein A from *Staphylococcus aureus*, the fusion protein can be recovered by simply passing the cellular extract through a column of immobilized IgG. Other schemes fuse products to glutathione S-transferase, which allows purification of the product with an affinity column (Box 3.9) of immobilized glutathione, or fuse them to a stretch of histidine residues, and then purify them by exploiting the metal complexation of histidine. Yet another example is fusion to the periplasmic maltose-binding protein of *E. coli*, which allows purification of fused proteins on a crosslinked amylose column. This approach has the added advantage that the fusion proteins may be secreted into the periplasm, the space that is located between the cytoplasmic membrane and the outer membrane and contains many fewer protein species than the cytoplasm.

Formation of Inclusion Bodies

When expressed at high levels in *E. coli* cytoplasm, many foreign proteins, especially those of eukaryotic origin, form insoluble aggregates called *inclusion bodies*. They are presumed to form where high concentrations of the overproduced, nascent proteins favor intermolecular interactions between the hydrophobic stretches of incompletely folded polypeptide chains, and they lead to aggregation and misfolding of these proteins (Figure 3.23).

These high, localized concentrations of nascent proteins are partly a consequence of the use of overexpression systems with their high gene dosage and powerful promoters. They are also partly due to the

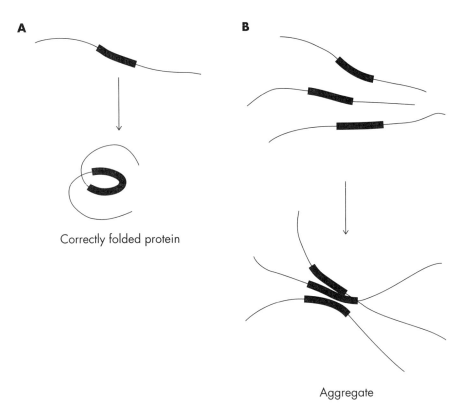

Figure 3.23
Presumed mechanism for the aggregation of overexpressed proteins. **A.** Normally, nascent polypeptides fold into a globular conformation, with hydrophobic stretches (thick line) hidden in the interior. **B.** However, when concentrations of nascent polypeptides are very high, there is increased likelihood that an exposed hydrophobic region on one molecule will interact with that on another molecule before the chains have a chance to fold properly. These intermolecular interactions between nascent chains result in aggregation and in an irreversible misfolding of the protein, producing inclusion bodies.

prokaryotic structure of *E. coli*. Under the typical eukaryotic conditions of synthesis, many nascent human and animal proteins would be sequestered into compartments separated from the cytosol, such as the lumen of the endoplasmic reticulum. In *E. coli*, however, the newly synthesized proteins must remain at large in the undifferentiated bacterial cytoplasm. Furthermore, several factors tend to retard the folding of foreign proteins in *E. coli*, thus increasing the chances of intermolecular association and aggregation: (1) The conditions in the *E. coli* cytoplasm —for example, pH, ionic strength, and redox potential—are different from the normal environment in which these proteins are folded into their final conformations. Many secreted proteins of eukaryotic origin cannot fold in the highly reducing cytoplasm of *E. coli* because disulfide bonds, which are normally formed in the oxidizing environment of the endoplasmic reticulum and help the folding process, are not produced. (2) The correct folding of many polypeptides is facilitated by various proteins (Box 3.10). These include prolyl *cis/trans* isomerase, which facilitates the interconversion of two forms of prolyl groups; protein disulfide isomerase, which catalyzes the exchange of disulfide linkages in the substrate protein, thereby facilitating the production of the form with correct disulfide pairs; and a group of *molecular chaperones*, or *chaperonins*, which also enhance the folding process or at least prevent the premature formation of aggregates of denatured proteins. The nature and the concentration of such helper proteins in *E. coli* obviously differ from those in the various compartments of the eukaryotic cells.

In some cases the formation of inclusion bodies can be avoided, as described in the next section. Even when this is difficult, however, we may be able to use inclusion bodies to advantage in the purification of recombinant proteins. The cells are broken, the extracts centrifuged,

 BOX 3.10   **Foldases and Molecular Chaperones**

Christian B. Anfinsen's group showed in 1957 that a completely denatured ribonuclease A can be spontaneously renatured into its native conformation. This famous discovery was interpreted by many to imply that all proteins become folded spontaneously, without assistance from any other cellular component. However, the fact that certain reactions occur spontaneously does not mean that cells do not use enzymes to facilitate those processes. Indeed, recent years have witnessed the discovery of two classes of proteins that assist the folding of newly made proteins. The members of one class of such proteins possess enzyme activities and are sometimes called "foldases." These include prolyl *cis-trans* isomerase and protein disulfide isomerase. The former enzyme helps in the folding process by facilitating the interconversion between *cis-* and *trans-* configurations of the peptide bond linking the nitrogen atom of proline and the carboxyl group of the preceding amino acid residue. Practically all of the peptide bonds in proteins have the *trans* configuration, but bonds involving proline are the exception. Spontaneous *cis-trans* isomerization of such bonds occurs slowly. Protein disulfide isomerase is important in the folding of proteins that contain disulfide bonds. If, in the course of folding, disulfide bonds are formed between incorrect pairs of cysteine residues, the protein is likely to be permanently misfolded. Protein disulfide isomerase, however, allows exchanges of disulfide bonds between cysteine residues and thus helps

the protein go through a trial-and-error process for reaching the correct combination of cysteine residues for disulfide bond formation.

The second class of proteins that assist in the folding process are molecular chaperones, or chaperonins, which appear to perform more subtle and complex functions. The structure of chaperonins has been conserved strongly during evolution, and most of them belong to the *heat shock proteins*—either to the Hsp70 or the Hsp60 class. In *E. coli*, the chaperonins DnaK and GroEL are representative of these two classes, respectively. They are called heat shock proteins because in many organisms they are overproduced when the organism experiences high temperatures; these proteins are thought to facilitate the unfolding and proper refolding of heat-denatured proteins. Hsp70 binds to partially denatured proteins and appears to release them upon ATP hydrolysis. One mechanism of action proposed for this protein suggests that repeated attachment and release help the folding process by sequestering the unfolded polypeptide and by preventing its interaction with other unfolded polypeptides (interaction that would result in irreversible aggregation and precipitation). Hsp60, in contrast, exists as a 14-mer in a double-doughnut configuration. In cooperation of Hsp10, a smaller protein that exists as heptamers, it catalyzes ATP hydrolysis and appears to help fold the polypeptide bound to its surface.

> **BOX 3.11   Renaturation of Proteins Containing Disulfide Bonds**
>
> If the protein contains disulfide bonds, denaturation requires their cleavage. This can be done by reduction (with either dithiothreitol or mercaptoethanol). In this case the solubilized denatured protein must be purified, always in the presence of reducing agents. Alternatively, the disulfide bond can be cleaved by converting cysteine sulfur atoms into S-sulfonates with the addition of sodium sulfite. S-sulfonates are stable at neutral or acidic pH, and thus the solubilized proteins can be conveniently purified if alkalinization of the samples is avoided. After purification, S-sulfonates can be reconverted to sulfhydryl groups by the addition of mercaptoethanol. In both procedures, the renaturation of the protein is accomplished by removal of the reducing agent and of the denaturing agent, and the oxidation of the cysteine residues into disulfides is accomplished by exposure to air.

and the inclusion bodies recovered as a sediment. Because the sediment also contains membrane fragments, it is customary to wash it by resuspension in detergent solutions (to dissolve and remove membrane components) and by recentrifugation. In this manner the complex and tedious process of protein purification can be almost completely bypassed. Finally, the inclusion bodies are solubilized with protein denaturants, such as 6 M urea or 8 M guanidinium hydrochloride, and the proteins are renatured by the gradual removal of denaturants (Box 3.11). Procedures of this type have been successfully used for the purification of many proteins. Although theoretical considerations indicate that *in vitro* renaturation should be done at low concentrations of the protein to minimize intermolecular interactions, in practice some systems tolerate fairly high concentrations, presumably because even these concentrations are quite low compared with those reached in overproducing cells.

Preventing the Formation of Inclusion Bodies

Although inclusion bodies are a convenient starting material for purification, the denaturation and the controlled renaturation steps are costly. It is especially problematic that the renaturation process usually works best at low protein concentrations, a requirement that increases cost and decreases yield. Careful cost analysis shows that the expense of the renaturation step is the main reason why the commercial production of large proteins such as tissue plasminogen activator and Factor VIII by recombinant DNA technology is carried out in animal cell cultures rather than in *E. coli*. In animal cell cultures, the recombinant protein folds spontaneously into the native conformation, and inclusion bodies are not formed.

Production costs for these proteins would be further reduced if they could be produced in the native conformation in a microbial host. Much effort has therefore been devoted to finding conditions that would decrease the extent of inclusion body formation in *E. coli*. So far, a technique that universally and drastically decreases inclusion body formation has not been discovered. Among the approaches tried, lowering of the growth temperature was most effective. Attempts have also been made, with some success, to co-clone eukaryotic chaperones and foldases to improve the correct folding of eukaryotic proteins in *E. coli*. In one study, overproduction of *E. coli* molecular chaperone DnaK (see Box 3.10) decreased the aggregation of human growth hormone in the *E. coli* cytoplasm. Another study took a different approach: Coding sequences for eukaryotic proteins were fused to the *E. coli* thioredoxin gene, and the fused proteins were found to be produced in a totally soluble form (Box 3.12). It may be that thioredoxin acts as a partner in disulfide exchange reactions, thereby facilitating the correct folding of the foreign polypeptide.

BOX 3.12   Thioredoxin Fusions

E. coli thioredoxin is a small protein (molecular weight 11,675) with two cysteine residues in close proximity. Its normal function is to serve as a carrier of reducing power in reactions such as the synthesis of deoxyribonucleoside diphosphates from ribonucleoside diphosphates. The thiols (—SH groups of cysteine residues) are oxidized to an intramolecular disulfide in the reaction catalyzed by ribonucleotide reductase.

$$\text{Ribonucleoside diphosphate} + \text{Thioredoxin}\begin{smallmatrix}\text{SH}\\\text{SH}\end{smallmatrix} \longrightarrow \text{Deoxyribonucleoside diphosphate} + \text{Thioredoxin}\begin{smallmatrix}\text{S}\\\text{S}\end{smallmatrix}$$

The oxidized thioredoxin is reduced at the expense of NADPH by thioredoxin reductase.

$$\text{Thioredoxin}\begin{smallmatrix}\text{S}\\\text{S}\end{smallmatrix} + \text{NADPH} + \text{H}^+ \longrightarrow \text{Thioredoxin}\begin{smallmatrix}\text{SH}\\\text{SH}\end{smallmatrix} + \text{NADP}^+$$

When overexpressed from plasmid vectors, *E. coli* thioredoxin can accumulate in soluble form to 40% of total cellular protein. Both the N- and the C-termini of thioredoxin are exposed on the surface of the molecule. In the fusion proteins, a DNA sequence encoding an amino acid sequence, with a site cleavable by the enzyme enterokinase (see Table 3.2), was introduced between *trxA* (the gene encoding thioredoxin) and the gene encoding the desired protein. Thioredoxin fusion proteins were found to accumulate to a high level in the *E. coli* cytoplasm without forming inclusion bodies.

Secretion Vectors

Whether the recombinant proteins form inclusion bodies in the cytoplasm of the host bacterium or remain in soluble form, their purification is always a challenge. One way to simplify the task of separating the recombinant protein from the myriad of host proteins, at least in principle, is to cause the recombinant proteins to be secreted into the culture medium. After that, purification becomes quite straightforward, because bacteria are usually grown in simple, protein-free media. Secreted proteins are also likely to fold properly, because their concentration in the extracellular space tends to be much lower than in the cytoplasm. For both these reasons, much effort has been spent on the construction of *secretion vectors.*

In both prokaryotes and eukaryotes, proteins destined to be secreted from the cell are synthesized with an extra sequence, a *leader* (or *signal*) *sequence*, of about two dozen residues at the N-terminus. This sequence guides the nascent protein to the secretory apparatus in the membrane and is split off by leader peptidase after the polypeptide is translocated across the membrane. The presence of the leader sequence is a necessary, but not always a sufficient, condition for secretion: Some artificial constructs composed of leader sequences fused to soluble, cytosolic proteins fail to be secreted, presumably because the mature part of the protein folds quickly to a stable, globular conformation and cannot be translocated in that condition. This suggests that the secretion vector strategy will work best when the products are fairly small peptides and thus are unlikely to fold into a tight, stable conformation.

Indeed, this strategy proved useful in the production of insulin-like growth factor I (IGF-1), a peptide composed of about 70 amino acid residues. The cDNA for IGF-1 was cloned behind the sequence coding for the leader sequence of protein A, a secreted, immunoglobulin (IgG)-binding protein from *Staphylococcus aureus.* In addition, two copies of the sequence coding for the IgG-binding domain of protein A were inserted between the leader sequence and the cDNA for IGF-1 to facilitate the purification and to inhibit proteolytic degradation (see page 128). An "affinity handle" such as this is important, because when proteins are secreted, they must be purified from the culture supernatant, with its very large volume. The affinity handle provides a way of rapidly and efficiently concentrating the desired product. In this case, the culture supernatant was passed through a column of IgG-Sepharose, which adsorbed all of the secreted proteins containing the IgG-binding protein A sequence. The fusion protein was then cleaved with hydroxylamine by taking advantage of the hydroxylamine-sensitive Asn-Gly sequence introduced just in front of the IGF-1 sequence. The IGF-1 peptide was then purified by conventional column chromatography methods.

Although secretion vectors have often proved useful, secretion is not yet a universally applicable approach. Some proteins fail to be secreted even when fused to a leader sequence, as mentioned earlier. In addition, the most frequently used bacterial host, *E. coli*, is protected by an outer membrane as well as by the cytoplasmic membrane. Secreted proteins usually cannot cross the outer membrane even when they are successfully secreted across the cytoplasmic membrane. Most of them can then be found in the periplasm, the space between the outer and cytoplasmic membranes. However, even when the proteins are in the periplasm, they tend to form inclusion bodies much less frequently than in the cytoplasm, partly because the periplasm is a more oxidizing environment than the cytoplasm, and partly because periplasm contains a disulfide isomerase. Periplasm contains only a few dozen proteins, so secretion into periplasm may make purification of the protein product significantly easier than from the cytoplasm. It is also possible to cause the secreted proteins to cross the outer membrane by using various outer-membrane-leaky strains or by exploiting the machinery that *E. coli* and its relatives use naturally to export proteins such as hemolysin and pullulanase into the medium. An alternative strategy is to use Gram-positive bacteria, such as *Bacillus subtilis*, as the host; these do not possess an outer membrane.

AN EXAMPLE: PRODUCTION OF CHYMOSIN (RENNIN) IN *E. COLI*

Chymosin is the major protease produced in the fourth stomach (abomasum) of calves. Its production is limited to the few weeks during which the calves are nourished by milk. Thereafter, an alternative protease, pepsin, becomes the major secreted enzyme. Chymosin is synthesized in the cytoplasm of mucosal cells as preprochymosin (containing the signal sequence, the "pro" sequence removed at the time of activation, and the mature chymosin sequence). The signal sequence of 16 amino acid residues is removed, and the protein is secreted, presumably through the endoplasmic reticulum–Golgi pathway (see Chapter 4). The secreted product, prochymosin (molecular weight 41,000), is an inactive zymogen that under acidic conditions becomes converted into the active enzyme chymosin (molecular weight 35,600) by autocatalytic cleavage of the N-terminal "pro" sequence of 27 amino acid residues.

Chymosin is an aspartyl protease. It coagulates milk very efficiently through the limited hydrolysis of κ-casein and is used extensively in the manufacture of cheese (Chapter 1, page 44). Chymosin is particularly useful for this purpose because it combines a very high milk-coagulating activity with a low general proteolytic activity. Because the production of cheese has increased rapidly in recent decades and the supply of suckling calves has declined, the availability of chymosin or chymosin substitutes has become an important issue in the dairy industry.

One major solution has been the commercialization of fungal enzymes from *Mucor* and *Endothia* as substitutes for chymosin. These enzymes are less expensive and are fairly effective, but they do not quite attain the high coagulation/proteolysis ratio of calf chymosin, and this results in subtle but real differences in the flavor of the cheese. Moreover, the fungal enzymes are more resistant to heat inactivation than calf chymosin is, so they are more likely to continue proteolysis on into the later phases of cheese production. A more satisfactory solution, therefore, would be to produce chymosin by cloning, if it can be done in a cost-effective manner.

Several laboratories succeeded in cloning chymosin cDNA in the early 1980s. In every case, the original template was mRNA from the mucosa of the calf abomasum. In one study, the RNA was extracted from frozen, pulverized abomasum tissue in the presence of bentonite, which was added to inhibit the strong ribonuclease activity present in the tissue. The polyanionic mucosal polysaccharides, which could interfere with RNA purification, were removed by precipitation with LiCl. PolyA-tailed mRNAs were isolated by taking advantage of their association with polyU-Sepharose. A large fraction of the mRNA obtained in this manner coded for prochymosin (and its precursor), as was discovered when *in vitro* translation of the mRNA preparation gave prochymosin as the major product (Figure 3.24). (Possibly the preprochymosin was processed to prochymosin because crude reticulocyte lysate was used as the translation system).

cDNA was prepared from this mRNA, in some cases after size fractionation in order to further enrich for (pre)prochymosin mRNA. The cDNA preparation was used in the primary cloning to isolate the sequence coding for chymosin (and its precursors). The cDNA was inserted into plasmid vectors such as pBR322, and the recombinant plasmids were screened. The screening was done differently in different laboratories. In one laboratory, a dodecanucleotide probe was synthesized on the basis of the known amino acid sequence of prochymosin and was used in a hybridization assay (see Figure 3.18). In another laboratory, fractionated mRNA was used in the colony hybridization assay (see Figure 3.18). Because such preparations also contain mRNAs that code for other proteins, positive hybridization serves only as a prospective signal of success. However, if a recombinant plasmid DNA contained the sequence coding for prochymosin, one of its strands would be complementary to the prochymosin mRNA. Thus denatured plasmid DNA was added to an *in vitro* translation system containing the fractionated mRNA and was shown indeed to inhibit specifically the synthesis of prochymosin, a result proving unequivocally that the recombinant DNA indeed contained the desired gene. Subcloning and

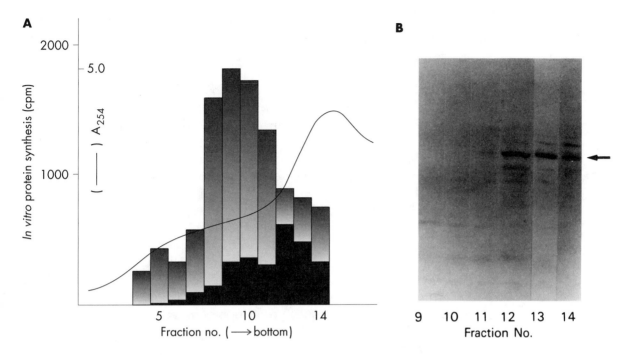

Figure 3.24
Products of *in vitro* translation of fractionated mRNA from calf abomasum.
A. mRNA was fractionated by size through sucrose-density gradient centrifugation. mRNA in each fraction was used to direct *in vitro* protein synthesis with an amino acid mixture containing [^{3}H] leucine. Open bars show the amounts of total protein synthesized; filled-in bars show the amount of prochymosin synthesized, as revealed by adsorption to the specific antibody.
B. Analysis by sodium dodecylsulfate-polyacrylamide gel electrophoresis of proteins produced by these mRNA fractions confirmed that larger mRNA (fractions 12 to 14) contained the mRNA for prochymosin (arrow). From Beppu, T. (1983), *Tr. Biotechnol.* 1:85, with permission.

sequencing then confirmed that the cloned cDNA contained the sequence coding for prochymosin. (Somewhat unexpectedly, it did not contain the entire sequence for preprochymosin, but a part of the sequence coding for the leader was apparently lost at some step during the cloning.)

The next step was the cloning in a suitable expression vector. In the calf abomasum, chymosin is made as a preprotein. Researchers had to decide in which form it should be expressed in *E. coli*. No attempt was made to express chymosin in its mature, processed form, because the production of such an active protease in the *E. coli* cytoplasm was expected to be harmful to host cells. Nor, in the initial efforts, was an attempt made to express the entire preprochymosin sequence because of concern that the eukaryotic signal sequence might not lead to efficient secretion in *E. coli*. In several laboratories, therefore, prochymosin was chosen as the form to be expressed.

Two methods were used. In one, the prochymosin sequence was fused to the N-terminal portion of LacZ or TrpE, and the protein was expressed as the fusion protein. In this case, the prokaryotic promoters and the RBS sequence present in front of these highly expressed prokaryotic genes were used to initiate transcription and translation efficiently. In another approach, the sequence coding for prochymosin was inserted directly behind a sequence containing a suitable prokaryotic promoter, RBS sequence, and ATG codon. Some adjustment of distance between the RBS sequence and ATG, as well as of the actual base sequence, was needed to optimize the expression in this case (Table 3.3). (Even subtle changes in the upstream sequence can profoundly affect the rate of expression in ways that are often difficult to predict. This is why the natural *E. coli* promoters, together with the 5′-terminal fragments of the *E. coli* genes, are still frequently used in expression vectors.) Both approaches led to the production of prochymosin at a level corresponding to up to 5% of total *E. coli* protein.

The overproduced prochymosin, however, accumulated in a denatured form as inclusion bodies. When attempts were made to purify the inclusion bodies and to renature prochymosin from this material by the procedure already described (pages 129–132), the yield of active prochymosin was disappointingly low, owing primarily to difficulties in the renaturation step. The increased production cost that would result might be tolerated if the product were a human therapeutic compound, but for agricultural products such as prochymosin, the cost was clearly prohibitive. Prochymosin has since been expressed more efficiently in yeasts (see Chapter 4).

Effect of Distance Between the RBS and the ATG Codons on Level of Prochymosin Expression

TABLE 3.3

Sequence*	Expression Level (% of cell protein)
AAGGGTATCG**ATG**	0.3
AAGGGTATCGAT**ATG**	9.8
AAGGGTATCGATAAGCTT**ATG**	3.2
AAGGGTATCGATAAGCTAGCTT**ATG**	0.8

* The putative RBS (AAGG) and the initiation codon (ATG) are shown in boldface.
SOURCE: Beppu, T., (1988) in Thomson, J.A., *Recombinant DNA and Bacterial Fermentation*, CRC Press.

SUMMARY

Some proteins and peptides of therapeutic value are difficult to purify in sufficient amounts from their human and animal sources. Recombinant DNA methods have had a revolutionary impact in the production of these compounds. Once the DNA sequences coding for these proteins and peptides have been cloned and amplified in microorganisms, the latter can continue to function as a living factory for the inexpensive production of such compounds. Bacteria, especially *Escherichia coli*, are used extensively as the host microorganism. Segments of "foreign DNA" coding for these products are first obtained either by cutting the genomic DNA or by synthesizing a DNA sequence complementary to the mRNA (*cDNA*) with reverse transcriptase. Such segments must first be inserted into vector DNA, which contains information that makes possible its replication in the bacterial host. In addition to plasmids, which are widely used as general-purpose cloning vectors, there are several types of vectors for cloning in bacteria. Lambda phage vectors and cosmids are useful for the cloning of large segments of DNA (20–40 kb), and single-strand DNA phage vectors are especially well suited for sequencing of DNA and for site-directed mutagenesis. The recombinant DNA—that is, the vector DNA containing the foreign DNA insert—is then introduced into the host cell by transformation or by injection from phagelike particles after it has been packaged into phage heads. The clone that contains the desired gene sequence is identified, sometimes among a vast majority of clones not containing this sequence, by using either the DNA sequence itself or the protein product of the gene as the marker. In some cases, however, PCR (polymerase chain reaction) enables one to bypass all the steps of primary cloning and screening by direct amplification of the DNA sequence *in vitro*.

Regardless of the source, the sequence coding for the desired product can then be inserted into expression vectors to maximize the synthesis of the product in bacteria. The overproduction of foreign proteins in bacteria, however, frequently results in misfolding and aggregation of these proteins. Several strategies for avoiding aggregation are available, but none of them appears to be universally applicable to all proteins. However, aggregation does not necessarily mean a total failure, because protein aggregates can be easily purified, totally denatured, and then renatured under controlled conditions.

An informative example of the practical problems involved in cloning of DNA coding for animal proteins is the work that has been done with prochymosin, a calf abomasum protease important in the production of cheese. The cDNA sequence coding for prochymosin was successfully cloned in plasmids, the clone containing the desired sequence was identified by screening, and the gene was inserted into expression vectors. Nevertheless, the overexpression of the gene produced insoluble aggregates of prochymosin, which turned out to be difficult to renature. Prochymosin is now produced by using fungal vectors.

SELECTED REFERENCES

General References on Recombinant DNA Methods

Watson, J.D., Gilman, M., Witkowski, J., and Zoller, M., 1992. *Recombinant DNA.* 2d ed. Scientific American Books.

Old, R.W., and Primrose, S.B., 1994. *Principles of Gene Manipulation: An Introduction to Genetic Engineering.* 5th ed. Blackwell.

Hackett, P.B., Fuchs, J.A., and Messing, J.W., 1988. *An Introduction to Recombinant DNA Technology.* Benjamin/Cummings.

Sambrook, J., Fritsch, E.F., and Maniatis, T., 1989. *Molecular Cloning: A Laboratory Manual.* 2d ed. Cold Spring Harbor Laboratory Press.

Vectors

Morales, V.M., and Bagdasarian, M., 1991. Cloning in bacteria: vectors and techniques. In Prokop, A., Bajpai, R.K., and Ho, C. (eds). *Recombinant DNA Technology and Applications*, pp. 3–28, McGraw-Hill.

Balbas, P., Soberon, X., Merino, E., Zurita, M., Lomeli, H., Valle, F., Flores, N., and Bolivar, F., 1986. Plasmid vector pBR322 and its special-purpose derivatives—A review. *Gene* 50:3–40.

Young, R.A., and Davis, R.W., 1983. Efficient isolation of genes by using antibody probes. *Proc. Natl. Acad. Sci. USA.* 80:1194–1198.

Rosenberg, M., Ho, Y.-S., and Shatzman, A., 1983. The use of pKC30 and its derivatives for controlled expression of genes. *Meth. Enzymol.* 101:123–138.

Proteolysis

Enfors, S.-O., 1992. Control of *in vivo* proteolysis in the production of recombinant proteins. *Tr. Biotechnol.* 10:310–315.

Protein Folding and Molecular Chaperones

Schein, C.H., 1989. Production of soluble recombinant proteins in bacteria. *Bio/Technology* 7:1141–1149.

Ellis, R.J., and van der Vies, S.M., 1990. Molecular chaperones. *Ann. Rev. Biochem.* 60:321–347.

Horwich, A.L., Neupert, W., and Hartl, F.-U., 1990. Protein-catalysed protein folding. *Tr. Biotechnol.* 8:126–131.

Langer, T., and Neupert, W., 1991. Heat shock proteins hsp60 and hsp70: Their roles in folding, assembly and membrane translocation of proteins. *Curr. Top. Microbiol. Immunol.* 167:3–30.

Blum, P., Velligan, M., Lin, N., and Matin, A., 1992. DnaK-mediated alterations in human growth hormone protein inclusion bodies. *Bio/Technology* 10:301–304.

LaVallie, E.R., DiBlasio, E.A., Kovacic, S., Grant, K.L., Schendel, P.F., and McCoy, J.M., 1993. A thioredoxin gene fusion expression system that circumvents inclusion body formation in the *E. coli* cytoplasm. *Bio/Technology* 11:187–193.

Protein Production in *E. coli* Versus That in Animal Cells

Datar, R.V., Cartwright, T., and Rosen, C.-G., 1993. Process economics of animal cell and bacterial fermentations: A case study analysis of tissue plasminogen activator. *Bio/Technology* 11:349–357.

Protein Secretion from Bacteria

Pugsley, A.P., 1993. The complete general secretory pathway in Gram-negative bacteria. *Microbiol. Rev.* 57:50–108.

Simonen, M., and Palva, I., 1993. Protein secretion in *Bacillus* species. *Microbiol. Rev.* 57:109–137.

Pérez-Pérez, S., Márquez, G., Barbero, J.-L., and Gutiérrez, J., 1994. Increasing the efficiency of protein export in *Escherichia coli. Bio/Technology* 12:178–180.

Prochymosin

Beppu, T. 1988. Production of chymosin (rennin) by recombinant DNA technology. In Thomson, J.A. (ed.), *Recombinant DNA and Bacterial Fermentation*, pp. 11–21. CRC Press.

Marston, F.A.O., Angal, S., Lowe, P.A., Chan, M., and Hill, C.R., 1988. Scale-up of the recovery and reactivation of recombinant proteins. *Biochem. Soc. Trans.* 16:112–115.

CHAPTER

4 Production of Proteins in Yeast

In Chapter 3 we described the cloning of foreign genes in bacteria, mostly in *E. coli*. In passing, we touched on the difficulties encountered when bacteria are used to clone and express genes from higher organisms—that is, from eukaryotes. For example, many eukaryotic proteins normally undergo one or more posttranslational modifications that are important to their functions or stability. Yeast has often been referred to as a model eukaryote, and in this chapter we will show how yeast cells are able to carry out many of the posttranslational modifications necessary to produce accurately synthesized proteins using the genes or cDNA of higher organisms.

Glycosylation—the addition of oligosaccharide units to a protein—is one of the most important posttranslational modifications that occur to the gene products of eukaryotic cells. Indeed, most secreted eukaryotic proteins are glycosylated (Figure 4.1). Glycosylation often helps ensure the correct folding of proteins and protects them from proteolytic enzymes. In some cases, specific receptors on animal cells recognize serum proteins whose N-linked oligosaccharides lack certain sugars and remove these proteins (usually "old" proteins) from circulation (Figure 4.2). Thus the presence of the correct oligosaccharides is very important in producing recombinant human proteins that work well and last for a long time *in vivo*. Glycosylation, and other modifications described in Box 4.1, do not occur if eukaryotic genes are expressed in bacteria such as *E. coli*.

141

A

$$\text{SA} \longrightarrow \text{Gal} \longrightarrow \text{GlcNAc} \longrightarrow \text{Man}$$
$$\text{SA} \longrightarrow \text{Gal} \longrightarrow \text{GlcNAc} \longrightarrow \text{Man}$$

Fuc (dashed arrow down to GlcNAc)

Man $\longrightarrow$ GlcNAc $\longrightarrow$ GlcNAc $\longrightarrow$ Protein

B

Man
Man $\longrightarrow$ Man
Man

Man $\longrightarrow$ GlcNAc $\longrightarrow$ GlcNAc $\longrightarrow$ Protein

C

Fuc (dashed arrow down to GlcNAc)

Gal $\longrightarrow$ GlcNAc $\longrightarrow$ Man
Gal $\longrightarrow$ GlcNAc $\longrightarrow$ Man

Fuc (dashed arrow up to lower GlcNAc)

Xyl (arrow down to Man)
Fuc (dashed arrow down to GlcNAc)

Man $\longrightarrow$ GlcNAc $\longrightarrow$ GlcNAc $\longrightarrow$ Protein

D

Man   Man
↓   ↓
Man   Man-Ⓟ
↓   ↓
Man   Man
↓   ↓
Man $\cdots$ Man $\longrightarrow$ Man $\longrightarrow$ Man
Man $\longrightarrow$ Man $\longrightarrow$ Man
Man $\longrightarrow$ Man $\longrightarrow$ Man $\longrightarrow$ Man
↑
Man

Man
Man

Man $\longrightarrow$ GlcNAc $\longrightarrow$ GlcNAc $\longrightarrow$ Protein

E

Man $\longrightarrow$ Man
Man $\longrightarrow$ Man $\longrightarrow$ Man
Man $\longrightarrow$ Man $\longrightarrow$ Man $\longrightarrow$ Man $\longrightarrow$ Man
Man

Man

Man $\longrightarrow$ GlcNAc $\longrightarrow$ GlcNAc $\longrightarrow$ Protein

F

$$\text{SA} \longrightarrow \text{Gal} \longrightarrow \text{GalNAc} \longrightarrow \text{Protein}$$

SA (dashed arrow up to GalNAc)

In eukaryotic cells, secretory proteins are synthesized by ribosomes associated with the membrane of the endoplasmic reticulum and are translocated across that membrane cotranslationally by a mechanism involving a *signal-recognition particle*. On entering the lumen of the endoplasmic reticulum, these proteins are immediately glycosylated (Box 4.2). By contrast, the prokaryotes' homologs of signal-recognition particles may not play a major role in the export of most proteins, and bacterial secretory proteins are almost never glycosylated.

Because glycosylation is of such importance in eukaryotes, it follows that eukaryotic microbes, such as yeasts, may be better hosts for the expression of genes of higher eukaryotes. Yeast cells do export many proteins using the endoplasmic reticulum-Golgi pathway, apparently with the participation of a signal-recognition particle, and glycosylate those proteins in the process. Yeast cells also carry out posttranslational N-acetylation and myristylation of proteins (see Box 4.1).

One might even expect that yeasts, as eukaryotes, would be able to carry out the correct splicing of nascent RNA transcripts of mammalian genes. However, yeasts contain few introns and therefore may fail to process mammalian intron sequences. The safest procedure when expression of a mammalian protein is desired is to utilize an intron-free gene—that is, to generate a cDNA copy of the mature mRNA for the gene of interest and use that as a template.

Yeasts can be grown to very high densities in simple, inexpensive media. More important, the components and metabolic products of

---

Figure 4.1 (*on facing page*)
Types of oligosaccharide groups found in glycoproteins. Many proteins acquire oligosaccharide substituents at asparagine residues via an *N*-glycosidic bond. In higher animals these "N-linked" oligosaccharides are typically of the complex type, in which additional branching can produce more complicated structures (**A**), although some carry high-mannose-type oligosaccharides (**B**). Glycoproteins of higher plants (**C**) lack the sialic acid residues that are commonly present in the animal glycoproteins. Yeast glycoproteins characteristically carry oligosaccharides containing very large numbers of mannose residues (**D**), except in the case of a mutant (*mnn*9) that cannot add outer-chain mannose residues (**E**). Other oligosaccharides can be linked to serine or threonine residues via an *O*-glycosidic bond. These "O-linked" oligosaccharides are generally much simpler in structure (**F**) than the N-linked oligosaccharides. Abbreviations: SA, sialic acid; Gal, D-galactose; GlcNAc, *N*-acetyl-D-glucosamine; Man, D-mannose; Fuc, L-fucose; Xyl, D-xylose; Ⓟ, phosphate; GalNAc, *N*-acetyl-D-galactosamine. Broken arrows indicate partial substitutions.

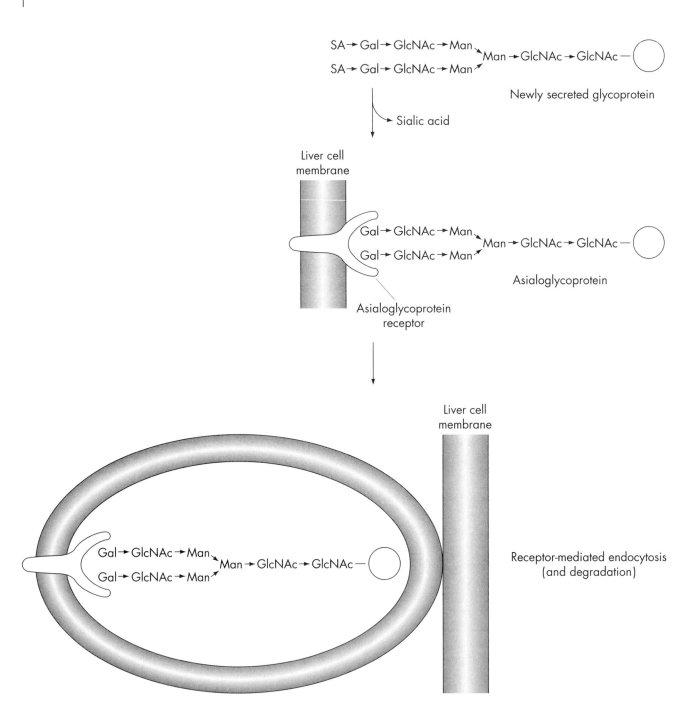

Figure 4.2

Removal of serum proteins lacking terminal sialic acid residues. Glycoproteins slowly lose their terminal sialic acid residues during circulation. Such *asialoglycoproteins* are bound by a specific receptor on liver cells and are removed from circulation by the process of receptor-mediated endocytosis. Glycoproteins with terminal mannose residues (for example, those carrying high-mannose-type oligosaccharides) are removed by the use of another receptor, mannose receptor, on liver endothelial cells. SA, sialic acid.

> ## BOX 4.1 Acylation of amino acid residues
>
> In many eukaryotic proteins, the amino group of the N-terminal amino acid residue is modified by acylation—that is, by the formation of an acyl (acetyl, myristyl, etc.) amide linkage. N-acetylation interferes with recognition of the protein by the intracellular proteolytic degradation machinery of eukaryotic cells and thus preserves the proteins for a longer period within the animal or human body. Another characteristic modification of the N-terminal residue is N-myristylation, which adds the 14-carbon, saturated fatty acid known as myristic acid onto the amino group. The myristylated proteins can bind to membranes at the fatty acid, thus becoming peripheral membrane proteins. Similar targeting of certain other proteins occurs by the covalent attachment of palmitic acid, a 16-carbon, saturated fatty acid. In these cases the fatty acid is attached to the sulfhydryl groups of internal (not N-terminal) cysteine residues.

yeast cells are not toxic to humans (remember that LPS, an integral component of the *E. coli* outer membrane, is a very toxic molecule also known as endotoxin). In the following pages, we will discuss how foreign DNA is introduced into yeast cells (most often *Saccharomyces cerevisiae*, or baker's yeast), the types of yeast cloning vectors that are used, some barriers to the efficient expression of foreign genes in yeast, and the use of secretion vectors. In the course of these discussions, the use of yeast to produce two proteins, hepatitis B virus surface antigen and prochymosin, will be described in some detail.

BOX 4.2   **Protein-secretion pathways in prokaryotic and eukaryotic cells**

In prokaryotes, secretory proteins are made with an N-terminal signal sequence and are secreted via the SecY-SecD-SecE-SecF protein complex found in the plasma membrane. SecA protein is thought to help bring the signal sequence to the export machinery. The signal sequence is cleaved when the junction between it and the mature sequence appears on the outer side of the cytoplasmic membrane (see Figure **A**).

Secreted proteins made by eukaryotic cells also have an N-terminal signal sequence. However, the signal sequence is recognized by a complex structure, the signal-recognition particle (SRP), which contains six proteins held together by a small RNA. SRP then binds to the membrane-associated SRP receptor, thus guiding the nascent protein to the export apparatus located specifically within the membrane of the *rough endoplasmic reticulum*. One of the proteins in SRP also arrests translation until this "docking" at the export apparatus takes place, thus preventing the protein's misfolding in the cytosolic environment. The protein passes across the membrane, presumably in an extended form, and enters the lumen of the rough endoplasmic reticulum. The signal sequence is split off soon after a partial translocation of the protein across the membrane. The environment in the lumen is less reducing than the cytosol, and folding of the protein is often followed by the formation of disulfide bonds. Because the creation of disulfide bonds between "wrong" pairs of cysteine residues might produce a permanently misfolded protein, the lumen contains disulfide isomerase, which splits and reforms disulfide bonds so as to encourage the correct conformation (see Box 3.10).

Even while the polypeptide is being extruded through the membrane, some sites within the secreted protein become glycosylated. Figure **B** shows the formation of a complex type of N-linked oligosaccharide of the simplest structure in animal cells (there can be many variations on the details of the pathway). Within the endoplasmic reticulum, a "core" oligosaccharide containing 2 proximal *N*-acetylglucosamine residues, 9 mannose residues, and 3 glucose residues is attached to an appropriate site on the protein; subse-

A

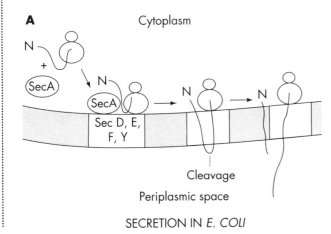

SECRETION IN *E. COLI*

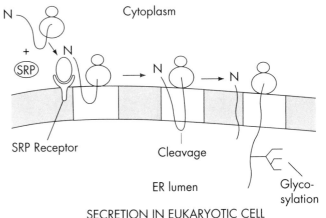

SECRETION IN EUKARYOTIC CELL

INTRODUCTION OF DNA INTO YEAST CELLS

DNA can be introduced into bacteria in a variety of ways: directly through transformation, through injection by phage particles, or through plasmid-transfer mechanisms involving cell-to-cell contact. With yeasts, however, transformation is the only practical means of introducing DNA. In one method, the yeast cell wall is removed by enzyme digestion, and the resulting "spheroplasts" (cells bounded essentially only by the cytoplasmic membrane) are incubated with DNA in the presence of Ca^{2+} and polyethyleneglycol. Both Ca^{2+} and polyethyleneglycol are agents that stimulate the membrane fusion process and

> **BOX** 4.2   (*continued*)

quently, all of the glucose residues and 1 mannose residue are "trimmed off." The glycoprotein is then transported, probably via small membrane vesicles, into the Golgi apparatus, another complex, membrane-bounded organelle: First it enters *cis* Golgi vesicles, where 3 more of the mannose residues are removed. In the next compartment, the lumen of the medial Golgi vesicles, more mannose residues are trimmed off, and 2 *N*-acetylglucosamine residues are added

on. Finally, in the *trans* Golgi compartment, 2 galactose residues are added to the *N*-acetylglucosamine residues, and sialic acid residues are added on top of them. The completed glycoprotein is then secreted from the cell by the fusion of glycoprotein-containing vesicles with the plasma membrane. As demonstrated in Figure 4.1, the N-linked oligosaccharides may have different structures in different proteins, and, especially, in different organisms.

B

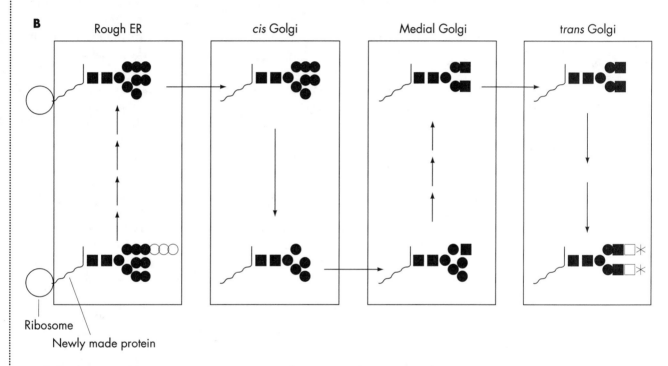

Symbols: ■, *N*-acetylglucosamine; ●, mannose; ○, glucose; □, galactose; ✳, sialic acid.

thereby enhance fusion between spheroplasts. DNA is probably taken up by yeast cells in the process of spheroplast fusion, but it is not yet clear whether the fusion is necessary for this uptake to occur. In another method, intact yeast cells (with cell wall in place) are treated with alkali metal ions such as Li⁺ and then incubated with DNA and polyethyleneglycol. The mechanism of DNA uptake remains obscure in this case, too. A third method is to apply transient high voltages to a suspension of cells. This process, called *electroporation*, creates transient holes in the walls and membranes, as described in Chapter 3. If pieces of DNA are present in the medium, they may be taken up by the cells, presumably by chance. Electroporation is a very useful method for making an entire culture of yeast cells competent to receive DNA.

Yeast Cloning Vectors

Several types of cloning vectors are used to manipulate recombinant DNA constructs in yeast. Most are "shuttle" vectors: vectors that can multiply in yeast as well as in *E. coli*. The reason shuttle vectors are preferred is that the basic recombinant DNA manipulations are more easily carried out in *E. coli*, but the resulting DNA constructs must be transferred to yeast to take advantage of the superior properties of this host, such as the expression of glycosylated proteins. Shuttle vectors can be moved between the two hosts, because they also contain the origin of replication recognized by *E. coli* and selection markers useful in *E. coli*, as well as features that enable them to survive in yeast cells. There are five major types of yeast cloning vectors: yeast integrative plasmids (YIp), yeast replicating plasmids (YRp), yeast episomal plasmids (YEp), yeast centromeric plasmids (YCp), and yeast artificial chromosomes (YAC).

Yeast Integrative Plasmids

YIps (Figure 4.3) are essentially bacterial plasmid vectors with an added marker that makes possible their genetic selection in yeast. As we have seen already, antibiotic resistance genes are commonly used as selection markers in bacterial vectors. However, not many antibiotics are effective against yeasts. Thus selection procedures in yeast are commonly designed to utilize a host strain that is defective in the biosynthesis of amino acids, purines, or pyrimidines and a vector that contains a yeast gene for the missing function. Some commonly used nutritional markers are the yeast genes LEU2 (a gene involved in leucine biosynthesis), URA3 (a gene involved in uracil biosynthesis), and HIS3 (a gene involved in histidine biosynthesis). For example, the leu2 host strain (in yeast genetics, a mutated, and hence usually functionally defective, allele is denoted in lower-case letters) will not grow in a minimal medium — that is, a medium containing only the requisite minerals and a carbon source — but the same strain harboring a LEU2-containing plasmid will grow, because it is able to synthesize leucine.

YIps lack an origin for replication that can be recognized by the yeast DNA synthesis machinery. Therefore, they can be maintained in yeast cells only when they become integrated into a yeast chromosome (usually by homologous recombination at the site of the yeast marker gene or one of the other yeast sequences present in the vector). Once integrated, they are inherited quite stably as a part of the yeast genome. However, integration is a rare event, so the frequency of transformation with plasmids of this type is extremely low (1–100 transformants per microgram DNA compared to the 100,000 transformants per microgram that can be obtained with *E. coli*). The frequency of integration can be enhanced somewhat by cutting the plasmid within the region of

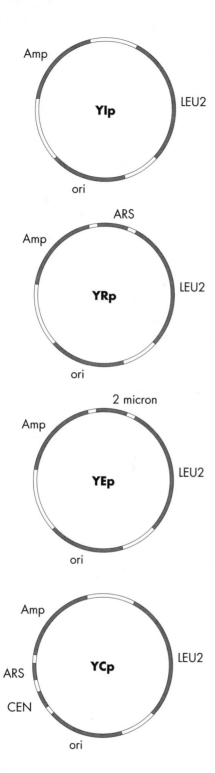

Figure 4.3
Plasmid vectors useful for cloning in yeast. Examples of four types of vectors are shown. Abbreviations: ori, *E. coli* origin of replication; Amp, ampicillin resistance gene for selection in *E. coli*; LEU2, a gene involved in leucine biosynthesis, for selection in yeast. Regions controlling replication and segregation in yeast, such as ARS, CEN, and 2-μm plasmid sequence, are described in the text.

yeast homology, a procedure that promotes homologous recombination to some degree (Box 4.3). Another drawback of these plasmids is their low copy number. Usually only one copy at most is integrated in one yeast cell, effectively limiting the level of expression of the cloned gene. One way to circumvent this problem is to design the plasmid to integrate into genes that exist in multiple copies in the yeast chromosomes. For example, there are more than 100 copies of the genes coding for rRNA in a yeast cell, so multiple integrations into these sites could create a cell genome with many copies of the cloned genes. Alterna-

BOX 4.3 Homologous Recombination Process

Homologous recombination begins with alignment of the homologous regions in two parental DNA duplexes. This is followed by "nicking" (single-stranded cleavage) of one parent-DNA helix and generation of single-stranded "whiskers" (step 1). A whisker wanders into the other duplex and forms Watson-Crick pairs with the complementary strand of the other DNA duplex (step 2). Finally, the end of the whisker is joined covalently to one of the strands of the other duplex, completing the process of crossing over (step 3). This mechanism requires an initial cut in one or both strands. Consequently, using plasmids that are already cut in the region of homology increases the frequency of recombination. More comprehensive schemes for the entire recombination pathway have been proposed [Orr-Weaver, T.L., Szostak, J.W., and Rothstein, R.J. (1981), Yeast transformation: A model system for the study of recombination, *Proc. Natl. Acad. Sci. USA* 78:6354–6358].

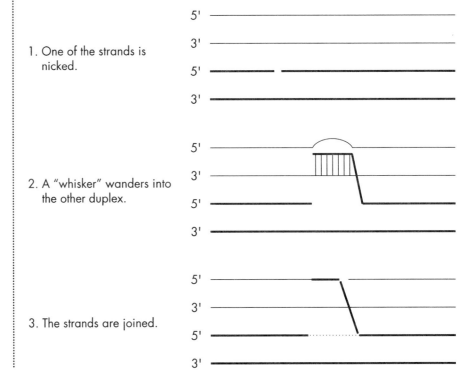

1. One of the strands is nicked.

2. A "whisker" wanders into the other duplex.

3. The strands are joined.

tively, one can use, as the selectable marker on the YIp vector, a gene that has to exist in a large number of copies for the yeast to survive under certain conditions. For example, if the plasmid contains the CUP1 gene, which codes for metallothionein, a protein that protects yeast cells by binding to heavy metals, yeast cells will survive in a medium containing Cu^{2+} only when a large number of copies of the CUP1 have been integrated into the genome—that is, when the gene has become "amplified." YIps are quite useful in spite of their typically low copy number, because plasmid stability can often become a major problem with other kinds of yeast vectors (see below).

Yeast Replicating Plasmids

In addition to selection markers useful in yeast, YRps (Figure 4.3) contain an origin of replication derived from the yeast chromosome and termed *ARS* (autonomously replicating sequence). With this origin, the plasmids can replicate without having to be integrated into the chromosome. However, yeast cells divide unequally by budding. In the process, only a disproportionately small fraction of the plasmids that were present in the mother cell are partitioned off into the buds, and many of the progeny cells are likely to lack plasmids entirely. Thus YRp plasmids are lost rapidly unless constant selection pressures are applied. Consequently they are not very useful for the reproducible expression of cloned genes.

Yeast Episomal Plasmids

Most strains of *S. cerevisiae* contain an endogenous, autonomously replicating, high-copy-number plasmid called 2-μm plasmid. The origin of this plasmid is added to YIps to produce YEps (Figure 4.3), which can exist in high copy numbers (30–50 copies per cell). (An *episome* is a genetic element that can exist either free—as a plasmid—or as a part of the cellular chromosome.) Like YRps, YEps are poorly segregated into daughter cells, but they are maintained more stably because of their higher copy number. If the entire 2-μm DNA (6.3 kb) is inserted into a YIp plasmid and introduced into yeast cells that lack an endogenous 2-μm plasmid, copy numbers in excess of 200 per cell can be achieved under certain conditions. Plasmids of this type are obviously most suitable when high-level expression of a foreign gene is desired. One vector of this type, pJDB219, contains an intact LEU2 gene but not its promoter. Because of the lack of promoter, this construction, called *leu2-d*, does not produce the full-scale expression of the LEU2 protein. Nevertheless, an extremely low level of expression does occur, presumably because of the nonspecific binding of the RNA polymerase or a very weak "read through" from upstream genes. This low-level expression produces a detectable phenotype because even a small amount of the

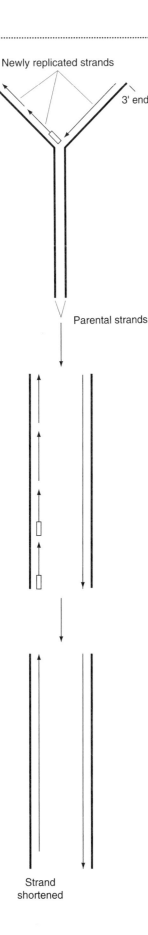

Newly replicated strands

5' end

3' end

Parental strands

Strand shortened

The ends of linear DNA duplexes, such as chromosomes and YACs, cannot be replicated faithfully, because DNA synthesis always occurs in the 5'-to-3' direction: One of the strands is thus replicated in short segments (Okazaki fragments) formed by the elongation of RNA primers (rectangles). When the RNA primer that is complementary to the very end of the DNA is degraded, there is no mechanism for synthesizing DNA to replace it. Consequently, such linear DNAs become shorter with each replication cycle. Eukaryotic chromosomes solve the problem by having repeated oligonucleotide sequences (telomeres) at their ends (for example, telomeres in human chromosomes have the sequence $[TTAGGG]_n$). When telomeres become too short, they are elongated by an enzymatic mechanism that does not require a template.

enzyme is enough to produce some leucine. But because the amount of the enzyme produced by a single plasmid is far from sufficient, the plasmid-carrying cells cannot grow at a reasonable rate in minimal medium unless the plasmid is present in very large numbers (200–300 per cell) in order to complement the completely defective $leu2^-$ allele of the host. In other words, this plasmid is designed so that its presence in high copy numbers will be favored.

Yeast Centromeric Plasmids

YCps (Figure 4.3) are YRps, or sometimes YEps, in which the sequence of a yeast centromere has been inserted. The centromeric sequence allows these plasmids to behave like regular chromosomes during the mitotic cell division, so YCps are faithfully distributed to daughter cells and are highly stable even without maintenance by selection. However, the "chromosomelike" behavior of these plasmids also means that their copy number is kept very low (1–3 per haploid cell). This is a potential disadvantage when the plasmids are used for the expression of cloned genes, although the expression can be increased by the use of highly inducible promoters.

Yeast Artificial Chromosomes

YACs are linear plasmids containing an ARS, a centromeric sequence, and, most important, a telomere (Box 4.4) at each end (Figure 4.4). These features allow the plasmids to behave exactly like chromosomes. Because the plasmid is linear, there is no limit to the amount of foreign DNA that can be cloned into it. This is the most important feature of YACs. Animal genes contain many introns and can exceed 100 kb in size. Such genes cannot be cloned in a single vector except in YACs. (Even cosmids can accommodate only up to 40 kb of foreign DNA; see Chapter 3). YACs, however, are not the first choice when the main objective is a high level of expression of foreign genes.

Figure 4.4
An example of a YAC vector. LEU2 gene is needed for selection in yeast, ARS for replication in yeast, and telomeres for stability in yeast.

ENHANCING THE EXPRESSION OF FOREIGN GENES IN YEAST

There are several points to consider when designing a system for expressing foreign-gene products in yeast cells.

Plasmid Copy Number

A high-copy-number plasmid of the YEp class is the best choice for maximal expression of any cloned gene. However, the expression of foreign proteins is often toxic for the yeast cells. Probably one of the main reasons is that some foreign proteins are likely to misfold in the cytoplasm, sequestering many of the chaperonin molecules needed for the correct folding and functioning of the yeast's own proteins (see Chapter 3). In this situation, low-copy-number YEps and, paradoxically, even YIps may produce higher sustainable yields than high-copy-number plasmids.

Another problem that may be important in commercial production runs is the instability of some of the plasmids. In commercial fermentation, the organism must last through a far higher number of generations than is usually necessary in a small, laboratory-scale experiment. Thus even a moderate degree of plasmid instability can cause a major problem.

Promoter Sequence

Because promoters of foreign origin are unlikely to be expressed efficiently in yeast cells, the coding sequence of a foreign gene is usually inserted behind a strong yeast promoter. Yeast promoters are quite different from bacterial promoters. Although both contain AT-rich recognition sequences for RNA polymerase [typically TATATAA for yeast, in contrast to the TATAAT consensus sequence (Box 4.5) for *E. coli*], the "TATA sequence" in yeast is located much further upstream (40–120 bases) from the mRNA initiation site than it is in *E. coli*, in which the "TATAAT," or Pribnow box, is typically located only 10 bases upstream of the transcription initiation site. In addition, yeast promoters usually require an upstream activator sequence (UAS), an

> ┄┄┄┄┄┄┄┄► **BOX** 4.5 **Consensus Sequence**
>
> Promoters of various genes share a common stretch of nucleotide sequence, with only minor variations. Such homologous sequences, and often their idealized versions, are called consensus sequences. Actual sequences may deviate significantly from idealized sequences: for example, the −10 sequences for *E. coli trp* operon (TTAACT) and *lac* operon (TATGTT) are similar to but not completely identical with the consensus TATAAT.

enhancer-like sequence located very far upstream (100–1,000 bases) from the transcription initiation site. Because of the location of the UAS, most yeast expression vectors contain a long "promoter sequence" (typically around 1 kb). Two frequently used promoters are the upstream sequences for an alcohol dehydrogenase gene (ADH1) and for a triose phosphate dehydrogenase gene (TDH3). ADH1 was thought to be expressed constitutively at a high level; and its use was popular at one time. However, we now know that this particular isozyme of alcohol dehydrogenase becomes repressed when the culture reaches a high density, so its use has fallen off. (In contrast, another isozyme of alcohol dehydrogenase, ADH2, becomes derepressed when glucose in the medium becomes exhausted. The promoter for this enzyme is often used as a regulatable promoter, as we shall see).

If the expression of the foreign protein inhibits the growth of the yeast cells, it becomes necessary to use regulatable promoters and to initiate expression of the foreign genes only when the culture has reached a high density. For example, the genes involved in galactose catabolism, GAL1, GAL7, and GAL10, have been extensively used as sources of regulatable promoters for cloned genes, because they are repressed in the presence of glucose but are induced by the addition of galactose to the medium. The regulation of these genes involves the binding of a positive activator, GAL4, to upstream sequences of GAL1, GAL7, and GAL10. Thus if the recombinant DNA containing the latter genes exists in multiple copies in a cell, and GAL4 is expressed from a single copy of the gene on the chromosome, the GAL4 protein in the cell might become exhausted by binding before all the recombinant genes are activated. However, this limitation can be removed if GAL4 is also introduced into the vector so that multiple copies of GAL4 are present in a cell. A drawback of this system is that it tends to increase the expression of the cloned gene even in the absence of galactose, so it is dangerous if the product is toxic to yeast cells. Other regulatable promoters that have been used include ADH2 (alcohol dehydrogenase, regulated by ethanol and glucose) and PHO5 (acid phosphatase, regulated by phosphate). Another attractive regulatable promoter is the one for CUP-1, which codes for metallothionein, a Cu^{2+}-binding protein, and which is induced by the addition of metal ions such as Cu^{2+} or Zn^{2+} to the medium. Several systems that are induced by elevated temperatures have been used successfully in the laboratory, but it may be difficult to get the temperature to change fast enough in a large fermentation tank.

Several hybrid promoters have also been used. These contain (1) a UAS from a regulatable promoter for controlling the level of expression of the gene and (2) the TATA box region from a strong, constitutive promoter for increasing the maximal level of expression. For example, a hybrid promoter containing the UAS sequence of ADH2 and the downstream sequences (containing the TATA box) from the TDH3 promoter has been effective in producing some foreign proteins at levels sometimes exceeding 10% of the total yeast protein.

Transcription Termination and Polyadenylylation of mRNA

In higher animals, termination of transcription usually occurs very far downstream from the coding sequence. Following termination, the nascent RNA transcript is cleaved at or near the cleavage signal, AAUAAA, present hundreds of nucleotides upstream from the transcript's 3′-terminus. This newly exposed 3′-terminus is then polyadenylylated—that is, a stretch of A is added. In yeast, this processing follows a rather different pattern, with polyadenylylation apparently occurring quite close to the 3′-end of the transcript. Unfortunately, the precise structure of a yeast terminator sequence is still not very clear. Because of this uncertainty, when recombinant DNA is constructed for gene expression in yeast, a large segment of "terminator" sequence, taken from the downstream sequence of yeast genes whose transcription is terminated efficiently, is usually placed downstream of the gene to be expressed.

Stability of mRNA

Yeast mRNAs differ greatly in their stability. The sequences that determine their degradation rates have been located in the 3′ untranslated regions and the coding regions of mRNA, but this knowledge does not appear to have been used successfully to increase the stability of foreign-gene transcripts in yeast.

If the gene is not followed by an effective terminator sequence, this usually produces unstable mRNA, presumably because it lacks the proper 3′-end that could become protected by polyadenylylation. Repeated experiments have shown that such a situation drastically decreases the yield of foreign proteins in yeast. Thus it is desirable to clone an efficient yeast terminator sequence downstream from the coding sequence of the foreign gene to be expressed, as described above.

Recognition of the AUG Initiation Codon

Efficient synthesis of mRNA, though necessary, is not sufficient in itself to ensure high-level production of a protein. Another basic requirement is efficient translation, for which the correct AUG codon must be readily recognized by initiation factors and by the ribosome machinery. In bacteria, this involves pairing of the Shine–Dalgarno sequence with the complementary sequence in 16S rRNA. There is no equivalent recognition sequence in eukaryotes, but the efficiency of translation initiation is known to depend on the sequences surrounding the AUG codon (such surrounding sequences are often called *context*). Analysis of gene sequences in yeast has shown that the consensus sequence (see Box 4.5) of the context is xxxAxA**AUG**UCU (where x is usually either A or U) (this is called Kozak's rule), quite different from the one in animal and plant mRNAs. However, efficient expression does not have an absolute requirement for this consensus sequence. In fact, although changing both of the −1 and −3 positions (relative to the initiation codon, AUG) to bases other than A had a rather drastic effect, changing the most important base at the −3 position from A (best) to U (worst) decreased the expression only by 50%.

If the 5′ untranslated segment of the mRNA tends to form base-paired loops, translation can be inhibited quite significantly. G occurs infrequently (taking about 5% of the positions) among the 20–40 bases immediately preceding the AUG codon; the presence of a large number of G residues in this region is known to inhibit translation initiation. These facts should be taken into consideration when designing the part of the DNA sequence that codes for the 5′-terminal portion of the mRNA.

Elongation of the Polypeptide

The genetic code is degenerate; in other words, several different codons may code for one amino acid. However, a given species will use certain of these codons more frequently than the others, and it will usually contain a more abundant supply of tRNAs for those frequently used codons than for the others.

Foreign genes often contain codons that are rarely used by yeast, and this may slow the translation process. Strings of rare codons occurring close together are especially detrimental. To enhance the expression of foreign genes in such cases, codons that yeast prefers have been substituted for those rarely used by yeast, usually by the site-directed mutagenesis procedure described in Chapter 3. The preferred yeast codons can be determined by analyzing the codons used in endogenous genes that are continuously expressed at high levels, such as genes for yeast glycolytic enzymes.

Folding of the Foreign Protein

Many foreign proteins have been expressed at a high level in yeast cells and have been shown to fold correctly. For example, the hepatitis B virus core protein, the P-28-1 protective antigen of the schistosome, and human superoxide dismutase have all been expressed to a level that corresponds to 20–40% of total yeast protein, and yet the proteins do not appear to misfold or to form intracellular aggregates. This is in a striking contrast to the nearly ubiquitous formation of aggregates or inclusion bodies when foreign proteins are expressed in the cytoplasm of bacteria (see Chapter 3). Although the formation of such aggregates has been reported in yeast, they are not found nearly so frequently. This could be due to the presence, in the yeast cell cytoplasm, of many kinds of molecular chaperones (the so-called heat-shock proteins) (discussed briefly in Chapter 3).

Proteolysis

Many proteins in eukaryotic cells are subject to degradation by the ubiquitin pathway. These proteins have at their N-terminus certain amino acids that are recognized by a small protein, called ubiquitin, that tags them for proteolytic degradation. All eukaryotic proteins are translated with methionine at the N-terminus, but subsequent removal of the N-terminal amino acid residues may expose one of the "destabilizing" amino acids and lead to destruction of the protein. If this problem exists in a cloning situation, one way of solving it is to alter the N-terminal amino acid sequence. Another recourse is to fuse the protein to the N-terminal segment of another protein, preferably of yeast origin, that is known not to be degraded by this pathway.

Glycosylation

Animal proteins secreted through the endoplasmic reticulum-Golgi pathway (see Box 4.2) are usually glycosylated in the process. As we have noted, this may help the proteins fold correctly and make them more resistant to proteases. Such posttranslational modifications do occur to foreign proteins cloned in yeast cells (if the proteins successfully enter the secretion pathway, see also below), but the yeast system can add only the high-mannose type of oligosaccharides, not the complex type most common in the glycoproteins of higher animals (see Figure 4.1). Sometimes this affects the folding and protease sensitivity of the protein and, more important, the half-life of the protein *in vivo* (see Figure 4.2).

Example: Hepatitis B Virus Surface Antigen

The commercial production of the hepatitis B virus surface antigen (HBsAg) in yeast, a process that led to the first recombinant DNA vaccine licensed in the United States (Chapter 5), illustrates several of the features we have discussed.

HBsAg is a major component of the envelope of the hepatitis B virus, and immunization with this protein was known to confer good protection against viral infection (such a substance is called a protective antigen). The coding sequence for this 226-residue protein was identified on the virus genome, and it was successfully inserted into YEp-type yeast cloning vectors in several laboratories in the early 1980s (Figure 4.5). Remarkably, HBsAg folded correctly in yeast and became assembled in the form of empty envelopes, or "22-nm particles," making the subsequent purification somewhat easier. Several years later, the production of HBsAg was commercialized by two companies, Merck, Sharpe & Dohme in the United States and Smith Kline-RIT in Belgium.

S. cerevisiae strains transformed with the first-generation recombinant plasmids produced only small amounts of HBsAg. For example, pHBS-16 (see Figure 4.5), the first plasmid reported to produce HBsAg in yeast, made no more than 25 μg of HBsAg per liter of culture. The subsequent development that led from this plasmid to establishment of the commercial production process at Merck, Sharpe & Dohme is unfortunately not documented in detail in the open literature. However, some of the improvements carried out at Smith Kline-RIT have been documented, so we can get a glimpse of what they entailed.

Plasmid Copy Number. YEp vectors appear to be the most suitable for high-level expression because of their high copy numbers, and they were used in the production of HBsAg. As we have said, it is possible to increase the copy number of the plasmids by replacing LEU2 with the promoterless *leu2-d*, so that only cells containing hundreds of copies of the plasmid can make enough leucine to survive. The Smith Kline-RIT group tried such an "improved" vector for HBsAg production but found that the cells rapidly lost the capacity to make HBsAg. It seems likely that the cells were losing the portion of the plasmid that coded for HBsAg. After all, when HBsAg is constitutively expressed (see below), a high level of this foreign protein is likely to be deleterious to the growth of the host cell. Therefore, progeny cells that inherit the *leu2-d*-containing part of the plasmid (which is essential for growth) but fail to inherit the gene for HBsAg are more likely to flourish. This example shows that one cannot blindly apply methods that are supposed to work better without testing and taking numerous factors into account. Production of foreign proteins is rarely neutral for the host, and one should always be alert to their possible toxic effects.

Promoter Sequence. In the first-generation plasmids pHBS-16 (Merck) and pRIT10764 (Smith Kline-RIT), the promoter sequences came from the alcohol dehydrogenase (ADH1) and ornithine carbamoyl-transferase (ARG3) genes, respectively (see Figure 4.5). In the improved production strains used at both Merck and Smith Kline-RIT, the promoter comes from the gene for glyceraldehyde 3-phosphate dehydro-

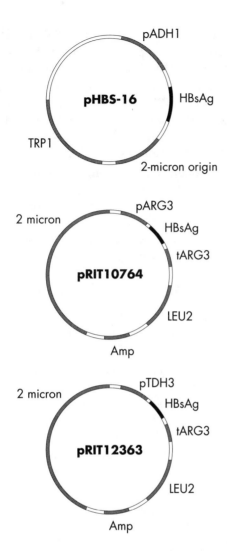

Figure 4.5
Recombinant plasmids used for the production of HBsAg. The "first-generation" plasmids include pHBS-16 and pRIT10764. These were further developed for commercial use by Merck, Sharpe & Dohme and Smith Kline-RIT,
respectively. pRIT12363 is an improved expression plasmid said to be in
use at Smith Kline-RIT. Here p indicates a promoter sequence, and t denotes a terminator sequence. TRP1 in pHBS-16 is a yeast gene coding for an
enzyme of the tryptophan biosynthetic pathway and serves as the selective
marker in yeast.

genase (TDH3). The TDH3 promoter is especially powerful, as one
might have predicted from the fact that the dehydrogenase expressed
from this promoter constitutes 5% of the total yeast protein. Clearly,
use of the TDH3 promoter was advantageous.

As mentioned above, ADH1 was later found to become repressed
toward the end of the growth cycle, so TDH3 remained preferable. In

pRIT10764, the scientists chose a host strain with a leaky mutation in arginine biosynthesis so that the cells would be starved for arginine and so that the expression of ARG3, a gene involved in arginine synthesis, could be sustained at a high level (for regulation of amino acid biosynthetic genes, see Chapter 12). However, the paucity of arginine slowed the growth of the culture, again creating a less favorable situation for commercial fermentation. In these cases, there are rational explanations why the use of TDH3 promoter was preferable, but we must emphasize that in general it is difficult to predict the levels of expression of foreign genes from the levels of expression of the endogenous yeast genes. There are many reasons why foreign genes may not be expressed as efficiently as host genes: instability of the mRNA, possible effects of the untranslated 5' sequences of mRNA on the efficiency of translation initiation (see the next section), and the possibility that the coding sequences of the yeast genes contained enhancerlike sequences that were absent in the cloned foreign gene.

Transcription Termination and Polyadenylylation of mRNA. Of the first-generation plasmids pHBS16 and pRIT10764, the latter was reported to produce a higher yield of HBsAg—about 200 μg/L of culture compared to the reported yield of less than 25 μg/L for pHBS16. Although much of this difference could be due to trivial factors such as the different quantitation methods used in various laboratories, there is an obvious difference between the two plasmids that could have contributed significantly to the higher yield of HBsAg in strains containing pRIT10764. In this plasmid, the HBsAg sequence is followed by a terminator sequence taken from the downstream sequence of the ARG3 gene, whereas no special terminator sequence is present in pHBS16.

Recognition of the AUG Initiation Codon. Another difference between pHBS16 and pRIT10764 is the relative content of G residues directly upstream of the AUG codon. We have noted that large numbers of G residues in this region inhibit the initiation of translation. Of 25 bases in this region in pHBS16, 9 are G residues (36%), far more than the proportion found in native yeast promoter sequences. In contrast, pRIT10764, which produced a higher reported yield, contains only 3 G residues, well within the range found in native yeast promoters.

Glycosylation, Folding, and Acetylation. HBsAg made in human cells is N-glycosylated. This suggests that it is exported to the cell surface via the endoplasmic reticulum–Golgi pathway, in spite of the fact that there is no typical, cleaved, signal sequence at its N-terminus. When HBsAg is made in yeast cells, it is not glycosylated, and the protein accumulates in the cytoplasm without entering the endoplasmic–Golgi pathway. Perhaps the cloned sequence is incomplete. In the hepatitis B virus, the HBsAg sequence is preceded by an upstream "preS" sequence. Transcription in human cells may start at the preS sequence, which may contain the export signal. (Analysis of RNA transcripts is difficult with hepatitis B virus, because it cannot be grown in cultured cells). When the HBsAg sequence is cloned and expressed together with the upstream extension, the product *is* glycosylated in yeast, a result that is consistent with the hypothesis that the preS sequence contains the export signal.

Despite the lack of glycosylation, HBsAg obviously folds correctly. This and the assembly of the protein into 22-nm particles presumably are important in achieving the desired overproduction of the antigen; if HBsAg were folded incorrectly to produce inclusion bodies in the cytoplasm, this would tie up foldases and chaperonins (Chapter 3) that are needed for the folding of essential proteins of yeast, thereby interfering with the growth of the host cells.

The N-terminus of HBsAg becomes acetylated when produced in human cells; in yeast, at least a fraction of the HBsAg molecules become acetylated.

Fermentation Conditions. Some seemingly minor improvements in the fermentation conditions can have major effects on the yield. With the Smith Kline-RIT strains, the initial recombinant plasmid pRIT10764 was reported to produce HBsAg to a level of 0.06% of total yeast protein. Two years later, investigators in the same company reported a yield of 0.4% with the identical plasmid—an improvement presumably caused by a fine-tuning of culture conditions. However, the use of the ARG3 promoter still limited the growth of yeast cells to about 1 g/L. In pRIT12363, which was used for the production strain, use of the TDH3 promoter increased expression to about 1% of the yeast cell protein. However, the major improvement seems to have resulted from the fact that whereas pRIT10764 necessitated the use of leaky arginine biosynthesis mutants as the host, with pRIT12363, prototrophic strains could be used as the host, resulting in a much higher final density of yeast cells: about 60–70 g/L.

EXPRESSION OF FOREIGN-GENE PRODUCTS IN A SECRETED FORM

As with bacterial hosts, it is advantageous in many ways if the yeast cell secretes the foreign-gene products into the medium. First, because *S. cerevisiae* does not naturally produce many extracellular proteins, purification of the products is much simpler: One does not have to start from a mixture containing many hundreds of other cytoplasmic proteins. Second, the secreted protein goes through the endoplasmic reticulum–Golgi pathway, where disulfide bonds—and hence, a stable protein—may be formed under optimal conditions with the help of protein disulfide isomerase, which is present in the lumen of the endoplasmic reticulum. Indeed, α-interferon secreted by yeast cells has been shown to have disulfide bonds at the same positions as α-interferon made by human cells. In contrast, when the same protein is made in the cytoplasm, a large fraction of it appears to become misfolded. Third, the proteins may become glycosylated during their passage through the endoplasmic reticulum and Golgi apparatus. Fourth, hormone precursors may be made into mature products by processing proteases during their secretion by yeast.

Proteins are brought into the endoplasmic reticulum–Golgi pathway when the components of the secretory pathway recognize their hydrophobic signal sequence (see Box 4.2). It is possible to design a recombinant plasmid so that the protein will enter this pathway, by fusing the DNA coding for an effective signal sequence to the coding sequence for the protein. Secretion vectors, which already contain DNA segments coding for the signal sequence, are useful in producing such recombinant plasmids. Signal sequences for secreted invertase (SUC2) and for secreted acid phosphatase (PHO5) have been used in this way and have resulted in the successful secretion of several animal proteins of interest. In some cases, however, a large fraction of the secreted foreign protein remains trapped within the yeast cell wall. This is reminiscent of certain secreted yeast proteins that do not seem to become freely dispersed in the culture medium. We can release these proteins into the medium by digestion of the cell wall, so it is clear that they are not anchored to the plasma membrane; rather, they appear to be trapped in the space between the cytoplasmic membrane and the cell wall. This problem has led to the exploration of yeast mechanisms that produce the genuine secretion of peptides into the surrounding medium.

In *S. cerevisiae*, one such system produces and excretes the mating factor α, a 13-residue peptide. The immediate product of its structural gene, MFα1, is a 165-residue polypeptide containing an N-terminal signal sequence and four copies of the α-factor sequence. The α-factor sequences are separated by a spacer with the sequence Lys-Arg-(Glu-Ala)$_n$ (n = 2 or 3) (Figure 4.6). After cleavage of the signal sequence in the lumen of the endoplasmic reticulum, the polypeptide undergoes further proteolytic processing in the later stages of the secretion process. First, KEX2 protease cleaves it between Arg and Glu. Then the peptide is shortened from both ends, KEX1 carboxypeptidase removing the Arg and Lys residues from the C-termini and STE13 dipeptidyl aminopeptidase removing Glu-Ala units from the N-termini. This complex processing scheme may prove useful to biotechnologists, because it may afford them some flexibility in the design of fusion joints. In many laboratory experiments, genes coding for animal and plant proteins have been fused to the N-terminal portion of the MFα1 gene. In most cases, the products were successfully secreted, although in some cases the processing with the STE13 enzyme was not complete, leaving extra Glu and Ala residues at the N-terminus of the secreted protein.

In yeast the Asn-linked core oligosaccharides that are attached to the *N*-glycosylation sites of foreign proteins sometimes become

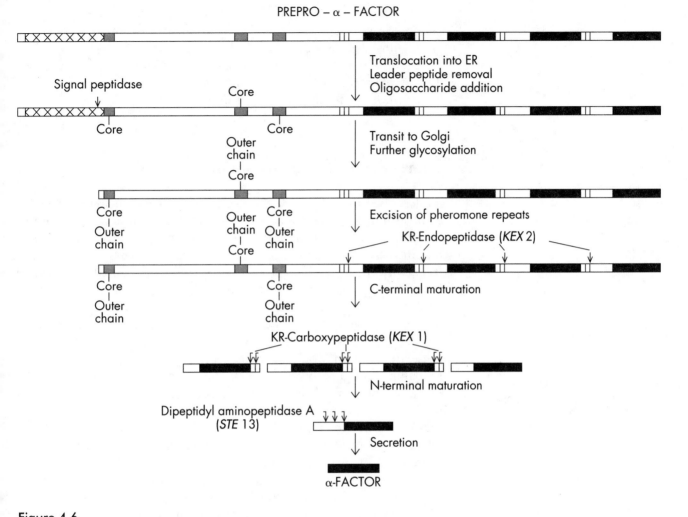

Figure 4.6
The processing of α-factor within the secretion pathway. From Fuller, R.S., Sterne, R.E., and Thorner, J., 1988. Enzymes required for yeast prohormone processing. *Ann. Rev. Physiol.* 50: 345–362. With permission from the Annual Reviews, Inc.

extended into large, outer chains, producing high-mannose-type oligosaccharides (see Figure 4.1). These enormous oligosaccharides may impair the proper folding and functioning of the animal-derived proteins. It may therefore be advantageous to use mutants, such as *mnn*9, that are defective in the addition of outer-chain mannose residues (see Figure 4.1). For example, human α-1-antitrypsin, secreted from a *S. cerevisiae mnn*9 mutant, carries three N-linked oligosaccharides similar in size to those attached to the human protein.

One general problem with yeast secretion systems is low yield. However, screening of mutagenized yeast cells that contain secretion plasmids has produced high-secretion mutants. In one case, the combination of two mutations produced a strain that secreted 80% of the prochymosin synthesized. Although such an increase in secretion does not occur with all secreted proteins, continued studies in this area may ultimately enable us to design optimal procedures for the synthesis and secretion of any foreign protein of interest.

Expression of Prochymosin in Yeast

As described in Chapter 3, the expression of calf prochymosin in the cytoplasm of *E. coli* resulted in the formation of inclusion bodies. It was thought that this problem might be overcome by expression of the protein in yeast cells, because inclusion bodies are apparently formed less frequently in yeast. The prochymosin gene was cloned into several YEp-type expression plasmids behind effective yeast promoters. However, the accumulated product was largely insoluble when overproduced, although lower-level expression produced some soluble protein.

Better results were expected with the yeast secretion vectors, because the protein would then be secreted through the endoplasmic reticulum–Golgi pathway, which is similar to that in animal cells. The recombinant plasmid that was tested contained (1) a strong yeast promoter, such as the one for the triose phosphate isomerase gene, (2) the DNA sequence coding for the signal sequence and several of the N-terminal amino acid residues of the mature invertase, a secreted yeast protein, and (3) the sequence for prochymosin fused to the invertase fragment. These YEp-type plasmids directed the secretion of prochymosin, but the fraction secreted was quite low, usually less than 5%. Mutagenesis and screening of the host strain have refined the system to the point where up to 80% of the synthesized fused protein is secreted from the yeast cell, as described above. However, the reported yield is still rather low—around 1 mg/g of total yeast protein—even in the best combinations of host with plasmid. A possible explanation is that *S. cerevisiae* normally secretes only a very small fraction of its cellular proteins across its cytoplasmic membrane; in wild-type strains, secreted invertase corresponds to much less than 0.1% of the total cellular protein. For this reason, the recent trend has been to explore other, more secretion-competent species of yeast. However, in some cases it is possible to obtain high-secretion strains from *S. cerevisiae*. A recent success combined chemical mutagenesis with an efficient selection for host cells that have a higher capacity for secretion. When plasmids with a powerful promoter (for yeast vacuolar endoprotease B) were introduced into these improved yeast strains, large amounts of mammalian proteins were secreted; for example, the secretion of 140 mg/liter of human albumin was achieved. To date, however, there have been no reports of successful prochymosin secretion using this system.

As we mentioned, there are yeast species that naturally secrete proteins more effectively than *S. cerevisiae*. In one experiment a recombinant plasmid was made in which the sequence coding for prochymosin was placed between a strong LAC4 promoter and the LAC4 terminator sequence from *Kluyveromyces lactis*, a lactose-utilizing yeast species. When this plasmid was linearized and integrated into the *Kluyveromyces* genome, there was only a low-level expression of prochymosin. But most of the prochymosin was secreted into the medium, even though the cloned DNA lacked the sequence coding for the signal sequence. When the prochymosin gene was cloned together with the sequence coding for its own signal sequence, the prochymosin production increased 50- to 70-fold, and 95% of the product appeared in the medium in a correctly processed form. The yield was reported as about 100 enzyme units/mL, which corresponds roughly to 1 g/L, or about 10% of the total cellular protein. Other yeast species that have been shown to produce (and secrete) foreign proteins at a higher level than *S. cerevisiae* include *Pichia pastoris* and *Hansenula polymorpha*, both methylotrophic yeasts (yeasts capable of using methanol as the carbon source), and *Yarrowia lipolytica*, which can grow on alkanes. *P. pastoris*, for example, produces HBsAg to a level of about 50% of the total cellular protein.

There are also nonyeast fungal species that are known to secrete very large amounts of proteins; for example, *Trichoderma reesei* and *Aspergillus awamori* secrete more than 20 g of protein per liter. These species were hypothesized to be even more proficient in catalyzing the export of large amounts of foreign proteins. Initial yields, obtained after cloning of the prochymosin gene in an expression vector, were not exceptional, but several optimization steps increased the yield significantly. These procedures included inactivation of the gene that encodes a prochymosin-inactivating protease and fusion of the prochymosin sequence to the 3′-end of a complete sequence coding for a glucoamylase, an enzyme secreted in very large amounts by *A. awamori*. Such modifications increased the yield to the range of 100 mg/L. Finally, random mutagenesis and screening of the host *A. awamori* strain increased the yield to about 1 g/L, apparently a level that would make production commercially profitable. Importantly, the high-secretion mutant strain selected on the basis of prochymosin production also secretes other foreign proteins at a higher efficiency.

With a variety of tools for solving the production problem—chief among them the approaches described above—several laboratories are now attempting to modify, by site-directed mutagenesis, the structure of the chymosin molecule itself (the active enzyme produced by autocatalytic cleavage of prochymosin) so as to improve further its properties as a specific protease. Because the three-dimensional structure of chymosin is known from X-ray crystallographic studies, it is possible to select, on a rational basis, the residues that would influence the specificity and catalytic throughput of this enzyme. Thus chymosin is a well-suited object for such an effort.

SUMMARY

Saccharomyces cerevisiae and other yeast species have considerable potential as host organisms for the production of foreign proteins, especially proteins of animal origin, by the use of recombinant DNA. Many different vectors are available, and most are shuttle vectors, which allow the recombinant DNA manipulations to be conveniently carried out in *E. coli* before the final recombinant product is introduced into yeast. The expression of cloned genes in yeast can be maximized by the use of proper promoter and terminator sequences and by optimization of codon usage and of the sequence surrounding the initiation codon AUG. To avoid triggering the ubiquitin pathway, which starts the proteolytic degradation of a protein, it may be necessary to alter the N-terminal sequence of the protein or to fuse another protein to the N-terminus.

One major advantage of expression in yeasts is that foreign proteins appear to have less tendency to become misfolded in yeast than in bacterial hosts, partly because yeast cells presumably contain more efficient chaperonins and foldases. Furthermore, proteins can become glycosylated in yeast cells if they can be introduced into the endoplasmic reticulum–Golgi apparatus protein-secretion pathway. Glycosylation not only facilitates the correct folding of some proteins but also makes them less susceptible to degradation in the animal body, thus prolonging their half life when they are administered as therapeutic agents. However, N-linked oligosaccharides synthesized in yeast cells are usually of the high-mannose type, not the complex type most frequently found in proteins synthesized by animal cells. Some proteins can also become acylated at their amino terminus when produced in yeast; this modification, which occurs frequently in animal cells, helps prevent the proteolytic degradation of the synthesized protein within the cell.

Hepatitis B surface antigen (HBsAg) and prochymosin are two proteins of animal origin that have been successfully produced in yeasts. HBsAg did not enter the secretion pathway and was not glycosylated, however, apparently because the cloned DNA fragment lacked the segment coding for the signal sequence. Nevertheless, it was folded correctly and assembled into a structure resembling the envelope of the virus. When the cDNA for prochymosin was cloned into a secretion vector that supplied the signal sequence and the recombinant plasmid was introduced into an *S. cerevisiae* host, the protein entered the endoplasmic reticulum–Golgi pathway, was folded correctly and glycosylated, and was secreted, although the yield remained low. When non-*Saccharomyces* yeasts and nonyeast fungi that physiologically secrete very large amounts of proteins were used as hosts, commercially acceptable yields of secreted prochymosin were achieved.

SELECTED REFERENCES

General

Barr, P., Brake, A., and Valenzuela, P. (eds.), 1989. *Yeast Genetic Engineering*. Butterworth.

Marino, M.H., 1991. *Expression of heterologous proteins in yeast*. In Prokop, A., Bajpai, R.K., and Ho, C. (eds.), *Recombinant DNA Technology and Applications*, pp. 29–65. McGraw-Hill.

Bitter, G.A., Egan, K.M., Koski, R.A., Jones, M.O., Elliott, S.G., and Griffin, G., 1987. Expression and secretion vectors for yeast. *Meth. Enzymol.* 153:516–544.

Cregg, J.M., Vedvick, T.S., and Rashke, W.C., 1993. Recent advances in the expression of foreign genes in *Pichia pastoris*. *Bio/Technology* 11:905–910.

Robinson, A.S., Hines, V., and Wittrup, K.D., 1994. Protein disulfide isomerase overexpression increases secretion of foreign proteins in *Saccharomyces cerevisiae*. *Bio/Technology* 12:381–384.

Protein Glycosylation

Elbein, A.D., 1991. The role of N-linked oligosaccharides in glycoprotein function. *Tr. Biotechnol.* 9:346–352.

Goochee, C.F., Gramer, M.J., Andersen, D.C., Bahr, J.B., and Rasmussen, J.R., 1991. The oligosaccharides of glycoproteins: Bioprocess factors affecting oligosaccharide structure and their effect on glycoprotein properties. *Bio/Technology* 9:1347–1355.

Hepatitis B Virus Surface Antigen

Valenzuela, P., Medina, A., Rutter, W.J., Ammerer, G., and Hall, B.D., 1982. Synthesis and assembly of hepatitis B virus surface antigen particles in yeast. *Nature* 298:347–350.

Harford, N., Cabezon, T., Crabeel, M., Simoen, E., Rutgers, A., and de Wilde, M., 1983. Expression of hepatitis B surface antigen in yeast. In *Second WHO/IABS Symposium on Viral Hepatitis: Standardization in Immunoprophylaxis of Infections by Hepatitis Viruses* (Developments in Biological Standardization, vol. 54), pp. 125–130. S. Karger.

de Wilde, M.J., 1990. Yeast as a host for the production of macromolecules of prophylactic or therapeutic interest in human health care. In Verachtert, H., and de Mot, R. (eds.), *Yeast: Biotechnology and Biocatalysis*, pp. 479–504. Marcel Dekker.

Prochymosin Production

Smith, R.A., Duncan, M.J., and Moir, D.T., 1985. Heterologous protein secretion from yeast. *Science* 229:1219-1224.

Sleep, D., et al, 1991. *Saccharomyces cerevisiae* strains that overexpress heterologous proteins. *Bio/Technology* 9:183–187.

van den Berg, J.A., et al., 1990. *Kluyveromyces* as a host for heterologous gene expression: expression and secretion of prochymosin. *Bio/Technology* 8:135–139.

Buckholz, R.G., and Gleeson, M.A.G., 1991. Yeast systems for the commercial production of heterologous proteins. *Bio/Technology* 9:1067–1072.

Harkki, A., Uusitalo, J., Bailey, M., Penttila, M., and Knowles, J.K.C., 1989. A novel fungal expression system: secretion of active calf chymosin from the filamentous fungus *Trichoderma reesei*. *Bio/Technology* 7:596–603.

Dunn-Coleman, N.S., et al. 1991. Commercial levels of chymosin production by *Aspergillus*. *Bio/Technology* 9:976–981.

Recombinant and Synthetic Vaccines

I n developing countries, infectious diseases still cause 30–50% of all deaths. For many of the diseases that plague these regions, effective chemotherapeutic agents simply do not exist, and those that do exist are often far too costly for much of the population. Vaccines thus become the most important tool for fighting infectious diseases in those parts of the world.

The situation is very different in developed countries, where infectious diseases account for only 4–8% of all deaths. However, this does not mean that vaccines are not important there. The low rate of infectious diseases is in fact largely due to the widespread use of vaccination in developed societies (Figure 5.1). In addition to the well-known example of the smallpox vaccine, which has succeeded in completely eradicating the disease, the use of several other vaccines has also resulted in dramatic decreases in the incidence of numerous grave diseases. For example, at the beginning of this century, diphtheria (caused by the bacterium *Corynebacterium diphtheriae*) infected about 3,000 children each year in developed countries, out of a then total population of 1 million. Because diphtheria targets young children, this incidence corresponds to several percent of children of the susceptible age, and nearly one-tenth of the infected children died. Now, thanks to a mass immunization program, diphtheria incidence in the United States is less than 0.2 per million, a decrease of more than a thousandfold. Another example is poliomyelitis (caused by an RNA-containing virus). As recently as 1955, the U.S. and Canadian incidence of polio was 200 per million of the population, but the development of vaccines has decreased

169

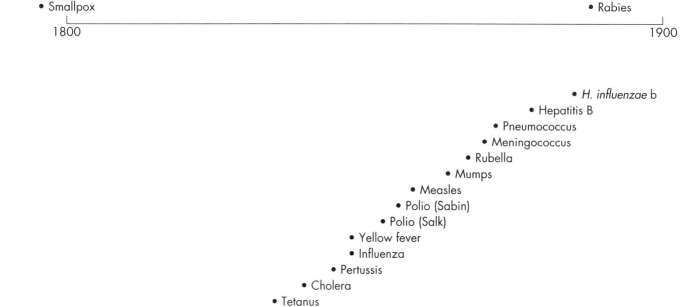

Figure 5.1
Vaccines since Jenner. From Warren, K.S. (1983), *Proc. Natl. Acad. Sci. USA* 83:9275–9277, with permission.

polio cases by more than 4000-fold, to less than 0.05 per million. Similar rapid decreases in incidence occurred for measles and rubella (German measles) after introduction of the vaccines in the 1950s and 1960s (see Figure 5.1).

A second reason that vaccines remain important in developed countries is efficiency. Vaccination is much less costly than treating people who are already sick. Modern antibiotics and other chemotherapeutic agents can be very expensive, not to mention the social cost of morbidity both in lost productivity and in increased allocation of resources to health care.

Finally, vaccines continue to play an important role in veterinary medicine, especially because cost pressures usually mean that farm animals are kept in tight quarters, a practice that enormously increases the chances of cross infection.

Effective vaccines can be made in traditional ways, and these are still used for prevention of many diseases. However, serious problems exist with many of these traditional vaccines. Recombinant DNA techniques and synthetic organic chemistry have had a major impact in overcoming these problems and in developing vaccines of a new type. Furthermore, these methods enable us to develop vaccines against diseases for which traditional vaccines do not exist.

PROBLEMS WITH TRADITIONAL VACCINES

Traditional vaccines are of two types, live and killed. *Live vaccines* include attenuated (weakened) viral or bacterial strains. Such attenuated strains have most often been obtained by totally empirical procedures, such as prolonged storage or cultivation under suboptimal conditions. *Killed vaccines* are either killed whole cells of bacteria or inactivated toxin proteins (called *toxoids*; Table 5.1). Many traditional vaccines are quite effective, but new ones are desperately needed—and new techniques for producing vaccines as well. We lack vaccines for many important diseases (Table 5.2), and certain of the traditional ones are ineffective or not entirely safe.

Foremost among the problems encountered with traditional live vaccines is the danger of reversion to the virulent state. For instance,

TABLE 5.1

▶ **Commonly Used Vaccines**

Live viral vaccines

Poliomyelitis
Mumps
Measles
Rubella
Yellow fever

Inactivated viral vaccine

Influenza

Viral subunit vaccine

Hepatitis B

Killed bacterial vaccines

Pertussis (whooping cough)
Typhoid fever

Inactivated bacterial toxins (toxoids)

Diphtheria
Tetanus

Some Diseases for Which Effective Vaccines Are Not Yet Available

Disease	Pathogen	Infections (millions/year)	Deaths (thousands/year)
AIDS	Virus	0.75	670
Diarrheal diseases	Usually bacteria	3000–5000	5000–10000
Malaria	Protozoa	800[a]	1200
Schistosomiasis	Trematodes	200	500–1000
Chagas disease	Protozoon	12	60
Leishmaniasis	Protozoon	12	5
African trypanosomiasis	Protozoon	1	5
Leprosy	Bacterium	(12)[b]	Low

TABLE 5.2

[a] This number may seem rather high, but only one infection out of five is estimated to produce recognizable disease.

[b] Leprosy is a chronic disease, so the number of new infections is difficult to estimate. The number given here is the estimated total number of patients.

SOURCES: The estimates of infection and death rates are from Walsh, J.A., and Warren, K.S. (1979), *New Engl. J. Med.* 301:967–974. The estimates on AIDS (for the period 1991–1992) are from Mann, J.M., Tarantola, D.J.M., and Netter, T.W. (eds.) (1992), *AIDS in the World* (Harvard University Press).

although the oral (Sabin) vaccine for poliomyelitis is generally considered quite safe, the nucleic acid sequence of a vaccine strain and that of the virulent parent strain differ by only two nucleotide substitutions. Mutant strains with such slight alterations do revert from time to time, and between 1973 and 1984, use of the Sabin vaccine produced an estimated 1 case of poliomyelitis for every 520,000 administrations of the first dose. Because most attenuated vaccine strains have been obtained by ill-defined, empirical procedures, the genetic alterations that make these strains nonpathogenic have not been defined. Many of these alterations may be as slight and as readily reversible as those in the Sabin vaccine. Another danger is that the viruses used in traditional vaccines have to be grown in tissue culture cells or in animals, which poses the risk of introducing cryptic (hidden) viruses from host cells. In one well-known case, a cell line used for propagation of the polio vaccine was found to contain a virus capable of producing tumors in experimental animals. Still another drawback is that even attenuated pathogens can produce severe disease in people with immune system deficiencies. This may be a serious problem in developing countries, where many malnourished children suffer from such deficiencies.

The chief problem with the traditional killed vaccines is that they themselves can cause severe reactions. For example, the typhoid, pertussis, and several other vaccines consist of whole killed cells of Gram-negative bacteria, whose outer membranes contain lipopolysaccharide (LPS; also called *endotoxin*; see Figure 2.3) as a major component. Even very small amounts of endotoxin may elicit strong host responses. In sensitive animals, such as rabbits, endotoxin in amounts as low as 1 ng/kg of body weight can produce a measurable increase in body temperature, and the consequences of administering endotoxin are not limited to fever. Crude killed-cell preparations may contain other toxic materials as well. Because of the widespread fear of side effects caused by such killed bacterial cell preparations, many governments have changed the status of pertussis vaccination of infants from compulsory (or highly recommended) to voluntary. A second problem is the direct risk run by the workers who cultivate dangerous pathogens in large amounts to manufacture the vaccines. Third is the possibility that the organism or the toxin in the vaccine may not be completely killed or inactivated. The killing or inactivation procedure is usually a mild one, designed to inactivate the organism or toxin without destroying its ability to produce specific immunity. In several widely publicized cases, mass infections and effects of toxins killed many who were "vaccinated," because the viral vaccines accidentally contained living viruses and the toxoid-based vaccines contained incompletely inactivated toxins. A final problem is that in some cases, producing a sufficient quantity of the infectious agent is extremely costly or even impossible. For example, to grow malaria parasites on a large scale using human blood cells would be prohibitively expensive, and there would be a significant risk of introducing contaminating viruses into the vaccine produced. An example of infectious agents that cannot be grown in tissue culture cells is the hepatitis B virus.

IMPACT OF BIOTECHNOLOGY ON VACCINE DEVELOPMENT

Developments in biotechnology have made it possible to produce entirely new kinds of vaccines. Some of these are directed at new targets; others are simply more effective or produce fewer side reactions than traditional vaccines.

Traditional vaccines are the products of pathogens, either the live or inactivated organisms or their secretions. In contrast, the new vaccines are produced by organisms we have come to regard as friends. Once researchers have identified a *molecule* that can produce specific immunity (among the thousands of components in a pathogenic microorganism), recombinant DNA methods can be used to produce this molecule, or *protective antigen*, in a safe nonpathogenic organism such as *E. coli* or yeast. This should make it possible to produce vaccines even when the pathogens are difficult or impossible to cultivate. Because the new vaccines contain only some of the molecules found in the original pathogen, they are often called *subunit vaccines*. The hepatitis B subunit vac-

An antigen is any molecule that elicits a specific immune response, either (1) the production of antibody proteins that have complementary binding sites or (2) the proliferation of lymphocytes (T effector cells) that have specific surface receptors again with complementary binding sites. In a narrower sense, an antigen is a molecule that binds to these complementary sites; such a molecule may be called an immunogen if it also elicits the immune response described above.

An antigen is described as protective if the immune response it elicits in an animal protects the animal from later infection by the pathogen containing the antigen.

cine, developed through recombinant DNA technology, is now licensed for human use, and several others are undergoing field trials.

A Subunit Vaccine for Hepatitis B

The hepatitis B virus, transmitted through contaminated needles and sexual contact, infects an estimated 200,000 Americans every year, 20,000 of whom become carriers. One in 5 carriers dies of cirrhosis of the liver, and 1 in 20 develops liver cancer. Yet the virus does not grow in tissue culture cells; until recently, it was available only from the plasma of carriers. Vaccines were made either by purifying the viral surface antigen or by inactivating living virus by chemical treatment (for example, with formaldehyde). However, this source of the virus was quite limited, and use of the killed vaccine always carried the risk that not all the particles were inactivated.

The surface antigen or the surface glycoprotein of the virus was known to be an effective vaccine, so the first step in producing a subunit vaccine was to clone the gene for this protein from the viral genome. The hepatitis B virus genome consists of a partially double-stranded DNA (Figure 5.2), which codes for a core protein as well as for the major surface protein (S protein); most of the protein subunits making up the viral envelope are S protein molecules (226 residues). The DNA coding for the S protein was inserted into a YEp plasmid vector behind an effective yeast promoter and before a terminator (the construction of the first-generation recombinant plasmid is shown in Figure 5.3; for plasmids made later, see Figure 4.5). Presumably, the expectation that yeast would glycosylate the envelope protein contributed to its choice as a host (see Chapter 4), but the protein produced in yeast was not glycosylated. Nevertheless, it seemed to have folded properly, self-assembling into a form that resembled an empty virus envelope 22 nm in diameter that was nearly indistinguishable from those found in the plasma of patients (Figure 5.4). (Maneuvers that increased the yield of the recombinant protein were already discussed in Chapter 4 on page 161). This yeast-produced vaccine, lacking the oligosaccharides, was as effective as the vaccine derived from human plasma, and in 1986 it became the first recombinant-DNA-based vaccine licensed for use in the United States. Before this, it took about 40 liters of infected human serum to produce a single dose of the hepatitis B vaccine; now we can obtain many doses of the recombinant vaccine from the same volume of yeast culture.

Potential Problems of Recombinant Subunit Vaccines

Recombinant DNA subunit vaccines, if they are effective, have many advantages over the traditional vaccines. They can be produced easily,

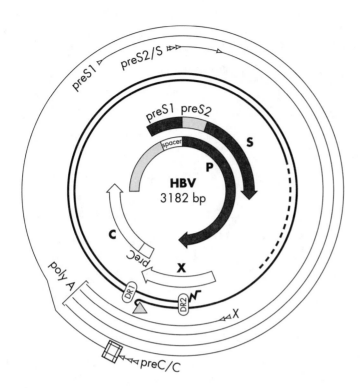

Figure 5.2
The genome of the hepatitis B virus. The figure shows the partially double-stranded genome (inner ring) and RNA transcripts (outer arcs). Open reading frames for four proteins [P, X, C, and S (with PreS2 and PreS1)] are also shown in the center. P protein is the DNA polymerase that synthesizes one strand of DNA by reverse transcription of the longest mRNA (preC/C in the figure) and then synthesizes the other strand by using the just-synthesized DNA strand as the template. The precise function of X protein, which is present in minute quantities, is not known. C protein is the major component of inner capsid, and S protein is the major surface (envelope) protein. Some transcripts cover only the S region, producing the HBsAg, but other transcripts also cover preS2 or both preS1 and preS2 regions, producing larger protein products. These latter, less abundant products containing PreS2 and PreS1-PreS2 are now known to comprise minor surface proteins of the virion. Although hepatitis B virus cannot infect tissue culture cells, its DNA can be introduced into cells by transfection, and this method made possible analysis of the transcription pattern of the genome. From Nassal, M., and Schaller, H. (1993), *Tr. Microbiol.* 1:221–228.

safely, and inexpensively and are devoid of all the extraneous components of the pathogen that often cause undesirable side effects. Another advantage is that there is absolutely no possibility that a living pathogen will be present in the subunit vaccine.

If the subunit vaccines produced by recombinant DNA technology are really so effective and so advantageous, why have they not yet replaced the traditional vaccines, especially those administered against

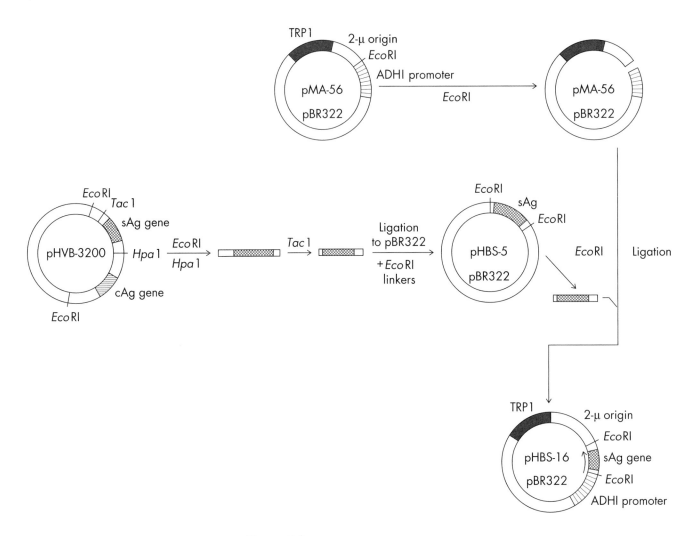

Figure 5.3
Construction of a plasmid expressing hepatitis B surface antigen in yeast.
From a plasmid (pHVB-3200) that contains both the surface antigen (sAg)
gene and the core antigen (cAg) gene of the hepatitis B virus, a clone containing only the sAg gene, pHBS-5, was constructed. The sAg gene was then
inserted behind an alcohol dehydrogenase (ADH1) promoter in the plasmid
pMA-56, to produce pHBS-16. Note that the final plasmid contains not only
the replication origin functional in *E. coli* (from plasmid pBR322) but also
the sequence that enables the plasmid to replicate in yeast cells (2-μm plasmid origin, from pMA-56). From Valenzuela, P., et al. (1982), *Nature*
298:347–350, with permission.

viral infections? The reason is that the recombinant DNA approach is
still beset by a number of technical problems. For one thing, the gene
may have a low level of expression. For another, the protein may fold
improperly—because it is in an environment different from that encountered in the mammalian host, because it is being produced in un-

A　　　　　　　　　　　　　**B**

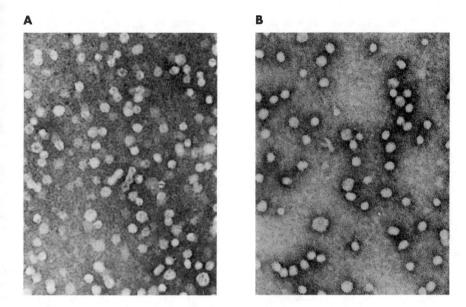

Figure 5.4
Negatively stained electron micrographs of (**A**) plasma-derived and (**B**) yeast-derived hepatitis B surface antigen vaccines. From Hilleman, M.R. (1988), *Vaccine* 6:175–179, with permission.

usually large amounts, or because of a lack of posttranslational processing (Chapters 3 and 4). Many viral proteins, including those of fowl plague, vesicular stomatitis, and herpes viruses, have been successfully expressed in *E. coli*, but improper folding apparently caused difficulties with many others.

Cultured mammalian cells have been used to express genes for protective antigens in the hope that in these cells the products would be properly modified and folded. This requires cell lines that multiply indefinitely, as tumor cells do. Indeed, cells of many of these lines are known to induce tumors when injected into appropriate hosts. In order to prevent the introduction of tumor-causing DNA material into vaccines when one is using such a system, it becomes mandatory to remove all of the host cell DNA from the vaccines, a very difficult and costly process. An alternative that is being investigated is the use of insect host cells and baculovirus vectors. The drawback to this approach is that we know very little about the posttranslational modification system, or indeed the nature of the intracellular environment, in insect cells.

An even bigger problem in some cases than incorrect folding is the weakness and short duration of the immunity produced by many of the subunit vaccines. Presumably our immune systems, which have evolved to react against natural pathogens and thus react well with the traditional vaccines that are similar to the pathogens, often respond only feebly to subunit vaccines that are radically different from the natural pathogens. Thus we need to develop a more thorough understanding of the mechanisms of immunity in order to improve the performance of subunit vaccines—a knowledge that was not necessary in the development of traditional vaccines.

MECHANISMS FOR PRODUCING IMMUNITY

In the vertebrate body, the first lines of defense against pathogenic microorganisms are nonspecific. The infecting organism may be killed by the antimicrobial substances in tissues, or it may be ingested by macrophages in tissue or polymorphonuclear leucocytes migrating into infected tissues from the bloodstream. Most infections are presumably arrested at this stage. However, when the pathogens survive this initial defense, the body's specific immune responses are activated.

In many cases, immunity is acquired through the production of *antibodies*, proteins with binding sites complementary to the structure of the immunizing foreign antigen (Figure 5.5). Many vaccines act by stimulating the synthesis of antibodies that bind to various components of the vaccines and the pathogens. When these antibodies bind to the protein toxins produced by pathogens, they very often succeed in inactivating (neutralizing) the toxins. This is the way the diphtheria and tetanus vaccines work. Protein toxins secreted by diphtheria and tetanus

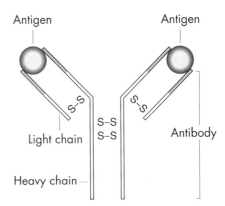

Figure 5.5
A schematic structure of an antibody of the IgG type. The antibody is composed of two heavy chains (longer polypeptides in the center) and two light chains (shorter polypeptides on the sides), linked via disulfide bridges. The two antigen-binding sites of the IgG molecule are made up from N-terminal ends of heavy and light chains, regions in which there is much variation in amino acid sequence among antibody molecules (*hypervariable region*, shaded in the figure).

bacteria cause the major symptoms of those diseases, but immunizing with inactivated toxin vaccines stimulates the synthesis of antibodies that bind to the toxins and neutralize them. Even when toxins do not play a major role in the development of a disease, antibodies may still be very effective in preventing it; when the antibodies bind to the surface of the invading pathogen, they are recognized by the phagocytic cells, which then ingest and kill the invader (Figure 5.6). Furthermore, the bound antibodies initiate the *complement cascade*, a series of reactions

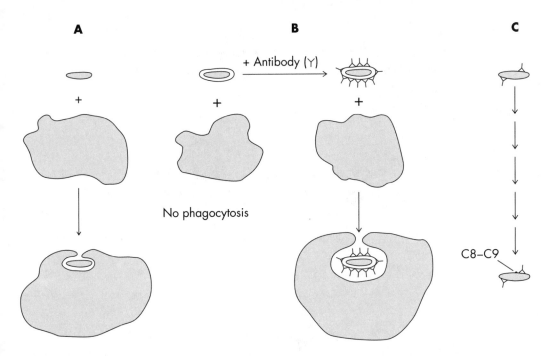

Figure 5.6
Killing of invading microorganisms. **A.** If the invading microorganisms have a cell surface that can be easily recognized as "foreign" (for example, a less hydrophilic surface), they are nonspecifically ingested by phagocytes such as macrophages and polymorphonuclear leucocytes and usually are killed in this process. **B.** Most successful pathogens have hydrophilic surface structures (for example, capsules) that enable them to evade the nonspecific phagocytosis. However, when antibodies bind to the surface of the pathogens, phagocytic cells recognize the Fc portion of the antibody and succeed in ingesting and killing the invading pathogen. (Fc is the nonspecific domain of the antibody composed of the C-termini of heavy chains; in Figure 5.5 it corresponds to the bottom part of the structure.) This function of the antibody is called opsonization. **C.** If the invading pathogen is a Gram-negative bacterium, it can be killed without phagocytosis. Antibody binding to the surface activates a series of reactions in the complement pathway, and the final components of this pathway, C8 and C9, form a *membrane attack complex* that inserts into the membranes of the pathogen and kills them directly.

involving many serum proteins, that leads to the migration of phago-cytes out of the bloodstream as well as other important effects, including the direct killing of Gram-negative bacterial invaders without the in-volvement of phagocytosis (see Figure 5.6).

Any antigen (see the box on page 173), including vaccines, stimu-lates antibody production through a process called *clonal selection*, in which the antigen first binds to an antibody on the surface of a particular lymphocyte (B cell), one of a preexisting assemblage of lymphocytes each of which produces a different antibody (Figure 5.7). The antibody has a combining site complementary to some portion of the antigen's structure. The antigen/antibody binding on the surface of the B cell stimulates that line of cells to proliferate, and the resulting clones dif-ferentiate into plasma cells that secrete large amounts of the specific antibody (see Figure 5.7). The result is immunization through vaccina-tion.

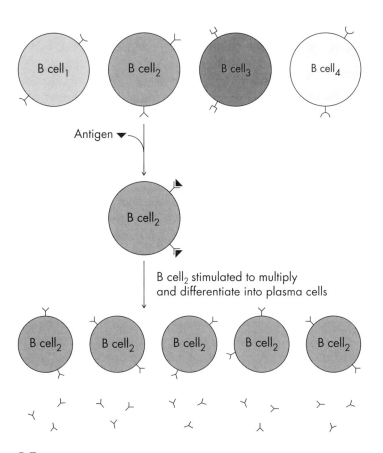

Figure 5.7
Clonal selection of B cells producing specific antibody. Antigen molecules encounter a spectrum of B cells producing different kinds of antibodies. Only the B cell whose antibody binds the antigen (in this case, B cell$_2$) multiplies and differentiates into antibody-producing plasma cells.

It is important to note that antibodies cannot bind to entire macro-molecular antigens. The antigen-binding site of antibodies can accommodate only structures of 20×30 Å—that is, structures containing 18–20 amino acids if the antigen is an α-helical protein. Thus the antibody actually binds to the *epitope*, a small portion of the antigen that determines the specificity of the particular antibody.

The mechanism of clonal selection is actually more complex than this simplified description suggests. In nearly all cases, another class of lymphocytes, called T cells, is also involved. T cells have receptors on their surface (T-cell receptors), which are antibodylike proteins that specifically bind to a part of an antigen. There are various subclasses of T cells that perform very different functions, and those that collaborate with B cells in the antibody response are called T helper cells, or T_H cells. For an effective antibody response, the antigen molecules must bind to the receptor on the surface of T_H cells (Figure 5.8). Usually, the part of the antigen recognized by the antibody on the B cells (the B-cell epitope) is different from the part of the antigen recognized by the T cells (the T-cell epitope). Effective production of antibodies occurs only if the vaccine contains both the B-cell and T-cell epitopes in close proximity.

Although an effective vaccine stimulates antibody production, the body generally does not produce the antibodies forever. Some vaccines do produce a long-term effect, however, even lifelong immunity. These are situations in which a successful clonal selection of immune cells leaves behind a small number of "memory cells" that persist and that can respond immediately if the individual is challenged again by the same antigen (pathogen). A good vaccine should induce the persistence of both B and T_H memory cells, which will respond effectively in such a secondary, or *anamnestic*, response.

The production of antibodies is not the only mechanism vertebrates use to fight invading pathogens. Antibodies are not effective against invaders that live *inside* host cells (such as all viruses and some bacteria, including *Mycobacterium*, *Brucella*, and some *Salmonella*), because they cannot enter the cytoplasm by diffusing across the plasma membrane. Instead, cellular immunity is the principal defense against these invaders. Two types of T cells, Tc and Td, are involved. These T cells have receptors that recognize specific antigens, such as viral proteins, which are expressed on the surface of the pathogen-infected cells. Recognition leads to the selection and proliferation of the antigen-specific Tc and Td cells, through a clonal mechanism similar to that outlined for the selective propagation of specific B cells (see Figure 5.7). Tc cells (c for cytotoxic) kill the target cell by direct contact with it, and the Td cells (d for delayed-type hypersensitivity) contribute to the host defense by inducing the migration of phagocytes into the infected area.

One important feature of the recognition of antigenic epitopes by T cells of all types is that they cannot recognize antigen molecules until

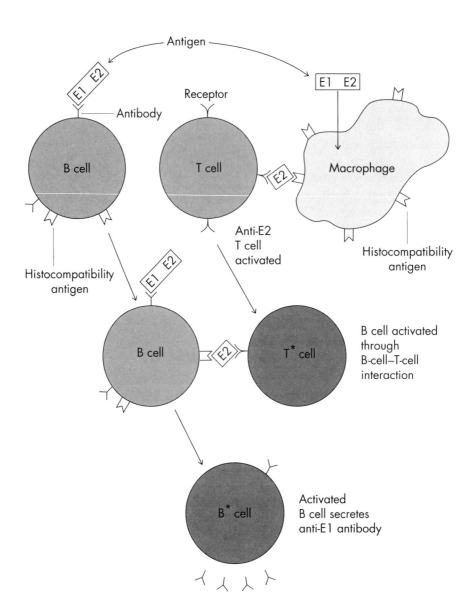

Figure 5.8

T-cell–B-cell cooperation in the production of antibody. In order to be effective in the production of antibody, an antigen usually must contain both a B-cell epitope (E1) and a T-cell epitope (E2) in the same molecule (rectangle labeled "E1 E2" at top left). The host animal contains many different types of B cells, each with antibody of a different specificity. Only the B cell that happens to have an antibody complementary to the E1 epitope becomes stimulated to proliferate (top left; see also Figure 5.7). At the same time, some of the antigen is "processed" (partially degraded) by macrophages, and the E2 epitope is "presented" in a form embedded within the class II histocompatibility antigen on the surface of the macrophage. Again, among a population of many T cells, only a small number of T cells with the correct receptor would fit with the particular E2 epitope and would become activated (top right). At stage II, some of the antigen molecules that were bound to B cells initially are also degraded, and the fragment containing the E2 epitope is presented (embedded in the histocompability antigen of B cells) for recognition by the activated T cell (shown as T* cell). This T-cell–B-cell interaction finally results in the full activation and differentiation of the B cell (shown as B* cell), making possible the secretion of antibodies (stage III). Modified from Davis, B.D., et al. (1990), *Microbiology*, 4th ed. (Lippincott).

they have been partially degraded or "processed" by other cells, such as macrophages or B cells. Moreover, the fragment (epitope) must be properly "presented": The T cell will not recognize it unless it is embedded in a special class of proteins, called histocompatibility antigens, that are present on the surface of most vertebrate cells (see Figure 5.8). Histocompatibility antigens differ from one individual to another and enable the immune cells to distinguish the body's own cells from foreign cells. The requirement for epitope presentation ensures, for example, that Tc cells become activated only when they are in contact with our own cells that are being invaded by a virus and have viral antigens on the cell surface—that is, with the cells that the Tc cells must kill. If there were no such requirement, Tc cells that are not in contact with a target cell would become indiscriminately and uselessly activated by viral antigens floating in the body fluids.

Cellular immunity, especially that mediated by Tc cells, may also be important in the body's early detection of malignant tumor cells. Tumor cells usually express abnormal antigens on their surface. Tc cells recognize the new antigens and destroy the cells carrying them. This *immune surveillance* is thought to eliminate most tumor cells that arise in the body.

IMPROVING THE EFFECTIVENESS OF SUBUNIT VACCINES

Knowledge on the mechanism of protective immunity has made it possible to design several approaches to improving the efficacy of recombinant-DNA-based subunit vaccines.

Strategies for Administering Antigen

As we have seen, the production of antibody requires the presence of both B-cell and T-cell epitopes in the vaccine. With the subunit vaccines, this is usually not a problem, because a protective antigen, being a macromolecule (often a protein), usually contains both. However, to generate a strong and long-lasting immunity, it may also be necessary to administer the antigen in a specific manner. Our immune response mechanisms have probably been refined during their evolution such that they mount an effective immune response against real pathogens while avoiding accidental reactions against similar-looking antigens derived from our own tissues. Perhaps it is because of this that the strongest responses usually occur (1) when the antigen molecules are present in a concentrated form, as on the surface of a virus particle or a bacterial cell, and (2) when, as in many cases, the immune system recognizes on the surface of a pathogen the presence of molecules that are characteristic of invading organisms in general and then uses these molecules as signals to stimulate the immune response. Such molecules are called adjuvants. Typical adjuvants of bacterial origin are fragments of the peptidoglycan cell wall, the smallest active unit being the *muramyl dipeptide* (MDP; Figure 5.9), and the LPS of Gram-negative outer membrane.

A consequence of these recognition mechanisms is that a pure subunit vaccine almost always produces a weaker response than does the whole pathogen. The surface antigen of the hepatitis B virus was a lucky exception to this rule, probably because it assembled into particles of a size similar to that of the virus particles themselves, with the same antigens exposed at high concentrations on its surface as on the virus. Thus the hepatitis vaccine was excellent in mimicking the natural virus and in satisfying at least the first condition above for provoking a strong immune response, a situation that was not achieved with most other subunit vaccines.

Recognition of the difficulties inherent in producing subunit vaccines has led to efforts to present the antigenic subunits in a more effective manner. To satisfy the first condition (concentration), various ways

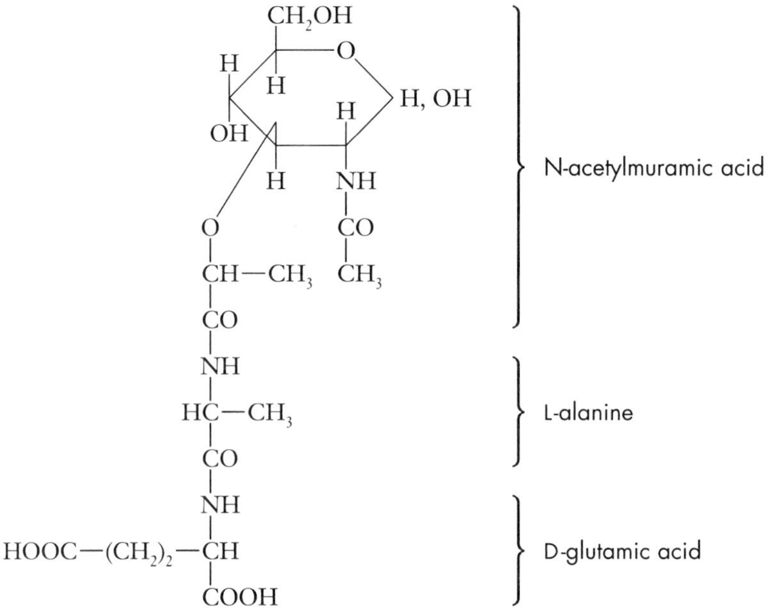

Figure 5.9
Structure of muramyldipeptide.

have been devised to fix a large number of antigenic protein molecules on the surface of a particulate carrier. One approach, for example, takes advantage of the self-assembling feature of the hepatitis B surface antigen and fuses the important parts of other subunit antigens to this protein by the recombinant DNA technique. To satisfy the second requirement (nonspecific stimulation), the immune response is enhanced most often by the use of adjuvants. In one promising approach, more than 130 synthetic analogs of the natural adjuvant MDP were tested, and one compound, in which the L-alanine residue of MDP was replaced by L-threonine, was found to have a potent stimulatory activity on the immune response without the unwanted side effects of MDP. When this MDP analog was used in combination with several viral antigens, produced by recombinant DNA methods, and dispersed on the surface of the hydrophobic microsphere of squalane by the action of the surface-active agent L121 (Figure 5.10), both requirements were satisfied, and very effective vaccination was obtained in animal models. The development of effective antigen delivery strategies thus seems crucial for future formulations of subunit vaccines.

Although the surface of the hepatitis B vaccine produced in yeast is very similar to that of the virus, as described above, it is not identical. In the virus, the region of DNA that precedes the portion coding for the surface antigen is transcribed occasionally, producing larger proteins that become assembled on the surface of the virus particle (see Figure 5.2). It has been reported that the presence of the larger proteins enhances the efficacy of the vaccine. Furthermore, the intact virus contains another major protein, the core antigen, and there is some experimental evidence that this antigen also contributes to the immunity. (In addition, cloning and expression of the core antigen have provided a useful tool for distinguishing people who were previously infected with hepatitis B virus—including carriers—from those who were simply immunized with the recombinant HBsAg vaccine. The former, but not the latter, have antibodies directed against the core antigen.) These results indicate that although one antigen may play a major role in protection, other antigens may enhance the immunity, and that we may lose some efficacy by limiting the vaccine to a single "pure" antigen.

Use of Live, Attenuated Vectors

In some cases, the requirements for a strong immune response that we have described can also be met by incorporating the subunit antigens into live, attenuated viruses or bacteria. Such strategies have both advantages and disadvantages.

Some highly effective traditional vaccines are live vaccines (see Table 5.1). Because they present the antigens to the body's defense

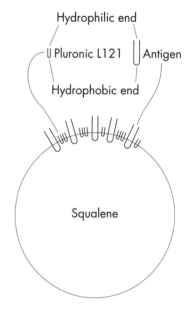

Figure 5.10
Antigenic protein molecules embedded in the squalane–Pluronic L121 system. Pluronic L121, being an amphiphilic molecule, inserts its hydrophobic ends into the surface of a droplet of squalane, a completely hydrophobic compound. This stabilizes the entire structure, because the surface of the complex is now covered only with the hydrophilic ends of Pluronic L121. Antigen molecules also insert themselves partly into the surface of this complex, thereby producing a two-dimensional dense array, a configuration that favors recognition by immune cells. Modified from Allison, A.C., and Byars, N.E. (1987), *Bio/Technology* 5:1041–1045.

mechanisms in a "natural" manner (that is, in a concentrated form, often accompanied by molecules that act as effective adjuvants), they very often confer stronger immunity, sometimes for a longer period. Furthermore, live vaccines can usually be administered in smaller dosages, because to a limited extent they may be able to multiply within the host. Another advantage of some live vaccines is that they do not have to be administered parenterally (by injection).

Introducing the gene for a protective antigen into a live vector creates a recombinant DNA vaccine that has all these advantages. In addition, with these vaccines there is no need to produce and purify the antigenic protein in a manufacturing plant, so they tend to be much less expensive than the subunit vaccines discussed previously. The live vectors, however, do present the dangers we noted before, such as the chance of their reverting to virulence or of their acting as virulent strains in hosts with weakened immune systems.

Viral Vectors

The vaccinia vector seems to be the most promising candidate for a viral vector, because it is reasonably safe and because its large genome can accommodate a fair amount of foreign DNA. The large vaccinia virus DNA is difficult to manipulate *in vitro*, but a series of clever techniques has been devised to overcome this obstacle. In a typical situation, the foreign genes are cloned into short stretches of vaccinia DNA in conventional plasmids, using *E. coli* as the vector. The plasmid DNA is then isolated and introduced into mammalian cells that are simultaneously infected with vaccinia DNA. The foreign DNA inserts into the vaccinia DNA through homologous recombination (Figure 5.11).

Approaches of this type were used to produce experimental vaccines in the vaccinia vector directed against rabies, hepatitis B, influenza, Friend murine leukemia, herpes simplex, and other diseases. Many proved highly efficacious in animal experiments, and some have been tested in field trials. Recombinant vaccinia virus containing the gene for a glycoprotein of rabies virus has been administered with baits for wild animals and has virtually eradicated rabies in a wide area. This is significant because live, attenuated rabies vaccine was known to cause disease in some species of wild animals and was also considered dangerous because it was known to revert to the virulent state.

The vaccinia vector thus appears quite promising. A dozen or more genes for foreign proteins could perhaps be inserted into its genome, making it possible (theoretically, as yet) to produce a vaccine that, in a single administration, would produce immunity against many different diseases. However, vaccinia virus is also known to produce severe side effects on occasion, including encephalitis. When vaccinia was used for

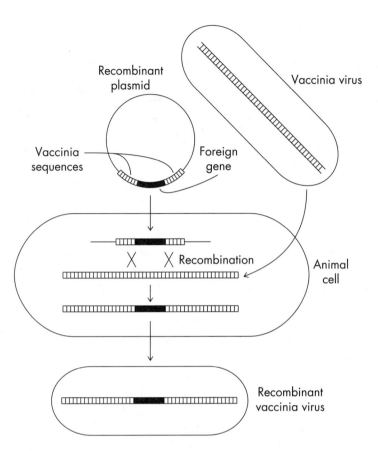

Figure 5.11
Cloning of foreign DNA into vaccinia virus DNA. The piece of foreign
DNA is first cloned in between vaccinia sequences in plasmids. The plasmid
is then introduced into cultured animal cells that are simultaneously infected
with vaccinia virus. Homologous recombination between viral DNA and
the vaccinia sequences in the plasmid leads to the creation of a hybrid vaccinia DNA, which when encapsulated becomes a recombinant vaccinia virus
particle.

immunization against smallpox, the public accepted this risk, partly because of the perceived gravity of the consequences of smallpox epidemics. It seems doubtful that the vaccinia vector would be accepted as
readily for the vaccination of human population against other diseases.
(There would not be such an obstacle for veterinary use, especially for
vaccination of wild animals, as in the rabies-vaccinia hybrid virus.) It
may be desirable to modify the vector so as to reduce the risks of undesirable side effects.

Bacterial Vectors

Usually, the only efficient protection against bacterial pathogens that cause infections of the gastrointestinal tract is to generate a localized mucosal immunity by oral administration of attenuated bacteria. For example, parenteral administration of killed *Salmonella* vaccines generates only moderate immunity against typhoid fever, and the toxicity of the endotoxin (LPS) very often causes significant side effects. In contrast, oral vaccination with live, attenuated strains appears to be more effective and to cause less severe side effects in both animals and people.

Several types of *Salmonella* mutants are being tested. One class lacks enzymes for the synthesis of aromatic compounds, including *p*-aminobenzoic acid. This mutation prevents multiplication of the bacteria in animal tissues, because salmonella, like other bacteria, has to synthesize an essential co-factor, folic acid, from *p*-aminobenzoic acid and cannot utilize the prefabricated folic acid found in animals. Another class has deletions in the genes for adenylcyclase and cAMP-binding protein; this makes the mutant avirulent, presumably because cAMP-dependent regulation is involved in the biosynthesis of various proteins the bacteria need, especially under conditions of starvation and stress. Another class lacks the enzyme for galactose synthesis, UDP-galactose 4-epimerase (*galE*). Because galactose is one of the major constituents of the LPS of many salmonella serotypes, including *S. typhi* and *S. typhimurium*, these mutants cannot synthesize the complete LPS, which is required for virulence. However, they can utilize the small amounts of galactose available in the host tissues and build up small numbers of complete LPS molecules, a feature that is thought to give them just enough ability to proliferate slowly in the host and provide a very effective immunity. A *galE* mutant strain of *S. typhi*, Ty21a, was made by chemical mutagenesis and has been examined extensively in field tests. It appears to be a safe vaccine without side effects. In the first field trial in Egypt it was reported to be very effective, but a subsequent trial in Chile yielded less convincing results.

Some problems still complicate the use of these live vaccine strains. For example, a double mutant of *S. typhi*, lacking the ability to synthesize *p*-aminobenzoic acid as well as purines, was very safe but was rather weak in terms of eliciting antibody response, presumably because the nutritional deficiencies were too effective at blocking bacterial growth. On the other hand, when *S. typhi* strains with only the *galE* gene defect were constructed by using recombinant DNA methods, they remained highly virulent in humans, which suggests that the lack of virulence in Ty21a was due to unknown mutations introduced presumably in the course of heavy chemical mutagenesis. This discovery leaves Ty21a open to the various criticisms that were marshaled against the traditional live vaccines.

Although these attenuated vaccine strains are not yet suitable for all applications, they are attractive systems for stimulating local, mucosal immunity. Efforts are therefore being made to use these strains as vectors—that is, to add protective antigens of other pathogens to them. These include the antigens of *Shigella*; those of *E. coli* strains that cause diarrhea, especially traveler's diarrhea; and even those of streptococci that are implicated in the generation of dental caries.

FRAGMENTS OF ANTIGEN SUBUNIT USED AS SYNTHETIC PEPTIDE VACCINES

Subunit vaccines use only one or a few macromolecular components of the pathogenic organism. Because only small parts (epitopes) of these macromolecules are involved in binding to the antibody or to the T-cell receptor, it is possible to extend this approach even further. Researchers predicted that an immunization response could be elicited with nothing more than the small peptide corresponding to the epitope. Immune response indeed occurs frequently when such a peptide is first attached to a macromolecular "carrier" protein and then administered to animals.

A *peptide vaccine* has several advantages. The most prominent is that peptides can be made by chemical synthesis, eliminating the purification steps necessary for the production of recombinant-DNA-based subunit vaccines. Such purification is often a difficult and expensive process, because all traces of host-cell DNA must be removed if animal cells are used for expression and because highly toxic components of bacterial cells, such as LPS, have to be removed if bacteria such as *E. coli* are used. For this reason, peptide vaccines tend to be less expensive, purer, and more stable than protein-containing subunit vaccines. In addition, using only a part of the antigenic protein means there will be no unwanted immunological reactions to other parts of the protein.

Identifying the Epitope

To produce a peptide vaccine, one first has to identify the antibody-binding epitopes on the surface of the antigenic protein. The identification itself is not so difficult; the difficulty is in selecting the *correct* epitope, which, when used as a vaccine, will protect the vaccinated human or animal against attack by the pathogenic organism.

Natural variation often provides a clue to the location of these epitopes. That is, the antigenic proteins of some virus species vary so much from strain to strain that infection by one does not necessarily build up immunity against the others. The influenza virus is a notorious example of this phenomenon. The proteins on the surface of this virus undergo such rapid variation that people infected by it one year have no immunity at all against the next year's epidemic. The coat proteins of the foot-and-mouth disease virus, an animal pathogen, show similar variation. In such cases, analyses of nucleic acid sequences usually show that there are several regions of high variability on the antigenic protein molecule (Figure 5.12). Because such regions produce variants differentiated by the body's immune response, it is clear that they react with antibodies and are able to stimulate the proliferation of a particular line of B cells; in other words, these regions are *immunogenic epitopes.*

Another strategy for identifying epitopes involves the *in vitro* creation of a special kind of antibody. *In vivo*, antibody diversity is generated by the random joining of many genes coding for segments of their polypeptide chains. Before an animal is exposed to an antigen, its body already contains more than a million different kinds of B cells, many of which produce antibodies that can bind to different epitopes of any particular antigen with different degrees of affinity. All of these latter cells are stimulated to mature and divide when that antigen is introduced and binds to them. Thus the antibodies produced in response to one antigen in an immunized human or animal are a heterogeneous mixture, coming from many independent clones of antibody-producing cells: They are "polyclonal." When polyclonal antibodies come in contact with the antigen, various antibodies bind to various parts of the antigen molecule, and the identification of epitopes is arduous and complex.

In the laboratory, on the other hand, individual B-cell clones can be "immortalized" by fusion with a tumor cell line. From each clone one can then isolate a homogeneous population of antibodies, *monoclonal antibodies,* that bind only to a single epitope with uniform affinity. Thus with each monoclonal antibody, one can identify, in molecular detail, a site (epitope) that is recognized by the antibody. When antibodies are generated by using proteins as immunogens, a large fraction of the antibodies (usually 50% or more) bind well only when the proteins are intact and properly folded. These antibodies are directed to epitopes that are three-dimensional sites formed by the juxtaposition of regions

---

Figure 5.12
Amino acid sequences of VP1 proteins from various strains of foot-and-mouth disease virus. Amino acid residues in strains A-10, O-1, and C-3 are listed only when they differ from those found in strain A-12. There are several major regions of high variability, including residues 42–51 and residues 137–158. The latter region (boxed) was used in making the peptide vaccines. From Rowlands, D.J., et al. (1984), in Chanock, R.M., and Lerner, R.A. (eds.), *Modern Approaches to Vaccines: Molecular and Biochemical Basis of Virus Virulence and Immunogenicity,* pp. 93–101 (Cold Spring Harbor Laboratory), with permission.

A-12	Thr	Thr	Ala	Thr	Gly	Glu	Ser	Ala	Asp	Pro	Val	Thr	Thr	Thr	Val	Glu	Asn	Tyr	Gly	Gly	
A-10			Thr																		
O-1			Ser	Ala																	
C-3			Thr																		20
A-12	Glu	Thr	Gln	Val	Gln	Arg	Arg	His	His	Thr	Asp	Val	Ser	Phe	Ile	Met	Asp	Arg	Phe	Val	
A-10	Asp												Gly								
O-1				Ile				Gln													
C-3				Ile									Ala		Val	Leu					40
A-12	Lys	Ile	Lys	Ser	Leu	Asn	Pro	Thr	His	Val	Ile	Asp	Leu	Met	Gln	Thr	His	Gln	His	Gly	
A-10			Asn			Ser												Lys			
O-1		Val	Thr	Pro	Gln		Gln	Ile	Asn	Ile	Leu					Ile	Pro	Ser		Thr	
C-3		Val	His	Val	Ser	Gly	Asn	Gln		Thr	Leu		Val			Val		Lys	Asp	Ser	60
A-12	Leu	Val	Gly	Ala	Leu	Leu	Arg	Ala	Ala	Thr	Tyr	Tyr	Phe	Ser	Asp	Leu	Glu	Ile	Val	Val	
A-10	Ile																				
O-1									Ser										Ala		
C-3	Ile																		Ala		80
A-12	Arg	His	Asp	Gly	Asn	Leu	Thr	Trp	Val	Pro	Asn	Gly	Ala	Pro	Glu	Ala	Ala	Leu	Ser	Asn	
A-10																					
O-1	Lys		Glu		Asp											Lys			Asp		
C-3	Thr		Thr		Lys										Val	Ser			Asp		100
A-12	Thr	Gly	Asn	Pro	Thr	Ala	Tyr	Asn	Lys	Ala	Pro	Phe	Thr	Arg	Leu	Ala	Leu	Pro	Tyr	Thr	
A-10		Ser																			
O-1		Thr										Leu									
C-3		Ala						His		Gly		Leu									120
A-12	Ala	Pro	His	Arg	Val	Leu	Ala	Thr	Val	Tyr	Asn	Gly	Thr	Asn	Lys	Tyr	Ser	Ala	Ser	Gly	
A-10											Asp									Asp	
O-1													Glu	Cys	Arg		Asn	Arg	Asn	Ala	
C-3									Thr		Thr			Thr	Ala		Thr			Ala	140
A-12	Ser	Gly	-	Val	Arg	Gly	Asp	Phe	Gly	Ser	Leu	Ala	Pro	Arg	Val	Ala	Arg	Gln	Leu	Pro	
A-10		-	Arg	Ser	-			Leu			Ile		Ala				Thr				
O-1	Val	Pro	Asn	Leu				Leu	Gln	Val			Gln	Lys				Thr			
C-3	-	-	Arg	-				Leu	Ala	His			Ala	Ala	His			His			160
A-12	Ala	Ser	Phe	Asn	Tyr	Gly	Ala	Ile	Lys	Ala	Glu	Thr	Ile	His	Glu	Leu	Leu	Val	Arg	Met	
A-10									Gln		Gln	Ala									
O-1	Thr										Thr	Arg	Val	Thr				Tyr			
C-3	Thr				Phe			Val				Thr		Thr							180
A-12	Lys	Arg	Ala	Glu	Leu	Tyr	Cys	Pro	Arg	Pro	Leu	Leu	Ala	Ile	Glu	Val	Ser	Ser	Gln	Asp	
A-10									Lys						Lys		Thr				
O-1					Thr										His	Pro	Thr	Glu	-	Ala	
C-3											Val		Pro	Val	Gln	Pro	Thr	Gly	-		200
A-12	Arg	His	Lys	Gln	Lys	Ile	Ile	Ala	Pro	Gly	Lys	Gln	Leu	-							
A-10		Tyr								Ala				Leu							
O-1							Val			Val			Thr	Leu							
C-3					Pro	Leu				Ala				Leu	214						

that are widely separated from each other in the protein's primary structure. Such sites are often called *assembled topographic sites* or *discontinuous epitopes*. In contrast, some monoclonal antibodies—those that define *continuous epitopes*—do bind well to certain, continuous fragments of the protein that was used in the immunization. When a continuous epitope is identified on a pathogen, a synthetic peptide can be made to correspond to it, and the hope is that when humans or animals are immunized with the synthetic peptide, they will generate antibodies that bind to that epitope and will protect them against the pathogen. (Monoclonal antibodies can also be used to discover genetic variations in epitope structure. If a given monoclonal antibody neutralizes a given virus population, for example, any of the virus that survives the attack by this antibody is likely to have a mutation that alters the epitope.)

It is far more difficult to use peptides to generate antibodies to a discontinuous epitope. However, there are possible approaches. Geysen and his coworkers, for example, reasoned that even in an assembled epitope, certain amino acid side-chains must be adjacent in space to some others (although the former may be far away from the latter in primary sequence) and therefore that an assembled epitope could possibly be mimicked by peptides, whose sequences do not necessarily correspond to the primary structure of the antigen. Thus a monoclonal antibody specific for a discontinuous epitope was incubated with a large number of peptides with random sequences, those peptides that bound to the antibody were identified, and their structure was refined by replacing each of the amino acid residues to maximize the fit between the peptide and the antigen-binding site of the antibody. However, it is not yet known whether the peptide selected in this manner is effective in generating high-affinity antibodies that react with the whole protein.

Predicting Epitopes from Primary Structure

The experimental determination of epitopes just described can be difficult and tedious, and the results are not always satisfactory. For example, not all peptides that correspond to continuous epitopes proved effective in eliciting antibody response when they were used in immunization. That is, some of the peptides were not *immunogenic*. Predictive strategies have therefore been developed to make the search for continuous epitopes more efficient; some pinpoint promising regions of primary sequence as presumptive epitopes, and others predict a peptide's immunogenicity.

Hydrophilicity and Segmental Mobility. It is often possible to recognize, on the basis of the primary structure alone, *short sequences* that act as good epitopes and are also immunogenic. Because epitopes must be on the surface of the protein in order to combine with the antigen-binding sites of the antibody, they must occur in exposed, hydrophilic regions of the protein. Hence *hydrophilicity plots*, which identify such regions along the amino acid sequence, are popular tools for prediction. (In some cases, however, the initial interaction between the antibody and the antigenic protein produces conformational alterations in the latter, so that some initially hidden residues of the antigenic protein come to participate in the binding process.)

Although hydrophilic and exposed segments tend to be good epitopes, actual experience shows that many of them are not very immunogenic when administered as short peptides. A plausible reason for this is the rapid fluctuation in the three-dimensional conformation of the peptides. Short peptides usually do not assume stable conformations in water, whereas corresponding short segments of the parent protein are likely to exist in a distinct conformation, stabilized by interaction with other parts of the protein. Thus most of the antibodies generated in response to the short peptides do not bind to the corresponding segment of the protein (Figure 5.13). This suggests that to be effective, a continuous epitope segment must have high mobility (flexibility) in the protein so that it can fit into the antigen-binding site of the antibody. Indeed, when the immunogenicity of peptides corresponding to various parts of the tobacco mosaic virus protein was tested, highly immunogenic regions were found to correspond to flexible parts of the protein, as shown by X-ray crystallography (Figure 5.14).

The idea outlined above has been advocated by R. A. Lerner and his associates and has been applied in many attempts to develop synthetic peptide vaccines. The results thus far have been rather disappointing. Although short peptides corresponding to hydrophilic, mobile loops are able to generate antibodies that combine with the corresponding areas of the parent protein (and are often useful as tools in the laboratory), the antibodies usually do not offer much clinical protection because their binding affinity tends to be too weak. In retrospect, this is not surprising. Because the peptides lack a finite secondary structure, the immunization presumably generates a wide range of antibodies, only some of which combine with the target protein when its mobile peptide segment happens to be in a conformation complementary to theirs.

Stable Secondary Structure. The correlation between immunogenicity and segmental mobility breaks down as the peptides get longer (14–20 residues). Surprisingly, the longer peptides that are more immunogenic and produce antibodies with stronger affinity appear to correspond to protein regions with a stable secondary structure. It is likely that a significant fraction of these peptides assume definite structures in water. A successful case of the production of an experimental peptide vaccine in this manner is described below. (Theoretically, the same effect should be obtainable even with short peptides, if they could be fixed in a stable

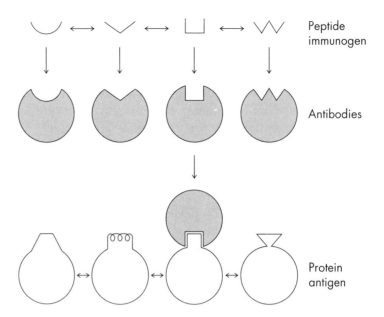

Figure 5.13

The reaction of antipeptide antibodies with a whole-protein antigen. A short peptide can take on many conformations in water (top). In animals and humans it therefore generates antibodies of many different types, each specific for one of the multiple conformations of the peptide (center). If the peptide corresponds to an exposed and flexible region of the whole protein, one of the possible conformations of that region of the protein may correspond to one of the statistically possible conformations of the peptide and therefore may bind some of the antipeptide antibody (bottom). Even in this case, only a small fraction of the antipeptide antibody will participate in the binding.

conformation by covalent linkages. This is one of the approaches being pursued.)

FMDV Vaccine: An Experimental Peptide Vaccine

In economic terms, foot-and-mouth disease is the most important disease of farm animals. It afflicts cattle, sheep, goat, and swine populations around the world, except in North America and Australia. Fear of the spread of the foot-and-mouth disease virus (FMDV) is one of the major reasons why the U.S. Department of Agriculture forbids the importation of uncooked meat products from various parts of the world.

Currently, veterinarians are trying to prevent the disease with a traditional vaccine containing killed virus particles. However, this vac-

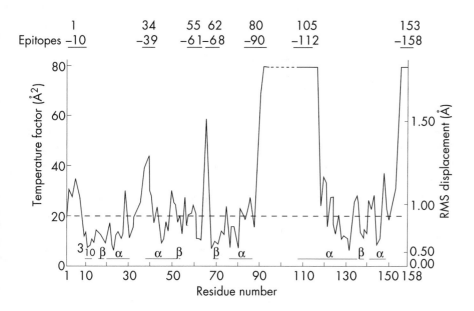

Figure 5.14
The continuous antigenic epitopes and flexibility of the regions of the to-bacco mosaic virus protein. At the top, seven continuous epitopes are shown. These correspond to areas that have high flexibility, which give diffuse scattering patterns in X-ray crystallography. Because such patterns are caused by thermal movement of the atoms, the diffuseness decreases at lower temperature. This is why the extent of flexibility can be estimated from the *temperature factor* observed in the X-ray crystallographic analysis. Secondary structures of various segments (α-helix, β-sheet, or 3_{10}-helix) are shown at the bottom. From Westhof, E., et al. (1984), *Nature* 311:123–126, with permission.

cine becomes inactivated rapidly if it is not kept constantly at refrigerator temperature. Presumably this is why vaccination has not reduced the incidence of the disease drastically except in Western Europe. Furthermore, killed or inactivated vaccines can be dangerous, as we noted earlier, because some of the viruses may survive the inactivation process. This indeed occurred with one batch of the FDMV vaccine, causing an outbreak of the disease in Western Europe in 1981. In addition to these problems, the production of killed vaccine is enormously expensive, because each year, about 2 billion doses of the vaccine have to be made by growing the extremely dangerous virus in cultured animal cells. These considerations argue persuasively for the development of a modern "engineered" vaccine to combat this virus.

The virus produces four major capsid proteins. Because the blood sera of animals who have survived foot-and-mouth disease contain antibodies that react with the capsid protein VP1, the first attempt to produce a subunit vaccine focused on producing VP1 with recombinant

DNA techniques. VP1 can be overproduced in *E. coli*, but unfortunately, it cannot be recovered in its native conformation. Thus vaccination with this material produces hardly any immunity to the disease.

Subsequent efforts centered on the development of potential peptide vaccines. Following an approach described earlier, several regions of high variability were discovered in the VP1 protein and identified as potential epitopes (see Figure 5.12). When peptide 141–160 was synthesized, coupled to a large carrier protein, and used in vaccination, high titers of antibodies were produced that successfully neutralized (inactivated) the virus particles and conferred very effective immunity against the disease, at least in guinea pigs. In later experiments, good immunity was even produced by injecting uncoupled peptides, as long as they were administered with some adjuvants. It should be noted that this peptide is fairly long and comes from a region that is likely to assume a stable secondary structure.

Unfortunately, there are still obstacles to overcome before the FMDV peptide vaccine is ready for use in the field. First, the vaccine works extremely well in guinea pigs, but only marginally in cattle. The different responses in different animal species may be due in part to a problem of recognition by T cells (discussed below). Second, in order to be of practical value, the vaccine will have to be made more powerful. Perhaps this can be achieved by arranging the antigenic molecules in closely spaced arrays or by improving the adjuvant. Cloning of the DNA sequence that codes for the VP1 sequence 140–161 into the gene for hepatitis B virus core antigen dramatically improved the vaccine's potency, presumably because hepatitis core antigen, like hepatitis surface antigen, self-assembles into a particulate structure.

Perhaps the most difficult problem—and one that is likely to plague other peptide vaccines as well—is the danger that mutant viruses will thrive in the vaccinated host. The antibodies generated in response to peptide vaccines are all directed to a single epitope and may not bind well to viruses with altered sequences in that region of the protein. One way of avoiding this problem would be to design a vaccine equipped with two or more epitopes from different regions of the antigenic protein. Some authors argue for incorporating at least three epitopes because the frequency of epitope alteration by mutation is so high. Such a feat would be difficult in many cases, because prominent continuous epitopes are not particularly common. It is also true that a similar selection for mutants occurs, to a limited extent, in nonvaccinated populations that have developed postinfection immunity, so the repertoire of epitope variations found in a given protein in different strains of viruses may already represent the full range of changes that can be made without affecting the function of the protein. According to this optimistic view, all that would be needed to produce an effective vaccine would be to include in it all known variant sequences of one single epitope.

Recruiting the Assistance of T-helper Cells

As described earlier, antibody production requires the presence within the antigen of both B-cell and T-cell epitopes (see Figure 5.8). Most large antigenic proteins contain both epitopes. However, when a peptide vaccine has been constructed by identification and cloning of the B-cell epitope only, there is no guarantee that the peptide will also contain a T-cell epitope. It is true that the peptide is usually conjugated to a large carrier protein that supplies T-cell epitopes, but the host that is immunized in this way may not be able to mount an effective immune response (secondary response; see page 181) when challenged with the real pathogen. This is because the host will not have memory T_H cells specific for the T-cell epitopes in the real antigen in the pathogenic organism, although it will have memory cells directed to the epitopes in the artificial carrier. The best approach therefore would be to incorporate the real T-cell epitope from the antigenic protein, as well as the B-cell epitope, into the synthetic peptide.

Another important consideration is that the response to T-cell epitopes varies greatly from one individual to another. Possibly because the T-cell receptor also has to recognize a specific histocompatibility antigen (see Figure 5.8), its "repertoire," in terms of antigen recognition, is rather limited. That is, T-cell receptors in any given individual, with a given histocompatibility antigen, can recognize only a subset of T-cell epitopes. This is not a problem when the entire pathogenic organism is being used in vaccination, because in any pathogen there are many proteins, each containing at least several T-cell epitopes, and any individual host will respond well to at least one of them. However, when a single peptide is used in immunization, the immune response depends greatly on the make-up of the host's histocompatibility antigen, and a given individual may not respond significantly to the T-cell epitope in the vaccine. A somewhat similar situation may explain the often pronounced differences with which various animal species react to a given peptide vaccine. (Recall that the experimental FMDV vaccine elicited an excellent response in guinea pigs, but not in cattle.)

These considerations suggest that researchers' initial optimism about the development of synthetic peptide vaccines may not have been warranted. As we have mentioned, the vertebrate immune system usually reacts very well when challenged by traditional vaccines, probably because it evolved through a long-standing interaction with the pathogens in question. However, the immune system tends to react poorly to the administration of peptide vaccines that contain only isolated epitopes of an antigen taken totally out of their natural context. If peptide vaccines are ever to be successful, researchers must exploit everything we know about stimulation of the immune response.

Peptides for Generating Cellular Immunity

Although many problems arise in using peptides to stimulate B cells, it turns out that peptides are excellent instruments for generating cellular immunity through the selection of T-cell clones. This is mainly because T-cell selection involves the accessory cells that "present" already processed (proteolytically cleaved) antigen to the T cells—the relevant structures are therefore short peptide strings. In other words, most T-cell epitopes are continuous ones, unlike B-cell epitopes, many of which are of the assembled type. However, these are theoretical considerations, and it remains to be seen whether useful vaccines that generate cellular immunity can actually be produced by the peptide immunization approach.

DIFFICULTIES IN MAKING VACCINES FOR CERTAIN DISEASES

Some pathogens have developed extraordinarily "clever" weapons against the immune defenses of the host, so producing an effective vaccine is far from simple. We will discuss only two examples, leprosy and malaria, but very similar considerations explain why effective vaccines are not yet available for several other important diseases.

Leprosy

Hansen's disease, or leprosy, is caused by *Mycobacterium leprae*, which grows extremely slowly within the phagocytes of infected individuals. Twelve million people in the world are now estimated to be suffering from the disease, to which a unique stigma has been attached throughout history. A combination of antibacterial agents—the antibiotic rifampicin and the chemotherapeutic agents dapsone and clofazimine—is very effective in treating the disease, but this therapy is still prohibitively expensive in some parts of the world, and there is fear that drug-resistant *M. leprae* strains may become more common. An effective vaccine would therefore be a great boon. Because *Homo sapiens* seems to be the only natural host of *M. leprae*, an effective vaccine might even eradicate this disease entirely, just as smallpox was eradicated from the human population through vaccination.

In practical terms, producing conventional vaccines against leprosy is all but impossible. To begin with, *M. leprae* cannot be grown *in vitro*. It can multiply in the armadillo but does not attain a high concentration within the tissues for several years.

These difficulties make the development of subunit vaccines based on recombinant DNA technology an ideal alternative. The World Health Organization recruited a team of very capable molecular biologists to clone and express *M. leprae* antigens in *E. coli*. Their strategy was as follows: (1) to produce a number of mouse monoclonal antibodies against the component macromolecules of *M. leprae*, (2) to identify the *M. leprae* proteins that bound to these antibodies, and (3) to produce these protein antigens by recombinant DNA techniques, in *E. coli*, for use as a vaccine. These proteins stimulated the mouse immune system and caused the production of antibodies, so it was expected that at least some of them would behave as *protective antigens*, bestowing immunity on vaccinated subjects. The results have so far been disappointing, however, at least in animal models.

One reason for this failure might be that *M. leprae* multiplies *inside* the host cells: Cellular immunity, rather than antibodies, may thus be more important to combat the pathogen. An important consideration in this regard is that the disease of leprosy has a very wide spectrum. Patients with *tuberculoid leprosy* have few lesions (granulomas), each containing very few *M. leprae* cells. This suggests that an effective machinery of cellular immunity is operating locally to combat the infection. The enormously high number of Tc cells in tuberculoid lesions supports this conclusion. In contrast, patients with *lepromatous leprosy* have many lesions, each containing enormous concentrations of *M. leprae* cells. It is clear that the cellular immunity is not operating properly in these patients. In fact, the lesions have been shown to be enriched in T cells of a special type (T-suppressor cells), which actually suppress the immune function. Interestingly, these patients usually have very high levels of antibodies that react with components of *M. leprae*.

Although it is not known what causes the patient to respond in one way or the other, the knowledge that the serum antibody levels are very high in patients with a more rapidly progressing type of disease seems to buttress the notion that stimulating antibody production cannot produce effective protection. The next step, then, is to try to identify epitopes that stimulate the proper types of T cells, the Tc and Td cells, without stimulating the T-suppressor cells, even though it is not yet totally clear that such epitopes exist. In another development, workers found that a high percentage of γ-δ T cells from leprosy patients reacted with *M. leprae* components. Most T cells (about 95%) produce receptors that are composed of two protein chains, called α and β. A minority have a receptor composed of γ and δ chains, and the functions of γ-δ T cells are not well understood. What is most interesting is that the γ-δ T cells appear to be enriched only in lesions that presumably correspond to the very beginning of the granulomatous response found in the tuberculoid form of leprosy, and thus there is some hope that they represent a decisive event that steers the course of immune response onto the desirable, tuberculoid path. In any case, producing an effective vaccine will not be easy, because many of the disease's symptoms appear to result from an improper response on the part of the host's own immune system.

Malaria

Malaria presents a major problem in the tropical and subtropical parts of the world. It is estimated that each year nearly 200 million new cases occur worldwide, which result in more than 1 million deaths. Increasing urbanization and other factors have significantly increased the incidence of malaria during the last few decades. Furthermore, species of *Plasmodium*, the causative organism, are becoming increasingly resistant to the most effective therapeutic agent, chloroquine. It is thus most important to develop effective vaccines against this disease.

Because it is difficult, demanding, and expensive to grow malaria parasites, subunit vaccines produced by recombinant DNA technology are again an obvious approach to prevention. Many laboratories around the world have been working intensively to develop vaccines of this type. Their progress has been impressive, but they still seem to be far from achieving a truly efficacious product.

Clearly, one of the major problems is the complexity of the causative organism, *Plasmodium*. Its life cycle consists of at least three completely separate stages within the vertebrate hosts (Figure 5.15), and the organism at each stage has a completely distinct antigenic make-up. As is well known, malaria infection begins with the bite of an *Anopheles* mosquito. The organism that enters the bloodstream through this route is in the *sporozoite* stage. Sporozoites enter liver cells and multiply there. After 1–2 weeks the organisms are released into the bloodstream again, this time as *merozoites*. Many of the symptoms of malaria are due to merozoites infecting red blood cells, multiplying, and being released after 2 or 3 days to reinfect fresh red blood cells (see Figure 5.15). A small fraction of merozoites eventually develop into sexual forms called *gametocytes*. When these enter the gut of a mosquito, they develop and mate, producing *ookinetes* that eventually produce sporozoites again. Vaccine development has targeted each of the stages that occur in the human host.

Sporozoite Vaccine

A vaccine for preventing infection by malaria parasites must produce an immunity directed against sporozoites. However, such an immunity must work perfectly; even a single sporozoite escaping and entering a liver cell will eventually cause the disease.

In animal models, killed sporozoites are effective vaccines. However, these preparations do not confer enough protection on humans.

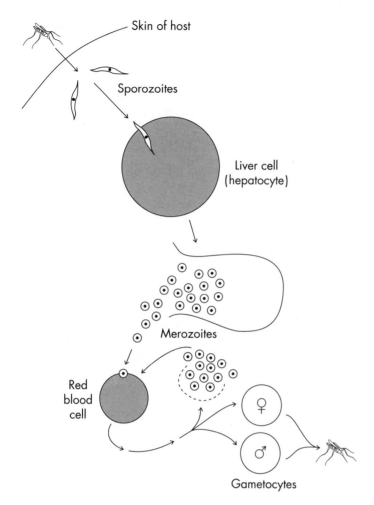

Figure 5.15
A simplified life cycle of the malaria parasite.

Thus we need stronger immunogenicity than can be obtained with the whole sporozoites—a difficult goal to attain with subunit vaccines, as we saw above.

Following the standard approach to producing subunit vaccines, scientists immunized an experimental host with killed whole sporozoites and then identified the antigen to which most of the resulting antibodies were directed. In this manner, the *circumsporozoite* (*CS*) protein, a protein component on the surface of the sporozoite, was identified as the predominant antigen. The central part of this protein consists of many repetitions (37 times in one strain and 43 times in another strain of *P. falciparum*) of a tetrapeptide, Asn-Ala-Asn-Pro, with a few repetitions of a similar tetrapeptide, Asn-Val-Asp-Pro (Figure 5.16). Because a high fraction of the antibodies produced upon immunization with the killed

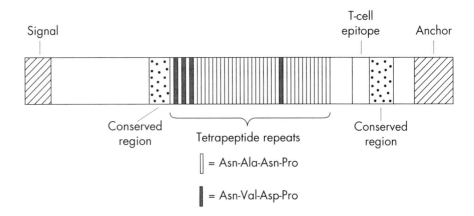

Figure 5.16
Schematic structure of CS antigen from a strain of *Plasmodium falciparum*. The N-terminus of the protein (left end in the figure) begins with a signal sequence for export, and the C-terminus (right end in the figure) ends with a hydrophobic sequence that presumably anchors the protein at the surface of the cell membrane. The middle portion of the protein consists of tetrapeptide repeats (37 repeats of Asn-Ala-Asn-Pro and 4 repeats of Asn-Val-Asp-Pro. The T-cell epitope is located on the C-terminal side of the repeat region. Based on Kemp, D.J., Coppel, R.L., and Anders, R.F. (1987), *Ann. Rev. Microbiol.* 41:181–208.

sporozoites are directed against this region, the first subunit vaccines consisted of a synthetic peptide containing three of these repeats conjugated to a protein carrier, and a recombinant-DNA-produced vaccine composed of 32 of these repeats fused to another protein. These vaccines were injected into human volunteers, and the subjects who developed the highest antibody titers were then infected by living sporozoites of *P. falciparum*. Only a small fraction of this group (one-third and one-sixth, respectively) was protected against the disease.

Later studies showed that the repetitive region did not contain any T-cell epitopes. As we noted earlier, for an effective immune response to occur, the antigen must contain both B-cell and T-cell epitopes. The predominant epitope recognized by T_H cells is located in the nonrepeating, C-terminal region of CS protein. When subunit vaccines that presumably contained T-cell epitope(s) were made by expressing the recombinant DNA in yeast, the immune response was superior in experimental animals. However, even if this approach produces a vaccine capable of protecting human recipients, problems can arise. Although the central repeat regions are similar in various strains of *P. falciparum*, the predominant T_H-cell epitope is located in a highly variable region of the protein. Thus using the vaccine may simply select out those variants of the pathogen that have a different T_H-cell epitope.

Also in question is whether the antibodies directed against CS antigen, and especially against its repeating units, are at all protective. In field studies conducted where malaria is prevalent, workers found no correlation between the levels of antibodies against the repeat region in a given group of people and the incidence of malarial infection. A pessimistic interpretation of this finding is that because the malarial parasite presents this highly repetitive and therefore highly immunogenic sequence to the host, most of the host's immune resources are directed toward producing useless antibodies, thereby preventing the production of antibodies that are directed to other (presumably more important) proteins and thus would truly be protective. If this hypothesis is valid, vaccination with repeat sequences will not produce any benefit. An alternative interpretation is based on the finding that similar but slightly different repeats exist in other proteins produced by the same parasite. Normally, prolonged exposure to a pathogen will select out B-cell clones with the highest affinity to the predominant epitopes on its surface. However, the presence of cross-reacting epitopes would mislead the B-cell clones and prevent the selection of clones that produce antibodies with truly high affinity, leaving the host with an assembly of less effective, lower-affinity, antibodies. If this is true, immunization with engineered homogeneous repeat sequences would increase the production of high-affinity antibodies, at least to those repeat sequences, by overcoming the "misdirection" from other cross-reacting proteins, a response that might be beneficial for the host. More data are needed before investigators can evaluate these competing hypotheses.

Blood-Stage Immunity

This is the stage in which the most extensive proliferation of malarial parasites occurs. Among natural populations of humans, this proliferation is inhibited in those individuals who are heterozygous for a mutant form of hemoglobin, sickle-cell hemoglobin. (Its association with protection from malaria is assumed to be the reason why this mutant allele is prevalent—up to 40% in some parts of Africa—even though the allele causes a severe disease, sickle-cell anemia, when present in a homozygous form.) If a vaccine could produce significant levels of antibody directed to merozoites, the merozoites could be attacked during their transition from one red blood cell to the other, and the symptoms of malaria might similarly be alleviated. (Antibodies cannot cross the cell membrane, so they cannot attack merozoites multiplying within the red blood cells). However, huge numbers of merozoites would always be waiting to be released from the liver cells, a location that antibodies cannot reach, so a vaccine against merozoites is not likely to be of much use in the complete prevention of the disease. Nevertheless, there are valid arguments for at least including antigens of this type in the future vaccine for malaria.

Several pertinent antigens have been studied. Pf155/RESA is a protein found on the surface of infected erythrocytes, and antibodies against this antigen are able to block merozoite invasion *in vitro*. Again, the immunodominant region of this antigen contains repeat sequences. Another antigen, called PMMSA or Pf195, is a precursor of the major merozoite surface antigens. Manuel Patarroyo and his associates in Bogotá, Colombia, have produced what appears to be the most promising vaccine. They synthesized a 45-residue peptide containing potential epitopes from three merozoite protein antigens and then polymerized this peptide into a macromolecule by using disulfide bonds. This vaccine, which delayed or suppressed parasitemia in volunteers, is being tested in a rigorous field trial in Africa under the sponsorship of World Health Organization. It is worth noting that this vaccine overcomes one weakness of the peptide vaccines in general by polymerizing the peptide; thus the antigenic epitopes are presented in a dense array (see pages 183–184).

Transmission-Blocking Immunity

Although such immunity would do little to help the immunized individual, it would benefit the community significantly by preventing the further spread of the disease. However, attempts thus far have not been encouraging, and there is speculation that gametocyte antigens may lack good T-helper epitopes.

Other Considerations

In addition to the problems presented by the complexity of the life cycle of the malarial parasite, some of the malarial antigens have apparently evolved structures that do not stimulate the immune system. For example, when overlapping peptides from the CS protein were tested for their ability to stimulate T cells, 40% of adults from West Africa (an area where malaria is very prevalent) showed no T-cell response whatsoever.

Finally, there is the problem of frequent antigenic variation, an example of which is the aforementioned variability of the predominant T-cell epitope of the CS antigen. Sporozoite proteins did perform effectively as immunogens in a mouse–malarial parasite model, but this result was probably a laboratory artifact. In the model system in question, the parasite strain was maintained by always infecting virgin animals (those not previously exposed to malarial parasite) in the laboratory. Thus the parasites multiplied without encountering many antibodies, and there was no environmental pressure for antigenic diversification. All of the parasites were therefore probably monoclonal, being covered by the same antigen molecule. The efficacy of vaccination in such a situation is likely to be very different from what it would be in an area where malaria is endemic and the parasites have undergone maximal diversification.

SUMMARY

Vaccines of the traditional type are live organisms of attenuated virulence, killed organisms, or inactivated toxins. Usually, only minor genetic alterations distinguish virulent organisms from the live vaccine strains, which therefore have the potential to revert to the virulent state. With killed or inactivated vaccines, there is always a danger that their inactivation might have been incomplete. Furthermore, they often produce undesirable side effects because they are likely to contain extraneous toxic components that are not needed to produce the protective immunity.

These and other shortcomings of the traditional vaccines can be overcome by subunit vaccines, which are made up only of the immunity-conferring "protective antigen," most often a single protein component of the pathogenic organism. Such an antigen may be produced safely and inexpensively by introducing, into harmless organisms such as *E. coli* or yeast, recombinant DNA molecules containing the appropriate gene. Recombinant hepatitis B vaccine, composed of the major capsid protein of the virus, is a successful commercial product licensed for human use.

The subunit vaccines, however, are not always as effective as the traditional vaccines. This is because the vertebrate immune system is optimized to recognize features of the whole pathogen, such as the repetitive presence of the same antigen on its surface, and the presence of components, such as LPS or peptidoglycan, that occur commonly in foreign, invading microorganisms but not in host cells. Therefore, obtaining better immune response often requires improvements in the administration strategy of subunit vaccines so that they mimic the appearance of the antigen in whole pathogens. In some cases, attenuated live pathogens, which can elicit a natural immune response, are used as an effective vector to carry the vaccine subunit: Vaccinia virus carrying the gene for the capsid glycoprotein of rabies virus is now successfully used to immunize wild animals against rabies. It is especially difficult to develop subunit vaccines for pathogens that go through complex life cycles (for example, malaria parasites) or interact with the immune system in unusual ways (for example, leprosy bacillus). Because subunit vaccines present antigens to animals and humans in an artificial context, the production of effective subunit vaccines requires much more in-depth knowledge of the microbes and of the functions of immune cells than the production of traditional vaccines.

Because only small parts of the antigenic protein are recognized by antibodies, it is also possible to generate immunity by injecting small peptides (usually conjugated to a carrier protein) instead of the whole subunit or antigen. Peptide vaccines are attractive because pure, stable preparations can be produced in large quantities by chemical synthesis. An experimental peptide vaccine for foot-and-mouth disease has been successfully produced. Nevertheless, effective peptide vaccines are difficult to produce because they must contain sequences recognized by B cells as well as those recognized by T cells.

SELECTED REFERENCES

General

Ada, G.L., 1982. Vaccines for the future? *Austral. J. Exp. Biol. Med.* 60:549–569.

Liew, F.Y., 1985. New aspects of vaccine development. *Clin. Exp. Immunol.* 62:225–241.

Hilleman, M.R., 1988. Hepatitis B and AIDS and the promise for their control by vaccines. *Vaccine* 6:175–179.

Warren, K.S., 1986. New scientific opportunities and old obstacles in vaccine development. *Proc. Natl. Acad. Sci. USA* 83:9275–9277.

Woodrow, C.C., and Levine, M.M. (eds.), 1990. *New Generation Vaccines*, Marcel Dekker.

Subunit Vaccines Made with Recombinant DNA Technology

Valenzuela, P., Medina, A., Rutter, W.J., Ammerer, G., and Hall, B.D., 1982. Synthesis and assembly of hepatitis B virus surface antigen particles in yeast. *Nature* 298:347–350.

Techniques for Delivery of Vaccines

Allison, A.C., and Byars, N.E., 1987. Vaccine technology: Developmental strategies; Adjuvants for increased efficiency. *Bio/Technology* 5:1038–1045.

Bachmann, M.F., et al., 1993. The influence of antigen organization on B cell responsiveness. *Science* 262:1448–1451.

Live, Attenuated Vectors

Smith, G.L., Mackett, M., and Moss, B., 1983. Infectious vaccinia virus recombinants that express hepatitis B virus surface antigen. *Nature* 302:490–495.

Germanier, R., and Füer, E., 1975. Isolation and characterization of *galE* mutant Ty 21a of *Salmonella typhi*: A candidate strain for a live, oral typhoid vaccine. *J. Inf. Dis.* 131:553–558.

Paoletti, E., Lipinskas, B.R., Samsonoff, C., Mercer, S., and Panicali, D., 1981. Construction of live vaccines using genetically engineered poxviruses: Biological activity of vaccinia virus recombinants expressing the hepatitis B virus surface antigen and the herpes simplex virus glycoprotein D. *Proc. Natl. Acad. Sci. USA* 81:193–197.

Brochier, B., et al., 1991. Large-scale eradication of rabies using recombinant vaccinia-rabies vaccine. *Nature* 354:520–522.

Synthetic Peptide Vaccines

Barteling, S.J., 1988. Possibilities and limitations of synthetic peptide vaccines. *Adv. Biotechnol. Processes* 10:25–60.

Synthetic Peptides as Antigens. Ciba Foundation Symposium, no. 119. Wiley, 1986.

Lerner, R.A., 1984. Antibodies of predetermined specificity in biology and medicine. *Adv. Immunol.* 36:1–44.

Bittle, J.L., et al., 1982. Protection against foot-and-mouth disease by immunization with a chemically synthesized peptide predicted from the viral nucleotide sequence. *Nature* 298:30–33.

Brown, F., 1988. Use of peptides for immunization against foot-and-mouth disease. *Vaccine* 6:180–182.

Rowlands, D.J., 1986. New advances in animal and human virus vaccines. *Advan. Biotech. Proc.* 6:253–285.

Westhof, E., et al., 1984. Correlation between segmental mobility and the location of antigenic determinants in proteins. *Nature* 311:123–126.

Al Moudallal, Z., Briand, J.P., and van Regenmortel, M.H.V., 1985. A major part of tobacco mosaic virus protein is antigenic. *EMBO J.* 4:1231–1235.

Satterthwait, A. C., et al., 1988. Conformational restriction of peptidyl immunogens with covalent replacements for hydrogen bond. *Vaccine* 6:99–103.

Geysen, H.M., Rodda, S.J., and Mason, T.J., 1986. The delineation of peptides able to mimic assembled epitopes. In *Synthetic Peptides as Antigens* (CIBA Foundation Symposium, no. 119), pp. 130–144. Wiley.

Celada, F. and Sercarz, E.E., 1988. Preferential pairing of T-B specificities in the same antigen: The concept of directional help. *Vaccine* 6:94–98.

Berzofsky, J.A., et al., 1986. Molecular features of class II MHC-restricted T-cell recognition of protein and peptide antigens: The importance of amphiphilic structures. *Curr. Top. Microbiol. Immunol.* 130:13–24.

Davis, M.M., and Bjorkman, P.J., 1988. T-cell antigen receptor genes and T-cell recognition. *Nature* 334:395–402.

Mechanism of Immune Response

Hood, L.E., Weissman, I.L., Wood, W.B., Wilson, J.H., 1984. *Immunology*, 2d ed. Benjamin/Cummings.

Paul, W.E. (ed.), 1993. *Fundamental Immunology*, 3d ed. Raven Press.

Kuby, J., 1994. *Immunology*, 2d ed. W.H. Freeman.

Leprosy

Gaylord, H., and Brennan, P.J., 1987. Leprosy and the leprosy bacillus: Recent developments in characterization of antigens and immunology of the disease. *Ann. Rev. Microbiol.* 41:645–675.

Malaria

Mitchell, G.H., 1989. An update on candidate malaria vaccines. *Parasitology* 98:S29–S47.

Good, M.F., Kumar, S., and Miller, L.H., 1988. The real difficulties for malaria sporozoite vaccine development: Nonresponsiveness and antigenic variation. *Immunol. Today* 9:321–355.

Good, M.F., and Miller, L.H., 1990. T-cell antigens and epitopes in malaria vaccine design. *Curr. Top. Microbiol. Immunol.* 155:65–78.

Kemp, O.J., Coppel, R.L., and Anders, R.F., 1987. Repetitive proteins and genes of malaria. *Ann. Rev. Microbiol.* 41:181–208.

Cattani, J.A., 1989. Malaria vaccines: Results of human trials and directions of current research. *Exp. Parasitol.* 68:242–247.

Valero, V., et al., 1993. Vaccination with SPf66, a chemically synthesised vaccine, against *Plasmodium falciparum* malaria in Colombia. *Lancet* 341:705–710.

Microbial Insecticides

> *The concerted effect of the exponentially increasing costs of insecticide development, the dwindling rate of commercialization of new materials, and the demonstration of cross or multiple resistance to new classes of insecticides almost before they are fully commercialized makes pest resistance the greatest single problem facing applied entomology. The only reasonable hope of delaying or avoiding pest resistance lies in integrated pest management programs that decrease the frequency and intensity of genetic selection by reduced reliance upon insecticides and alternatively rely upon multiple interventions in insect population control by natural enemies, insect diseases, cultural manipulations, and host-plant resistance."* Metcalf, R.L. (1980), Changing role of insecticides in crop protection, *Ann. Rev. Entomol.* 25:219–256.

The competition for crops between humans and insects is as old as agriculture, but chemical warfare against insects has a much shorter history. Farmers began to use chemical substances to control pests in the mid-1800s. Not surprisingly, the development of insecticides paralleled the development of chemistry. Early insecticides were in the main inorganic and organic arsenic compounds. They were followed by organochlorine compounds, organophosphates, carbamates, pyrethroids, and formamidines, many of which are in use today. Global sales of chemical insecticides have now reached about $5 billion a year.

There are disadvantages to relying exclusively on chemical pesticides. Foremost among these problems is that the widespread use of single chemical compounds confers a selective evolutionary advantage on the progeny of pests that have acquired resistance to such substances. For example, housefly strains (*Musca domestica*) worldwide have developed resistance to virtually every insecticide used against them. A second problem is that some pesticides affect nontarget species, with disastrous results. Unintentional elimination of desirable predator insects has resulted in explosive multiplication of secondary pests. A third concern is the environmental persistence and toxicity of many pesticides, which have led to the abandonment of many chemical pesticides and increased the cost of developing new and safer ones. Cumulatively, these disadvantages provide a strong incentive to find alternative ways of controlling pests.

Like all living things, insects are susceptible to infection by pathogenic microorganisms (bacteria, fungi, and protozoa) and viruses. Many of these biological agents have a narrow host range and so do not cause

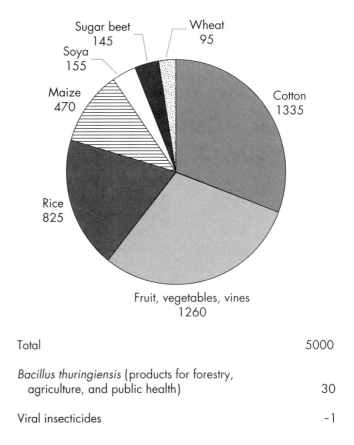

Total	5000
Bacillus thuringiensis (products for forestry, agriculture, and public health)	30
Viral insecticides	~1

Figure 6.1
Global insecticide sales and their distribution among the world's major crops. Numbers represent millions of U.S. dollars at their 1985 value. Data from Jutsum, A.R. (1988), Commercial application of biological control: Status and prospects, *Phil. Trans. R. Soc. Lond.* B 318:357–373.

random destruction of beneficial insects and are not toxic to vertebrates. In spite of this very attractive feature, microbial pest control agents represent less than 1% of total insecticide sales. *Bacillus thuringiensis* has been used for pest control since the 1920s and still accounts for over 90% of the minuscule share of the insecticide market that biological control agents have captured (Figure 6.1). The unusual biological properties of *Bacillus thuringiensis* are central to its role as a microbial insecticide. What scientists are learning about this particular family of pathogenic microorganisms bodes well for more successful and ecologically sound control of insect pests in the near future. Hence, current knowledge about this bacterium is presented in depth in this chapter. Two other members of the genus *Bacillus* are briefly discussed.

Insect viruses (baculoviruses) have achieved some modest commercial success as pest control agents (see Figure 6.1). In recent years, recombinant baculoviruses, introduced into large-scale suspension cultures of insect cells, have successfully expressed bulk amounts of heterologous eukaryotic proteins. Further genetic manipulation of these agents may well increase the utility of these viruses as pesticides. Therefore, baculoviruses are also discussed in some detail here.

BACILLUS THURINGIENSIS

The discovery of *Bacillus thuringiensis* is credited to S. Ishiwata. In Japan in 1901 he isolated the organism responsible for flacherie, a disease of silkworm larvae (*Bombyx mori*) and named it *Bacillus sotto*. (Larvae dying of this disease become soft and flaccid and eventually turn black; *sotto* is a Japanese word roughly equivalent to *limp* in English.) In 1911, a similar bacillus was isolated by E. Berliner from diseased larvae of the Mediterranean flour moth (*Anagasta kühniella*). Berliner named his isolate *Bacillus thuringiensis* after the province of Thüringen, in which it had been discovered. Numerous strains of *B. thuringiensis* have been described since that time and each has its own distinct spectrum of pathogenic effects on its host insects.

Up to 1976, only *B. thuringiensis* strains pathogenic to Lepidoptera (butterflies and moths) were known. These strains showed poor larvicidal activity against blackflies, mosquitoes, and beetles. Recent surveys, however, have led to the isolation of strains pathogenic to dipteran (flies, midges, and mosquitoes) and coleopteran (beetles) pests. These pathogens are neither rare nor difficult to isolate. The greater than 60-year gap between the discovery of the lepidopteran pathotype and that of the dipteran and coleopteran pathotypes was due solely to the lack of a strong incentive to search. This incentive came from the eventual recognition of the urgent need to find new biological agents to control disease-causing insect pests. Joel Margalit has provided a vivid account of the discovery of the first dipteran pathotype, *B. thuringiensis* var. *israelensis*:

> *In 1975–76, Drs. Tahori and Margalit conducted a survey in Israel for biocontrol agents against mosquitoes. During this survey (August 1976) the senior author of this paper came across a small pond in a dried-out river-bed in the northcentral Negev Desert, near Kibbutz Zeelim. This mosquito breeding site, 15 × 60 m, with a maximum depth of 30 cm, contained brackish water with an approximate salinity of 900 mg Cl/ liter and a heavy load of decomposing organic material. A very dense population of exclusively* Culex pipiens *[a common species of mosquitoes] complex dead and dying larvae was found as a "thick carpet" on the surface in an epizootic situation [an epizootic is a disease that affects many animals of one kind at the same time; it corresponds to* epidemic *as applied to diseases of humans]. In addition, pupae and sunk adults attempting to emerge from their pupal cases were floating on the surface.*
>
> *A sample collected from the edge of the pool, containing dead and decomposing larvae, water and silty mud was taken to the laboratory and refrigerated. Bacteria were isolated from this sample in the lab, in association with Mr. L.H. Goldberg, and purified to single colonies. Thus, from a single colony designated ONR 60A, were derived all known cultures of* B.t.i. *now in use.* Margalit, J., and Dean, D. (1985), The story of *Bacillus thuringiensis* var. *israelensis (B.t.i.)*, *J. Am. Mosq. Control Assoc.* 1:1–7.

The discovery of *B. thuringiensis* var. *israelensis* was of immediate practical importance. Bloodsucking dipteran insects, such as mosquitoes and blackflies, transmit a broad spectrum of animal diseases. The blood-borne pathogens include viruses, bacteria, protozoa, and helminths. Mosquitoes, for example, are the vectors that spread the protozoan that causes malaria, with its annual incidence of 200–300 million cases. Insecticidal preparations from *B. thuringiensis* strains toxic to mosquitoes and blackflies are now used successfully as biological control agents in virtually every country where such pests are a severe problem.

B. thuringiensis var. *tenebrionis*, a pathotype effective against the larvae of Coleoptera, was described in 1983. This strain can infect the Colorado potato beetle, the most damaging pest of potatoes in Europe and North America. Populations of this beetle can reach a density of hundreds of insects on a single plant. It has developed resistance to many chemical insecticides and is extremely difficult to control. Fortunately, among 850 strains of *B. thuringiensis* isolated from a wide variety of locations in the United States in 1988, 55 were active against Coleoptera.

B. thuringiensis is a Gram-positive soil bacterium that can grow either by digesting organic matter derived from dead organisms (*saprophytic metabolism*) or by colonization within living insects (*parasitic metabolism*). The bacterium has been isolated from the dust inside silkworm-rearing houses in Japan and from the soils surrounding them. Outbreaks of infection with *B. thuringiensis* are found in insectaries that rear pink bollworm and among larvae that inhabit grain-storage bins. Although *B. thuringiensis* is commonly found inside insects, epizootics, such as the one described by Margalit, are rare in nature. The organism appears to have a low capacity to spread through insect populations. In one study, healthy larvae of *A. kühniella* and *Pieris brassicae* were confined with larvae of the same species that had been infected with *B. thuringiensis*. Most of the diseased larvae died within a few days, and their carcasses were left in the cages. None of the initially healthy *A. kühniella* and only 3 of 180 *P. brassicae* larvae developed *B. thuringiensis* infections. Thus *B. thuringiensis* acts more like a chemical insecticide than an infectious agent.

B. thuringiensis strains currently are classified into different serotypes, or varieties (subspecies), on the basis of their flagellar antigens. This taxonomy is important in identifying the particular strains, but it has little value for predicting the specificity and potency of the insecticidal proteins (see below) produced by a particular strain.

Strains of *B. thuringiensis* are also classified into five pathotypes on the basis of their insecticidal range: lepidopteran-specific (e.g., var. *berliner*); dipteran-specific (e.g., var. *israelensis*); coleopteran-specific (e.g., var. *tenebrionis*); active against both Lepidoptera and Diptera (e.g., var. *aizawai*); and no known toxicity in insects (e.g., var. *dakota*). Even within each of these pathotypes, the various strains differ markedly in potency and in specificity against different insects.

Crystalline Inclusion Bodies

The early studies on flacherie, in 1915, found that only those *B. thuringiensis* cultures that had undergone sporulation were toxic to silkworm larvae. The importance of this observation was not appreciated for over 40 years, though it seems obvious now that the toxic agent must therefore be some molecular species produced specifically during the sporulation process. Once this connection was made in the 1950s, the toxic substance was not hard to identify. Unlike nearly all other *Bacillus* species, *B. thuringiensis* produces crystalline inclusion bodies during sporulation, and these are readily visible under the light microscope. The stages in the sporulation of *B. thuringiensis* var. *kurstaki* are shown schematically in Figure 6.2. Approximately 8 h into the sporulation process, a large bipyramidal crystal and a smaller cuboidal crystalline inclusion develop within the vegetative cell. Chemical analysis has shown the inclusion bodies to consist of proteins that exhibit highly specific

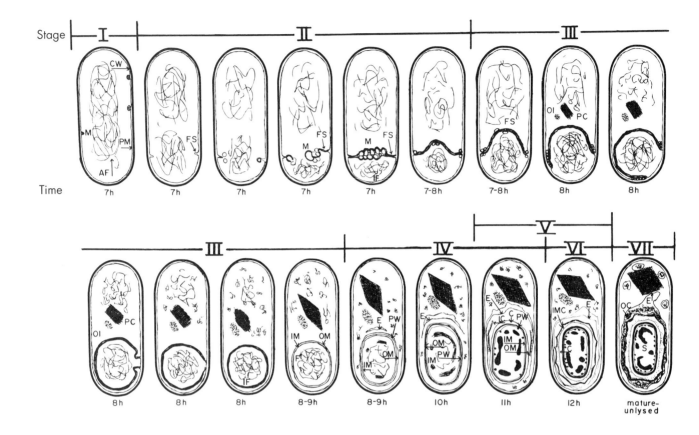

insecticidal activity. Actively growing cells lack the crystalline inclusions and thus are not toxic to insects.

δ-Endotoxins and *cry* Genes

To summarize, during sporulation *B. thuringiensis* strains produce a spore and one or more large protein-containing crystalline inclusions. Ingestion of the protein crystals by susceptible insects leads to their death. The proteinaceous crystals consist of inactive protoxin molecules known as δ-endotoxins. When certain insect larvae ingest them, the crystals are dissolved by the larva's alkaline midgut juices. Next the protoxins are cleaved by gut proteases to generate the active protein toxins. The mature toxins bind to specific receptors on the plasma membrane of larval gut epithelial cells and then insert into the membrane to form cation-conducting pores, 10–20 Å in diameter, thereby destroying the permeability barrier to ions and protons. An influx of water accompanies the entrance of ions into the intestinal cells and

Figure 6.2
Diagrammatic scheme of sporulation in *Bacillus thuringiensis*. Abbreviations: M, mesosome; CW, cell wall; PM, plasma membrane; AF, axial filament; FS, forespore septum; IF, incipient forespore; OI, ovoid inclusion; PC, parasporal crystal; F, forespore; IM, inner membrane; OM, outer membrane; PW, primordial cell wall; E, exosporium; LC, lamellar spore coat; C, cortex; UC, undercoat; OC, outer spore coat; S, mature spore in an unlysed mother cell. Reproduced with permission from Faust, R.M., and Bulla, L.A., Jr., (1982), Bacteria and their toxins as insecticides, in Kurstak, E. (ed.), *Microbial and Viral Pesticides*, p. 88 (Marcel Dekker).

Figure 6.3
A. Scanning electron micrograph of a healthy midgut epithelium of a large white butterfly (*Pieris brassicae*) larva. **B.** Scanning electron micrograph of the midgut epithelium of a larva fed 5 μg of δ-endotoxin and dissected 15 min after endotoxin ingestion. Note the resulting disappearance of microvilli from the epithelium. Courtesy of Dr. Peter Lüthy.

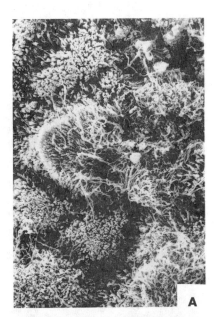

A

B

causes them to swell and lyse (Figures 6.3 and 6.4). The lack of ion regulation also causes paralysis of the muscles of the gut and mouth parts. As a result, feeding stops and death ultimately ensues.

During the active phase of vegetative growth, certain varieties of *B. thuringiensis* produce a low-molecular weight, heat-stable toxin called β exotoxin. This toxin has a nucleotidelike structure (Figure 6.5) and inhibits the activity of DNA-dependent RNA polymerase of both bacterial and mammalian cells. It has been used in Russia to control various cattle fly larvae, and its potential to control the Colorado potato beetle has been examined in the United States, but its nucleotide analog structure, teratogenic effects on insects, and toxicity when injected into mammals have discouraged its use as a biological control agent. Currently, the use of *B. thuringiensis* preparations that contain β-exotoxin is forbidden in North America and Western Europe.

The major crystal protein genes of *B. thuringiensis* (*cry* genes) fall into at least four classes based on similarity of nucleotide sequence and on the insecticidal spectra of the proteins they encode (Table 6.1). The four classes are Lepidoptera-specific (CryI), Lepidoptera- and Diptera-specific (CryII), Coleoptera-specific (CryIII), and Diptera-specific (CryIV) genes. In *B. thuringiensis* var. *israelensis*, one crystal constituent is a small protein (designated CytA) that exhibits cytolytic activity against a variety of cells from invertebrates and vertebrates, but it is totally unrelated in sequence to the *cry* gene products. The pathotype of a given *B. thuringiensis* strain is a reflection of the particular endotoxin gene or genes the strain expresses.

Some *B. thuringiensis* strains produce a single δ-endotoxin; others produce several δ-endotoxins of different specificity. *B. thuringiensis* var. *kurstaki* strain HD-1 is the basic strain used in most preparations currently sold for control of caterpillars. This strain produces two crystalline inclusions—a large bipyramidyl structure and a small cuboidal inclusion—often located at one apex of the bipyramidyl crystal (Figures 6.6 and 6.7). The bipyramidyl crystal contains three protoxin proteins (designated P1 proteins), one of 135 and two of 140 kDa. These protoxins have insecticidal activity against Lepidoptera. The cuboidal crystal contains a single protoxin (designated P2 protein) of 65 kDa that exhibits activity against both Lepidoptera and Diptera.

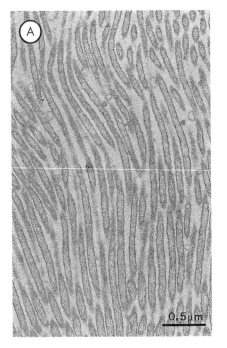
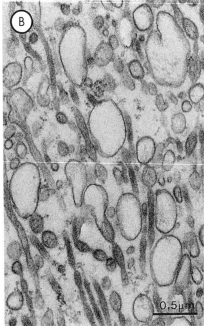
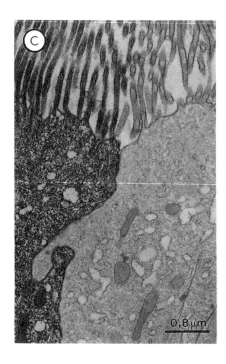

Figure 6.4
A. Intact microvilli of a columnar cell from the gut epithelium of *Pieris brassicae* (control). **B.** Appearance of microvilli 10 min after δ-endotoxin ingestion. **C.** Within 10 min of ingestion of δ-endotoxin, the cells of the midgut epithelium begin to lose the ability to control permeability and become permeable to the indicator stain ruthenium red. The cell on the left has taken up the stain, whereas the cell on the right still retains control of permeability. Reproduced with permission from Lüthy, P., and Ebersold, H.R. 1981, *Bacillus thuringiensis* delta-endotoxin: Histopathology and molecular mode of action, in Davidson, E.W. (ed.), *Pathogenesis of Invertebrate Microbial Diseases*, p. 244 (Allanheld, Osmun).

Figure 6.5
β-Exotoxin of *B. thuringiensis*.

Properties of Representative Insecticidal Crystal Protein Genes of *B. thuringiensis*

**T
A
B
L
E**

6.1

Gene Type	Host Range*	Number of Amino Acids Encoded	Predicted Molecular Mass (kDa)
cryIA(a)	Lepidoptera	1176	133.2
cryIA(b)	Lepidoptera	1155	131.0
cryIA(c)	Lepidoptera	1178	133.3
cryIC	Lepidoptera	1189	134.8
cryID	Lepidoptera	1165	132.5
cryIIA	Lepidoptera and diptera	633	70.9
cryIIB	Lepidoptera	633	70.8
cryIIIA	Coleoptera	644	73.1
cryIVA	Diptera	1180	134.4
cryIVB	Diptera	1136	127.8
cryIVC	Diptera	675	77.8
cryIVD	Diptera	643	72.4
cytA	Diptera/cytolytic	248	27.4

* Lepidopterans ("scale wings") are butterflies and moths; dipterans ("two wings") are flies, midges, and mosquitoes; coleopterans ("sheath wings") are beetles.

SOURCE: Höfte, H., and Whiteley, H.R. (1989), Insecticidal crystal proteins of *Bacillus thuringiensis*, *Microbiol. Rev.* 53:242–255.

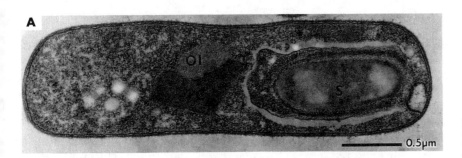

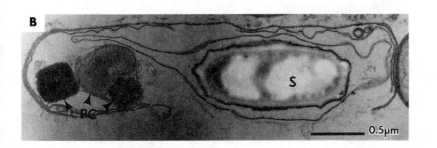

Figure 6.6
Electron micrographs of sporulated cells of *Bacillus thuringiensis* var. *kurstaki* (**A**) and var. *israelensis* (**B**). Abbreviations: S, spore; OI, ovoid inclusion (cuboidal crystal); PC, parasporal crystal (bipyramidal crystal). ×25,370. Reproduced with permission from Faust, R.M., and Bulla, L.A., Jr. (1982), Bacteria and their toxins as insecticides, in Kurstak, E. (ed.), *Microbial and Viral Pesticides*, p. 91 (Marcel Dekker).

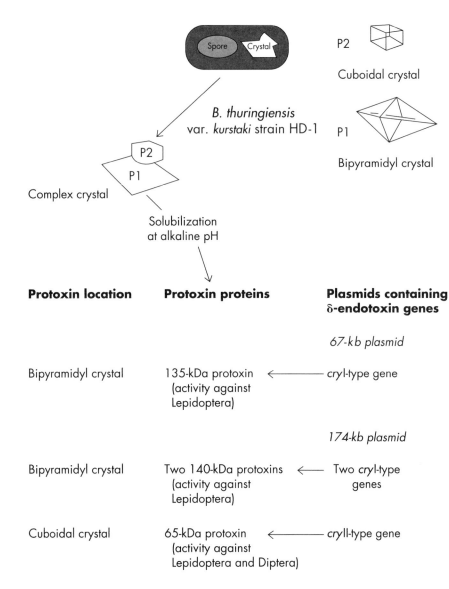

Figure 6.7
δ-Endotoxin genes and insecticidal crystal proteins of *Bacillus thuringiensis* var. *kurstaki* strain HD-1. Based on Carlton, B.C. (1988), Development of genetically improved strains of *Bacillus thuringiensis*, in P.A. Hedin, J.J. Menn, and R.M. Hollingworth (eds.), *Biotechnology for Crop Protection*, Fig. 1 (American Chemical Society).

The insecticidal protoxins of strain HD-1 are encoded by four genes. The gene for the 135-kDa P1 protein resides on a 67-kb plasmid, whereas the two genes that encode two similar 140-kDa P1 proteins and the gene for the 65-kDa P2 protein reside on a 174-kb plasmid (see Figure 6.7). The 67-kb plasmid is readily transmitted from strain HD-1

B. *thuringiensis* var. *berliner* δ-endotoxin (Lepidoptera)

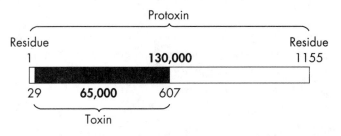

B. *thuringiensis* var. *israelensis* δ-endotoxin (Diptera)

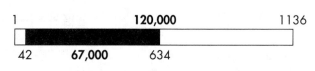

B. *thuringiensis* var. *tenebrionis* δ-endotoxin (Coleoptera)

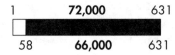

Figure 6.8
Structural features of *B. thuringiensis* insecticidal crystal proteins. Boldface numbers above the bars give the molecular weights of the protoxins. Those below the bars give the molecular weights of the toxins.

to other *B. thuringiensis* strains. The 174-kb plasmid is not self-transmissible.

Regardless of the molecular weight of the protoxin or its ultimate insecticidal specificity, proteolysis in the gut generates active toxin fragments of similar size, 60–70 kDa. Deletion mapping of the different gene types (see Table 6.1) for various 130–145-kDa protoxins showed in each case that the mature toxin resides within the amino-terminal 62–70-kDa portion of the protoxin. One exception is the CryIVD protein, in which the toxic fragment is reported to be about 30 kDa. Depending on the particular protoxin, the amino-terminal border of the active fragment ranges from residue 29 to 39, and the carboxyl-terminal border from residue 607 to 677 (Figure 6.8). The *cryIIIA* gene, encoding a Coleoptera-specific protoxin, directs the synthesis of a 73-kDa protein (see Table 6.1). This protein is converted into a 66-kDa toxin by spore-associated proteases that remove 57 amino-terminal residues (see Figure 6.8). The *cryIIIA* gene is homologous to the toxin-encoding domain of the genes that specify the larger 130–145-kDa protoxins and lacks a region corresponding to the 3′ portion of these genes. Deletion analysis confirmed these conclusions: Truncation of the *cryIIIA* gene at its 3′-end led to loss of toxic activity.

The 3′-portions of the large *cry* genes, encoding the part of the protoxin sequence that starts at about residue 700, show a high degree of sequence homology. This finding suggests that the carboxyl-terminal portion of such protoxins may play a role in the formation of the crystalline protoxin inclusions.

Structure–Function Relationships in the Insecticidal Crystal Proteins

The *cry* genes were cloned in *E. coli* and then sequenced. Deletion mapping was performed to define the domain required for toxic activity in each type of protoxin (Figure 6.8). Separate expression of each gene made it possible to address several important questions.

- *Can a single δ-endotoxin protein assemble into a crystalline inclusion morphologically indistinguishable from that seen in the* B. thuringiensis *strain from which it is derived?* B. thuringiensis subsp. *aizawai* IPL7 synthesizes two protoxins of 130 kDa and 135 kDa, respectively. The amino acid sequence of these two proteins differs at about 10% of the positions out of a total of over 1100 residues. When cloned individually in *E. coli* expression plasmids, each insecticidal protein was able to form a bipyramidyl crystal, virtually free of other components and of the same shape as those observed in the parental *B. thuringiensis* strain (Figure 6.9). The information for the assembly of the bipyramidyl crystals resides in the structure of a single protoxin, and no other *B. thuringiensis* components are required for crystal formation.

- *Does a single δ-endotoxin crystal exhibit larvicidal activity of a potency and specificity that match those of the intact crystals purified from its parent strain?* The genes encoding δ-endotoxins from strains representative of three different *B. thuringiensis* pathotypes, var. *berliner*, var. *israelensis*, and var. *tenebrionis* (specific for Lepidoptera, Diptera, and Coleoptera, respectively), have been cloned, sequenced, and expressed in *E. coli*. The recombinant var. *berliner* crystal protein was a 130-kDa polypeptide, as was the crystal protein from var. *israelensis*. The recombinant crystal protein cloned from var. *tenebrionis* was a 72-kDa polypeptide. In var. *tenebrionis* this protein is cleaved to a 66-kDa protoxin before assembly into crystals, but the cleavage of the protein in *E. coli* was incomplete. The recombinant proteins were purified, and their toxicity was compared on a weight basis to that of

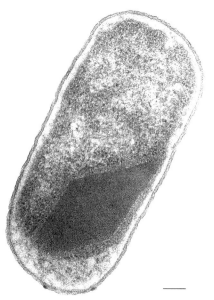

Figure 6.9
Electron micrograph of a cell of *E. coli* JM103 (pTB1) containing a crystal of 130-Kda insecticidal protein of *B. thuringiensis* subsp. *aizawai* IPL7. Bar: 100 nm. Reproduced with permission from Oeda, K., et al. (1989), Formation of crystals of the insecticidal proteins of *Bacillus thuringiensis* subsp. *aizawai* IPL7 in *Escherichia coli*, *J. Bacteriol.* 171:3568–3571.

Comparison of the Toxicity of *B. thuringiensis* Crystals and Purified Recombinant Crystal Proteins

	Target Insect		
	Manduca sexta *Tobacco hornworm* *(Lepidoptera)* LC_{50} *(ng/cm²)*	Aedes aegypti *Mosquito* *(Diptera)* LC_{50} *(ng/ml)*	Leptinotarsa decemlineata *Colorado potato beetle* *(Coleoptera)* LD_{50} *(ng/larva)*
B. thuringiensis *Strain*	*1st instar*	*3rd instar*	*1st instar*
var. *berliner*			
Purified crystal	7.5	—	—
Recombinant protoxin	6	—	—
var. *israelensis*			
Purified crystal	—	9	—
Recombinant protoxin	—	25	—
var *tenebrionis*			
Purified crystal	—	—	13
Recombinant protoxin	—	—	20

TABLE 6.2

Note: LC_{50} is the concentration at which 50% of the exposed insects die; LD_{50} is the administered dose at which 50% of the exposed insects die. A higher number indicates a lower potency. "—" indicates absence of activity. *Instar* is a form an insect assumes between successive molts. The larval stage (in which several molts occur) is represented by several instars.

SOURCE: Höfte H., Chungjatupornchai, W., Van Rie, J., Jansens, S., and Vaeck, M. (1988), Molecular organization and applications of *Bacillus thuringiensis* delta endotoxins, in Fehrenbach, F.J., et al. (eds.), *Bacterial Protein Toxins* pp. 435–440 (Gustav Fischer).

the crystals of the parental *B. thuringiensis* strains. For the proteins of all three pathotypes, the host specificity of the recombinant protein matched that of the crystalline inclusions purified from the parental *B. thuringiensis* strain (Table 6.2). For var. *berliner* and var. *tenebrionis*, the specific activities of the recombinant protein and the purified crystal were similar. These results indicate that in each case the larvicidal specificity and potency are properties of the protoxin *per se* and that potentiation by other components in the parental *B. thuringiensis* strains need not be invoked. However, for var. *israelensis*, the recombinant protein is significantly less active than the native crystals. This raises another question. In cases where a *B. thuringiensis* strain forms crystalline inclusions that contain several protoxins, do these components cooperate in the larvicidal action?

B. thuringiensis var. *israelensis* crystal inclusions contain four insecticidal proteins, one of 27.4 kDa, one of 65 kDa, and two of 130 kDa. The 27.4-kDa protein, encoded by *cytA* (see Table 6.1), is not related structurally to the two 130-kDa proteins or to the 65-kDa protein, which are encoded by *cry* genes. The 27.4-kDa protein displays hemolytic activity toward mammalian erythrocytes as well as cytolytic activity toward a number of mammalian, dipteran, and lepidopteran cell lines. In addition, it has significant mosquitocidal activity though not as strong as that of the 130-kDa and 65-kDa proteins. The lower toxic activity of the single recombinant *israelensis* 130-kDa crystal protein relative to that of the intact crystals (see Table 6.2) indicates that there is either additivity or synergism among the four crystal proteins of var. *israelensis*.

- *Do protoxins bind with high affinity to specific receptors in the larval midgut epithelium, and do they compete for the same receptor?* The protocol of a study directed at this question is given in some detail in Box 6.1. The binding of three Lepidoptera-specific toxins (Bt2, Bt3, and Bt73) to the midgut membranes of two target insects, *M. sexta* and *H. virescens* was analyzed, with the following conclusions.

  Saturable high-affinity toxin binding sites for Bt2, Bt3, and Bt73 (with dissociation constants in the 10^{-9} M range) are present on brush border membrane vesicles of *M. sexta* and *H. virescens.* Such sites do not exist on corresponding vesicles from nonsusceptible insects. δ-Endotoxins from *B. thuringiensis* strains that are not toxic to *M. sexta* did not bind to the brush border membrane vesicles from that organism. These results suggest that toxicity does correlate with the presence of specific receptors.

  The three toxins had similar larvicidal activity in *M. sexta.* However, they differed in their toxicity against *H. virescens.* Judging by the LC_{50} values, Bt3 is approximately 20 times less active than Bt2, which in turn is about 3-fold less active than Bt73 (see Box 6.1). Measurement of the *binding site concentration* on the midgut membranes for each of these toxins showed differences that paralleled the differences in *in vivo* toxicity. Competition binding experiments with *H. virescens* vesicles indicated the presence of at least three distinct types of high-affinity toxin-binding sites. One population binds all three toxins with similar affinity. A second population binds Bt2 and Bt73 toxins but not Bt3 toxin, and a third is accessible only to Bt73, the most toxic of the three δ-endotoxins.

These studies demonstrate that two important determinants of the effectiveness of a particular insecticidal crystal protein in a target insect

6.1 Binding specificity of *B. thuringiensis* toxins to brush border membrane of the midgut of target insects.

STEP 1: Three Lepidoptera-specific δ-protoxins were obtained as follows:

Protoxin   Organismal Origin and Protein Source

Bt2   Product of a *B. thuringiensis* var. *berliner* gene cloned and expressed in *E. coli*

Bt3   Product of a *B. thuringiensis* var. *aizawai* HD-68 gene cloned and expressed in *E. coli*

Bt73   Purified from *B. thuringiensis* var. *aizawai* HD-73. This strain produces only one insecticidal crystal protein.

STEP 2: The protoxins were converted to toxins.
The protoxins were converted to toxins by treatment with trypsin. This produced toxic fragments of about 60 kDa with boundaries similar to those created by the action of the alkaline proteolytic enzymes of the insect midgut. (It had previously been shown that activation by trypsin or by insect gut proteases results in a similar toxicity pattern.)

STEP 3: Toxicity assays were performed.
The larvae of two lepidopteran target insects were used to assay the toxins, the tobacco hornworm (*Manduca sexta*) and the large white butterfly (*Pieris brassicae*). First-instar larvae were fed an artificial diet coated with known amounts of the toxins. The results are tabulated below. LC_{50} values are concentrations required to kill 50% of the insects tested.

LC_{50} (ng toxin/cm²)		
Toxin	*M. sexta*	*H. virescens*
Bt2	20	7
Bt3	20	157
Bt73	9	2

STEP 4: Labeled toxins and brush border membrane vesicles were prepared.
For binding measurements, activated Bt2, Bt3, and Bt73 toxins were labeled covalently with the radioactive iodine isotope, ^{125}I. Brush border membrane vesicles were prepared from *M. sexta* and *H. virescens* midgut membranes and purified by a differential centrifugation method.

STEP 5: Binding assays were performed.
Samples of ^{125}I-labeled toxin, either alone or in combination with varying amounts of other unlabeled toxins, were incubated with brush border membrane vesicles under appropriate conditions. At the end of the incubation period, the vesicles were separated from free toxin by filtration through glass-fiber filters.

SOURCE: Van Rie, J., et al. (1989), Specificity of *Bacillus thuringiensis* δ-endotoxins: Importance of specific receptors on the brush border membrane of the mid-gut of target insects. *Eur. J. Biochem.* 186:239–247.

are the *presence of specific high-affinity receptors* for the toxin in the midgut epithelium of the larva and the *concentration* of such receptors. The importance of these factors emerges clearly in an examination of the basis of insect resistance to particular toxins.

Resistance to Cry Proteins

Repeated exposure of an insect population to a pathogen exerts a strong selective pressure for the proliferation of resistant mutants that may arise in the population. Such selective pressure has been applied in the laboratory to the Indian meal moth (*Plodia interpunctella*), the almond moth (*Cadra cautella*), and the tobacco mealworm (*Heliothis virescens*). Each experiment resulted in the evolution of strains that were less sensitive to the *B. thuringiensis* used to infect them.

The first report of such resistance occurring in the field, in 1985, noted that colonies of *P. interpunctella* taken from grain storage bins that had routinely been treated with *B. thuringiensis* var. *kurstaki* were less susceptible than insects taken from untreated bins. In laboratory-reared *P. interpunctella* strains raised on a *B. thuringiensis*-containing diet, the median lethal dose increased some 250-fold relative to that of the sensitive parent strains.

The biochemical basis of the increased resistance was investigated in *P. interpunctella* strain 343, selected for a high level of resistance against "Dipel," a commercial formulation of a mixture of crystals and spores from *B. thuringiensis* var. *kurstaki*. Figure 6.7 lists the insecticidal crystal proteins present in Dipel. These proteins are encoded by genes that belong to the *cryIA* and *cryII* families (see Table 6.1).

The LD_{50} value (the dose required to kill 50% of the insects tested) for the resistant strain (R strain) and for the sensitive parent strain (S strain) was determined for Dipel, for recombinant CryIA(b) protoxin and toxin from *B. thuringiensis* var. *berliner*, and for recombinant CryIC protoxin and toxin from *B. thuringiensis* var. *entomocidus* HD110. The S and R strains responded similarly to Dipel and to the CryIA(b) protoxin, which is closely related to one of the insecticidal crystal proteins in Dipel. The R strain showed a similar resistance level to the CryIA(b) protoxin and the CryIA(b) toxin prepared from it by *in vitro* activation (Table 6.3). This indicated that resistance was not due to a lack of proteolytic activation of the protoxin in the midgut of the R strain. Other studies showed that the resistance was not due to destruction of the stable toxic fragment by enhanced proteolytic degradation in the R strain.

Receptor-binding studies with ^{125}I-labeled CryIA(b) toxin were performed with brush border membrane vesicles derived from larval midguts of the S and R strains in the manner described in Box 6.1. Although the concentrations of the receptors in the S and R strains were similar, the data showed that the apparent affinity of the receptors from the R strain was 50-fold lower than for the S strain (Table 6.4). These results suggest that resistance to CryIA(b) toxin is due to a decrease in receptor affinity.

The R strain shows no enhanced resistance to a different insecticidal protein toxin, CryIC, that is structurally unrelated to the CryIA(b) toxin (see Table 6.3). The results of binding experiments with the

T A B L E

6.3

> **Toxicity of *B. thuringiensis* Insecticidal Crystal Proteins to Sensitive and Resistant Strains of the Indian Meal Moth *P. interpunctella***

| Insecticidal Crystal Protein | LD_{50} of *P. interpunctella* strain (microgram/larva) | |
	Sensitive	Resistant
Dipel	1.21	> 30
CryIA(b) protoxin	0.12	> 12.8
CryIA(b) toxin	0.03	26.3
CryIC protoxin	0.20	0.02
CryIC toxin	0.11	0.03

SOURCE: Van Rie J., et al. (1990), Mechanism of insect resistance to the microbial insecticide *Bacillus thuringiensis*, Science 247:72–74.

CryIC toxin were more complicated than those with the CryIA(b) toxin. In the S strain, two kinds of binding sites for CryIC were detected with dissociation constants, designated K_{d1} and K_{d2}, of 0.31 and 154 nM, respectively. In the R strain, only the binding site characterized by K_{d1} was detected (see Table 6.4). Competition experiments comparing the

T A B L E

6.4

> **Binding of *B. thuringiensis* Insecticidal Crystal Proteins to Brush Border Membrane Vesicles from S and R Strains of *P. interpunctella****

Strain	Insecticidal Crystal Protein	K_{d1} (nM)	REC_1 (pmol/mg protein)	K_{d2} (nM)	REC_2 (pmol/mg protein)
Sensitive	CryI(A)(b) toxin	0.72	1.44		
	CryIC toxin	0.31	0.38	154	6.17
Resistant	CryI(A)(b) toxin	36.3	1.77		
	CryIC toxin	0.18	1.15		

* K_d values are the dissociation constants of toxin–membrane receptor complexes. For the CryI(A)(b) toxin, the binding data showed a single population of saturable high-affinity sites in both the S and R strains. For the CryIC toxin, there were two populations of sites. The CryIC toxin bound to the strong sites with an affinity some 500-fold higher than to the second population of sites (see values of K_{d1} and K_{d2}). REC values represent the binding site (receptor) concentrations.

SOURCE: Van Rie et al. (1990), Mechanism of insect resistance to the microbial insecticide *Bacillus thuringiensis*, Science 247:72–74.

activities of CryIC and CryIA(b) in the S strain showed that CryIC and CryIA(b) compete for the sites characterized by K_{d2} and that these sites correspond to the strong binding sites for CryIA(b). CryIA(b) does not compete for the strong binding sites for CryIC. Moreover, the affinity of CryIC for its own strong sites is the same, within experimental error, in the S and R strains (see Table 6.4).

To summarize, the midgut epithelium of *P. interpunctella* has different high-affinity receptor sites for the distantly related insecticidal crystal proteins CryIA(b) and CryIC. CryIC binds strongly to its own receptor and about 50-fold more weakly to the CryIA(b) receptor. CryIA(b) binds strongly only to its own receptor. The presence on the midgut epithelium of different receptors recognized specifically by different toxins has practical implications. Treating the target insect strain with a mixture of toxins that recognize different receptors would greatly reduce the likelihood that the insect will develop resistance to the toxins.

B. thuringiensis as a Biological Insecticide Today

As illustrated by the partial listing in Table 6.5, numerous serious insect pests are sensitive to insecticidal preparations made from naturally occurring strains of *B. thuringiensis*. However, for several reasons, the use of *B. thuringiensis* preparations as insecticides has grown very slowly.

Microbial pesticides are generally more expensive to produce than many chemical pesticides. Because epizootic infections are rare, large quantities of toxin have to be applied to a crop to ensure that each larva will ingest a lethal dose. Nevertheless, the cost may well decrease as the demand increases. In Canada, where *B. thuringiensis* has been used in large amounts in preference to chemical pesticides for the control of forest pests, the average cost of the bacterial preparation was halved between 1980 and 1983.

The inherent specificity of different *B. thuringiensis* strains is both a strength and a weakness. The majority of *B. thuringiensis* products, produced from *B. thuringiensis* var. *kurstaki* strain HD-1, have excellent activity against some insect pests, such as gypsy moth larvae and the cabbage loopers, but poor activity against other pests, such as the cotton bollworm. Pesticide users who are accustomed to the broad spectrum of toxicity of many chemical pesticides may consider a product with a narrower range to be less convenient. Moreover, the slower speed with which the microbial pesticides kill their insect targets contributes to a perception that they are less effective than the traditional chemical agents.

At the same time, *B. thuringiensis* preparations have some particularly favorable characteristics. These include stability during several years of storage and resistance of the crystals to inactivation by the

Main Target Insects Susceptible to *B. thuringiensis* Preparations

TABLE 6.5

Insect	Crop or Product
Pieris brassicae (large white butterfly)	Vegetables
Pieris rapae (imported cabbage looper)	Vegetables
Plutella maculipennis (diamondback moth)	Vegetables
Trichoplusia ni (cabbage looper)	Vegetables, tobacco, soybeans, cotton
Heliothis virescens (tobacco budworm)	Cotton, soybeans, tobacco
Heliothis zea (cotton bollworm)	Cotton, soybeans, tobacco
Ostrinia nubilalis (European corn borer)	Corn
Manduca sexta (tobacco hornworm)	Tobacco
Leptinotarsa decemlineata (Colorado potato beetle)	Potatoes
Lymantia dispar (gypsy moth)	Forests
Choristoneura fumiferana (spruce budworm)	Forests
Malacosoma disstria (tent caterpillar)	Forests
Denrolimus sibiricus (Siberian silkworm)	Forests
Plodia interpunctella (Indian meal moth)	Stored products
Anagasta kühniella (Mediterranean flour moth)	Stored products
Ephestia cautella (almond moth)	Stored products
Insect Vectors of Vertebrate Diseases	*Disease*
Culex spp., *Aedes* spp., *Anopheles* spp. (mosquito species)	Disseminate malaria, yellow fever, and Dengue fever
Simulium spp. (black fly species)	Disseminate onchoceriasis (African river blindness)

SOURCE: This table is based primarily on Table 3 in Lüthy, P., Cordier, J.-L., and Fischer, H.-M. (1982), *Bacillus thuringiensis* as a bacterial insecticide: Basic considerations and application. Kurstak, E. (ed.), *Microbial and Viral Pesticides*, pp. 35–74 (Marcel Dekker).

ultraviolet rays of the sun. The most important factor in favor of expanded future use is the demonstrated safety of *B. thuringiensis* preparations in over 30 years of field application around the world.

Implications of the Diversity of B. thuringiensis cry *Genes in Nature*

Studies such as those we have discussed have shown that the pathotype of specific *B. thuringiensis* strains is a result of the recognition by its toxin(s) of specific receptors on the midgut epithelium of the host insects. The toxicity of a particular insecticidal crystal protein against different insects belonging to the same class, such as Lepidoptera, can vary by 100-fold or more.

To attack a specific insect pest, it would therefore be appropriate to use a strain of *B. thuringiensis* whose insecticidal crystal proteins were expressed at a high level and had high affinity for receptors in the midgut epithelium of the target insect. In naturally occurring strains of *B. thuringiensis*, there are dozens—perhaps hundreds—of insecticidal crystal protein genes encoding protoxins that differ from each other in specificity. This situation, combined with improved understanding of the structure–function relationships in the toxin–receptor interaction that X-ray crystallographic and other structural studies have yielded, should soon make it possible to engineer new genes that encode toxins with high affinity for the receptors of specific pests.

One might imagine that in the future, different insects will be screened for susceptibility to a panel of insecticidal crystal proteins, much the way bacteria are screened for sensitivity to various antibiotics. Preparations for field use against an insect could then be created by using one or more of the toxins found to have greatest potency against that pest. Such a practice would exploit the highly selective toxicity of biological insecticides much as medicine now exploits the selectivity of antibiotics against specific infectious microorganisms.

Genetic Improvement of B. thuringiensis *and Other Organisms*

Useful new *B. thuringiensis* strains have been generated by genetic methods which do not involve the use of recombinant DNA.

Selective Plasmid Curing. In many *B. thuringiensis* strains, the individual bacterium contains multiple plasmids, several of which may carry one or more of the various *cry* genes. The level of expression of any one of the *cry* genes (as measured by the amount of the gene-encoded crystal protein in the cell) is influenced by the expression of the other *cry* genes. Sometimes, one of the *cry* genes encodes a protoxin with high activity against a specific target pest, while the other *cry* genes may encode protoxins with low activity against the pest. In such cases, it would be advantageous to obtain more of the high-activity protoxin per cell. This can be done by eliminating the genes that encode the low-activity protoxins.

As an example, let us examine *B. thuringiensis* var. *kurstaki* strain HD-1. This strain contains a 67-kb plasmid, which carries a gene encoding a 135-kDa Lepidoptera-specific crystal protein, and a 174-kb plasmid, which carries genes that encode other crystal proteins (see Figure 6.7). Deliberate inducement of heat shock or repeated transfer of cultures can be used to cause the plasmids to mis-segregate during cell division, creating variants that have lost some or all of their plasmids. A variant of *B. thuringiensis* var. *kurstaki* strain HD-1, which had lost the 174-kb plasmid, expressed its single remaining 135-kDa protoxin gene from the 67-kb plasmid at a higher level.

Construction of New Strains by Conjugation. Certain of the plasmids in *B. thuringiensis* strains are self-transmissible. Such plasmids can be used to construct new strains by conjugation, as in the following example.

As we saw earlier, the Colorado potato beetle (Coleoptera) is the most destructive insect pest of potatoes in North America. But certain lepidopteran insects, such as the European corn borer and the potato tuber worm, can also significantly damage this crop. No single strain of *B. thuringiensis* isolated from nature is toxic to all of these insects. However, a strain effective against all three pests was constructed as follows: A *B. thuringiensis* strain active against the Colorado potato beetle was isolated, and its *cry* gene(s) were found to be on a transmissible plasmid. This strain was mated with a recipient *B. thuringiensis* strain that was active against the European corn borer. The transconjugant strain was active against both insects and in field tests showed promise as a general-purpose potato insecticide.

Transfer of the cry *Genes to Other Organisms.* By means of genetic engineering, any organism can be endowed with and made to express the *cry* genes, or modified versions thereof. In fact, *cry* genes have already been transferred into various bacteria and plants and expressed at a high level. The generation of insect-resistant transgenic crops that express *B. thuringiensis* crystal proteins is discussed in Chapter 9. The transfer of *cry* genes to cyanobacteria (blue-green algae), which flourish in aquatic environments, has been considered as a way to maintain high levels of toxin in mosquito-infested environments. Gene transfers of this sort open the way to new approaches for the protection of crops and the suppression of disease-bearing insects.

BACILLUS SPHAERICUS

B. sphaericus, an aerobic spore-forming bacterium common around the world in soil and aquatic environments, is more active than *B. thuringiensis* against the larvae of certain important mosquito species (*Culex* and *Anopheles*). The insecticidal activity of *B. sphaericus* strains appears to be almost exclusively limited to mosquito larvae. Numerous other insects are unaffected by the presence of these organisms.

Some strains of *B. sphaericus* are able to grow on decomposing organic matter in heavily polluted water. This is noteworthy because it is the habitat favored by the insect host, and the *B. sphaericus* strains do well there whether the mosquitos are present or not. Treatment of cesspits and latrines in Tanzania with a dry-powder preparation of a *B. sphaericus* strain at 10 g/m² provided control of mosquito larvae for 6–10 weeks. In contrast, *B. thuringiensis* is unable to grow saprophytically in this environment, and its spores and crystals do not persist long in polluted water. The World Health Organization is strongly encouraging further evaluation and development of *B. sphaericus* as a microbial larvicide.

Like *B. thuringiensis*, certain *B. sphaericus* strains produce a crystal-line protein toxin whose mode of action on susceptible mosquito larvae shares both similarities with and differences from those of *B. thuringiensis* toxins. The major components of the *B. sphaericus* crystal are two proteins of 51- and 42-kDa, respectively; neither protein is toxic to larvae, and both are required for toxicity. Hence the toxin is described as a *binary* toxin. Following ingestion of crystal-containing sporulated cells of *B. sphaericus*, the crystal is solubilized in the midgut by the alkaline pH, and the 51- and 42-kDa proteins are processed to 43- and 39-kDa proteins, respectively. The toxin proteins bind to the midgut epithelial cells. This interaction leads to feeding inhibition and ultimately to the death of the larvae. The toxin may exert its effect by ADP-ribosylation of specific proteins in the cells of susceptible larvae. Such a toxic mechanism has been established for the binary *Clostridium botulinum* C2 toxin, where one protein is involved in binding to the cell membrane and the other exerts a toxic effect by ADP-ribosylation of nonmuscle actin, blocking its polymerization and leading to disruption of the intracellular cytoskeleton network.

BACILLUS POPILLIAE

B. popilliae is a pathogen of various scarabeid beetles and was first isolated from the Japanese beetle *Popillia japonica*. Unlike *B. thuringiensis* and *sphaericus*, *B. popilliae* does not produce toxins. Instead, after being ingested by the larvae, the bacteria of this species invade the hemocoel (*coel* is from the Greek *koilos*, "hollow"; in arthropods, the body cavity formed by expansion of the blood-vascular system). They multiply rapidly there for about 2 days, killing the larvae in the process. The bacteria pass from the vegetative to the sporulating phase as the larva dies. A single larval cadaver may be packed with as many as 10^{10} spores, and these have proven to be very persistent in soil.

A constraint on the use of *B. popilliae* is that the organism is an obligate pathogen, so the only way to obtain spores for commercial use is to grow the bacteria in Japanese beetle larvae. Spores are produced commercially in this way, on a small scale, for control of the Japanese beetle on ornamental plants and in lawns.

BACULOVIRUSES

Viral infections are common in insects. About 1600 virus isolates — grouped into more than 10 families on the basis of their morphological and biochemical properties — have been reported to cause disease in some 1100 species of insects and mites. One of these families, the

Principal Baculovirus Candidates for Insect Pest Control

Virus	Target Host	Crop
Nuclear polyhedrosis viruses		
Anticarsia gemmatalis NPV (soybean looper)	*A. gemmatalis*	Soybeans
Autographa californica NPV (alfalfa semilooper)	*Orgyia pseudotsuga* *Trichoplusia ni*	Forests Cabbage
Gilpinia hercyniae NPV (spruce sawfly)	*G. hercyniae*	Forests
Heliothis spp. NPV (cotton bollworm)	*Heliothis* spp.	Cotton, maize, sorghum
Lymantria dispar NPV (gypsy moth)	*L. dispar*	Forests
Mamestra brassicae NPV (cabbage moth)	*Mamestra, Heliothis,* and *Diparopsis* spp.	Cotton, cabbage, and other vegetables
Neodiprion sertifer NPV (pine sawfly)	*N. sertifer*	Forests
Neodiprion lecontei NPV (redheaded pine sawfly)	*N. lecontei*	Forests
Orgyia pseudotsugata NPV (Douglas fir tussock moth)	*O. pseudotsuga*	Forests
Spodoptera littoralis NPV (cotton leafworm)	*S. littoralis*	Cotton
Trichoplusia ni NPV (cabbage looper)	*T. ni*	Cabbage
Granulosis viruses		
Cydia pomonella GV (codling moth)	*C. pomonella*	Orchards
Plodia interpunctella GV (Indian meal moth)	*P. interpunctella*	Stored grain
Nonoccluded baculoviruses		
Oryctes rhinoceros virus (coconut rhinoceros beetle)	*O. rhinoceros*	Coconut and oil palm

Note: With the exception of *G. hercyniae* NPV and *P. interpunctella* GV, each of these viruses is either being produced for field use or has been the subject of a commercial trial.

SOURCE: Payne, C.C. (1988), Pathogens for the control of insects: Where next? *Phil. Trans. Roy. Soc. Lond.* B. 318:225–248.

Baculoviridae (baculoviruses), are restricted in their host range to a small number of arthropod insects, some crustaceans, and mites. However, the susceptible insects include many important pests (Table 6.6). Baculoviruses account for more than 60% of all viruses known to infect insects. They are not pathogenic to vertebrates or plants.

The baculoviruses (Figure 6.10) are classified, on the basis of morphological properties, into three groups: the nuclear polyhedrosis viruses (NPVs), the granulosis viruses (GVs), and the *Oryctes*-type viruses.

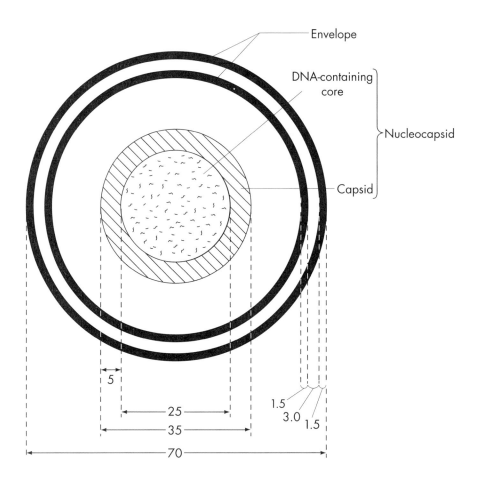

Figure 6.10

Diagrammatic representation of a cross section of a budded baculovirus virion; all dimensions are in nanometers. The core of the virus contains the viral DNA associated with a small, strongly basic protein and lesser amounts of other tightly bound proteins. This nucleoprotein core is surrounded by the *capsid*, a rod-shaped protein shell. Baculovirus capsids are about 35–50 nm in diameter and 200–400 nm in length. The virus capsid containing the nucleoprotein core is called the *nucleocapsid*. Membrane-enveloped nucleocapsids are referred to as *virus particles* or *virions*. If the membrane is derived from the plasma membrane in the process of budding, the enveloped virion is referred to as the *budded virus* (BV). Virus particles that obtain their envelope within the nucleus are usually also embedded in *occlusion bodies* (*polyhedra*) within a crystalline protein matrix formed by the protein *polyhedrin*. Such virions are termed *occluded virions*.

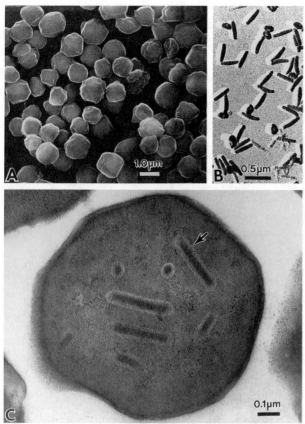

Figure 6.11
A. Scanning electron micrograph of the corn earworm, *Heliothis zea*, nuclear polyhedrosis virus (NPV) polyhedra, or viral occlusions. **B.** Virions of *H. zea* NPV released from viral occlusion bodies at alkaline pH. **C.** Section of *H. zea* viral occlusion body with single nuclear polyhedrosis viruses embedded in a paracrystalline protein matrix. The virions contain a single nucleocapsid per envelope (arrow). Courtesy of Dr. J.R. Adams.

In the NPVs and GVs, the genome codes for a protein called polyhedrin, which forms aggregates, or *occlusion bodies*, around the virions (Figure 6.11). The NPV occlusion body is polyhedral in shape (0.8–5 μm in diameter) and holds many virions. The GV occlusion body is generally ellipsoidal (0.3–0.5 μm long). The *Oryctes*-type viruses are not occluded. NPVs and GVs infect host larvae; *Oryctes*-type viruses can be infectious in both the larval and the adult stages of the host.

Biology of Baculoviruses

The sequence of events associated with an NPV infection is illustrated in Figure 6.12. NPVs are the only viruses known to use two phenotypes, *budded virus* and *occluded virion*, to complete their life cycle in nature (see Figure 6.10 for the definition of these terms). The two virus forms have identical nucleocapsids, but their membranes differ in protein composition. Infection of a susceptible larva results when it ingests food contaminated with viral occlusion bodies. The occlusion body disaggregates in the alkaline contents of the gut of the larval host, and virus particles are released to enter the gut epithelial cells by fusion. The coat surrounding the viral DNA is removed at the nuclear pores or in the nucleus. The single baculovirus DNA molecule is double-stranded, circular, and supercoiled, and is up to about 130 kb in length. It begins to replicate about 6 h after entering the cell. About 10 h after that, virus is

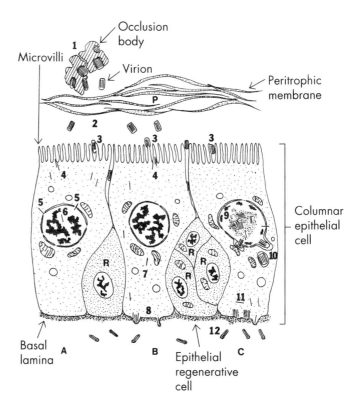

Figure 6.12
Pathways of infection by NPV in insect larvae. (1) Ingested viral occlusion bodies dissolve in the midgut of the larvae, releasing embedded virions. (2) Released virions pass through the peritrophic membrane lining the gut lumen and associate with columnar epithelial cells of the midgut wall. (3) Virions gain entry into these cells by fusion with the plasma membrane of their microvilli or lateral surfaces. Once inside the cell, a virion may follow one of two routes: initiation of replication (cell A) or translocation to the basal membrane (cell B). In cell A (4), nucleocapsids are transported through the cytoplasm to the nucleus, probably through association with microtubules, and (5) enter the nucleus, probably through nuclear pores. (6) Uncoating of the viral genome and initiation of viral replication occurs in the nucleus. In cell B (7), some nucleocapsids bypass the nucleus and migrate directly to the basal membrane of the columnar epithelial cells.

(8) These nucleocapsids can bud through the basal membrane and basal lamina into the hemolymph, acquiring a cell-membrane-derived envelope in the process. In cell C (9), after step (6) a diffuse virogenic stroma is formed in the infected nuclei of the columnar epithelial cells in which nucleocapsid assembly takes place. No viral occlusions are formed in columnar epithelial cells. (10) Nucleocapsids exit the nucleus by budding through the nuclear envelope, forming transport vacuoles. (11) Transport vacuoles are lost during transit of nucleocapsids to the basal membrane where budding of progeny virus takes place. (12) Extracellular virus in the hemolymph transmits the disease throughout the larval body. Reproduced with permission from Fraser, M.J. (1986), Transposon-mediated mutagenesis of baculoviruses: Transposon shuttling and implications for speciation, *Ann. Entomol. Soc. Amer.* 79:773–783.

released from the cell by budding into the hemolymph. This extracellular virus (budded virus) transmits the disease to other tissues of the host. Very few, if any, viral occlusion bodies are formed in the columnar epithelial cells. However, massive production of occlusion bodies takes place in cells of other infected tissues (Figure 6.13). Whereas extracellular virus levels reach a maximum approximately 2 days after infection, occlusion bodies continue to accumulate in the nucleus of infected cells for 4–6 days, until the cells lyse (Figure 6.14).

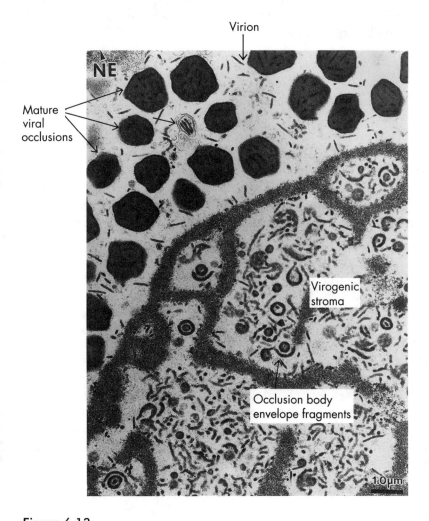

Figure 6.13
Late stage of viral replication (80–100 h after infection) in the *Heliothis zea* fat body. The fat body is a very abundant tissue in lepidopteran larvae. This section reproduced here shows mature viral occlusions between the virogenic stroma and the nuclear envelope (NE). The virogenic stroma contains occlusion body envelope fragments, nucleocapsids, and virions. The unlabeled arrow points to a bundle of virions among the viral occlusion bodies. Courtesy of Dr. J.R. Adams.

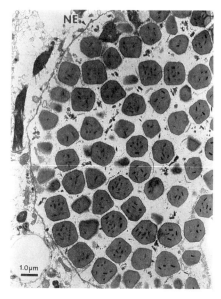

Figure 6.14
Late stage of infection of a cell in the fat body of the beet army worm, *Spodoptera exigua*, by *Autographa californica* NPV. The section shows a late stage of maturation of the viral occlusions (the nucleus is filled with mature viral occlusions). NE, nuclear envelope. Courtesy of Dr. J.R. Adams.

Unfortunately, the infected insect continues to feed almost to the end. Even 2 or 3 days after infection, when the majority of the cells in the epidermis and the fat body are infected, the midgut is surprisingly free of virus. This temporary recovery of the midgut is accounted for, at least in part, by the fact that when insect larvae molt, a large portion of the midgut epithelial cells are sloughed off and replaced. Hence the insect is able to feed until a very late stage of the disease.

After death and decomposition of the host, the occlusion bodies protect the virions from loss of infectivity for long periods of time. The disease spreads within the insect population when the occlusion bodies are ingested by other larvae.

Baculoviruses as Insecticides

Commercial preparations of *Neodiprion sertifer* NPV and *Mamestra brassicae* NPV are used to control the pine sawfly and the cabbage moth, respectively. *Oryctes rhinoceros* virus has been successfully used in many countries in the South Pacific region to control the rhinoceros beetle, a pest of coconut and oil palms. In spite of these successes, several factors have limited the wider use of baculoviruses as pest control agents.

First, the viral pesticides are more expensive than chemical pesticides. Whereas many baculoviruses are highly host-specific and are therefore ideal choices for selective pest control, most crops are attacked simultaneously by several different lepidopteran pests. There has been little enthusiasm for the notion of producing and marketing a different virus for each of these pest species. Furthermore, baculovirus infection kills too slowly to satisfy most farmers. As we noted earlier, 4–6 days elapse from the time of ingestion until the death of the host, and the larva continues to feed on the crop for much of that time. In addition, when baculoviruses are applied to the surface of plants by conventional spraying techniques, they are rapidly inactivated by the ultraviolet rays of the sun. Finally, in most cases, application of baculoviruses has not led to total elimination and long-lasting control of target pests. Adequate control has been achieved only when these viruses have been applied repeatedly.

Future Possibilities for Baculovirus Insecticides

In principle, it is possible to overcome each of the major obstacles to increasing the use of baculoviruses as insecticides. To begin with, it is likely that an increase in the use of baculoviruses will lead to a decrease in production costs. In addition, it may become possible, by genetic manipulation, to produce strains with the desired broad host range. A highly selective host range is not an invariable property of baculoviruses. *Autographa californica*, an NPV from the alfalfa semilooper, has a known range of 43 species of Lepidoptera in 11 families, many of which are known pests. The factors that determine the host range and virulence of baculoviruses are under investigation.

Recombinant DNA technologies might also increase the speed with which baculoviruses kill their hosts. Because foreign genes can be incorporated into baculovirus genomes and expressed in insect hosts, it might be possible to insert genes that encode a hormone that affects feeding behavior or a neuropeptide that paralyzes the larva.

Such recombinant viruses would still have to be grown in insect cell cultures or in larvae. Consequently, the recombinant virus would have to be engineered in such a way that the toxic gene is not expressed during commercial production of the virus. One way to achieve this would be to create a temperature-sensitive mutant of the virus that could express the toxic gene only under specific temperature conditions.

Different approaches could be taken to decrease the sensitivity of the virus to ultraviolet light. Spraying the virus onto the underleaf surfaces of plants would help somewhat, as would the engineering of more effective DNA-repair systems for UV-induced damage.

In an integrated pest management approach, total kill of the target insect is not essential. A primary contribution of the baculovirus would be to destroy the population that includes the insecticide-resistant mutants that might arise in response to a chemical insecticide used on the crop, before they could give rise to resistant progeny.

SUMMARY

Many chemical pesticides have an unfavorable environmental impact, and there is pressure for decreased reliance on such agents and greater regulatory control of their use. Moreover, many pests have acquired resistance to widely used chemical pesticides. The trend is toward *integrated pest management programs* that decrease the frequency and intensity of the genetic selection of resistant insect mutants by employing in concert different means of insect population control: insecticides, microbial or viral pathogens, other natural enemies, and resistance of the plant host.

Many bacteria and viruses cause disease and death in insects. Three *Bacillus* species are used as bacterial insecticides: *B. thuringiensis*, *B. sphaericus*, and *B. popilliae*. *B. thuringiensis*, which can be cultured *in vitro*, synthesizes insecticidal crystal proteins (δ-endotoxins) during sporulation. The δ-endotoxins represent up to 30% of the dry weight of sporulating cells. Particular *B. thuringiensis* strains produce toxins active against Lepidoptera (butterflies and moths), Diptera (flies, midges, and mosquitoes), or Coleoptera (beetles). After ingestion by larvae, the δ-endotoxins are processed to active toxins. The latter bind to specific receptors on the plasma membrane of larval gut epithelial cells, cause small pores permeant to cations to form in the membrane, and thereby destroy the permeability barrier to ions. The destruction of the larval gut epithelium leads to cessation of feeding, and death ultimately ensues. *B. sphaericus* strains are facultative pathogens. Their insecticidal activity appears to be limited almost exclusively to mosquito larvae. Numerous other insects are unaffected by the presence of these organisms. The *B. sphaericus* crystal, like that of *B. thuringiensis*, forms upon sporulation. The major components of this crystal are two proteins of 51 and 42 kDa, respectively; neither protein is toxic to larvae, and both are required for toxicity. Hence the toxin is described as a *binary* toxin. Following ingestion of crystal-containing sporulated *B. sphaericus* cells, the crystal is solubilized in the midgut by the alkaline juices, and the 51- and 42-kDa proteins are processed to 43- and 39-kDa proteins, respectively. The toxin proteins bind to the midgut epithelial cells. This interaction leads to feeding inhibition and ultimately to the death of the larvae. The binary toxin may exert its effect by ADP-ribosylation of specific proteins in the cells of susceptible larvae. *B. popilliae* strains are obligate pathogens and are grown on live larvae of the Japanese beetle (*Popillia japonica*). *B. popilliae* find limited use as insecticides against certain coleopteran larvae.

Many viruses pathogenic to insects have been identified. Over 100 different insects are susceptible to *baculoviruses*. Baculoviruses are desirable biological insecticides because they are unrelated to known vertebrate or plant viruses, have a restricted host range (certain insects), and are harmless to nontarget insects or vertebrates. Baculoviruses have a large (up to 200-kb), double-stranded, circular DNA genome enclosed within a rod-shaped nucleocapsid. In the nuclei of cells infected with nuclear polyhedrosis viruses, one or more nucleocapsids are *occluded* within a paracrystalline proteinaceous matrix to form an inclusion body or *polyhedron*. At a late stage in infection, the nuclei are filled with inclusion bodies. When baculoviruses are used as an insecticide, occluded virus preparations are sprayed on foliage. After ingestion by target larvae, the inclusion bodies dissolve in the alkaline juices of the insect gut and release the embedded infectious virions. The virions initiate infection by passing through the membrane that lines the gut lumen and entering the epithelial cells of the midgut wall.

Application of genetic engineering techniques to the production and modification of these bioinsecticides is opening new horizons in the use of microbial agents in insect pest control.

SELECTED REFERENCES

General

Hansen, M., 1988. A history of pest control: From traditional agriculture through the pesticide era. In *Escape from the Pesticide Treadmill: Alternatives to Pesticides in Developing Countries*, pp. 9–29. Institute for Consumer Policy Research, Consumers Union.

Metcalf, R.L., 1980. Changing role of insecticides in crop protection. *Ann. Rev. Entomol.* 25:219–256.

Microbial Insecticides

Angus, T.A., 1984. Inspiration, sweat and serendipity: The proof of *Bacillus thuringiensis* in biological control. In *Comparative Pathobiology*, vol. 7, *Pathogens of Invertebrates: Application in Biological Control and Transmission Mechanisms*, ed. T.C. Cheng, pp. 35–45. Plenum Press.

Lacey, L.A., and Undeen, A.H., 1986. Microbial control of black flies and mosquitoes. *Ann. Rev. Entomol.* 31:265–296.

Laird, M., Lacey, L.A., and Davidson, E.W. (eds.), 1989. *Safety of Microbial Insecticides.* CRC Press.

Currier, T.C., 1990. Commercial development of *Bacillus thuringiensis* bioinsecticide products. In Nakas, J.P., and Hagedorn, C. (eds), *Biotechnology of Plant–Microbe Interactions*, pp. 111–143. McGraw-Hill.

Maramorosch, K. (ed.), 1991. *Biotechnology for Biological Control of Pests and Vectors.* CRC Press.

Stone, R., 1992. Researchers score victory over pesticides—and pests—in Asia. *Science* 256:1272–1273.

Moffat, A.S., 1992. Improving plant disease resistance. *Science* 257:482–483.

Microbial and Viral Diseases of Insects

Steinhaus, E.A., 1975. *Disease in a Minor Chord. Being a Semihistorical and Semibiographical Account of a Period in Science When One Could be Happily Yet Seriously Concerned with the Diseases of Lowly Animals without Backbones, Especially the Insects.* Ohio State University Press.

Davidson, E.W. (ed.), 1981. *Pathogenesis of Invertebrate Microbial Diseases.* Allanheld, Osmun.

Kurstak, E. (ed.), 1982. *Microbial and Viral Pesticides.* Marcel Dekker.

Wood, R.K.S., and Way, M.J. (eds.), 1988. *Biological Control of Pests, Pathogens, and Weeds: Developments and Prospects.* The Royal Society.

Aronson, A.I., Beckman, W., and Dunn, P., 1986. *Bacillus thuringiensis* and related insect pathogens. *Microbiol. Rev.* 50:1–24.

Bacillus thuringiensis Strains with Activity Against Diptera and Coleoptera

de Barjac, H., and Sutherland, D.J. (eds.), 1990. *Bacterial Control of Mosquitoes and Black Flies. Biochemistry, Genetics and Applications of* Bacillus thuringiensis israelensis *and* Bacillus sphaericus. Rutgers University Press.

Krieg, V.A., Huger, A.M., Langenbruch, G.A., and Schnetter, W., 1983. *Bacillus thuringiensis* var. *tenebrionis*, a new pathotype effective against larvae of Coleoptera. *Z. Angew. Entomol.* 96:500–508.

Talbot, H.W., et al. 1989. Unique strains of *Bacillus thuringiensis* with activity against Coleoptera. In Demain, A.L., Somkuti, G.A., Hunter-Cevera, J.C., and Roosmoore, H.W. (eds.), *Novel Microbial Products for Medicine and Agriculture*, pp. 213–218. Society for Industrial Microbiology.

Insecticidal Crystal Proteins (δ-Endotoxins)

Whiteley, H.R., and Schnepf, H.E., 1986. The molecular biology of parasporal crystal body formation in *Bacillus thuringiensis*. *Ann. Rev. Microbiol.* 40:549–576.

Höfte, H., and Whiteley, H.R., 1988. Insecticidal crystal proteins of *Bacillus thuringiensis*. *Microbiol. Rev.* 53:242–255.

Brousseau, R., and Masson, L., 1988. *Bacillus thuringiensis* insecticidal crystal toxins: Gene structure and mode of action. *Biotech. Adv.* 6:697–724.

Fehrenbach, F.J., et al. (eds.), 1988. *Bacterial Protein Toxins.* Gustav Fischer.

Carlton, B.C., 1988. Development of genetically improved strains of *Bacillus thuringiensis*: A biological insecticide. In Hedin, P.A., Menn, J.J., and Hollingworth, R.M. (eds.), *Biotechnology for Crop Protection*, pp. 260–279. American Chemical Society.

Van Rie, J., Jansens, S., Höfte, H., Degheele, D., and Van Mellaert, H., 1989. Specificity of *Bacillus thuringiensis* δ-endotoxins: Importance of specific receptors on the brush border membrane of the mid-gut of target insects. *Eur. J. Biochem.* 186:239–247.

Oeda, K., et al., 1989. Formation of crystals of the insecticidal proteins of *Bacillus thuringiensis* subsp. *aizawai* IPL7 in *Escherichia coli. J. Bacteriol.* 171:3568–3571.

Mechanism of δ-Endotoxin Action

Knowles, B.H., and Ellar, D.J., 1987. Colloid-osmotic lysis is a general feature of the mechanism of action of *Bacillus thuringiensis* δ-endotoxins with different insect specificities. *Biochim. Biophys. Acta* 924:509–518.

Li, J., Carroll, J., and Ellar, D.J., 1991. Crystal structure of insecticidal δ-endotoxin from *Bacillus thuringiensis* at 2.5 Å resolution. *Nature* 353:815–821.

Wolfersberger, M.G., 1992. V-ATPase-energized epithelia and biological insect control. *J. Exp. Biol.* 172:377–386.

Chen, X.J., Lee, M.K., and Dean, D.H., 1993. Site-directed mutations in a highly conserved region of *Bacillus thuringiensis* δ-endotoxin affect inhibition of short circuit current across *Bombyx mori* midguts. *Proc. Natl. Acad. Sci. USA* 90:9041–9045.

Resistance to Insecticidal Crystal Proteins

Van Rie, J., McGaughey, W.H., Johnson, D.E., Barnett, B.D., and Van Mellaert, H., 1990. Mechanism of insect resistance to the microbial insecticide *Bacillus thuringiensis*. *Science* 247:72–74.

Gibbons, A., 1991. Moths take the field against biopesticide. *Science* 254:646.

McGaughey, W.H., and Whalon, M.E., 1992. Managing insect resistance to *Bacillus thuringiensis* toxins. *Science* 258:1451–1455.

Bacillus sphaericus

Baumann, P., Clark, M.A., Baumann, L., and Broadwell, A.H., 1991. *Bacillus sphaericus* as a mosquito pathogen: Properties of the organism and its toxins. *Microbiol. Rev.* 55:425–436.

Thanabalu, T., Berry, C., and Hindley, J., 1993. Cytotoxicity and ADP-ribosylating activity of the mosquitocidal toxin from *Bacillus sphaericus* SSII-1: Possible roles of the 27- and 70-kilodalton peptides. *J. Bacteriol.* 175:2314–2320.

Baculoviruses

Adams, J.R., and Bonami, J.R. (eds.), 1991. *Atlas of Invertebrate Viruses*. CRC Press.

Granados, B.R., and Federici, B.A., 1986. *The Biology of the Baculoviruses*. Vol. I, *Biological Properties and Molecular Biology*. Vol. II, *Practical Application for Insect Control*. CRC Press.

Miller, L.K., 1987. Expression of foreign genes in insect cells. In Maramorosch, K. (ed.), *Biotechnology in Invertebrate Pathology and Cell Culture*, pp. 295–303. Academic Press.

Lucknow, V.A., and Summers, M.D., 1988. Trends in the development of baculovirus expression vectors. *Bio/Technology* 6:47–55.

Greer, F., Ignoffo, C.M., and Anderson, R.F., 1990. The first viral pesticide. *Chemtech* 347:606–611.

Engelhard, E.K., Kam-Morgan, L.N., Washburn, J.O., and Volkman, L.E., 1994. The insect tracheal system: A conduit for the systemic spread of *Autographa californica* M nuclear polyhedrosis virus. *Proc. Natl. Acad. Sci. USA* 91:3224–3227.

Microbial Enzymes

The dairy, baking, and brewing industries, and other major food industries as well, have long depended on enzymes from animals, plants, and microorganisms for producing cheese, bread, and malt beverages, for clarifying fruit and vegetable juices, and for tenderizing meats. Today the dominant uses of partially purified or pure *microbial* enzymes are in the production of glucose and fructose syrups, in washing powders and detergents, and in the manufacture of textiles. The enzymes used in the greatest amounts are listed in Table 7.1, and estimates of their annual sales value are given in Table 7.2. With the exception of glucose isomerase and invertase, all these enzymes are extracellular; that is, they are excreted by the microorganisms.

This chapter describes some of the industrial processes that employ bacterial and fungal enzymes on a large scale. All are applications involving the hydrolysis of biological polymers—either polysaccharides or polypeptides. Microbial enzymes have many other important applications, notably in organic syntheses, diagnostics, and research. These applications, which generally require much more modest quantities of enzymes, are the subject of Chapter 14.

LEGAL REGULATION OF ENZYMES USED IN FOODS

Fewer than 50 bacteria and fungi produce enzymes for the food industry. The most prominent are the *Bacillus* species, such as *B.*

Microbial Enzymes with Industrial-Scale Applications and Some of Their Sources

T A B L E

7.1

Enzyme	Source	Action	Applications
α-Amylase	Bacillus subtilis Bacillus licheniformis Aspergillus oryzae	Endo-hydrolysis of α-1,4-glucosidic linkages	Starch processing
Glucoamylase	Aspergillus oryzae Aspergillus niger Rhizopus oryzae	Removes glucose from nonreducing end of starch, also splits α-1,6-linkages at branch points but more slowly	Starch processing; brewers' and distillers' mashes
Pullulanase	Klebsiella aerogenes	Splits α-1,6-glycosidic linkages in pullulan and amylopectin	Starch processing
Glucose isomerase	Bacillus coagulans Streptomyces albus	Converts D-glucose to D-fructose. This enzyme is actually a xylose isomerase that converts D-xylose to D-xylulose	Production of high-fructose syrups
β-Glucanase	Bacillus subtilis Aspergillus niger Penicillium emersonii	Degrades β-glucan by cleaving β-1,3 (4)-glucosidic linkages	Brewing
Invertase	Saccharomyces cerevisiae	Splits sucrose to glucose and fructose	Confectionery industry; baking
Lactase	Saccharomyces lactis A. oryzae, A. niger, Rhizopus oryzae	Splits lactose to glucose and galactose	Dairy industry (treatment of milk and whey)
Pectinase	A. oryzae, A. niger, Rhizopus oryzae	Degrades pectin, α-1,4-linked anhydrogalacturonic acid with some of the carboxyl groups esterified as the methyl esters	Clarification of fruit juices and wines
Neutral protease	Bacillus subtilis Aspergillus oryzae	Hydrolyzes peptide bonds in proteins	Flavoring of meat and cheese; baking
Alkaline protease	Bacillus licheniformis	Hydrolyzes peptide bonds in proteins	Laundry detergents
Rennin	Mucor miehei spp. Recombinant enzyme produced in E. coli and fungi	Hydrolyzes a specific bond in κ-casein, leading to coagulation of milk proteins	Cheesemaking
Lipase	A. oryzae, A. niger, Rhizopus oryzae	Hydrolyzes ester linkages in fats	Dairy industry; detergents

T
A
B
L
E

7.2

Worldwide Sales of Microbial Enzymes (estimate for 1989)

Class of enzymes	Sales (U.S. $ million)
Proteolytic enzymes	
Alkaline proteases	150
Neutral proteases	70
Rennins	60
Carbohydrases	
Amylases	100
Isomerases	45
Pectinases	40
Lipases	20

SOURCE: Arbige, M.V., and Pitcher, W.H. (1988), Industrial enzymology: A look towards the future, *Tr. Biotechnol.* 7:330–335.

subtilis and *B. licheniformis,* the yeast *Saccharomyces cerevisiae,* and the filamentous fungi *Aspergillus niger* and *A. oryzae* (Table 7.1). These organisms are nonpathogenic, produce no known toxins, and have a well-established record of safety. *Lactobacillus* and lactic *Streptococcus* species, as well as certain of their enzymes, have long been used in the dairy industry for preparation of fermented foods and beverages and are also considered safe.

If an organism is *not* on the regulatory agencies' GRAS ("generally recognized as safe") list for food manufacture or processing, it may take several years for a manufacturer to obtain permission to use an enzyme from that source. Petitions for approval must include the following information:

- Description of the enzyme and its source organism.

- The methods of enzyme preparation: (1) complete composition of the medium, (2) equipment used in production, and (3) fermentation conditions (aeration, agitation, temperature, and pH).

- Composition of the enzyme preparation: (1) single enzyme, enzyme mixture, or whole cell; (2) impurities such as heavy metals, traces of solvents, or other chemicals from the manufacturing process (an example is the cross-linking agent glutaraldehyde, which is frequently used in enzyme-immobilization procedures); and (3) details of the analytical methods used to obtain the information on the composition and amounts of impurities.

"There are many requirements for the producing organism. High enzyme production capacity is not the only criterion. The most efficient producer is not necessarily useful for the industry at all. If the organism is not on the GRAS list (Generally Recognized as Safe) you must spend a lot of time and money to obtain acceptance. Moreover, it is also necessary to take into account the fact that the organism will perhaps never be accepted for production of food enzymes. One alternative is to forget about the whole thing, which may be a wise decision." Linko, M. (1989), Enzymes in the forefront of food and feed industries, *Food Biotechnology* 3:1–9.

- Toxicological data on the enzyme or microorganism: (1) The organism must be nonpathogenic. (2) It must not produce mycotoxins or other toxic chemicals. (3) The organism must not produce antibiotics.

Because enzyme preparations are good substrates for microbial growth, specifications for enzyme preparations used in food processing always require an assessment of microbial contamination. This includes values for the total count of viable bacteria and for specific kinds of bacteria such as coliform bacteria, *E. coli* (to detect contamination of fecal origin), and *Salmonella* (because of its pathogenicity).

The advent of recombinant enzymes in food industry applications may result in a significant acceleration of the approval process. Recombinant DNA technology uses a limited variety of microorganisms and plasmids to produce a wide range of enzymes. Consequently, it will not be necessary to redetermine the safety of the host and plasmid system every time approval is sought for a new recombinant enzyme.

PRODUCTION OF MICROBIAL ENZYMES

The first step in enzyme manufacture is to identify microbial strains that produce an enzyme with both the appropriate catalytic specificity and the desired physical properties. Modification of the producer microbial strain is generally required in order to increase greatly the amount of the enzyme produced per cell. Composition of the culture medium and the fermentation conditions are important determinants of the yield and cost of an enzyme preparation. All of these points are considered in the paragraphs that follow.

Strain Selection and Development

So that many wild-type and mutant organisms can be examined quickly, screening for organisms suitable for the industrial production of a particular enzyme must include methods for enrichment of desirable strains and for approximate assay of the level of enzyme production. The search for an enzyme with particular physical properties is aided by general correlations observed between the optimal conditions for growth of various microorganisms and the properties of their enzymes.

Table 7.3 illustrates the positive correlation that exists between the growth temperature of an organism and the stability of its extracellular proteases. Such correlations help narrow the range of environments to be searched for microorganisms whose *extra*cellular enzymes have the

T
A
B
L
E

7.3

Thermostability of Proteases from Mesophilic and Thermophilic Microorganisms

Source	Growth temperature (°C)	In vitro enzyme incubation temperature (°C)	Half-life* (min)
Bacillus licheniformis	37	70	10
B. stearothermophilus NCIB 8924	55	74	15
B. thermoproteolyticus	55	80	60
Thermus aquaticus T351	75	80	1800

* Time required for 50% loss of enzymatic activity.

SOURCE: Cowan, D., Daniel, R., and Morgan H. (1985), Thermophilic proteases: Properties and potential applications, *Tr. Biotechnol.* 3:68–72.

desired properties. On the other hand, although the *intra*cellular enzymes of thermophiles (organisms with an optimal growth temperature greater than 45°C) are generally more stable than those of mesophiles (organisms with an optimal growth temperature range of 20–45°C), environmental temperature conditions cannot be counted on to predict the properties of intracellular enzymes.

Different microorganisms often produce enzymes that catalyze the same reaction, but each type of microorganism produces a different amount of the enzyme. Moreover, such *isofunctional* enzymes frequently have very different compositions and properties. For example, proteolytic enzymes from *Bacillus* species growing in alkaline environments (alkalophilic strains) have pH optima several units higher than proteases from *Bacillus* species growing in environments with a neutral pH.

When a prospective producer microorganism has been identified, several steps are required to convert it into a strain that is suitable for commercial use. First, a good industrial strain produces a high yield of enzyme and in high concentration. As a rule, the wild-type organism produces the desired enzyme in amounts too low for commercial use. One major impediment to high levels of enzyme production is the phenomenon of *catabolite repression*, in which the presence of readily metabolized carbon sources such as glucose represses the biosynthesis of many degradative enzymes. Catabolite repression can be minimized either by manipulating the medium or by genetically modifying the organism. For example, selection for mutants resistant to the presence of nonmetabolizable analogs of glucose makes it possible to isolate strains resistant to carbon catabolite repression. Resistance to 2-deoxyglucose has been used to isolate mutants of *Trichoderma* sp. that overproduce cellulases. Deregulated mutants of *B. licheniformis* that overproduce α-amylase have also been isolated.

Inducers for Some Commercial Enzymes

TABLE 7.4

Enzyme	Substrate	Inducer
α-Amylase	Starch	Starch or maltodextrins
Glucoamylase	Starch	Maltose or isomaltose
Invertase	Sucrose	Sucrose
Pullulanase	Pullulan	Pullulan or maltose
Xylose isomerase	Xylose	Xylan or xylose

Another difficulty is that, as a rule, degradative enzymes are synthesized in significant quantity only when a specific inducer is present in the medium (Table 7.4). The following negative regulatory mechanism operates in many of these cases. In the absence of the inducer, a repressor protein prevents the structural genes for degradative enzymes from being expressed. It is the formation of a complex between the inducer (or some product derived from it) and the repressor that releases the structural genes from inhibition. Some inducers are too expensive to be used in large-scale enzyme production. However, the need for an inducer can be eliminated by mutations either of the structural gene that encodes the repressor or of the regulatory site(s) on the chromosome where the repressor binds. Such *constitutive* mutants produce the normally inducible enzyme in the absence of the inducer.

More gene products are required in the production of extracellular than of intracellular enzymes, and this must be taken into account in the development of commercial enzyme-producing strains. The polypeptides that are precursors to extracellular enzymes are synthesized inside the cell or at sites on the cell membrane and are then secreted into the medium. Frequently, these processes are accompanied by posttranslational modifications such as proteolytic processing of precursor polypeptides to the mature form as well as (in fungi) glycosylation.

The development of a *B. subtilis* hyperproducer strain for α-amylase (Figure 7.1 and Table 7.4) demonstrates the general way in which industrial strains are generated by classical genetic approaches. In the development of this strain, mutant alleles of regulatory genes responsible for elevated production of α-amylase in *B. subtilis* were introduced either by transformation of cells with DNA extracted from naturally occurring α-amylase hyperproducing strains or by treatment of the bacterial cells with N-methyl-N'-nitro-N-nitrosoguanidine (NTG), a mutagenic chemical. The customary procedure is to adjust the extent of exposure to the mutagen such that a large fraction of the population is killed. This ensures that the surviving population will include a high percentage of genetically modified individuals. As shown in Figure 7.1, *B. subtilis* strain T2N26, into which six regulatory alleles (*amy R3, amyS, tmrA*, *papS*, C-108, and N-26) were introduced stepwise, produces 1600 times as much α-amylase as the original *B. subtilis* strain 6160. Mutations affecting the integrity of the cell wall have a particularly important influence on the level of secretion of extracellular enzymes.

Fermentation Processes and Composition of the Medium

There are two types of commercial processes for using microbes in enzyme production (Box 7.1). In the more traditional method, the koji process (or *solid-state fermentation*), microorganisms are grown on solid or semisolid media in trays, a setup that makes it difficult to prevent contamination and to achieve uniform control of temperature, aeration, and humidity. Thus the second method, the use of fermenters, now predominates. In modern fermenters, parameters such as pH, temperature, concentration of dissolved oxygen, rate of oxygen uptake, rate of carbon dioxide evolution, aeration, back-pressure, foaming, and agitation are both monitored and adjusted by computer-controlled instruments. Cell density and enzyme content of the broth (the liquid growth medium) are determined by frequent sampling.

Substrate expenses can easily account for up to 80% of the cost of an enzyme fermentation process. It is therefore important that the producer microorganism be able to grow rapidly on a cheap medium. It must also be able to grow on a concentrated medium in order to yield high amounts of enzyme per fermenter batch. The typical carbohydrate constituents of inexpensive media are molasses, barley, corn, wheat, starch hydrolyzate, and lactose. Soybean, peanut, and cottonseed meal, corn steep liquor, yeast hydrolyzate, whey, gluten, and gelatin are used as nitrogen sources. Appropriate inorganic salts provide supplementary sources of nitrogen, phosphorus, sulfur, or calcium. Trace metals either come from the tap water used for preparation of the media or are present in the main nutrients mentioned above. Generally, phosphate buffers are used for pH control.

LARGE-SCALE APPLICATIONS OF MICROBIAL ENZYMES

The most extensive uses of microbial enzymes are in the manufacture of sweeteners, in laundry detergents, and in the manufacture of cheeses.

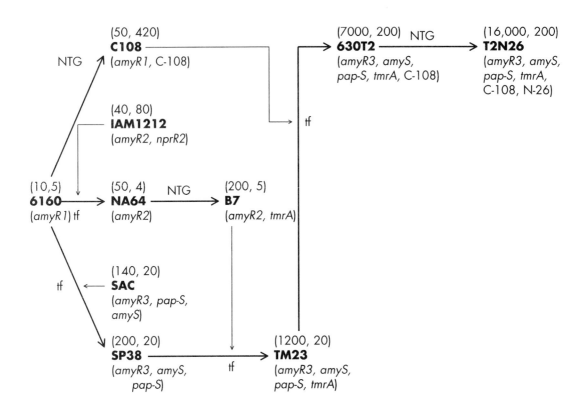

Construction of a Strain of *B. subtilis* That Overproduces α-Amylase

Allele	Source	Stimulation of α-Amylase Production	Comments
amyE⁺	*B. subtilis* 6160		Structural gene for α-amylase
amyR1	*B. subtilis* 6160	Basal level of production (10 units/ml)	Chromosome location adjacent to *amyE*
amyR2	*B. natto* IAM1212	4–5 fold	
amyR3	*B. subtilis* SAC	4–5 fold	
amyS	*B. subtilis* SAC	2–3 fold	Distal to *amyE*. AmyS was not functional in the donor strain.
papS	*B. subtilis* SAC	2–3 fold	Also stimulates neutral protease secretion
tmrA	*B. subtilis* B7	5–7 fold	Selected as an NTG mutant resistant to tunicamycin[a]
"C-108"	*B. subtilis* 108	2–3 fold	Selected as an NTG mutant resistant to cycloserine[b]
"N-26"	*B. subtilis* T2N26	2–3 fold	Selected as NTG mutant resistant to ampicillin[b]

[a] Tunicamycin inhibits glycosylation of glycoproteins in eukaryotic cells and the biosynthesis of peptidoglycan and teichoic acid in bacteria.

[b] Cycloserine and ampicillin interfere with peptidoglycan biosynthesis.

Starch Processing

Starch is a mixture of linear and branched homopolymers of D-glucose. It is formed as a carbohydrate reserve in plants and is present in considerable amounts in potato tubers, cassava, and tapioca, and in the seeds of wheat, corn, barley, and sorghum. The linear component, *amylose*, consists of chains of α-1,4-D-glucopyranose ranging in degree of polymerization from about 10^2 to 4×10^5. In the branched component, *amylopectin*, shorter chains (17–23 units long) of α-1,4-D-glucopyranose join together by α-1,6 bonds to form a branched structure with a degree of polymerization between 10^4 and 4×10^7 (Figure 7.2).

Starch is readily hydrolyzed by mineral acids, such as hydrochloric acid; and in dilute suspensions, acid hydrolysis leads to near-total conversion of the starch to D-glucose. In fact, acid hydrolysis of starch was the chief method for preparing glucose syrups for much of the early twentieth century. It was not feasible to use dilute suspensions for commercial purposes, however, so acid hydrolysis was usually carried out with suspensions containing 20% starch at pH 2 and 140°C. Under such conditions reversion products are formed, as well as repolymerized polyglucoses, which limit the yield of glucose to about 50% of theoretical. The hydrolyzate was neutralized with sodium carbonate (with the consequent formation of salt) and then refined to a glucose syrup. Among the disadvantages of acid hydrolysis were the low glucose yields, the formation of large amounts of salt, and the need to use corrosion-resistant equipment.

Glucose has only approximately 75% of the sweetness of sucrose, whereas its isomer fructose has about twice the sweetness of sucrose (Figure 7.3). Consequently, fructose is the preferred sweetener in low-calorie foods, providing twice the sweetness of sucrose at half the weight. It is possible to convert glucose to fructose chemically by exposing glucose to alkaline conditions at high temperature. However, those

---

Figure 7.1

Construction of *B. subtilis* strain T2N26, an α-amylase overproducer. Steps resulting from mutagenesis with N-methyl-N'-nitro-N-nitrosoguanidine are indicated by NTG, and those resulting from transformation by tf. Strain designations are in boldface type. The two numbers in parentheses above each strain designation indicate units of enzyme activity per milliliter secreted into the medium for α-amylase and neutral protease, respectively. The relevant genotypes are indicated in parentheses under each strain designation. The table provides information on the individual regulatory alleles, their organismal source, and their influence on the level of α-amylase production, chromosomal location, and other features. Based on Maruo, B., and Yoshikawa, H. (eds.) (1989), *Bacillus subtilis: Molecular biology and industrial application*, Fig. 8.2.2 (Elsevier).

BOX 7.1 Fermentation Processes and Fermenters (Bioreactors)

Solid-state fermentation. In this process, microorganisms grow on porous solid substrates, generally in the absence of free water. The water adsorbed on the substrate is sufficient to support the growth of the cells. The growth and secretion of product occur both on the surface of the solid support and within the support matrix. Support substrates include wheat bran, wheat straw, brewer's spent grain, rice hulls, sugar beet pulp, and bagasse. The majority of solid-state fermentations employ fungi to produce proteases, lipases, cellulases, and oxidases.

Bioreactors. There are four common types of bioreactors: the stirred-tank reactor, the bubble column, the air lift, and the packed bed (or fixed bed). Simplified schematic views of these reactors are given below.

Stirred-tank reactor. This is the most widely used type of bioreactor. It is particularly well suited for fermentations that involve abundant fungal mycelia and for the production of biopolymers where high viscosity is encountered.

Bubble column. The agitation and aeration in this reactor are provided by a bottom sparger. To ensure even agitation, the sparger nozzles must be distributed uniformly over the cross section of the bottom. Either a ring with regularly spaced holes, a small number of parallel pipes, or a starlike arrangement of pipes is used.

Air lift. This tubular, tower reactor consists of two pipes connected at the top and bottom. The air is introduced at the bottom into one of the pipes (the *riser*). The air rises and escapes at the top. Therefore, there is little air in the other pipe, the *downcomer*. The introduction of the air results in circulation of the reactor contents between the riser and the downcomer. The circulation is induced by the difference in density between the aerated liquid in the riser and the liquid in the downcomer.

Packed bed. This bioreactor consists of a tubular column packed with a support that retains (or entraps) microbial and fungal cells. The reactor can be operated in either the upflow or the downflow mode; that is, the liquid containing the substrate can be introduced at either the top or the bottom of the reactor. Such reactors are not used for the production of proteins. They are commonly used for small-molecule bioconversions.

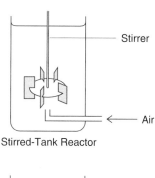

Stirred-Tank Reactor

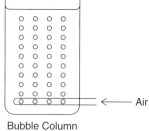

Bubble Column

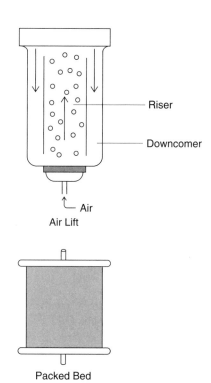

Air Lift

Packed Bed

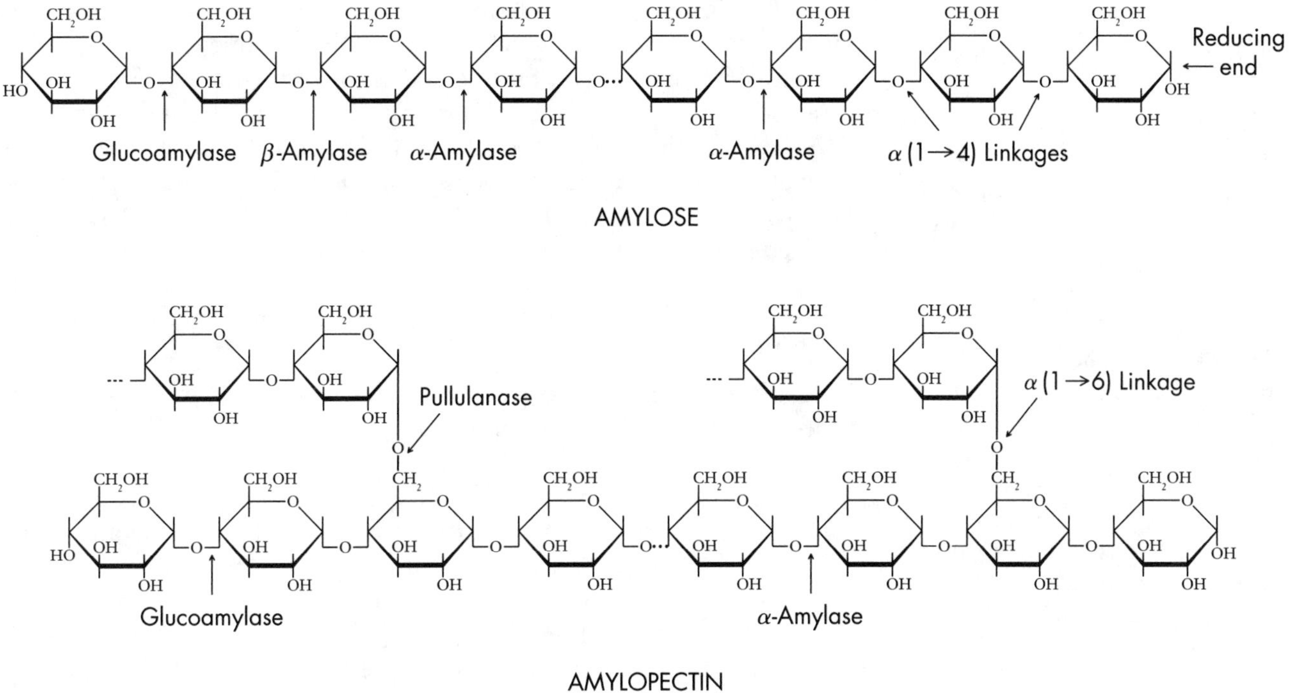

Figure 7.2
Structure of starch, indicating points of cleavage by α- and β-amylase, glucoamylase, and pullulanase.

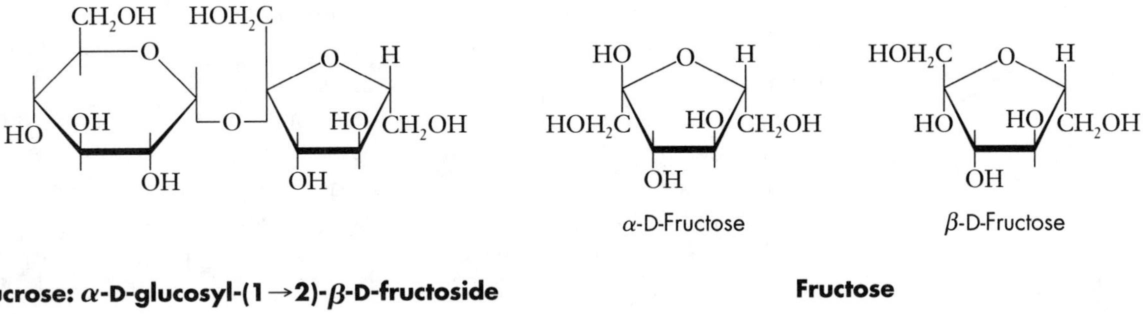

Sucrose: α-D-glucosyl-(1→2)-β-D-fructoside

Fructose

Figure 7.3
Structure of sucrose and of the α and β anomers of fructose.

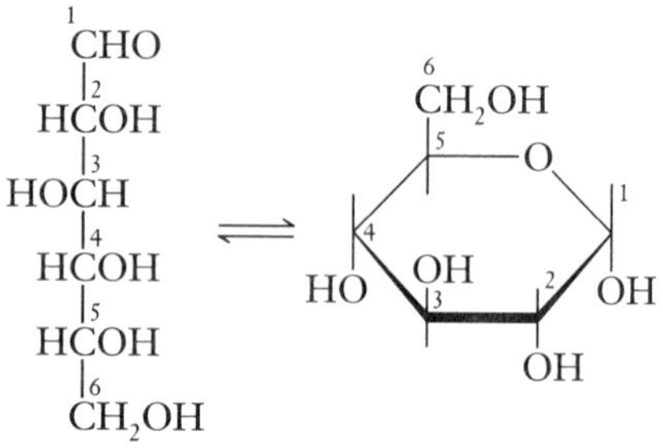

Figure 7.4
The equilibrium between the straight-chain structure and the ring structure in D-glucose. In water, the predominant form of glucose is the cyclic intramolecular *hemiacetal* shown at the right. This *pyranose* ring form is in equilibrium with the open-chain form containing a free aldehyde group, which is also called the *reducing group*.

conditions lead to the formation of undesirable by-products, such as the monosaccharide psicose, which is not metabolizable by humans.

The problems inherent in the chemical conversion of starch to glucose and fructose are avoided by using processes that depend entirely on enzymes, and the world production of high-fructose corn syrups from starch by enzyme-catalyzed processes reached 4.25 million tons in 1983. High-fructose corn syrups account for over 30% of total nutritive sweetener consumption in the United States.

Terminology

Several technical terms are in common use in descriptions of starch processing and of the resulting products.

- *Dextrose equivalent (DE).* Hydrolysis of glucosidic bonds leads to the exposure of reducing aldehyde groups (Figures 7.2 and 7.4). The hydrolysis of starch to a mixture of sugar monomers and oligomers is monitored by measuring their specific chemical reducing power relative to that of the same amount of pure glucose (dextrose); the latter is assigned a DE value of 100. A fully hydrolyzed sample of starch has a DE value of 100.

- *High-fructose corn syrup (HFCS).* HFCS is a sugar syrup with high fructose content. Other accepted terms are *high-fructose glucose syrup (HFGS)* or *isoglucose (IG)*.

- *Degree of isomerization (DI).* The degree of isomerization (DI) is the percentage of sugars in a syrup, dry basis, by weight of fructose. For example, a 42-DI HFCS contains 42% fructose, 52% glucose, and 6% disaccharides and higher sugars. Such a product would be described as HFCS-42.

Enzymatic Conversion of Starch to Dextrins and Sugars

Approximately one-third of all starch production is used in the paper and textile industries and the balance in the food industry. Variations in the extent of starch hydrolysis create a range of products that differ in osmolarity, viscosity, and sweetening power (Table 7.5). Three classes of starch-degrading enzymes are used commercially in large amounts. α-Amylases (α-1,4 glucan-4-glucanohydrolases) are enzymes that hydrolyze internal α-1,4 linkages and bypass α-1,6 linkages. Glucoamylases (amyloglucosidases; α-1,4 glucan glycohydrolases) hydrolyze α-1,4 and α-1,6 bonds to produce glucose as the major product of hydrolysis (see Figure 7.2 and Table 7.2). Pullulanases and other debranching enzymes hydrolyze only α-1,6 linkages.

A process for the manufacture of glucose and fructose syrups from starch is described in Box 7.2. This process is inexpensive, and the

Properties and Industrial Applications of Hydrolyzed Starch Products

TABLE 7.5

Type of Syrup	DE	Composition (%)	Properties	Applications
Low-DE maltodextrins	15–30	1–20 D-glucose 4–13 maltose 6–22 maltotriose 50–80 higher oligomers	Low osmolarity	Clinical feed formulations, raw materials for enzymic saccharification, thickeners, fillers, stabilizers, glues, pastes
Maltose syrups	40–45	16–20 D-glucose 41–44 maltose 36–43 higher oligomers	High viscosity, reduced crystallization, moderately sweet	Confectionary, soft drinks, brewing and fermentation, jams, jellies, ice cream, sauces
High-maltose syrups	48–55	2–9 D-glucose 48–55 maltose 15–16 maltotriose	Increased maltose content	Hard confectionary, brewing and fermentation
High-DE syrups	56–68	25–35 D-glucose 40–48 maltose	Increased moisture holding, increased sweetness, reduced viscosity, higher fermentability	Confectionary, soft drinks, brewing and fermentation, jams, sauces
Glucose syrups	96–98	95–98 D-glucose 1–2 maltose 0.5–2 isomaltose	Commercial liquid "dextrose"	Soft drinks, caramel, baking, brewing and fermentation, raw materials
Fructose syrups	98	48 D-glucose 52 D-fructose	Alternative industrial sweeteners to sucrose	Soft drinks, jams, sauces, yoghurts, canned fruits

SOURCE: Kennedy, J.F., Cabalda, V.M., and White, C.A. (1988), Enzymic starch utilization and genetic engineering, *Tr. Biotechnol.* 6:184–189.

products compete successfully in price with sucrose. The first step, *liquefaction*, involves the conversion of starch to low-DE maltodextrins. This step is performed at near-neutral pH for a relatively short period of time at temperatures ranging from 95-107°C. Under the stabilizing influence of very low concentrations of Ca^{2+} ions and of the substrate (starch), *Bacillus licheniformis* α-amylase is able to withstand these extreme temperatures. Conversion of the low-DE maltodextrins to glucose, *saccharification*, is performed with a mixture of two enzymes. Glucoamylase from the fungus *Aspergillus niger* rapidly splits α-1,4 linkages with stepwise release of glucose molecules from the nonreducing ends of the starch chains. This enzyme also splits α-1,6 linkages, but

> **BOX** 7.2 Manufacture of Glucose and
> Fructose Syrups from Starch

Conversion of starch slurry to low-DE maltodextrins. In a typical process, starch (1000 kg) is slurried with water (2000 liters) at pH 6.0–6.5, and *Bacillus licheniformis* α-amylase (0.5–2.0 kg) is added. The uniform slurry is heated at 105–107°C for 3–5 min. Above 100°C, the starch goes through a solid-to-gel transition and is liquefied by degradation by the α-amylase. To produce further dextrins, the incubation is continued at 95–97°C. The length of the incubation depends on the nature of the desired product. The product obtained after a 1–2 h incubation has a DE of 5–15 (*low-DE maltodextrins*).

If the foregoing treatment produces too much dextrinization, the glucoamylase used in the next step will have a low affinity for the short dextrins that make up the bulk of the mixture, and saccharification rates (hydrolysis to glucose) are lowered. On the other hand, if the degree of starch dextrinization is too low, the high-molecular-weight dextrins form micelles, which are attacked poorly by glucoamylase.

Saccharification of low-DE maltodextrins to glucose syrup. This is accomplished by the addition of *Aspergillus niger* glucoamylase and a *Bacillus* species pullulanase (debranching enzyme; see Table 7.1), both at an enzyme/substrate weight ratio of about 0.03% at pH 4.5 and 60°C. At the end of a 48–60 h incubation, the final DE is 97–98.5, and the yield of glucose is 95–97.5%.

To obtain a high-purity glucose syrup, the mixture is subjected consecutively to filtration, decolorization with activated carbon, filtration, ion exchange treatment (passage through cation, anion, and mixed-bed resins) to remove salts, treatment with activated carbon, filtration, and evaporation. The resulting syrup can be converted to fine powdered or crystalline glucose.

Conversion of glucose syrup to fructose syrup. This step is performed with regenerable immobilized preparations of D-glucose isomerases. There are many variants of this process. We describe one [Antrim, R.L., and Auterinen, A.-L. (1986), A new regenerable immobilized glucose isomerase, *Starch* 38:132–137]. Highly purified glucose isomerase from *Streptomyces rubiginosus* is adsorbed to granular solid carrier particles made up of fibrous diethylaminoethyl (DEAE)-cellulose (30%), food-grade polystyrene (50%), and titanium dioxide (20%). The enzyme is bound to this cationic carrier primarily by electrostatic interaction. Like other glucose isomerases, the enzyme requires Mg^{2+} for activity and is both compatible with and stabilized by bisulfite (2 mM), which is added as a preservative. Glucose syrup containing 40–50% by weight solids is pumped into an enzyme reactor at pH 7.6 at 55–57°C. Mg^{2+} (1.5 mM) is added both as a required activator and to compete with Ca^{2+}, which is present in the substrate solution and inhibits the enzyme. The enzyme converts the glucose syrup into a syrup containing 42% fructose. In commercial production, more than 9 metric tons of 42% fructose syrup solids are obtained per kilogram of immobilized enzyme. Moreover, the enzyme can be desorbed from the carrier, repurified, and reimmobilized several times. The savings in processing costs made possible by introducing immobilized preparations of glucose isomerase have been very important in reducing the manufacturing cost of high-fructose syrups.

slowly; however, the second enzyme, the pullulanase, secreted by *Bacillus* species, splits them rapidly. Saccharification is performed under mildly acidic conditions and at a lower temperature, to prevent the formation of psicose (which is favored by alkaline pH) and to avoid forming colored products by caramelization of glucose at high temperatures.

The development of techniques for the large-scale *isomerization* of glucose to fructose offered several challenges. The well-known metabolic route for conversion of glucose to fructose proceeds by several steps: Glucose is phosphorylated to glucose-6-phosphate, glucose-6-phosphate is isomerized to fructose-6-phosphate, and the latter is dephosphorylated to fructose. A microbial enzyme capable of converting glucose directly to fructose was first reported by Richard O. Marshall and Earl Kooi in 1957. Marshall had observed that the bacterium *Aerobacter cloacae* grown on xylose was able to convert glucose to fructose in the presence of arsenate and magnesium chloride. This conversion was catalyzed by xylose isomerase, whose metabolic role is to convert D-xylose to D-xylulose (Figure 7.5). Numerous disadvantages prevented commercial exploitation of this initial discovery, however, including the high cost of xylose, the low affinity of xylose isomerase for glucose, fructose yields of only 33%, and long reaction times. Moreover, the use of arsenate was particularly undesirable in a step leading to a food product.

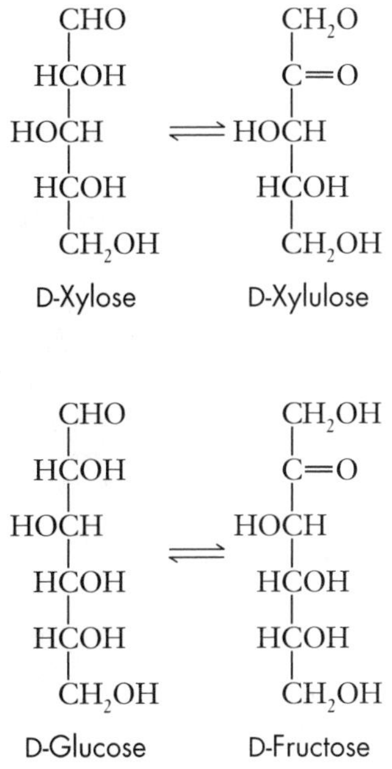

A search for xylose isomerases in other organisms led to the discovery that certain *Streptomyces* species produce xylose isomerases that do not require arsenate. Investigation of streptomycetes also solved the problem of the high expense of using xylose as a substrate for growth. Several *Streptomyces* species produce an extracellular xylanase as well as intracellular xylose isomerase (henceforth referred to as glucose isomerase). Xylanase degrades polymeric xylan to xylose, and xylan can be obtained cheaply from straw or wood. Thus the substrate costs were reduced dramatically. The yield of fructose from glucose achieved with the *Streptomyces* enzyme was in the 40–50% range. This improvement was important because the sweetness of syrups containing 42% fructose and 58% glucose is equivalent to the sweetness of syrups containing sucrose.

The early commercial processes for enzymic conversion of glucose to fructose in batches included the use of stirred-tank reactors. Crude glucose isomerase preparations were added to high-DE sugar in large quantities to compensate for the low substrate affinity of the enzyme. However, the enzyme preparations were expensive, and the subsequent refining step destroyed them. Later modifications in the process sought to conserve the enzymes. One solution was to use continuous-flow reactors containing immobilized bacterial cells. These later gave way to reactors in which pure enzyme was bound either covalently or noncovalently to a solid support, as in the method described in Box 7.2. The

Figure 7.5
Isomerization reactions catalyzed by xylose isomerase. This enzyme is frequently referred to as glucose isomerase.

> Many bacteria utilize D-xylose as an energy source. Active transport systems in the cytoplasmic membrane bring the sugar into the cell, where xylose isomerase converts it to D-xylulose. D-Xylulose is phosphorylated by xylulose kinase to form D-xylulose-5-phosphate. This phosphorylated sugar is then metabolized in the pentose phosphate and the glycolytic pathways.

reactor used in that example made possible the commercial production of over 9 metric tons of 42% fructose syrup solids per kilogram of immobilized enzyme.

The equilibrium for the glucose isomerization reaction lies at a fructose concentration of about 52%. In practice, the conversion is carried out only to about 42% fructose, because attainment of equilibrium is slow. Syrups with higher fructose levels can be obtained by cation exchange chromatography under appropriate conditions; fructose is more strongly adsorbed than glucose, and a syrup containing 42% fructose can be enriched to 85% fructose by a single pass through a column.

Textile Desizing

In fabric manufacture, the warp is highly stressed mechanically during weaving. Application of starch paste, or *sizing*, reinforces the warp and prevents loss of string by friction, as well as generation of static electricity on the string. However, for subsequent processing of the cloth (dyeing, bleaching, and finishing), the starch has to be completely removed. Continuous desizing procedures that depend on the stability of bacterial α-amylases at high temperatures are a principal approach to the removal of sizing. After being prewashed in hot water at 90°C for about 10 s, the cloth is passed one or more times through a bath containing a 0.5–1% wt/vol solution of *Bacillus subtilis* α-amylase, a 0.05% wt/vol solution of calcium chloride (to stabilize the enzyme), and a wetting agent (0.2–0.5% wt/vol nonionic surfactant) at temperatures between 65 and 80°C for about 20 s each time. The cloth is then rinsed.

Enzyme-Containing Detergents

Several million pounds of microbial proteolytic enzymes are produced annually for use as additives to laundry detergents. During the washing of clothes, the proteolytic enzymes degrade the proteins present in stains from foodstuffs, blood, and so on. Dirt adhering strongly to the stained fibers is much more readily removed after the protein components of the stains have been degraded. About 80% of laundry detergents on the market contain proteolytic enzymes at 0.015–0.025% by weight.

Properties of Subtilisins

To ensure the degradation of a wide variety of proteins to small peptides, the enzymes used in laundry detergents must combine high proteolytic activity with low substrate specificity. For the enzymes to remain active for prolonged periods of time under the harsh conditions of the washing process, they must also exhibit all of the following properties: (1) stability at elevated temperatures (up to 70°C); (2) resistance to denaturation by nonionic detergents; (3) high activity in the alkaline pH range (pH 8–11); (4) no metal-ion requirement for catalytic activity (all detergents contain chelating agents, such as ethylenediamine tetraacetate or tripolyphosphates, which would make many metal ions unavailable); and (5) resistance to oxidizing agents (such as hypochlorite and perborate) that are used as bleaches.

These stringent requirements are best satisfied by the *subtilisins*, a family of proteolytic enzymes secreted at the onset of sporulation by *Bacillus* species. The subtilisin of *B. licheniformis* has particularly favorable properties and is widely used in laundry detergents. It has a broad substrate specificity and a pH optimum of 9, and it readily converts proteins to a mixture of water-soluble peptides. It also has adequate thermostability and is active in the presence of most nonionic and anionic detergents at temperatures up to 65°C.

Subtilisin is a member of the "serine protease" class of proteolytic enzymes. Its catalytic mechanism is illustrated in Figure 7.6. There is a triad of residues involved in the catalysis. Ser-221 acts as a catalytic nucleophile, His-64 as a general base, and Asp-32 as a stabilizer for the correct tautomeric form of His-64 in the transition state. There is no requirement for a metal ion in the catalysis, so the activity of the enzyme is not affected by the presence of chelating agents.

Genetic Engineering of Subtilisin

The enzymatic activity of subtilisin decreases only slowly in the presence of perborate. However, the enzyme is very sensitive to oxidation by hydrogen peroxide, which is liberated from perborates at temperatures higher than 60°C. Oxidation of Met-222 (adjacent to the active-site residue Ser-221) to the sulfoxide results in a 90% loss of enzyme activity.

$$NaBO_3 + H_2O \rightarrow NaBO_2 + H_2O_2$$

| Sodium perborate | Sodium metaborate | Hydrogen peroxide |

Met-222 + H_2O_2 → Met-222 sulfoxide + H_2O

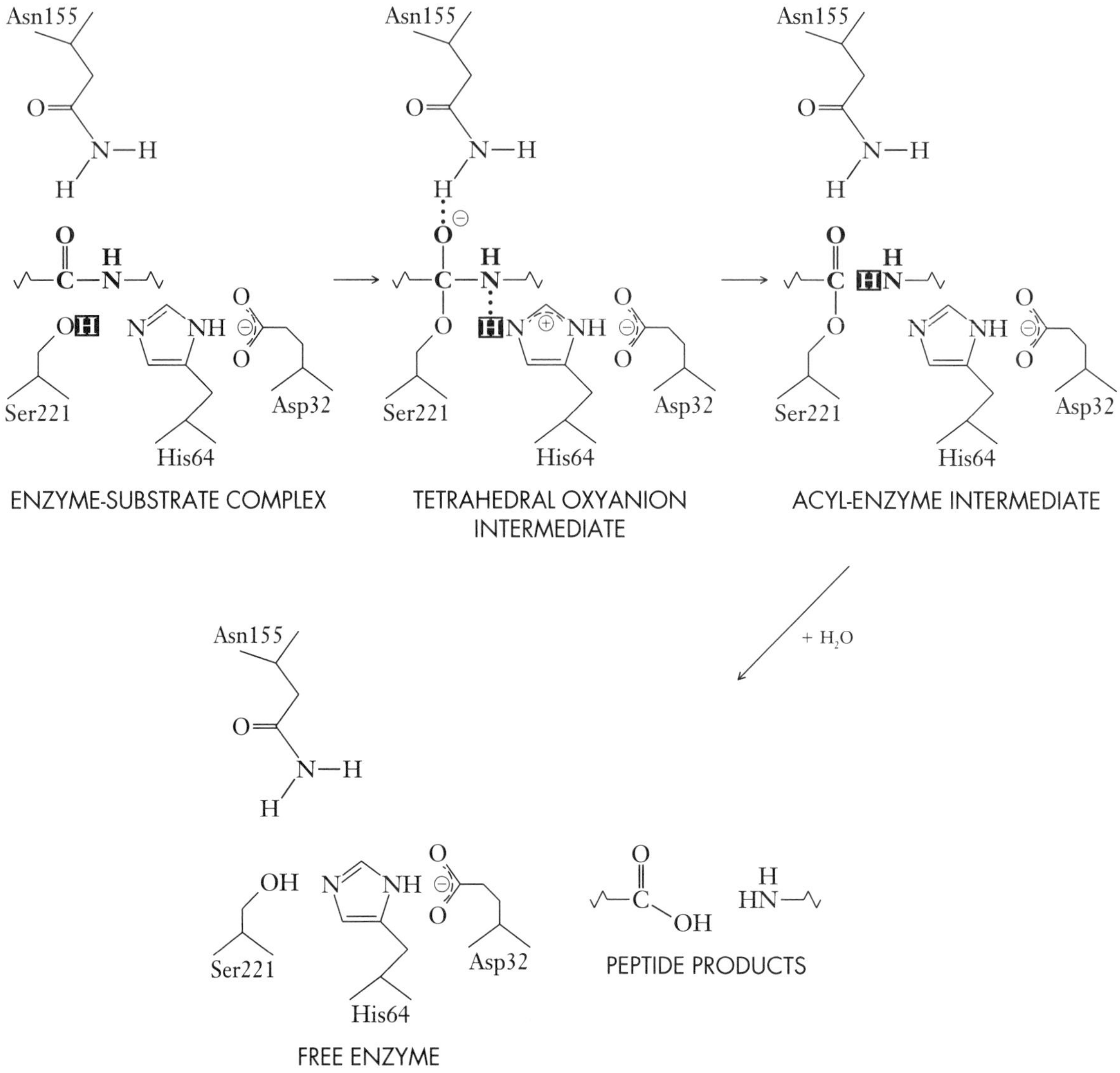

Figure 7.6
Mechanism of hydrolysis of peptide bonds by subtilisin.

This inactivation of subtilisin by H_2O_2 imposes an inconvenient upper limit on the temperature at which laundry detergents containing this enzyme can be used. However, genetic engineering has led to the creation of subtilisins that show improved stability to oxidants.

The solution to the oxidative damage problem lies in site-specific modification of subtilisin's structure. Because Met-222 is not directly involved in catalysis, it can be replaced by a different amino acid residue without destroying the enzyme's activity. Ala-222 subtilisin retains 53% of the catalytic activity of the wild-type enzyme and is totally resistant to exposure to 1 M H_2O_2 for an hour.

Genetic engineering has also improved the activity of subtilisin at the strongly alkaline pH of laundry detergents. The activity of *B. amyloliquefaciens* subtilisin BPN' normally decreases between pH 9.5 and 11 with an apparent pK_a value of about 10.5. This decrease in activity appears to be attributable to the ionization of one or both of two tyrosine residues, Tyr-104 and Tyr-217, that lie in the substrate-binding region of subtilisin BPN'. Replacement of Tyr-104 by phenylalanine via site-specific mutagenesis significantly increased the activity of the enzyme at high pH.

Un-ionized and ionized forms of a tyrosyl residue

Phenylalanyl residue

At pH 10.6, for example, the k_{cat}/K_m ratio of the Tyr104 → Phe enzyme was increased 1.2-fold relative to its k_{cat}/K_m ratio at pH 8.6, whereas the k_{cat}/K_m for wild-type subtilisin BPN' was halved.

An empirical approach was applied successfully to generate subtilisin variants with improved thermostability. *B. amyloliquefaciens* colonies that express randomly mutated subtilisin genes were grown on filters that bound the subtilisin secreted from each colony. The filters were transferred to assay plates of agar rendered opaque by inclusion of a protein substrate. After incubation at high temperature, the stability of the various mutant subtilisins was assessed by comparing the size of the "halos" (the zones cleared by proteolysis) produced by each enzyme variant. One mutant, Asp-218 → Ser, identified by means of this screen was more thermostable and also showed a 2-fold increase in k_{cat}/K_m in degrading a synthetic peptide substrate.

> The K_m (Michaelis constant) is the concentration of substrate at which the velocity of an enzyme-catalyzed reaction is half-maximal. k_{cat} is the "turnover number" of the enzyme. It represents the maximal number of substrate molecules converted to products per active site per unit time. The ratio k_{cat}/K_m is an apparent second-order rate constant for the enzyme-catalyzed reaction at substrate concentrations that are low relative to the K_m.

Cheese Making

In the manufacture of cheese, which was described briefly in Chapter 1 (page 44), the curdling of milk is achieved by the addition of calcium ions and rennet. Rennet is an aqueous extract from the fourth stomach of unweaned calves, and the component responsible for its curdling action is the proteolytic enzyme chymosin (also called rennin). Chymosin is synthesized as a 381-residue precursor called preprochymosin, which is processed during secretion to a 365-residue molecule, prochymosin. The latter has no enzymatic activity, but at low pH values it undergoes a multistep autocatalytic proteolytic conversion to a 323-residue proteolytic enzyme, chymosin. The chymosin in rennet is actually a mixture of prochymosin and chymosin.

The Action of Chymosin

To explain the action of chymosin, we must first describe the structure of casein micelles. The term *casein* encompasses a family of proteins, and milk contains about 25 g of them per liter. The major constituents $\alpha_s 1$-, $\alpha_s 2$-, β-, and κ-casein represent approximately 36, 12, 36, and 12%, respectively, of the total. The molecular weight of α-casein and of β-casein is approximately 24,000, and that of κ-casein is 19,000. $\alpha_s 1$-Casein contains eight phosphoserine residues and β-casein five. κ-Casein contains only one phosphoserine residue, at Ser-149, but it carries a glycosyl moiety, [*N*-acetylneuraminyl-(2 $\rightarrow$ 3(6))-galactosyl-(1 $\rightarrow$ 6)-*N*-acetylgalactosaminyl $\rightarrow$ Thr], on Thr-131.

The caseins form complexes called *casein submicelles*. In milk, these submicelles are crosslinked by calcium phosphate clusters to form *casein micelles*. The calcium phosphate clusters interact with the phosphoserine residues, and their formation may in fact be nucleated by these residues. The hydrophobic amino-terminal portion of κ-casein lies within the micelle. The hydrophilic portion of the molecule projects from the surface of the micelle and prevents the micelles from aggregating (Figure 7.7).

κ-Casein is a protein of 169 residues. At pH 6.6, the pH of milk, chymosin cleaves κ-casein specifically between Phe-105 and Met-106, removing the glycosylated polypeptide (residues 106–169) from the micelle. In the presence of calcium ions, the modified micelles then aggregate to form a three-dimensional network called *paracasein*. This is the phenomenon commonly referred to as milk coagulation, and it is the first step in the production of cheese.

Periodic shortages of authentic calf chymosin have led to the use of fungal aspartyl proteases from *Endothia parasitica*, *Mucor miehei*, and *M. pusillus* as substitutes for calf rennin. The taste and consistency of older cheese prepared with these enzymes are said to compare unfavorably with those of cheese prepared with rennin.

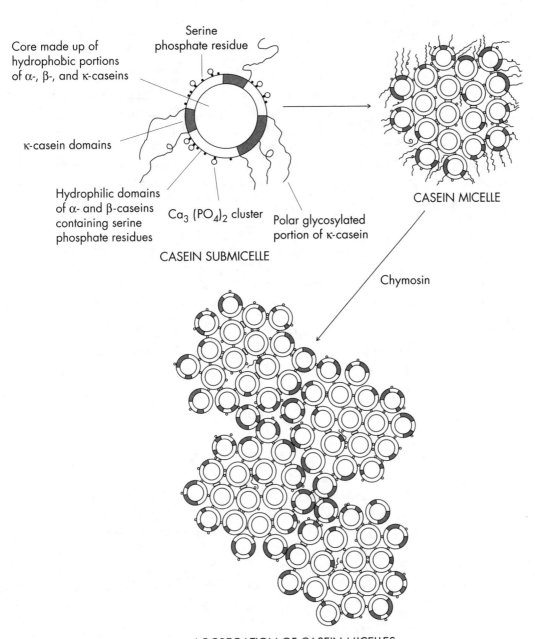

Figure 7.7
Casein submicelles are bridged by calcium phosphate complexes to form casein micelles. Detachment of the polar "tails" of the κ-casein molecules by site-specific proteolysis with chymosin in the presence of calcium ions leads to aggregation of the micelles, with the ultimate formation of a three-dimensional network called paracasein. Based on Stadhouders, J., and van den Berg, G. (1988), The technology of cheesemaking. *Endeavour, New Series* 12:107–112, Fig. 4.

Recombinant Chymosin

Commercial calf chymosin preparations vary in purity, and their price has risen sharply in recent years. Worldwide, the dairy industry spends over $100 million a year on chymosin. This situation has provided an incentive to produce a recombinant enzyme on a large scale. On March 23, 1990, the U.S. Food and Drug Administration approved the use, in dairy products, of bovine chymosin produced in *E. coli* by recombinant DNA technology. Chymosin thus became the first genetically engineered protein approved for human consumption.

We have seen that prochymosin expressed in *E. coli* forms inclusion bodies (see Chapter 3) and that the preparation of pure active enzyme then requires several additional steps: isolation of the inclusion bodies, dissolution of the inclusion bodies under denaturing conditions, renaturation of the protein, and purification. Much effort has therefore been devoted to developing expression systems for chymosin that secrete the recombinant chymosin into the medium, where it can be recovered directly as an active enzyme (see Chapter 4). Such systems have been successfully constructed in yeasts (*S. cerevisiae* and *Kluyveromyces*) and in two filamentous fungi, *Aspergillus nidulans* and *Trichoderma reesei*. The *T. reesei* system gave the highest yield of active chymosin.

SUMMARY

Bacterial and fungal enzymes are used on a large scale in the production of nutritional sweeteners, as ingredients in laundry detergents, and in the manufacture of cheeses. The key microbial enzymes in the conversion of starch to high-glucose and high-fructose syrups are α-amylase, glucoamylase, pullulanase, and glucose isomerase. Subtilisins, *Bacillus subtilis* proteolytic enzymes, are included in laundry detergents to remove proteinaceous food stains. Bovine chymosin catalyzes the first step in the manufacture of cheese from milk. In 1990, recombinant chymosin, produced in *E. coli* for use in dairy products, became the first genetically engineered protein to be approved for human consumption. The generation of a *Bacillus subtilis* hyperproducer strain for α-amylase illustrates the semi-empirical approach used for the selection and development of industrial producer strains.

Mutant selection and site-specific modification have been successfully exploited to improve the properties of industrial enzymes. These approaches have led to the production of subtilisins that exhibit greatly enhanced stability to oxidants, higher activity at alkaline pH, and improved thermostability.

SELECTED REFERENCES

General

Peppler, H.J., and Perlman, D. (eds.), 1979. *Microbial Technology: Microbial Processes*. 2d ed., vol. I. Academic Press.

Kennedy, J.F. (ed.), 1987. *Enzyme Technology*. Rehm, H.-J., and Reed, G., (eds.), *Biotechnology*, vol. 7a. VCH.

Finn, R.K., et al. (eds.), 1988. *Biotechnology Focus 1: Fundamentals. Applications. Information*. Hanser.

Amylase Research Society of Japan, 1988. *Handbook of Amylases and Related Enzymes: Their Sources, Isolation Methods, Properties and Applications*. Pergamon Press.

Jacobson, G.K., and Jolly, S.O., 1989. *Gene Technology*. Rehm, H.-J., and Reed, G., (eds.), *Biotechnology*, vol. 7b. VCH.

Roland, J.F., 1981. Regulation of food enzymes. *Enzyme Microb. Technol.* 3:105–110.

Volesky, B., and Luong, J.H.T., 1985. Microbial enzymes: Production, purification, and isolation. *CRC Crit. Rev. Biotechnol.* 2:119–146.

Cowan, D., Daniel, R., and Morgan, H., 1985. Thermophilic proteases: Properties and potential applications. *Tr. Biotechnol.* 3:68–72.

Kennedy, J.F., Cabalda, V.M., and White, C.A., 1988. Enzymic starch utilization and genetic engineering. *Tr. Biotechnol.* 6:184–189.

Wells, J.A., and Estell, D.A., 1988. Subtilisin—an enzyme designed to be engineered. *Tr. Biochem. Sci.* 13:291–297.

Linko, S., 1989. Novel approaches in microbial enzyme production. *Food Biotechnol.* 3:31–43.

Arbige, M.V., and Pitcher, W.H., 1989. Industrial enzymology: A look towards the future. *Tr. Biotechnol.* 7:330–335.

Chymosin

Stadhouders, J., and van den Berg, G., 1988. The technology of cheesemaking. *Endeavour, New Series* 12:107–112.

Beppu, T., 1988. Production of chymosin (rennin) by recombinant DNA technology. In Thomson, J.A. (ed.), *Recombinant DNA and bacterial fermentation*, pp. 11–21. CRC Press.

Harrki, A., Uusitalo, J., Bailey, M., Penttilä, M., and Knowles, J.K.C., 1989. A novel fungal expression system: Secretion of active calf chymosin from the filamentous fungus *Trichoderma reesei*. *Bio/Technology* 7:596–603.

Fermentation Processes

Mudgett, R.E., 1986. Solid-state fermentations. In Demain. A.L., and Solomon, N.A. (eds.), *Manual of Industrial Microbiology and Biotechnology*, pp. 66–83. American Society for Microbiology.

Chisti, Y., and Moo-Young, M., 1991. Fermentation technology, bioprocessing, scale-up and manufacture. In Moses, V., and Cape, R.E. (eds.), *Biotechnology: The science and the business*, pp. 167–209. Harwood Academic Publishers.

Reisman, H.B., 1993. Problems in scale-up of biotechnology production processes. *CRC Crit. Rev. Biotechnol.* 13:195–253.

Microbial Polysaccharides and Polyesters

This chapter deals with two classes of biopolymers: polysaccharides and polyesters. Polysaccharides include some of the most abundant carbon compounds in the biosphere, the plant polysaccharides cellulose and hemicelluloses (discussed in Chapter 10), as well as the much less abundant but useful algal polymers such as agar and carrageenan. Bacteria and fungi also produce many different types of polysaccharides, some in amounts well in excess of 50% of cell dry weight. Polyesters are produced exclusively by bacteria and until very recently were of interest only to students of microbial physiology.

Polysaccharides are used to modify the flow characteristics of fluids, to stabilize suspensions, to flocculate particles, to encapsulate materials, and to produce emulsions. Among many other examples are the use of polysaccharides as ion-exchange agents, as molecular sieves, and (in aqueous solution) as hosts for hydrophobic molecules. Polysaccharides are also used in enhanced oil recovery and as drag-reducing agents for ships.

The discovery that many bacteria synthesize large amounts of *biodegradable* polyester polymers of high molecular weight, which can be used to manufacture plastics, has aroused considerable interest. There are hundreds of varieties of synthetic plastics; their uses are too many to enumerate. Current annual production of these materials in the United States alone exceeds 30 billion pounds. Plastics are manufactured from petrochemicals and decompose extremely slowly in the natural environment.

In this chapter, we discuss the structures of some of the microbial polysaccharides and polyesters, their biosynthesis and functions, and certain uses of these compounds.

POLYSACCHARIDES

Microorganisms and plants produce polysaccharides of widely varying composition and structure. Cellulose and starch from plants are the most abundant, important, and familiar of these biopolymers, but many other naturally occurring polysaccharides have interesting and useful properties. Agar, a mixture of polysaccharides extracted from marine red algae, has been manufactured in Japan since about the year 1760 and is one of the most effective gelling agents known. Gel formation occurs at agar concentrations as low as 0.04%. In her book *The Algae and their Life Relations*, published in 1935, Josephine E. Tilden described the many uses of agar.

> *It is prepared as a food by being boiled and allowed to cool to form a jelly. It is likewise an ingredient of soups and sauces and many desserts. It is used also for purifying sake, for sizing textiles, for stiffening the warp of silks, and for clarifying beverages.*
>
> *In Europe and North America agar is used in making such foods as ice cream, jellies, candies, or pastries; in canning fish to prevent soft canned fish from being shaken to pieces during transportation; in clarifying liquors; in sizing material; and in dressing wounds. Its chief use in Europe and America, however, is in the preparation of bacteriological media for scientific purposes.*

Dried agar suspended in water melts in the temperature range of 60–90°C (the precise temperature varies with the algal source) and, at about 1–1.5% by weight, sets between 32 and 39°C to form a firm gel. This large temperature hysteresis (see page 276) is a particularly valuable property of agar that it shares with few other polysaccharides. Among polysaccharides produced by bacteria and fungi, some have properties resembling those of agar and others have distinctive rheological (flow) properties that are valuable in certain pharmaceutical or industrial applications.

Only one microbial polysaccharide, xanthan, ranks among the ten industrial polysaccharides consumed in the largest amounts (Table 8.1). Much of this chapter is devoted to xanthan. The other microbial exopolysaccharides have found only modest commercial use, and only a few are produced on a large scale (Table 8.2). However, the enormous range of polysaccharides synthesized by microorganisms has yet to be adequately explored.

Major Food and Industrial Polysaccharides in Order of Decreasing Consumption

Cornstarch and derivatives
Cellulose and derivatives
Guar gum and derivatives
Gum arabic
Xanthan*
Alginate
Pectin
Carrageenan
Locust bean gum
Ghatti

T A B L E **8.1**

* By weight, the consumption of xanthan represents less than 1.0% that of starches and derivatives.

Bacterial and Fungal Polysaccharides with Commercial Uses

T A B L E **8.2**

Polysaccharide	Organism(s)	Composition
Bacteria		
Xanthan	*Xanthomonas campestris*	The repeating unit is a pentasaccharide containing glucose, mannose, glucuronic acid, and acetyl and pyruvate substituents
Dextran	*Acetobacter* sp. *Leuconostoc mesenteroides* *Streptococcus mutans*	Polyglucose linked by α-1,6-glycosidic bonds; some 1,2-, 1,3-, or 1,4-bonds are also present in some dextrans
Alginate	*Pseudomonas aeruginosa* *Azotobacter vinelandii*	Blocks of β-1,4-linked D-mannuronic acid residues, blocks of α-1,4-linked L-guluronic acid residues, and blocks that contain both uronic acids in either random or alternating sequence
Curdlan	*Alcaligenes faecalis*	β-1,3 glucan (polyglucose)
Gellan	*Pseudomonas elodea*	Partially O-acetylated polymer of glucose, rhamnose, and glucuronic acid
Fungi		
Scleroglucan	*Sclerotium glutanicum*	Glucose units primarily β-1,3-linked with occasional β-1,6-glycosidic bonds
Pullulan	*Aureobasidium pullulans*	Glucose units primarily α-1,4-linked with occasional α-1,6-glycosidic bonds

Peptidoglycan layer

Outer membrane

Capsule

Cytoplasmic membrane

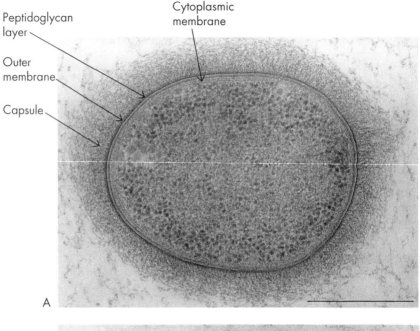

A

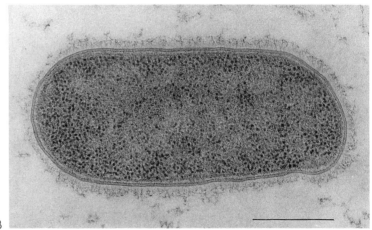

B

Figure 8.1

A. Thin section of the bacterium *Klebsiella pneumoniae*, showing the fine structure of the polysaccharide capsule. The capsule is approximately 160 nm thick and appears to be made up of two layers. The inner layer is formed by a palisade of thick bundles of fibers extending outward at right angles to the surface of the outer membrane. In the second layer, thin fibers spread from the ends of the bundles, forming a fine network. **B.** Thin section of the bacterium *E. coli* K1, showing a capsule that is much thinner than that of *K. pneumoniae*, but has a similar morphology. Staining was with uranyl acetate and lead citrate. Bars, 0.5 μm. Reproduced with permission from Amako, K., Meno, Y., and Takade, A. (1988), Fine structure of the capsules of *Klebsiella pneumoniae* and *Escherichia coli* K1. *J. Bacteriol.* 170:4960–4962.

Bacterial Polysaccharides

Under the proper culture conditions, the majority of bacterial species secrete mucoid substances of high molecular weight. When these viscous materials remain associated with the cell, they are called variously *capsules, sheaths,* or *slime layers* (Figure 8.1). Whereas capsules and sheaths are well-defined layers external to the cell wall, slime accumulates in large quantities outside the cell wall and diffuses into the medium. The slime formers may produce such copious amounts of viscous slime when they are grown in a liquid medium that the culture flask can be inverted and the culture will remain in place! Mucoid colonies grown on solid media usually have a moist, glistening surface.

Capsules have long been a subject of great interest because of their role as important virulence factors in bacteria that cause invasive infections. Capsules protect bacteria from phagocytosis and impart resistance against bactericidal effects of serum (see Figure 5.6). Frequently, pathogens cultured in the laboratory spontaneously produce unencapsulated mutants; these mutants are no longer pathogenic (for an example, see Figure 3.1).

Most of the extracellular polymers produced by bacteria are polysaccharides, although a few bacteria produce capsules made up of polypeptides of D-amino acid residues. The precise structure of the capsular polysaccharides of many pathogenic organisms is strain-specific. Such individual "coding" of its outermost layer protects the bacterial cell from being attacked by an immune system that may have developed antibodies in response to an earlier, perhaps closely related, invader.

In contrast, the structures of the extracellular polysaccharides of many slime-forming organisms are simpler in composition and less homogeneous in size. The molecular weight and even the composition of these extracellular polysaccharides (exopolysaccharides) may vary depending on the culture conditions.

Aspects of Polysaccharide Structure

The glycosidic bond between two monosaccharides is formed between the hydroxyl group on the anomeric carbon of one monosaccharide and any hydroxyl group on the other monosaccharide (for example, see sucrose in Figure 7.3). Hence the formation of a disaccharide from two identical hexopyranose ring structures of the D-series can result in 11 different isomers. In 8 of the isomers the glycosidic linkage is between C-1, in either the α or the β anomeric configuration, and C-2, C-3, C-4, or C-6 of the other pyranose residue. These are designated α-D-$(1 \rightarrow 2)$ linkages, α-D-$(1 \rightarrow 3)$ linkages, etc., where α and β refer to the anomeric configuration at C-1. The other 3 isomers are created by acetal formation between both C-1 atoms through the glycosidic oxygen atom in the α,α, the α,β, or the β,β configuration. A similar series of 11 isomers results if the two identical hexopyranose residues are of the L-series. The number of possible isomers is even higher if furanose forms (5-membered rings) are included. When the two monosaccharides are not identical, the number of possible structures increases, because either carbohydrate residue can occupy the first or the second position—that is, can be either the reducing or the nonreducing residue (see Figure 8.2). Each additional carbohydrate residue brings a large increase in the number of possible isomers. For complex polysaccharides, the number of theoretically possible structures is astronomical.

The meaning of the terms *primary* and *secondary structure* in reference to the structure of polysaccharides is similar to their meaning in reference to proteins. The primary structure of a polysaccharide consists of the identity, sequence, linkage(s), and anomeric configurations of all its monosaccharide residues, and the nature and position of any other substituents. *The individual sugar rings in a polysaccharide are essentially rigid.* The secondary structure of the polysaccharide is thus determined by the relative orientations of the component monosaccharide residues about the glycosidic linkage. As illustrated in Figure 8.2, two rotational angles (ϕ and ψ) specify the glycosidic bond between two carbohydrate residues, except for $(1 \rightarrow 6)$-linked residues, where

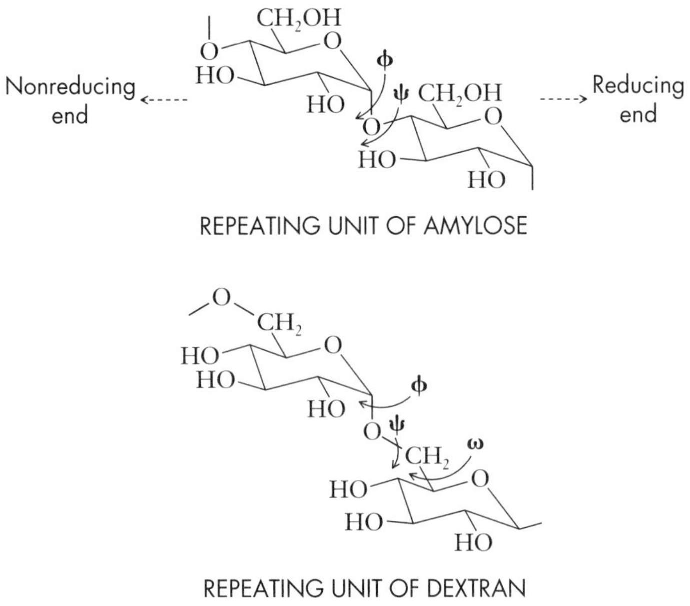

REPEATING UNIT OF AMYLOSE

REPEATING UNIT OF DEXTRAN

Figure 8.2
The relative orientations of adjacent carbohydrate residues joined by glycosidic linkage to a ring hydroxyl are defined by two torsion (rotational) angles, φ and ψ, except for $(1 \rightarrow 6)$-linked polysaccharides, where a third angle, ω, specifies the rotation around the exocyclic carbon–carbon bond.

specification of a third rotational angle (ω) is required. The range of permitted values for the rotational angles ϕ, ψ, and ω is limited by steric constraints, including the geometric relations within each residue of the polysaccharide and, to a lesser degree, the interactions between adjacent residues in the chain. Depending on the nature of such constraints, polysaccharides with a primary structure made up of identical repeating units exist in solution as relatively stiff extended or crumpled ribbons, loose helical coils, or [for $(1 \rightarrow 6)$-linked polysaccharides] flexible "random" coils. For example, curdlan, a $(1 \rightarrow 3)$-β-D-glucan, exists in solution as a flexible helix with 7-fold symmetry. For charged polysaccharides, interaction with ions can be very important. Alginate, a block copolymer of $(1 \rightarrow 4)$-β-D-mannuronic acid and $(1 \rightarrow 4)$-α-L-guluronic acid (see Table 8.2), exists as a stiff random coil in solution, but on binding Ca^{2+} ions, the poly-L-guluronate sequences undergo dimerization as shown in Figure 8.3, a step leading to gel formation.

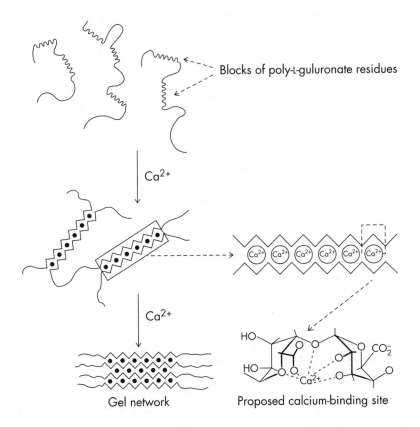

Blocks of poly-L-guluronate residues

Ca^{2+}

Ca^{2+}

Gel network

Proposed calcium-binding site

Figure 8.3
Interaction of Ca^{2+} ions with the poly-L-guluronate sequences in alginate causes lateral association of these chain segments, leading to the formation of a gel network. The proposed structure of the calcium-binding site is shown on the right.

Noncovalent aggregation of polysaccharide chains to form higher-order structures is analogous to subunit aggregation and the formation of *quaternary structure* in proteins.

The wide range of physical properties seen among microbial polysaccharides is a direct result of the differences in the nature and proportion of their monosaccharide building blocks, of substituents (such as acetyl groups) on the monosaccharides, of the linkages between the building blocks, and, ultimately, of the stereochemical consequences of all these factors. It follows that modulation of the composition of microbial polymers, through genetic modification or manipulation of culture conditions, can enhance their existing desirable properties or generate entirely new polymers with novel characteristics.

Roles of Microbial Polysaccharides in Nature

Organisms that produce extracellular polysaccharides have been isolated from a wide variety of environments. The functions of these polysaccharides are equally varied. We have already mentioned the capsular polysaccharides that protect pathogenic microorganisms from immune system defenses. Capsules may also serve as physical barriers to infection by bacteriophage. Moreover, capsules and sheaths retain water and in some settings play an important role in preventing dehydration of the cells. A common function of extracellular polysaccharides of water and soil microorganisms is to bind cells to surfaces, such as soil particles or rocks, as well as to each other. Likewise, it has been proposed that the extracellular polysaccharides of some plant pathogens play a role in attaching the bacteria to the surfaces of host plants.

XANTHAN GUM

Xanthan gum was discovered in the mid-1950s, during a systematic search for useful biopolymers. In a screen of its large microbial culture collection, the Northern Utilization Research and Development Division of the U.S. Department of Agriculture discovered that the bacterium *Xanthomonas campestris* (originally isolated from the rutabaga plant) produced a polysaccharide that had potentially valuable physical properties. Substantial commercial production of xanthan gum began in 1964, and in 1969 the U.S. Food and Drug Administration authorized its use in food. Today, xanthan gum has numerous uses in the food industry and elsewhere (Box 8.1). The total U.S. consumption of xanthan in 1990 is estimated to have been 40 million pounds.

> **BOX** 8.1   Applications of Xanthan Gum
>
> **Applications in Food**
> *As a stabilizer, thickener, gelling, or suspending agent* in dressings, sauces, gravies, syrups, toppings, desserts, ice cream, shakes, processed cheese spread, cottage cheese, cakes, confectionery, etc.
>
> **Applications in Products Other than Foods**
> *As a suspending agent* in thixotropic paints, in textile printing and dyeing to prevent migration of dye, in clay coatings for paper, in ceramic glazes, in insecticide sprays, etc.; *as a water-thickening polymer* (to provide viscosity control) in abrasives, in drilling muds (used for displacement of oil in enhanced oil recovery), in adhesives, etc.; *as a gelling agent* in explosives, detergents, and deodorants.

Xanthomonas, a Plant Pathogen

The genus *Xanthomonas* consists exclusively of plant pathogens. It is a relative of the genus *Pseudomonas*, a group of bacteria discussed in this book in the context of the degradation of xenobiotics (Chapter 15).

Bacteria of the genus *Xanthomonas* are yellow-pigmented, motile, aerobic, Gram-negative rods. *X. campestris* pathovar (abbreviated pv.) *campestris*, the producer of xanthan gum, causes black rot, one of the most serious diseases of plants in the genus *Brassica*, which includes such vegetables as cabbage, cauliflower, Brussels sprouts, broccoli, rutabaga, and turnip. The infection is spread by seeds contaminated with bacteria. As the seedling emerges and grows, the bacteria colonize the surface of the plant. In the case of *X. campestris* pv. *campestris*, the epiphytic bacteria (living on the surfaces of plants) enter the internal tissues through the *hydathodes*, structures on the leaf margin that enable the plant to excrete water. They then migrate through the vascular system and progressively cause chlorosis (destruction of chlorophyll), vein blackening (deposition of melaninlike pigments), and ultimately rotting. At least 30 genes contribute to the pathogenicity of *X. campestris*.

Structure of Xanthan Gum

The primary structure of xanthan gum is shown in Figure 8.4. As in cellulose, the xanthan backbone consists of β-(1 $\rightarrow$ 4)-linked D-glucose; however, the 3-position of alternating glucose monomer units carries a trisaccharide side chain that contains one glucuronic acid and two mannose residues. The nonterminal D-mannose unit carries an acetyl group at the 6-position, and a pyruvate is attached to the terminal D-mannose residue by a ketal linkage to the 4- and 6-positions. The degree of acetylation and the pyruvate ketal content vary with culture conditions and from one strain of the microorganism to another. Xanthan gum also binds cations.

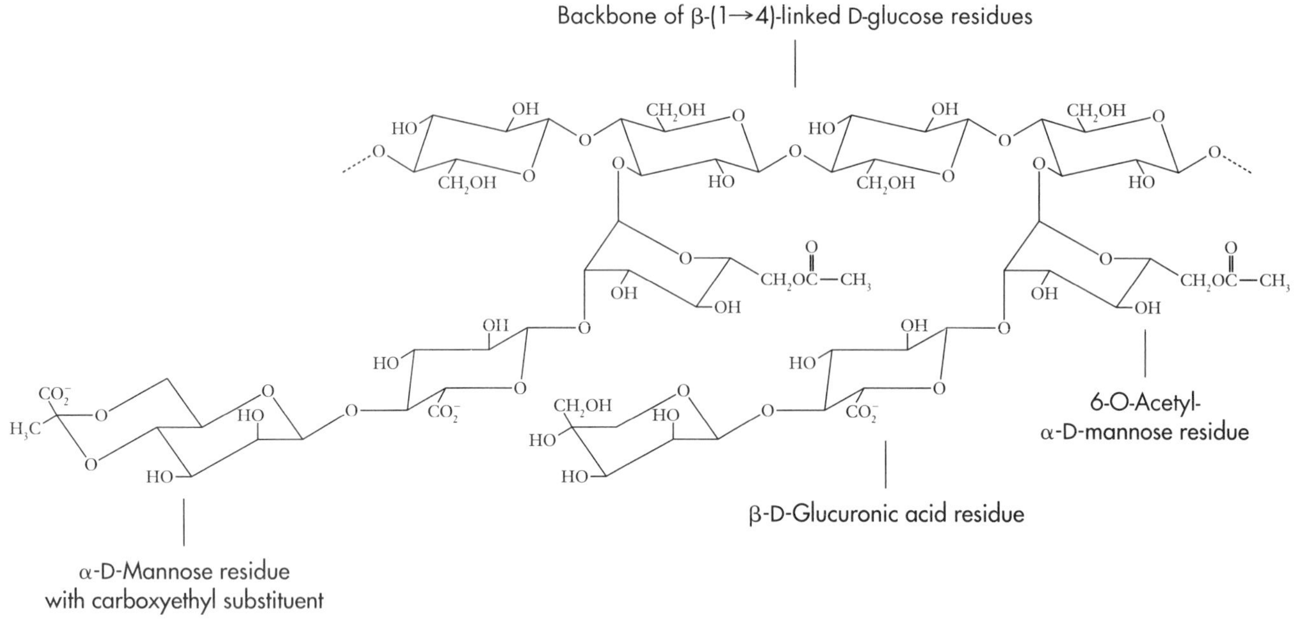

Figure 8.4
Primary structure of xanthan gum. As shown, the carboxyethylation of the terminal mannose residues is incomplete.

In solution, xanthan forms a stiff right-handed double helix with 5-fold symmetry and a pitch of 4.7 nm; the two chains most likely run antiparallel to each other as shown in Figure 8.5. The trisaccharide branches are closely aligned with the polymer backbone. The double-helical molecules interact side by side to form body-centered lattices, which in turn form microcrystalline fibrils. Estimates of the molecular weight of xanthan in solution range from 2 to 15 million. This wide spread in estimated values reflects the difficulty of evaluating the contribution of intermolecular association to the apparent molecular weight.

Properties of Xanthan Gum

Xanthan gum displays an extraordinary combination of physical properties that make this exopolysaccharide ideal for a wide range of applications.

Dilute xanthan solutions (0.5–1.5% by weight) have a high viscosity, and when the concentration of salt in a solution is 0.1% or higher, the viscosity stays uniform over a temperature range of 0–100°C. As a consequence, gravies and sauces containing xanthan maintain their thickness even at 80°C. The high viscosity of xanthan solutions is due to

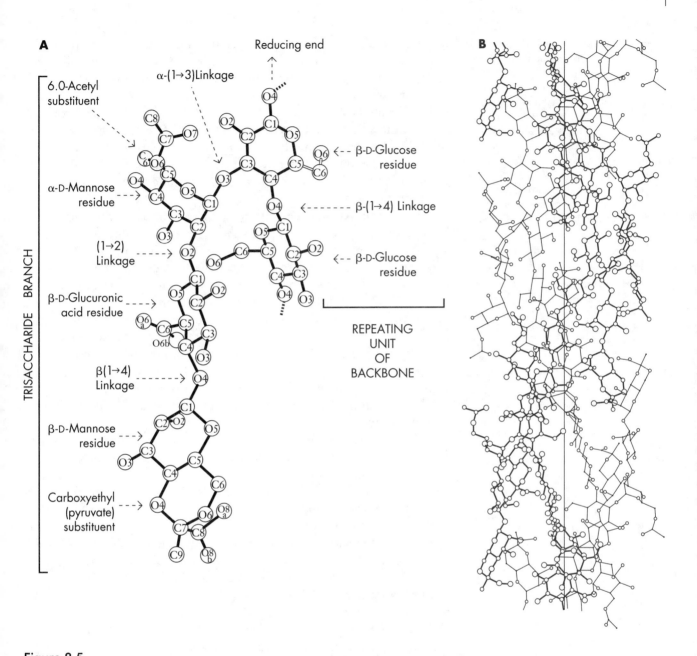

Figure 8.5

Structure of xanthan exopolysaccharide. **A.** The conformation of the penta-saccharide repeating unit of xanthan. **B.** The 5_1 antiparallel xanthan double helix viewed perpendicular to the helix axis. Based on Okuyama, K., et al. (1980), Fiber diffraction studies of bacterial polysaccharides, in French, A.D., and Gardner, K.H. (eds.), *Fiber Diffraction Methods*, ACS Symposium Series, vol. 141, pp. 411–427, Figs. 3 and 4 (American Chemical Society).

> **Hysteresis** is a phenomenon in which the relationship between two or more physical quantities depends on their prior history. More specifically, the response **B** takes on different values for an increasing input **A** than for a decreasing **A**.
>
> For example, when solid agar is heated, it does not dissolve until a temperature of 60°C or greater is reached. Once dissolved, however, agar remains in solution until the temperature decreases to about 39°C or lower.

> **Rheology** is the study of the deformation and flow of matter. It is that part of mechanics that deals with the relationship between force and deformation in material bodies. The term *rheological behavior* is applied to materials that show nonlinear and time-dependent deformation in response to shear stress.

the very stable two-chain helical, rigid, and rodlike structure of this branched exopolysaccharide (see Figure 8.5). The small amounts of salt added to these solutions shield the negative charges on the trisaccharide branches and minimize electrostatic repulsion. Often salt is already present in the materials to which xanthan is added; if not, the amount of salt that must be added does not present a problem in most products. In the complete absence of salt, the electrostatic repulsion destabilizes the two-chain structure sufficiently that the individual strands dissociate at elevated temperatures. Such separation can immediately be reversed by the addition of salt.

A particularly important property of xanthan is its rheological behavior. In the absence of shear stress, a xanthan solution is viscous. When a shear stress is applied beyond a certain low minimal value, however, the viscosity decreases steeply with shear rate. The term *shear thinning* describes this behavior. Moreover, xanthan solutions show no hysteresis: shear thinning and recovery are instantaneous. This property has considerable practical significance. For example, paints are formulated to be shear-thinning: A xanthan-containing paint maintains a high viscosity at low shear rates, it will not drip from the brush, and a film of the paint freshly applied to a vertical wall will not sag. However, a brushing motion applies a shear stress, and because shear thinning of xanthan solutions takes place at modest shear rates, the paint will thin and can be applied without undue exertion.

The shear-thinning and particle-suspending properties of xanthan solutions have led to the inclusion of this polymer as a component of drilling muds used in drilling oil wells. Drilling muds suspend the sand particles from oil wells and carry them to the surface; they also act as lubricants for the drill bit. (At the tip of the drill bit, the shear rate is very high and the viscosity of xanthan solutions very low. The reverse is true in the drill shaft, where the shear rate is low and the solution viscosity high.) Settling tests show that the suspending ability of xanthan gum surpasses that of every other polymer used in drilling fluids. Although xanthan gum is more expensive than guar gum, the most commonly used plant polysaccharide in the oil industry, the exceptional suspending ability of xanthan gum solutions at low polymer concentration favors its use where transportation costs are high.

The molecular explanation of the shear-thinning behavior of xanthan solutions is found in the way the two-chain xanthan molecules associate to form an entangled network of stiff molecules. In the absence of shear or at low shear rates, this network accounts for the effective suspending properties exhibited by xanthan solutions. As shear rates increase, the network is progressively disrupted, with an attendant decrease in viscosity.

In the presence of 0.1% (or higher) salt, xanthan shows high solubility, uniform viscosity, and excellent chemical stability over a pH range from 1 to 13. This chemical stability is particularly impressive. It is reported that aqueous solutions of xanthan in 5% sulfuric or nitric acid, 25% phosphoric acid, or 5% sodium hydroxide are reasonably stable for several months at room temperature. The unusual chemical stability of xanthan at high temperature or in strong acid or base is a direct outcome of the protective effect of the trisaccharide branches. In the presence of salt, the branches of xanthan interact with the main chain and shield the labile glycosidic linkages in the cellulose backbone from hydrolytic cleavage.

Biosynthesis of Xanthan Gum

The biosynthesis of xanthan gum is similar to the biochemical pathways that produce such common bacterial extracellular polymers as peptidoglycan and lipopolysaccharide. In all these cases, the challenge is to transfer a largely hydrophilic macromolecule across a lipid bilayer, the cell membrane. Microorganisms accomplish this task by utilizing a C_{55} isoprenoid lipid carrier, bactoprenol (undecaprenyl alcohol; Figure 8.6), in the stepwise construction of the polysaccharide and its transfer across the cytoplasmic membrane. As shown in detail in Figure 8.7, the 5-sugar repeating unit of xanthan is assembled by consecutive transfers of sugars from sugar nucleoside diphosphates. Each step is catalyzed by a specific glycosyl transferase. The site-specific acetylation of the repeating unit is catalyzed by a specific acetylase with acetyl coenzyme A as acetyl donor. Pyruvoylation (carboxyethylation) of the terminal mannose is catalyzed by a specific ketalase with phospho*enol*pyruvate as co-substrate. The acetylation and carboxyethylation reactions need not be consecutive. The completed xanthan building block, still linked through a pyrophosphoryl linkage to the bactoprenol, is transferred to another lipid-linked repeating unit by "tail-to-head" polymerization. The carrier bactoprenyl pyrophosphate, released from the growing xanthan molecule in the addition step, is hydrolyzed to the monophosphate (reaction 9), contributing favorably to the energetics of the polymerization reaction (reaction 8) and regenerating lipid carrier for the synthesis of another repeating unit. The lipid carrier attached to the growing xanthan chain remains behind when the polysaccharide is ultimately released into the medium.

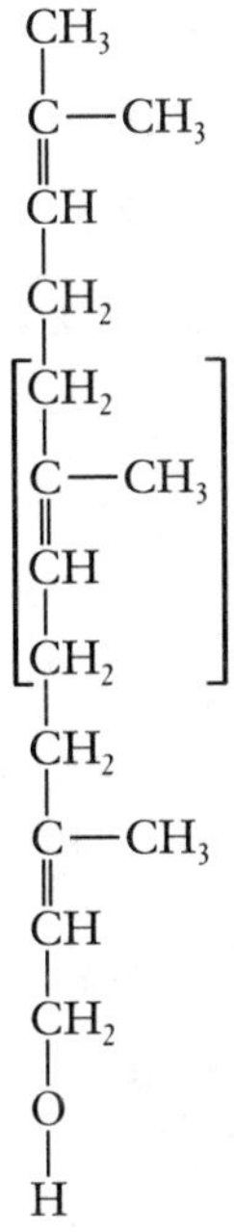

Production of Xanthan Gum

The yield of microbial polymers is often strongly influenced by the composition of the medium and by the nature of the growth-limiting

Figure 8.6
Bactoprenol (undecaprenyl alcohol).

1. Bactoprenol-**P** + UDP-Glc → Bactoprenol-**P-P**-Glc + UMP

2. Bactoprenol-**P-P**-Glc + UDP-Glc → Bactoprenol-**P-P**-Glc-(4←1)-β-D-Glc + UDP

3. Bactoprenol-**P-P**-Glc-(4←1)-β-D-Glc + GDP-Man → Bactoprenol-**P-P**-Glc-(4←1)-β-D-Glc + GDP

                                                      3

                                                      ↑

                                                      1

                                         α-D-Man

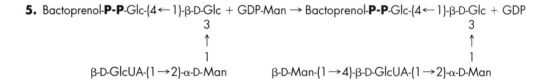

4. Bactoprenol-**P-P**-Glc-(4←1)-β-D-Glc + UDP-GlcUA → Bactoprenol-**P-P**-Glc-(4←1)-β-D-Glc + UDP

             3   3

             ↑   ↑

             1   1

          α-D-Man   β-D-GlcUA-(1→2)-α-D-Man

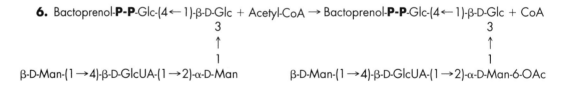

5. Bactoprenol-**P-P**-Glc-(4←1)-β-D-Glc + GDP-Man → Bactoprenol-**P-P**-Glc-(4←1)-β-D-Glc + GDP

             3   3

             ↑   ↑

             1   1

 β-D-GlcUA-(1→2)-α-D-Man   β-D-Man-(1→4)-β-D-GlcUA-(1→2)-α-D-Man

6. Bactoprenol-**P-P**-Glc-(4←1)-β-D-Glc + Acetyl-CoA → Bactoprenol-**P-P**-Glc-(4←1)-β-D-Glc + CoA

             3   3

             ↑   ↑

             1   1

β-D-Man-(1→4)-β-D-GlcUA-(1→2)-α-D-Man   β-D-Man-(1→4)-β-D-GlcUA-(1→2)-α-D-Man-6-OAc

Figure 8.7
Biosynthetic pathway for the formation of bactoprenol-linked xanthan polymer. Steps 1–5 are catalyzed by specific glycosyltransferases; step 6 is catalyzed by an acetylase, and step 7 by a ketalase. Step 8 is catalyzed by a polymerase. The abbreviations used are Glc, glucose; Man, mannose; GlcUA, glucuronic acid; Ac, acetyl; P, phosphate; UDP-Glc, uridine-5′-diphosphoglucose; GDP-Man, guanosine-5′-diphosphomannose; PEP, phospho*enol*pyruvate.

nutrient. Optimization of culture conditions is performed by growing cells continuously in a chemostat, where the cell growth is limited by a selected nutrient.

Chemostat experiments, performed at a constant dilution rate, have shown that the yield of xanthan from glucose in *X. campestris* is strongly

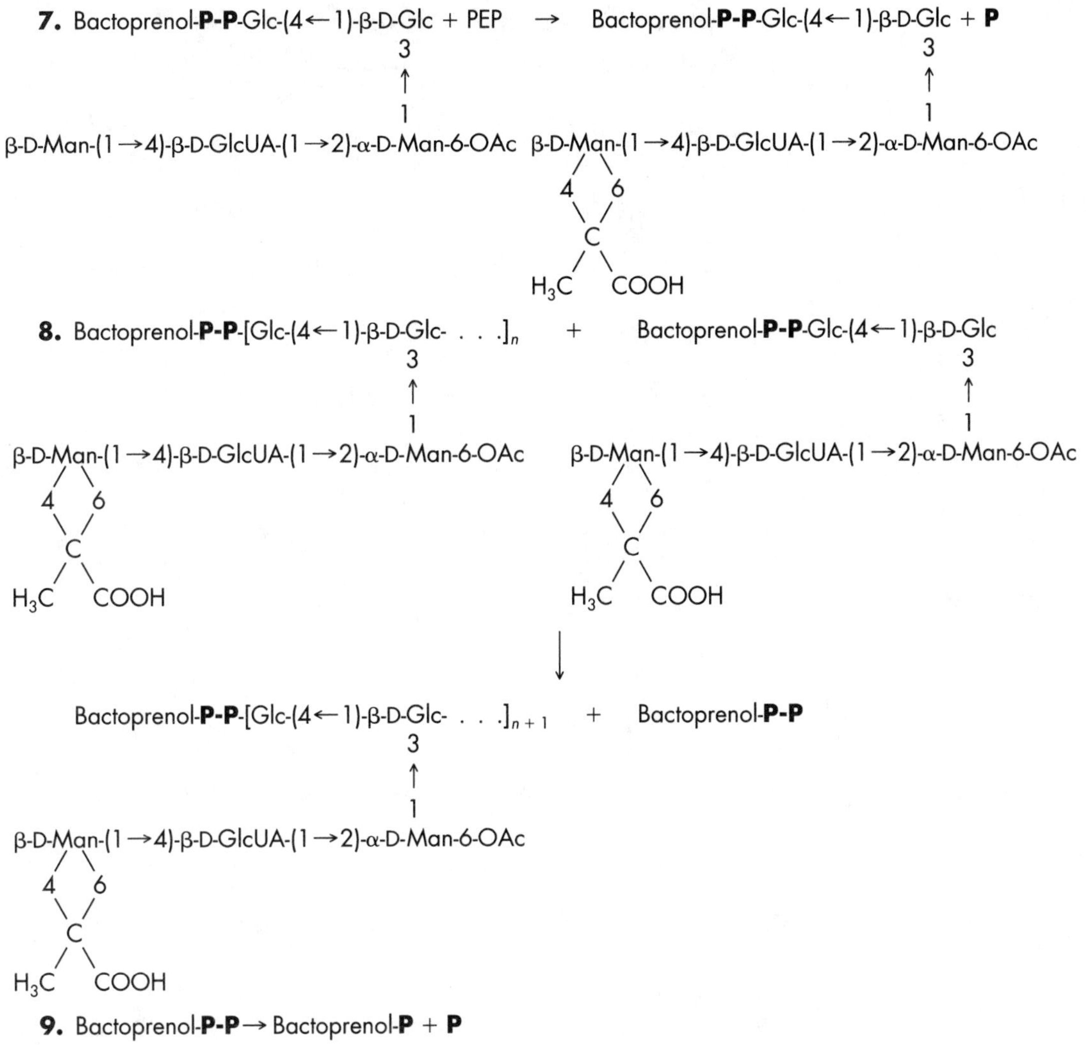

7. Bactoprenol-**P**-**P**-Glc-(4←1)-β-D-Glc + PEP → Bactoprenol-**P**-**P**-Glc-(4←1)-β-D-Glc + **P**

β-D-Man-(1→4)-β-D-GlcUA-(1→2)-α-D-Man-6-OAc   β-D-Man-(1→4)-β-D-GlcUA-(1→2)-α-D-Man-6-OAc

8. Bactoprenol-**P**-**P**-[Glc-(4←1)-β-D-Glc- . . .]$_n$ +   Bactoprenol-**P**-**P**-Glc-(4←1)-β-D-Glc

β-D-Man-(1→4)-β-D-GlcUA-(1→2)-α-D-Man-6-OAc   β-D-Man-(1→4)-β-D-GlcUA-(1→2)-α-D-Man-6-OAc

Bactoprenol-**P**-**P**-[Glc-(4←1)-β-D-Glc- . . .]$_{n+1}$ +   Bactoprenol-**P**-**P**

β-D-Man-(1→4)-β-D-GlcUA-(1→2)-α-D-Man-6-OAc

9. Bactoprenol-**P**-**P** → Bactoprenol-**P** + **P**

Figure 8.7 (*continued*)

influenced by the choice of the growth-limiting nutrient, the highest yield being seen under conditions of nitrogen limitation (Table 8.3). Temperature is an important variable. Cell growth is most rapid between 24 and 27°C, whereas the yield of xanthan is highest between 30 and 33°C. Commercial fermentations are run at about 28°C. Xanthan production invariably lags behind cell growth (Figure 8.8). It is possible that this lag reflects competition for bactoprenol among three synthetic pathways—those for peptidoglycan, lipopolysaccharide, and xanthan —during the period of exponential cell growth.

> The **chemostat** is an apparatus for the continuous culture of microorganisms. Sterile medium is fed into the culture vessel at a constant rate. The volume of the culture is kept constant by removing cells and spent medium at the same rate at which fresh medium is added. The chemostat controls the growth rate of the culture by limiting the supply of one essential nutrient while providing other essential nutrients in excess.
>
> The dilution rate is the volume of fresh medium added per hour and is expressed as a fraction of the volume of the vessel.

Effect of Growth-Limiting Nutrient on the Production of *X. campestris* Exopolysaccharide

Limiting Nutrient	Yield of Xanthan*
Glucose	0.54
NH_4^+	0.60
SO_4^{2-}	0.53
Mg^{2+}	0.55
K^+	0.42
PO_4^{3-}	0.31

TABLE 8.3

* The value given is grams of xanthan produced per gram of glucose consumed. A theoretical yield of approximately 0.85 is calculated for the acid form of xanthan containing 1 mole of acetate and 0.5 mole of pyruvate per repeating unit at an energy efficiency ratio (P/O ratio) of 2.

SOURCE: Davidson, I.W. (1978), Production of polysaccharide by *Xanthomonas campestris* in continuous culture, *FEMS Microbiol. Lett.* 3:347–349. Yields calculated by Margaritis, A., and Pace, G.W. (1985), Microbial polysaccharides, in Blanch, H.W., Drew, S., and Wang, D.I.C. (eds.), *Comprehensive Biotechnology*, vol. 3, pp. 1020–1021.

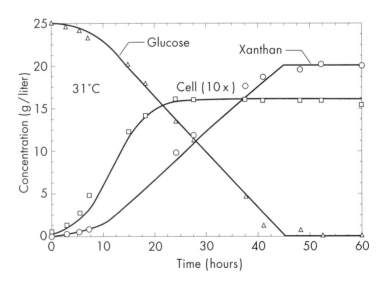

Figure 8.8
Consumption of glucose, cell growth, and xanthan gum formation in a submerged aerobic batch fermentation of *X. campestris*. Based on Shu, C.-H., and Yang, S.-T. (1990), Effects of temperature on cell growth and xanthan production in batch cultures of *Xanthomonas campestris*, *Biotechnol. Bioeng.* 35:454–468, Fig. 1a.

Medium for Batch Culture of *X. campestris*

Glucose	25 g
Yeast extract	3 g
K_2HPO_4	2 g
$MgSO_4 \cdot 7H_2O$	0.1 g
Antifoam emulsion	300 ppm
Tap water*	to 1000 ml
Adjusted to pH 7 with HCl	

* Source of required trace metals.

SOURCE: Shu, C-H., and Yang, S-T. (1990) Effects of temperature on cell growth and xanthan production in batch cultures of *Xanthomonas campestris*, *Biotechnol. Bioeng.* 35:454–468.

Xanthan gum is produced commercially by submerged aerobic batch fermentation of a pure *X. campestris* culture, using glucose, sucrose, or starch as a carbon source in a medium such as that described in Table 8.4. Acids are produced during the fermentation, and the pH is maintained near neutrality by periodic addition of sodium hydroxide. This adjustment of the pH is necessary because production of xanthan gum ceases below pH 5. In the production of food-grade gum, fermentation is stopped when the carbohydrate in the medium is exhausted, and the broth is pasteurized at near-boiling temperature. Isopropyl alcohol is added to the broth to precipitate the xanthan gum, and then the precipitate is dried and milled.

Modification of Xanthan Structure

Xanthans with altered structure have been obtained in several different ways. To begin with, different strains of *X. campestris* synthesize xanthans with widely differing numbers of acetyl and pyruvate groups. Modifications in the culture media and environmental conditions also lead to xanthans that vary widely with respect to these substituents. Insertional mutagenesis with transposons (see Box 3.8) has been used to generate mutant strains of *X. campestris* that produce xanthan polymers with truncated side branches.

The composition of the carbohydrate structure of xathan is unaffected by changes in the carbon source supplied to *X. campestris*. However, the modification of the branch sugars is strongly influenced by the choice of carbon substrate. For example, carboxyethyl groups represent

8% of the dry weight of xanthan when glycerol is used as the carbon source but represent only 5% when the carbon source is alanine (Table 8.5).

The acetyl and carboxyethyl groups of xanthan influence the conformational properties and the association behavior of xanthan in solution. The acetyl groups stabilize the ordered conformation of xanthan, which can be induced by specifically increasing the ionic strength of dilute aqueous solutions (by adding calcium ions). Carboxyethyl groups have been shown to have a strong destabilizing effect on the ordered conformation. This is probably the result of an unfavorable electrostatic contribution due to repulsion between these negatively charged groups. The optical rotation of xanthan solutions depends in part on the xanthan's conformation. One can exploit this dependence in order to follow the transition between the ordered and random coil conformations of xanthan in solution and to assess the effect of the acetyl and carboxyethyl substituents on the stability of the ordered conformation. The midpoint temperature (T_m) for the optical rotation change in deionized water for xanthan containing, by weight, 4.5% of acetyl groups and 4.4% of carboxyethyl groups is 44°C, but it rises to 54.5°C for polymer containing 7.7% acetyl and 1.7% carboxyethyl groups.

Thus xanthan gum offers an excellent example of how the interplay between structure and physical and chemical properties can endow a particular exopolysaccharide with considerable industrial value. Similar accounts could be given of many other microbial exopolysaccharides, but at this time there is only modest demand for these materials. However, studies on the exopolysaccharides listed in Table 8.2, and to a lesser extent on other microbial exopolysaccharides, have provided extensive information about the relationships between structure and properties in such polymers. This information, coupled with the ability to genetically engineer microbial strains so that they produce high yields of modified polysaccharides, will lead to new biopolymers with valuable properties.

Effect of Carbon Substrate on the Percentage (Dry Weight) Composition of *X. campestris* Exopolysaccharide

TABLE 8.5

Substrate	Glucose	Mannose	Glucuronic Acid	Acetate	Pyruvate
Glucose	30.1	27.3	14.9	6.5	7.1
Alanine	29.1	30.2	14.3	6.0	5.0
Glycerol	30.0	28.2	14.3	4.2	8.0
Pyruvate	29.6	29.8	13.8	6.4	10.6

SOURCE: Sutherland, I.W. (1987), Polysaccharide modification—physiological approach, in Yalpani, M. (ed.), *Industrial Polysaccharides: Genetic Engineering, Structure/property Relations and Applications* (Elsevier).

POLYESTERS

Bacteria differ from one another in the type of *reserve material* they accumulate when they are grown with unbalanced supplies of nutrients. The type of metabolite accumulated depends on the genotype of the organism and the kind of limitation. If the carbon-to-nitrogen ratio is high, and if nitrogen or oxygen is limiting, many bacteria accumulate glycogen and/or aliphatic polyesters, poly(3-hydroxyalkanoates), in amounts ranging from 30 to 80% of their cellular dry weight.

Polyhydroxyalkanoates produced from naturally occurring organic substrates have the general structure

$$\left[\begin{array}{c} CH_3 \\ | \\ (CH_2)x \qquad O \\ | \qquad\quad \| \\ -O-C^*H-CH_2-C- \end{array}\right]_n$$

where $x = 0-8$ or higher. The asterisk indicates an asymmetric center. In 1926, M. Lemoigne described the discovery, in *Bacillus megaterium*, of the first compound of this class, poly-(R)(3-hydroxybutyrate) (PHB), where x is zero. Many years later, studies showed that polyhydroxyalkanoates are thermoplastic materials. Thermoplastic polymers melt when heated to a certain temperature but harden again as they cool. This cycle can be repeated many times. Familiar thermoplastic materials, such as polyethylene, polypropylene, polyvinyl chloride, polystyrene, polycarbonate, and nylon, are used in the manufacture of hundreds of different types of plastic objects used in everyday life. In contrast to such nonbiodegradable plastics, those manufactured from polyhydroxyalkanoates were anticipated to be biodegradable. Applications for patents covering uses of poly-(R)(3-hydroxyalkanoates) were filed in 1962, but industrial production of these biopolymers did not occur until 1982.

Poly(3-hydroxybutyrate) shows similarities in its molecular structure and physical properties to polypropylene, a widely used polymer. (Polypropylene is used for packaging, rope, wire insulation, pipe and fittings, bottles, and appliance parts.) Both polymers are *isotactic*; that is, in each polymer the methyl group attached to the backbone is present in a single configuration throughout the chain. Some of the physical properties of poly(3-hydroxybutyrate) and of polypropylene are compared in Table 8.6. The most notable difference between the two materials is in biodegradability. Polypropylene is highly resistant to biodegradation, whereas poly(3-hydroxybutyrate) is ultimately completely degraded in a

Comparison of Some Properties of Polyethylene and Poly(3-hydroxybutyrate)

T
A
B
L
E

8.6

Property	Polypropylene	Poly(3-hydroxybutyrate)
Molecular weight	$(2.2-7) \times 10^5$	$(1-8) \times 10^5$
Melting point (°C)	171–186	171–182
Crystallinity (%)	65–70	65–80
Density (g · cm^{-3})	0.905–0.94	1.23–1.25
Flexural modulus (GPa)[a]	1.7	3.5–4
Tensile strength (MPa)[b]	39	40
Extension at break (%)	400	6–8
Resistance to ultraviolet irradiation	poor	good
Biodegradability	very poor	good[c]

[a] Flexural modulus is a measure of the elastic stiffness of a material. It relates the strain to the applied stress. GPa = 10^9 pascals or newtons per square meter.

[b] Tensile strength of a material is a measure of the maximal stress a material can withstand. MPa = 10^6 pascals.

[c] See Table 8.7.

SOURCE: Brandl, H., Gross, R.A., Lenz, R.W., and Fuller, R.C. (1990), Plastics from bacteria and for bacteria: Poly(β-hydroxyalkanoates) as natural, biocompatible, and biodegradable polyesters, *Adv. Biochem. Engin./Biotechnol.* 41:77–93, Table 6.

variety of environments (Table 8.7). The density difference between polypropylene and poly(3-hydroxybutyrate) plays a role in this regard. Polypropylene is less dense than water. Consequently, when discarded in rivers or oceans, objects made of polypropylene float. In contrast, the much higher density of poly(3-hydroxybutyrate) causes objects made of this material to sink to the bottom sediment layers, where they degrade.

Biodegradable plastics have many obvious applications in the manufacture of packaging containers, bottles, wrapping films, bags, and the like. They have important medical uses as well, serving as surgical pins, staples, wound dressings, bone replacements and plates, and biodegradable carriers for the long-term release of medicines.

Occurrence of Polyhydroxyalkanoates in Nature

Poly(3-hydroxyalkanoates) from different bacteria, extracted by means of organic solvents, have molecular weights of up to 2×10^6, corresponding to a degree of polymerization of about 20,000. The polymer accumulates in discrete granules in the cytoplasm of more than 50

Rate of Degradation of Random Copolymers of 3-Hydroxybutyrate and 3-Hydroxyvalerate in Various Environments

TABLE 8.7

Environment	Weeks Required for 100% Weight Loss[a]
Anaerobic sewage	6
Estuarine sediment	40
Aerobic sewage	60
Soil	75
Sea water	350

[a] Degradation of a molded film 1 mm thick.

SOURCE: Luzier, W.D. (1992), Materials derived from biomass/biodegradable materials, *Proc. Natl. Acad. Sci. USA* 89:839–842.

bacterial genera, each granule containing several thousand polymer chains. These include aerobic and anaerobic heterotrophic bacteria (such as *Azotobacter beijerinckii* and *Zoogloea ramigera*), many methylotrophs, aerobic and anaerobic phototrophs (*Chlorogloea fritschii*, *Rhodospirillum rubrum*, and *Chromatium okenii*), and archaebacteria (such as *Haloferax mediterranei*). Table 8.8 gives the poly(3-hydroxyalkanoate) content of a variety of bacteria grown on a wide range of carbon sources.

Biosynthesis and Biodegradation of Poly(3-hydroxyalkanoates)

The biochemical pathway leading to poly(3-hydroxybutyrate) branches from the central pathways of metabolism at acetyl-CoA, which is transformed to poly(3-hydroxybutyrate) by a sequence of three reactions. These are catalyzed by a 3-ketothiolase (Figure 8.9, reaction 1), by a NADPH-linked acetoacetyl-CoA reductase (reaction 2), and by a poly(3-hydroxyalkanoate) polymerase (reaction 3), respectively. Figure 8.9 shows the metabolic pathways in *Alcaligenes eutrophus* that lead to poly(3-hydroxybutyrate) from fructose or from butyrate.

When the carbon sources in the growth medium are exhausted, the polymer is degraded by an intracellular poly(3-hydroxyalkanoate) depolymerase to form D-(−)-hydroxybutyric acid. It is believed that the depolymerase is an *exo*-type hydrolase. D-(−)-Hydroxybutyric acid is then oxidized to acetoacetate by an NAD-specific dehydrogenase. The acetoacetate is converted to acetoacetyl-CoA by acetoacetyl-CoA synthetase. Thus acetoacetyl-CoA is an intermediate common to both the

> **Accumulation of Poly(3-hydroxyalkanoates) in Various Bacteria**

Genus	Description	Substrate	Maximal Poly(β-hydroxyalkanoate) (% dry weight)
Azotobacter	Aerobic rod	Glucose	25
Bacillus	Endospore-forming rod	Glucose	25
Beggiatoa	Gliding bacterium	Acetate	57
Caulobacter	Stalked bacterium	Glucose/glutamate	36
Chromatium	Phototrophic anaerobe	Acetate	20
Halobacterium	Archaebacterium	Glucose	38
Leptothrix	Sheathed bacterium	Pyruvate	67
Methylocystis	Methylotroph	Methane	70
Pseudomonas	Aerobic rod	Methanol	67
Rhizobium	Aerobic rod	Mannitol	57
Rhodobacter	Phototrophic anaerobe	Acetate	80
Spirulina platensis	Oxygenic phototroph	CO_2	6

SOURCE: Brandl, H., Gross, R.A., Lenz, R.W., and Fuller, R.C. (1990), Plastics from bacteria and for bacteria: Poly(β-hydroxyalkanoates) as natural, biocompatible, and biodegradable polyesters, *Adv. Biochem. Engin./Biotechnol.* 41:77–93, Table 1. Data for *Spirulina platensis* from Campbell, J., III, Stevens, S.E., Jr., and Balkwill, D.L. (1982), Accumulation of poly-β-hydroxybutyrate in *Spirulina platensis*, *J. Bacteriol.* 149:361–363.

biosynthesis and the degradation of poly(3-hydroxybutyrate). Note that in the reaction sequence leading from butyrate to poly(3-hydroxybutyrate), the (S)-3-hydroxyacyl-CoA intermediate in the fatty acid β oxidation pathway is inverted to form the (R)-3-hydroxyacyl-CoA precursor of poly(3-hydroxybutyrate) (see Figure 8.9).

The enzymes involved in the biosynthesis and degradation of poly(3-hydroxyalkanoates) are homologous to those just described for the metabolism of poly(3-hydroxybutyrate).

Biosynthesis of Random Copolymers of 3-Hydroxybutyrate and 3-Hydroxyvalerate

Alcaligenes eutrophus produces poly(3-hydroxybutyrate) when grown under appropriate conditions with glucose as the sole carbon source. However, when propionic acid is added in controlled amounts, a random copolymer of 3-hydroxybutyrate and 3-hydroxyvalerate (3-HB-co-3-HV copolymer) is formed, containing a predictable fraction of randomly distributed 3-HV units. The pathways leading to the formation of these copolymers are shown in Figure 8.10.

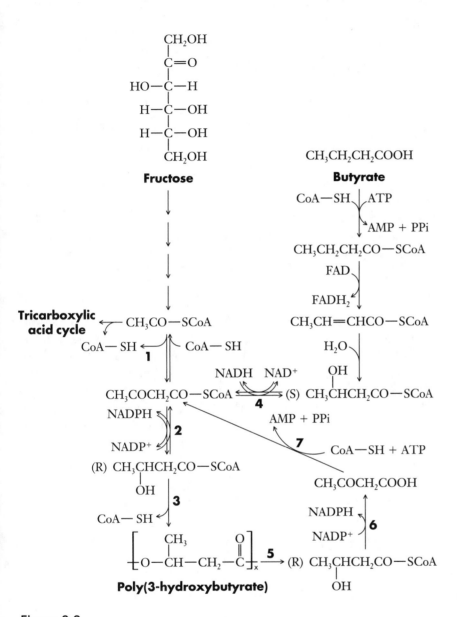

Figure 8.9

Pathways of biosynthesis and degradation of poly(3-hydroxybutyrate) in *Alcaligenes eutrophus*. The enzymes that catalyze reactions 1–7 are **(1)** 3-ketothiolase, **(2)** NADPH-dependent acetoacetyl-CoA reductase, **(3)** polyhydroxyalkanoate polymerase, **(4)** NADH-dependent acetoacetyl-CoA reductase, **(5)** polyhydroxyalkanoate depolymerase, **(6)** (R)-3-hydroxybutyrate dehydrogenase, and **(7)** acetoacetyl-CoA synthetase. Modified from Figure 1 in Doi, Y., *et al.* (1992), Synthesis and degradation of polyhydroxyalkanoates in *Alcaligenes eutrophus*, *FEMS Microbiol. Rev.* 103:103–108.

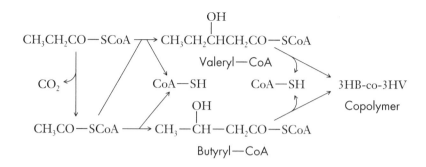

Figure 8.10
Biosynthesis of 3HB-co-3HV copolymers in *Alcaligenes eutrophus* grown on propionate as a sole carbon source. Modified from Anderson, A.J., and Dawes, E.A. (1992), Occurrence, metabolism, metabolic role, and industrial uses of bacterial polyhydroxyalkanoates, *Microbiol. Rev.* 54:450–472.

Alone, poly-(3-hydroxybutyrate) forms a brittle plastic that is difficult to process because it decomposes at about 10°C above its melting point of 177°C. However, plastics composed of 3-HB-co-3-HV copolymers have more favorable characteristics. The physical properties of 3-HB-co-3-HV copolymers are very sensitive to the mole percent of 3-HV. As this increases (Figure 8.11), the decomposition temperature remains unchanged but the melting temperature drops progressively.

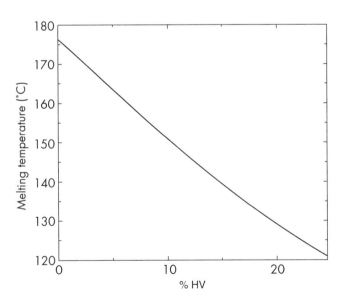

Figure 8.11
Influence of the 3-HV content on the melting point of 3-HB-co-3-HV copolymers. Data from Luzier, W.D. (1992), Materials derived from biomass/biodegradable materials, *Proc. Natl. Acad. Sci. USA* 89:839–842.

The percent crystallinity of the plastic decreases from over 80% for poly(3-hydroxybutyrate) to under 30% at a mole fraction of 25% 3-HV. Likewise, the copolymers show several-fold impovements in flexibility and toughness. This example highlights the importance of being able to make copolymers that vary in the nature and proportion of their building blocks.

Biosynthesis of Copolyesters from Single Substrates

Alcaligenes eutrophus requires two substrates—glucose and propionate —to synthesize 3-HB-co-3-HV copolymers. Other microorganisms synthesize copolymers with varying ratios of 3-HB to 3-HV when grown on a single substrate. This is illustrated in Table 8.9 for *Rhodococcus ruber* grown on different carbon sources. The 3-HV content of the polymer ranges from above 99% for *R. ruber* grown on valerate (pentanoate) to about 57% for cells grown on malate.

Manipulating Growth Conditions to Produce Novel Bacterial Polyesters—An Example

In nature, *Pseudomonas oleovorans* utilizes *n*-alkanes and *n*-alkanoic acids as sole carbon sources for growth. Under nutrient-limiting conditions but with excess carbon source, *P. oleovorans* forms large amounts of

TABLE 8.9 Composition and Yield of 3HB-co-3HV Copolymers Formed by *Rhodococcus ruber* Grown on Different Substrates

Substrate	Molar ratio 3HV:3HB	Yield of Polyhydroxyalkanoate as % of Cell Dry Weight
Malate	1.3	7.1
Acetate	2.5	40.4
Pyruvate	3.6	9.0
Glucose	3.8	31.1
Lactate	5.1	32.2
Succinate	12.0	7.1

SOURCE: Anderson, A.J., Williams, R.D., Taidi, B., Dawes, E.A., and Ewing, D.F. (1992), Studies on copolyester synthesis by *Rhodococcus rubber* and factors influencing the molecular mass of polyhydroxybutyrate accumulated by *Methylobacterium extorquens* and *Alcaligenes eutrophus*, FEMS Microbiol. Rev. 103:93–102.

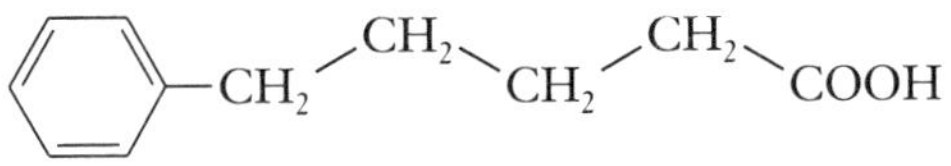

5-Phenylvaleric acid
(5-phenylpentanoic acid)

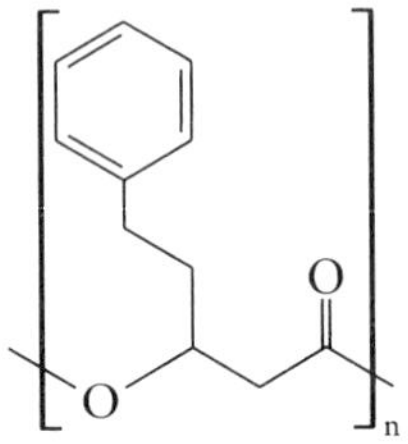

Poly(3-hydroxy-5-phenylpentanoate)

Figure 8.12
Structures of 5-phenylvalerate and poly(3-hydroxy-5-phenylvalerate).

poly(3-hydroxyalkanoates) from these substrates. *P. oleovorans* can use the "unnatural" substrate, 5-phenylvalerate (Figure 8.12), as the only carbon source. With this substrate in excess under conditions of nitrogen deprivation, *P. oleovorans* forms a pure polyester, poly(3-hydroxy-5-phenylpentanoate).

Biosynthesis of Novel Bacterial Polyesters through Cometabolism

A phenomenon known as *cometabolism* also adds greatly to the variety of copolymers that bacteria are able to synthesize. Cometabolism was first reported in 1960 in studies of methane-oxidizing bacteria. When these bacteria metabolize a growth-supporting hydrocarbon substrate, such as methane, they can also incorporate and oxidize other hydrocarbons, even though these hydrocarbons would not be utilized if they were present alone. We will encounter this poorly understood phenomenon again in Chapter 15 in the context of environmental degradation of foreign organic compounds.

Examples of cometabolism are seen in poly(3-hydroxyalkanoate) metabolism in *P. oleovorans*. When *P. oleovorans* grows on a hydrocarbon substrate, which itself can supply monomer units for poly(3-hydroxyalkanoate) formation, it is able to incorporate either a non-polymer-producing substrate or a non-growth-producing substrate into the polymer. Thus these bacteria can form polyesters from compounds not found in nature. The proportion of each building block in the

Composition and Yield of Copolymers Produced by *P. oleovorans* Grown with Mixtures of Nonanoic Acid and 11-Cyanoundecanoic Acid

TABLE 8.10

Molar Ratio of Nonanoic to 11-Cyanoundecanoic Acid in Growth Medium	Yield of Copolymer as % of Biomass	Mole % in Copolymer of Units Derived from 11-Cyanoundecanoic Acid
1:1	19.6	32
7:5	30.5	25
2:1	36.3	17

SOURCE: Lenz, R.W., Kim, Y.B., and Fuller, R.C. (1992), Production of unusual bacterial polyesters by *Pseudomonas oleovorans* through cometabolism, *FEMS Microbiol. Rev.* 103:207–214.

copolymers produced by cometabolism is determined by the ratio of the starting materials in the culture medium. Table 8.10 shows data for a copolymer of nonanoic acid (a natural substrate) and 11-cyanoundecanoic acid (which is able to support growth of *P. oleovorans* but not polymer synthesis on its own).

Genetic Engineering of Microorganisms and Plants for the Production of Poly(3-hydroxyalkanoates)

To permit large-scale use of microbial polyesters in the manufacture of plastics, these polymers need to be produced inexpensively in massive amounts. Genetic engineering offers approaches to these objectives.

Genetically Engineered E. coli. The three genes encoding the enzymes for the *Alcaligenes eutrophus* poly(3-hydroxybutyrate) biosynthetic pathway were introduced into *E. coli* on a multicopy plasmid. The *E. coli* strains harboring this plasmid accumulated poly(3-hydroxybutyrate) to over 90% of the cell dry weight when grown on glucose.

A special lysis strain of *E. coli* was engineered to facilitate recovery of the poly(3-hydroxybutyrate) granules. In addition to the plasmid carrying the poly(3-hydroxybutyrate) biosynthetic genes, these cells contained a separate plasmid that carried the bacteriophage T7 lysozyme gene, which was expressed at a low level throughout the cell cycle. The presence of the bacteriophage lysozyme did not interfere with cell growth or with poly(3-hydroxybutyrate) production. At the end of the poly(3-hydroxybutyrate) accumulation phase, the cells were collected by centrifugation and resuspended in a buffered solution of the chelating agent ethylenediamine tetraacetate (EDTA). The EDTA disrupted the integrity of the cytoplasmic membrane, allowing the lysozyme to reach and degrade the peptidoglycan layer of the cell wall. Subsequently, addition of a low concentration of a nonionic detergent (Triton X-100) sufficed to lyse the cells and allow quantitative release of the poly(3-hydroxybutyrate) granules.

Genetically Engineered Plants. In Chapter 1, we alluded to visions of "amber waves of plastics" and "plastic potatoes." For bacterial polyesters to compete in price with petroleum-derived plastics such as polyethylene, these visions may need to become reality. Bacterial biomass is much more expensive to produce than plant biomass. To appreciate the potential of plants as producers of polymers, one need only consider starch. The yield from a field of potatoes is about 20,000 kg of starch per hectare (1 hectare = 10,000 m^2).

Transgenic *Arabidopsis thaliana* plants (see Box 9.2) expressing *Alcaligenes eutrophus* genes for acetoacetyl-CoA reductase and poly(3-hydroxybutyrate) synthase produce poly(3-hydroxybutyrate) granules that are indistinguishable from those accumulated in bacteria. These transgenic plants grow more slowly and produce fewer seeds than wild-type plants do, because the diversion of a large fraction of acetyl-CoA and acetoacetyl-CoA from pathways leading to mevalonate and malonate slows the production of essential products such as flavonoids, lipids, phytohormones, carotenoids, sterols, and quinones (Figure 8.13).

These preliminary studies suggest a number of ways in which the production of poly(3-hydroxybutyrate) in plants might be achieved with less impact on the plant. For example, because plants efficiently accumulate starch and lipids in specific tissues, it may be possible, through genetic manipulation, to create a plant that expresses the poly(3-hydroxybutyrate) biosynthetic pathway *selectively in these tissues* while simultaneously reducing starch and oil synthesis there so as to minimize competition for common precursors. Alternatively, it might be possible to engineer a plant that expresses poly(3-hydroxybutyrate) in tissues where the storage carbon is not essential for the growth of the plant—tissues such as the tuber of potato (rich in starch) and the root of sugarbeet (rich in sucrose).

These and other possibilities all require extensive manipulation of plant metabolism. The controlled synthesis of complex copolymers in plants would offer additional challenges. Thus bacterial fermentation is likely to remain the source of poly-(R)(3-hydroxyalkanoates) for the foreseeable future.

SUMMARY

Plants and algae are the sources of the most commonly used polysaccharides, such as starch, cellulose, guar gum, gum arabic, agar, and carrageenan. Bacteria and fungi also produce polysaccharides—under

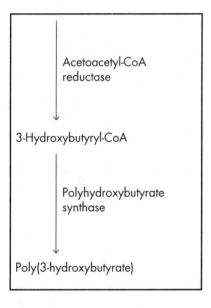

Acetyl-CoA $\longrightarrow$ Malonyl-CoA - - -> Pathways leading to lipids and flavonoids

3-Ketothiolase

Acetoacetyl-CoA $\longrightarrow$ Mevalonate - - -> Pathways leading to phytohormones, carotenoids, sterols, and quinones

Acetoacetyl-CoA reductase

3-Hydroxybutyryl-CoA

Polyhydroxybutyrate synthase

Poly(3-hydroxybutyrate)

Figure 8.13

Pathway of poly(3-hydroxybutyrate) synthesis in transgenic *Arabidopsis thaliana* expressing *Alcaligenes eutrophus* genes *phbB* and *phbC*, which encode acetoacetyl-CoA reductase and poly(3-hydroxybutyrate) synthase, respectively. The portion of the poly(3-hydroxybutyrate) biosynthetic pathway catalyzed by the *A. eutrophus* genes is boxed. The major end products of plant metabolic pathways that utilize acetyl-CoA and acetoacetyl-CoA as precursors are shown on the right. Modified from Poirier, Y., Dennis, D., Klomparens, K., Nawrath, C., and Somerville, C. (1992), Perspectives on the production of polyhydroxyalkanoates in plants, *FEMS Microbiol. Rev.* 103:237–246, Fig. 1.

appropriate conditions in amounts well in excess of 50% of cell dry weight. Of the many microbial polysaccharides, which vary greatly in composition, structure, and properties, only xanthan is in widespread use. Xanthan is produced by a plant pathogenic bacterium, *Xanthomonas campestris*. The xanthan backbone consists of β-(1 $\rightarrow$ 4)-linked D-glucose, and the 3-position of alternating glucose monomer units carries a trisaccharide side chain that contains one glucuronic acid and two mannose residues. The D-mannose units carry acetyl and carboxyethyl substituents. In solution, xanthan forms a stiff, right-handed double helix with 5-fold symmetry, the two chains running antiparallel to each other. Xanthan has extraordinary chemical stability: Aqueous solutions of xanthan in strong acid or base are stable for several months at room temperature. The high viscosity of dilute xanthan solutions stays uniform over a temperature range of $0-100°C$. Xanthan displays unusual flow behavior. In the absence of shear stress, a xanthan solution is viscous, but when a shear stress is applied beyond a certain low minimal value, the viscosity decreases sharply with shear rate. Because of this unique behavior, xanthan is used widely as a stabilizer, thickener, or gelling or suspending agent in many foods, as a suspending agent in paints, and as a water-thickening polymer in drilling muds (used to displace oil in enhanced oil recovery).

Many bacteria, when grown with unbalanced supplies of nutrients, accumulate massive amounts of polyhydroxyalkanoate polymer granules. Homopolyesters, copolyesters, or both are formed, depending on the nutrients supplied in the growth medium. Cometabolism makes possible the formation of copolymers that include unnatural building blocks. For example, when *Pseudomonas oleovorans* is grown on nonanoic acid (a natural substrate) and 11-cyanoundecanoic acid (a carbon source that on its own can support growth of *P. oleovorans* but not polymer synthesis), a copolyester accumulates whose yield and composition are determined by the molar ratio of nonanoic to 11-cyanoundecanoic acid in the medium.

The microbial polyesters are thermoplastic. Thermoplastic polymers melt when heated to a certain temperature but harden again as they cool. Thus the biodegradable polyhydroxyalkanoates are potential replacements for familiar nonbiodegradable thermoplastic materials (such as polyethylene, polypropylene, polyvinyl chloride, polystyrene, polycarbonate, and nylon) that are used in the manufacture of a multitude of plastic objects. The future of microbial polyhydroxyalkanoates will depend on our developing the ability to produce these materials cheaply in massive amounts. Appropriately genetically engineered *E. coli* strains produce high levels of poly(3-hydroxybutyrate) and allow easy isolation of the polymer granules. The genes for the pathway of poly(3-hydroxybutyrate) production have also been successfully expressed in plants.

SELECTED REFERENCES

General

Kennedy, J.F. (ed.), 1988. *Carbohydrate Chemistry*. Clarendon Press.

Yalpani, M. (ed.), 1987. *Industrial Polysaccharides: Genetic Engineering, Structure/Property Relations and Applications*. Progress in Biotechnology, vol. 3. Elsevier.

Yalpani, M., 1988. *Polysaccharides: Syntheses, Modifications and Structure/Property Relations*. Elsevier.

Bacterial Capsules

Jann, K., and Jann, B. (eds.), 1990. Bacterial capsules. *Current Topics in Microbiology and Immunology*, vol. 150. Springer-Verlag.

Microbial Polysaccharides

Blanshard, J.M.V., and Mitchell, J.R. (eds.), 1979. *Polysaccharides in Food*. Butterworth.

Berkeley, R.C.W., Gooday, G.W., and Ellwood, D.C. (eds.), 1979. *Microbial Polysaccharides and Polysaccharases*. Academic Press.

Sutherland, I.W., 1982. Biosynthesis of microbial exopolysaccharides. *Adv. Microb. Physiol.* 23:79–150.

Sutherland, I.W., 1983. Extracellular polysaccharides. In *Biotechnology*, vol. 3, *Biomass, Microorganisms for Special Applications, Microbial Products I, Energy from Renewable Resources* (ed. H. Dellweg), pp. 531–574. VCH.

Bushell, M.E. (ed.), 1983. Microbial polysaccharides. *Prog. Ind. Microbiol.*, vol. 18. Elsevier.

Margaritis, A., and Pace, G.W., 1985. Microbial polysaccharides. In *Comprehensive Biotechnology*, vol. 3, *The Practice of Biotechnology: Current Commodity Products* (ed. H.W. Blanch, S. Drew, and D.I.C. Wang), pp. 1005–1044. Pergamon Press.

The Genus *Xanthomonas*

Starr, M.P., 1981. The genus *Xanthomonas*. In Starr, M.P., et al. (eds.), *The Prokaryotes*, vol. 1, pp. 742–763. Springer-Verlag.

Palleroni, N.J., 1985. Biology of *Pseudomonas* and *Xanthomonas*. In Demain, A.L., and Solomon, N.A. (eds.), *Biology of Industrial Microorganisms*, pp. 27–56. Benjamin/Cummings.

Daniels, M.J., 1989. Pathogenicity of *Xanthomonas* and related bacteria towards plants. In Hopwood, D.A., and Chater, K.F. (eds.), *Genetics of Bacterial Diversity*, pp. 353–371. Academic Press.

Xanthan Gum

Okuyama, K., et al., 1980. Fiber diffraction studies of bacterial polysaccharides. In French, A.D., and Gardner, K.H. (eds), *Fiber Diffraction Methods*. ACS Symposium Series, vol. 141, pp. 411–427. American Chemical Society.

Kennedy, J.F., and Bradshaw, I.J., 1984. Production, properties, and applications of xanthan. *Prog. Ind. Microbiol.* 19:319–371.

Tait, M.I., Sutherland, I.W., and Clarke-Sturman, A.J., 1986. Effect of growth conditions on the production, composition and viscosity of *Xanthomonas campestris* exopolysaccharide. *J. Gen. Microbiol.* 312:1483–1492.

Polyhydroxyalkanoates

Lemoigne, M., 1926. Produits de déshydratation et de polymérisation de l'acide β oxybutyrique. *Bull. Soc. Chim. Biol.* 8:770–782.

Brandl, H., Gross, R.A., Lenz, R.W., and Fuller, R.C., 1990. Plastics from bacteria and for bacteria: Poly(β-hydroxyalkanoates) as natural, biocompatible, and biodegradable polyesters. *Adv. Biochem. Engin./Biotechnol.* 41:77–93.

Anderson, A.J., and Dawes, E.A., 1990. Occurrence, metabolism, metabolic role, and industrial uses of bacterial polyhydroxyalkanoates. *Microbiol. Rev.* 54:450–472.

Huisman, G.W., et al., 1991. Metabolism of poly(3-hydroxyalkanoates) (PHAs) by *Pseudomonas oleovorans*: Identification and sequences of genes and function of the encoded proteins in the synthesis and degradation of PHA. *J. Biol. Chem.* 266:2191–2198.

Luzier, W.D., 1992. Materials derived from biomass/biodegradable materials. *Proc. Natl. Acad. Sci. USA* 89:839–842.

Poirier, Y., Dennis, E., Klomparens, K., and Somerville, C., 1992. Polyhydroxybutyrate, a biodegradable thermoplastic, produced in transgenic plants. *Science* 256:520–523.

Schlegel, H.G., and Steinbüchel, A. (eds.), 1992. Papers presented at the International Symposium on Bacterial Polyhydroxyalkanoates, Göttingen, FRG, 1–5 June 1992. *FEMS Microbiol. Rev.* 103:91–376.

MICROORGANISMS IN PLANT BIOTECHNOLOGY

The nitrogen-fixing nodules on bean roots formed by association with the bacterium *Rhizobium*. From Galston, A.W. (1993), *Life Processes of Plants* (Scientific American Library). Copyright 1994 by Scientific American Library. All rights reserved.

Plant–Microbe Interactions

The production of high-quality agricultural products in quantities sufficient to feed the world's people is clearly a matter of vital importance. Equally clear is agriculture's importance in the global *economic structure*, although a precise estimate of its monetary value is notoriously difficult to make. Table 9.1 provides an estimate of sorts, but for many reasons (including the scarcity of reliable information), the figures are inexact. We cite the export/import prices for meat, for example, but it is likely that these prices are different from domestic prices, because the quality of exported meat may be different from that of meat earmarked for domestic consumption, and because prices may be affected by government regulations on exports. Nevertheless, even these imprecise estimates demonstrate that agriculture is one of humankind's major economic activities. Any general improvement in agricultural practices can therefore have a tremendous economic impact, and the methods of biotechnology suggest at least two important strategies for bringing about widespread improvements.

- *Improvement through the use of symbiotic or potentially pathogenic microorganisms* Plants live in intimate association with many microorganisms. Leaf surfaces are covered by layers of symbiotic and sometimes pathogenic bacteria and yeasts, and the soil close to the root area is populated by bacterial species very different from the species found in soil away from plants. There have been a number of attempts to modify symbiotic microorganisms so as to make them even more

Worldwide Agricultural Products (1986)

	Production per Year (millions of tons)	Estimated Value (billions of dollars)
Cereals	1867	270
Root crops[a,b]	592	51
Vegetables[c,d]	414	192
Fruits[e]	326	205
Oil crops[f]	130	27
Sugar (raw)	100	13
Coffee (green)	5.2	23
Cocoa bean	2.0	4.4
Tea	2.3	3.9
Vegetable fibers[g]	21	36
Tobacco	6.1	22
Rubber	4.4	3.7
Meat	155	306
Milk	521	175
Eggs	31	25

TABLE 9.1

The production figures are from *FAO* (Food and Agriculture Organization of the United Nations) *Production Yearbook 1986*. The unit prices were obtained from the *FAO Trade Yearbook 1986* and are thus averaged international import prices.

[a] Potato makes up about half of this category (by weight).

[b] Unit price was calculated by assuming that that of nonpotato root crops was 50% of the unit price of potato.

[c] Vegetable production in small fields is not included in many countries. This is estimated to correspond to as much as 40% of the total production in some countries.

[d] Statistics on 14 individual categories of vegetables are available in FAO yearbooks, but they still comprise only half of the total vegetable production. Because of the extreme diversity and complexity of this category, it is impossible to calculate a precise value. Thus we oversimplified the situation and took, as the unit price, the average of the unit prices of tomatoes and onions (the two vegetables grown in largest amounts). This figure is therefore only a guess.

[e] Apples, bananas, citrus fruits, and grapes account for about two-thirds of the total fruit production. The unit price was calculated by averaging the prices of these items.

[f] Soybean comprises the major part of this category.

[g] 75% of this category is cotton. About 20% is jute.

beneficial to plants. Examples mentioned in Chapter 1 include the use of ice nucleation-defective *Pseudomonas syringae* to reduce frost damage and the use of modified *Rhizobium* species to improve the nitrogen nutrition of cultivated plants.

- *Improvement through the production of transgenic plants*
Plants now being cultivated represent the end result of many years (sometimes centuries) of painstaking efforts at improve-

ment. Traditional plant-breeding programs rely on two methods: the selection of advantageous spontaneous mutations and the introduction of desirable traits from closely related species by crossbreeding. The first method is extremely inefficient and slow; the second is limited by the paucity of usable sources. Recombinant DNA technology is bringing revolutionary change to these endeavors. It enables scientists to select precisely defined genes with well-characterized properties, clone them, and transfer them to a given plant. These genes can be taken from organisms unrelated to the target plant—even from bacteria or animals.

Transferring cloned genes from prokaryotes (bacteria) into higher eukaryotes, such as plants, is difficult. However, there are two species of bacteria, *Agrobacterium tumefaciens* and *A. rhizogenes*, that transfer a small piece of DNA into plants as part of their normal life cycle. To date, this natural DNA transfer system has been central in the construction of plant varieties with superior properties. The current role of *Agrobacterium* in the genetic engineering of plants underscores the importance of studying the physiology of other naturally occurring bacteria to discover what contributions *they* may make to the new technology.

Several of the ways in which agriculture could benefit from genetic engineering are surveyed in this chapter. The obvious goal of increased crop yields can be achieved either by direct manipulation of the plant or indirectly by importing foreign genes that encode disease- or insect-resistance traits. One attractive long-range goal is to incorporate into cultivated plants the capacity to fix nitrogen. Another is to improve the quality of the product. Some strategies can be defined at the molecular level. For example, for making a superior bread, wheat flour must contain a protein (high-molecular-weight glutenin) that provides strong elasticity, apparently through the entanglement of polypeptide chains containing many turns. It should be quite feasible to change the properties of this protein through site-directed mutagenesis and then to insert the allele providing the desired level of elasticity into an otherwise preferred cultivar of wheat. Another highly desirable long-range goal is to change the amino acid composition of cereal proteins. Seed grains are the main source of protein for most of the world's population, yet they are poor in certain essential amino acids, such as methionine and lysine. Perhaps the storage proteins of these grains can be "engineered" to alter their composition and enhance their nutritional value. As another example, by changing the levels of expression of fatty acid desaturase, one might produce vegetable oil with more desirable properties. Finally, improvements that would result in easier harvesting, resistance to decay during transportation, and longer shelf life have very important economic potential.

Use of symbionts and pathogens

As we shall see later, much of the current agricultural research effort is directed at introducing potentially beneficial foreign genes into plant stocks. However, because there are many symbiotic bacteria normally associated with specific organs of various plants, a simpler plan might be to modify such bacteria and then use their interactions with the plants to introduce the modified traits at the appropriate locations. This alternative approach is technically easier, because the engineering of bacterial DNA through recombinant DNA methods is now routine, whereas the manipulation of plant DNA has yet to be perfected.

Protection of Plants from Frost Damage via Engineered Symbiotic Bacteria

One of the earliest examples of the successful modification of a symbiotic bacterium was performed in *Pseudomonas syringae*, which is found at high concentrations on the leaves of many plants. Many strains of this bacterium produce an ice nucleation protein that is apparently located on the surface of the bacterial cell, and the presence of this protein causes the formation of ice at temperatures only a little below 0°C, inflicting significant frost damage on important crop plants and so facilitating subsequent invasion of plant tissues by the bacteria. Steven Lindow and his associates first cloned the ice nucleation gene of *P. syringae* by using cosmid vectors. *E. coli* cells that received recombinant DNA molecules were screened for nucleation of ice formation at −9°C. A deletion was made in this gene by a recombinant DNA method, and the deletion was put back into the *P. syringae* chromosome by a homologous recombination process. The resulting Ice⁻ strain was identical to the parent Ice⁺ strain in all other properties, and heavy application of the mutant onto the leaves of strawberries, for example, led to a colonization that competed successfully with the wild-type bacteria, thereby protecting the plants from frost damage.

Use of Nitrogen-Fixing Bacteria to Improve Crop Yields

Another focus of current research—one that has a potentially much wider impact—is the process of nitrogen fixation. All animals and plants and most bacteria depend on the availability in their environment of some form of "combined nitrogen": nitrate (NO_3^-), ammonia (NH_3), or such nitrogen-containing organic compounds as amino acids. The huge amounts of N_2 that exist in the atmosphere are unavailable to the biological world except through the process of nitrogen fixation.

Nitrogen has been "fixed," or converted into fertilizers (mostly ammonium salts), by industrial processes since the beginning of this century. Because N_2 is an exceptionally stable compound, its conversion to NH_3 requires very harsh conditions—temperature higher than 500°C and pressure exceeding 200 atm. Thus the manufacture of chemical fertilizers consumes a significant portion of the energy expended globally. In addition, a considerable amount of the fertilizer applied to fields escapes into streams, ponds, and eventually the ocean, causing significant water pollution, including the growth of unwanted microalgae and other microorganisms.

In contrast, the biological process of nitrogen fixation, carried out by a small number of prokaryotic species, does not require the consumption of fossil fuels or electricity and, because it is regulated by the need for nitrogen in a given environment, does not produce environmental pollution. Thus increasing the extent of biological nitrogen fixation has been an important goal for biotechnology. Interestingly, even given the enormous amounts of chemical fertilizers currently in use, it is estimated that much more atmospheric nitrogen is fixed by nitrogen-fixing organisms. According to one estimate, the biological process fixes 6 times more nitrogen (24×10^7 tons/year) than is converted into chemical fertilizers by industrial processes. Thus even a small improvement in exploiting the biological process would have a global impact.

The biological process of nitrogen fixation is complex and consumes a large number of ATP molecules, because the enzymes involved in it must overcome the same huge activation energy barrier that faces the chemical synthesis of ammonia. Two enzymes are required: component I (nitrogenase) and component II (nitrogenase reductase). After component II is reduced by a strong biological reductant (ferredoxin or flavodoxin), 16 molecules (it is believed) of ATP are hydrolyzed to accomplish the reduction of component I. Reduced component II alone is not capable of overcoming the activation energy barrier. Reduced component I finally reduces N_2 to two molecules of NH_3 (Figure 9.1). Nitrogen fixation is a strongly reductive reaction, and the enzymes involved are usually irreversibly inactivated when they are exposed to oxygen. This oxygen sensitivity is important in understanding the biology of the N_2-fixing microorganisms, described below.

The ability to fix nitrogen is found in scattered members of the eubacterial kingdom. Some of the nitrogen-fixing genera are only very distantly related to others, which suggests that this function was probably transferred "laterally"—that is, between different organisms—during evolution. Several groups fix nitrogen as free-living organisms. Among those, *Clostridium* and *Klebsiella* fix nitrogen only under anaerobic conditions, an observation that is consistent with the oxygen sensitivity of the enzymes. Other free-living bacteria, however, can fix

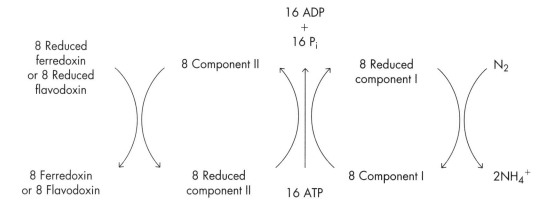

Figure 9.1
The enzymatic mechanism of nitrogen fixation.

nitrogen even under aerobic conditions, because each of these organisms has developed a complex machinery to protect the nitrogen-fixing apparatus from oxygen. Thus the cyanobacteria, which carry out oxygen-evolving photosynthesis, perform nitrogen fixation only in specialized cells called heterocysts, which do not produce oxygen. *Azotobacter* consumes oxygen at an extremely high rate, and in doing so apparently protects its nitrogen-fixing machinery. Another group of bacteria fix nitrogen only when they are in a symbiotic relationship with plants. The best-studied symbiotic nitrogen fixer is *Rhizobium*, which invades the root tissues of leguminous plants, such as alfalfa, pea, clover, and soybean, and lives in intracellular vacuoles. The vacuoles become filled with an oxygen-binding protein, leghemoglobin, produced by the plants. This creates an environment low in free oxygen, where *Rhizobium* is able to carry out nitrogen fixation.

The organization of genes involved in nitrogen fixation was first elucidated in *Klebsiella*. Remarkably, in that genus a very large number of genes are organized into a single *nif* gene cluster (Figure 9.2). This finding led, in the early 1970s, to the idea that cloning this cluster and putting the clones into desired crop plants might produce plant stocks that did not require chemical fertilizers—a possibility that, if realized, would revolutionize agriculture. Of course, the situation is much more complicated. First, if nitrogen-fixation machinery were produced in plant cells that were not also supplied with the necessary protective mechanisms, it would rapidly be inactivated by oxygen. Second, the large amounts of ATP needed for the process must be supplied as well. The importance of an ATP supply becomes clear when we compare the amounts of nitrogen fixed by symbionts and by free-living bacteria. Soil covered with red clover, carrying its symbiotic partner *Rhizobium*, fixes about 300 kg N/hectare/year. In contrast, free-living organisms such as *Azotobacter*, in the same area of soil, are thought to fix only negligible

K. pneumoniae
J H D K T Y E N X U S V W Z M F L A B Q
A. vinelandii
H D K T Y E N X U S V W Z M F A B Q
B. japonicum
nod and
fix genes
D K E N X U' S B H A
Pyruvate flavodoxin oxidoreductase
Subunits of nitrogenase (complex I and complex II)
Genes involved in synthesis of FeMo cofactor
Flavodoxin (reductant of complex II)
Regulatory genes

Figure 9.2
The organization of nitrogen fixation (*nif*) genes in *Klebsiella pneumoniae*, *Azotobacter vinelandii*, and *Bradyrhizobium japonicum*. For the latter two organisms, dotted lines represent regions where other genes and open reading frames are present, not drawn to scale. In *Bradyrhizobium*, the *nod* genes are involved in the formation of symbiotic root nodules; the *fix* genes are nitrogen fixation genes that do not have homologs in free-living *K. pneumoniae* or *A. vinelandii*. Because mutants of many *fix* genes still allow slow fixation of nitrogen in plant-free cultures but no fixation in plants, they are likely to function in the utilization of reductants provided by plants. After Dean, D.R., and Jacobson, M.R. (1992), Biochemical genetics of nitrogenase, in Stacey, G., Burris, B.H., and Evans, H.J. (eds.), *Biological Nitrogen Fixation*, pp. 763–834 (Chapman and Hall).

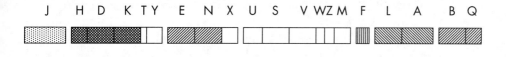

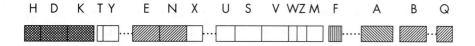

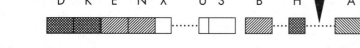

amounts (about 1 kg N/hectare/year). Among free-living organisms, cyanobacteria contribute more, but the amount of N_2 fixed by those bacteria under the best of conditions is still about one order of magnitude lower than the activity of the *Rhizobium*–clover combination. This difference is almost entirely due to the extent of the energy supply. Leguminous plants have evolved together with *Rhizobium* and contribute heavily to the successful symbiosis by expressing more than 20 genes specifically for that purpose. One of the contributions of these host plants is to supply a steady stream of compounds, such as dicarboxylic acids, that serve as the energy source for the bacteria. N_2 fixation by free-living organisms is usually constrained by a limited supply of energy (this explains the relative success of the cyanobacteria, which are phototrophic). Thus simply introducing *nif* genes does not bestow on a plant the ability to fix nitrogen.

These considerations led scientists to try more modest approaches directed at improving symbiotic N_2-fixing bacteria, such as *Rhizobium*. One possible target for improvement is their rate of nitrogen fixation. In *Rhizobium* species, all the genes involved in the nitrogen-fixation process itself, as well as most of the genes involved in its interaction with plants, are known to be located on plasmids ("*Sym* plasmids"). These range in size from 200–300 kb in *R. leguminosarum* to 1200–1500 kb in *R. meliloti*. Altering the expression of the regulatory protein for these genes, or simply increasing the number of copies of the genes, was found to produce a small but significant increase in the amount of nitrogen fixed in the appropriate host plant.

Another area with potential for improvement is the host-bacteria interaction. Although the interaction between *Rhizobium* and its hosts is extremely complex, many of the relevant genes have been identified. *Rhizobium* not only recognizes a given plant as a host but also induces a whole set of reactions in it, causing the root hair to curl, infection thread to form, and the thread to develop into a membrane that envelops the bacterium. It also induces the plant to secrete leghemoglobin, filling the space around the bacterium, and to supply constantly a large amount of energy source (such as dicarboxylic acids) to the bacterium, as described earlier. For example, the *Rhizobium nodD* gene product responds to specific flavonoid compounds produced by plants and activates the other genes involved in nodulation. By altering the sequence of *nodD*, it has been possible to change (and sometimes broaden) the host specificity of a given *Rhizobium* strain. In a later step of the nodulation process, the *nodH* and *nodQ* gene products of *Rhizobium* synthesize a low-molecular-weight signaling molecule that is recognized by a specific host plant, which then responds with curling of the root hair and so on. Substituting genes from a different species of *Rhizobium* for these genes resulted in successful alteration of the host range.

These results are impressive, but no one has yet produced an "improved" strain that performs effectively under field conditions. One major problem is that strains introduced from external sources are usually not competitive in the natural soil. In fields where leguminous plants are grown on a regular basis, the soils tend to contain a wealth of *Rhizobium* strains that are especially well suited to surviving in that particular environment, even though their N_2-fixing efficiency may not rival that of the newly engineered strains. Studies have shown that when strains of *Rhizobium* that supposedly fix N_2 more efficiently are introduced into such fields, they rarely survive the pressures of competing with the indigenous strains. Any hope of introducing a genetically engineered *Rhizobium* strain hinges on the production of better survivors and better colonizers—unfortunately, an area where investigators as yet know very little.

The effort to expand the host range of *Rhizobium* to include nonleguminous plants has been mentioned. However, in view of the complex nature of the interaction between *Rhizobium* and its host plants, this will not be easy. Accordingly, much work is being done on "associative" N_2 fixers, whose symbiotic relationship with plants is much less intimate. *Azospirillum* species, for example, which grow in association with important monocotyledonous crop plants such as sugar cane, associate with these host plants only loosely, most of the time by colonizing the surface of the roots. There is a price to pay for the loose nature of the interaction, however. Because the plants cannot supply nutrients rapidly to the bacteria under such conditions, the efficiency of N_2 fixation in such a system cannot be very high.

PRODUCTION OF TRANSGENIC PLANTS

As we saw earlier, an important application of biotechnology in agriculture is in the production of improved plant stocks. However, introducing cloned genes with desirable traits into plants is not a trivial matter. Plant cells are surrounded by a thick and rigid cell wall, and DNA cannot usually be inserted unless that wall is first removed. Even when pieces of foreign DNA are successfully transported into the plant cell, they are not likely to be replicated in succeeding generations. As we saw in Chapters 3 and 4, the standard approach to overcoming this difficulty is to put the cloned DNA segment into plasmids, which *will* replicate indefinitely in certain hosts. Lower eukaryotes such as yeast sometimes contain plasmids, which are used in the construction of shuttle vectors. However, most plant cells are not known to contain any plasmid DNA. An alternative, then, is to integrate the cloned DNA into the host chromosome, but this is an uncertain event.

To produce transgenic plants, therefore, the crucial steps are the efficient introduction of the cloned genetic material into the plant cell nucleus and the facilitated integration of the cloned gene into the plant

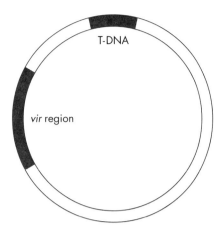

Figure 9.3
The structure of the Ti plasmid, with its *vir* region and T-DNA. In addition, the plasmid contains an origin of replication and several genes that assist in the colonization of host plants, such as the genes for the enzymes that degrade octopines and nopalines.

chromosome. It is interesting that the best method for doing so uses a system that already exists in nature, the system by which the plant-pathogenic bacterium *Agrobacterium tumefaciens* injects a portion of its plasmid DNA into plants and inserts it into the plant genome. (This underscores again the practical importance of studies of the "natural history" of bacteria-plant interactions.)

Introduction of Cloned Genes into Plants by the Use of *Agrobacterium tumefaciens*

A. tumefaciens is a Gram-negative bacterium that causes uncontrolled multiplication of "transformed," "tumorlike" cells—the disease known as crown gall—in host plants. Interestingly, rRNA homology has shown *A. tumefaciens* to be closely related to *Rhizobium*, but *A. tumefaciens* also contains a large (200 kb or even larger) "tumor-inducing" plasmid, the Ti plasmid (Figure 9.3). A small region of the plasmid, the *vir* (virulence) region, contains about two dozen genes that are involved in the infection of plants and in the transfer of a small part of the plasmid, T-DNA, into plant cells (Figure 9.4). Recent years have seen impressive progress in the understanding of this very complex process.

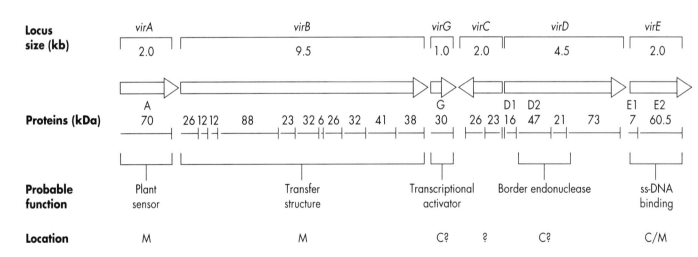

Figure 9.4
The organization and functions of the virulence region of the Ti plasmid. The large open arrows indicate the direction of transcription. C and M denote the cytoplasmic and the membrane location of the gene products, respectively. Based on Zambryski, P. (1988), Basic processes underlying *Agrobacterium*-mediated DNA transfer to plant cells, *Ann. Rev. Genet.* 22:1–30; and Kuldau, G.A., et al. (1990), The *Vir B* operon of *Agrobacterium tumefaciens* pTiC58 encodes 11 open reading frames, *Mol. Gen. Genet.* 221:256–266.

The genes of the *vir* region become activated in response to substances that exude from injured plant tissues. These substances thus serve as "signals" that tell an *A. tumefaciens* cell that it is next to a plant that is wounded and vulnerable to invasion. The proteins that recognize and respond to these signals, VirA and VirG, belong to a large family of prokaryotic regulatory proteins, often called two-component systems, that enable various organisms to respond adaptively to changes in osmolarity, availability of nitrogen sources, presence of chemoattractants, and so on (Figure 9.5). Such two-component systems typically contain a sensor protein located in the cytoplasmic membrane. The presence of signal molecules activates the protein kinase function of this sensor, with the consequent phosphorylation of the second component, the response regulator or transducer protein. This phosphorylation

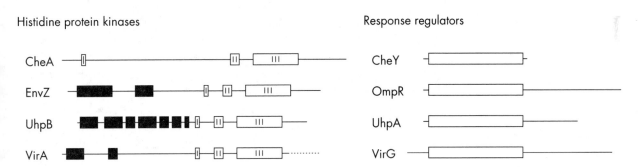

Figure 9.5

Two-component regulatory systems of bacteria. A system is composed of two proteins, a histidine kinase and a response regulator. There are now more than two dozen examples known. All histidine kinases are related in sharing three regions of homology (shown as I, II, and III), and all response regulators share similar sequences in their N-terminal region (shown as a box). In many cases, the histidine kinase is a membrane protein (the putative transmembrane domain is shown as a black box) and senses the nature of the external environment—for example, osmotic pressure (EnvZ), the presence of hexose phosphate (UhpB), or the presence of plant injury signals such as acetosyringone (VirA). Activation of the histidine kinase results in the phosphorylation of a conserved histidine residue in region I of these proteins, and then the phosphate is transferred to a conserved aspartate residue on the response regulator protein. This activates the response regulator, which typically affects the transcription rates of relevant genes (OmpR, phosphorylated by EnvZ, regulates porin genes; UhpA, phosphorylated by UhpB, regulates the expression of hexose phosphate transport genes; and VirG, phosphorylated by VirA, stimulates the transcription of *vir* genes). Some systems, however, have cytosolic histidine kinases, for example CheA that functions in chemotaxis by phosphorylating CheY, which acts to determine the sense of flagellar rotation. For a recent review on two-component systems, see Parkinson, J.S., and Kofoid, E.C. (1992), *Ann. Rev. Genet.* 26:71–112.

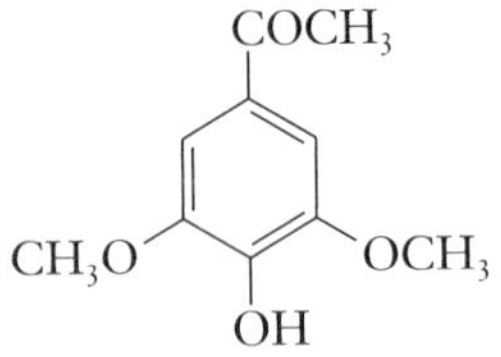

Figure 9.6
Structure of acetosyringone.

activates the response regulator, and it can, for example, activate the transcription of pertinent genes.

In the VirA–VirG system, VirA, a membrane protein, apparently acts as the sensor. It is activated synergistically by the presence of phenolic compounds, such as acetosyringone (Figure 9.6), leaking out of damaged plant tissues and by the presence of D-glucose, D-galactose, L-arabinose, or other sugars commonly found in plants. The sugars bind to a binding protein in the periplasm, and then the binding protein interacts with the periplasmic domain of VirA. The mechanism of interaction of acetosyringone is not yet clear. The activated VirA phosphorylates its own cytoplasmic domain, and the phosphate group is transferred to VirG, which is in the cytoplasm. The phosphorylated VirG then binds to the promoter regions of other *vir* genes and activates their transcription.

The next stage in the process is the nicking of the Ti DNA at specific points by "border nucleases" encoded by the *virD1* and *virD2* genes. A single-stranded DNA fragment of about 22 kb, called the T-strand, is released by an unwinding reaction, and at the same time a replacement strand is synthesized (Figure 9.7). This is accompanied by an orderly transfer of the T-strand, beginning with the "right" border, into the plant cell. The process—the nicking of the double-stranded DNA and the unwinding and injection of the single-stranded fragment —is remarkably similar to the events that occur in bacterial conjugation, and the two processes are thought to have a common evolutionary origin.

The product of the *virE2* gene, a single-stranded DNA binding protein, binds to the T-strand, presumably to facilitate the unwinding and injection. The *virD2* protein remains attached to the 5′-end of the T-strand and is thought to function as a "pilot" that leads the DNA into the plant cell.

The injection itself is catalyzed by the products of *virB* genes. The *virB* locus contains 11 genes, a fact that apparently reflects the complexity of the transport process. Some of the *virB* gene products are found in the cytoplasmic membrane, where they probably act as components of the channel. At least one of them contains a nucleotide-binding fold and is hypothesized to couple ATP hydrolysis to the transport process.

After it is injected into a plant cell, the T-strand has to enter the cell nucleus, a complementary strand has to be synthesized, and the double-stranded product, the "T-DNA," must become integrated into the plant genome. These processes, however, remain largely unidentified.

Once the T-DNA has been integrated into one of the plant chromosomes, various genes in the fragment begin to be expressed. These genes code for the synthesis of opines and of plant hormones. The opines (amino acid derivatives that can be used only by fellow *Agrobacterium* cells; see Figure 9.8) encourage the invasion of the plant by more *Agrobacterium* cells, and the hormones (auxin: indoleacetic acid, cyto-

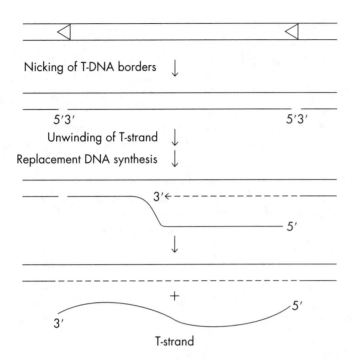

Figure 9.7
Nicking and transfer of T-DNA. The open arrowheads in the top diagram represent the 25-bp border sequences, which are cleaved by the specific endonuclease. Synthesis of the replacement strand (broken line) results in the release of the T-DNA as a single strand (T strand), which is injected into plant cells. From Zambryski, P., et al. (1989), Transfer and function of T-DNA genes from *Agrobacterium* Ti and Ri plasmids in plants, *Cell* 56:193–201, with permission.

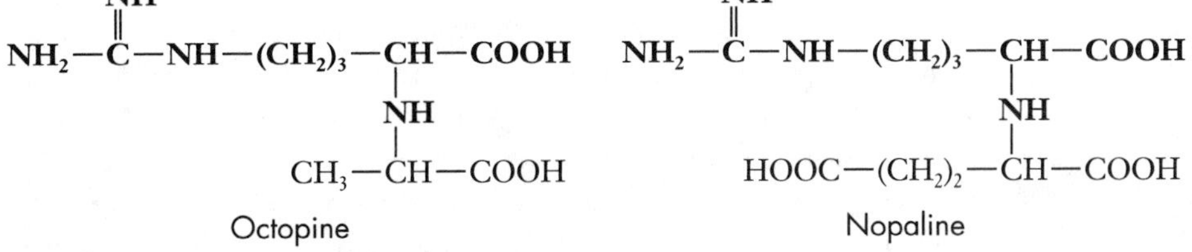

$$NH_2-\underset{\underset{NH}{\|}}{C}-NH-(CH_2)_3-\underset{\underset{NH}{|}}{CH}-COOH$$

$$CH_3-CH-COOH$$

Octopine

$$NH_2-\underset{\underset{NH}{\|}}{C}-NH-(CH_2)_3-\underset{\underset{NH}{|}}{CH}-COOH$$

$$HOOC-(CH_2)_2-CH-COOH$$

Nopaline

Figure 9.8
Structure of the opines, octopine and nopaline. These compounds are made of arginine residues (bold letters) joined to an acidic compound. *Agrobacterium* also produces families of similar compounds in which the arginine residue is replaced by other basic amino acids.

kinin: N^6-isopentenyladenine) stimulate the division and growth of plant cells, producing the characteristic tumor or gall.

The *Agrobacterium* Ti system seems almost tailor-made for exploitation by biotechnologists. The transfer of T-DNA is determined essentially by the 25-bp "border" repeats. The DNA between those repeats is transferred and integrated regardless of its sequence. Thus the insertion of extraneous genetic material into the middle of the T-DNA segment of the plasmid has no negative impact on the transfer and integration of that segment into a host cell.

Unfortunately, the Ti plasmid is too large to be conveniently manipulated *in vitro*. Therefore, the piece of foreign DNA to be cloned is almost always introduced into a smaller vector plasmid first. Then one of two strategies is used to transfer the cloned sequence into plants.

- *Use of a cointegrate intermediate.* In this, the more frequently used method, the foreign gene is inserted into a small vector, such as a derivative of pBR322 (so often used for cloning in *E. coli*), by the usual *in vitro* methods of recombinant DNA technology. Then *A. tumefaciens* cells that contain modified (non-tumor-producing) Ti plasmids are transformed with this recombinant plasmid. The modified Ti plasmids also contain a stretch of pBR322 sequence between the left and right borders of their T-DNA region. Because of this homology with the pBR322 replicon, recombination can take place to generate a cointegrate plasmid containing the entire sequence of the smaller plasmid between its left and right T-DNA borders (Figure 9.9). If the population is selected for the presence of a marker gene, such as a drug resistance gene on the pBR322 plasmid, only the cells that contain the cointegrate survive the selection. The cointegrate then injects the sequence between the two borders, containing the foreign gene, into plants.

  The cointegrate method has some drawbacks, however. The piece of DNA that is injected is relatively large, containing much extraneous information, so it is difficult to control the gene transfer process with precision. Often, only portions of this DNA become integrated into the plant genome. Furthermore, antibiotic resistance "marker" genes may get transferred into the Ti plasmid by mechanisms other than the homologous recombination, and thus there is no guarantee that the donor *Agrobacterium* cell contains the desired cointegrate. Finally, ascertaining the structure of the cointegrate by the usual *in vitro* methods is difficult because of its large size. Consideration of these points prompted the development of the binary plasmid approach.

- *Use of binary vectors.* This method takes advantage of the fact that the *vir* genes on one plasmid can catalyze the excision and transfer of a T-DNA sequence located on another plasmid; that

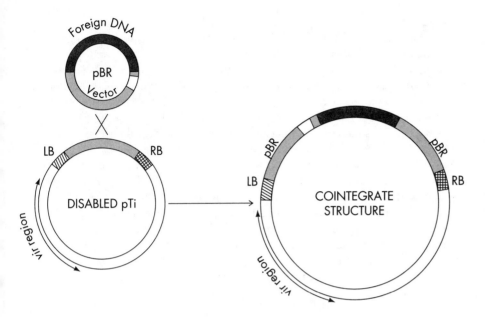

Figure 9.9

Transfer of T-DNA via the formation of a plasmid cointegrate. The pBR plasmid contains a marker selectable in *A. tumefaciens*, such as an appropriate antibiotic resistance marker (open box). The foreign DNA is cloned in the pBR plasmid, and the composite plasmid is introduced into *A. tumefaciens*, which already contains a Ti plasmid that had been "disarmed" by removing genes responsible for tumor production and had been further modified by incorporating a small fragment of the pBR sequence in the middle of the T-DNA. The origin of replication in the pBR plasmid functions well in *E. coli* but not at all in *A. tumefaciens*. Thus the antibiotic selection enriches only the cells that contain cointegrates, which were formed by the homologous recombination, utilizing the homologous pBR sequences, between the Ti plasmid and the pBR plasmid. Because the cointegrate contains all the *vir* genes, the foreign gene will be transferred into plants as a part of the modified T-DNA. From Lurquin, P.F. (1987), Foreign gene expression in plant cells, *Prog. Nucl. Acid Res.* 34:143–188, with permission.

is, these genes can act in *trans*. The binary plasmid approach consists of cloning the DNA fragment of interest into the T-DNA sequence of a plasmid vector with a broad host range that is capable of replicating in both *E. coli* and *A. tumefaciens*. The plasmid DNA is then introduced into *A. tumefaciens* cells that contain a "disarmed" Ti plasmid with *vir* genes but no T-DNA sequence. The *vir* genes of the disarmed plasmid effect the transfer of the T-DNA from the other plasmid without the formation of a cointegrate intermediate (Figure 9.10). In this method, only the piece of DNA that had been inserted between the left and right borders of the smaller plasmid is transferred into plants, allowing a more precise control of the process.

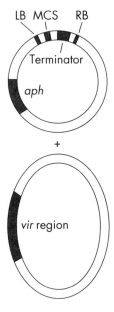

Figure 9.10

Transfer of T-DNA by the binary vector system. The foreign DNA is cloned into the middle of T-DNA in the plasmid, shown at the top, that contains one or more origins of replication that function both in *E. coli* and *A. tumefaciens*. In the particular vector shown, the original T-DNA sequence had been replaced by a multiple cloning site (MCS), followed by a terminator sequence functional in plants, located between the left border (LB) and right border (RB) sequences. The plasmid also contains an antibiotic resistance gene (in this case an aminoglycoside phosphoryltransferase, *aph*, to make possible selection of the plasmid-containing cells on aminoglycoside-containing plates). This composite plasmid is introduced into *A. tumefaciens* cells that contain another plasmid, shown at the bottom. This larger plasmid, a derivative of the Ti plasmid, contains only the *vir* region and the origin of replication and is totally devoid of the T-DNA region. Because the two plasmids have no common sequences, there is no recombination and cointegrate formation. Nevertheless, the products of the *vir* genes on the larger plasmid can mediate the transfer, into plants, of the T-DNA sequence of the other plasmid.

Both of these methods require several modifications in the Ti plasmid. In the Ti plasmid used for the cointegrate method, tumor-producing genes are either inactivated or removed; otherwise, transformed plant cells would become tumors, not healthy "transgenic plants." In addition, a pBR sequence is inserted into the T-DNA region, as shown in Figure 9.9. In the Ti plasmid used for the binary plasmid strategy, the entire T-DNA region is removed (see Figure 9.10); otherwise, T-DNA from the Ti plasmid would compete with the injection of T-DNA from the smaller plasmid.

Regardless of which strategy is used, the site of insertion of the foreign gene is usually sandwiched between a promoter sequence that functions effectively in plants and a terminator sequence. The 35S protein promoter from the cauliflower mosaic virus (CaMV) is popular because of its reliably high expression in a variety of plants. A popular terminator sequence is the one for nopaline synthetase. (An effective terminator ensures that the 3'-ends of mRNA are processed and polyadenylated so that the mRNA achieves a reasonable degree of stability.)

In addition, the region to be introduced into plants must contain a good marker so that plant cells that have received and integrated the foreign genes can be recognized easily. β-Glucuronidase is a marker whose activity can be detected readily in plant tissue. Even more useful are markers that enable one to select, not just screen, for plant cells with integrated T-DNA. Currently the most popular marker of this kind is neomycin (aminoglycoside) phosphotransferase II gene, which effectively inactivates aminoglycoside antibiotics that have detectable activity against plant cells, such as neomycin and kanamycin.

▶ Only three groups of DNA viruses are currently known to infect plants. Cauliflower mosaic virus is a member of one of these groups, the caulomoviruses. The host range of CaMV is restricted largely to the Cruciferae, the mustard family of plants. *Arabidopsis* belongs to this family.

Examples of Transgenic Plants
We Owe to *A. tumefaciens*

The *Agrobacterium* system is an excellent system for introducing foreign genes into the chromosomes of plants. The transfer into intact plant cells occurs at high frequency, the T-DNA is usually integrated into the plant chromosomes at high frequencies without undergoing structural alterations, and the cells that have received the T-DNA can be selected easily by using antibiotic resistance such as the neomycin resistance marker. Finally, the transgenic plants produced in this manner are quite stable for at least several generations. Consequently, almost all existing transgenic plants with potentially desirable traits have been obtained by this method.

In the simplest case, the desirable traits arise in transgenic plants through the continuous expression of the foreign genes. Below we list a few examples of transgenic plants of this type.

Herbicide-Resistant Plants

The advantages of making crop plants resistant to herbicides are obvious. Although there is a fear that such resistance may eventually increase the use of herbicide chemicals, there are also reasons to expect that these transgenic plants will promote the use of safer, more biodegradable herbicides, perhaps in smaller amounts. One example involves the herbicide glyphosate, which inhibits 5-*enol*pyruvylshikimate 3-phosphate synthase—an enzyme involved in the biosynthesis of aromatic amino acids—by acting as a structural analog of phospho*enol*pyruvate (Figure 9.11). This enzyme has been purified from crop plants and sequenced, and DNA probes corresponding to its amino acid sequence have been synthesized. These probes were used to isolate cDNA for the enzyme from the cDNA library of a plant cell line known to overproduce 5-*enol*pyruvylshikimate 3-phosphate synthase. The cDNA was then cloned behind the strong CaMV promoter, and the promoter-gene complex was introduced into plant cells (for example, petunia) via a disarmed Ti plasmid vector. The transgenic plants produce a much higher level of the target enzyme and thus are significantly more resistant to glyphosate (Figure 9.12). These results are encouraging because glyphosate has very low toxicity to animals and is rapidly degraded in soil.

A variation of this strategy is the introduction of mutant genes of a target enzyme. Sulfonylurea compounds act on plants by inhibiting acetolactate synthase, an enzyme involved in branched-chain amino acid synthesis. Plant mutants that showed resistance to this herbicide were selected in *Arabidopsis*. The small size and short generation time (Box 9.1) of *Arabidopsis* makes possible mutant selection of this type at

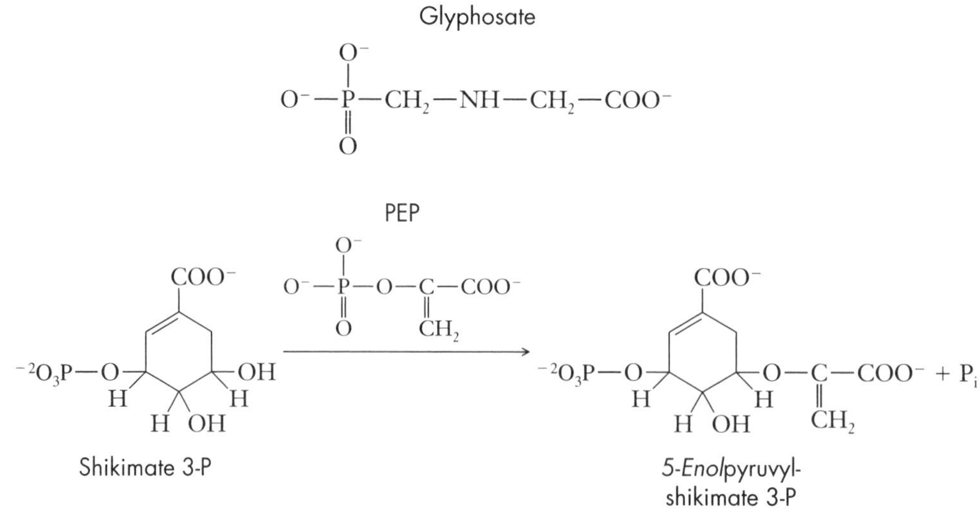

Glyphosate

$$O^- - \overset{\overset{\displaystyle O^-}{|}}{\underset{\underset{\displaystyle O}{\|}}{P}} - CH_2 - NH - CH_2 - COO^-$$

PEP

$$O^- - \overset{\overset{\displaystyle O^-}{|}}{\underset{\underset{\displaystyle O}{\|}}{P}} - O - \overset{\overset{\displaystyle}{}}{\underset{\underset{\displaystyle CH_2}{}}{C}} - COO^-$$

Shikimate 3-P

$$\longrightarrow$$

5-Enolpyruvyl-shikimate 3-P

$+ \ P_i$

Figure 9.11

Glyphosate and its mode of action. Glyphosate, acting as an analog of phospho*enol*pyruvate, inhibits the formation of 5-*enol*pyruvylshikimate 3-phosphate, a precursor of aromatic amino acids.

Figure 9.12

Production of glyphosate-resistant plants. The gene for 5-*enol*pyruvylshikimate 3-phosphate synthase, cloned behind a strong promoter, was introduced into petunia plants as described in the text. When these transgenic petunia plants (top) and the unaltered control plants (bottom) were sprayed with Roundup (a pesticide containing glyphosate), after 3 weeks the control plants were dead but the transgenic, resistant plants were completely healthy. From Shah, D.M., et al. (1986), Engineering herbicide tolerance in transgenic plants, *Science* 233:478–481, with permission.

> **BOX** 9.1 *Arabidopsis*
>
> *Arabidopsis* is a small cruciferous plant that is rapidly becoming the species of choice for genetic and recombinant DNA studies in plants—it is assuming something like the position of *E. coli* among prokaryotes. Its many advantages include a very small genome (70,000 kb, only 5 times the size of the yeast genome and only 10% of the size of the genome of typical crop plants) with an exceptionally small amount of repetitive DNA. It also has a rapid reproduction cycle (seeds can be obtained 6 weeks after germination); it is self-fertile, so mutant strains can be maintained easily; and it is susceptible to *Agrobacterium* transformation, so genetic material can be introduced by recombinant DNA methodology. For review, See Estelle, M.A., and Somerville, C.R. (1989), The mutants of *Arabidopsis*, *Tr. Genet.* 2:89–93; and Meyerowitz, E.M. (1989), *Arabidopsis*, a useful weed, *Cell* 56:263–269.

the whole-plant level. The acetolactate synthase gene from the mutant *Arabidopsis* was cloned by utilizing sequence homology with the known yeast enzyme and was shown to differ by a single base from the wild type, resulting in the change of a single amino acid residue. When this mutant gene was introduced into tobacco plants, the transgenic plants were nearly 100-fold more resistant to the sulfonylurea chlorosulfuron.

Altering the levels or nature of the enzymes targeted by herbicides is thus an effective approach to producing herbicide-resistant plants. However, this approach has several limitations. First, it yields only moderate degrees of resistance, because greater amounts of herbicide will still inhibit an overproduced or less sensitive target enzyme. Second, the overproduction of one of the enzymes in a complex pathway may disturb the regulation of such a pathway and could slow the growth of the plant. Third, it is difficult to alter an enzyme so that it will bind a toxic analog less well yet continue to bind its true substrate in a more or less normal manner. Considerations such as these have prompted the exploration of alternative methods for developing herbicide resistance in plants.

One highly successful technique for producing resistant transgenic plants is the introduction of genes coding for herbicide-detoxifying enzymes. An interesting example involves phosphinothricin, an analog of glutamic acid that inhibits glutamine synthetase (Figure 9.13). Although phosphinothricin is chemically synthesized and is sold commercially as an herbicide, it was originally discovered as the active moiety of an antibiotic, bialaphos, produced by a *Streptomyces* strain. As described in Chapter 13, antibiotic-secreting microorganisms frequently produce enzymes that detoxify the antibiotic to protect themselves. The bialaphos-producing *Streptomyces* was not an exception, and it produced an enzyme that inactivated both bialaphos and phosphinothricin by acetylation. The gene coding for this enzyme was cloned from the *Streptomyces*, put behind the CaMV 35S promoter, and introduced into crop

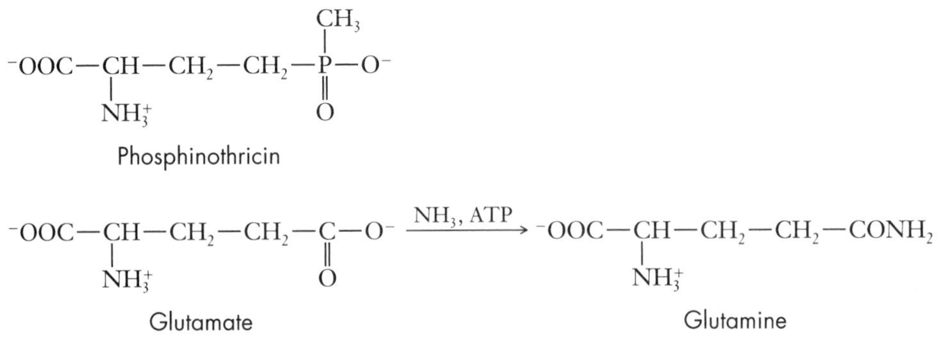

$$^-\text{OOC}-\underset{\underset{\text{NH}_3^+}{|}}{\text{CH}}-\text{CH}_2-\text{CH}_2-\underset{\underset{\text{O}}{\|}}{\overset{\overset{\text{CH}_3}{|}}{\text{P}}}-\text{O}^-$$

Phosphinothricin

$$^-\text{OOC}-\underset{\underset{\text{NH}_3^+}{|}}{\text{CH}}-\text{CH}_2-\text{CH}_2-\underset{\underset{\text{O}}{\|}}{\text{C}}-\text{O}^- \xrightarrow{\text{NH}_3,\ \text{ATP}}\ ^-\text{OOC}-\underset{\underset{\text{NH}_3^+}{|}}{\text{CH}}-\text{CH}_2-\text{CH}_2-\text{CONH}_2$$

Glutamate   Glutamine

Figure 9.13
Phosphinothricin and its mode of action. Phosphinothricin is an analog of glutamate. It binds to glutamine synthetase to inhibit the synthesis of glutamine. Bialaphos, an antibiotic produced by a *Streptomyces* species, is a tripeptide (phosphinothricinyl-alanyl-alanine) that is converted into phosphinothricin by plant peptidases.

plants, such as potato. The transgenic plants showed strong resistance to phosphinothricin, even in tests under field conditions.

Insect-Resistant Plants

Bacillus thuringiensis is used in the biological control of caterpillars because its sporulating cells contain toxic proteins (Chapter 6). The gene for one toxic protein was cloned behind promoters that are effectively expressed in plants and was introduced into plant cells via a Ti plasmid vector, thus producing plants toxic to caterpillars. The major problem with this approach has been the low level of expression of the toxin protein in plants, which is presumably related to the fact that the gene comes from a bacterium. Nevertheless, the method has produced tomato and tobacco plants that proved quite resistant to caterpillars in field tests. Cotton, however, is often attacked by insects that are more resistant to the *B. thuringiensis* toxin, and the low level of expression of this toxin in transgenic cotton plants provided the plants with little if any protection. Recently, the coding sequence of the toxin was altered extensively to replace codons that are rarely used in plants as well as to preclude the formation of a strong secondary structure in mRNA. When cotton plants were provided with this modified gene (placed behind the CaMV promoter with a duplicated enhancer region), their production of the bacterial toxin increased 100-fold, and they showed impressive resistance to common lepidopteran insects that damage unmodified plants.

Some seeds contain high concentrations of protease inhibitors, which are thought to interfere with the digestive process in insects. The cloning of cowpea trypsin inhibitor genes, for example, and their transfer to tobacco plants have resulted in good resistance to a wide variety of leaf-eating insects. In this case, there is no problem with the expression of the protein. It reaches levels as high as 1% of the total plant protein, presumably because the cloned gene is of plant origin. Also in contrast to *B. thuringiensis* toxin, the protection covers a broader range of insects.

Virus-Resistant Plants

The method most frequently used for producing virus-resistant plants sprang from the observation that, oftentimes, plants infected by a nearly avirulent virus are thereafter resistant to superinfection by a related, highly virulent one. Thus, deliberate infection with avirulent virus strains has been used to produce protection in crop plants. The method is not totally safe, however, because the avirulent strains may mutate to produce strains that are significantly pathogenic. Most plant viruses are positive-strand-RNA viruses covered by coat protein subunits (Figure 9.14). When the virus enters an injured plant cell, the replication process begins, starting with the progressive uncoating of the virus from the 5'-end of the RNA. Various observations suggested that the cross resistance between viruses occurs because the presence of virus coat proteins in the previously infected plant cells interferes with this uncoating process. Indeed, transgenic plants, whose genomes contain introduced tobacco mosaic virus (TMV) coat protein genes and which continuously synthesize the coat proteins, show resistance to virus infection. The phenotype of these plants is consistent with the idea that resistance is a result of interference with the uncoating of the virus particle: Although the plant cells are resistant to intact TMV particles, they remain sensitive to the TMV RNA or to partially uncoated TMV particles.

The coat protein gene is usually put behind a strong promoter such as the 35S promoter of CaMV and is introduced into a plant's genome via the *Agrobacterium* Ti system. Transgenic plants showing significant resistance to TMV, alfalfa mosaic virus, cucumber mosaic virus, tobacco streak virus, and tobacco rattle virus have already been produced in this manner. In a recent experiment, two virus coat protein genes, from potato virus X and potato virus Y, were introduced simultaneously into a commercially important potato cultivar, and one of the resulting transgenic plants proved quite resistant to both viruses under field test conditions. In most systems, the resistance level more or less parallels the level of coat protein produced.

Another approach to creating virus-resistant plants involves the cloning of satellite RNA genes. Populations of some plant viruses include a subpopulation of virus-like particles that contain smaller RNA molecules called *satellite RNA*. Satellite RNA does not seem to code for

Figure 9.14
Structure of TMV (tobacco mosaic virus). This drawing shows only 1/20 of the length of the rod-shaped virus particle. The single-stranded RNA is wound between the helices formed by the protein subunits. For the virus RNA to replicate, it must first be "uncoated" by the removal of proteins. This process begins from the 5'-end of the RNA and is driven by the translation process using this viral RNA as the message. After Klug, A., and Caspar, D.L.D. (1960), The structure of small viruses, *Adv. Virus Res.* 7:225–325, with permission.

any protein, but it is replicated by the enzymes coded for by the normal virus RNA and is then packaged into virus-like particles with the virus coat protein. What is interesting is that some satellite RNA *inhibits* virus replication strongly, presumably because it has a very high affinity to the viral replication machinery. Thus when cDNA made from the satellite RNA was inserted between the CaMV 35S promoter and an appropriate terminator and was transferred into plants via the *Agrobacterium* system, many of the transgenic plants showed significant resistance to tobacco ringspot virus and to cucumber mosaic virus. A strong promoter is necessary because satellite RNA must be produced by the transcription of this "gene." Compared with the resistance imparted by the overproduction of coat proteins, this type of resistance is reportedly more difficult to overcome with a higher dose of viruses.

In the future, it may also be possible to take advantage of the "hypersensitive response" with which plants react to infection by viruses, bacteria, and fungi. One aspect of this response is the production of low-molecular-weight secondary metabolites, such as *phytoalexins* (Box 9.2). It is likely that genetic enhancement of this reaction would produce plants with a generalized resistance to a broad range of infectious agents. Such an approach is technically quite difficult, however, because the genes involved are regulated in a complex manner.

Modification of Endogenous Plant Genes

In most of the examples we have discussed, crop plants have been improved by the introduction of genes taken from distantly related organisms. The transgenic plants have been designed to express these genes

---

BOX 9.2   Phytoalexins

Unlike higher animals, plants cannot produce specific antibodies to fight invading microorganisms. Many plants instead produce, as a response to microbial infection, low-molecular-weight secondary metabolites—phytoalexins—that inhibit the growth of invading microorganisms. The structure of phytoalexins is specific to the producing plant species. Here we show two examples, phaseollin (an isoflavonoid compound produced by green beans) and rishitin (a norsesquiterpene produced by potato).

Phaseollin   Rishitin

constitutively, often at quite high levels. This is not the only way to produce improved stocks, however. Many improvements can be introduced by using recombinant DNA technology to alter the species' endogenous genes.

One area in which some success has already been attained through this means is the modification of flower color. The biosynthetic pathways of flower pigments are known in detail, so it has been possible to bring in genes from different species of plants to produce "unnatural" colors that dramatically change a flower's appearance. In one instance, a maize gene was introduced into petunias (through the agency of the *Agrobacterium* system) to produce flowers with a brick-red color.

Another activity of great interest is the effort to delay the softening of fruits, such as tomatoes. Slowing the softening process would obviously lengthen the storage life of fruits and facilitate their transportation. Evidence suggested that hydrolysis of polygalacturonate, a polysaccharide component of the fruit cell wall, by the enzyme polygalacturonase was involved in the softening of tomatoes. The gene coding for this enzyme was cloned, and it was put behind the strong CaMV 35S promoter in an *inverted* orientation. This construct was then introduced into tomato plants via the *Agrobacterium* Ti system. The resulting transgenic plants produced much lower levels of polygalacturonase, because the "antisense mRNA" produced by the reading of the gene in reverse orientation interfered with translation of the normal mRNA, possibly by annealing to it. The ultimate effect of lowering the expression of polygalacturonase, however, is controversial; some tests did not show any differences either in the softening process or in the hydrolysis of cell wall pectin. Nevertheless, reducing the level of this enzyme appears to inhibit the degradation of pectin into small fragments, and it is reported to decrease damage during transport. One version of these engineered tomatoes is currently being marketed in the United States. Although there has been concern over the presence, in such products, of antibiotic-resistance genes used to select the transgenic plant cells, there are now methods available for the removal of such marker genes from plants.

One obvious area for future exploitation is modification of the nucleotide sequences of the genes for cereal proteins either to enhance their nutritional value or to improve their processing qualities, as mentioned at the beginning of this chapter. Some other examples have also been described. Efforts to improve the yields of crop plants or to change their properties in a major way, however, usually require complicated and subtle maneuvers. After all, plants are complex organisms containing highly differentiated organs and tissues that are regulated in a time- and environment-dependent manner. The challenge will be to devise a way of regulating endogenous genes in a carefully coordinated fashion. It is a difficult goal, but perhaps an attainable one, considering the solid progress being made in our understanding of how the plant genes are regulated.

The *Agrobacterium* system has been used in producing nearly all examples of transgenic plants mentioned so far. *Agrobacterium* has a very wide host specificity and in principle can infect most dicotyledonous plants. However, after *Agrobacterium*-mediated gene transfer has taken place, the subsequent steps leading to the production of transgenic plants have sometimes been hard to accomplish. In most cases this has been due to the difficulty of regenerating complete plants out of cells capable of receiving T-DNA. The use of embryonic tissues (such as epicotyl or hypocotyl) as the recipients of T-DNA is now being explored, because of their high regeneration potential.

Direct Introduction of DNA Fragments

In nature, *Agrobacterium* does not infect monocotyledonous plants (monocots), the subclass of plants that includes all the cereal crops. Apparently, this is largely due to the very weak wound response found in the tissues of cereal plants, and until very recently there have been no reproducibly successful methods for introducing and integrating T-DNA into these species. Although an *Agrobacterium*-mediated transfer of genes into embryonic rice tissues was reported very recently, it remains to be seen whether such an approach can be reproduced in other species and varieties.

This difficulty with the *Agrobacterium*-mediated system has led to a number of attempts to introduce DNA directly into plant cells by the following methods:

- *Incubation of plant protoplasts with DNA.* Protoplasts take up DNA in the presence of polyethyleneglycol (Chapter 4). When the protoplasts are made from embryogenic cells, it is often possible to regenerate the transgenic plants.

- *Fusion of DNA-containing liposomes with protoplasts.* This method will work for plants that can be regenerated from protoplasts. However, inasmuch as DNA can be introduced efficiently by using polyethyleneglycol, this method does not seem to offer any advantages.

- *Introduction of DNA into protoplasts by electroporation.* Transient application of high electric voltage across a protoplast membrane will produce large pores, through which DNA in the medium diffuses spontaneously (Chapter 4). Again, this approach offers no great advantage over the first method described.

- *Bombardment of plant cells with DNA-coated microprojectiles.* Pellets of microscopic size are coated with solutions of DNA and are literally "shot" into tissues and cells. This method, sometimes called biolistics, has great potential because it does not involve the often difficult process of regenerating cells and tissues out of protoplasts. Nevertheless, there are a number of drawbacks. The most important limitation is the very low frequency of successful introduction and integration of the foreign DNA, which means that one must examine a vast number of progeny cells to find the few that have acquired the new gene.

None of these methods has achieved the efficiency and reliability of the *Agrobacterium*-mediated gene transfer methods used in dicotyledonous plants. Nevertheless, we now know that the direct introduction of foreign DNA into plant cells and protoplasts does result in the integration of the DNA into the plant genome, sometimes with reasonably high efficiency, even though the plant cells are not supplied with specific exogenous systems for catalyzing the integration reactions. These methods, especially the direct introduction of DNA into protoplasts with polyethyleneglycol or by electroporation, have been used successfully to produce transgenic plants in economically important monocots such as rice, wheat, and corn.

SUMMARY

Many species of microorganisms interact with plants either as symbionts or as pathogens, and scientists have taken advantage of this intimate relationship in their efforts to improve agricultural production through biotechnology. For example, plant-pathogenic bacteria, inactivated through the deliberate deletion of genes essential for pathogenesis (such as the ice nucleation gene) have been used successfully as competitors against natural pathogens. Similar efforts are being made to improve the properties—for example, the host range—of beneficial symbionts such as nitrogen-fixing bacteria. In most cases, however, the focus of improvement has been the genetic constitution of a given plant—that is, the production of transgenic plants. But even these cases have exploited the natural capacity of the plant pathogen *Agrobacterium tumefaciens* to introduce a portion of its plasmid DNA, T-DNA, into plant cells. The introduction of exogenous genes from other plants or microorganisms in this manner has resulted in the production of plants that are resistant to herbicides, insect pests, or viruses. Perhaps an even more significant strategy under study is that of improving the quality of an agricultural product through manipulation of endogenous plant genes. Fruits and vegetables with improved shelf life and flowers with new and unexpected colors have been successfully produced. Even the production of crop proteins with higher nutritional values, or at much higher yields, may not be out of reach. A better understanding of the regulation of plant genes, however, is likely to be required before these goals can be achieved.

SELECTED REFERENCES

General

Grierson, D. (ed.), 1991. *Plant Genetic Engineering.* Chapman and Hall.

Use of Symbiotic Microorganisms

Lindow, S.E., Panopoulos, N.J., and McFarland, B.L., 1989. Genetic engineering of bacteria from managed and natural habitats. *Science* 244:1300–1305.

Stacey, G., Burris, R.H., and Evans, H.J. (eds.), 1992. *Nitrogen Fixation.* Chapman and Hall.

Long, S.R., 1989. Rhizobium-legume nodulation: Life together in the underground. *Cell* 56:203–214.

Martinez, E., Romero, D., and Palacios, R., 1990. The *Rhizobium* genome. *CRC Crit. Rev. Plant Sci.* 9:59–93.

Paau, A.S., 1991. Improvement of *Rhizobium* inoculants by mutation, genetic engineering and formulation. *Biotechnol. Adv.* 9:173–181.

Quispel, A., 1991. A critical evaluation of the prospects for nitrogen fixation with non-legumes. *Plants Soil* 137:1–11.

The *Agrobacterium* Ti System

Zambryski, P., Tempe, J., and Schell, J., 1989. Transfer and function of T-DNA genes from *Agrobacterium* Ti and Ri plasmids in plants. *Cell* 56:193–201.

Jin, S.G., Prusti, R.K., Roitsch, T., Ankenbauer, R.G., and Nester, E.W., 1990. Phosphorylation of the VirG protein of *Agrobacterium tumefaciens* by the autophosphorylated VirA protein: Essential role in biological activity of VirG. *J. Bacteriol.* 172:4945–50.

Shimoda, N., et al., 1990. Control of expression of *Agrobacterium vir* genes by synergistic actions of phenolic signal molecules and monosaccharides. *Proc. Natl. Acad. Sci. USA.* 87:6684–6688.

Cangelosi, G.A., Ankenbauer, R.G., and Nester, E.W., 1990. Sugars induce the *Agrobacterium* virulence genes through a periplasmic binding protein and a transmembrane signal protein. *Proc. Natl. Acad. Sci. USA.* 87:6708–12.

Zambryski, P.C., 1992. Chronicles from the *Agrobacterium*-plant cell DNA transfer story. *Ann. Rev. Plant Physiol. Plant Mol. Biol.* 43:465–490.

Production of Transgenic Plants by *Agrobacterium*-mediated Transformation

Gasser, C.S., and Fraley, R.T., 1989. Genetically engineered plants for crop improvement. *Science* 244:1293–1299.

Oxtoby, E., and Hughes, M.A., 1990. Engineering herbicide tolerance into crops. *Tr. Biotechnol.* 8:61–65.

Barton, K.A., Whiteley, H.R., and Yang, N.-S., 1987. *Bacillus thuringiensis* delta-endotoxin expressed in transgenic *Nicotiana tabacum* provides resistance to lepidopteran insects. *Plant Physiol.* 85:1103–1109.

Register, J.C., III, and Beachy, R.N., 1988. Resistance to TMV in transgenic plants results from interference with an early event in infection. *Virology* 166:524–532.

Kaniewski, W., et al., 1990. Field resistance of transgenic Russet Burbank potato to effects of infection by potato virus X and potato virus Y. *Bio/Technology* 8:750–754.

Lamb, C.J., Lawton, M.A., Dron, M., and Dixon, R.A., 1989. Signals and transduction mechanisms for activation of plant defenses against microbial attack. *Cell* 56:215–224.

Raineri, D.M., Bottino, P., Gordon, M.P., and Nester, E.W., 1990. *Agrobacterium*-mediated transformation of rice (*Oryza sativa* L.). *Bio/Technology* 8:33–38.

Tucker, G.A., 1990. Genetic manipulation of fruit ripening. *Biotechnol. Genet. Eng. Rev.* 8:133–159.

Visser, R.G.F., and Jacobsen, E., 1993. Towards modifying plants for altered starch content and composition. *Tr. Biotechnol.* 11:63–68.

Beck, C.I., and Ulrich, T., 1993. Biotechnology in the food industry. *Bio/Technology* 11:895–902.

Goldsbrough, A.P., Lastrella, C.N., and Yoder, J.I., 1993. Transposition mediated re-positioning and subsequent elimination of marker genes from transgenic tomato. *Bio/Technology* 11:1286–1292.

Transformation of Plant Cells by Direct Introduction of DNA

Potrykus, I., 1990. Gene transfer to cereals: An assessment. *Bio/Technology* 8:535–542.

Oard, J.H., 1991. Physical methods for the transformation of plant cells. *Biotech. Adv.* 9:1–11.

FROM BIOMASS
TO FUELS

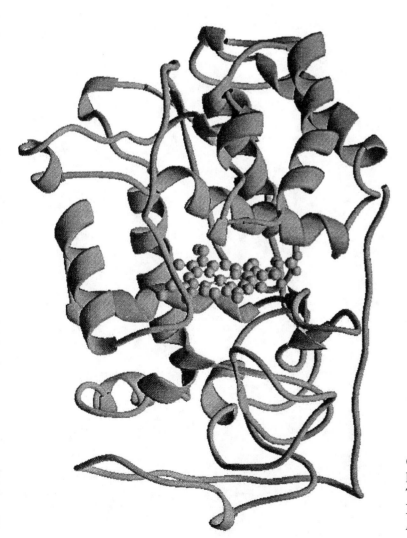

Crystallographic refinement of
lignin peroxidase at 2Å. Poulos,
T.L., Edwards, S.L., Wariishi,
H., and Gold, M.H. (1993), *J.
Biol. Chem.* 268:4429–4440.

Biomass

We all have dreamed of producing an abundance of useful food, fuel, and chemical products from the cellulose in urban trash and the residues remaining from forestry, agricultural, and food-processing operations. Such processes potentially could: (1) help solve modern waste-disposal problems; (2) diminish pollution of the environment; (3) help alleviate shortages of food and animal feeds; (4) diminish man's dependence on fossil fuels by providing a convenient and renewable source of energy in the form of ethanol; (5) help improve the management of forests and range lands by providing a market for low-quality hardwoods and the other "green junk" that develops on poorly managed lands; (6) aid in the development of life-support systems for deep space and submarine vehicles; and (7) increase the standard of living—especially of those who develop the technology to do the job!

At present, all of these aspirations are frustrated by two major features of natural cellulosic materials, crystallinity and lignification. Cowling, E.B., and Kirk, T.K. (1976), Properties of cellulose and lignocellulosic materials as substrates for enzymatic conversion processes, *Biotechnology and Bioengineering Symposium* 6:95–123.

Biomass can have broader definitions, but in the context of biotechnology, it is generally taken to mean "all organic matter that grows by the photosynthetic conversion of solar energy." The sun, either directly or indirectly, is the principal source of energy on Earth, its power converted to a usable organic form—biomass—by green plants, algae, and photosynthetic bacteria. The biomass produced *annually*, by photosynthesis on land and in the oceans, contains an estimated 3×10^{21} joules of energy, some 10 times the yearly worldwide human consumption.

Oil, coal, and natural gas represent the concentrated energy resource capital of the Earth, the accumulated residue of billions of years of photosynthesis. Projections of world rates of petroleum consumption and estimates of recoverable crude oil reserves indicate that the petroleum supply will be exhausted some time in the twenty-first century. Known coal and oil shale reserves, meanwhile, are expected to be adequate for centuries into the future, and processes for the conversion of coal to liquid hydrocarbon fuels are well established. However, exploitation of coal and oil shale deposits has a serious negative environmental impact, and massive use of fossil fuels adds to the carbon dioxide in the atmosphere, increasing the possibility of global warming—the "greenhouse effect." The use of biomass is promoted as a partial alternative, a large-scale renewable source of liquid fuels and chemical industry feedstocks. As long as biomass regeneration matched biomass use, there would be no net increase in the atmospheric content of carbon dioxide from this source.

How large is the store of world biospheric organic carbon compounds, and where is it found? *Forests*, which cover some 10% of total land area and account for about 43% of the net carbon fixed in the biosphere each year, contain about 90% of the biomass carbon of the Earth. Tropical forests make the largest contribution to this total. *Cultivated land* occupies a similar portion of total land area (about 9%) but accounts for less than 6% of mean annual net carbon fixation. As shown in Table 10.1, marine sources, savannah, and grasslands follow forests as large contributors to standing carbon reserves and carbon fixation in the biosphere.

Estimates of the Distribution of World Biospheric Carbon

TABLE 10.1

	Area		Net carbon production		Standing carbon	
	$10^6 km^2$	%	$10^9 tons/year$	%	$10^9 tons$	%
Marine sources	361	70.8	24.6	31.8	4.5	0.5
Forests	48.5	9.5	33.3	42.9	744	89.3
Savanna and grasslands	24	4.7	8.5	11.0	33.5	4.0
Swamp and marsh	2.0	0.4	2.7	3.5	14.0	1.7
Remaining terrestrial sources	74.5	14.6	8.4	10.8	37.5	4.5

SOURCE: Klass, D.L. (1983), Energy and synthetic fuels from biomass and wastes, in Meyers, R.A. (ed), *Handbook of Energy Technology and Economics*, pp. 712–786 (Wiley).

Current uses of biomass—food, fuel, fibers, building materials, and many other products—account for only a small fraction of the Earth's annual production. Forests and tree plantations are particularly rich sources of excess biomass, and in some parts of the world, agriculture can produce more food than is consumed locally or exported. This surplus capacity makes possible the cultivation of "energy crops" whose constituents can be converted to alcohol fuels or industrial chemicals. Energy crops include plants high in sugar (such as sugar cane, sugar beet, and sweet sorghum) or starch (cassava) content, those with high cellulose content (kenaf, elephant grass), and those high in hydrocarbon (jojoba, milkweed, and vegetable oilseeds such as sunflower and rapeseed). Anaerobic bacterial digestion of aquatic plants, such as marine and freshwater algae and the fast-growing water hyacinth, produces high yields of methane, making these plants energy crops as well.

Current uses of biomass also generate large amounts of organic wastes: agricultural wastes, such as wheat straw, corn cobs, oat hulls, and sugar cane bagasse (a fibrous residue left after extraction of juice); residues from logging and timber milling, such as wood chips and sawdust; spoiled produce and food-processing wastes; and urban solid waste such as paper, cardboard, and kitchen and garden refuse. Note that bioconversion of these waste products to fuels or protein does not decrease food production.

The conversion of biomass to fuel alcohol is the central topic of Chapter 11. We set the stage for it here with an examination of lignocellulose, the only major, nearly universal component of biomass. Lignocellulose makes up about half of all matter produced through photosynthesis. It consists of three types of polymers: cellulose, hemicellulose, and lignin. Each is intimately associated with the others by physical and chemical linkages, and all are degraded in the natural environment by bacteria and fungi.

MAJOR COMPONENTS OF PLANT BIOMASS

In the cell walls of the vascular tissues of higher land plants, cellulose fibrils are embedded in an amorphous matrix of lignin and hemicelluloses. These three kinds of polymers bind strongly to each other, by noncovalent forces as well as by covalent crosslinks, making a composite material that is known as lignocellulose, which represents over 90% of the dry weight of a plant cell. The quantity of each of the polymers varies with the species and age of a plant and from one part of the plant to another. Softwoods usually have a higher content of lignin than hardwoods. Hemicellulose content is highest in the grasses. On average, lignocellulose consists of 45% cellulose, 30% hemicelluloses, and 25% lignin (Table 10.2).

Source	Cellulose	Hemicellulose	Lignin
Grasses	25–40	25–50	10–30
Softwoods	45–50	25–35	25–35
Hardwoods	45–55	24–40	18–25

TABLE 10.2 Percentage Composition of Lignocellulose in Vascular Plants

Cellulose

Cellulose is the most abundant organic compound on Earth. *In situ*, a cellulose polymer is a linear chain of thousands of glucose molecules linked by β(1,4)-glycosidic bonds. The basic repeating unit is *cellobiose* (Figure 10.1). Consecutive glucose units in cellulose are rotated through

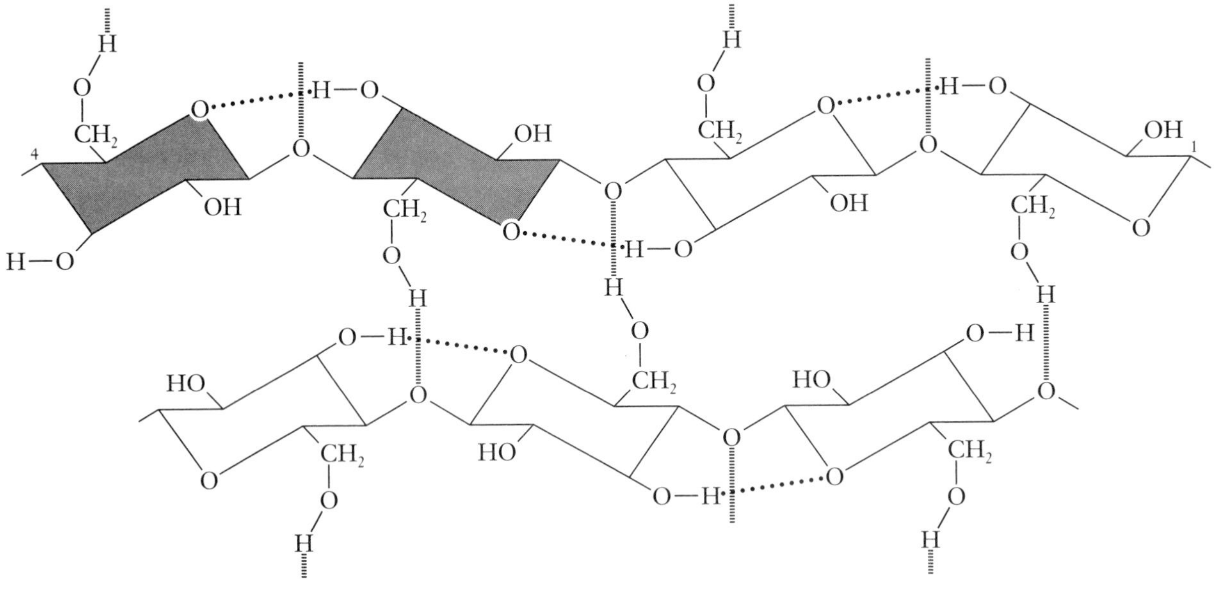

Figure 10.1
Cellulose chains showing the β-1,4-linked glucose residues rotated through 180° with respect to their neighbors in the chain. The conformation shown is stabilized by intramolecular hydrogen bonds (dots) within each chain. Intermolecular hydrogen bonds contribute to the interaction of adjacent chains within a microfibril. A cellobiose unit, the basic repeating disaccharide unit of the cellulose chain, is indicated by the shading within its pyranose rings.

180° with respect to their neighbors along the axis of the chain, and the terminal cellobiose can thus appear in one of two stereochemically different forms. The cellulose polymer chain has a flat, ribbonlike structure stabilized by internal hydrogen bonds. Other hydrogen bonds between adjacent chains cause them to interact strongly with one another in parallel arrays of many chains that all have the same polarity. The resulting very long, largely crystalline aggregates are called *microfibrils*. The microfibrils (250 Å wide) are combined to form larger fibrils. These are then organized in thin layers (lamellae) to form the framework of the various layers of the plant cell wall. Cellulose fibrils have regions of high order, *crystalline regions*, and regions of lesser order, *amorphous regions*. The fraction of crystalline cellulose varies with the source and with the way the material is prepared. Cellulose is water-insoluble, has a high tensile strength, and is much more resistant to degradation than other glucose polymers such as starch.

Hemicelluloses

Whereas cellulose is a linear homopolymer with little variation in structure from one species to another, hemicelluloses are highly branched, generally noncrystalline heteropolysaccharides. The sugar residues found in the hemicelluloses include pentoses (D-xylose, L-arabinose), hexoses (D-galactose, L-galactose, D-mannose, L-rhamnose, L-fucose), and uronic acids (D-glucuronic acid). These residues are variously modified by acetylation or methylation. Hemicelluloses show a much lower degree of polymerization (fewer than 200 sugar residues) than cellulose. Simplified structures of the three most common types of hemicelluloses are shown in Figure 10.2.

The name *hemicellulose* was introduced by E. Schultze in 1891 to describe the fraction of plant cell wall polysaccharides that are extractable with dilute alkali. Unfortunately, solubility behavior is a wholly inadequate criterion for assessing the hemicellulose content of various woods. Dilute alkali solubilizes both polysaccharides with a β-(1,4)-linked xylan backbone and galactoglucomannans, but glucomannans, which are a quantitatively important component of softwood hemicelluloses (Table 10.3), are insoluble under this condition and thus are left behind, still tightly associated with the cellulose fibers (Figure 10.3). The hemicelluloses are better defined as those polysaccharides that are noncovalently associated with cellulose.

Softwood Hemicelluloses. There are three major softwood hemicelluloses: glucomannan, galactoglucomannan, and arabinoglucuronoxylan. The two mannose-containing polymers differ greatly in galactose content. Their approximate sugar compositions (galactose:glucose:mannose) are 0.1:1:4 and 1:1:3, respectively. Their backbone consists of

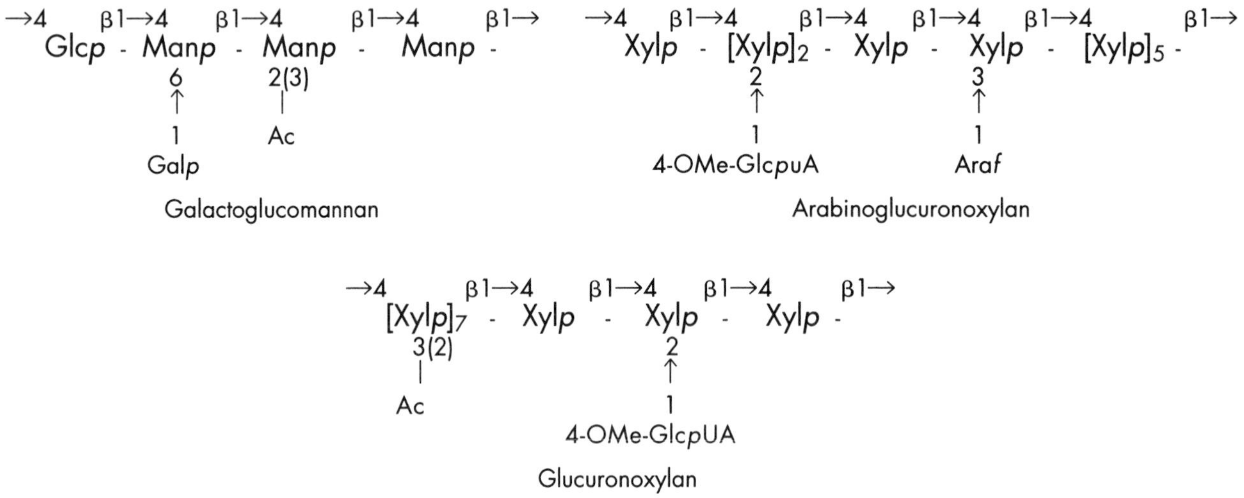

→4   β1→4   β1→4   β1→4   β1→   →4   β1→4   β1→4   β1→4   β1→4   β1→
Glc*p* - Man*p* - Man*p* - Man*p* -   Xyl*p* - [Xyl*p*]₂ - Xyl*p* - Xyl*p* - [Xyl*p*]₅ -
      6   2(3)   2   3
      ↑ | ↑   ↑
      1   Ac   1   1
Gal*p*   4-OMe-Glc*p*uA   Ara*f*

Galactoglucomannan   Arabinoglucuronoxylan

→4   β1→4   β1→4   β1→4   β1→
[Xyl*p*]₇ - Xyl*p* - Xyl*p* - Xyl*p* -
3(2)   2
 | ↑
Ac   1
              4-OMe-Glc*p*UA

Glucuronoxylan

Figure 10.2
Common types of hemicelluloses. Galactoglucomannans and arabinoglucur-
onoxylans are the principal hemicelluloses in softwoods. Glucuronoxylan
is a major hemicellulose in hardwoods. Abbreviations: Ac, acetyl; Ara*f*,
L-arabinofuranose; Gal*p*, D-galactopyranose; Glc*p*, D-glucopyranose;
Glc*p*UA, D-glucuronopyranose; Man*p*, D-mannopyranose; OMe, O-methyl;
Xyl*p*, D-xylopyranose.

Composition of the Lignocellulose of Various Wood Species[a]

TABLE 10.3

				Hemicellulose	
Species	*Common Name*	*Lignin*	*Cellulose*	*Glucomannan*[b]	*Glucuronoxylan*[c]
Softwoods					
Pseudotsuga menziesii	Douglas fir	29.3	38.8	17.5	5.4
Tsuga canadensis	Eastern hemlock	30.5	37.7	18.5	6.5
Juniperus communis	Common juniper	32.1	33.0	16.4	10.7
Pinus radiata	Monterey pine	27.2	37.4	20.4	8.5
Picea abies	Norway spruce	27.4	41.7	16.3	8.6
Larix sibirica	Siberian larch	26.8	41.4	14.1	6.8
Hardwoods					
Acer saccharum	Sugar maple	25.2	40.7	3.7	23.6
Fagus sylvatica	Common beech	24.8	39.4	1.3	27.8
Betula verrucosa	Silver birch	22.0	41.0	2.3	27.5
Alnus incana	Gray alder	24.8	38.3	2.8	25.8
Eucalyptus globulus	Blue gum	21.9	51.3	1.4	19.9
Ochroma lagopus	Balsa	21.5	47.7	3.0	21.7

[a] All values are given as percent of the wood dry weight.

[b] Includes galactose and acetyl substituents in softwoods (see Figure 10.2).

[c] Includes arabinose and acetyl substituents in hardwoods (see Figure 10.2).

SOURCE: Sjöström, E. (1981), *Wood Chemistry: Fundamentals and Applications*, Appendix (Academic Press).

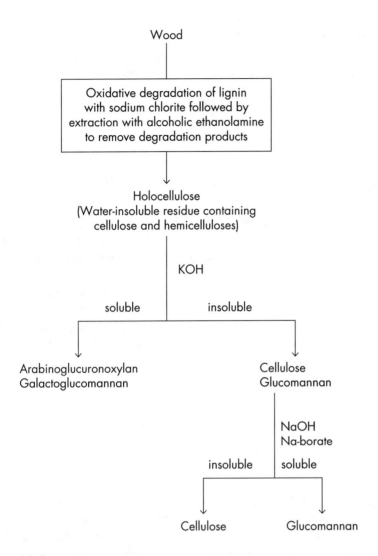

Figure 10.3
A procedure for the separation of the polysaccharide components of soft-wood. The dilute alkali extraction method of Schultze (see the text) leaves glucomannan hemicelluloses in the insoluble fraction.

1,4-linked β-D-glucopyranose and β-D-mannopyranose units. α-D-Galactopyranose units are linked to the backbone by 1,6 bonds. The backbone sugars are acetylated at C-2 or C-3 with one acetyl group for every 3–4 units (see Figure 10.2).

Arabinoglucuronoxylan has a backbone of 1,4-linked β-D-xylopyranose units partially substituted at C-2 by 4-O-methyl-α-D-glucuronic acid residues and at C-3 with α-L-arabinofuranose units (see Figure 10.2).

Hardwood Hemicellulose. The major hardwood hemicellulose is glucuronoxylan. This polymer's backbone consists of 1,4-linked β-D-xylopyranose units, the majority of which are acetylated at C-2 or C-3. About one 4-O-methyl-α-D-glucuronic acid, attached to the backbone at C-2, is present for every 10 xylose units (see Figure 10.2).

Lignin

Lignin is found in the cell walls of higher plants (gymnosperms and angiosperms), ferns, and club mosses, predominantly in the vascular tissues specialized for liquid transport. It is not found in mosses, lichens, and algae that have no tracheids (long tubelike cells peculiar to xylem). The increased mechanical strength that *lignification* confers on woody tissues enables huge trees several hundred feet tall to remain upright. Lignification is the process whereby growing lignin molecules fill up the spaces between the preformed cellulose fibrils and hemicellulose chains of the cell wall.

Building Blocks of Lignin

Lignin, the most abundant aromatic polymer on Earth, is a random copolymer built up of phenylpropane (C_9) units. The direct precursors of lignin are three alcohols—the *lignols*—derived from *p*-hydroxycinnamic acid: coniferyl, sinapyl, and *p*-coumaryl alcohols (Figure 10.4). Lignins are described as softwood lignin, hardwood lignin, or grass lignin on the basis of the relative amounts of these building blocks. A typical softwood (gymnosperm) lignin contains building blocks originating mainly from coniferyl alcohol, some from *p*-coumaryl alcohol, but none from sinapyl alcohol. Hardwood (angiosperm) lignin is composed of equal amounts (46% each) of coniferyl and sinapyl units and a minor amount (8%) of *p*-hydroxyphenylpropane units (derived from *p*-coumaryl alcohol). Grass lignin is composed of coniferyl, sinapyl, and *p*-hydroxyphenylpropane units, with *p*-coumaric acid (5–10% of the lignin) mainly esterified with the terminal hydroxyl group of *p*-coumaryl alcohol side chains.

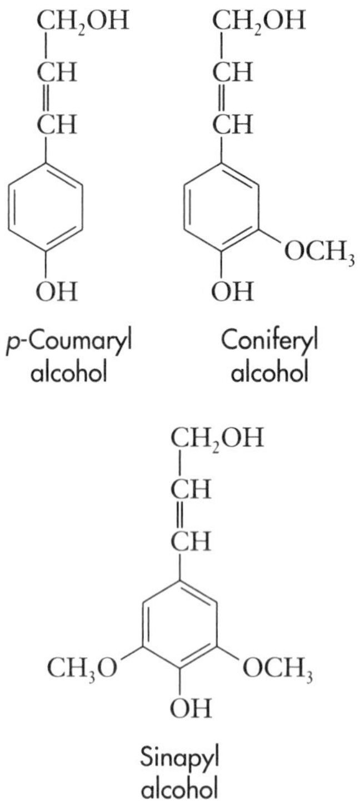

Figure 10.4
Three *p*-hydroxycinnamyl alcohols (monolignols) are direct biosynthetic precursors of lignin.

Biosynthesis of Lignin

The biosynthesis of lignin deserves particular attention because both the formation of this extraordinary polymer and its biodegradation depend on an extensive array of free-radical-mediated reactions.

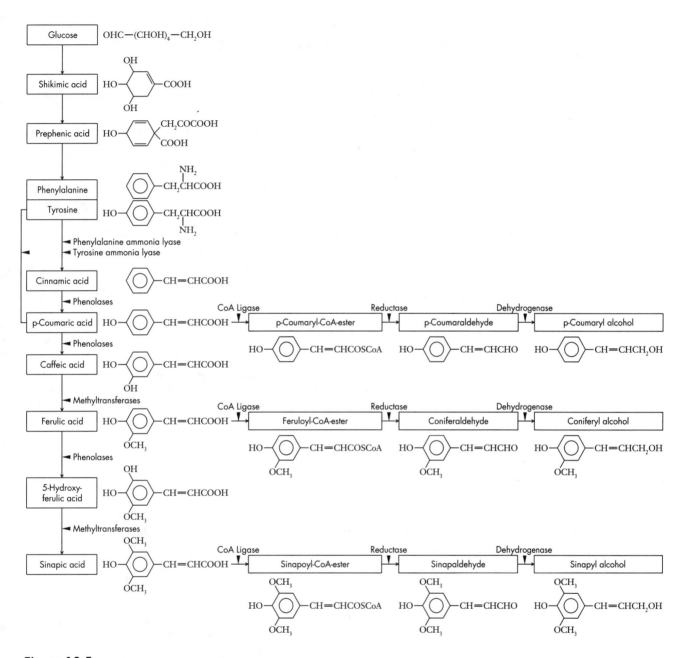

Figure 10.5
The shikimate pathway of biosynthesis of aromatic amino acids in plants leads to phenylalanine and tyrosine and thence to *p*-coumaryl, coniferyl, and sinapyl alcohols.

The synthesis of the building blocks of lignin utilizes the *shikimic acid pathway* of aromatic amino acid biosynthesis. In fact phenylalanine, and to a lesser extent tyrosine, are precursors in the biosynthesis of the lignols. The biosynthetic pathway is illustrated in Figure 10.5.

The lignin precursors are synthesized in the Golgi apparatus and/or the endoplasmic reticulum of plant cells and transported to the cell wall by vesicles. The precursors are then polymerized to lignin by a series of dehydrogenative reactions catalyzed by a peroxidase bound to the cell walls. Complex H_2O_2-requiring, peroxidase-catalyzed reactions convert the coniferyl, sinapyl, and p-coumaryl alcohols to phenoxy radicals, each of which can form a number of highly reactive resonance structures. Random coupling of such radicals leads to various quinone methide derivatives, which are converted to dilignols by the addition of water (Figure 10.6) or by intramolecular nucleophilic attack by primary alcohol or quinone groups on the benzyl carbons. Peroxidase-catalyzed dehydrogenation of the dilignols in turn leads to generation of radicals that form the lignin polymer by radical couplings followed by nucleophilic attack on the benzyl carbons of the oligomeric quinone methides by water and by aliphatic and phenolic hydroxyl groups of lignols. In parallel, nucleophilic attack by hydroxyl groups of sugar residues on cell wall polysaccharides forms lignin-hemicellulose crosslinks (Figure 10.7). There is roughly 1 bond to carbohydrate for every 40 phenylpropane units. Lignin also forms covalent bonds with the glycoproteins of the cell wall.

The principal modes of linkage between monomeric phenylpropane units, and their relative abundance in softwood lignin, are shown in Figure 10.8. A simplified qualitative structural model of softwood lignin appears in Figure 10.9.

Architecture and Composition of the Wood Cell Wall

The wood cell wall (Figure 10.10) is made up of several morphologically distinct concentric layers all of which contain cellulose, hemicelluloses, and lignin in varying proportions. The *middle lamella*, 0.2–1.0 μm in thickness, fills the space between cells and serves to bind them together. In mature wood (*latewood*), the middle lamella is highly lignified. The *primary wall*, on the outside of the cell, is a layer only 0.1–0.2 μm thick made up of cellulose, hemicelluloses, pectin, and proteins completely embedded in lignin. Just inside that is the *secondary wall*, which consists of three layers designated S_1 (the *outer layer*, 0.2–0.3 μm thick), S_2 (the *middle layer*, 1–5 μm thick), and S_3 (the *inner layer*, about 0.1 μm thick). These layers of the secondary wall, made up of variously oriented cellulose microfibrils embedded in a matrix of lignin and hemicelluloses,

Figure 10.6
Coupling of free radicals generated from p-coniferyl alcohol during lignin biosynthesis to form dilignols.

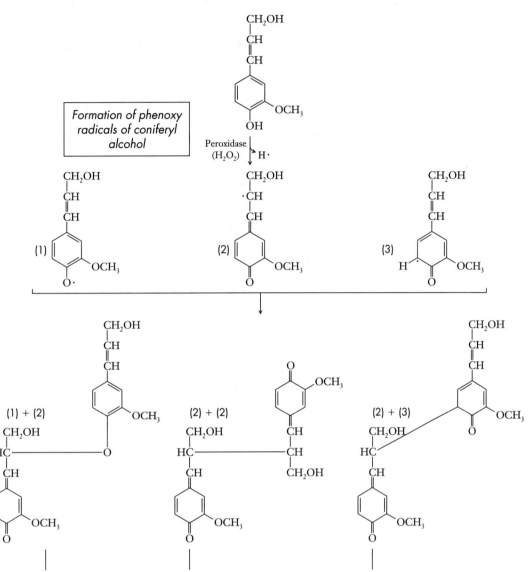

Formation of phenoxy radicals of coniferyl alcohol

Formation of dilignols by addition of water or by intramolecular nucleophilic attack by primary alcohol groups on the benzyl carbons

Peroxidase-mediated coupling of dilignols ⟶ LIGNIN POLYMER

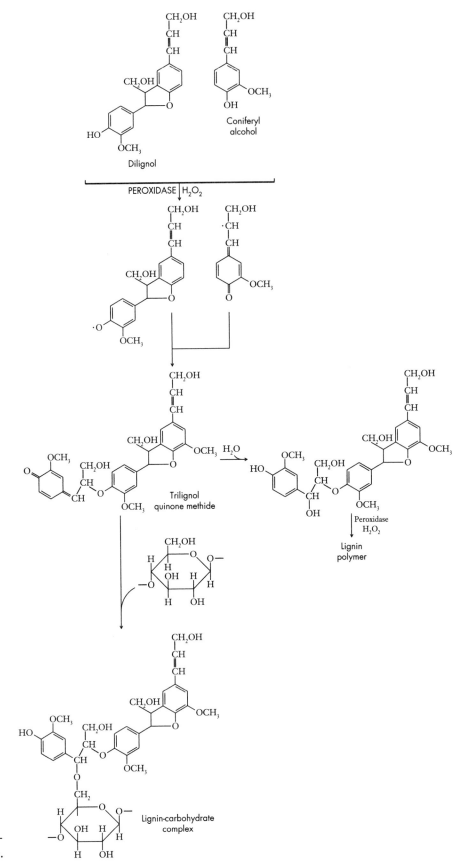

Figure 10.7
Peroxidase-catalyzed reactions of dilignols leading to the formation of the lignin polymer and lignin–carbohydrate crosslinks via an oligo-lignol quinone methide intermediate.

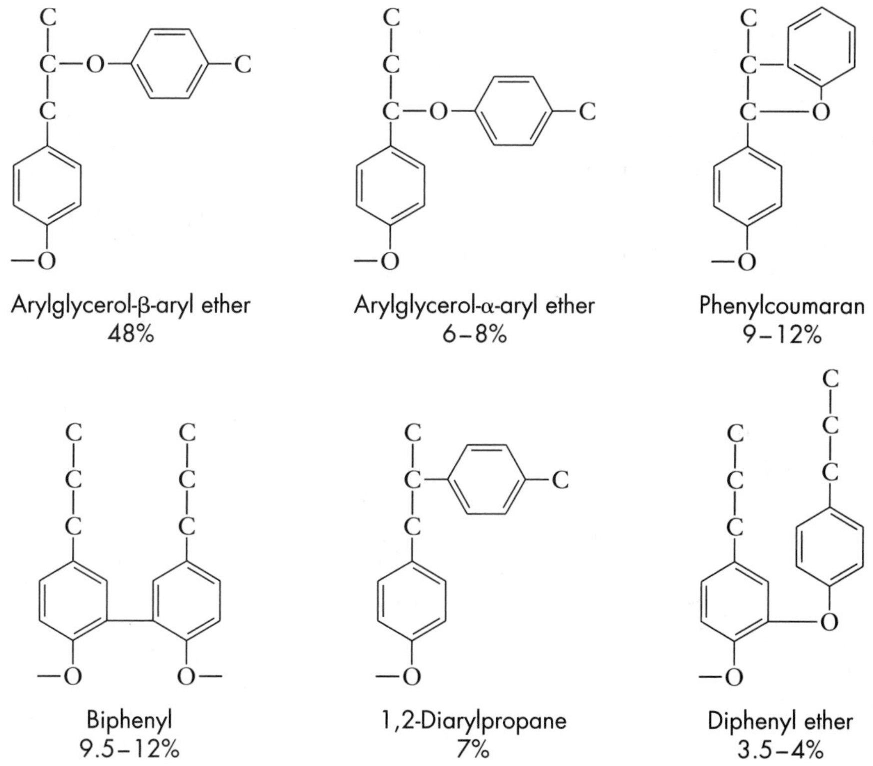

Figure 10.8

Major types of bonds between monomeric phenylpropane units in a softwood (spruce) lignin. See Betts, W.B., et al. (1991), Biosynthesis and structure of lignocellulose, in Betts, W.B. (ed.), *Biodegradation: Natural and Synthetic Materials*, pp. 140–155 (Springer-Verlag).

represent phases in cellulose synthesis and localization within the original protoplast. As long as the native structure of the plant cell wall is preserved, the lignin in it prevents degradative enzymes from reaching the cellulose microfibrils.

DEGRADATION OF LIGNOCELLULOSE BY FUNGI AND BACTERIA

Because plant tissues are the main repository of organic matter in the biosphere, it is not surprising that the ability to decompose and obtain carbon and energy from lignin, cellulose, and hemicellulose is widespread among fungi and bacteria.

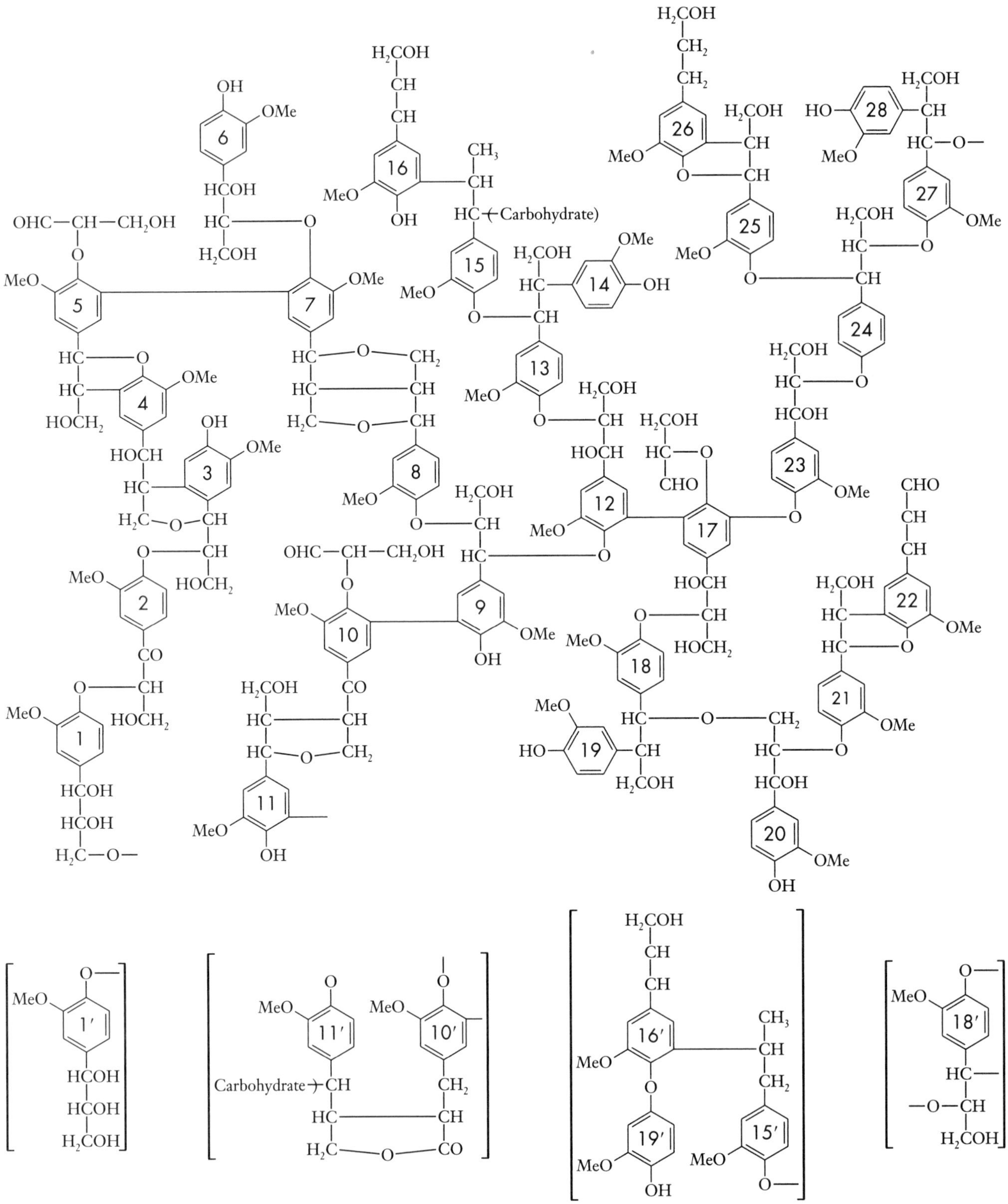

Figure 10.9

A simplified qualitative model of softwood lignin. Additional units that may be found within lignin are indicated in parentheses in the lower part of the figure. From Sakakibara, A. (1983), Chemical structure of lignin related mainly to degradation products, in Higuchi, T., Chang, H.-M., and Kirk, T.K. (eds.), *Recent Advances in Lignin Biodegradation Research*, pp. 12–33 (UNI).

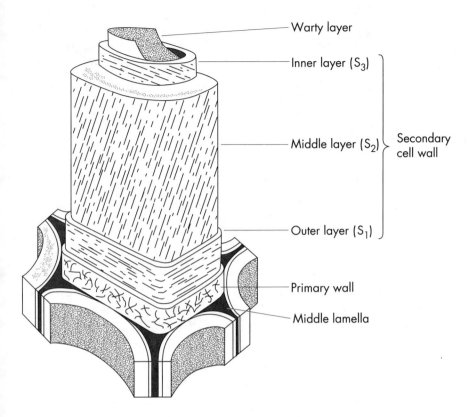

- Warty layer
- Inner layer (S₃)
- Middle layer (S₂)
- Secondary cell wall
- Outer layer (S₁)
- Primary wall
- Middle lamella

Figure 10.10
A simplified representation of the cell wall layers of a wood cell.

Degradation of Wood by Fungi

The wood-decaying fungi are the primary contributors to the degradation of wood in nature, secreting extracellular enzymes that degrade the polymeric components of the wood cell walls. They are classified as *white-rot, brown-rot,* and *soft-rot fungi* on the basis of morphological aspects of the decay. Of the more than 2000 species of wood-rotting fungi known, over 90% are white-rot fungi. Once a fungus has managed to invade the wood, it expands by growth of its hyphae in the lumina of the parenchyma and vascular cells. The hyphae penetrate from one cell to the next either through pits or through cell walls. The hyphae of some fungi grow within the middle lamellae or the secondary walls.

White-rots and brown-rots are filamentous fungi that belong to the subdivision Basidiomycetes. White-rots typically cause bleaching of the wood, giving it a fibrous or spongy consistency. The white-rot fungi degrade the lignin and polysaccharides at a similar rate, and most of them prefer hardwoods. Brown-rots preferentially degrade the polysaccharide components of the wood and cause little degradation of lignin. The decaying wood becomes brown and brittle. Most brown-rot fungi attack softwoods.

Soft-rots, a group of fungi belonging to the Ascomycetes and the Fungi imperfecti, are a major cause of decay in wood that is exposed to moisture. Soft-rots degrade both polysaccharides and lignin, but at different rates that depend on the fungal species. Soft-rots are found in both softwoods and hardwoods, and their action leads to a softening of the wood.

Action of Bacteria on Lignocellulose

Bacteria are poor degraders of lignin in the presence of oxygen, and no bacterial degradation is seen under anaerobic conditions. For example, aspen wood, in which the lignin was selectively labeled biosynthetically with ^{14}C, was not significantly degraded during 6 months of incubation in various soils. However, many bacteria produce cellulases as well as enzymes capable of cleaving the various bonds in hemicelluloses. Once these polysaccharides are exposed by fungal delignification, further degradation is accomplished by fungi and bacteria together.

Degradation of Lignin

We have seen that lignin is a complex polymer of irregular structure, formed by free-radical-mediated condensation reactions and lacking in readily hydrolyzable bonds (see Figure 10.9). The degradation of this extraordinary polymer likewise exploits free radical chemistry. The mechanism has been most extensively studied in the white-rot fungus *Phanerochaete chrysosporium*.

Wood-decaying fungi are generally mesophilic, with an optimal temperature for growth between 20 and 30°C. However, *P. chrysosporium*, which grows in environments such as self-heating piles of wood chips, has an optimal temperature for growth of 40°C and continues to grow at up to 50°C. Lignin degradation by *P. chrysosporium* in culture is triggered by nitrogen limitation. Such nutrient limitation mimics the situation in nature, because wood is low in organic nitrogen compounds. *P. chrysosporium* is an exceptionally efficient lignin decomposer capable of degrading the lignin in wood pulp at a rate that approaches 3 g of lignin per gram of fungal cell protein per day.

Lignin Peroxidase

A breakthrough in the study of lignin degradation was the discovery in 1983 of *P. chrysosporium* lignin peroxidase. This hydrogen-peroxide-requiring extracellular enzyme with an acidic pH optimum catalyzes the oxidative breakdown of lignin and of model compounds designed to represent portions of the lignin structure. Lignin peroxidase catalyzes the cleavage of the arylpropane side chains, ether bond cleavage, aromatic ring opening, and hydroxylation. All of these reactions can be adequately explained by a mechanism that begins with the catalyzed abstraction of an electron from aromatic nuclei in lignin to form unstable cation radicals. Subsequently, various products are formed by nonenzymatic reaction of the radical cations with water, other nucleophiles, and oxygen (Figures 10.11 and 10.12).

The catalytic cycle of lignin peroxidase explains its ability to catalyze 1-electron oxidations (Figure 10.13). The native enzyme contains a protoporphyrin prosthetic group with a high-spin ferric iron, Fe(III). Oxidation by hydrogen peroxide removes 2 electrons and converts the prosthetic group of the enzyme to an oxo-iron(IV) porphyrin radical cation, an enzyme form designated compound I. A 1-electron reduction by abstraction of an electron from a donor molecule (such as an aromatic nucleus) produces an enzyme with an oxo-iron(IV) porphyrin, a form designated compound II. A second 1-electron reduction completes the catalytic cycle by regenerating the native ferric enzyme.

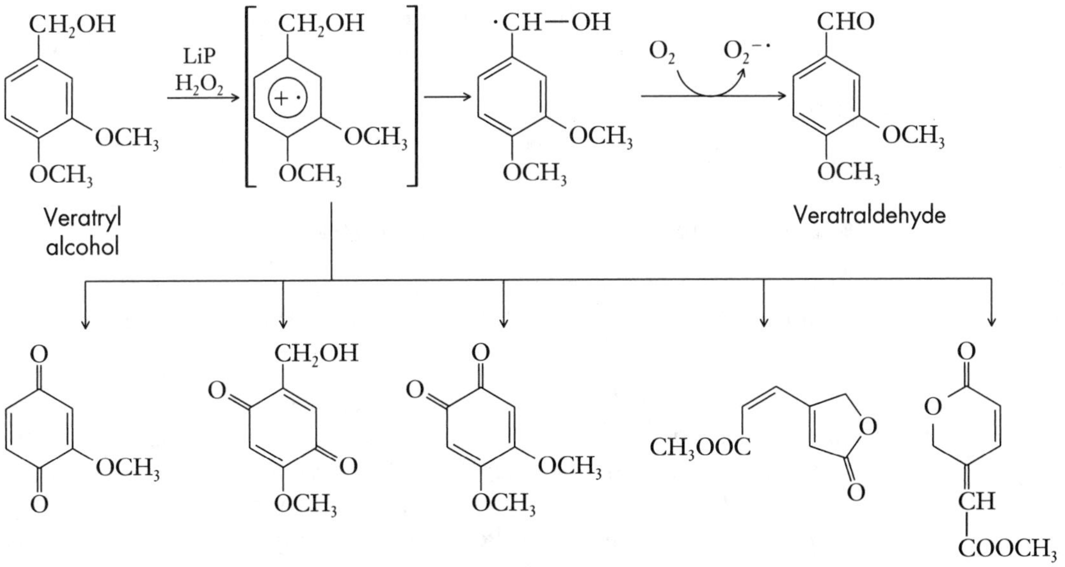

Figure 10.11
Oxidation of veratryl alcohol by lignin peroxidase (LiP) in the presence of H_2O_2—a model for lignin degradation. Veratryl alcohol is synthesized by various white-rot fungi and may be the normal inducer of the ligninolytic system. Veratryl alcohol is also a substrate for LiP. It is oxidized primarily to veratraldehyde, but also to a variety of other products, as illustrated here.

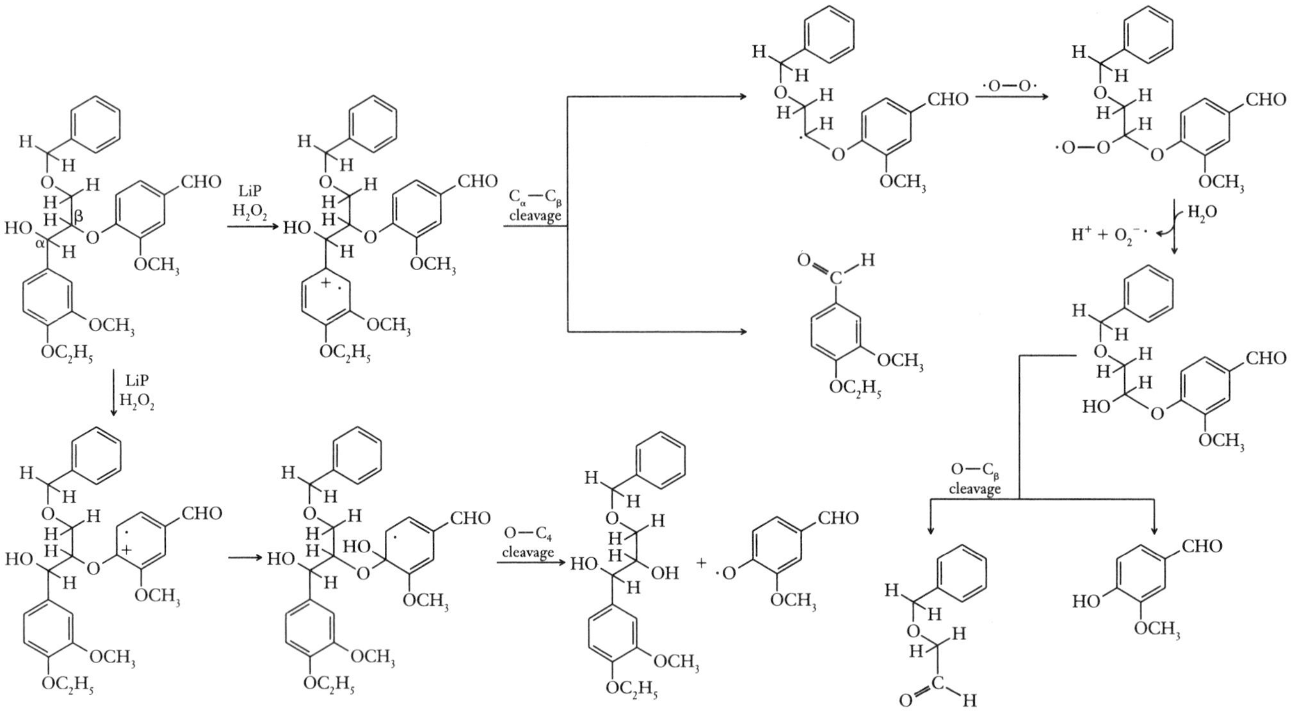

Figure 10.12
Pathways and products of degradation of a model compound for a component of lignin substructure, 4-ethoxy-3-methoxyphenylglycerol-β-vanillin-γ-benzyl diether, by *P. chrysosporium* lignin peroxidase (LiP). See Higuchi, T. (1990), *Wood Sci. Technol.* 24:23–63.

Other Enzymes with Roles in Lignin Breakdown

In addition to lignin peroxidase, ligninolytic cultures of *P. chrysosporium* produce other enzymes that function either in the breakdown of lignin or in modification of the breakdown products.

Mn(II)-dependent Peroxidases. Successful degradation of lignin requires attacks on both the nonphenolic and the phenolic lignin components. The extracellular Mn(II)-dependent peroxidases oxidize phenolic components of lignin (an example is shown in Figure 10.14), but they cannot oxidize the nonphenolic substrates of lignin peroxidase such as veratryl alcohol or the nonphenolic model compounds of lignin substructure. The prosthetic group of this family of isoenzymes, a proto-heme IX with high-spin ferric iron, is the same as that of lignin peroxidase. However, in addition to H_2O_2, these enzymes require Mn(II) and an organic acid such as malonate or oxalate. Mn(II) serves as a reductant for compound II (see Figure 10.13). The resulting Mn(III) is chelated by the organic acid, diffuses into the lignin, and oxidizes phenolic moieties.

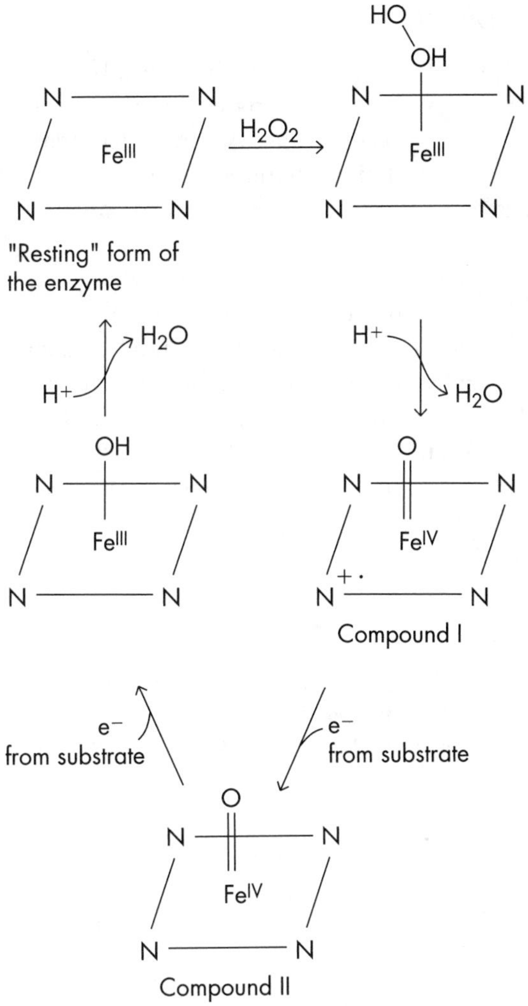

Figure 10.13
The catalytic cycle of lignin peroxidase. Compound I is an iron(IV) porphyrin π cation radical, where one of the electrons has been removed from the iron and one from the porphyrin ligand.

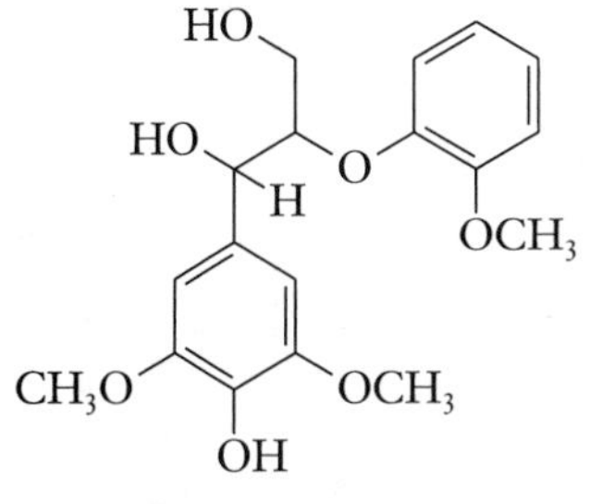

Quinone Reductases. Quinones are among the products of lignin degradation by the peroxidases, and *P. chrysosporium* produces both intracellular and extracellular quinone reductases. The extracellular cellobiose-quinone oxidoreductase is active only in the presence of cellulose. This enzyme uses cellobiose as a hydrogen donor to reduce quinones to hydroquinones. The intracellular quinone reductases use NAD(P)H as cofactor. The hydroquinones are rapidly metabolized by the fungus. Perhaps the role of the quinone reductases is to remove products of lignin degradation that might otherwise rapidly repolymerize.

Syringylglycerol-β-guaiacyl ether

Figure 10.14
A model compound for study of the degradation of the phenolic β-O-4 components of lignin such as the one shown in Figure 10.12.

Metabolism of Lignin

The major reactions leading to lignin biodegradation are summarized in Figure 10.15. Degradation is initiated by lignin peroxidase and the phenol oxidases, which catalyze the oxidative degradation of the polymer to oligolignols and thence to mono- and dilignols. Mediators produced by the action of these enzymes may also contribute to the oxidation of lignin.

Mediators are reactive low-molecular-weight species, such as chelated Mn(III) derived from the Mn(II)-dependent peroxidase, and the veratryl alcohol cation radical produced by lignin peroxidase. They may act as diffusible 1-electron oxidants, capable of penetrating and reacting within internal regions of the lignin polymer. Further oxidative cleavage of mono-, di-, and trilignols catalyzed by the peroxidases leads to the production of an array of low-molecular-weight compounds: C_1, C_2,

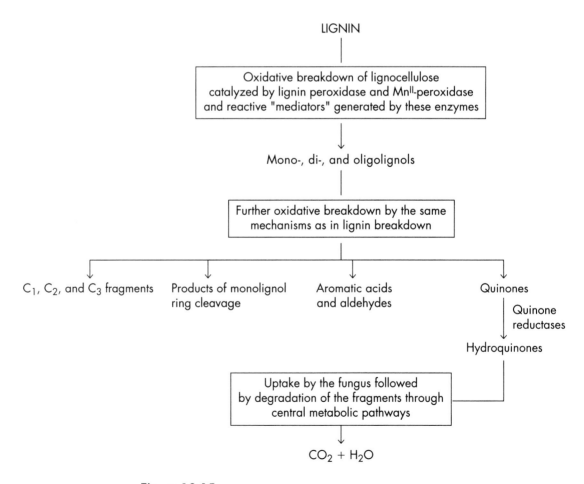

Figure 10.15
Schematic representation of the major steps in lignin degradation by ligninolytic cultures of the fungus *Phanerochaete chrysosporium*.

and C_3 fragments derived from the propane side chains of the phenyl-propane building blocks, aromatic acids and aromatic aldehydes, products of aromatic ring cleavage, and quinones. Metabolism of these compounds by combinations of reduction and oxidation reactions leads to their complete degradation to carbon dioxide and water. The C_2 and C_3 fragments serve as substrates for *glyoxal oxidase*. Their degradation by this extracellular enzyme is accompanied by the generation of H_2O_2, which is needed for the peroxidase-mediated depolymerization of lignin. Aromatic acids (such as vanillic and syringic acid) derived from lignin breakdown are taken up by the mycelium and degraded further by dioxygenases that function in the metabolism of simple aromatic acids. Quinones are rapidly reduced to hydroquinones by cellobiose-quinone oxidoreductase, with the concomitant formation of cellobionic acid. Hydroquinones are rapidly taken up by the fungus and degraded.

In the initial steps of lignin breakdown, the various lignols could undergo a reverse reaction, peroxidase-induced polymerization, to give a lignin polymer of "unnatural" structure. Such reactions must take place to some extent. Presumably, the rapid uptake, reduction, and degradation of smaller fragments by the fungal cells ensures that the equilibrium strongly favors degradation.

Degradation of Cellulose

Hundreds of species of fungi and bacteria are able to degrade cellulose. These organisms include aerobes and anaerobes, mesophiles and thermophiles. They are both widespread and abundant in the natural environment. However, although many microorganisms can grow on cellulose or produce enzymes that can degrade amorphous cellulose, relatively few produce the entire complement of extracellular cellulases able to degrade crystalline cellulose *in vitro*. Among the latter organisms, the most extensively studied sources of cellulolytic enzymes have been the fungi *Trichoderma* and *Phanerochaete* and the bacteria *Cellulomonas* (an aerobe) and *Clostridium thermocellum* (an anaerobe).

A Rotting Cartridge Belt

Scientific study of the degradation of cellulose became a practical necessity during World War II, when the clothing and equipment of American military units stationed in the jungles of the South Pacific were found to rot at an alarming rate. All cotton gear—tents, uniforms, webbing, and knapsacks—deteriorated rapidly. Replenishment of this equipment usurped most of the cargo space needed for military supplies. It was easy to see that the damage was being done by fungi, but the particular culprits had to be identified.

In 1944, the Army set up a basic research program at the U.S. Army Natick Development Center in Massachusetts to determine the nature of the deterioration processes, to identify the organisms responsible and their mechanism of action, and to develop methods of control that did not require the use of fungicides. Because cotton consists almost entirely of cellulose, a central focus of this effort was to develop an understanding of cellulose degradation. Over 10,000 strains of fungi in soil samples or damaged materials were shipped to Natick for study. The screening of these microorganisms was laborious. Thousands of microorganisms were grown on tens of thousands of textile strips, and each strip was tested for its loss of tensile strength. Those fungi that were found most active by this test were then grown in shake flasks with cellulose (ground filter paper) as a substrate and were compared in terms of their ability to produce the extracellular enzymes responsible for the degradation of cotton.

This screen uncovered a powerful cellulose-destroying strain of *Trichoderma viride*, which was isolated from a rotting cartridge belt found in New Guinea. The culture filtrates from this strain were the most active of all tested in their ability to convert crystalline cellulose to glucose. This discovery focused attention on *Trichoderma* strains. To this day, mutants derived from *Trichoderma reesei* QM6a (formerly *Trichoderma viride* QM6a), a strain isolated from a Bougainville Island cotton duck shelter, are among the most efficient known producers of extracellular cellulases.

The studies at Natick also revealed that the cellulolytic activities of a particular organism in the laboratory were not predictive of the importance of the organism to the destruction of cotton in the field. Samples of deteriorated cotton materials were sent to Natick from 24 military bases in South Pacific and Southwest Pacific areas extending from Hawaii to New Guinea and from localities in China, Burma, and India. Among 4500 strains of fungi isolated from these exposed cotton fabrics, *Trichoderma* strains numbered 385 (8.6%) and were found in samples from most locations. Yet direct examination of many hundreds of samples of decaying cotton fabrics from all these areas did not show any *fruiting structures* of *T. viride*. This observation indicates that although *T. viride* was not *growing* on them, it was commonly present (probably as spores) on these fabrics. There is no evidence that it played any role in the deterioration of these materials in the field. The ecological factors that govern how much the various cellulolytic fungi contribute to the rotting of cotton fabrics in the field are still not understood.

Trichoderma *Cellulases*

The cellulolytic system of *Trichoderma reesei* is typical of such systems among the filamentous fungi. *T. reesei* produces three types of cellulolytic enzymes that cooperate in the degradation of cellulose: endoglu-

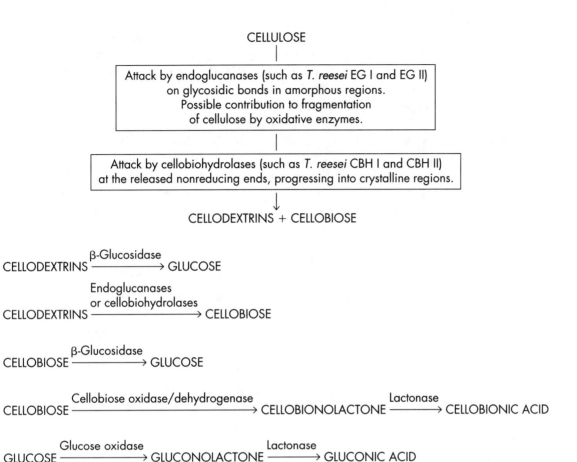

Figure 10.16
Proposed mechanism of lignin degradation by fungi. See Coughlan, M.P. (1990), in Fogarty, W.M., and Kelly, C.T. (eds.), *Microbial Enzymes and Biotechnology*, 2d ed., pp. 1–36 (Elsevier).

canases, cellobiohydrolases, and β-glucosidases. The endoglucanases are believed to hydrolyze internal bonds in disordered regions along cellulose fibers. The ends generated in this manner are then attacked by the cellobiohydrolases, with release of the disaccharide cellobiose. As the hydrolysis proceeds, the cellobiohydrolases also, apparently, begin to disrupt chain-chain interactions in the crystalline regions of the cellulose fiber. Finally, cellobiose is hydrolyzed to glucose by β-glucosidase (Figure 10.16).

Relationship of Structure to Function in the Fungal Cellulases

Four genes, coding for the two endoglucanases (EG I and EG II) and the two cellobiohydrolases (CBH I and CBH II), have been isolated from *T. reesei*. Each of the enzymes encoded by these genes is composed of two distinct domains joined by an extended, flexible "hinge" region (Figure 10.17). Note that the order of the domains in CBH I and EG I is the reverse of that seen in CBH II and EG II. The larger domain, composed of about 500 amino acids, includes the active site. The hinge region, a heavily glycosylated sequence 24–44 amino acids long and rich in prolyl and hydroxyamino acid residues, joins the catalytic domain to a 33-residue cellulose-binding domain.

The flexible hinge enables the cellulose-binding domain to attach the enzymes to the cellulose fiber with little restriction on the interaction of the catalytic domain with the cellulose. In this manner, the cellulose-binding domain contributes to maintaining a high concentra-

	Number of Amino Acid Residues			
	1. Signal sequence	2. Catalytic domain	3. Hinge ("linker")	4. Cellulose-binding domain
CBH I	17	425	26	33
CBH II	24	385	44	33
EG I	22	363	29	33
EG II	21	327	34	33

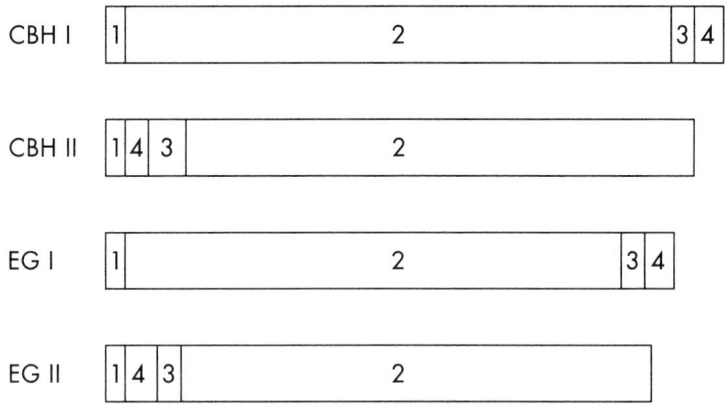

Figure 10.17
The table shows the lengths of the various domains of the preprotein forms of *Trichoderma reesei* cellobiohydrolases I and II (CBH I and CBH II) and endoglucanases I and II (EG I and EG II). The schematic shows a representation of those lengths and of the domains' relative location. Note that the order of the cellulose-binding domain, the hinge ("linker"), and the catalytic domain in CBH I and EG I is opposite of that in CBH II and EG II. Data from Gilkes, N.R., et al. (1991), *Microbiol. Rev.* 55:303–315; and Abuja, P.M., et al. (1988), *Eur. Biophys. J.* 15:339–342.

tion of cellulase on the substrate. Studies on the three-dimensional structure of the cellulose-binding domain indicate that it interacts preferentially with the crystalline regions in cellulose. The catalytic domain, on the other hand, has a high affinity for amorphous regions and a low affinity for the crystalline regions. In the intact enzyme, the low affinity of the catalytic domain for the crystalline regions is offset by the high affinity of the binding domain, with the result that many more intact enzyme molecules are bound in the crystalline regions, increasing the probability of glycosidic bond cleavage in these regions.

When the small terminal domain is removed from CBH I and CBH II by proteolysis, both enzymes' affinity for and activity toward insoluble cellulose is significantly reduced, whereas their activity toward small soluble substrates is unchanged. This suggests that the cellulose-binding domain contributes both to binding the cellulase to the cellulose surface and to altering the susceptibility of the substrate to degradation. The interaction of the cellulose-binding domain with crystalline cellulose may lead to local destabilization of interactions between chains, a disruption that in turn may lead to enhanced activity of the intact enzyme toward crystalline cellulose.

The classical approach to the study of enzyme specificity is to examine the enzymes' action on small synthetic substrates. As illustrated in Figure 10.18, however, such small substrates do not differentiate clearly between the specificities of exocellulases and endocellulases. Comparative biochemical studies of many cellulolytic enzymes, combined with examination of high-resolution crystal structures of the catalytic domain of *T. reesei* CBH II and a bacterial cellulase, have provided more specific information about the molecular determinants of exo- versus endocellulolytic cleavage. These studies indicate that in the exocellulases, the active site is in an enclosed tunnel through which a cellulose chain threads. Two aspartyl residues, believed to be the catalytic residues, are located in the middle of the tunnel. The architecture of the active site is such that productive substrate binding is predicted to lead to the release of disaccharide units (cellobiose). In contrast, the active site in the endocellulases lies in an open groove where there is no restriction on the position of cleavage of the β-1,4-glycosidic linkage in a cellulose chain.

Bacterial Cellulases

The ability to degrade crystalline cellulose is widespread among both aerobic and anaerobic bacteria. The bacteria either secrete their cellulases as soluble extracellular enzymes or assemble them into large complexes, called *cellulosomes*, that are attached to the bacterial cell surface. The extracellular endo- and exoglucanases produced by a variety of bacteria (for example, the soil bacterium *Cellulomonas fimi*) show the same general structural features as those of *T. reesei*.

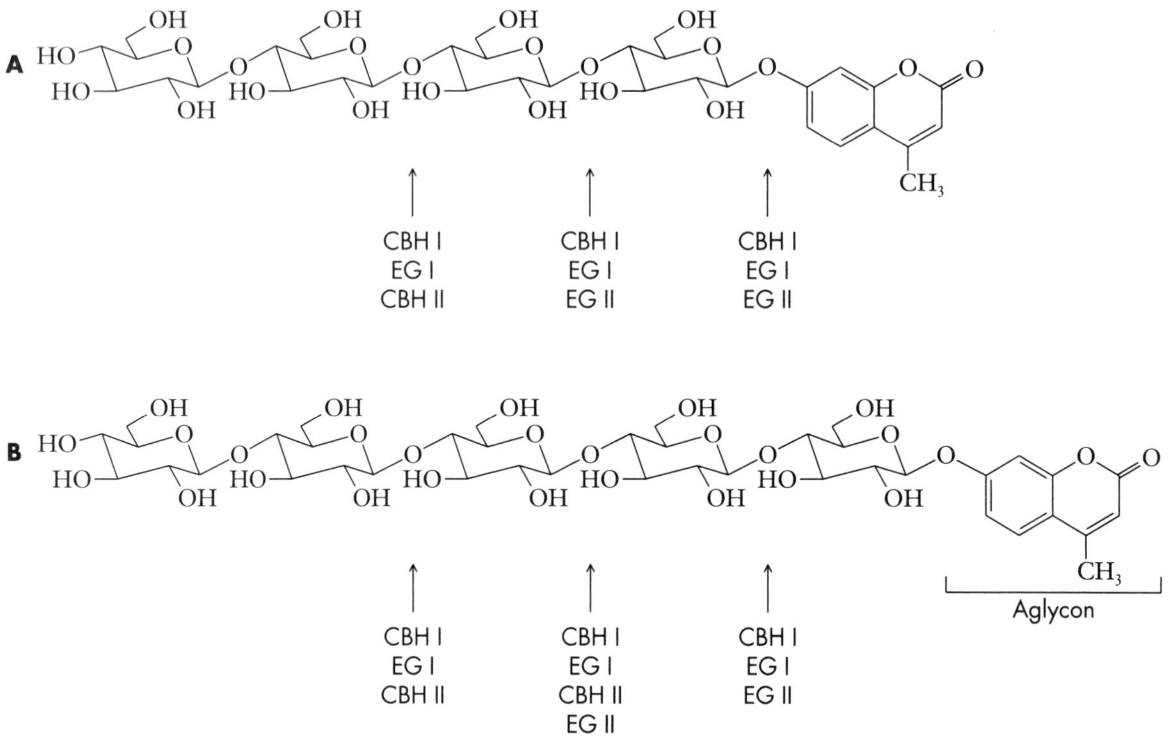

Figure 10.18
Cleavage of β-glycosidic bonds in cellotetraose (**A**) and cellopentaose (**B**) by *T. reesei* endoglucanases (EG I and EG II) and cellobiohydrolases (CBH I and CBH II). For each of the substrates, abbreviations below the arrows specify which enzymes cleaved a particular bond. The aglycon, 4′-methylumbelliferyl-, was chosen to facilitate detection of substrates and degradation products by detecting the released 4′-methylumbelliferone (4-methyl-7-hydroxycoumarin) via absorption or fluorescence spectroscopy. Data from Clayessens, M., and Henrissat, B. (1992), *Prot. Sci.* 1:1293–1297.

The best-studied cellulosome, that of the thermophilic anaerobic bacterium *Clostridium thermocellum*, is made up of 14–18 polypeptides and has a mass of about 2×10^6 daltons. This cellulosome contains several endoglucanases, as well as exoglucanases and xylanases. Cellulose-binding domains are absent from *C. thermocellum* glucanases. The cellulosome also contains polypeptides with no enzymatic activity that function in the organization of the cellulosome or help attach it to the cell surface and to cellulose. The latter ability bestows a competitive advantage on this bacterium in environments where numerous organisms may compete for the soluble products of cellulose degradation.

Architecture of the Cellulosome

The cellulosome is a highly active cellulase system capable of completely degrading cellulose. It has the unusual property of being more active against crystalline cellulose than against disordered (amorphous) cellulose. During the exponential phase of bacterial growth, the cellulosomes are primarily found attached to the cell surface. As the cells enter their stationary phase, a large percentage of these complexes is released into the medium.

A detailed model for the cell-bound cellulosome, compatible with what is currently known about the amino acid sequence of many of the cellulosome components and with the results of biochemical studies, is shown in Figure 10.19.

Cells of *C. thermocellum* are bounded by three layers: the cytoplasmic membrane, the peptidoglycan layer, and a surface protein layer (S-layer). CipA, a protein associated with the outer surface is the scaffold for the assembly of the other cellulosome polypeptides. The amino acid sequence of CipA contains nine domains, each of about 146 residues, separated by Pro- and Thr-rich segments of 17 to 19 residues. These domains recognize a 22-residue duplicated repeat present in each of the catalytic polypeptides (endoglucanases, cellobiohydrolases, β-glucosidases, and xylanases) of the cellulosome. Through their interactions with CipA, the catalytic components of the cellulosome are assembled into a regular array on this polypeptide. CipA and certain of the catalytic components of the cellulosome also possess cellulose-binding domains that attach the cellulosome to the surface of crystalline cellulose. It seems likely that the spacing of the hydrolytic enzyme sites on

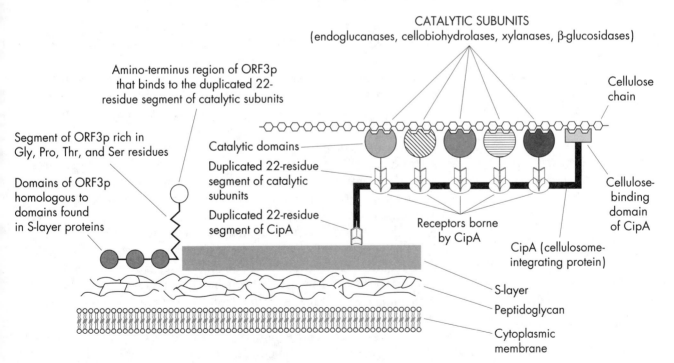

Figure 10.19

A schematic model proposed for the structure of the *Clostridium thermocellum* cell-bound cellulosome. The details of this model are discussed in the text. Based on Fujino, T., Béguin, P., and Aubert, J.-P. (1993). Organization of a *Clostridium thermocellum* gene cluster encoding the cellulosomal scaffolding protein CipA and a protein possibly involved in attachment of the cellulosome to the cell surface, *J. Bacteriol.* 175:1891–1899, Fig. 6.

the cellulosome is such as to allow near-simultaneous multisite cleavage of bonds in the highly ordered crystalline cellulose substrate. This would account for the cellulosome's preference for crystalline over amorphous cellulose.

Additional sites for attachment of degradative enzymes may be provided by a polypeptide designated ORF3p located on the *C. thermocellum* outer surface. The carboxyl-terminus of ORF3p has three sequence repeats (indicated by spheres in Figure 10.19), with strong homology to sequences in S-layer proteins. These repeats are thought to integrate ORF3p into the surface protein layer. A long region of sequence rich in Gly, Pro, Thr, and Ser residues links the amino- and carboxyl-terminal domains of ORF3p. A prediction from the amino acid sequence is that this central region of ORF3p will assume an extended conformation. In analogy to the situation seen with CipA, the amino-terminus of ORF3p also contains a domain that recognizes the 22-residue duplicated repeats in each of the catalytic polypeptides. Thus ORF3p may function in anchoring individual polysaccharide-degrading enzymes to the cell surface.

Degradation of Hemicelluloses

Most studies of enzymatic hemicellulose degradation have focused on xylans, to the virtual exclusion of galactoglucomannans and glucomannans. Xylans are the major hemicelluloses in wood from angiosperms (flowering plants), where they account for 15–30% of the total dry weight. In the gymnosperms (the ferns, the conifers and their allies), however, xylans contribute only 7–12% of the total dry weight.

Because of the high content of xylans in lignocellulose, economically feasible conversion of biomass to alcohol requires a practicable method for their fermentation. The issue of xylan degradation arises in two other significant contexts as well. Paper manufacture generates effluents from wood pulping and pulp processing that contain large amounts of xylans. These wastes frequently pollute streams and rivers. Agricultural residues also contain appreciable amounts of xylans.

The structure of xylans is complex, and their complete biodegradation requires the concerted action of several enzymes. Bacteria and fungi isolated from many different habitats contain various xylanases. The domain type of organization described for *T. reesei* cellulases is seen in xylanases as well.

The complete conversion of a glucuronoxylan into its building blocks requires the combined actions of an endo-β-1,4-xylanase and a β-xylosidase, as well as an α-glucuronidase, an α-L-arabinofuranosidase, and an acetylxylan esterase (Figure 10.20). Examples of microorganisms that are able both to degrade xylan and to convert the resulting xylose to ethanol and other products are given in Table 10.4.

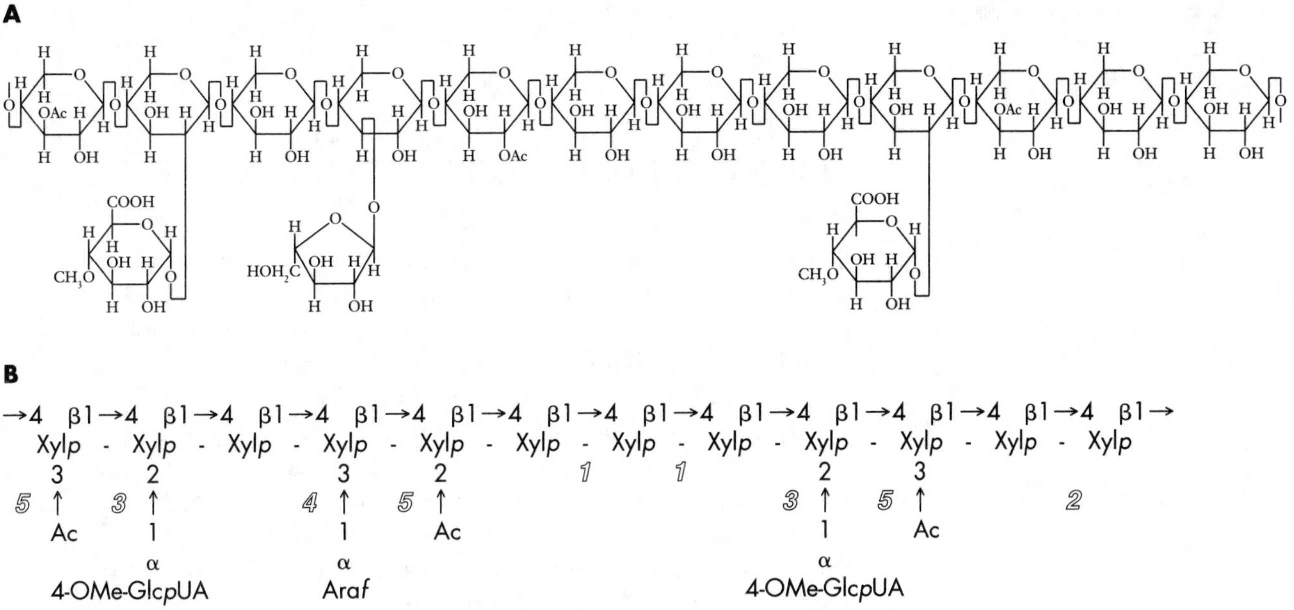

A

B

→4 β1→4 β1→4 β1→4 β1→4 β1→4 β1→4 β1→4 β1→4 β1→4 β1→4 β1→4 β1→
 Xylp - Xylp - Xylp - Xylp - Xylp - Xylp - Xylp - Xylp - Xylp - Xylp - Xylp - Xylp
  3   2   3   2   2   3
5 ↑   3 ↑   4 ↑   5 ↑   3 ↑   5 ↑   2
 Ac   1   1   Ac   1   Ac
       α   α   α
4-OMe-GlcpUA   Araf   4-OMe-GlcpUA

Figure 10.20
Sites of attack on a fragment of an arabinoglucuronoxylan by microbial xylanolytic enzymes. **A.** Chemical structure of arabinoglucuronoxylan. **B.** Representation of the structure using the conventional abbreviations defined in the legend to Figure 10.2. Outlined numbers indicate which type of enzyme cleaves a particular bond: 1, endo-β-1,4-xylanase; 2, β-xylosidase; 3, α-glucuronidase; 4, α-L-arabinofuranosidase; 5, acetylxylan esterase.

TABLE 10.4

Various Bacteria and Fungi Able to Ferment Xylan to Ethanol

Microorganism	Products
Bacteria	
Clostridium thermocellum	Ethanol, acetic acid, lactic acid
Clostridium thermohydrosulfuricum	Ethanol
Clostridium thermosaccharolyticum	Ethanol
Thermoanaerobacter ethanolicus	Ethanol, acetic acid, lactic acid
Thermoanaerobium brockii	Ethanol, acetic acid, lactic acid
Thermobacteroides acetoethylicus	Ethanol, acetic acid, lactic acid
Fungi	
Neurospora crassa	Ethanol
Pichia stipitis	Ethanol

THE PROMISE OF ENZYMATIC LIGNOCELLULOSE BIODEGRADATION

The first step in the utilization of biomass for fuel, chemical feedstocks, or food is to release the carbohydrate and aromatic building blocks locked up in the structure of lignocellulose. This can be achieved by chemical means, but only at the expense of producing undesirable by-products and chemical waste at the same time.

Fortunately, many fungi and bacteria possess enzyme systems that are capable of degrading the polymers of lignocellulose and utilizing the breakdown products as sources of cell carbon and energy. Large amounts of these enzymes are readily available through cloning and overexpression in bacterial and fungal host cells. Organisms that over-produce particular enzymes or that produce a mix of enzymes appropriate for a particular application are also available. For example, conventional mutation and strain selection has created strains of *T. reesei* that produce up to 40 g per liter of extracellular protein, mostly components of the cellulase system, in relative proportions similar to those of the parent wild-type strain; and genetic engineering has been used to produce *T. reesei* strains with modified amounts and proportions of EG I, EG II, CBH I, and II.

The availability of cloned genes for lignin peroxidases, glucanases, and xylanases from many different microorganisms opens the way to genetic engineering of these enzymes in order to improve their functional and physical properties—for example, to minimize product inhibition and increase the enzymes' thermostability. Such abilities will in turn make it possible to convert more of the lignocellulose into useful products and to do so faster and more economically.

SUMMARY

Wood contains three types of biopolymers: cellulose, hemicelluloses, and lignin. These polymers are the major storage form both of energy trapped by photosynthesis and of organic matter in the biosphere. Cellulose is a linear polymer of thousands of glucose molecules. The basic repeating unit is cellobiose, glucose-β-1,4-glucose. Hemicelluloses are highly branched heteropolysaccharides of fewer than 200 sugar residues. The sugars in hemicellulose include pentoses, hexoses, and uronic acids. The sugars in hemicellulose are variously modified by acetylation or methylation. Lignin is a very high-molecular-weight irregular polymer made up of oxygenated C_9 (phenylpropane) units linked in many different ways. Lignin is formed by peroxidase-catalyzed generation and subsequent reaction of phenoxy radicals of (methoxy-) substituted *p*-hydroxycinnamyl alcohols: *p*-coumaryl, coniferyl, and sinapyl alcohols.

White-rot, brown-rot, and soft-rot fungi are the primary wood degraders in nature. They secrete extracellular enzymes—lignin peroxidases—that mediate lignin depolymerization. Both fungi and bacteria produce extracellular enzymes that degrade cellulose and hemicelluloses. The enzymatic machinery for cellulose and hemicellulose degradation by certain clostridia is packaged in intricate complexes known as cellulosomes that are bound to the cell surfaces of these bacteria.

The genes for lignin peroxidases, cellulases, and various hemicellulose-degrading enzymes have been cloned and the enzymes expressed in fungi and *E. coli*. Moreover, genetically engineered strains producing modified amounts and proportions of the degradative enzymes have been constructed. These advances hold promise for the efficient utilization of lignocellulose as an abundant source of sugar substrates for fermentation processes such as the production of ethanol.

SELECTED REFERENCES

General

Siu, R.G.H., 1951. *Microbial Decomposition of Cellulose.* Reinhold.

Sjöström, E., 1981. *Wood Chemistry. Fundamentals and Applications.* Academic Press.

Fengel, D., and Wegener, G., 1984. *Wood: Chemistry, Ultrastructure, Reactions.* Walter de Gruyter.

Hartley, B.S., Broda, P.M.A., and Senior, P.J., 1987. Technology in the 1990s: Utilization of lignocellulosic wastes. *Phil. Trans. Roy. Soc. Lond. A* 321:403–568.

Fry, S.C., 1988. *The Growing Plant Cell Wall: Chemical and Metabolic Analysis.* John Wiley.

Rayner, A.D.M., and Boddy, L., 1988. *Fungal Decomposition of Wood: Its Biology and Ecology.* Wiley.

Wood, W.A., and Kellogg, S.T. (eds.), 1988. Biomass. Part A: Cellulose and Hemicellulose. *Methods in Enzymology*, vol. 160. Academic Press.

Wood, W.A., and Kellogg, S.T. (eds.), 1988. Biomass. *Part B: Lignin, Pectin, and Chitin. Methods in Enzymology*, vol. 161. Academic Press.

Betts, W.B. (ed.), 1991. *Biodegradation: Natural and Synthetic Materials.* Springer-Verlag.

Lignin

Kirk, K.T., and Farrell, R.L., 1987. Enzymatic "combustion": The microbial degradation of lignin. *Ann. Rev. Microbiol.* 41:465–505.

Higuchi, T., 1990. Lignin biochemistry: Biosynthesis and degradation. *Wood Sci. Technol.* 24:23–63.

Schoemaker, H.E., 1990. On the chemistry of lignin biodegradation. *Recl. Trav. Chim. Pays-Bas* 109:255–272.

Wariishi, H., Valli, K., and Gold, M.H., 1992. Manganese (II) oxidation by manganese peroxidase from the basidiomycete *Phanerochaete chrysosporium*: Kinetic mechanism and role of chelators. *J. Biol. Chem.* 267:23688–23695.

Poulos, T.L., and Fenna, R.E., 1994. Peroxidases: Structure, function, and engineering. In Sigel, H., and Sigel., A. (eds.), *Metal Ions in Biological Systems*, vol. 30, pp. 25–75. Marcel Dekker.

Cellulose

Coughlan, M.P., 1985. The properties of fungal and bacterial cellulases with comment on their production and application. *Biotechnol. Genet. Eng. Revs* 3:39–109.

Coughlan, M.P., 1990. Cellulose degradation by fungi. In Fogarty, W.M., and Kelly, C.T. (eds.), *Microbial Enzymes and Biotechnology*, 2d ed., pp. 1–36. Elsevier Applied Science.

Stutzenberger, F., 1990. Bacterial cellulases. In Fogarty, W.M., and Kelly, C.T. (eds.), *Microbial Enzymes and Biotechnology*, 2d ed., pp. 37–70. Elsevier Applied Science.

Gilkes, N.R., Henrissat, B., Kilburn, D.G., Miller, R.C., Jr., and Warren, R.A.J., 1991. Domains in microbial β-1,4-glycanases: Sequence conservation, function, and enzyme families. *Microbiol. Rev.* 55:303–315.

Claeyssens, M., and Henrissat, B., 1992. Specificity mapping of cellulolytic enzymes: Classification into families of structurally related proteins confirmed by biochemical analysis. *Prot. Sci.* 1:1293–1297.

Stålberg, J., Johansson, G., and Pettersson, G., 1991. A new model for enzymatic hydrolysis of cellulose based on the two-domain structure of cellobiohydrolase I. *Bio/Technology* 9:286–290.

Teeri, T.T., Pentillä, M., Keränen, S., Nevalainen, H., and Knowles, J.K.C., 1992. Structure, function, and genetics of cellulases. In Finkelstein, D.B., and Ball, C. (eds.), *Biotechnology of Filamentous Fungi: Technology and Products*, pp. 417–445. Butterworth-Heinemann.

Cellulosome

Lamed, R., and Bayer, E.A., 1988. The cellulosome concept: Exocellular/extracellular enzyme reactor centers for efficient binding and cellulolysis. In Aubert, J.-P., Béguin, P., and Millet, J. (eds.), *Biochemistry and Genetics of Cellulose Degradation*, FEMS Symposium No. 43, pp. 101–116. Academic Press.

Wu, J.H.D., 1993. *Clostridium thermocellum* cellulosome. New mechanistic concept for cellulose degradation. In Himmel, M.E., and Georgiou, G. (eds.), *Biocatalyst Design for Stability and Specificity*, ACS Symposium Series 516, pp. 251–264. American Chemical Society.

Wang, W.K., Kruus, K., and Wu, J.H.D., 1993. Cloning and DNA sequence of the gene coding for *Clostridium thermocellum* cellulase S_s (CelS), a major cellulosome component. *J. Bacteriol.* 175:1293–1302.

Salamitou, S., Raynaud, O., Lemaire, M., Coughlan, M., Béguin, P., and Aubert, J.-P., 1994. Recognition specificity of the duplicated segments present in *Clostridium thermocellum* endoglucanase CelD and in the cellulosome-integrating protein CipA. *J. Bacteriol.* 176:2822–2827.

Salamitou, S., Lemaire, M., Fujino, T. Ohayon, H., Gounon, P., Béguin, P., and Aubert, J.-P., 1994. Subcellular localization of *Clostridium thermocellum* ORF3p, a protein carrying a receptor for the docking sequence borne by the catalytic components of the cellulosome. *J. Bacteriol.* 176:2828–2834.

Felix, C.R., and Ljungdahl, L.G., 1993. The cellulosome —the exocellular organelle of *Clostridium*. *Ann. Rev. Microbiol.* 47:791–819.

Hemicelluloses

Timell, T.E., 1964. Wood hemicelluloses: Part I. *Advan. Carbohyd. Chem.* 19:247–302.

Timell, T.E., 1964. Wood hemicelluloses: Part II. *Advan. Carbohyd. Chem.* 19:409–483.

Woodward, J., 1984. Xylanases: Functions, properties and applications. In Wiseman, A. (ed.), *Topics in Enzyme and Fermentation Biotechnology*, vol. 8, pp. 9–30. Horwood.

Biely, P., 1985. Microbial xylanolytic systems. *Tr. Biotechnol.* 3:286–290.

Wong, K.K.Y., Tan, L.U.L., and Saddler, J.N., 1988. Multiplicity of β-1,4-xylanase in microorganisms: Functions and applications. *Microbiol. Rev.* 52:305–317.

Törrönen, A., et al., 1992. Two major xylanases from *Trichoderma reesei*: Characterization of both enzymes and genes. *Bio/Technology* 10:1461–1465.

CHAPTER

11

Ethanol

Barring a legislative mandate to use biofuels or reduce fossil fuel consumption, I believe that an open energy market will be one of the major determinants, and that crude oil prices will be the near- to mid-term driving force of changes in renewable energy consumption. Directly related to crude oil price is the complacency of energy users. Interest in establishing a long-term alternative fuels policy for the nation will decline when the price of crude stabilizes at levels that consumers judge to be acceptable. But ultimately, when the economics are right, growth in consumption of renewables and consequent fuel displacement will occur. Klass, D.L. (1990), The U.S. biofuels industry, *Chemtech*, December, pp. 720–731.

Chapter 10 described the major components of plant biomass—cellulose, hemicelluloses, and lignin—and their natural pathways of biodegradation. Many view the sugars locked up in cellulose and in the hemicelluloses as an immense storehouse of renewable feedstocks for the fermentative production of fuel alcohol. This chapter begins with a discussion of the conversion of such sugars to ethanol, or ethyl alcohol, the characteristic component of alcoholic beverages, and ends with an assessment of the future impact of fermentation alcohol as a fuel.

The manufacture of alcoholic beverages by microbiological fermentation of sugars originated over 5000 years ago and remains popular today. Wine was a source of both pleasure and relative safety in the ancient world. Water was frequently impure, whereas fermented fruit juices, protected from spoilage by their high alcohol content, were generally safe to drink.

359

In microbiology, *fermentation* is defined as a metabolic process leading to the generation of ATP and in which degradation products of organic compounds serve as hydrogen donors as well as hydrogen acceptors. Oxygen is not a reactant in fermentation processes. In the words of Louis Pasteur, "[l]a fermentation est la vie sans l'air": "Fermentation is life without air." The long history of brewing and wine making has produced highly refined technologies for large-scale fermentation and for the recovery of ethyl alcohol. And in addition to its role as a beverage, ethanol can serve as a fuel and as a starting material for the manufacture of such chemicals as acetic acid, acetaldehyde, butanol, and ethylene (itself a key intermediate in the petrochemical industry).

Although the industries that currently produce fuels and organic chemicals require fossil feedstocks (petroleum and natural gas) as raw materials, the dramatic rise in the price of oil that occurred in the 1970s has prompted assessment of alternative sources. Ethanol seems a particularly promising alternative. Anhydrous ethanol was already in use as a fuel in internal combustion engines by the late nineteenth century. In fact, in an attempt to assist the mechanization of farms, the United States Congress in 1906 removed the tax on alcohol to encourage farmers to produce their own engine fuels.

With appropriate pretreatment, various forms of biomass can serve as substrates for alcohol fermentation. Since 1975, the Brazilian National Alcohol Program—the most determined effort yet to replace gasoline with alcohol—has used sucrose, obtained directly from sugar cane, as the substrate for fermentation. Over 50 billion liters of ethanol were produced during the program's first decade of operation; in 1989 about 40% of Brazil's 14 million vehicles were operating on ethanol (95% ethanol, 5% water) and 60% on a blend of 78% gasoline and 22% ethanol.

The United States is the world's second largest alcohol producer. In 1980, the government passed the Energy Security Act, which included a Biomass Energy and Alcohol Fuels Act, designed to provide loans and loan guarantees to alcohol and other biomass energy projects. The objective of this program is to encourage the addition of 10% alcohol to gasoline, creating "gasohol." Gasohol is a lead-free fuel whose combustion produces lower amounts of nitrogen oxides and carbon monoxide than the combustion of regular gasoline. During 1988 about 840 million gallons of ethanol, prepared by fermenting corn and other starch-rich grains, were sold in gasohol blends, primarily in the midwestern United States. These blends accounted for some 7% of total nationwide gasoline sales. The cost-competitiveness of ethanol in both Brazil and the United States depends on government subsidies, however.

Figure 11.1 presents a flowchart describing the conversion of various feedstocks to alcohol. In the first stage, the polymeric substrates are broken down to monosaccharides through physical, chemical, or

STAGE I: CONVERSION OF BIOMASS TO FERMENTABLE SUGARS

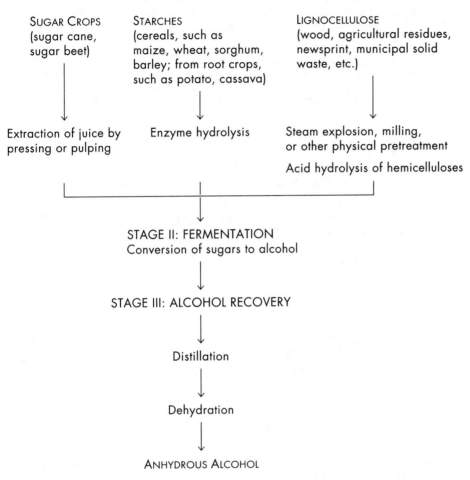

Figure 11.1
Stages in the conversion of biomass to ethanol.

enzymatic techniques, as appropriate. In the second stage, microbial (generally yeast) fermentation converts the sugars to alcohol. In the third stage, alcohol is recovered by distillation (as a constant-boiling mixture of 95.6% ethanol and 4.4% water, by volume). Further distillation procedures are needed to obtain anhydrous ethanol.

If alcohol is to be competitive with other sources of fuels and chemical feedstocks, all three stages in its production must be simple and inexpensive, regardless of the starting substrate. In this chapter, we examine relevant aspects of the biochemistry, microbiology, and technology of the first two stages: the breakdown of sugars to monosaccharides and their conversion to alcohol. Methodological improvements in those stages also have an impact on the third stage, ethanol recovery, but the techniques employed in that stage belong largely to the realm of process engineering.

STAGE I: FROM FEEDSTOCKS TO FERMENTABLE SUGARS

As indicated in Figure 11.1, in stage I, the carbohydrate-containing raw materials are pretreated in ways that make the sugars they contain readily available to microorganisms. The principal biomass substrates for the production of alcohol by microbial fermentation are sugars, starches, and cellulose. The chemical structures of sugars and starches are described in Chapter 7, and that of cellulose in Chapter 10.

Sugars

Sucrose (α-glucose-1,2-β-fructose; see Figure 7.3) is the sweetener most commonly used for human consumption. Sugar cane and sugar beet contain up to 20% sucrose by weight, the other major components being water (about 75%), cellulose (5%), and inorganic salts (about 1%). The substrate for fermentation is obtained by extraction of sucrose with water after mechanically crushing the cane or after stripping and pulping the beet. Sucrose from sugar cane (the Brazilian approach) is a particularly favorable substrate for the production of alcohol by yeast fermentation. Yeast produces the enzyme invertase in both a cytoplasmic and a secreted form, and this enzyme hydrolyzes sucrose to glucose and fructose, which are then fermented by the yeast cells. The conversion of other substrates to alcohol involves additional pretreatment. *Saccharomyces cerevisiae* and related yeasts can take up and metabolize many sugars, including glucose, fructose, galactose, mannose, maltose (glucose-α-1,4-glucose), and maltotriose.

Starches

In the United States, cornstarch is the major feedstock for the production of fuel alcohol. It consists of a water-soluble fraction called amylose (20%) and a water-insoluble higher-molecular-weight fraction called amylopectin (80%). The structures of these glucose polymers are shown in Figure 7.2. To obtain cornstarch, dry corn is milled, water is added, and the slurry (watery suspension) is sent to a cooker. Heating the slurry solubilizes the starch and makes it vulnerable to enzyme hydrolysis. In the next stage (fermentation), amylases and amyloglucosidase are added to convert the starch polymers to glucose (see Box 7.2, page 254).

Cellulose

Cellulose is obtained from lignocellulose (see Chapter 10). Pretreatment of the lignocellulose makes the cellulose more accessible to hydrolytic enzymes and begins to disrupt, at least in part, the highly crystalline structure of the cellulose fibers. Numerous raw material pretreatment processes have been designed to achieve these objectives. We will consider, as examples, processes developed by the Iotech Corporation Limited of Ottawa, Canada (the Techtrol/Iotech process) and by the United States Army Natick Laboratories (the Natick process).

In the Techtrol/Iotech process, small wood chips are charged with steam in a heated pressure vessel to about 500°F and maintained at that temperature for about 20 s, at which point the vessel is rapidly decompressed. Pressure in the vessel reaches 600 p.s.i. before release. After the explosive decompression, the cellulose in the wood is susceptible to enzyme hydrolysis.

This "steam explosion" treatment has the further effect of solubilizing the hemicellulose in the wood so that it can be washed away with water. Then the cellulose and lignin can be separated from one another in either of two ways. In one method, the lignin is extracted in high yield with methanol or with dilute sodium hydroxide before the cellulose is degraded. In the other method, the cellulose is converted to glucose by hydrolytic enzymes, while the lignin remains insoluble and is subsequently removed by filtration. The glucose solution is then fermented to alcohol.

The process developed at the Natick Laboratories consists of extensive physical disruption of the wood followed by enzymatic degradation of cellulose. In this "ball milling" process, the lignocellulose is fragmented by milling and suspended in water. A mixture of *Trichoderma reesei* cellulases (see Chapter 10, page 348) is isolated from large amounts of culture broth of *T. reesei* grown on cellulosic materials. Adding the enzymes to the slurry of milled biomass results in a 45% conversion of cellulose to glucose. The volumes are adjusted to obtain a 10% solution of glucose, which is then transferred to a fermentation vessel for conversion to alcohol.

STAGE II: FROM SUGARS TO ALCOHOL

The second stage utilizes the simple sugars released from polysaccharides in stage I as substrates in microbial fermentations to produce alcohol.

Yeasts

Near-quantitative conversion of glucose to alcohol is carried out by many yeasts but by few bacteria. Industrial processes use primarily yeasts in the genus *Saccharomyces*. Although yeasts have many of the attributes of an ideal ethanol producer, they have significant limitations, such as a narrow substrate range and limited tolerance to alcohol. Below, we consider the various facets of yeast fermentations.

Substrate Range

The greatest constraint in employing yeasts as agents of fermentation is the limited range of substrates they are able to use (Table 11.1). Most oligosaccharides formed during the hydrolysis of starch, for instance, are not fermented by yeasts. These resistant compounds include malto-dextrins longer than maltotriose, and isomaltose (an α-1,6-linked dimer of glucose). Yeasts thus require the addition of glucoamylases (see Figure 7.2 and Table 7.1) to utilize starch completely. Nor can yeast cells utilize cellulose, hemicellulose, cellobiose, or most pentoses. Their inability to ferment a diversity of cheaper and readily available substrates is the major obstacle to lowering the cost of alcohol production. The search for ways to use more substrates is a major focus of research into improving ethanol fermentations.

Saccharomyces cerevisiae ferment a number of common substrates, including the disaccharides sucrose and maltose. The substrates are handled by one of two mechanisms: Either the disaccharides are hydrolyzed by extracellular enzymes and the monosaccharides transported into the cell, or the disaccharides are also transported into the cell and then hydrolyzed by intracellular enzymes. The uptake and metabolism of various substrates in a mixture occurs in an order determined by regulatory mechanisms at the level of gene expression. For example, glucose is the preferred substrate. If it is present, the permeases for the other substrates, such as maltose and maltotriose, are not induced until the glucose disappears. As a result, these substances are fermented sequentially rather than simultaneously. Di- and trisaccharides, once internalized, are hydrolyzed by an α-glucosidase. Fermentations must be run long enough to allow induction of the enzyme systems and full use of the various substrates. In this case, they generally go to completion with little fermentable substrate left at the end.

Saccharomyces strains are responsible for almost all the current industrial production of alcohol by fermentation. They convert glucose by the glycolytic pathway to high yields of ethanol and carbon dioxide (Figure 11.2). Only 2 moles of ATP are produced per mole of glucose metabolized, and the yeast cells use them for growth. Ethanol is recovered in 90-95% of theoretical yield.

Some Yeasts and Bacteria That Produce Significant Quantities of Ethanol, and the Major Carbohydrates Utilized as Substrates

T
A
B
L
E

11.1

	Substrates
Yeasts	
Saccharomyces spp.	
S. cerevisiae	Glucose, fructose, galactose, maltose, maltotriose, and xylulose
S. carlsbergensis	Glucose, fructose, galactose, maltose, maltotriose, and xylulose
S. rouxii (osmophilic)	Glucose, fructose, maltose, and sucrose
Kluyveromyces spp.	
K. fragilis	Glucose, galactose, and lactose
K. lactis	Glucose, galactose, and lactose
Candida spp.	
C. pseudotropicalis	Glucose, galactose, and lactose
C. tropicalis	Glucose, xylose, and xylulose
Bacteria	
Zymomonas mobilis	Glucose, fructose, and sucrose
Clostridium spp.	
C. thermocellum (thermophilic)	Glucose, cellobiose, and cellulose
C. thermohydrosulfuricum (thermophilic)	Glucose, xylose, sucrose, cellobiose, and starch
Thermoanaerobium brockii (thermophilic)	Glucose, sucrose, and cellobiose
Thermobacteroides acetoethylicus (thermophilic)	Glucose, sucrose, and cellobiose

Substrate Utilization

An understanding of the arithmetic of ethanol production begins with the equation established by Gay-Lussac in 1810 for the fermentative conversion of glucose to ethanol by yeast:

$$C_6H_{12}O_6 \longrightarrow 2\ C_2H_5OH + 2\ CO_2$$

180 g ⟶ 92 g 88 g

From this equation, the theoretical yield of alcohol from glucose is calculated to be 51.1% by weight. However, the production of alcohol

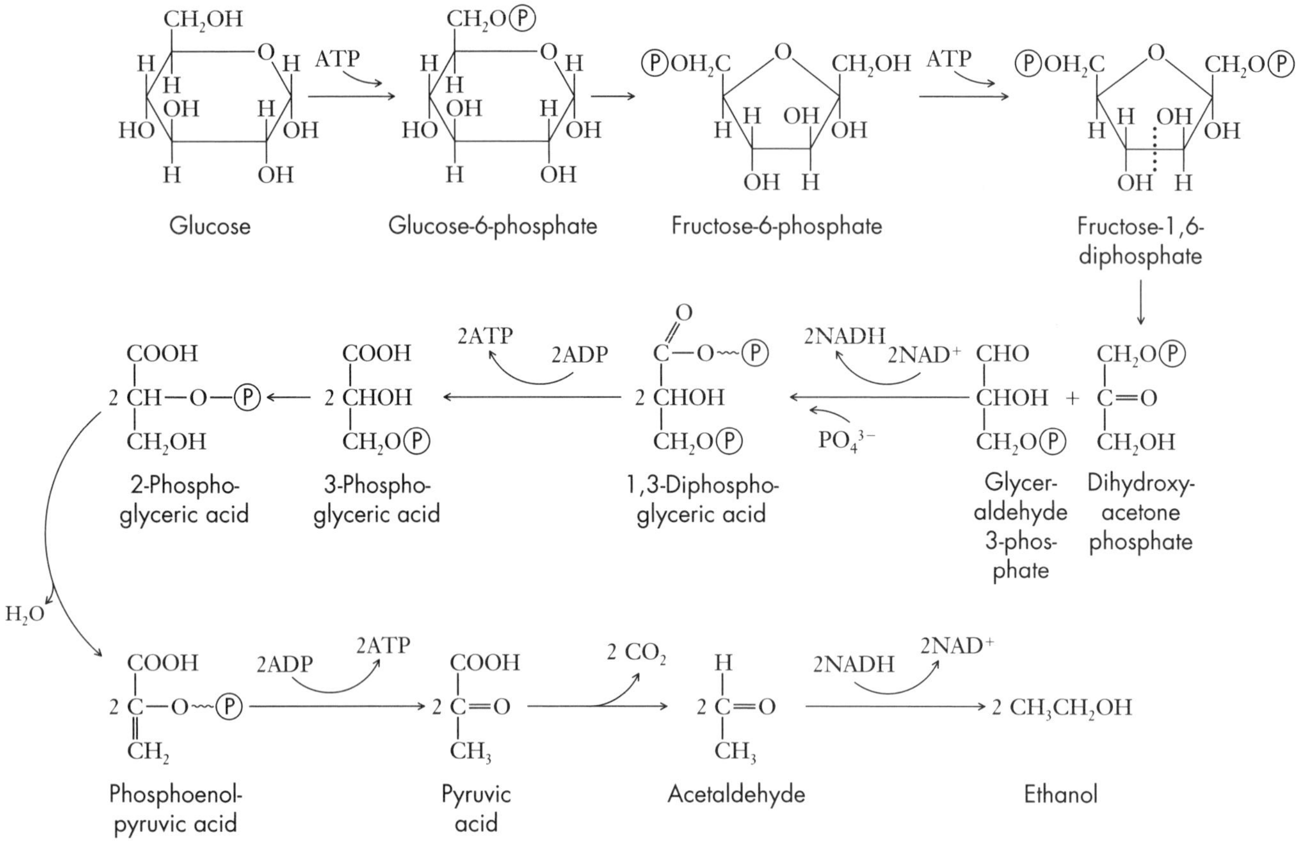

Figure 11.2
Formation of ethanol and carbon dioxide from glucose by the glycolytic (Embden–Meyerhof) pathway.

by yeast is actually a by-product of the yeast's growth, and some of the substrate is utilized to produce more cells: Rapidly growing, fermenting microorganisms produce about 10 g dry weight of cells for each mole of ATP they synthesize. As we have seen, each mole of glucose fermented to ethanol produces 2 moles of ATP, so the theoretical yield of cells is 20 g dry weight. The carbon content of these microbial cells is close to 50%. Because glucose is the sole carbon source in the fermentation medium, and the carbon content of glucose is 40%, 25 g glucose is needed to provide 10 g cell carbon. Therefore, allowing for cell growth, the maximum yield of alcohol is expected to be about 86%. How then do yeast fermentations result in higher yields of ethanol, 90–95% of theoretical?

The reason for the extremely high yield of ethanol is that not all the ATP is used to produce new cells. Some energy goes for other cellular functions (called maintenance, for want of a better term) regardless of the rate of cell growth. This proportion is smaller during rapid growth, but when growth slows, as it does during inhibition by ethanol or nutrient limitation, the maintenance energy requirement does not decrease. In fact, because the presence of ethanol may cause ion leakage across cell membranes, some of the cells' maintenance energy requirements may actually increase at high ethanol concentrations. Thus, as ethanol builds up during batch fermentations, the cells use an increasing share of the ATP for maintenance at the expense of reproduction, and the fraction of glucose carbon in the fermenter represented by cells decreases from 14 to 5% or less of the starting glucose, with a corresponding increase in the yield of ethanol.

Substrate represents the largest component of the cost of ethanol production. Small improvements in the efficiency of conversion can have a significant impact on costs. Increasing yield from 90 to 92% can lower the product cost by 1% or more.

As indicated by the Gay-Lussac equation, carbon dioxide and alcohol are produced from glucose in equimolar amounts. Additional reactions also take place in the fermenter, leading to small amounts of such minor by-products as glycerol, fusel oils, acetic acid, lactic acid, succinic acid, acetaldehyde, furfural, and 2,3-butanediol.

Of these, glycerol accumulates in the largest amount. Industrial fermentations produce up to 5 g glycerol for every 100 g ethanol. Glycerol is formed by reduction of the glycolytic intermediate, dihydroxyacetone phosphate (see Figure 11.2), to glycerol-3-phosphate, which is dephosphorylated to glycerol (Figure 11.3). *Saccharomyces* synthesizes glycerol as an osmoregulatory metabolite in response to the high osmotic pressure of the sugar solution in the fermenter. Osmoregulatory metabolites are organic compounds produced and accumulated by many living organisms to regulate their internal osmotic pressure in response to changes in extracellular water activity.

The choice of glycerol as an osmoregulatory metabolite in *Saccharomyces* may be influenced by the fact that, like the Embden–Meyerhof pathway, the pathway leading to glycerol directly regenerates NAD^+. When *Saccharomyces* are growing in media high in sugar content, the highest proportion of glycerol relative to ethanol is produced early in

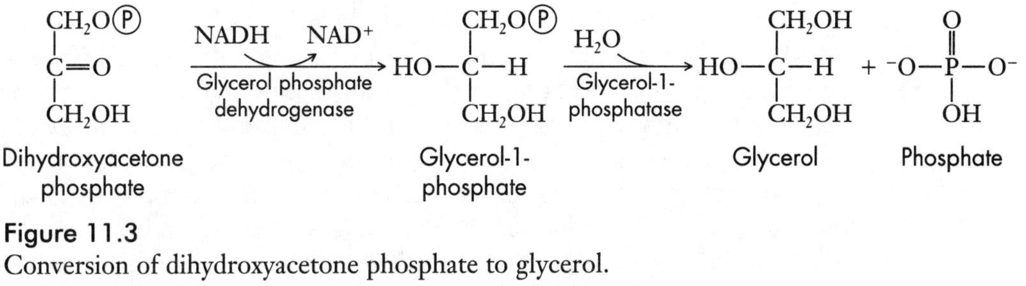

Figure 11.3
Conversion of dihydroxyacetone phosphate to glycerol.

the fermentation, when osmotic pressure and fermentation rate are at their peak. At high glycolysis rates, the terminal reductive steps in glycolysis (see Figure 11.2) appear to be rate-limiting, and the high concentrations of NADH that accumulate under these conditions favor the formation of glycerol. Even though glycerol is produced in significant amounts and is a valuable chemical, its recovery from the residues left at the end of the process in pure form is not economically feasible. In simultaneous saccharification and fermentation processes (see pages 383–384), where the steady-state concentration of sugars is relatively low, and in other processes that result in lower rates of cell growth, there should be lower amounts of glycerol formed at the expense of ethanol.

Fusel oils are a mixture of higher alcohols, primarily amyl and butyl alcohols. These compounds are produced from the degradation of amino acids, which in turn come from proteolysis of the proteins in the feedstock. When feedstocks are low in protein—sugar cane juice is an example—less fusel oil is produced. In contrast, fusel oils may represent as much as 0.5% of the crude distillate from starch feedstocks derived from grains.

Yeasts are little affected by pH in the range of 4–6. However, if the pH in the fermenter is allowed to rise above 5.0, the conditions favor the growth of *Lactobacillus*. These bacteria ferment glucose to produce lactate and acetate as well as ethanol. To avoid contamination, the pH in the fermenter is maintained below 5.0 by the addition of small amounts of acid.

Ethanol Tolerance

Because the separation of ethanol from water (during stage III, alcohol recovery) accounts for much of the energy used in the overall production process, the higher the concentration of ethanol in the fermenter, the lower the distillation cost per liter of product. However, ethanol is toxic to yeast cells at concentrations ranging between 8 and 18% by weight, depending on the strain of the yeast and the metabolic state of the culture (Figure 11.4). Yeast fermentation is totally inhibited by ethanol concentrations of about 11% by volume. To understand why, we must consider the properties of cytoplasmic membranes.

Structure and Function of Cytoplasmic Membranes. The cytoplasmic membranes of yeast and bacteria consist of a lipid bilayer with embedded protein complexes. Some of these transmembrane proteins serve as channels through which the interior of the cell interacts with the external milieu. Others are electron transport complexes responsible for setting up proton and ion gradients between the interior and exterior of the cytoplasmic membrane. An example is the ubiquitous F_0–F_1 ATPase, the enzyme complex that utilizes a proton gradient across the cytoplasmic membrane for the synthesis or hydrolysis of ATP.

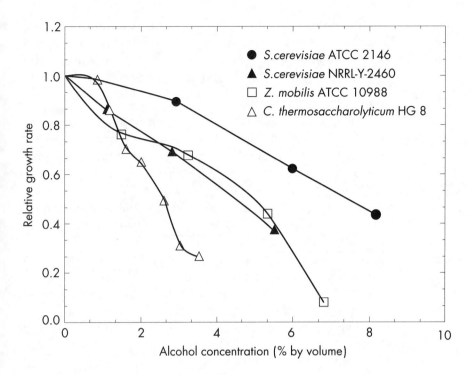

Figure 11.4

Comparison of the ethanol tolerances of *Saccharomyces cerevisiae* strains and of the bacteria *Zymomonas mobilis* and *Clostridium thermosaccharolyticum*. Growth data are for steady-state continuous cultures to which ethanol has been added at the concentration indicated. Data from Hogsett, D.A., Ahn, H.-J., Bernandez, T.D., South, C.R., and Lynd, L.R. (1992), Direct microbial conversion: Prospects, progress, and obstacles, *Appl. Biochem. Biotechnol.* 34/35:527– 541.

The transport of nutrients into the cell is dependent on a class of transmembrane proteins or protein complexes called the permeases, which mediate the passage of hydrophilic molecules through the cell membrane. Glucose permease, for example, facilitates the transport of glucose into the cell down a concentration gradient by a "facilitated diffusion" process. It depends on the enzyme hexokinase rapidly phosphorylating any free glucose inside the cells and thus keeping intracellular glucose levels very low. Other permeases catalyze the active transport of nutrients *against* internal nutrient concentrations that are orders of magnitude higher than those in the extracellular medium. Many of these active-transport proteins use the pH and ion concentration gradient ("proton motive force") across the cytoplasmic membrane to provide the energy for active transport. Other permeases use ATP as the energy source. The transport of maltose, amino acids, and ammonium ion into yeast cells depends on active transport by membrane-embedded protein transporters.

The cytoplasmic membrane provides a barrier to the diffusion of protons, other ions, and small polar molecules from the cell to the outside and vice versa. Without such a highly efficient boundary, the internal homeostasis essential to the survival of living cells could not be maintained.

Effect of Ethanol on Membrane Structure and Function. Although the cytoplasmic membrane's lipid bilayer is an efficient barrier to hydrophilic molecules, it does allow small amphiphilic molecules such as ethanol to pass through freely without the need for a specific permease. Consequently, the concentration of alcohol within the cell is the same as in the surrounding medium. As the alcohol concentration increases, it disrupts the structure of water, so the entropic contribution to the stabilization of lipid bilayer membranes is lower in alcohol-water mixtures than in water alone. Moreover, as alcohol partitions into the interior of the membrane, it disturbs lipid–lipid and lipid–protein interactions. Therefore, as alcohol levels increase, the membrane becomes progressively more and more leaky. The ion gradients that give rise to the proton motive force across the membrane gradually collapse, and small molecules leak out of the cell. In some yeast strains, a 50% decrease in the rate of uptake of sugars, ions, and amino acids is seen in the presence of ethanol concentrations as low as 4% by volume.

Influence of Membrane Composition on Resistance to Ethanol. Some yeast strains show greater resistance than others to increased concentrations of ethanol. In some cases, the membrane lipid bilayer structure is stabilized by the presence of longer hydrocarbon chains in the lipid molecules. This increases the interaction between neighboring chains. In other cases, leakage is minimized because of a high sterol content in the membrane; sterols decrease the nonspecific permeability of phospholipid bilayers.

Most yeast cells grown in the presence of ethanol show a small but significant increase in the average chain length of their membrane fatty acids. However, membranes rich in lipids with longer-chain *saturated* fatty acids tend to "freeze" and become rigid at the growth temperature of yeasts. Thus the longer-chain acids are produced as *unsaturated* fatty acids. For example, the membrane lipids of yeast grown in the presence of 7.5% ethanol contain 34% oleic acid ($18:1^{\Delta 9}$; Figure 11.5), whereas the lipids of cells grown without added ethanol contain only 17%. Much of this increase occurs at the expense of the shorter-chain, saturated fatty acid palmitate (16:0). Oxygen is required for ethanol tolerance. Because yeasts, like higher eukaryotes, make unsaturated fatty acids by using O_2 and NADPH, yeast cells cannot be grown completely anaerobically in the presence of ethanol. Oxygen is also required for the production of ergosterol, a cytoplasmic membrane sterol that contributes to membrane stability. Consequently, growth under anaerobic conditions in the presence of ethanol requires addition to the medium of unsaturated, longer-chain fatty acids as well as ergosterol (see below).

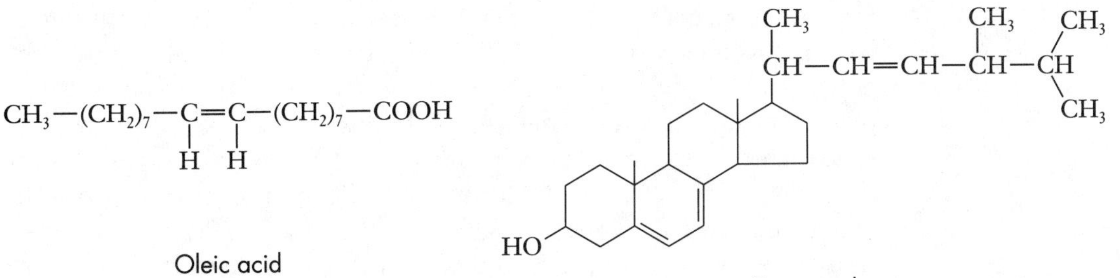

$$CH_3-(CH_2)_7-\underset{\underset{H}{|}}{C}=\underset{\underset{H}{|}}{C}-(CH_2)_7-COOH$$

Oleic acid
(*cis*-9-octadecenoic acid)

Ergosterol

Figure 11.5
Oleic acid and ergosterol.

Microaerophilic growth in the presence of ethanol also leads to an increase in the membrane content of ergosterol (see Figure 11.5). Ethanol induces the production of cytochrome P-450, a component of a monooxygenase system responsible for the demethylation of lanosterol to ergosterol in yeasts (and also for the conversion of saturated to unsaturated fatty acids). Thus it is likely that the differences in the level of tolerance to ethanol between different strains of microorganisms can be attributed in the main to the make-up of their cytoplasmic membranes and to their ability to vary their membrane composition in response to increasing ethanol concentration.

Temperature

The conversion of glucose to ethanol and CO_2 is an exothermic reaction: The complete fermentation of an 18%-by-weight glucose solution would raise the temperature of the medium by more than 20°C. Every 5°C increase in temperature multiplies the evaporative loss of ethanol by 1.5. Yeast metabolism rates also increase with temperature up to an optimum at 35°C; they then decrease gradually between 35 and 43°C and drop abruptly above 43°C. Therefore, the fermenter operating temperature must be maintained below 35°C.

Flocculence and "Cell Recycle"

The objectives of a fermentation process are to convert substrates to alcohol as rapidly as possible, to minimize the cost of alcohol recovery, and to decrease the amount of yeast cells produced as a by-product of the process. Exploitation of cell recycling and of the tendency of yeast cells to *flocculate* (clump) contributes to the attainment of these objectives.

Cells can be collected at the end of one fermentation to be used as the inoculum for the next. This procedure is termed *cell recycle*. By using large amounts of yeast cells recovered in this manner, it is possible to raise cell concentrations in the fermenter from a few grams dry weight per liter to tens of grams per liter. Because of this increase in the cell concentration, cell recycle may increase the amount of alcohol produced per unit volume even when inhibition by ethanol decreases the specific productivity of individual cells. One constraint, however, is the need to keep the cells viable during the collection process by providing adequate nutrition. Collecting cells by centrifugation or filtration costs more in equipment and attention than any savings that result from such recycling, so the key has been finding inexpensive ways to separate yeasts from the fermentation broth. Flocculation provides a partial solution.

Flocculation is a property in yeasts that depends on the gene *flo1*. Cells with this property stick together because they possess a wall protein, encoded by *flo1*, that binds in a calcium-ion-dependent manner to the wall mannans of other cells. As a result, the cells form clumps that sediment rapidly and are easily removed from the fermentation mixture for recycling. Nonflocculating strains, which do not form clumps, are called powdery. Although synthesis of the Flo1 protein is normally repressed by anaerobic growth, mutants expressing the protein during fermentation are readily found.

Continuous rapid fermentation mixtures require a great deal of agitation to ensure uniform suspension of cells. Moreover, a high rate of CO_2 production in the fermentation mixture in itself creates a great deal of agitation. Consequently, even flocculent strains are not necessarily easy to separate without additional equipment such as centrifuges or cross-membrane filters. Any marginal improvements in productivity resulting from cell recycle and the use of flocculent strains are usually negated by the extra expense of carrying out the cell separation. However, this approach does increase the amount of alcohol produced relative to the amount of yeast biomass generated.

Stillage

Stillage is the residue from the first distillation of fermented substrate (corn mash, sugar cane juice, etc.; see Figure 11.1). With sugar cane, about 12 liters of stillage are produced for each liter of ethanol. Such stillage contains 40–65 g organic matter per liter (Table 11.2). Depending on what is done with it, stillage is either a serious water-polluting waste or a source of valuable by-products. Questions of stillage disposal will be taken up later in this chapter. Processes that yield lower ratios of stillage to ethanol volumes entail lower costs.

Zymomonas mobilis, an Alternative Ethanol Producer

The specifications for an ideal fermentation alcohol producer would include the following important characteristics:

Major Components of Stillage Remaining after Alcohol Distillation from Fermented Sugar Cane Juice

TABLE 11.2

Component	Quantity (g/liter)
Organic matter	40–65
Nitrogen	0.7–1
Phosphorus	0.1–0.2
Potassium	4.5–8

- The ability to ferment a broad range of carbohydrate substrates rapidly
- Ethanol tolerance and the ability to produce high concentrations of ethanol
- Low levels of by-products, such as acids and glycerol
- Osmotolerance (ability to withstand the high osmotic pressures encountered at high concentrations of sugar substrates)
- Temperature tolerance
- High cell viability for repeated recycling
- Appropriate flocculation and sedimentation characteristics to facilitate cell recycle

For all their shortcomings, which we have described, *Saccharomyces* strains come closer to meeting these specifications than any other organisms known to produce ethanol. Only two other ethanol producers have attracted serious attention: *Zymomonas mobilis* and certain thermophilic clostridia.

Bacteria of the genus *Zymomonas* attracted the notice of microbiologists in 1912 as contributors to "cider sickness"—spoilage of fermented apple juice. Subsequently, *Zymomonas* was isolated from other fermenting sugar-rich plant juices: agave sap in Mexico, palm saps in various part of Africa and Asia, and sugar cane juice in Brazil. These fermentations produce alcoholic beverages such as pulque (from agave sap) and palm wines. Zymomonads are anaerobic, Gram-negative rods 2–6 μm in length and 1–1.5 μm in width. They are flagellated but lack spores or capsules.

Zymomonas mobilis takes up glucose and produces ethanol some 3–4 times more rapidly than yeast, with ethanol yields of up to 97% of the theoretical maximum. Moreover, unlike yeast, *Zymomonas* requires no oxygen for growth. The organism grows in minimal medium with no organic compound requirements. Many *Z. mobilis* strains grow at 38–40°C.

Zymomonads have high osmotolerance, and most strains grow in solutions containing 40% by weight glucose, but their salt tolerance is low. No strains are able to grow in 2% NaCl, whereas many yeasts can tolerate even higher salt concentrations. *Z. mobilis* strains are alcohol-tolerant, with fermentation yields of up to 13% alcohol by volume at 30°C. Few bacteria are able to survive such high levels of alcohol.

In spite of these favorable characteristics, *Z. mobilis* has not displaced yeast as a large-scale alcohol producer. Some of the reasons for this failure emerge from a detailed consideration of carbohydrate metabolism in *Zymomonas*.

Carbohydrate Utilization in Zymomonas

Zymomonas can utilize only three carbohydrates: glucose, fructose, and sucrose. The metabolism of each of these sugars has distinctive features. Therefore, after discussing the overall picture of glucose fermentation in *Zymomonas*, we will consider the distinctive reactions that come into play when fructose or sucrose is the substrate. The pathways of carbohydrate metabolism in *Zymomonas* are charted in Figure 11.6.

Entry of Glucose and Its Fermentation by the Entner-Doudoroff Pathway

Glucose enters the *Zymomonas* cell by means of a stereospecific, low-affinity, high-velocity facilitated-diffusion transport system. A constitutive glucokinase then converts the glucose to glucose-6-phosphate, after which step *Zymomonas* departs from the glycolytic pathway characteristic of other glucose fermenters, such as yeasts, and instead utilizes the Entner–Doudoroff pathway, shown in Figures 11.6 and 11.7.

Figure 11.6

Metabolism of sucrose, glucose, and fructose in *Zymomonas mobilis*. Enzymes (indicated by *circled* numbers) are as follows: 1, levansucrase; 2, invertase; 3, mannitol dehydrogenase; 4, glucose-fructose oxidoreductase; 5, fructokinase; 6, glucose-6-phosphate isomerase; 7, glucose dehydrogenase; 8, gluconolactonase; 9, gluconate kinase; 10, glucokinase; 11, glucose-6-phosphate dehydrogenase; 12, 6-phosphogluconolactonase; 13, 6-phosphogluconate dehydratase; 14, keto-deoxy-phosphogluconate aldolase; 15, glyceraldehyde 3-phosphate dehydrogenase; 16, phosphoglycerate kinase; 17, phosphoglycerate mutase; 18, enolase; 19, pyruvate kinase; 20, lactate dehydrogenase; 21, triose-phosphate isomerase; 22, phosphatase; 23, glycerol 3-phosphate dehydrogenase; 24, phosphatase; 25, pyruvate decarboxylase; 26, alcohol dehydrogenase. REFERENCE: Bringer-Meyer, S., and Sahm, H. (1988), Metabolic shifts in *Zymomonas mobilis* in response to growth conditions, *FEMS Microbiol. Rev.* 54:131–142.

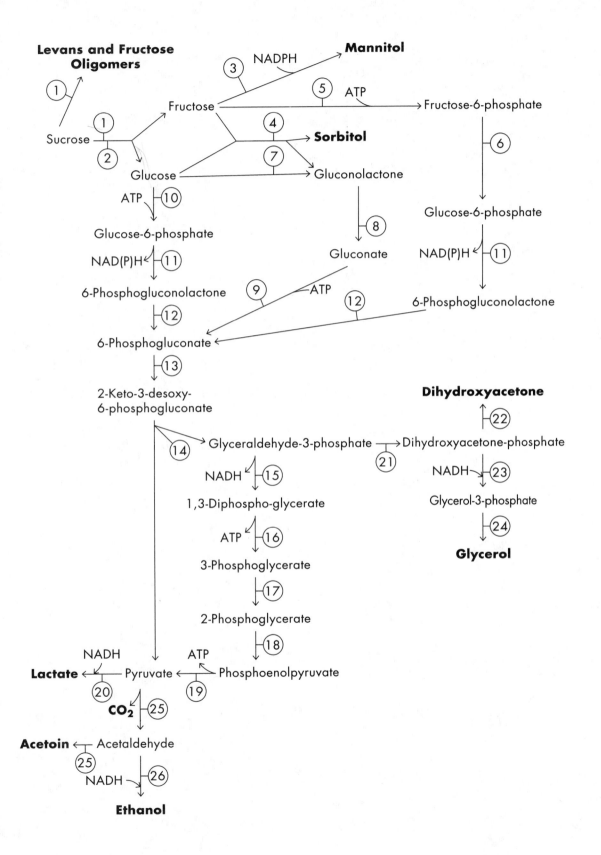

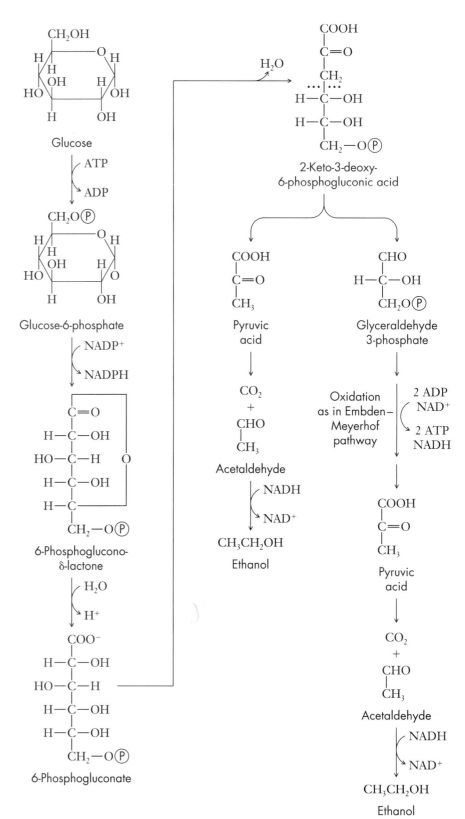

Figure 11.7
Formation of ethanol from glucose by *Zymomonas mobilis* via a fermentative version of the Entner-Doudoroff pathway.

In this pathway, glucose-6-phosphate dehydrogenase catalyzes the conversion of glucose-6-phosphate to 6-phosphoglucono-γ-lactone, with the concomitant reduction of NAD^+ to NADH. A lactonase with a very high catalytic activity rapidly hydrolyzes the lactone and 6-phosphogluconate dehydratase, then converts the resulting 6-phosphogluconate to 2-keto-3-deoxy-6-phosphogluconate, the unique intermediate of the Entner–Doudoroff pathway. This compound is cleaved by a specific aldolase to pyruvate and glyceraldehyde-3-phosphate, the latter being converted to a second molecule of pyruvate by a series of reactions that are part of the common glycolytic pathway (see Figure 11.2). The pyruvate is then converted to acetaldehyde and carbon dioxide by an unusual pyruvate decarboxylase that, unlike the enzyme in yeast, does not require the cofactor thiamine pyrophosphate for its catalytic activity. Finally, two alcohol dehydrogenases (ADH I and ADH II) reduce the acetaldehyde to ethanol, with the concomitant stoichiometric oxidation of NADH. ADH I is a tetrameric enzyme with zinc at the active site, like most of the commonly encountered alcohol dehydrogenases. ADH II is unusual in that it contains iron in its active site. At low alcohol concentrations, the V_{max} value of ADH I is about twice as high for acetaldehyde reduction as for alcohol oxidation, whereas ADH II shows a much higher rate of alcohol oxidation. In the absence of ethanol, both contribute equally to the enzymatic catalysis of the reduction of acetaldehyde. At high ethanol concentrations ADH I is strongly inhibited, but ethanol production continues because the rate of acetaldehyde reduction by ADH II is significantly increased under these conditions.

The Entner–Doudoroff pathway produces 2 moles of NADH for each mole of glucose consumed: 1 mole in the reaction catalyzed by glucose-6-phosphate dehydrogenase and 1 mole in the reaction catalyzed by glyceraldehyde-3-phosphate dehydrogenase. The 2 moles of pyruvate produced from each mole of glucose are converted to 2 moles of acetaldehyde. Reduction of the acetaldehyde to ethanol, which is catalyzed by the alcohol dehydrogenases, stoichiometrically regenerates NAD^+.

A key difference between the glycolytic and Entner–Doudoroff *Embden – Meyerhof* pathways is that the glycolytic pathway results in the net production of 2 moles of ATP per mole of glucose fermented, and the Entner–Doudoroff pathway produces only 1. In spite of its dependence on such an inefficient pathway for ATP generation, *Zymomonas* competes successfully with other microorganisms in the natural environment. It manages so well because of the very high levels of glycolytic and ethanologenic (pyruvate decarboxylase, alcohol dehydrogenase) enzymes it produces, which ensure a rapid flux of substrate through the pathway. Together, these enzymes represent about half the mass of the cytoplasmic proteins in *Zymomonas mobilis*.

Fructose Metabolism. Zymomonas takes up fructose like glucose, by a constitutive facilitated-diffusion transport system. The fructose is then phosphorylated to fructose-6-phosphate by a constitutive kinase that is highly specific for fructose and ATP and strongly inhibited by glucose (K_i = 0.14 mM). Glucose phosphate isomerase converts fructose-6-phosphate to glucose-6-phosphate, at which point this pathway merges with that for the metabolism of glucose (see Figure 11.7).

When glucose is the carbon source, the loss of carbon to the formation of by-products is insignificant. When fructose is the carbon source, the results are very different. Experiments using the same *Zymomonas* strain under similar conditions of batch fermentation show that the yield of ethanol from glucose is 95% of theoretical and from fructose only 90% of theoretical. The cell yield is also lower, showing that the decrease in the yield of ethanol does not reflect greater utilization of fructose for cell growth. A number of side reactions that occur in the presence of fructose are responsible for the difference.

Table 11.3 lists the products of a batch fermentation of fructose, at an original concentration of 15%. The major by-products are dihydroxyacetone, mannitol, and glycerol. *Zymomonas* contains an NADPH-dependent mannitol dehydrogenase with a high K_m value (0.17 mM) for fructose. At high fructose concentrations, this enzyme catalyzes the formation of mannitol from fructose at the expense of the NADPH generated when glucose-6-phosphate dehydrogenase (which utilizes NADP$^+$ in preference to NAD$^+$) oxidizes glucose-6-phosphate to 6-phosphogluconolactone. The resulting depletion of NAD(P)H

Products of the Fermentation of Fructose by *Zymomonas mobilis* Strain VTT-E-78082

TABLE 11.3

Product	Percent Yield (by weight)*
Ethanol	45.0
Cells	0.9
Dihydroxyacetone	4.0
Mannitol	2.5
Glycerol	1.7
Acetic acid	0.4
Sorbitol	0.3
Acetoin	0.3
Acetaldehyde	0.2
Lactic acid	0.1

* Starting fructose concentration was 148 g/liter. Final alcohol concentration was 66.7 g/liter.

SOURCE: Viikari, L. (1988), Carbohydrate metabolism in *Zymomonas, CRC Crit. Rev. Biotechnol.* 7:237–261.

(particularly acute in the early stages of the fermentation) leads to a build-up of acetaldehyde and, indirectly, to accumulation of intermediates at the triose phosphate level. With the build-up of these intermediates, other competing reactions come significantly into play. The equilibrium between glyceraldehyde-3-phosphate and dihydroxyacetone phosphate favors the latter compound, which in turn is converted to dihydroxyacetone by dephosphorylation or to glycerol phosphate by reduction. Glycerol phosphate is dephosphorylated to glycerol. Because fermentation through the Entner–Doudoroff pathway generates only 1 ATP per hexose, each of these side reactions, leading to the formation of dihydroxyacetone or glycerol, wastes all the energy produced in the reaction cycle. Moreover, both dihydroxyacetone and acetaldehyde inhibit cell growth. Such energy losses and inhibitory effects probably account for the lowered cell yield in fructose batch fermentations.

Sucrose Metabolism. Zymomonas produces an enzyme, levansucrase (sucrose: β-2,6-fructan fructosyltransferase), that hydrolyzes sucrose to glucose and fructose. This enzyme has been detected both in the culture medium and within the cells. In addition to its hydrolytic activity, levansucrase possesses a transfructosylating activity that leads to the formation of high-molecular-weight sugar polymers (up to 10^7 daltons) called *levans* when *Zymomonas* is grown on sucrose (see Figure 11.6). Substantial amounts of low-molecular-weight fructooligosaccharides are also formed (Figure 11.8). The main fructofuranosyl linkages in these compounds are $2 \rightarrow 6$ and $2 \rightarrow 1$. The formation of levans and

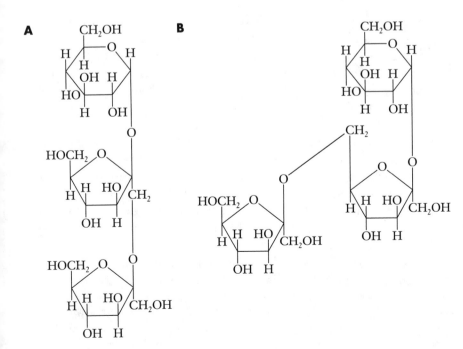

Figure 11.8
Structures of oligosaccharides formed by *Z. mobilis* during growth on sucrose. **(A)** 1^F-β-fructosylsucrose [O-α-D-glucopyranosyl-$(1 \rightarrow 2)$-O-β-D-fructofuranosyl-$(1 \leftarrow 2)$-β-D-fructofuranoside]; **(B)** 6^F-β-fructosylsucrose [O-α-D-glucopyranosyl-$(1 \rightarrow 2)$-O-β-D-fructofuranosyl-$(6 \leftarrow 2)$-β-D-fructofuranoside].

fructose oligomers competes with the fermentation of fructose to ethanol. Fortunately, levan formation is greatly diminished at higher temperatures (37°C), so the competition can be minimized.

Sorbitol Formation. Sorbitol is another by-product that lowers the yield of ethanol when *Zymomonas* ferments sucrose instead of glucose as a carbon source. It is the product of an abundant cytosolic enzyme, *glucose–fructose oxidoreductase*, which converts fructose to sorbitol using glucose as the reductant (see Figure 11.6). The enzyme contains tightly bound NADP$^+$ and does not require added cofactors. The second product of the reaction catalyzed by glucose–fructose oxidoreductase, gluconolactone (see Figure 11.6), is rapidly hydrolyzed by gluconolactonase to gluconate and then converted by gluconate kinase into 6-phosphogluconate, an intermediate of the Entner–Doudoroff pathway. When a mixture of glucose and fructose is added to the fermentation medium, or when they appear in the medium as a result of sucrose hydrolysis, *Z. mobilis* converts as much as 11% of the initial carbon sources to sorbitol, which cannot be used as a carbon source and merely accumulates in the medium.

Tolerance to Ethanol

As the ethanol concentration increases in the growth medium, most microorganisms begin to experience some impairment of membrane integrity. In *Z. mobilis*, however, unusual features of the cell membrane composition enable the organism to tolerate high levels of ethanol in the medium.

Although the *Z. mobilis* cell membrane contains the usual assortment of phospholipids, phosphatidylethanolamine being the most abundant, these phospholipids are exceptionally rich (up to 70%) in the monounsaturated fatty acid *cis*-vaccenic acid (18:1; see Figure 11.9). Moreover, the average hydrocarbon chain length in the membrane is greater by about one —CH$_2$— group than in most other Gram-negative bacteria.

Z. mobilis also profits from the presence of compounds known as hopanoids (Figure 11.10; Box 11.1) in the cell membrane. These pentacyclic triterpenoids, found in various prokaryotes, are functional analogs of sterols (Figure 11.11), but the biosynthesis of sterols involves oxygen-dependent reactions, and that of hopanoids does not. Sterols in the cell membrane contribute to the alcohol tolerance of yeasts (pages 370–371).

The fraction of cell lipid represented by hopanoids is strongly influenced by ethanol concentration. When a culture has been grown at different constant alcohol concentrations, bacteriohopanetetrol represents 2.5% of the total lipid at 0.5% ethanol, 21% at 6.3% ethanol, and 36.5% at 16% ethanol. The levels of the other hopanoids follow the

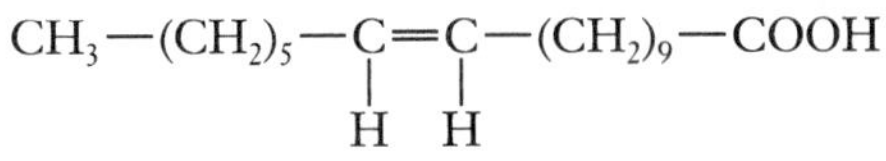

$$CH_3-(CH_2)_5-\underset{\underset{H}{|}}{C}=\underset{\underset{H}{|}}{C}-(CH_2)_9-COOH$$

Figure 11.9

Vaccenic acid (*cis*-11-octadecenoic acid).

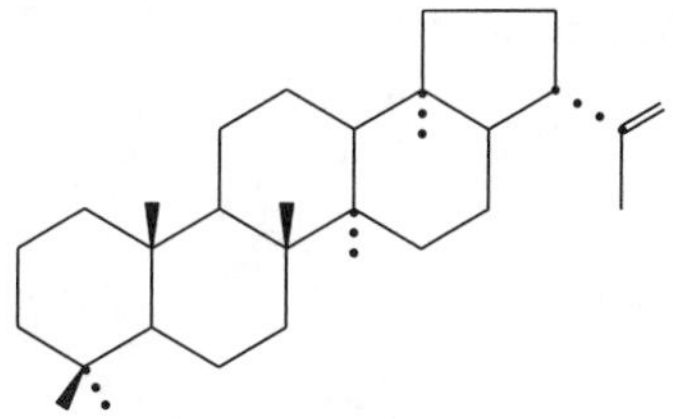

Hopane 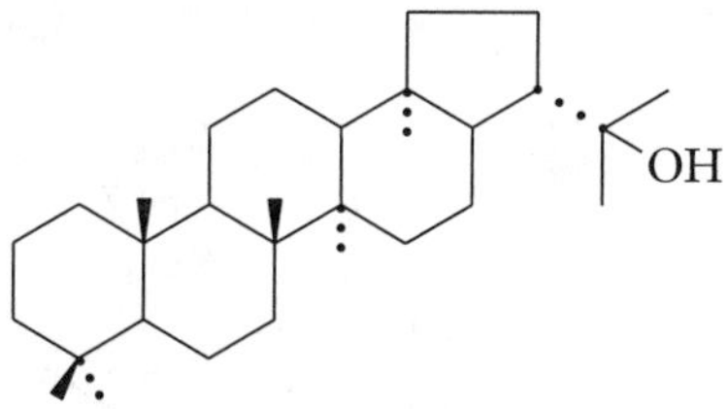   Diplopterol 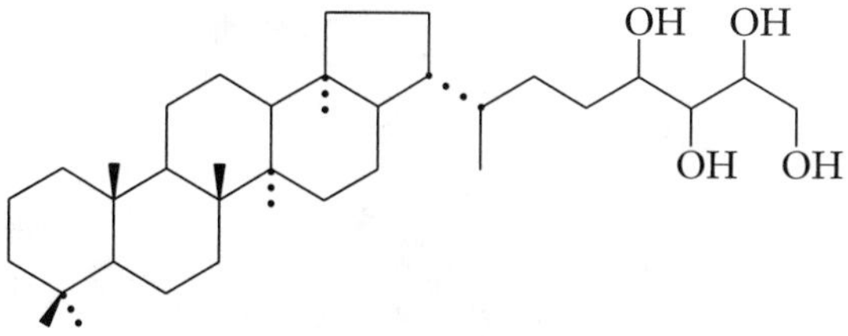   Bacteriohopanetetrol

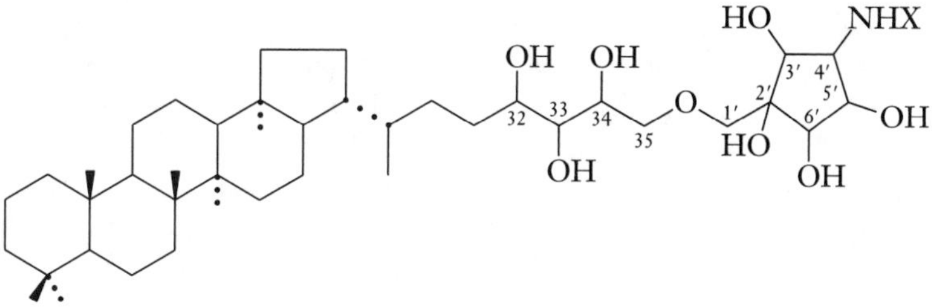

Bacteriohopanetetrol ether

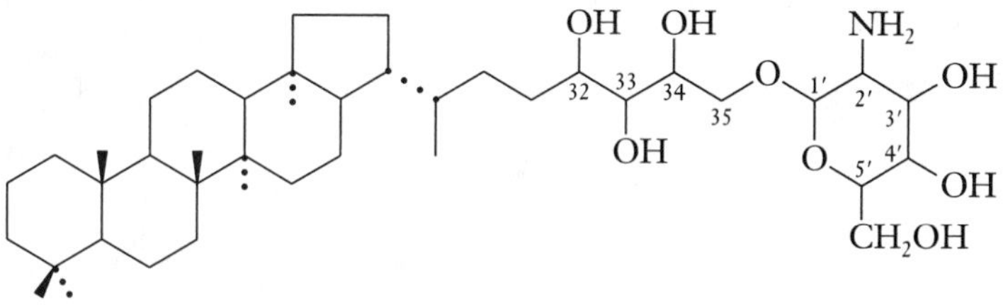

Glucosaminyl–bacteriohopanetetrol

Figure 11.10
Hopanoids found in *Zymomonas mobilis*.

BOX 11.1   Hopanoids

The existence of the hopanoid family of bacterial lipids was unsuspected until their geochemical transformation products (*geohopanoids*) were discovered in the 1960s as universal constituents of geological sediments, young and old. The chemical diversity of geohopanoids is considerable, and unrelated sediments usually contain different sets of geohopanoids. Thus "fingerprinting" of sediments by their geohopanoid patterns is useful in oil exploration. *The total amount of geohopanoids is on the order of 10^{12} tons, an amount similar to that estimated for the total mass of organic carbon in all currently living organisms.* Surveys of microorganisms for the presence of hopanoids have shown these polyterpenes to be widely distributed among prokaryotes.

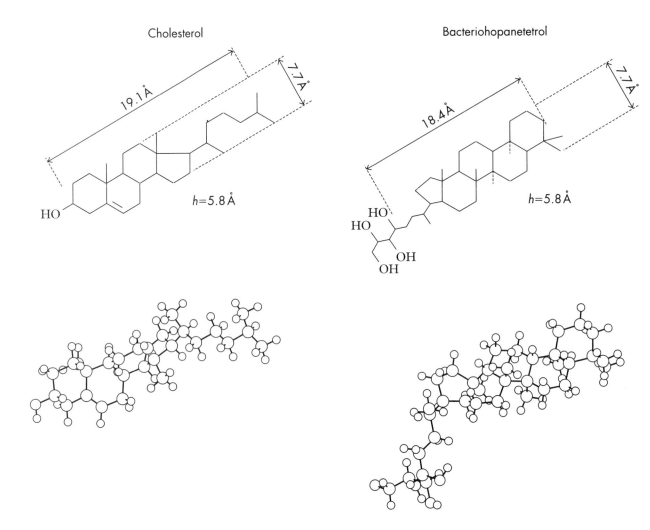

Figure 11.11
Structures of cholesterol and bacteriohopanetetrol. Based on Ourisson, G., and Albrecht, P. (1992), Hopanoids. 2. Biohopanoids: A novel class of bacterial lipids, *Acc. Chem. Res.* 25:403–408, Figure 2.

same trend. The hopanoids substitute for a portion of the phospholipids in the cell membrane; the absolute amount of phospholipids decreases as the amount of hopanoids increases. The hopanoids, just like the sterols, decrease membrane fluidity, and, presumably, in this manner counteract the permeabilizing influence of ethanol.

This interpretation is reinforced by the observation that raising the temperature of the medium causes changes in the composition of the *Z. mobilis* cell membrane that parallel those induced by increasing the ethanol concentration. A change in growth temperature from 30 to 37°C leads to an increase in the relative hopanoid content that is comparable to the increase seen at high ethanol concentrations. The membrane protein concentration also increases in response to an increase in the alcohol concentration or growth temperature, which suggests that ethanol tolerance in *Z. mobilis* is achieved by coordinated shifts in the relative amounts of cell membrane phospholipids, hopanoids, and proteins, all of which contribute to membrane stability.

SIMULTANEOUS SACCHARIFICATION AND FERMENTATION

Stages I and II Combined

In simultaneous saccharification and fermentation (SSF) processes, stages I and II, the hydrolysis of the substrate and the fermentation of monosaccharides to ethanol, are carried out in a single vessel. The hydrolytic enzyme(s) are produced in a separate reactor and introduced into the fermenter along with the substrate and yeast. For example, when partially hydrolyzed corn syrup is the substrate, *Aspergillus* sp. glucoamylase is added to the fermenter with the syrup and yeast cells. Starch hydrolysis produces glucose, which is immediately converted by the yeast to ethanol. With low steady-state concentrations of glucose, much smaller amounts of enzyme are able to achieve adequate rates of substrate hydrolysis, because product inhibition of the glucoamylase reaction is modest at low concentrations of glucose. When cellulose is the substrate, a mixture of *T. reesei* cellulases is used in place of glucoamylase. When lignocellulose is the substrate of an SSF process, pentose (xylose, arabinose) fermentation must be incorporated to maximize ethanol yields by utilizing both cellulose and hemicellulose. SSF-type processes have drastically reduced ethanol production costs.

Clostridial Fermentations

The direct conversion of cellulosic biomass to ethanol by anaerobic bacteria is potentially cheaper than a process that combines the actions of fungal cellulases and alcohol-producing yeasts. Three clostridial strains, *Clostridium thermocellum*, *C. thermosaccharolyticum*, and *C. thermohydrosulfuricum*, are regarded as candidate organisms for one-step conversion processes. These bacteria are thermophilic, Gram-positive, and strict anaerobes.

C. thermocellum was isolated first in 1926 from manure and subsequently from many other anaerobic environments rich in organic nutrients. It has an optimal growth temperature of 60°C. The cellulose-degrading complexes of *C. thermocellum*, cellulosomes, which form when the bacteria grow on cellulose, are described in Chapter 10. The cellulase activity levels found in culture supernatants of *C. thermocellum* match those produced by *Trichoderma reesei*, long considered to be the most efficient cellulase producer. Strains of *C. thermocellum* are able to ferment cellulose, xylans, and monomeric sugars (such as glucose, mannose, arabinose, and xylose). The products are ethanol, acetic acid, lactic acid, CO_2, and H_2.

C. thermosaccharolyticum has an optimal growth temperature of 55–60°C, with an upper limit of 67°C. It ferments diverse carbohydrates, including glycogen, starch, and pectin, to produce ethanol, acetate, butyrate, lactate, CO_2, and H_2. This organism has been studied thoroughly because it causes spoilage of canned foods: Contaminated cans blow open under the pressure of the gases produced by the fermentation. Batch cultures of resting cells of *C. thermosaccharolyticum* at pH 7.0 produce up to 1.2 moles of ethanol per mole of glucose.

C. thermohydrosulfuricum was isolated first in 1965 from the extraction juices of a beet sugar factory in Austria and subsequently from many other locations. This organism ferments a range of carbohydrates very similar to that of *C. thermosaccharolyticum*, but at an optimal temperature about 10°C higher and with an upper limit for cell growth at 74–76°C. Under appropriate conditions, *C. thermohydrosulfuricum* fermentation yields up to 1.5 moles of ethanol per mole of glucose, along with acetate, lactate, CO_2, and H_2.

Clostridial strains have the advantage over yeasts and *Zymomonas* of being excellent cellulase producers, thus allowing the conversion of cellulose and hemicelluloses to ethanol by a single organism. This eliminates the need for a second organism (one to produce cellulase and one to carry out the fermentation) and the feedstock needed to support it. However, clostridia are unlikely to become commercially important ethanol producers in the foreseeable future. The by-products of clostridial fermentation include large amounts of organic acids. The highest reported molar ratio of ethanol to acids has been about 2.3. These organisms also produce some H_2S from sulfur-containing amino acids. Furthermore, their ethanol tolerance is lower than that of yeasts and *Zymomonas*.

The clostridial cellulosome is a unique, complicated macromolecular complex of more than a dozen enzymes that work together in the efficient degradation of cellulose and hemicelluloses (see Figure 10.19). In view of this, the clostridia may become more important in the future as sources of enzymes for polysaccharide degradation than as direct ethanol producers.

FUEL ETHANOL FROM BIOMASS: AN ASSESSMENT

Energy can be extracted from biomass either by direct combustion or by first converting the biomass to another fuel (ethanol, methanol, or methane) and then combusting it. Most commonly, biomass is simply burned to generate heat; of the 2.8 billion tons of biomass consumed annually worldwide as material for burning, wood represents about 50%, crop residues some 33%, and dung most of the remainder. The fraction of biomass converted to other fuels, such as ethanol, is small.

As described in this and the preceding chapter, microbial fermentation of sucrose from sugar cane and beet, of corn starch, or of appropriately pretreated cellulosic biomass leads to the production of ethanol in high yield. Ethanol can be used as a fuel either "neat" or as a blend of 85% ethanol and 15% gasoline. The 3 billion gallons of fuel ethanol used annually in Brazil represent some 20% of the liquid fuel used in that country. At this time, ethanol's ability to compete with petroleum-based fuels depends on subsidies. However, future increases in the price of fossil fuels and improvements in the production of ethanol by fermentation may make it cost-competitive as a fuel. Does this mean that alcohol could then take the place of much of the fossil fuel used today?

An answer to this question must take into account (1) the average productivity of the biosphere, (2) the ratio of the output of energy in the form of fuels (specifically ethanol) to the input of energy in the manufacture of such fuels from different biomass sources, and (3) the level of energy use in high-technology societies. This type of analysis highlights the limitations of biomass as a replacement for fossil fuels and identifies the feedstocks and fermentation processes that offer the most favorable energy output/input ratio.

The average productivity of the terrestrial ecosystem (heat of combustion of biomass produced per unit time divided by the land surface of the Earth) is estimated to range from 0.05 to 0.10 W/m^2 (1 $W \cdot s$ = 0.239 cal; 0.29 W/m^2 = 5.988 $kcal/m^2/day$). Each form of biomass contributes differently.

Corn, the crop with the highest yield of food calories per hectare, represents about 0.6 W/m^2 of biomass, of which 0.34 W/m^2 is in the form of grain usable for food. The energy spent in producing the corn is approximately 30% of that output (Table 11.4). Consequently, the net yield of food energy is 0.24 W/m^2. In the most favorable case, where corn cobs and stalks are collected and burned to replace the fossil fuel requirement, the energy output/input ratio in the conversion of corn to alcohol is about 1.3 (Table 11.5). With bagasse (the crushed remains of sugar cane) used as fuel, the corresponding ratio for the conversion of sugar cane to ethanol is about 2 (see Table 11.5).

Energy Inputs per Hectare for Corn and Sugar Cane Production in the United States

TABLE 11.4

Input	Corn (10^3 kcal)	Sugar cane (10^3 kcal)
Labor	7	21
Machinery	1,485	1,944
Fuel	1,255	3,788
N (ammonium)	3,192	3,318
P (phosphate)	473	611
KCl	240	373
Limestone	134	353
Seed	520	802
Insecticides	150	250
Herbicides	200	620
Electricity	100	—
Transport	89	146
Total input	7,845	12,226
Output (crop yield)	26,000[a]	24,618[b]
Output/input ratio	3.31	2.01

[a] 6,500 kg corn = 26,000,000 kcal
[b] 88,000 kg sugar cane = 24,618,000 kcal

SOURCE: Pimentel, D., et al. (1988), Food versus biomass fuel: Socioeconomic and environmental impacts in the United States, Brazil, India and Kenya, *Advances in Food Research* 32:185–238.

Commercial forest land has a mean annual biomass production of less than 2500 kg dry weight per hectare (about 0.23 W/m²). However, short-rotation forestry of hardwoods produces 10-fold higher dry-weight yields of biomass (over 2 W/m²). For wood as a combustible fuel, the energy output/input ratio ranges from 7 to 15, depending on what one assumes the energy inputs for cultivation, cutting, and transportation to be. For stage II (see Figure 11.1), in which both the hexoses and the pentoses in hardwood are converted to ethanol, the presently attainable energy output/input ratio is about 5. As a result, absent the political value of farm subsidies, one would not choose corn over lignocellulose as a biomass feedstock suitable for conversion to alcohol.

How does the Earth's rate of generation of biomass energy compare with the rate of energy consumption by humans? Energy consumption is estimated to be 2.03 W/m² in Japan, 0.97 W/m² in Northern Europe, and 0.32 W/m² in the United States—estimates that encompass both the use of energy per person and the population density. In these technologically advanced parts of the world, the rate of energy consumption exceeds the average production of energy in the biosphere and is also greater than the rate of generation of energy in the food crops produced

Energy Inputs per 1000 Liters of Ethanol from Corn or Sugar Cane in the United States

Input	Corn (10^3 kcal)	Sugar cane (10^3 kcal)
Corn (2,700 kg)	3,259	
Sugar cane (14,000 kg)	—	1,945
Transport	325	400
Water	90	70
Stainless steel	89	45
Steel	139	46
Cement	60	15
Coal	4,617[a]	—
Bagasse	—	7,600[a]
Total input	8,579	10,121
Output	5,130[b]	5,130[b]
Output/input ratio	0.6	0.51[a]

[a] If corn cobs and stalks are collected to replace the fossil fuel requirement, then the output/input ratio becomes 5,130,000/3,962,000 = 1.29. If the energy cost of bagasse, a by-product of the manufacture of ethanol from sugar cane, is subtracted, then the output/input ratio becomes 5,130,000/2,521,000 = 2.03.
[b] 1000 liters of ethanol = 5,130,000 kcal

SOURCE: Pimentel, D., et al. (1988), Food versus biomass fuel: Socioeconomic and environmental impacts in the United States, Brazil, India and Kenya, *Advances in Food Research* 32:185–238.

by modern agriculture. With increases in human population and improvements in standards of living, the rate of energy use in the other parts of the world is also rising above the rate of energy production in the form of biomass. To replace the current world oil consumption with ethanol from corn would require an area approximately 4.5 times the total land area of the United States, which is about twice the amount of total arable land used for food crop production worldwide.

The necessary conclusion—that fermentation alcohol is unlikely to emerge as a major fuel replacement for gasoline—should not discourage those who are attempting to develop more efficient processes for the fermentation of lignocellulose sugars to alcohol. Paper (over 60% by weight cellulose) and other forms of lignocellulose (for example, from yard waste) represent over half the weight of municipal solid waste (Table 11.6), and we have seen that the energy output/input ratio for wood is highly favorable. Thus the wood components of municipal solid wastes, which are currently disposed of through a combination of incineration, composting, anaerobic digestion, and recycling, might be

> ▶ **Weight Percent Composition of Discarded Municipal Solid Waste**

Category	Percentage
Paper*	37.4
Yard waste	13.9
Glass	20.0
Metal	9.8
Wood	8.4
Textiles	3.1
Leather and rubber	2.2
Plastics	1.2
Miscellaneous	3.4

TABLE 11.6

* Newspaper is 61% cellulose, 16% hemicellulose, and 21% lignin.

SOURCE: Ackerson, D.M., Clausen, E.C., and Gaddy, J.L. (1993). Recovery of ethanol from municipal solid waste, in Khan, M.R. (ed.), *Clean Energy from Waste and Coal*, pp. 28–41 (ACS Symposium Series 515).

transformed into ethanol once efficient processes for its fermentation and recovery have been developed.

In one scenario for ethanol recovery from refuse, feedstock preparation consists of the removal of plastics, metal, and glass from solid wastes, after which the residue is shredded, ground, and conveyed to a reactor, where the lignocellulose is degraded with concentrated acid (sulfuric or hydrochloric acid at about 40°C). The concentrated-acid treatment results in a nearly complete conversion of hemicellulose and cellulose to monomeric sugars, with minimal creation of such undesirable degradation products as furfural (from xylose) and 5-hydroxymethyl-furfural (from glucose), which would otherwise inhibit subsequent fermentation. An acid recovery step separates the acid from the sugars, which are then converted to ethanol by fermentation with yeasts or bacteria capable of metabolizing both hexoses and pentoses. The advantage of acid hydrolysis is its efficient, rapid release of sugars from a particulate feedstock that is not likely to be easily degraded by enzymes. Its serious disadvantage is the difficulty of using and recovering large volumes of concentrated acids. It is possible that polysaccharide-degrading enzymes from extreme thermophiles, acting at temperatures above 100°C, could be used instead of the concentrated acids for the initial lignocellulose degradation.

Finally, in addition to the questions we have examined, it is important to note the problems that have arisen in countries such as Brazil and Argentina as a result of mass production there of fermentation alcohol. In Brazil, the expansion of ethanol production has had an adverse impact on the size of other agricultural crops and on forests, although the severity of this impact is disputed. The production of fermentation alcohol also creates massive amounts of wastewater. Indeed, the BOD (see Box 1.2) of wastewater from Brazilian alcohol plants is two-thirds that of the waste produced by the *total* human population in Brazil. The proper treatment of stillage is thus mandatory. Inadequate or nonexistent treatment in parts of Brazil and Argentina has led to severe degradation of water quality in rivers and streams, causing diarrhea and other diseases in the human population and the disappearance from polluted rivers of fish, a valuable source of protein. There are ways of lessening the wastewater problem: Feasibility studies show that the BOD of stillage can be reduced by some 75% through the microbial anaerobic digestion procedures used in sewage treatment plants (Chapter 15). Furthermore, the methane produced in the anaerobic digestion can be put to use as a fuel by the distilleries or elsewhere.

Fermentation alcohol will continue to play a role as an alternative fuel, in spite of its considerable limitations. Energy production is not a zero-sum process, and where *efficient* conversion of biomass is feasible, such as in solid waste fermentation, further development holds real promise. But concerning ethanol's contribution as an energy source and chemical feedstock in the more distant future, there are still more questions than answers.

SUMMARY

Cellulose, hemicelluloses, and starches are a vast renewable source of sugars convertible to ethanol by microbial fermentation. Production of ethanol from the polysaccharides of biomass proceeds in three stages: degradation of polysaccharides to fermentable sugars, fermentation, and alcohol recovery. Disruption of the physical structure of lignocellulose to make cellulose and hemicelluloses accessible to enzymatic attack is achieved either by steam explosion or by ball milling. *Saccharomyces* strains are responsible for nearly all the current industrial production of alcohol by fermentation. These yeasts produce high concentrations of alcohol with low levels of by-products and have the high cell viability and flocculation characteristics needed for repeated cell recycling. A serious limitation is that the yeasts ferment only a narrow range of carbohydrate substrates.

Zymomonas mobilis, a bacterium isolated from fermenting sugar-rich plant juices, takes up glucose and produces ethanol some 4-fold faster than yeast, with alcohol yields of up to 97% of the theoretical maximum. However, *Zymomonas* utilizes only three substrates: glucose, fructose, and sucrose. Growth on fructose and sucrose leads to conversion of 10% or more of the substrate to products other than alcohol, such as dihydroxyacetone, mannitol, and glycerol.

Simultaneous saccharification and fermentation (SSF) processes combine stages I and II. In SSF processes, enzymatic hydrolysis of the polysaccharide substrate and the fermentation of monosaccharides to alcohol are carried out in a single vessel. With partially hydrolyzed starch, for example, *Aspergillus* sp. glucoamylase is added to the fermenter along with yeast cells. One-step clostridial conversion of cellulose to alcohol is another version of an SSF process. Thermophilic clostridia that produce extracellular cellulases are able to convert cellulose all the way to alcohol. Unfortunately, clostridial fermentations also produce large amounts of organic acids and some hydrogen sulfide.

The future of ethyl alcohol as an alternative fuel depends on further improvements in the efficiency and economics of the conversion of lignocellulose to alcohol.

SELECTED REFERENCES

General

Henry, J.-F., 1979. The silvicultural energy farm in perspective. In Sarkanen, K.V., and Tillman, D.A. (eds.), *Progress in Biomass Conversion*, vol. 1, pp. 215–255. Academic Press.

Final Report. U.S. National Alcohol Fuels Commission, 1981. *Fuel Alcohol: An Energy Alternative for the 1980s.* U.S. Government Printing Office.

Klass, D.L., 1983. Energy and synthetic fuels from biomass and wastes. In Meyers, R.A. (ed.), *Handbook of Energy Technology and Economics*, pp. 712–786. Wiley.

Klass, D.L., 1990. The U.S. biofuels industry. *Chemtech*, December, pp. 720–731.

Pereira, A., 1986. *Ethanol, Employment and Development: Lessons from Brazil.* ILO Publications.

Sineriz, F., 1987. Alcohol production: Anaerobic treatment of process waste water and social considerations. In DaSilva, E.J., Dommergues, Y.R., Nyns, E.J., and Ratledge, C. (eds), *Microbial Technology in the Developing World*, pp. 226–237. Oxford University Press.

Rosillo-Callé, F., 1989. A reassessment of the Brazilian national alcohol programme (PNA). In Greenshields, R. (ed.), *Resources and Applications of Biotechnology: The New Wave*, pp. 332–345. Stockton Press.

Lynd, L.R., Cushman, J.H., Nichols, R.J., and Wyman, C.E., 1991. Fuel alcohol from biomass. *Science* 251:1318–1323.

Society of Automotive Engineers, 1991. *Alternative Liquid Fuels in Transportation.* SP-889. Society of Automotive Engineers.

Microbial Tolerance to Alcohols

Ingram, L.O., 1986. Microbial tolerance to alcohols: Role of the cell membrane. *Tr. Biotechnol.* 4:40–44.

Curtain, C.C., 1986. Understanding and avoiding alcohol inhibition. *Tr. Biotechnol.* 4:110.

Yeasts

Stewart, G.G., Panchal, C.J., Russell, I., and Sills, A.M., 1984. Biology of ethanol-producing microorganisms. *CRC Crit Rev. Biotechnol.* 1:161–188.

Tubb, R.S., 1986. Amylolytic yeasts for commercial applications. *Tr. Biotechnol.* 4:98–104.

Nagashima, M., 1990. Progress in ethanol production with yeasts. In Verachtert, H., and de Mot, R. (eds.), *Yeast: Biotechnology and Biocatalysis.* pp. 57–84. Marcel Dekker.

Stratford, M., 1993. Yeast flocculation: Flocculation onset and receptor availability. *Yeast* 9:85–94.

Zymomonas

Swings, J., and De Ley, J., 1977. The biology of *Zymomonas. Bacteriol. Rev.* 41:1–46.

Montenecourt, B.S., 1985. *Zymomonas,* a unique genus of bacteria. In Demain, A.L., and Solomon, N.A. (eds.), *Biology of Industrial Microorganisms.* pp. 261–289. Benjamin/Cummings.

Viikari, L., 1988. Carbohydrate metabolism in *Zymomonas*. *CRC Crit. Rev. Biotechnol.* 7:237–261.

Bringer-Meyer, S., and Sahm, H., 1988. Metabolic shifts in *Zymomonas mobilis* in response to growth conditions. *FEMS Microbiol. Rev.* 54:131–142.

Mejia, H.P., et al., 1992. Coordination of expression of *Zymomonas mobilis* glycolytic and fermentative enzymes: A simple hypothesis based on mRNA stability. *J. Bacteriol.* 174:6438–6443.

Hopanoids

Ourisson, G., and Albrecht, P., 1992. Hopanoids. 1. Geohopanoids: The most abundant natural products on Earth? *Acc. Chem. Res.* 25:398–402.

Ourisson, G., and Albrecht, P., 1992. Hopanoids. 2. Biohopanoids: A novel class of bacterial lipids. *Acc. Chem. Res.* 25:403–408.

Clostridia

Wiegel, J., 1992. The obligately anaerobic thermophilic bacteria. In Kristjansson, J.K. (ed.), *Thermophilic Bacteria*. pp. 105–184, CRC Press.

Sato, K., et al., 1992. Effect of yeast extract and vitamin B_{12} on ethanol production from cellulose by *Clostridium thermocellum* I-1-B. *Appl. Environ. Microbiol.* 58:734–736.

Hogsett, D.A., Ahn, H.-J., Bernandez, T.D., South, C.R., and Lynd, L.R., 1992. Direct microbial conversion: Prospects, progress, and obstacles. *Appl. Biochem. Biotechnol.* 34/35:527–541.

Lynd, L.R., 1989. Production of ethanol from lignocellulosic materials using thermophilic bacteria: Critical evaluation of potential and review. *Adv. Biochem. Eng. Biotechnol.* 38:1–52.

METABOLITES FROM MICROORGANISMS

PART

five

Colonies of *Streptomyces cinne-monensis*, producer of monensin, a polyketide antibiotic widely used in veterinary medicine. Courtesy of Mr. Frederick P. Mertz and Dr. Ronald E. Chance, Lilly Research Laboratories, Indianapolis.

12 Amino Acids

The production of amino acids on an industrial scale dates back to 1908, when the Japanese agricultural chemist K. Ikeda discovered that L-glutamate was responsible for the characteristic taste, much appreciated in Japan, of foods cooked with dried kelp (konbu). For the first 50 years, monosodium L-glutamate (MSG) was manufactured by expensive chemical processes based largely on the acid hydrolysis of proteins. This hydrolysis process was costly, because glutamate had to be separated from all other amino acids in the hydrolyzate. Significant amounts of MSG were also made by chemical synthesis. This process was expensive too, because it produced a mixture of D- and L-glutamate that had to be resolved to eliminate the D-isomer, which is tasteless.

A revolutionary change was introduced in 1957, when scientists at Kyowa Hakko Co. discovered a soil bacterium that excreted large amounts of L-glutamate into the medium. Similar bacteria were soon isolated by several other companies, ushering in a new industrial technology: amino acid fermentation, the production of amino acids by microorganisms. Except for ethanol, some other organic solvents, and a few vitamins, glutamate was the first organic compound produced on an industrial scale by a microbial fermentation technique.

Today many amino acids are manufactured either by fermentation or by a related process that uses immobilized enzyme columns (Table 12.1). In 1979, the total value of amino acid production was estimated at $1.9 billion a year. Amino acids are thus among the most important products of microbial biotechnology, ranking behind only ethanol and antibiotics (Chapter 1).

Industrial Production of Amino Acids

T
A
B
L
E

12.1

Amino Acid	World Annual Production (metric tons)	Methods[a]	Use
L-Alanine	130	1, 3c	Flavor enhancer
DL-Alanine	700	2	Flavor enhancer
L-Arginine	1,000	1, 3a	Infusion, cosmetics
L-Aspartate	4,000	1, 3c	Flavor enhancer, aspartame production
L-Asparagine	50	1, 2	Therapeutic
L-Cysteine	700	1	Bread additive, antioxidant
L-Glutamate	370,000	3a	Flavor enhancer
L-Glutamine	500	3a	Therapeutic
Glycine	6,000	2	Organic synthesis
L-Histidine	200	3a, 1	Therapeutic
L-Isoleucine	150	3a	Infusions
L-Leucine	150	1, 3a	Infusions
L-Lysine	70,000	3a, 3c	Feed additive, infusions
DL-Methionine	70,000	2	Feed additive
L-Methionine	150	3c	Therapeutic
L-Ornithine	50	3a, 3c	Therapeutic
L-Phenylalanine	3,000	3a, 3c	Infusions, therapeutic, aspartame production
L-Proline	100	3a	Infusions
L-Serine	50	3a, 3b	Cosmetics
L-Threonine	160	3a	Feed additive
L-Tryptophan	200	3a, 3c	Infusions, therapy
L-Tyrosine	100	1, 3c	Infusions, L-DOPA synthesis
L-Valine	150	3a, 3c	Infusions

[a] Production methods: 1, Hydrolysis of proteins; 2, chemical synthesis; 3a, direct fermentation; 3b, microbial transformation of precursors; 3c, use of enzymes or immobilized cells.

SOURCE: Soda, K., Tanaka, H., and Esaki, N. (1983), Amino acids, in Rehm, H.-J., and Reed, G. (eds.), *Biotechnology*, vol. 3, pp. 479–530 (Verlag Chemie).

Amino acids are used in a variety of ways, most of them associated with food. Their major use is as a flavor enhancer, and here L-glutamic acid is by far the most important (see Table 12.1). Amino acids are also used as food additives and feed additives. As is well known, most higher animals cannot synthesize lysine, methionine, tryptophan, leucine, isoleucine, valine, phenylalanine, threonine, and arginine—the "essential amino acids"—but must obtain them from the proteins in their diet. However, the less expensive, more abundant sources of food proteins, the seeds of crop plants, are rather deficient in some of the essential amino acids, particularly lysine, methionine, and tryptophan. For example, the proteins in corn contain only 0.2% lysine, whereas the proteins in animal meat contain 2.6% lysine. The nutritional value of plant seed proteins can be increased significantly if the seeds can be fortified with the deficient amino acids. A third important use of amino acids is as a starting material for the production of other compounds. For example, the sweetener aspartame is produced from L-phenylalanine and L-aspartate. Finally, some amino acids have medical uses, whether for specific therapeutic effects or as components of intravenous infusions given to patients who have difficulty taking in enough nutrients from food.

This chapter will discuss how amino acids are produced by fermentation, both with mutant and wild-type strains and by enzymatic processes.

FERMENTATION WITH MUTANT STRAINS

The fermentation processes that produce amino acids differ significantly from those that produce organic chemicals such as ethanol (described in Chapter 11). The latter compounds are by-products of catabolic fermentation pathways, whose physiological function in the microorganism is to generate ATP; these compounds are simply waste products and thus are excreted in large amounts into the medium. In contrast, the amino acids are building blocks that the cells synthesize—at a net expense of energy—and ultimately use in making proteins. It is to the cells' advantage not to waste energy by producing more amino acids than they need. Consequently, amino acid production is usually effectively regulated.

Our good understanding of such regulatory processes has come primarily through studies of *E. coli*, which lives in the intestinal tract of higher vertebrates and thus leads a life of "feast or famine." It can predict neither the nature of the amino acids that will become available in its environment nor when they will arrive. Consequently, the regulatory mechanisms of *E. coli* are much more sophisticated than those of many soil microorganisms, for example, whose environments are less complex and more stable.

In *E. coli*, the biosynthesis of most amino acids is regulated at two different levels: (1) by control of the *activity of pre-existing enzymes*, and (2) by control of the *synthesis of new enzyme molecules*. The former governs the response of *E. coli* to the sudden appearance of amino acids in the environment. The biosynthesis of amino acids consumes a large amount of chemical energy. Thus when a supply of exogenous amino acids becomes available, it is advantageous for the bacterium to shut down its own biosynthetic pathway and start utilizing the prefabricated amino acids. However, the cells already contain a full complement of

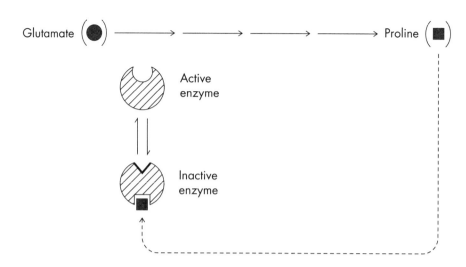

Figure 12.1

Regulation of proline biosynthesis by feedback inhibition. In this example, the biosynthesis of the amino acid proline is regulated by altering the activity of the first enzyme of the pathway, glutamate kinase. This enzyme is an allosteric enzyme, with a binding site for the substrate, glutamate, as well as a separate binding site for the inhibitor, proline. Whenever excess proline is present, it binds to the regulatory site of glutamate kinase, thereby converting the enzyme into an inactive form. This shuts down the endogenous proline synthesis instantaneously.

the enzymes that participate in the formation of amino acids. Under these circumstances, the best response for the cell is to lower the activity of pre-existing enzymes. This is usually achieved by *feedback inhibition*, a process in which an excess of end product, in this case an amino acid, inhibits the activity of the first enzyme of the biosynthetic pathway via an allosteric mechanism (Figure 12.1).

Feedback inhibition is an ideal mechanism for coping with rapid fluctuations in the supply of amino acids in the environment, but it is inefficient for long-term adaptation. If a population of bacteria continues to live in an environment in which tryptophan, for example, is always available at high concentrations, it makes no sense for the bacteria to synthesize all the enzymes needed for tryptophan biosynthesis, using much energy in the process, and then to keep all of these enzymes from functioning by a system of feedback inhibition. It is much more economical to shut down the synthesis of the unneeded enzymes. Thus regulation at the level of enzyme synthesis is also important, although it does not help the organism adjust to sudden changes in its environment.

Enzyme synthesis can be regulated by one of two mechanisms. In *repression*, the amino acid end product of the pathway binds to a specific

⋯⋯⋯▶ An *allosteric enzyme* is an enzyme whose activity is regulated by molecules other than its substrates. It contains two sites: The catalytic site binds the substrate, and the allosteric ("other position") or regulatory site binds the regulator, which often has a structure totally unrelated to that of the substrate. The occupancy of the regulatory site affects the catalytic activity of the enzyme through alteration of the conformation of the enzyme.

repressor protein as a *corepressor*, altering its conformation. The unliganded repressor has no effect on enzyme synthesis, but the corepressor-repressor complex binds to an upstream sequence of the genes and operons coding for the biosynthesis of that particular amino acid, preventing transcription of the messenger RNA (Figure 12.2). *Attenuation*, the other mechanism, works by controlling the frequency of RNA chain termination during mRNA transcription. In this mechanism, the 5′-end of the mRNA can form one of the two alternative stem-loop (or hairpin) structures (Figure 12.3). When segments 3 and 4 form a hydrogen-bonded stem loop, that structure acts as a rho-independent termination signal for RNA polymerase (see Box 3.6), because its GC-rich stem is followed by a continuous stretch of U residues. However, the stem loop formed when segment 2 becomes aligned with segment 3 does not act as a termination signal, because there is no stretch of U right after this hairpin. Attenuation works because segment 1 contains multiple codons

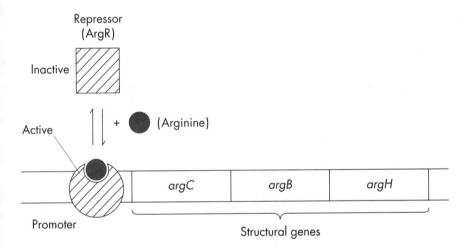

Figure 12.2
Regulation of amino acid biosynthesis by repression. In *E. coli* K12, the transcription by RNA polymerase of genes for arginine biosynthesis (here *argC*, *argB*, and *argH*, which form an operon) is regulated by the repressor protein (ArgR). When no free arginine is present in the cytoplasm, ArgR does not bind to the upstream region of the operon, and transcription occurs. When arginine is present, it binds to ArgR, and the liganded form of ArgR binds to the region of DNA just behind the promoter, presumably preventing the RNA polymerase from initiating the transcription process. Note that repression affects the synthesis of *all* the enzymes of the pathway (other genes of arginine synthesis are located elsewhere on the chromosome but are similarly regulated by ArgR). In contrast, feedback inhibition usually affects the activity of only the first enzyme of the pathway (or of the branch in a branched pathway).

A

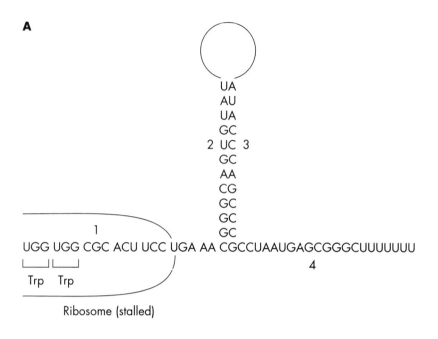

B

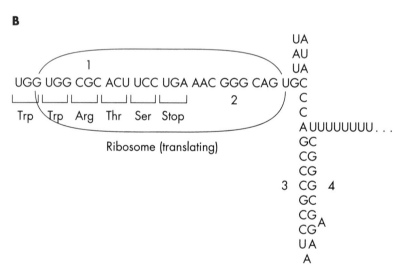

Figure 12.3

Regulation of amino acid biosynthesis by attenuation. In *E. coli*, the levels of the enzymes responsible for tryptophan biosynthesis are regulated predominantly by attenuation. The 5′-terminal part of the mRNA for *trp* operon can form two alternative looped structures, (**A**) one involving the pairing of regions 2 and 3, and (**B**) one involving the pairing of regions 3 and 4. Close to the beginning of this stretch of mRNA, there are two consecutive codons for tryptophan. When tryptophan (and therefore tryptophanyl-tRNA) is scarce, translation is stalled at the position of the tryptophan codons. In this state (**A**), region 1 is entirely covered by the ribosome, the 2-3 stem loop is formed, and transcription continues. When tryptophan is readily available, the ribosome moves past the tryptophan codons, inhibiting formation of the 2-3 stem loop, and structure (**B**) forms instead. The stem-loop structure created by the pairing of regions 3 and 4 is a typical rho-independent terminator, with a loop followed by stretches of U (see Box 3.6). In its presence, mRNA transcription is terminated, and tryptophan biosynthetic enzymes are not produced. REFERENCE: Yanofsky, C. (1981), Attenuation in the control of expression of bacterial operons, *Nature* 289:751–758.

for the particular amino acid whose synthesis is to be regulated—for example, tryptophan. If no tryptophanyl-tRNA molecules are present, then translation becomes stalled in segment 1. Segment 2 then forms a hairpin structure with segment 3, and transcription continues. In contrast, if tryptophanyl-tRNA is abundant, the continued translation inhibits the formation of a stem structure between segments 2 and 3, and the chain-terminating hairpin between 3 and 4 is formed instead, halting production of the enzymes needed to synthesize tryptophan.

In *E. coli*, the biosynthesis of many amino acids is regulated at the level of enzyme synthesis by both the repression and the attenuation mechanisms. This redundancy apparently expands the range of amino acid concentrations to which *E. coli* can respond. There is evidence that repression plays the predominant regulatory role when the amino acid concentrations in the medium are rather high and that attenuation becomes more important when the amino acids are scarce.

Production of Biosynthetic Intermediates by Auxotrophic Mutants

The foregoing discussion of the multiple regulatory mechanisms in *E. coli* may give the impression that to create mutant strains that overproduce amino acids is a hopelessly complex task. This is true to a certain extent, as we will see. Much more practicable, however, is the creation of strains that overproduce biosynthetic intermediates of amino acids. All the regulatory mechanisms—feedback inhibition, repression, and attenuation—require the end product in order to function. If one of the steps in amino acid biosynthesis is blocked, none of the regulatory machineries will work, and the cell will overproduce the intermediate that lies just before the blocked step. This type of mutant is termed *auxotrophic* (requiring growth factors, such as amino acids, purines, pyrimidines, or vitamins, for nutrition) because it cannot survive without the amino acid whose biosynthesis is blocked. These auxotrophic mutants are useful if the intermediate itself is a useful commodity or if it can be converted easily into the desired amino acid by a chemical process.

To isolate mutants that are defective in amino acid biosynthesis, one must create an environment in which they have the survival advantage. A classical technique is *penicillin selection*. Assume that the parent strain can synthesize all of the amino acids; that is, it is "prototrophic" and can grow in a mineral medium containing a single organic compound as its major source of cellular carbon and energy. In order to isolate mutants that are unable to synthesize L-arginine ("arginine auxotrophs"), one usually exposes the population of wild-type cells to mutagens so that the frequency of mutants will increase. Then the bacteria are grown in the same mineral medium with the carbon source, but this time it is fortified with L-arginine as well. When the population has reached a suitable density, the cells are harvested, washed, and transferred to a medium exactly like the last one except that it lacks L-arginine. When penicillin is added to this culture, only arginine auxotrophs survive. Penicillin kills growing bacteria, because it inhibits the crosslinking of the newly synthesized cell walls and eventually causes lysis of the cells. However, it is totally without effect on nongrowing bacteria, which do not possess new peptidoglycan material. The prototrophic cells are killed, because they are capable of growth in the arginine-free medium. The mutant cells, which are incapable of growth, survive. The arginine auxotrophs can then be recovered by plating the mixture on a medium containing L-arginine but no penicillin.

L-Ornithine, which has medicinal applications, is an example of a useful biosynthetic intermediate. Because L-arginine is made from L-citrulline, which in turn is derived from L-ornithine, L-ornithine is overproduced by mutants that are blocked in the arginine biosynthetic pathway at the step converting L-ornithine to L-citrulline. Mutants blocked at earlier or later steps are useless. To find L-ornithine-producing mutants, researchers isolated arginine auxotrophs and then selected those that were able to grow after the addition of citrulline to the medium but were unable to grow after the addition of ornithine. Such mutants are defective in ornithine transcarbamylase, which catalyzes the conversion of ornithine to citrulline, and they accumulate large amounts of ornithine because there is no arginine present to trigger negative regulation. During 1957, the first year of the modern era of amino acid fermentation, researchers isolated such a mutant belonging to the newly discovered glutamate-producing bacterial species *Corynebacterium glutamicum*; we will describe it shortly.

Auxotrophic mutations are also useful in the production of certain amino acids that are synthesized by branched pathways. If several amino acids—say A, B, C, and D—are produced as end products of a branched pathway, the effect, in some organisms, of cutting off the branches leading to B, C, and D is an increase in the production of A. This is primarily because the negative regulatory effects of B, C, and D are eliminated and secondarily because the flow of carbon to the cut off branches is decreased. These cases will be considered in more detail in our discussion of regulatory mutants of branched pathways.

A major drawback to the use of auxotrophic mutants is that the amino acid that the auxotrophic strain cannot synthesize must be added to the medium. If too much is added, the excess exerts negative regulatory effects on the biosynthetic pathway, the very effect the mutant was designed to avoid. Thus the mutant strains must be grown in a defined medium (such media tend to be expensive), and the required amino acids must be added in carefully controlled amounts (this process is called fed-batch fermentation). As a result, auxotrophic mutants are usually not the ideal strains for amino acid production.

Production of Amino Acids by Regulatory Mutants of Unbranched Pathways

Mutants that are defective in the negative regulation of amino acid biosynthesis overproduce the amino acid. Unlike auxotrophic mutants, these *regulatory mutants* can be grown in inexpensive, complex media, and they do not require careful control of growth conditions.

In almost all cases, these regulatory mutants are isolated through the use of amino acid analogs that inhibit the growth of the parent strain in a minimal medium. It has been known for many years that some analogs of amino acids are toxic. Earlier, the prevailing view was that these analogs inhibited growth by becoming misincorporated into proteins, thereby producing nonfunctional proteins. This may well occur in some instances, but in a vast majority of cases the major cause of inhibition appears to be that the analogs mimic the way the amino acid regulates its own production. Thus an analog may bind to the allosteric site of the first enzyme of the synthetic pathway or may bind effectively to the repressor, in either case shutting off the synthesis of that particular amino acid. Growth is inhibited because the cells now become starved of that amino acid. Researchers take advantage of this phenomenon to select for regulatory mutants by synthesizing a wide variety of analogs of an amino acid, selecting those analogs that effectively inhibit growth of the wild-type strain in a minimal medium, and then selecting for mutants that are able to grow in the presence of these analogs. In many such mutants, either the repressor or the first enzyme of the pathway is altered, which is why the amino acid synthesis proceeds even in the presence of the analog and will do so in the presence of the amino acid itself.

This approach does not work well when a biosynthetic pathway is regulated at more than one level. Mutation is a rare event, so the probability of isolating a strain that has simultaneously acquired two desirable mutations is miniscule. This is a major concern, because in *E. coli* many pathways are regulated at *three* levels: feedback inhibition, repression, and attenuation. Fortunately, many aquatic and soil microorganisms regulate amino acid biosynthesis more simply than *E. coli*, presumably because, unlike *E. coli*, they live in a more or less constant environment that is always poor in amino acids. In these organisms, the major function of the regulation of amino acid biosynthesis is not to enable the organism to adjust to a varying supply of the amino acids in the environment but to control the rate of amino acid biosynthesis so that it will meet the demands of the cell's protein synthesis machinery.

The foregoing points are well illustrated by certain mutant strains of a water and soil inhabitant, *Serratia marcescens*, that overproduce proline. Proline is the end product of an unbranched pathway (Figure 12.4). In *Serratia marcescens*, its synthesis is controlled almost entirely by feedback inhibition, making this a system ideally suited to the use of amino acid analogs. A few of the proline analogs tested, such as thia-

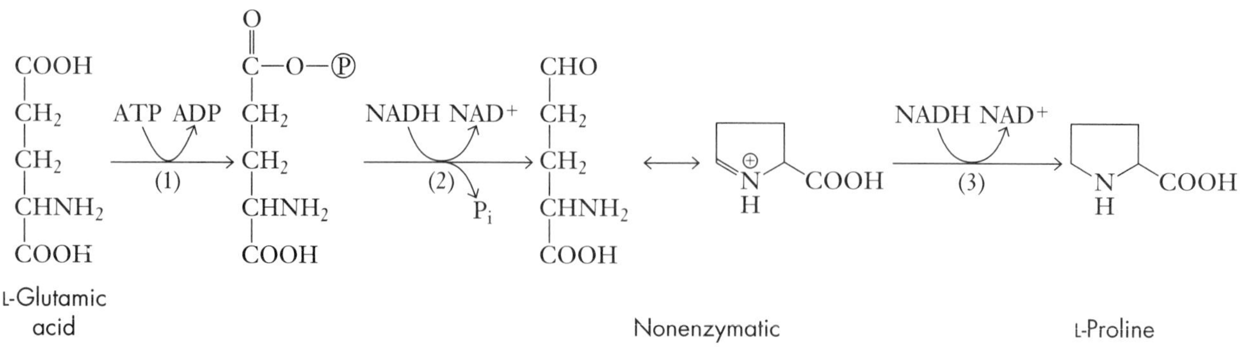

Figure 12.4
Proline biosynthetic pathway. The enzymes are (1) glutamate kinase, (2) glutamic γ-semialdehyde dehydrogenase, and (3) Δ¹-pyrroline 5-carboxylate reductase.

Proline

Thiazolidine-4-carboxylic acid

3,4-Dehydroproline

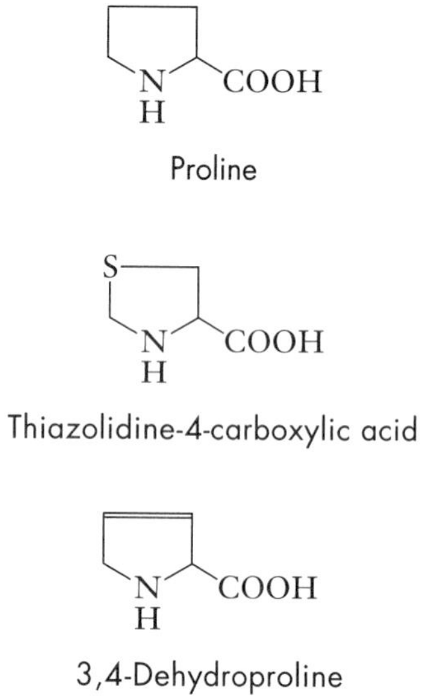

Figure 12.5
L-Proline and two toxic analogs.

zolidine-4-carboxylic acid and 3,4-dehydroproline (Figure 12.5), selected mutants in which the first enzyme of the pathway, glutamate kinase, was altered in such a way that it was no longer susceptible to feedback inhibition by proline. These mutants overproduced proline, as expected, but maximizing the production required the introduction of two additional changes. First, many wild-type strains of *Serratia marcescens* produce an enzyme that degrades proline. This enzyme enables the organism to use proline as a source of carbon and nitrogen and, in so doing, decreases the yield of proline by breaking some of it down after it has been synthesized. Thus it was necessary to inactivate the gene coding for this enzyme, proline oxidase. Second, proline functions as an osmoprotectant solute. When bacterial cells are forced to grow in a high-osmolarity medium, they try to avoid the injurious effect of high salt concentration in the cytoplasm by accumulating proline, both by overproducing it and by transporting it from the medium. Because of this additional regulatory mechanism, the yield of proline increases when the bacteria are grown in high-salt media. With these two improvements, yields of 60–75 g/liter of L-proline have been reported.

In a later attempt to increase further the production of proline by *Serratia marcescens*, the genes involved in proline biosynthesis were cloned from the overproducing mutant strain, inserted into multicopy plasmids, and reintroduced into *Serratia* strains defective in proline degradation. The new strains produced approximately 50% more proline, an amount that might seem modest until one considers that the mutant was already converting more than 20% of the carbon source into the final product.

When a pathway is regulated at more than one level, a production strain must contain a mutation at each level. Mutants selected with an analog are usually altered in only one of the regulatory mechanisms.

However, mutations can be combined via the classical methods of *E. coli* genetics, especially in organisms such as *Serratia* that are phylogenetically related to *E. coli*. This principle is nicely illustrated by the construction of a histidine-producing mutant of *S. marcescens*. Because *Serratia* uses histidine as a source of carbon and nitrogen, the first step was to create mutants in which the degradative enzyme, histidase, had been inactivated. Starting from this mutant, investigators then selected strains capable of growing in the presence of the toxic histidine analogs 2-methylhistidine and 1,2,4-triazole-3-alanine (Figure 12.6), respectively. As summarized in Table 12.2, one of the triazole-3-alanine (TRA)-resistant mutants, 142, overproduced enzymes of the histidine biosynthetic pathway (Figure 12.7), here exemplified by the much higher specific activity of one such enzyme, histidinol dehydrogenase, but the first enzyme of the pathway, phosphoribosylphosphate adenyltransferase, was unaltered in its susceptibility to the feedback inhibition. In contrast, one of the 2-methylhistidine (2MH)-resistant mutants, 581, produced an adenyltransferase that is remarkably resistant to feedback

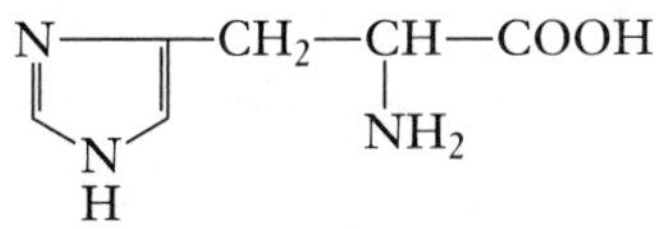

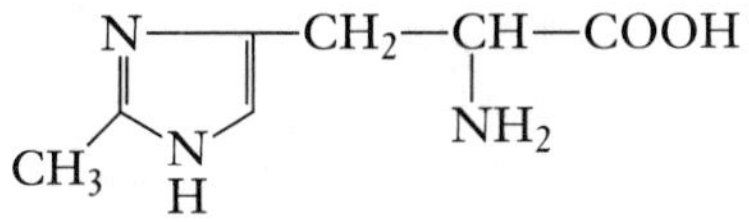

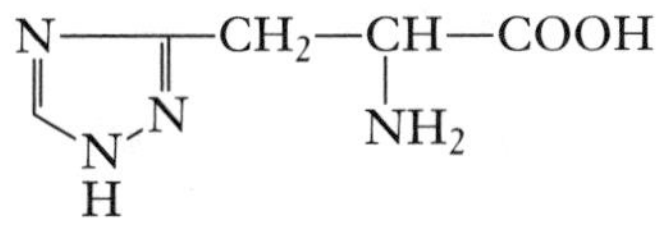

Figure 12.6
L-Histidine and two toxic analogs.

Histidine Biosynthesis in *Serratia* Strains[a]

Strain	1 *Extent of Repression* (specific activity of histidinol dehydrogenase, units/mg protein)	2 *Extent of Feedback Inhibition of the First Enzyme* (inhibition of PRPP adenylyltransferase by 10 mM histidine, %)	3 *Histidine Production* (g/liter)
Sr41 (wild type)	1.1	100	0
Hd-16 (histidase⁻)	0.9	100	0
581 (2MHʳ)	1.6	0	0.8
142 (TRAʳ)	12.4	94	1.3
2604 (142 × 581)	14.7	0	12.9

TABLE 12.2

[a] This table shows the effectiveness of the two regulatory mechanisms of histidine biosynthesis in various strains of *S. marcescens*. In column 1, the specific activity of one of the enzymes of the pathway is shown. The activity is low in the strains in which repression is inhibiting the expression of the genes, but it becomes higher in strains in which this regulatory mechanism has been altered. In column 2, the efficiency of feedback control is indicated. In the wild-type strain, the first enzyme of the pathway, phosphoribosylpyrophosphate (PRPP) adenyltransferase, is completely inhibited by the end product, histidine, but in mutant 581 this regulatory mechanism is practically eliminated.

Abbreviations: 2MH, 2-methylhistidine; TRA, 1,2,4-triazole-3-alanine.

SOURCE: Kisumi, M., Komatsubara, S., Sugiura, M., and Takagi, T. (1987), *CRC Crit. Rev. Biotechnol.* 6:233–252.

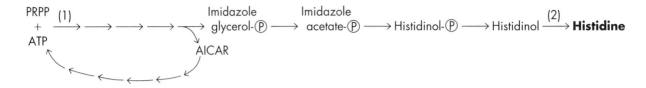

Figure 12.7
Histidine biosynthetic pathway. Abbreviations: PRPP, 5-phosphoribosyl pyrophosphate; AICAR, 5-aminoimidazole-4-carboxamide ribonucleotide. The steps (1) and (2) are catalyzed by PRPP adenyl transferase and histidinol dehydrogenase, respectively.

inhibition by histidine. Because each mutant was altered in only one regulatory mechanism, neither mutant produced large amounts of histidine (see column 3 of Table 12.2). However, scientists at Tanabe, a pharmaceutical company in Japan, were able to combine the two regulatory mutations by using transduction. The transductant, 2604 (see Table 12.2), was desensitized in both repression and feedback inhibition processes, and, as expected, it produced large amounts of histidine.

Creating a system that overproduces metabolites is usually more complicated than creating one that overproduces primary gene products. The increased flux through metabolic pathways may become limited in unexpected ways. This is one of the reasons why straightforward applications of recombinant DNA technology have not always produced spectacular results in this field. In the foregoing example, strain 2604 and similar histidine-production strains developed adenine deficiency that is presumably related to the consumption of ATP in the first step of the histidine pathway. Although an adenine moiety is expected to be regenerated from 5-aminoimidazole-4-carboxamide ribonucleotide (AICAR; see Figure 12.7), this process probably cannot keep pace with the extremely rapid rate of histidine biosynthesis. Tanabe scientists solved the problem by putting the mutant strain through yet another selection cycle, this time using a toxic adenine analog, 6-methylpurine. The result was a resistant strain that overproduces adenine and hence is able to produce histidine at an increased level (23 g/liter) without requiring the addition of adenine to the medium.

Production of Amino Acids by Regulatory Mutants of Branched Pathways

Proline and histidine, which we have described, are the products of unbranched biosynthetic pathways, but many other amino acids are produced by branched pathways. Members of the aspartate family of amino acids (lysine, methionine, threonine, and isoleucine) are produced by one branched pathway, those of the pyruvate family (valine and leucine) by another branched pathway, and those belonging to the aromatic family (tryptophan, phenylalanine, and tyrosine) by still another. The regulation of these pathways is much more complex, because it requires a system in which an excess of one product does not accidentally shut off the entire pathway. The regulatory mechanisms in the branched pathways of *E. coli* are very complicated indeed. For example, in the aspartate pathway (Figure 12.8), each product inhibits and/or represses the first enzyme of the common pathway, aspartate kinase. However, so that one product does not shut down all of the kinase activity, three different isozymes (different enzymes that catalyze the same chemical reaction) of aspartate kinase are produced: one sensitive to lysine inhibition, one sensitive to threonine inhibition, and one repressed mainly by an excess of methionine. In addition, *E. coli* is able to adjust to the presence of an unbalanced mixture of the products of the pathway. For example, if it encounters an excess of methionine but low levels of the other products (lysine, isoleucine, and threonine), regulating the aspartate kinase alone will not lead to the proper ratio of methionine to the other three amino acids. Thus each amino

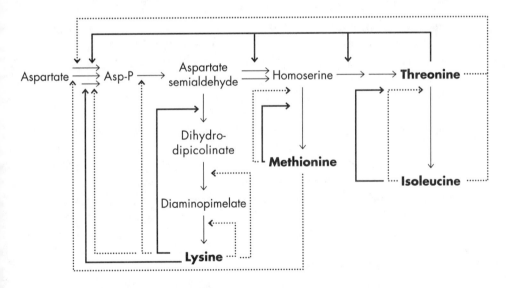

Figure 12.8
Regulation of synthesis of the aspartate family amino acids in *E. coli*. The pathways are shown in abbreviated form; for example, a single arrow from threonine to isoleucine represents five consecutive enzyme-catalyzed steps. Continuous thick arrows indicate regulation by feedback inhibition, and dotted arrows represent repression or attenuation regulation. Repression and attenuation affect all enzymes of a branch, but here only the effect on the first enzyme of the branch is shown. Step 1 is catalyzed by three isozymes of aspartate kinase. Modified from Tosaka, O., and Takinami, K. (1986), in Aida, K., et al. (eds.), *Biotechnology of Amino Acid Production*, pp. 152–172 (Kodansha).

acid usually controls the first enzyme of its own particular branch. Methionine and isoleucine regulate the pathway through both feedback inhibition and repression mechanisms, while lysine and threonine participate in more complex regulatory patterns.

Eliminating various types of controls in a system like this is a major endeavor. Such a maneuver was attempted with *E. coli* for the production of tryptophan. But even after the *tour de force* achievement of combining ten mutations, each with defined effects on specific aspects of the regulatory mechanism (Box 12.1), the yield was less than 1.5 g/liter, far below the commercially viable level. (More recent work, involving gene amplification with recombinant plasmids, has produced a more respectable yield in this organism.)

Microorganisms with Less Complex Regulatory Mechanisms

Luckily for us, numerous microorganisms have much simpler regulatory mechanisms, presumably because these organisms do not often encounter mixtures of amino acids, sometimes of unbalanced composition, in their environment. If the organism usually lives in an environment that is poor in amino acids, the major function of the regulation of amino acid biosynthesis is to adjust its rate in response to the growth rate of the organism. Thus there is no need to adjust the ratios of the various amino acids produced. When such an organism is producing excess lysine (for example, because its growth has slowed), it is also likely to be producing excess methionine, threonine, and isoleucine. The major regulation of the branched pathway can then be achieved by a system in which only one or a few of the products inhibit the first common enzyme. This type of simple regulatory scheme is indeed found in the soil bacterium *Brevibacterium flavum* (Figure 12.9), a member of the "coryneform cluster" of amino acid–producing organisms, which includes the glutamic acid producer *Corynebacterium glutamicum*.

Lysine production was extensively studied in these organisms, because lysine is one of the amino acids that occur only in small amounts in cereal proteins. Nutritionists evaluate food proteins by calculating the ratio between the weight the animal gains and the weight of the protein it is fed: the *protein efficiency ratio (PER)*. Corn proteins have a low PER of 0.85 because of their low lysine and tryptophan content, but this value can be increased about 3-fold, to 2.55, by adding 0.4% lysine and 0.07% tryptophan to the corn. The large impact of these small additions illustrates the importance of lysine as a food and feed additive.

Production by Auxotrophic Mutants

The simplicity of the regulatory schemes in the *Brevibacterium–Corynebacterium* group makes it possible to obtain a large amount of lysine simply by cutting off the branches leading to other amino acids.

BOX 12.1 Achieving Overproduction of
Tryptophan in *E. coli*

Tryptophan synthesis in *E. coli* is regulated by a complex set of mechanisms. By transducing each mutation one at a time, researchers combined a long list of alterations to these mechanisms within a single strain, creating a strain of *E. coli* capable of overproducing tryptophan. The first step of the aromatic pathway, the conversion of erythrose 4-phosphate and phospho*enol*pyruvate to 3-deoxy-D-*arabino*-heptulosonate 7-phosphate (DHAP), is catalyzed by three isofunctional enzymes regulated by tyrosine, tryptophan, and phenylalanine, respectively. Researchers simplified the system by mutationally inactivating the genes coding for tryptophan- and phenylalanine-regulated DHAP synthases (*aroG* and *aroH*) and by making the tyrosine-regulated enzyme insensitive to feedback inhibition, again through mutation (*aroF394*). The repression of this enzyme by tyrosine was prevented by inactivation of the repressor gene (*tyrR*). Other changes included cutting off the branches leading to tyrosine and phenylalanine (*tyrA* and *pheA*) so that the flow of carbon would not be diverted into other products of the aromatic pathway, and so that these endproducts could not act as inhibitors; inactivating the gene for tryptophanase (*tna*), a catabolic enzyme, to prevent the possible breakdown of the synthesized tryptophan; alleviating the feedback regulation of the tryptophan branch by making anthranilate synthetase insensitive to tryptophan (*trpE382*); removing the repression control on tryptophan pathway enzymes by inactivating the tryptophan repressor (*trpR*); and destroying the cell's attenuation control by mutating the gene for tryptophanyl-tRNA synthetase (*trpS*). The figure shows the functions of catalytic gene products modified (those of purely regulatory components such as TyrR, TrpR, and TrpS are not shown). Data from Tribe, D.E., and Pittard, J. (1979), *Appl. Environ. Microbiol.* 38:181–190.

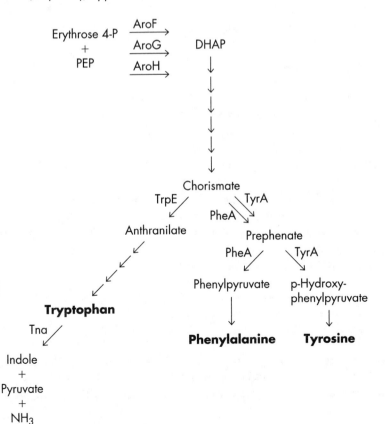

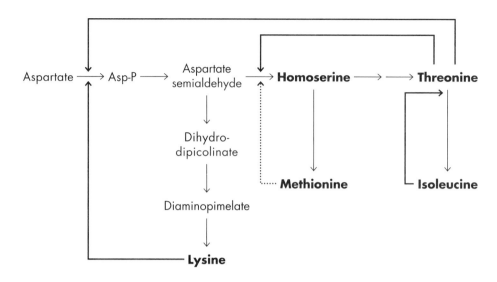

Figure 12.9
Regulation of synthesis of the aspartate family amino acids in *Brevibacterium flavum*. Thick continuous arrows show regulation by feedback inhibition, and dotted arrows represent repression control. After Tosaka, O., and Takinami, K. (1986), in Aida, K., et al. (eds.), *Biotechnology of Amino Acid Production*, pp. 152–172 (Kodansha).

For example, a yield of 34 g lysine per liter of medium was obtained from a mutant of *Brevibacterium flavum* that was defective in the branches leading to methionine and to threonine and isoleucine. The diversion of metabolites that would normally go to create those other amino acids contributed to the overproduction of lysine, but the primary reason it occurred is that efficient feedback inhibition of aspartate kinase requires both lysine and threonine (see Figure 12.9). When lysine and threonine are present simultaneously at 1 mM, they inhibit aspartate kinase by 94%, whereas each amino acid, when it is present alone at 1 mM, produces only marginal (12–20%) inhibition. However, lysine-producing auxotrophic mutants of this type must be fed methionine, threonine, and isoleucine continuously and in carefully measured amounts so that they are never present in large excess. Consequently, regulatory mutants are much more desirable for the commercial production of lysine.

$$
\begin{array}{cc}
NH_2 & NH_2 \\
| & | \\
CH_2 & CH_2 \\
| & | \\
CH_2 & CH_2 \\
| & | \\
CH_2 & S \\
| & | \\
CH_2 & CH_2 \\
| & | \\
CHNH_2 & CHNH_2 \\
| & | \\
COOH & COOH \\
\text{Lysine} & \textit{S}\text{-Aminoethylcysteine}
\end{array}
$$

Figure 12.10
Structures of lysine and *S*-aminoethylcysteine.

Production by Regulatory Mutants

Mutant strains of *Brevibacterium*, altered in the regulation of lysine biosynthesis, have typically been obtained by using the lysine analog *S*-aminoethylcysteine (AEC) (Figure 12.10). Like lysine, this analog inhibits the activity of aspartate kinase and hence inhibits the growth of the wild-type bacteria. Thus AEC-resistant mutants are likely to have an alteration in their aspartate kinase such that the altered allosteric, regulatory site of the enzyme has a lower affinity for AEC. These mutant enzymes are also likely to bind lysine with lower affinity and to be defective in the feedback regulation of lysine synthesis. A yield of 32 g lysine per liter was reported for such a mutant; additional fine-tuning that presumably modified the minor regulatory mechanisms increased the yield to 60 g/liter.

To isolate analog-resistant mutants, it is often necessary first to mutagenize the parent strain in order to increase the frequency of mutation. Because chemical and physical mutagenesis procedures are nonspecific, however, every mutant strain isolated tends to contain many other unrelated mutations, some of which may have adverse effects on the organism, such as slowing down the growth of the culture or making the cell more fragile. These unwanted mutations can be removed by backcrossing a mutant strain with a wild-type strain. (*Backcrossing* means crossing an individual with its wild-type parent). In organisms of the *Brevibacterium–Corynebacterium* group, conjugational crosses are difficult to carry out, so crossing is most easily done by protoplast fusion (Box 12.2). Figure 12.11 shows an example in which "vigor" was restored in a lysine-overproducing strain by such a backcrossing process. Faster growth means that the fermentation tank can be reutilized many more times, so this improvement is important in the commercial production of relatively inexpensive compounds such as amino acids.

Concept of Preferential Synthesis

Isamu Shiio and associates developed the notion of *preferential synthesis* to explain the regulation of the enzymes of the aspartate family. Shiio argued that the simplest mechanism for regulating a branched pathway

---

> **BOX** 12.2 **Protoplast Fusion**
>
> Natural mechanisms for genetic exchange are found in several species of bacteria (see Chapter 3), but for the vast majority of bacteria, such mechanisms either are unknown or do not exist. A technique known as protoplast fusion can be used to force such bacteria to exchange genetic material. The first step is to remove the cell wall, usually by exposing the bacterial cell to digestive enzymes such as lysozyme. If this is done in an osmotically protective medium, protoplasts (cells bounded by cytoplasmic membrane only) are generated. Application of fusogenic agents, such as polyethylene glycol, causes the protoplasts to fuse with each other. During the outgrowth of these protoplasts to generate progeny cells, the extra genome is lost, but in many cases the chromosomes will have exchanged parts and recombined beforehand. Protoplast fusion has been successfully used not only with bacteria (including *Streptomyces*) but also with fungi (including yeasts) and even plant cells.

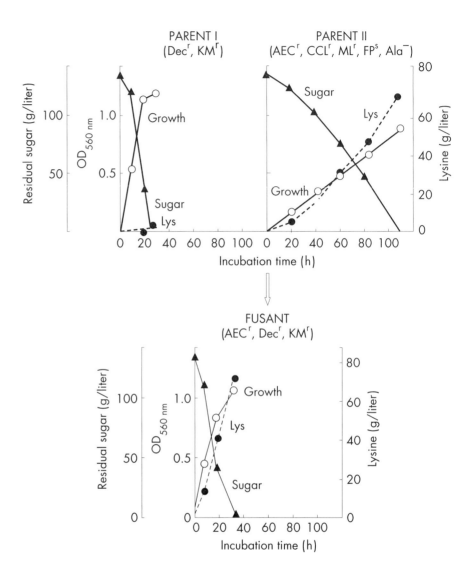

Figure 12.11

Restoration of strain vigor by backcrossing. Parent I is essentially a wild-type strain that grows rapidly but does not accumulate any lysine. Parent II was produced by successive mutagenesis and selection; it accumulates a large amount of lysine but grows very slowly. Crossing these two strains produced a recombinant ("fusant") that grew rapidly yet accumulated high concentrations of lysine. The ordinate shows growth (in OD_{560}—that is, optical density at 560 nm), the amount of sugar (carbon source) remaining in the medium, and the amount of lysine produced. Abbreviations: Dec, decoinine; KM, ketomalonate; CCL, γ-chlorocaprolactam; ML, γ-methyllysine; FP, β-fluoropyruvate. Modified from Tosaka, O., et al. (1982), Abstracts, Internat. Symp. Genet. Industrial Microorganisms, p. 61.

is to have very unbalanced enzyme activities at the branch point (Figure 12.12). For example, when the activities of various enzymes of the aspartate pathway (Figure 12.13) were assayed in cell extracts, the first enzyme on the homoserine branch, homoserine dehydrogenase, showed a specific activity 15 times higher than that of the first enzyme of the lysine branch, dihydropicolinate acid synthetase. A similar type of imbalance was found at every other branch. At the point where the methionine and threonine biosyntheses diverge, the first enzyme of the methionine branch, homoserine transacetylase, has a V_{max} value three times higher than that of the first enzyme of the threonine branch, homoserine kinase. When the activities were compared at the fairly low substrate concentrations likely to be found in the cytoplasm (for example 0.1 mM homoserine), the difference in activity was approximately 10-fold.

Because a typical protein does not contain such unbalanced amounts of different amino acids, the enzymes after the branch point must be

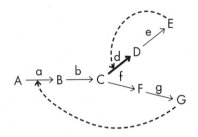

Figure 12.12
Principle of regulation by preferential synthesis. At every branch, one pathway is greatly favored over the other (here pathway "d" is favored over pathway "f "). The entire pathway is regulated efficiently if the end product of the favored branch (product "E") inhibits the first enzyme on pathway "d" after the branch and if the end product of the less-favored branch (product "G") inhibits the first enzyme of the unbranched part of the pathway. Based on Shiio, I. (1982), in Krumphanzl, K., Sikyta, B., and Vanek, Z. (eds.), *Overproduction of Microbial Products*, pp. 463–472 (Academic Press).

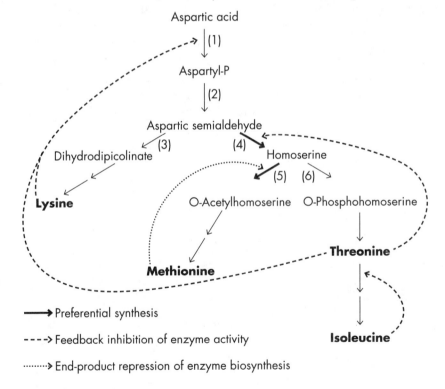

Figure 12.13
Regulation in *Brevibacterium flavum* of the synthesis of the amino acids of the aspartate family, as an example of preferential synthesis. The enzymes catalyzing the numbered steps are (1) aspartate kinase, (2) aspartic semialdehyde dehydrogenase, (3) dihydropicolinate synthetase, (4) homoserine dehydrogenase, (5) homoserine transacetylase, and (6) homoserine kinase. Based on Shiio, I. (1982), in Krumphanzl, K., Sikyta, B., and Vanek, Z. (eds.), in *Overproduction of Microbial Products*, pp. 463–472 (Academic Press).

regulated, either by activation of the weaker enzyme or by inhibition/ repression of the stronger enzyme. Bacteria generally utilize the latter mechanism. The stronger enzyme at the branch point is regulated by negative feedback by the end product of that branch (Figures 12.13 and 12.14). In intact cells, this mechanism enables the weaker enzymes after the branch point to operate at a rate similar to that of the stronger enzymes. Because the branch containing the weaker enzyme is regulated not after the branch point but at the start of the common pathway (Figure 12.12), the level of the end product of the branch reflects the level of activity of the common part of the pathway. Thus it is possible to control the activity of the common pathway by a feedback mechanism in which the end product of the weaker-enzyme branch regulates the first enzyme of the entire pathway (Figures 12.12 and 12.13). This enables the cell to achieve a balanced synthesis of all the end products of a branched pathway with a minimal number of regulatory mechanisms. As proposed by Shiio, this scheme is likely to be the basic mechanism regulating most branched pathways; similar regulatory patterns are in fact found in many "simple" soil bacteria used in amino acid production. The one drawback of this mechanism is that it cannot cope with an unbalanced supply of amino acids from the outside. For example, if the bacterium in Figure 12.12 encounters an amino acid mixture that is rich in "G" but contains almost no "E," it will not be able to grow because the entire pathway will be shut down by "G." Species that are likely to

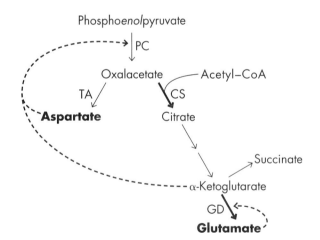

Figure 12.14
Aspartate, the starting material on the lysine pathway, is itself a product of a branched pathway. Thick arrows represent the favored branch, and open, curved arrows show the feedback inhibition control operating in *Brevibacterium flavum*. Abbreviations: PC, pyruvate carboxylase; CS, citrate synthase; TA, aspartate aminotransferase; GD, glutamate dehydrogenase. Based on Shiio, I. 1982, in Krumphanzl, K., Sikyta, B., and Vanek, Z. (eds.), *Overproduction of Microbial Products*, pp. 463–472 (Academic Press).

encounter complex mixtures of exogenous amino acids need a more complex regulatory scheme similar to that seen in *E. coli*.

This concept of preferential synthesis suggested a strategy for further increasing the yield of some amino acids. We have already seen that a *B. flavum* mutant with a feedback-insensitive aspartate kinase overproduces lysine. However, the starting material of the lysine pathway, aspartate, can in its turn be considered an end product of a branched pathway, as shown in Figure 12.14. The branch that leads to aspartate is a less favored branch, and aspartate regulates the first enzyme in the common part of the pathway, pyruvate carboxylase, by feedback inhibition. Therefore, a feedback-insensitive mutant of pyruvate carboxylase should increase the production of aspartate and, in turn, of lysine. Introduction of this alteration (together with a defect in citrate synthase to cut off the favored branch; see Figure 12.14) indeed increased the yield of lysine nearly 3-fold, from 20 to about 60 g/liter.

Amplification of Biosynthetic Genes

When the yield of a given amino acid is only marginal, amplification of the biosynthetic genes by recombinant DNA methods may produce significant improvements. This technique was used to construct a threonine-overproducing strain of *E. coli*. As shown in Figure 12.8, threonine exerts feedback inhibition on several enzymes of the aspartate pathway in this organism. In the first step, the researchers isolated a mutant resistant to the toxic threonine analog α-amino-β-hydroxyvaleric acid (Figure 12.15). This mutant produced a feedback-insensitive homoserine dehydrogenase, and its threonine-sensitive aspartate kinase was also altered. The simultaneous loss of sensitivity occurred because both activities depend on a single enzyme protein and apparently are regulated from a single allosteric site. The yield of threonine was still very low, but this was not surprising because there were other regulatory mechanisms left operating (see Figure 12.8). To stem the flow of materials into other branches and to remove some of the other regulatory mechanisms, the strain was made auxotrophic for methionine and isoleucine. (One consequence was that the absence of isoleucine made the attenuation of the pathway less efficient, because the leader sequence contains 4 isoleucine codons in addition to 8 threonine codons.) Though helpful, these changes increased the yield to only about 3 g/liter, still quite low. However, when the genes coding for the threonine pathway (*thrA*, aspartate kinase I and homoserine dehydrogenase I; *thrB*, aspartic semialdehyde dehydrogenase; *thrC*, threonine synthase) were cloned from this mutant strain, put onto a multicopy plasmid, and transformed back into the strain, the levels of the enzymes increased 5-fold, and the yield of threonine increased to 12 g/liter. Further improvement of the fermentation conditions, especially an increase in the oxygen partial pressure, raised the yield to 65 g/liter, a level commonly

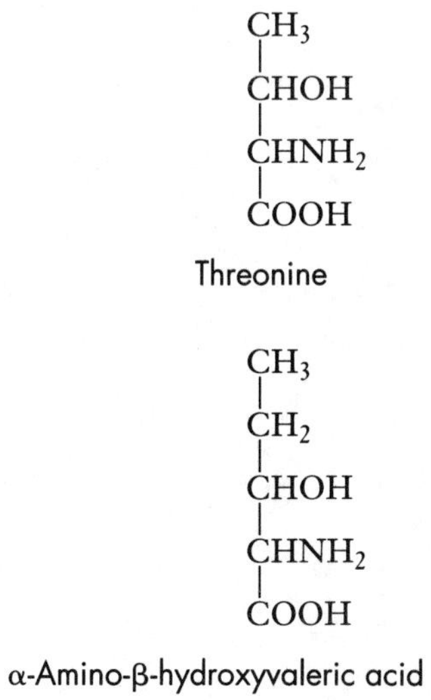

$$
\begin{array}{c}
CH_3 \\
| \\
CHOH \\
| \\
CHNH_2 \\
| \\
COOH
\end{array}
$$

Threonine

$$
\begin{array}{c}
CH_3 \\
| \\
CH_2 \\
| \\
CHOH \\
| \\
CHNH_2 \\
| \\
COOH
\end{array}
$$

α-Amino-β-hydroxyvaleric acid

Figure 12.15
Structures of threonine and α-amino-β-hydroxyvaleric acid.

> ┈┈┈┈┈┈▶ **BOX** 12.3 **Streptomycin Dependence Mutation**
>
> Streptomycin is an aminoglycoside antibiotic (Chapter 13) that kills bacterial cells by inhibiting the function of the 30S ribosomal subunit. Mutants that are highly resistant to this antibiotic have an altered ribosomal protein S12. Some of these mutants not only are resistant to streptomycin but also require the presence of streptomycin for growth, a property called streptomycin dependence.

seen in producing strains derived from *Brevibacterium* or *Corynebacterium*.

Plasmid vectors suitable for use in *Brevibacterium* and *Corynebacterium* are now available and have been utilized to improve the yield of amino acids in the laboratory. Because production strains go through a great many generations in fermentation runs in large tanks, the major problem with using plasmid-containing strains on an industrial scale is preventing loss of the plasmids. Solutions to this problem have been described, but it is not clear whether all of them can be used routinely in production strains. For example, a streptomycin-dependence mutation (Box 12.3) was put into the chromosome of the threonine-overproducing *E. coli* strain we have described, and the wild-type, streptomycin-independence marker was put into its plasmid. Streptomycin was not present in the medium, so when the plasmid was lost from the strain, the cells lost their ability to grow. In the laboratory, at least, this genetic manipulation prevents overgrowth by cells lacking plasmids.

FERMENTATION WITH WILD-TYPE STRAINS

As we have noted, amino acid fermentation technologies developed from the surprising observation that certain wild-type strains of soil bacteria excrete large amounts of glutamic acid into the medium. Since then, many unusual features related to this process have been discovered in the producing organisms.

Puzzle of Membrane Permeability

All glutamic acid overproducers are soil organisms belonging to the *Brevibacterium–Corynebacterium* group. The discovery of glutamic acid fermentation was a lucky accident that occurred because researchers did not know that all these overproducers require biotin for growth. Because biotin is required in very small amounts, the strains grew in the laboratory media, initially by utilizing biotin that was available as a contaminant; at the later stage of growth, however, they developed a biotin deficiency and began to excrete glutamic acid. When the strains were

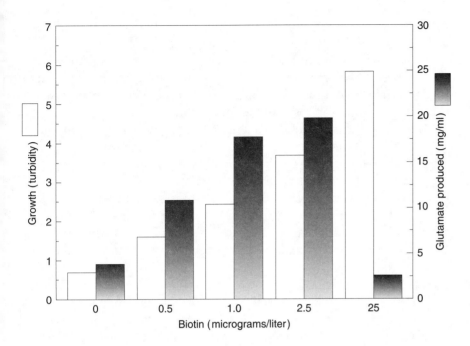

Figure 12.16
Influence of biotin concentration on the production of glutamic acid by
C. glutamicum. Based on Tanaka, K., Iwasaki, T., and Kinoshita, S. (1960),
Nihon Nogeikagakukai Zasshi 34:593–600.

grown with sufficient amounts of added biotin, there was almost no
production of glutamic acid (Figure 12.16).

Why is biotin starvation necessary for the excretion of glutamic
acid? The major function of biotin is to serve as a prosthetic group in
acetyl-CoA carboxylase, the first enzyme of fatty acid synthesis, so it was
hypothesized that the excretion might be related to an increased permeability of the cell membrane, caused by an insufficiency of fatty acids.
This notion was supported by the finding that other treatments affecting the integrity of the cell membrane—for example, adding to the
medium a detergent, Tween 60, or a low concentration of β-lactam
antibiotics (the latter treatment strains the cell membrane by weakening
the peptidoglycan cell wall) or limiting the amount of exogenously fed
oleic acid in unsaturated fatty acid auxotrophs—also increased the yield
of glutamic acid, even in the presence of excess biotin.

These observations were of practical importance, because they suggested ways to stimulate glutamic acid excretion while feeding the bacteria with inexpensive carbon sources (such as molasses) that happen to
be rich in biotin. During the period when petroleum was inexpensive,
petroleum hydrocarbons were also tried as a carbon source. When scientists found that unsaturated fatty acid auxotrophs do not produce
glutamic acid when petroleum hydrocarbons are present (presumably

because they are easily converted to unsaturated fatty acids by the organism), they tried an alternative way of weakening the cell membrane: They constructed conditional auxotrophs for glycerophosphate so that the synthesis of phospholipids from glycerophosphate became deficient under certain conditions. Such mutants also excreted glutamate when the supply of glycerophosphate became limiting.

The results described above lend strong support to the hypothesis that a leaky cell membrane is responsible for the excretion of glutamic acid. Several questions remain to be answered, however. Principally, how do the cells remain alive if the membrane becomes so leaky? In other words, what prevents the leakage of essential metabolites and ions? And why is this treatment needed only for the production of glutamic acid and not for that of other amino acids? In short, we still do not have a complete understanding of the mechanism of amino acid excretion.

It has recently been shown that the *Brevibacterium–Corynebacterium* group produces specific transporters that function solely to excrete glutamic acid as well as certain other amino acids that they also overproduce. These "efflux transporters" are thought to be useful when the cytoplasm becomes flooded with amino acids, possibly as a consequence of the rapid uptake of peptides from the medium. However, for glutamic acid to be excreted in large amounts, additional alterations that are caused by biotin deficiency are required, and the nature of these alterations is not clear. The addition of biotin to biotin-starved cells immediately stops glutamate excretion, whereas it takes a generation or so to return the lipid content of the cells to a normal level. In spite of these uncertainties about the molecular mechanisms of the excretion process, there is no doubt that the effective excretion of amino acids by these organisms, and that of glutamic acid in particular, contributes significantly to the overproduction, because excretion dramatically decreases the intracellular concentrations of amino acids, and hence any regulatory effects they may have on their own synthesis.

Need for a Functionally Truncated Citric Acid Cycle

Many *Brevibacterium–Corynebacterium* strains excrete very large amounts of glutamate into the medium, close to 100 g/liter of medium under certain conditions. This extreme overproduction obviously cannot be accounted for solely by leakage through the cell membrane; the metabolic machinery of these organisms must have unusual features that lead to the specific overproduction of glutamate. The search for these features has pointed once again to particular conditions applied in the cultivation of these microorganisms that as a fortuitous result trigger glutamate production.

The biochemical precursor of glutamic acid is α-ketoglutaric acid, an intermediate in the citric acid cycle. When the enzymes of the citric acid cycle were first examined, glutamic acid producers of the *Brevibacterium–Corynebacterium* group were reported to contain all the enzymes of the cycle except α-ketoglutarate dehydrogenase. More recently, α-ketoglutarate dehydrogenase was shown to be present, but in a labile form. Its activity appears low, although unfortunately no report comparing the activities of all enzymes of the citric acid cycle in one experiment has been published. In any case, the citric acid cycle appears to be essentially truncated because of a low activity of this enzyme in intact cells (Figure 12.17), and this causes the accumulation of α-ketoglutarate, the intermediate before the block.

The citric acid cycle is the pathway responsible for the complete oxidation of small molecules into CO_2 in all organisms. It was thus unexpected to find a truncated citric acid cycle in wild-type bacteria. How could these bacteria compete against other aerobic bacteria that have a complete, functioning citric acid cycle? It turns out that when oxygen is abundant, the glutamate producers *can* oxidize small molecules com-

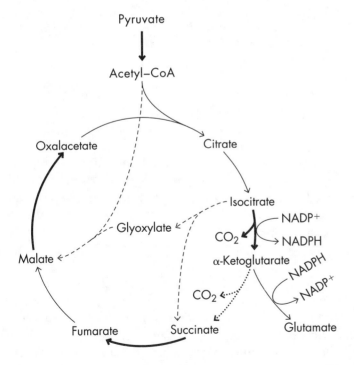

Figure 12.17
Citric acid cycle and glyoxylate shunt in *C. glutamicum*. The activity of α-ketoglutarate dehydrogenase (dotted arrow) is thought to be lower than those of the other enzymes. The reactions of the glyoxylate shunt are shown as dashed arrows. Thick arrows show oxidative steps that require either oxygen or other oxidants (such as pyridine coenzymes). See the text for details.

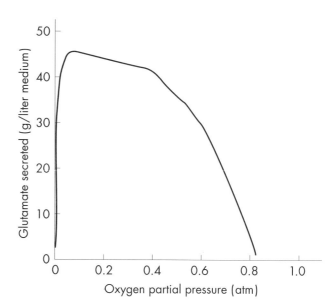

Figure 12.18
Effect of oxygen partial pressure (P_L) on the yield of L-glutamate. Modified from Hirose, Y., and Okada, H. (1979), in Peppler, H.J., and Perlman, D. (eds.), *Microbial Technology: Microbial Processes*, 2d ed., vol. 1, pp. 211–240 (Academic Press).

pletely by utilizing the glyoxylate shunt (see Figure 12.17)—that is, by converting the isocitrate into glyoxylate and succinate. Thus under these conditions, there is no accumulation of α-ketoglutarate and no excretion of glutamate (Figure 12.18). Likewise, little glutamic acid is generated when oxygen partial pressure is very low (see Figure 12.18); under these conditions, the end products of glycolysis presumably are not converted into acetyl-CoA (a step that is catalyzed by pyruvate dehydrogenase and that ultimately requires oxygen or another terminal acceptor of electrons) (see Figure 12.17). In contrast, when there is only a limited supply of oxygen and the complete oxidation of intermediates is not possible, as often happens when bacteria are cultivated at high density in laboratory media, α-ketoglutarate accumulates and is then removed by conversion into L-glutamate in the presence of excess ammonia (see Figure 12.18). The production of glutamate probably serves a physiological purpose for these organisms, regenerating the oxidized form of the pyridine nucleotide coenzymes, a role reminiscent of the production of ethanol by yeast (see Chapter 11). The reduced metabolites (ethanol or glutamate, depending on the organism) are then excreted as waste products.

In this interpretation, the metabolism of glutamate producers can be classified as a case of *incomplete oxidation*. A classic example of incomplete oxidation is the production of acetic acid by *Acetobacter*, which contains a functional citric acid cycle yet converts ethanol only as far as acetate when the oxygen supply is insufficient. (When excess oxygen is available, ethanol is oxidized completely into CO_2, as are carbon sources for glutamate producers). Although incomplete oxidation produces less ATP than complete oxidation, it is nevertheless more advantageous than anaerobic metabolism, because glutamate producers will manufacture additional ATP through the oxidation of pyruvate to acetyl-CoA, and *Acetobacter* will get its only ATP through the oxidation of ethanol to acetate when oxygen partial pressures are low.

It thus appears that the *Brevibacterium–Corynebacterium* group produces glutamate as a waste product of catabolism. This suggests that the enzyme responsible for the process, glutamate dehydrogenase, may not be regulated as tightly as the obligatorily biosynthetic enzymes. Evidence that such is indeed the case is described below.

Regulation of the Enzymes Involved in Glutamate Synthesis

In most microorganisms, the major pathway for glutamate synthesis is the GS-GOGAT pathway.

(1)   L-Glutamate + NH_3 + ATP $\longrightarrow$ L-glutamine + ADP + P_i

   (GS: glutamine synthetase)

(2)   L-Glutamine + α-ketoglutarate + NADPH + H^+ $\longrightarrow$

   2 L-glutamate + $NADP^+$

   (GOGAT: glutamine: oxoglutarate aminotransferase)

Sum (1) + (2):

α-ketoglutarate + NH_3 + ATP + NADPH + H^+ $\longrightarrow$

   L-glutamate + ADP + P_i + $NADP^+$

Although this pathway consumes one ATP for every molecule of L-glutamate synthesized, it is preferred to the alternative pathway described below because it has a low K_m (much lower than 1 mM) for NH_3 and thus can efficiently take up ammonia.

The second pathway is the glutamate dehydrogenase pathway:

α-Ketoglutarate + NH_3 + NAD(P)H + H^+ $\longrightarrow$ L-glutamate + $NAD(P)^+$ + H_2O

   (glutamate dehydrogenase)

Because of the rather high K_m value (usually around 1 mM) for NH_3, this pathway becomes significant only when NH_3 concentrations are high. In most organisms, mutational inactivation of this enzyme does not make the mutants auxotrophic for glutamate, which confirms the relatively minor role of this pathway for glutamate synthesis in most species. Because the GS-GOGAT pathway is so much more important, its enzymes are strongly regulated, as is typical of obligatorily biosynthetic enzymes; on the other hand, regulation of the glutamate dehydrogenase activity tends to be rather loose.

In the *Brevibacterium–Corynebacterium* group, the regulation of glutamate dehydrogenase is exceptionally loose; 50% inhibition of the enzyme requires the addition of 0.2 M glutamate to the cell extract, whereas similar inhibition of *Alcaligenes eutrophus* glutamate dehydrogenase requires only 20 mM. The gene for *Corynebacterium glutamicum* glutamate dehydrogenase has been cloned and sequenced, but the sequence is about 50% identical with the *E. coli* enzyme, and no striking peculiarity was noticed.

These considerations suggest that the overproduction of glutamic acid in the *Brevibacterium–Corynebacterium* group is due to a fortuitous combination of several factors. Over evolutionary time, the organisms of this group have developed a peculiar combination of enzymes, presumably equipping them to deal with ecological conditions that we do not yet fully understand. And it appears that the fermentation conditions used in the laboratory and in industry are precisely those that led to stimulation of the synthesis, and excretion, of glutamic acid.

AMINO ACID FERMENTATION AND RECOMBINANT DNA TECHNOLOGY

We have seen that amino acid fermentation processes are very different from processes designed to generate primary gene products via recombinant DNA methods. In fact, recombinant DNA technology has played only a minor role in this area, although it is important for improving the yields of amino acids such as threonine and tryptophan, whose overproduction is difficult to achieve by conventional methods. There are several reasons why this should be so.

1. Recombinant DNA technology is essential for expressing, in convenient microbial hosts, the genes of distantly related organisms (such as human genes) or genes for protective antigens of pathogens. Amino acids, in contrast, are ubiquitous components of all living cells, and the genes for amino acid synthesis already exist in almost every prototrophic microorganism.

2. Because amino acid biosynthesis is usually negatively regulated, the potential uninhibited (or derepressed) enzyme activity is usually very high. Thus it is often possible to achieve a good yield without amplifying the genes involved.

3. In the case of primary gene products, cloning the gene behind a strong promoter is virtually all that is needed to achieve their overproduction. In contrast, amino acids are low-molecular-weight metabolites synthesized through the cooperation of a large number of primary gene products (enzymes). Expressing the necessary numbers of genes in a coordinated manner can be difficult with recombinant DNA technology.

4. In successful cases, cultures with a cell density of about 20 g/liter produce amino acids in amounts of more than 60 g/liter. Thus the major portion of the metabolite flow is directed to amino acid synthesis. In these conditions, amplifying the enzymes of amino acid biosynthesis does not produce major improvements in yield, because the yield is likely to be limited by the availability of starting materials (page 404).

Regardless of whether recombinant DNA methods are used, industrial production of amino acids may sometimes pose unexpected problems. A recent example was the emergence, beginning in the late 1980s, of eosinophilia-myalgia syndrome, a new disorder that resulted in the death of more than two dozen people and caused severe illness in several hundred more. The syndrome was traced to the consumption of large amounts of L-tryptophan (an average of 1.5 grams per day) from particular batches made by Showa Denko in Japan. This company had been producing L-tryptophan by a fermentative process with *Bacillus amyloliquefaciens*, but the strain was altered just before the incident, via recombinant DNA technology, such that the starting material for the aromatic pathway, phosphoribosylpyrophosphate, was overproduced. When the implicated batch of L-tryptophan was analyzed, it was found to be contaminated by trace levels of two compounds related to tryptophan, 1,1'-ethylidene-*bis*[tryptophan] (EBT) and the ditryptophan aminal of acetaldehyde (DTAA), as well as by the unrelated compound 3-anilino-L-alanine (Figure 12.19). It appears that EBT may be generated by a nucleophilic attack by the indole nitrogen of tryptophan on acetaldehyde, during a final drying step in the tryptophan purification. At acid pH (such as that in the stomach), EBT is transformed to 1-methyl-1,2,3,4-tetrahydro-β-carboline-3-carboxylic acid (see Figure 12.19). Experiments with rats showed that administering mixtures of EBT and L-tryptophan to rats produced some, though not all, of the physiological changes associated with eosinophilia-myalgia syndrome. However, 3-anilino-L-alanine may also have played a significant role. In retrospect, researchers realized that a large outbreak of eosinophilia-myalgia syndrome had occurred in Spain in 1981. That outbreak was linked to ingestion of oil mixtures containing oil that had been treated with aniline. The speculation then was that a toxic product was generated by the reaction of aniline with some compound in the oil, because aniline itself did not cause eosinophilia–myalgia syndrome (which at the

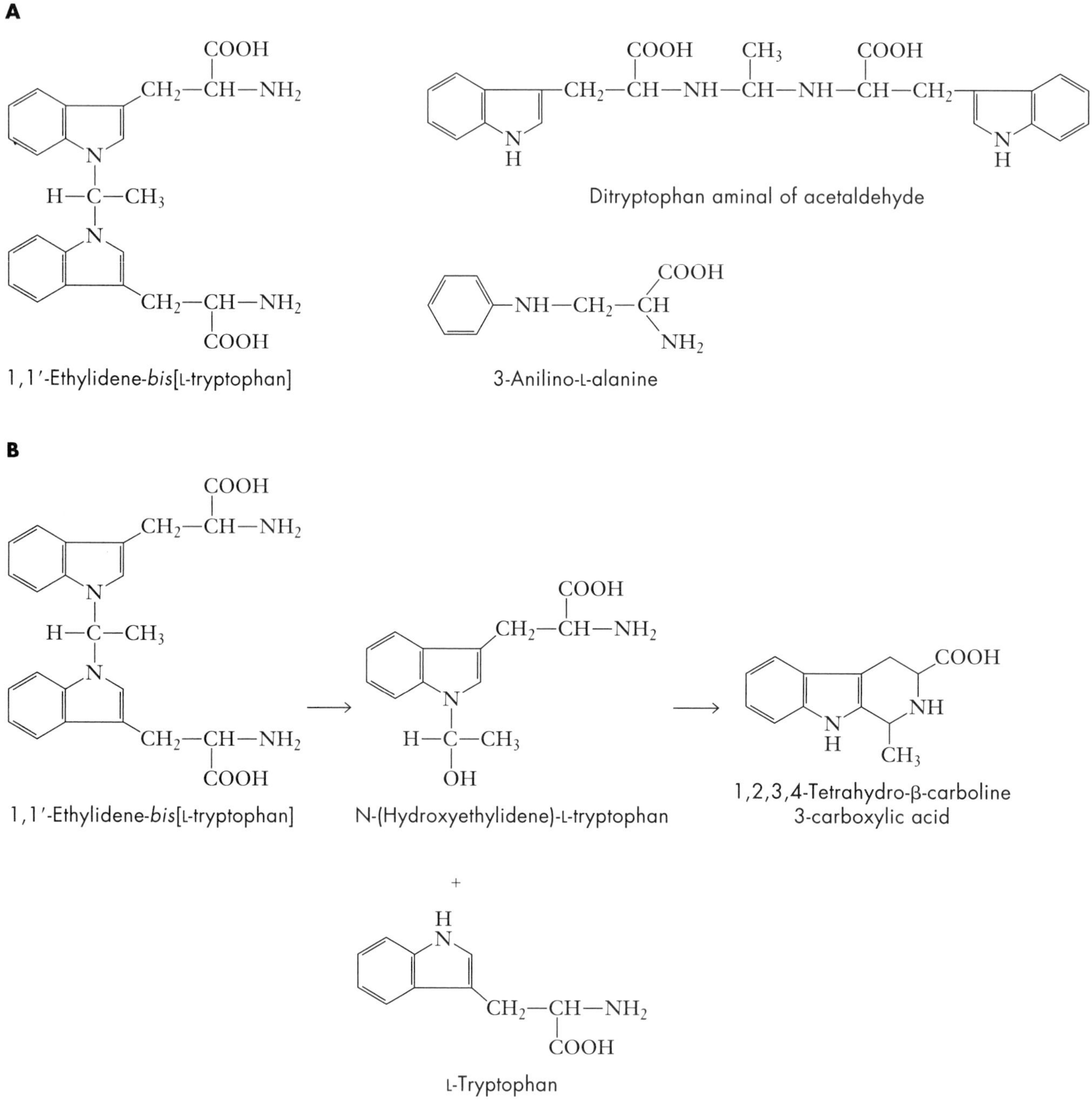

A

1,1'-Ethylidene-*bis*[L-tryptophan]

Ditryptophan aminal of acetaldehyde

3-Anilino-L-alanine

B

1,1'-Ethylidene-*bis*[L-tryptophan]

N-(Hydroxyethylidene)-L-tryptophan

1,2,3,4-Tetrahydro-β-carboline 3-carboxylic acid

L-Tryptophan

Figure 12.19
(**A**) Structures of impurities in batches of L-tryptophan associated with eosin-ophilia-myalgia syndrome. (**B**) The pathway of breakdown of the 1,1'-ethyli-dene-*bis*[L-tryptophan] in simulated human gastric juice.

time was designated "toxic oil syndrome"). The precise roles of the various contaminants will have to be clarified by further toxicological and epidemiological studies.

Because the Showa Denko tryptophan was made by fermentation, it is not surprising that the culture supernatant contained many metabolic waste products, including acetaldehyde. However, no problems were reported before October 1988. How did the "toxic" lots of Showa Denko L-tryptophan, made between October 1988 and June 1989, differ from previous batches? As we have noted, these lots were made with a new genetically engineered strain of *B. amyloliquefaciens* that increased the production of L-tryptophan several-fold. Consequently, the L-tryptophan concentration in certain production steps was much higher than before. Moreover, a reverse-osmosis membrane filter, intended to remove chemicals with molecular weights of more than 1,000 from the fermentation broth, was partially bypassed. Finally, the amount of activated charcoal, used in one of the purification steps to adsorb out impurities, was reduced by 50% during the time in which the contaminated batches were produced.

There is no basis for the claim, advanced by some people, that the use of recombinant DNA caused the eosinophilia–myalgia disaster. However, this example shows that utmost care is needed in producing amino acids by fermentation, to prevent trace amounts of reactive metabolites, such as aldehydes, from being carried through the purification process.

Amino acid production with enzymes

Although many amino acids are produced by fermentation processes, it is sometimes advantageous to use an enzymatic process instead. A good example is the production of L-aspartic acid, which is used as a starting material for the synthesis of aspartame. L-Aspartic acid has been produced in Japan since 1969, primarily by exploiting the *E. coli* enzyme aspartase, which catalyzes the conversion of L-aspartate into fumarate and NH_3.

$$\text{L-aspartic acid} \xrightarrow{\text{aspartase}} \text{fumaric acid} + NH_3$$

Aspartase is a catabolic enzyme used by *E. coli* to degrade exogenous aspartic acid into readily utilizable sources of carbon and nitrogen for growth. Under physiological conditions, the reaction equilibrium lies far toward the right. More precisely, because the K_{eq} is 20 mM, if the initial concentration of aspartate is 1 mM, about 95% of it will be converted into fumarate and NH_3. Thus at first, the enzyme appears to be a poor choice for *synthesizing* aspartate. However, the law of mass action favors aspartate synthesis when the reactant concentrations are high—that is, under conditions that would be used for industrial production of an amino acid. Based on the K_{eq} value cited above, we can calculate that if we start with 2 M fumarate and 2 M NH_3 the reaction will proceed far toward the left and will come to equilibrium when 90% of the reactants are converted into L-aspartate. Thus aspartase, although it is a degradative enzyme, is ideal for the industrial synthesis of aspartate. In contrast, most enzymes that catalyze biosynthetic reactions in intact cells are not suitable for the industrial production of amino acids, either because they use complex, expensive substrates or because they require an expensive cofactor, such as NADPH or ATP.

When amino acids *can* be produced by a simple enzyme reaction, such a process has many advantages.

1. The amino acids are produced in high concentrations (in the case of aspartate, about 1 M), making their recovery easy. In most such processes, the product amino acids spontaneously precipitate out of solution.

2. Because the solution contains only a few reactants, purification of the product is simple. In contrast, the supernatant in a fermentation run always contains some of the growth medium, the ingredients of which are usually quite complex and difficult to eliminate.

3. Compared with fermentation processes, the rate of production per unit volume is much higher. Fermentation requires huge tanks and much material for growth media, whereas enzymatic production can be done in a small space. Consequently, the required capital investment is much smaller. Of course, enzymatic processes do not make much economic sense when the reactants are expensive chemicals. Because amino acids are relatively inexpensive products, production cost is the more important issue.

Although enzymatic production can be carried out conveniently with soluble enzymes, it is much more efficient when the enzymes are immobilized by attachment to a supporting matrix. Separation and recovery of an immobilized enzyme are much simpler, so the enzyme can be reused economically. In addition, immobilization sometimes increases the stability of the enzyme. The immobilized enzyme preparations are usually packed into columns, a space- and time-saving format that makes it possible to maintain the reaction conditions necessary to drive the reaction in the desired direction.

High aspartic acid yields have been achieved with columns of immobilized aspartase. In the early years, the aspartase was purified from *E. coli* and attached covalently to an insoluble matrix. It was soon discovered that the purification is unnecessary and that attaching intact *E.*

coli cells to the matrix produces a very effective column. Because aspartase is an inducible enzyme, the *E. coli* must be grown in the presence of aspartate. Interestingly, the activity of the enzyme column increases nearly 10-fold during the first 1–2 days of incubation at 37°C. Apparently this is due to destruction of the membrane permeability barrier by autolysis. Even an intermediate method, in which *E. coli* cells were immobilized by entrapment in polyacrylamide gel, resulted in a cost reduction of 40% over the older batch method. This decrease was due to the increased production rate we have cited and to elimination of the cell-removal step associated with the batch process. In 1978, Tanabe scientists introduced a better immobilization medium, κ-carrageenan (a seaweed polysaccharide). Production with a properly treated column of this material is 170% higher than production with a polyacrylamide column. Moreover, the stability of the enzyme in this column is excellent; the half-life at 37°C is about 2 years. Adding a fresh mixture of fumarate and ammonia continuously to a small column 1 liter in volume results in a yield of several kilograms per day of L-aspartate. To obtain yields of this magnitude by fermentation would require tanks of at least several hundred liters in volume. As a further improvement, the gene for aspartase was cloned and inserted into a multicopy plasmid, increasing the aspartase production by the gene dosage effect.

Other amino acids, including L-alanine, L-lysine, and L-phenylalanine, have been produced successfully via enzymatic synthesis on a commercial scale. For example, L-alanine is produced by the enzymatic decarboxylation of L-aspartate, which can be generated inexpensively by the process described above.

$$\text{L-aspartate} \longrightarrow \text{L-alanine} + CO_2$$

A wild-type strain of *Pseudomonas dacunhae* is used as the source of the decarboxylase, and a batch process employing immobilized whole cells is reported to produce a yield of 400 g/liter.

Although L-lysine is usually produced by fermentation, it is also produced by the following ingenious enzymatic process:

$$\text{DL-}\alpha\text{-amino-}\varepsilon\text{-caprolactam} \xrightarrow{\text{hydrolase}} \text{L-lysine} + \text{D-}\alpha\text{-amino-}\varepsilon\text{-caprolactam}$$

$$\text{D-}\alpha\text{-amino-}\varepsilon\text{-caprolactam} \xrightarrow{\text{racemase}} \text{L-}\alpha\text{-amino-}\varepsilon\text{-caprolactam}$$

This process uses the inexpensive starting material DL-aminocaprolactam, produced by organic synthesis, and exploits the stereospecificity of the hydrolase, which is found in the yeast *Cryptococcus laurentii*, to produce L-lysine. The remaining D-isomer of the caprolactam is brought back into the production pathway by a racemase found in the bacterium *Achromobacter obae*. Batchwise production with intact cells of these organisms is reported to produce nearly complete conversion of 100 g/liter of aminocaprolactam into L-lysine.

There was an attempt to produce L-phenylalanine by utilizing phenylalanine aminolyase, which under normal, physiological conditions would function in the direction of phenylalanine degradation.

$$\text{L-phenylalanine} \longrightarrow \textit{trans}\text{-cinnamic acid} + NH_3$$

As in the case of L-aspartic acid synthesis, the law of mass action favors the synthesis of phenylalanine when the substrates are present at high concentrations. Yields of 80% or more, with a final L-phenylalanine concentration of 50–60 g/liter, have been produced in column reactors containing immobilized *Rhodotorula* yeast cells. Thus there is no problem with the process itself. Nevertheless, it has not been a commercial success, because the starting material, *trans*-cinnamic acid, is almost as expensive as the product.

In contrast, some attempts to commercialize an enzymatic process have been frustrated by problems inherent in the process. For example, the enzymes were not stable enough, or the reaction required cofactor(s) that were expensive or labile, or the reaction was relatively slow.

SUMMARY

Amino acids, some of the most important products of microbial biotechnology, are frequently produced by fermentation. As end products of biosynthetic pathways, their overproduction is normally prevented by various kinds of negative regulation. Such mechanisms can be complicated in organisms such as *E. coli* that live in a complex and fluctuating environment. In contrast, regulatory mechanisms are relatively simple in microorganisms that live in simple and constant environments. Such species, many of which are members of the *Brevibacterium–Corynebacterium* cluster, have been utilized widely in amino acid fermentation, though even with these organisms it is usually necessary to prevent negative regulation. This has been achieved by inactivating, through mutation, one of the enzymes of the biosynthetic pathway in order to eliminate negative control by the amino acid end product and thereby cause accumulation of the intermediate before the block. Alternatively, amino acids, rather than the intermediates, can be overproduced by altering the machinery of negative regulation; this is usually achieved by selecting mutants that grow in the presence of amino acid analogs, which mimic the amino acids in either the feedback inhibition or the repression process. In one exceptional case, wild-type strains in the *Brevibacterium–Corynebacterium* group produce large amounts of glutamic acid. This occurs only under special conditions, and apparently the cells use α-ketoglutarate, which accumulates when oxygen is low, to reoxidize the accumulated NADPH, generating glutamate as a waste product of catabolism. Glutamate excretion is enhanced by biotin deficiency through a poorly understood mechanism.

Recombinant DNA technology has not played a major role in the fermentative production of amino acids. This is primarily because it would require the coordinated overproduction of many enzymes and because it causes a major drain on the metabolism of the producing organism even without the use of recombinant DNA. However, with the construction of suitable vectors, recombinant DNA methods are likely to be introduced into an increasing number of systems.

Some amino acids are produced by the direct, enzymatic conversion of precursor compounds. Physiologically, these enzymes usually function to *degrade* amino acids, but they can also be made to *synthesize* amino acids when the reactants are present in high concentrations. The efficiency of the operation and the stability of the enzymes are improved when the enzymes (or, in some cases, cells containing the enzymes) are immobilized in enzyme columns.

SELECTED REFERENCES

General

Aida, K., Chibata, K., Nakayama, K., and Esaki, N. (eds.), 1986. *Biotechnology of Amino Acid Production*, vol. 24, Progress in Industrial Microbiology. Kodansha.

Yoshinaga, F., and Nakamori, S., 1983. Production of amino acids, In Hermann, K.M., and Somerville, R.L. (eds), *Amino Acids: Biosynthesis and Genetic Regulation*, pp. 405–429. Addison-Wesley.

Soda, K., Tanaka, H., and Esaki, N., 1983. Amino acids. In Rehm, H.-J., and Reed, G. (eds.), *Biotechnology*, vol. 3, pp. 479–530. Verlag Chemie.

Enei, H., and Hirose, Y., 1984. Recent research on the development of microbial strains for amino-acid production. In Russell, G.E. (ed), *Biotechnology and Genetic Engineering Reviews*, vol. 2, pp. 101–120. Intercept.

Production by Mutant Strains

Shiio, I., 1982. Metabolic regulation and over-production of amino acids. In Krumphanzl, K., Sikyta, B., and Vanek, Z. (eds.), *Overproduction of Microbial Products*, pp. 463–472. Academic Press.

Kisumi, M., Komatsubara, S., Sugiura, M., and Takagi, T., 1987. Transductional construction of amino acid-hyperproducing strains of *Serratia marcescens. CRC Crit. Rev. Biotechnol.* 6:233–252.

Omori, K., Suzuki, S.-I., Imai, Y., and Komatsubara, S., 1992. Analysis of the mutant *proBA* operon from a proline-producing strain of *Serratia marcescens. J. Gen. Microbiol.* 138:638–699.

Production by Wild-Type Strains

Kinoshita, S., 1985. Glutamic acid bacteria. In Demain, A.K., and Solomon, N.A. (eds.), *Biology of Industrial Microorganisms*, pp. 115–142. Benjamin/Cummings.

Hoischen, C., and Kramer, R., 1990. Membrane alteration is necessary but not sufficient for effective glutamate secretion in *Corynebacterium glutamicum. J. Bacteriol.* 172:3409–3416.

Broer, S., and Kramer, R., 1991. Lysine excretion by *Corynebacterium glutamicum*. 1. Identification of a specific secretion carrier system. *Eur. J. Biochem.* 202:131–135.

Bormann, E. R., Eikmanns, B. J., and Sahm, H., 1992. Molecular analysis of the *Corynebacterium glutamicum gdh* gene encoding glutamate dehydrogenase. *Mol. Microbiol.* 6:317–326.

Eikmanns, B. J., Kleinertz, E., Liebl, W., and Sahm, H., 1991. A family of *Corynebacterium glutamicum/Escherichia coli* shuttle vectors for cloning, controlled gene expression, and promoter probing. *Gene* 102:93–98.

Eosinophilia–Myalgia Syndrome and Tryptophan

Belongia, E. A., et al., 1990. An investigation of the cause of the eosinophilia-myalgia syndrome associated with tryptophan use. *New Engl. J. Med.* 323:357–365.

Roufs, J.B., 1992. Review of L-tryptophan and eosinophilia–myalgia syndrome. *J. Am. Dietetic Assn.* 92:844–850.

Driskell, W.J., et al., 1992. Identification of decomposition products of 1,1′-ethylidene*bis*[L-tryptophan], a compound associated with eosinophilia-myalgia syndrome. *Bull. Environ. Contam. Toxicol.* 48:679–687.

Goda, Y., Suzuki, J., Maitani, T., Yoshihira, K., Takeda, M., and Uchiyama, M., 1992. 3-Anilino-L-alanine, structural determination of UV-5, a contaminant in EMS-associated L-tryptophan samples. *Chem. Pharm. Bull.* 40:2236–238.

Enzymatic Production of Amino Acids

Hsiao, H.-Y., Walter, J.F., Anderson, D.M., and Hamilton, B.K., 1988. Enzymatic production of amino acids. In Russell, G.E. (ed.), *Biotechnology and Genetic Engineering Reviews*, vol. 6, pp. 179–219. Intercept.

13 Antibiotics

In terms of monetary value, antibiotics are currently the most important products of microbial biotechnology, apart from such "traditional" products as alcoholic beverages and cheese. Worldwide antibiotic production was estimated at a value of about $8 billion in 1981 (Chapter 1) and has doubled since that time. The primary use of antibiotics is in the treatment of human infectious diseases, although a significant number have agricultural and veterinary uses.

For much of our history, humanity did not have a truly effective way of combating pathogenic microorganisms. Medical practice consisted largely of weak palliative measures aimed at alleviating the symptoms rather than attacking the source of infectious diseases. This situation was radically altered with the invention of chemotherapy by Paul Ehrlich, a German physician. Ehrlich was impressed by the ability of certain dyes to stain specific parts of a cell in histological specimens. He reasoned that it might be possible to find a chemical that could specifically bind to a pathogenic microorganism but not to human cells. His effort to find such a "magic bullet" was finally rewarded when his assistant, Sahachiro Hata, discovered in 1910 that the 606th compound they tested was active against the causative agent of syphilis, which is transmitted through sexual contact and causes devastating mental symptoms and death in its later stages. Although this compound, salvarsan, had significant toxicity for humans, it was a major leap forward—the first effective treatment for an infectious disease.

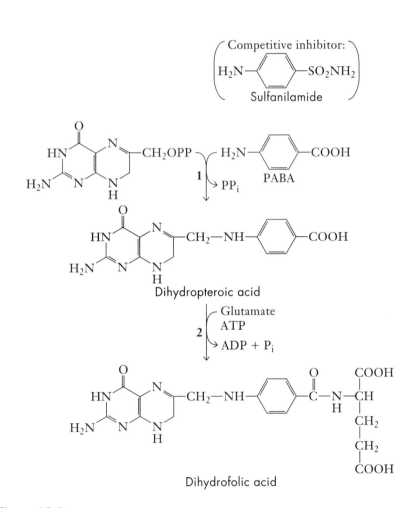

Figure 13.1

Structure of Prontosil. The part enclosed by dotted lines is the sulfanilamide structure, which has been shown to be active in antibacterial chemotherapy.

Figure 13.2

Folic acid biosynthesis and the sites of action of sulfa drugs. The step labeled 1 in the figure is catalyzed by the enzyme dihydropteroate synthase. This reaction is inhibited by sulfa drugs, which act as competitive inhibitors of p-aminobenzoic acid (PABA), the physiological substrate of the enzyme. Because higher animals ingest prefabricated folic acid as a vitamin, rather than synthesizing it within their cells, their metabolism is not inhibited by sulfa drugs.

The next major advance in chemotherapy did not occur until nearly 30 years later. In the early 1930s, Gerhardt Domagk, working in a German company involved in the manufacture of dyes and perhaps influenced by Ehrlich's earlier ideas, tested various synthetic dyes as antimicrobial agents. He found that the red dye Prontosil (Figure 13.1) was effective in the treatment of experimental streptococcal infections in mice. A few years later, French workers discovered that the active moiety in Prontosil was the colorless compound sulfanilamide. This realization led to the development of various sulfanilamide derivatives, the so-called sulfa drugs.

The sulfa drugs were the first truly successful chemotherapeutic agents because they were active against most bacteria, many of them could be taken orally, and they were remarkably nontoxic. In 1940 Donald D. Woods concluded that sulfa drugs act by mimicking a natural compound, p-aminobenzoic acid, and thus by competitively inhibiting the enzyme dihydropteroate synthetase, which catalyzes a step in the de novo biosynthesis of the vitamin folic acid (Figure 13.2). The drug is not toxic to humans because we do not synthesize folic acid but rather acquire it from our diet. Thus the "selective toxicity" that was Paul Ehrlich's ultimate goal is due in this case to the different nutritional requirements of bacteria and human beings.

Sulfa drugs are still used today in a few limited situations. For example, dapsone (Figure 13.3) is a drug of choice for treating Hansen's disease. Sulfa drugs are also used as general-purpose antibacterial agents in combination with another chemotherapeutic agent, trimethoprim, which inhibits dihydrofolate reductase and thereby blocks transformation of dihydrofolate (see Figure 13.2) to tetrahydrofolate. In many industrialized countries, however, sulfa drugs have been almost completely supplanted by the *antibiotics*—compounds produced by living organisms, usually microorganisms.

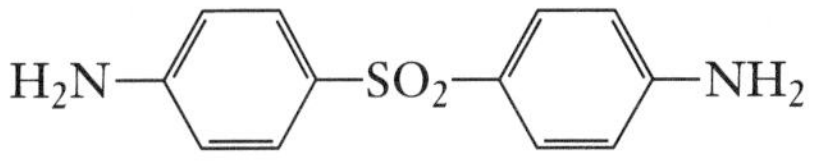

The tale of the discovery of the first antibiotic, penicillin (Figure 13.4), by Alexander Fleming is well known. Fleming is supposed to have kept a rather untidy laboratory. Tradition has it that he returned from vacation to find that mold had contaminated one of his bacterial cultures on a petri dish and that the bacterial colonies adjacent to the mold colony were lysed. This story is often cited as an example of the importance of serendipity in science. But it fails to take account of the fact that in an effort to find agents that could be used in the treatment of bacterial infections, Fleming had dedicated his entire career to the search for natural products that lyse bacterial cells. He had actually discovered the enzyme lysozyme several years earlier but then was disappointed to find that most human pathogens are intrinsically resistant to its lytic action. Because lysozyme is present in most of the tissues and body fluids of higher animals, pathogenic bacteria presumably have been subjected to evolutionary selection for such resistance.

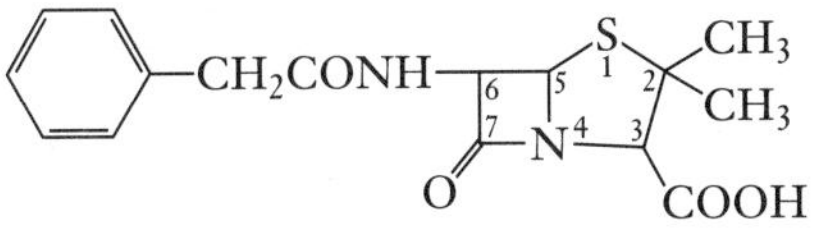

A key aspect of Fleming's discovery of penicillin is that he tried to isolate the active substance and use it to treat infections. Dozens of

Figure 13.3
Dapsone.

Figure 13.4
Penicillin G.

scientists before Fleming had published reports of either lysis or growth inhibition of bacteria caused by products of molds and other microorganisms. Yet nothing came of those findings because the authors operated as classical "naturalists," reporting "curious" phenomena. The eventual mass production and practical application of penicillin came about only after great effort by many outstanding chemists and a wartime collaboration among pharmaceutical companies. Seen in this light, Fleming's early efforts are all the more impressive: In his initial report of 1928, he showed the efficacy of the culture filtrate of his mold as a local therapeutic agent in an animal infection model.

The major reason why antibiotics have replaced synthetic chemotherapeutic agents in the treatment of human bacterial infections is the relative efficacy of these agents, as reflected by their *minimum inhibitory concentrations* (MICs). Sulfonamides typically have MIC values in the range of 10–100 μg/ml. Drugs are continuously excreted by patients, so maintaining a minimal tissue level of 100 μg/ml requires the administration of 10 g or more per day, a barely tolerable dose. In contrast, penicillin kills sensitive bacteria, such as *Streptococcus pneumoniae*, the causative agent for "classical" pneumonia, at 0.01 μg/ml or even lower. In addition to this much higher antibacterial activity, penicillin is remarkably nontoxic, even when administered in amounts that result in a tissue concentration of 100 μg/ml. In this way, penicillin and many other antibiotics give physicians a tremendous margin of safety in critical situations.

After the mode of action of sulfonamide was elucidated, much effort was invested in attempts to design other inhibitors of metabolism that might act as chemotherapeutic agents. In spite of these exertions, the number of agents that were successfully developed for antibacterial chemotherapy is not very large. Perhaps the lack of success was due in part to faulty strategy. Competitive inhibitors, such as sulfonamides, are inherently less effective than other agents, because the inhibition of enzyme activity leads to accumulation of the natural substrate, which then acts to overcome the inhibitory action of the drug. However, the failure was also due to the fact that the design of chemotherapeutic agents was often a matter of guesswork. In marked contrast, the antibiotics have already undergone precise structural refinement during billions of years of evolution and are extremely efficacious in attacking their targets. On the other hand, most of the new antibiotics that are introduced nowadays are semisynthetic; that is, they are based on microbial products but have been modified by synthetic organic chemistry. Thus, as we shall see, progress in the field is again dependent on the creative power of medicinal chemists. Furthermore, recombinant DNA methodology—the ability to clone, sequence, overproduce, and crystallize enzymes that carry out functions unique to specific types of microbial cells—should make it possible to design tight-binding transition-state analogs and irreversible inhibitors and may lead to the development of synthetic chemotherapeutic agents that are as efficacious as some of the natural or semisynthetic antibiotics.

CLASSES OF ANTIBIOTICS

Most commercially important antibiotics are produced either by fungi or by bacteria. The overwhelming majority of agents from bacterial sources are produced by *Streptomyces* or other organisms of the actinomycete line (Box 13.1); others are the products of various *Bacillus* species.

What is the function of an antibiotic in the natural environment of these organisms? The simplest hypothesis is that an antibiotic benefits the producing organism by inhibiting the growth of competing organisms in the environment; however, certain findings appear to conflict with this idea. First, strains isolated from nature but grown under laboratory conditions usually produce antibiotics at vanishingly low concentrations. This criticism may be countered by pointing out that most producing microorganisms are soil dwellers and secrete their antibiotics in a small confined space, such as the cracks in a soil particle. Furthermore, antibiotic production is usually inhibited by the presence of excess nutrients and thus is likely to be much more vigorous in a nutritionally limited natural environment. A second argument against the simplest hypothesis is that organisms under study also produce compounds that do not seem to have a biological function in the environment, such as compounds with antitumor activity. However, this argument appears to arise because our definition of antibiotics is anthropocentric. We classify as an "antibiotic" any small molecule that is produced by living organisms and shows an activity that is useful for us

> **BOX 13.1   Antibiotic Producers of the Actinomycete Line**
>
> Most antibiotics of prokaryote origin are produced by organisms belonging to the actinomycete line, and among them the majority come from *Streptomyces* species, which grow as a branched mycelium and bear spores at the ends of aerial hyphae (Chapter 2). The classification of these actinomycetes has been fraught with difficulties not uncommon in the classification of bacteria in general. Because of the economic importance of antibiotic production, there has been (and still is) a tendency to divide these organisms into too many taxonomic units, sometimes on the basis of rather insignificant detail. This tendency has contributed to the creation of many hundreds of "species" within the genus *Streptomyces*. On the other hand, analysis of 16S rRNA sequences (see Chapter 2) has shown that there is indeed unexpectedly wide diversity even among those actinomycetes that show similar morphological patterns of differentiation. Thus the genus *Micromonospora*, whose members are producers of some important antibiotics including gentamicin, is only distantly related to *Streptomyces*, even though they both produce spores at the tips of hyphae.

humans. Antibiotics in a broad sense thus include not only antimicrobial agents but also antitumor agents, inhibitors of specific enzymes, compounds that bind to receptors on human cells, and so on. This classification is strictly an expedient one for humans, and there is no reason why all the compounds classified as antibiotics should serve the same ecological function for the producing organisms. Although it seems likely that many of the "classical" antibiotics with strong antimicrobial activity do indeed function as antimicrobials in the natural environment of the producing organisms, we do not have many clues about the physiological functions of the compounds that show weak antimicrobial activity or no such activity. It has been proposed that some of the "antibiotics" play a regulatory, or even hormone-like, role within the *population* of the producing microbes (as opposed to within an individual cell).

Antibacterial Agents

Most of the antibiotics that are now in commercial production are compounds active against bacteria. It is important to distinguish between the agents that act against Gram-positive bacteria only and those that are also active against Gram-negative bacteria (Chapter 2).

The cells of Gram-negative organisms have the added protection of an outer membrane (see Figure 2.3). Small solutes (<1000 daltons) cross this permeability barrier primarily through the water-filled channels of special proteins, called porins. The porin channels can be quite narrow, with a size of only 0.7×1 nm in *E. coli* and its relatives and with a diameter estimated to be only about 2 nm in the bacteria with the largest channels. For this reason, the diffusion of water-soluble antibacterial agents larger than 1000 daltons through the outer membrane is severely limited. Not surprisingly, the permeability of these water-filled channels to lipophilic solutes is also poor. Furthermore, the lipid bilayer region of the outer membrane has an unusually low permeability toward lipophilic molecules, apparently because its outer leaflet is composed exclusively of an unusual lipid molecule, the lipopolysaccharide (LPS; see Figure 2.3). Consequently, the larger and more lipophilic antibiotics tend to have significant activity against Gram-positive bacteria only.

β-Lactams. Penicillin (benzylpenicillin, penicillin G; see Figure 13.4) is the classic example of a β-lactam antibiotic. The β-lactams inhibit synthesis of the eubacterial cell wall, which is composed of peptidoglycan, a polymer unique to true bacteria. Peptidoglycan is a network of polysaccharide (or glycan) chains composed of alternating N-acetylglucosamine and N-acetylmuramic acid residues. The polysaccharide chains are crosslinked to each other via short peptide chains that include some D-amino acids and are attached to the N-acetylmuramic acid residues (Figure 13.5; see also Figure 2.1). This structure gives peptidoglycan

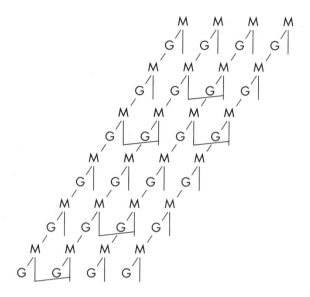

Figure 13.5
Structure of peptidoglycan from Gram-negative bacteria. Polysaccharide chains are made of alternating *N*-acetylmuramic acid units (M) and *N*-acetyl-glucosamine units (G). Short peptides (vertical lines) are attached to the *N*-acetylmuramic acid residues, and the polysaccharide chains are connected via peptide crosslinks (nearly horizontal lines).

exceptional chemical stability, mechanical strength, and rigidity. To add new glycan chains to the cell wall in growing cells, the peptide side chain of a new unit is crosslinked to the pre-existing peptidoglycan structure (Figure 13.6). The crosslinking reaction is catalyzed by a DD-transpeptidase. This enzyme cleaves the peptide bond between the two D-alanine residues in the side chain of a newly made glycan chain and transfers the glycan–peptidyl complex to the free amino group of the diamino acid residue in the side chain of a pre-existing peptidoglycan, thereby crosslinking different chains.

As first pointed out by Donald Tipper and Jack Strominger, the β-lactam ring system structurally resembles the D-alanyl–D-alanine of the nascent side chain (Figure 13.7). They demonstrated that the interaction between transpeptidase and penicillin generates a covalent penicilloyl enzyme and irreversibly inactivates the transpeptidase. (Because of the formation of stable penicilloyl enzymes, these transpeptidases are referred to as *penicillin-binding proteins*, or *PBPs*). Thus a β-lactam is a "suicide inhibitor": When it interacts with its target enzyme like a substrate, the two molecules undergo a chemical reaction that inactivates both of them permanently. Suicide inhibitors are more effective antimicrobial agents than competitive inhibitors (such as sulfonamides) are, because they achieve complete inhibition of the target.

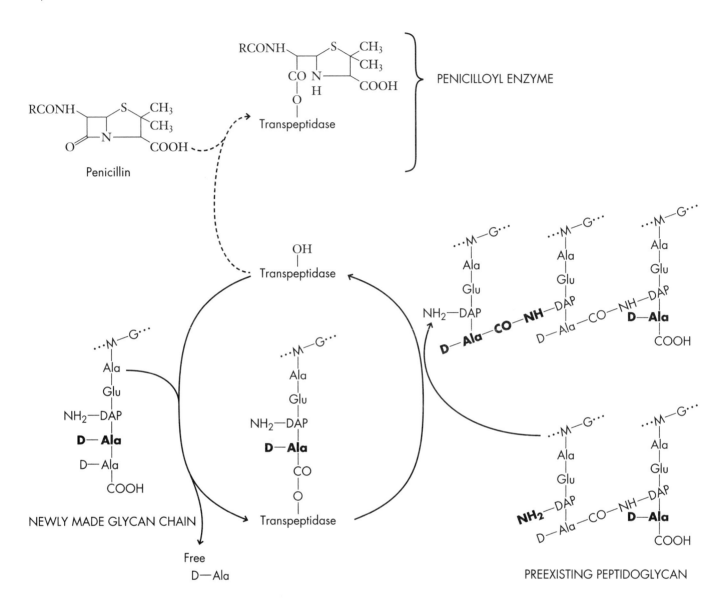

Figure 13.6
Addition of new material to the existing peptidoglycan and the mechanism of action of penicillin. The newly made, still uncrosslinked peptidoglycan (lower left) becomes covalently linked to transpeptidase (center below) by a reaction involving the splitting of the D-alanyl–D-alanine bond of the peptide. The new peptidoglycan material is then transferred onto the amino group of the diaminopimelic acid residue in a pre-existing peptidoglycan complex (right), regenerating the transpeptidase enzyme (center). Penicillin also produces a covalent complex with the transpeptidase (penicilloyl enzyme, top center), but this complex is stable and the enzyme becomes permanently inactivated (broken arrows).

Penicillin G is very effective at killing most Gram-positive bacteria but is ineffective against most Gram-negative bacteria because of its lipophilicity. However, the benzyl side chain at the 6-position (see Figure 13.4) can be chemically replaced with other substituents. Some of the resulting compounds show significant activity against Gram-negative bacteria and thus are broad-spectrum antibiotics (we discuss them below). Many years after the discovery of penicillin, antibiotics with a slightly different ring structure—cephalosporins—were discovered. Our discussion of cephalosporins begins on page 475. Although penicillins and cephalosporins account for most of the β-lactam antibiotics used today, compounds with other ring systems are also being developed (see below).

Macrolides. The structure of macrolide antibiotics is based on a large lactone ring (Figure 13.8). *Streptomyces* species (some now classified as *Saccharopolyspora*) synthesize them through head-to-tail condensations of several C3 units. Macrolides are therefore examples of polyketide compounds, whose synthesis is discussed later in this chapter (page 506).

Two inhibitors of prokaryotic protein synthesis, erythromycin and oleandomycin, belong to this class. Because they are large and hydrophobic, their inhibitory action is mostly limited to Gram-positive bacteria. However, some semisynthetic macrolides (such as azithromycin) have recently been developed that show significant anti-Gram-negative activity.

Ansamycins. Ansamycins also have a macrocyclic structure, but the ring differs from that of the macrolides because it contains an aromatic chromophore as well as an amide (lactam) linkage (Figure 13.9). These compounds are isolated from *Nocardia* (now *Amycolatopsis*) species, which like *Streptomyces* are members of the actinomycete branch of eubacteria (see Chapter 2). Rifamycins inhibit prokaryotic RNA polymer-

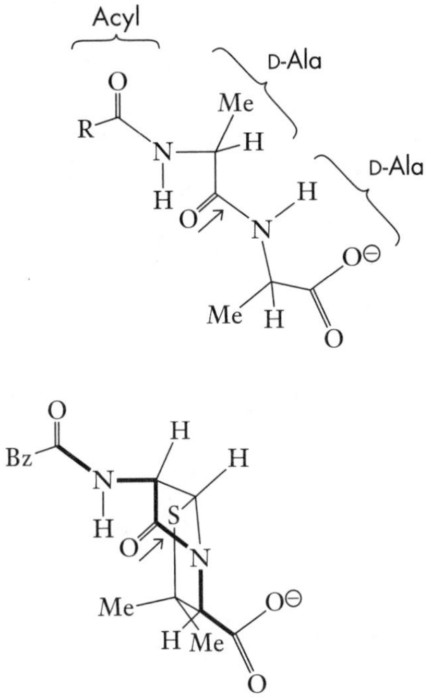

Figure 13.7
Structural similarity between the acyl–D-alanyl–D-alanine part of the peptidoglycan (upper right)—that is, the natural substrate of transpeptidase—and penicillin G (lower left). Bz in the penicillin structure denotes benzyl moiety. The bonds in penicillin G that correspond to the peptide backbone in acyl–alanyl–alanine are drawn in heavy lines. The bonds that are cleaved by the peptidoglycan transpeptidase are indicated by arrows. The drawing is based on Tipper, D.J., and Strominger, J.L. (1965), *Proc. Nat. Acad. Sci. USA.* 54:1133–1141, but it is turned around and simplified by representing the methyl group as Me.

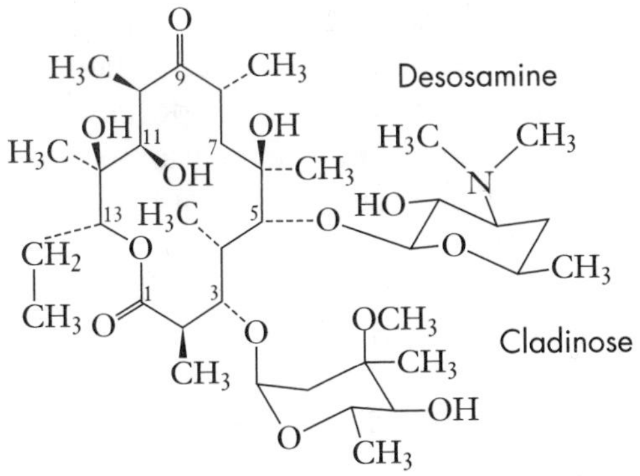

Figure 13.8
Structure of erythromycin A. The macrocyclic rings in most macrolides are substituted by sugar residues.

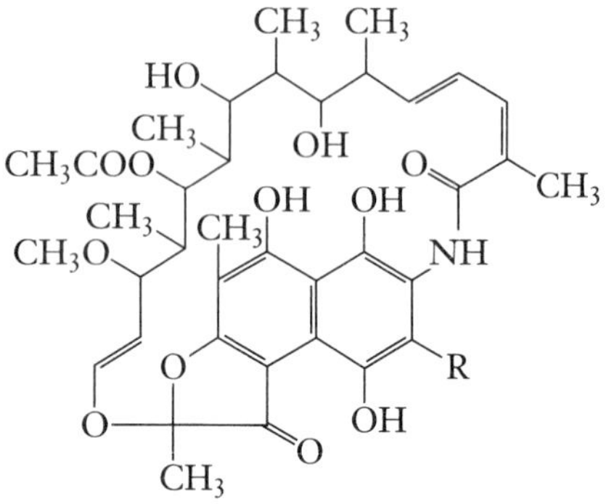

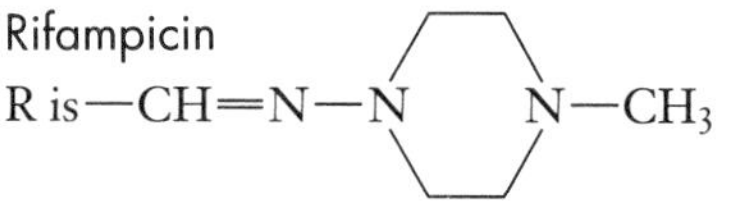

Figure 13.9
Structure of rifamycins.

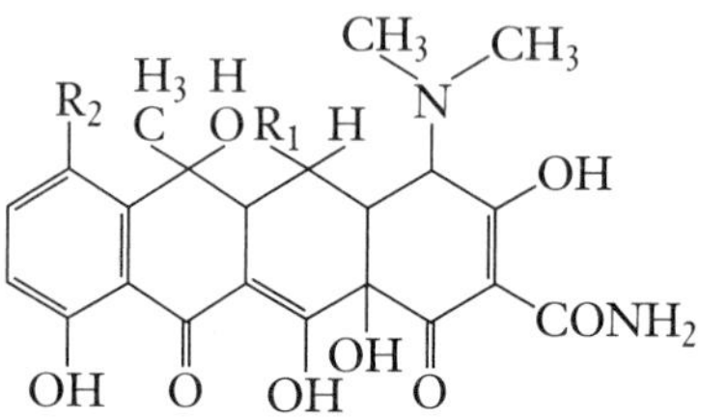

Figure 13.10
Structure of tetracyclines.

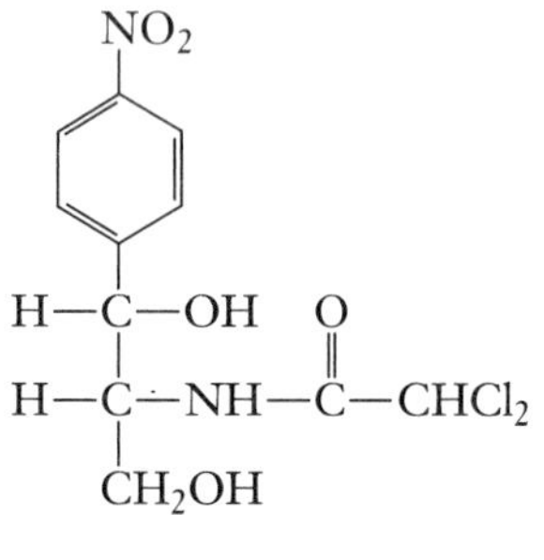

	R_1	R_2
Tetracycline	H	H
Chlortetracycline	H	Cl
Oxytetracycline	OH	H

Figure 13.11
Structure of chloramphenicol.

ase. Natural rifamycins, such as rifamycin SV (see Figure 13.9), have significant activity against Gram-positive bacteria only, as their hydrophobic character and large size lead us to expect. When a positively charged group is introduced chemically to make the molecule more polar, as in rifampicin (see Figure 13.9), some activity against Gram-negative bacteria becomes detectable. Perhaps more important, derivatives such as rifampicin can be administered orally. These compounds are used widely to combat the Gram-positive pathogens *Mycobacterium tuberculosis* (the cause of tuberculosis) and *M. leprae* (the cause of Hansen's disease).

Tetracyclines. Tetracyclines, produced by several species of *Streptomyces*, contain a fused four-ring system (Figure 13.10). These compounds are inhibitors of eubacterial protein synthesis. They are relatively hydrophilic because of the presence of several hydroxyl groups, an amide moiety, and a tertiary amine substituent, so they can cross the outer membrane of Gram-negative bacteria efficiently. Tetracyclines are thus broad-spectrum antibiotics active against both Gram-positive and Gram-negative bacteria.

In the past, adding low concentrations of tetracyclines to animal feed was a common practice all over the world; it tended to increase the rate of weight gain and decrease the incidence of infectious disease in livestock. However, in England, *Salmonella* containing tetracycline-resistance plasmids has caused several epidemics in calves that later spread to the human population. In most European countries, giving tetracycline to animals without veterinary prescription is now prohibited. The same is true of other antibiotics that are commonly used to treat human bacterial infections.

Chloramphenicol. Chloramphenicol (Figure 13.11) was originally isolated from the filtrate of *Streptomyces venezuelae* cultures. This compound or close analogs of it are also produced in large amounts by other members of the actinomycete family (Chapter 2). However, because chloramphenicol is a small molecule with a simple structure, it is more economically produced by chemical synthesis than by fermentation.

Chloramphenicol inhibits eubacterial protein synthesis. It penetrates through the outer membrane channels of Gram-negative bacteria easily because of its small size and is therefore another broad-spectrum antibiotic. However, chloramphenicol can also enter eukaryotic cells and inhibit mitochondrial protein synthesis. Thus its use can prevent the growth of rapidly proliferating animal cells and can lead, for example, to aplastic anemia in humans. Because of this side effect, chloramphenicol is not widely used today, except in treating diseases that involve the intracellular colonization of human cells, for example by *Salmonella typhi* (typhoid fever).

Peptide Antibiotics. Several peptide antibiotics (Figure 13.12), many the products of *Bacillus* species, are now in commercial production. These

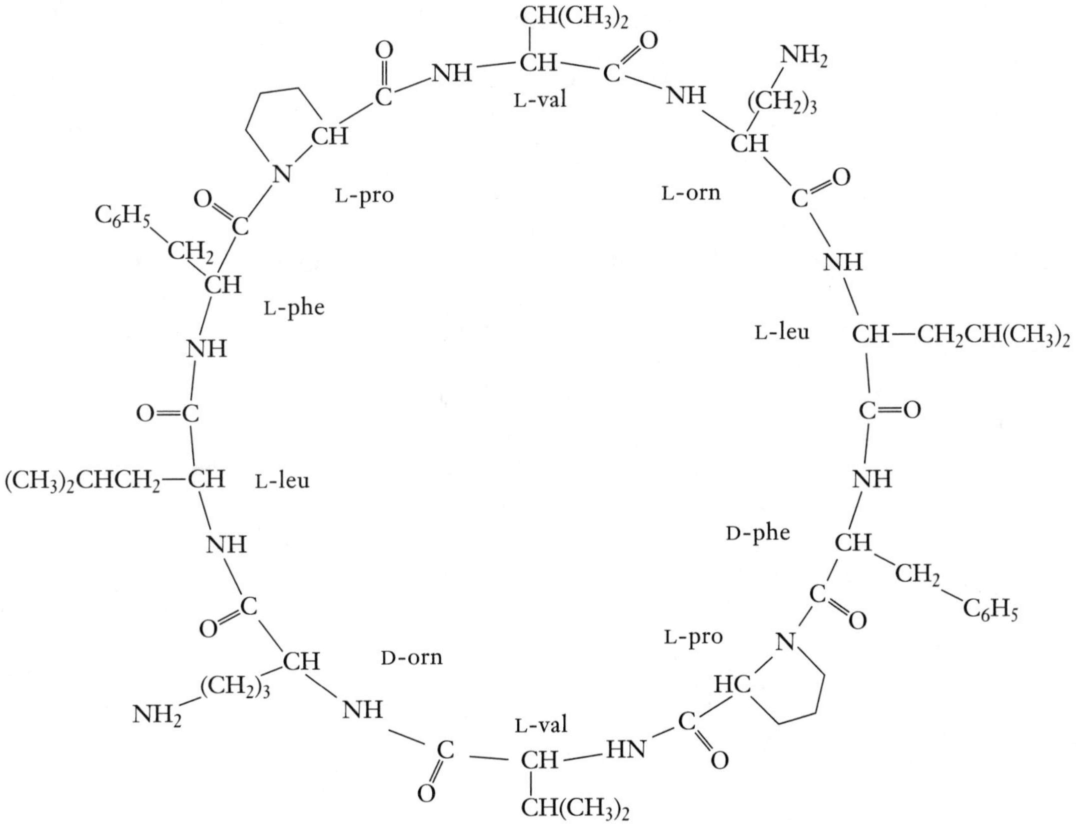

Gramicidin S

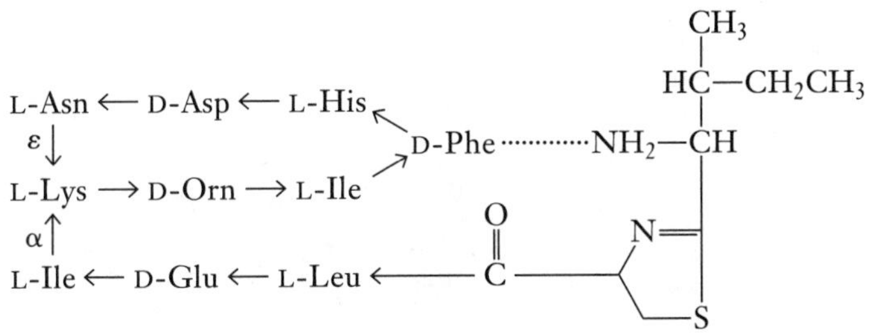

Bacitracin A

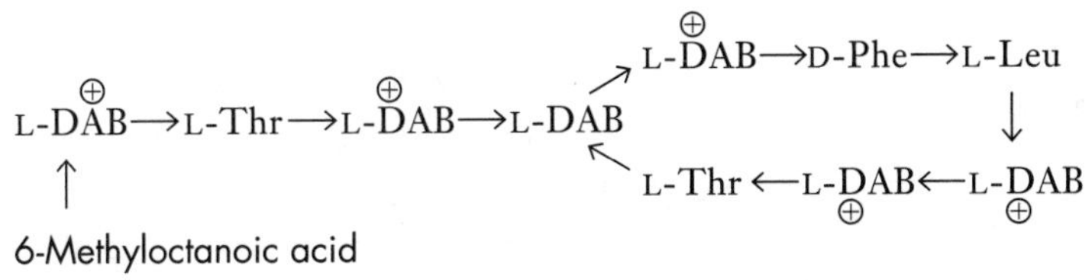

Polymyxin B_1

Figure 13.12
Some examples of peptide antibiotics. Arrows indicate the direction of peptide bonds (—CO → NH—). In the structure of polymyxin, DAB indicates 2,4-diaminobutyric acid. The positive charge due to the free 4-amino group of this residue is also indicated.

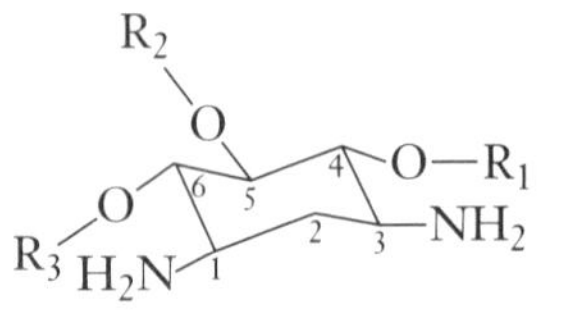

Figure 13.13
A general structure of aminoglycosides containing 2-deoxystrepta-
mine. The class with substituents at
the 4- and 6-positions (class I) includes kanamycins, tobramycin,
gentamicins, and sisomicin. The
class with substituents at the 4-
and 5-positions, here called class II,
includes neomycins, paromomycins,
lividomycins, and ribostamycin.
Class III contains substituents only
on the 4-position and includes apra-
mycin. Class IV contains substituents only on the 5-position and
includes destomycins and hygromycin B.

Figure 13.14
Polyoxin B.

peptides contain unusual amino acids, such as D-amino acids, ornithine, and diaminobutyric acid, and have significant toxicity in humans when ingested or injected. Thus they are useful for topical applications only. Some of these agents are used in Europe as feed additives because they are not utilized in human therapy (see the foregoing discussion on tet-racycline).

Many of these antibiotics are too bulky and hydrophobic to cross the outer membrane barrier of Gram-negative organisms, and their efficacy is therefore limited to Gram-positive bacteria. An interesting exception is polymyxin, which has a two-step mode of attack on Gram-negative bacteria. This polycationic antibiotic (see Figure 13.12) apparently binds to the highly negatively charged LPS molecules in the outer membrane of many Gram-negative bacteria and disrupts the molecular organization of the outer membrane bilayer. Once the integrity of the outer membrane is destroyed, polymyxin can bind to and penetrate the cytoplasmic membrane with its hydrophobic fatty acid tail, thus killing the bacterial cell by permeabilizing this membrane. Polymyxin shows strong activity against organisms such as *Pseudomonas aeruginosa*, which is resistant to almost all other antibiotics because of the unusually low permeability of its outer membrane. Hence, in spite of its significant toxicity, polymyxin is used sometimes to treat life-threatening infec-tions by *P. aeruginosa*.

Aminoglycosides (Aminocyclitols). Aminoglycosides, which are discussed in detail later in this chapter, consist of an aminocyclitol moiety (strep-tidine, a derivative of streptamine, in streptomycin, 2-deoxystreptamine in many other aminoglycosides; Figure 13.13) to which amino sugars are attached in various ways. All of these compounds are produced by the prokaryotes of the actinomycete line (*Streptomyces* in most cases, *Micromonospora* for gentamicin). These antibiotics are inhibitors of pro-karyotic protein synthesis. Because they are quite hydrophilic and suffi-ciently small, aminoglycosides can cross the Gram-negative outer membrane through the porin channels, so these compounds are equally effective against Gram-positive and Gram-negative bacteria.

Antifungal Agents

Fungi are eukaryotes and contain the same type of machinery for pro-tein and nucleic acid synthesis as the cells of higher animals. It is there-fore more difficult to find compounds that selectively inhibit fungal metabolism while presenting no toxicity to humans.

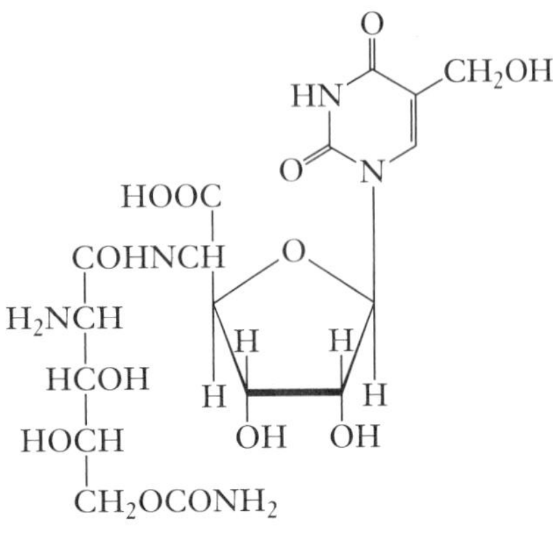

Polyoxin D (Figure 13.14), a product of *Streptomyces* sp., has a struc-ture reminiscent of UDP-*N*-acetylglucosamine and inhibits the synthe-sis of chitin, a polysaccharide found uniquely in fungal cell walls. It is used in agriculture as a fungicide. Griseofulvin (Figure 13.15), produced by the mold *Penicillium griseofulvum*, also inhibits fungi. It binds to proteins involved in the assembly of tubulin in microtubules and inhibits mitosis in fungi. It is used in human therapy as well as in agriculture.

Polyenes (Figure 13.16) have macrocyclic lactone structures similar to those in macrolides; however, the ring structure is larger, involving at least 26 atoms and at least 3 conjugated double bonds. Polyenes, produced by *Streptomyces* spp., form complexes with sterols in the fungal cell membranes, thereby perturbing the membrane and increasing its nonspecific permeability. Because animal cell membranes also contain a sterol (cholesterol), most polyenes are rather toxic to animals and are therefore used predominantly in topical applications.

Recently, several inhibitors of the biosynthesis of fungal cell wall polysaccharides have been identified among microbial products and have been developed into therapeutic compounds.

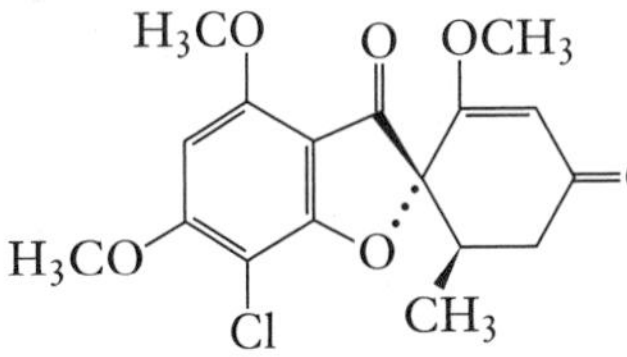

Antitumor Antibiotics

Certain antibiotics are used in chemotherapy against malignant tumors, some because they have polycyclic rings that intercalate between the bases in double-helical DNA. Daunomycin and adriamycin contain a naphthacene ring (four fused benzene rings; Figure 13.17), and dactinomycin and mithramycin contain three fused rings (Figure 13.18). In contrast, bleomycin, which also has a high affinity for DNA, has a totally different structure and destroys DNA by generating oxygen free radicals (catalyzed by a ferrous ion, which is coordinated by this antibiotic; Figure 13.19). All of these compounds are products of *Streptomyces* species.

Other Microbial Products with Pharmacological Activity

Culture filtrates of microorganisms are screened not only for antimicrobial activity but also for the presence of other activities, such as inhibition of particular animal enzymes or binding to specific receptors on animal cell surfaces. By this means, a number of compounds with

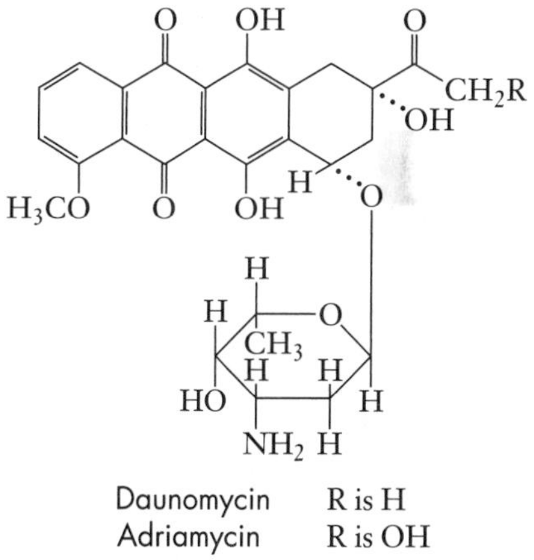

Figure 13.15
Griseofulvin.

Daunomycin   R is H
Adriamycin   R is OH

Figure 13.17
Daunomycin and adriamycin. These may be said to belong to the anthracyclines, because they appear to contain the anthracene ring (if we disregard the saturated ring to the extreme right). More accurately, they may be said to contain a tetrahydro derivative of the naphthacene ring, in which four benzene rings are fused together.

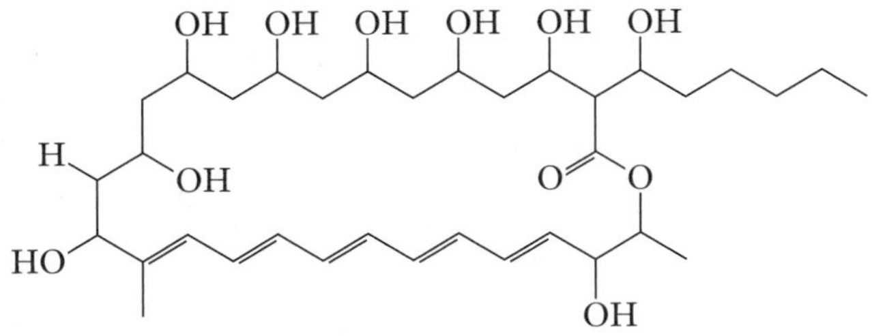

Figure 13.16
Filipin, an example of a polyene antibiotic.

Figure 13.18
Dactinomycin and mithramycin. Arrows idicate the direction of peptide bonds (—CO → NH—). Abbreviations: Sar, Sarosine; MeVal, N-methylvaline.

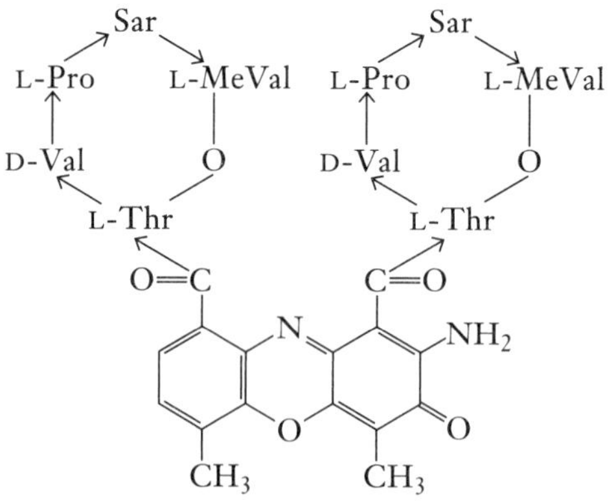

Dactinomycin

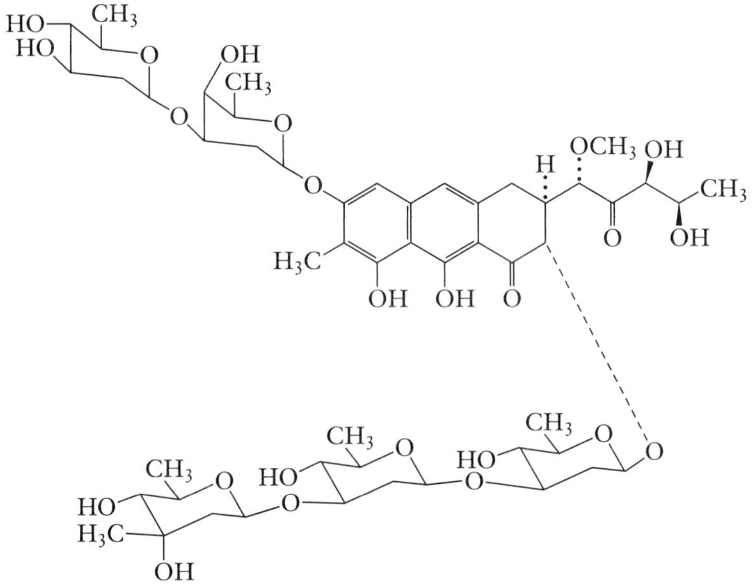

Mithramycin

Figure 13.19
Bleomycin.

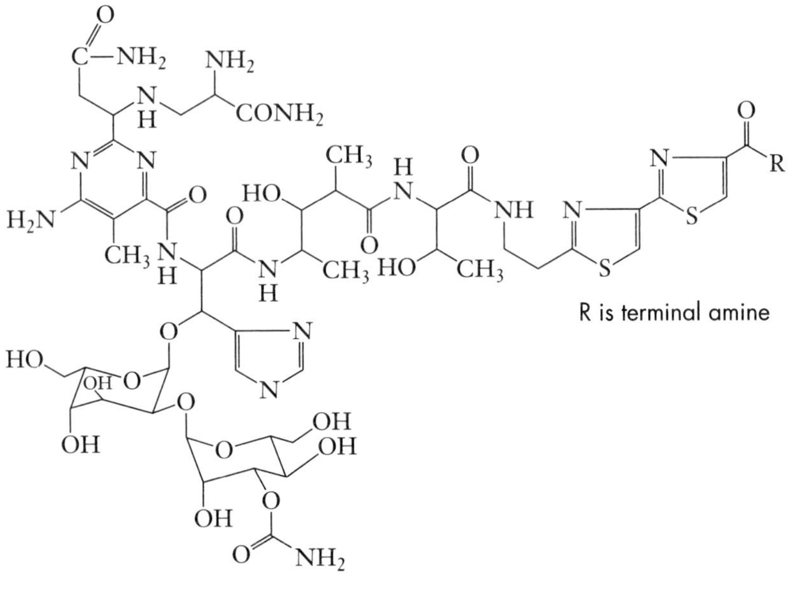

R is terminal amine

extremely interesting pharmacological action have been discovered. This is perhaps the most significant reason to screen for secondary metabolites of microorganisms nowadays. A few examples serve to illustrate the range of targets for which compounds have been found—compounds that do not fit the traditional, narrower definition of "antibiotics."

Protease/Peptidase Inhibitors. In the late 1960s, Hamao Umezawa had the remarkable foresight to screen culture filtrates for microbial secondary products that would inhibit proteases. Proteases, essential for many normal physiological functions (for example, the blood-clotting cascade), are also implicated in many pathological states (for example, elastase plays a role in the etiology of emphysema). Furthermore, the proliferation of many animal viruses requires proteolytic processing of viral polyprotein precursors by virally encoded proteases (an example is the aspartyl protease of the AIDS virus). Protease inhibitors are therefore expected to have important medical applications. A search for such inhibitors in culture filtrates of *Streptomyces* and related species led rapidly to the discovery of several extremely potent inhibitors of various types of proteases. Some of these molecules, such as leupeptin, antipain, and pepstatin, are very useful reagents in the research laboratory (Figure 13.20). Another prominent example of an inhibitor obtained through screening is an agent that blocks the angiotensin-converting enzyme, a key enzyme in the renin–angiotensin system that controls blood pressure.

Inhibitor of Cholesterol Biosynthesis. Limiting dietary intake of cholesterol-rich foods lowers blood cholesterol levels significantly in some individuals, but in others high cholesterol levels result from an elevated endogenous synthesis of cholesterol rather than from diet. In the 1970s, scientists at Sankyo Co. in Tokyo isolated several novel products from cultures of *Penicillium* species, initially because of their antifungal activity. Further study established that these compounds, one of which was compactin (Figure 13.21), competitively inhibited the enzyme that catalyzes the first unique step of cholesterol synthesis, hydroxymethylglutaryl–CoA reductase. Compactin was also effective at lowering serum cholesterol levels in experimental animals and in humans. In 1980 scientists at Merck, Sharpe & Dohme isolated a closely related compound from a culture filtrate of another fungus, *Aspergillus*. This compound is now marketed as lovastatin.

Some Other Inhibitors. There are other antibiotics that were not originally isolated by a screen for enzyme inhibition but that nevertheless turned out to be exceptionally strong and specific inhibitors of important enzyme reactions. Two such agents that have contributed enormously to experimental studies in biochemistry and cell biology are cerulenin (Figure 13.22), a fungal analog of a fatty acid derivative, which inhibits fatty acid synthesis in other organisms, and tunicamycin (Figure 13.23), which inhibits transfer of the N-acetylglucosamine-1-phosphate

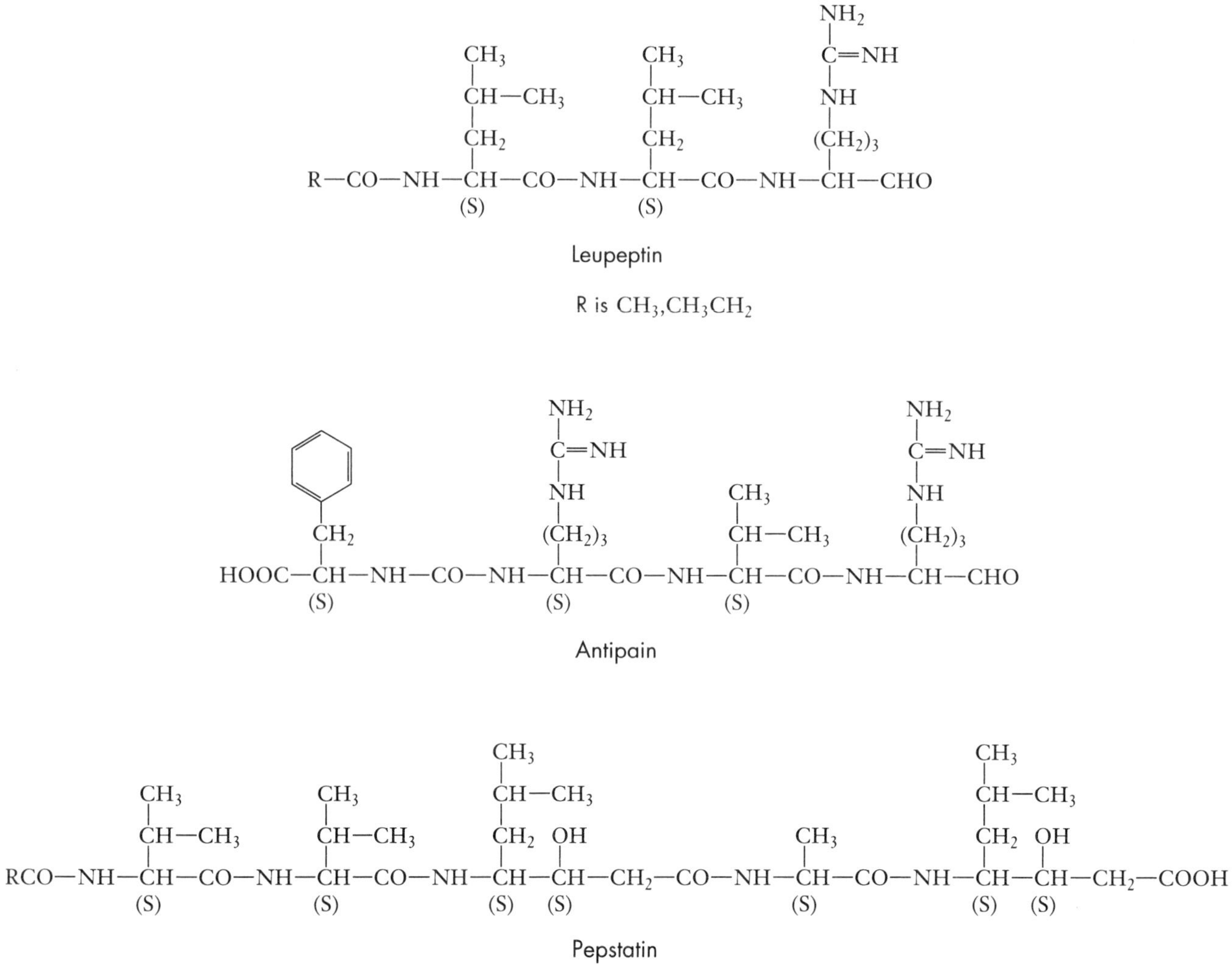

$$\begin{array}{c} & & & & & & NH_2 \\ & CH_3 & & CH_3 & & C{=}NH \\ & | & & | & & | \\ & CH{-}CH_3 & & CH{-}CH_3 & & NH \\ & | & & | & & | \\ & CH_2 & & CH_2 & & (CH_2)_3 \\ & | & & | & & | \\ R{-}CO{-}NH{-}CH{-}CO{-}NH{-}CH{-}CO{-}NH{-}CH{-}CHO \\ & (S) & & (S) & \end{array}$$

Leupeptin

R is CH_3, CH_3CH_2

$$\begin{array}{c} & & NH_2 & & & NH_2 \\ & & C{=}NH & & & C{=}NH \\ & & | & & & | \\ & & NH & CH_3 & & NH \\ & CH_2 & | & | & & | \\ & | & (CH_2)_3 & CH{-}CH_3 & & (CH_2)_3 \\ & | & | & | & & | \\ HOOC{-}CH{-}NH{-}CO{-}NH{-}CH{-}CO{-}NH{-}CH{-}CO{-}NH{-}CH{-}CHO \\ & (S) & & (S) & & (S) \end{array}$$

Antipain

$$\begin{array}{c} & & & & CH_3 & & & CH_3 \\ & CH_3 & & CH_3 & CH{-}CH_3 & & & CH{-}CH_3 \\ & | & & | & | & & CH_3 & | \\ & CH{-}CH_3 & & CH{-}CH_3 & CH_2\ OH & & | & CH_2\ OH \\ & | & & | & | & | & CH{-}CH_3 & | & | \\ RCO{-}NH{-}CH{-}CO{-}NH{-}CH{-}CO{-}NH{-}CH{-}CH{-}CH_2{-}CO{-}NH{-}CH{-}CO{-}NH{-}CH{-}CH{-}CH_2{-}COOH \\ & (S) & & (S) & (S)\ (S) & & (S) & (S)\ (S) \end{array}$$

Pepstatin

Figure 13.20
Some protease inhibitors of microbial origin. (R) and (S) indicate the stereo-chemistry of chiral centers (see Box 14.1).

moiety onto the lipid carrier in the biosynthesis of the asparagine-linked carbohydrate chains of glycoproteins.

Immunosuppressants. Some microbial products act as powerful immu-nosuppressants and have been useful in decreasing the incidence of al-lograft rejection in organ transplant patients. The best known of these compounds is cyclosporin (Figure 13.24A), a cyclic peptide produced by a fungus. Interestingly, this compound was originally caught in the screening net because of its antifungal activity. Cyclosporin forms a complex with a *cis–trans* peptidyl prolyl isomerase, cyclophilin, in eu-karyotic cytoplasm. The cyclosporin–cyclophilin complex then binds to and inhibits the action of a phosphoprotein phosphatase known as

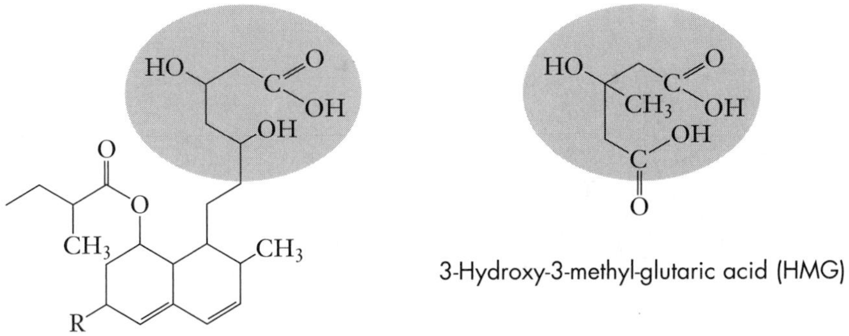

Compactin R is H
Mevinolin R is CH_3

3-Hydroxy-3-methyl-glutaric acid (HMG)

Figure 13.21
Compactin, mevinolin (lovastatin), and 3-hydroxy-3-methylglutaric acid. The antibiotics are drawn in their acid, rather than lactone, form because lactones are hydrolyzed rapidly into acids in the human body. Thus in the case of mevinolin, the structure would more accurately be called mevinolic acid. Shaded areas indicate regions of structural similarity.

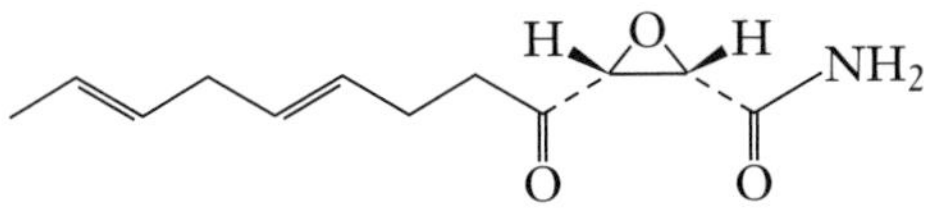

Figure 13.22
Cerulenin.

calcineurin, thus preventing a dephosphorylation event that leads finally to the expression, in T cells, of an autocrine lymphokine known as interleukin 2. As a result, the T cells involved in graft rejection are not activated.

More recently, FK-506 (Figure 13.24B) was isolated when microbial products were screened for the activity of suppressing interleukin 2 production.

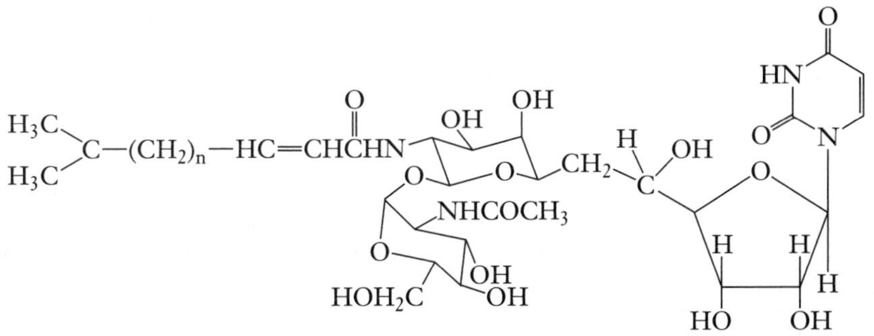

Figure 13.23
Tunicamycin.

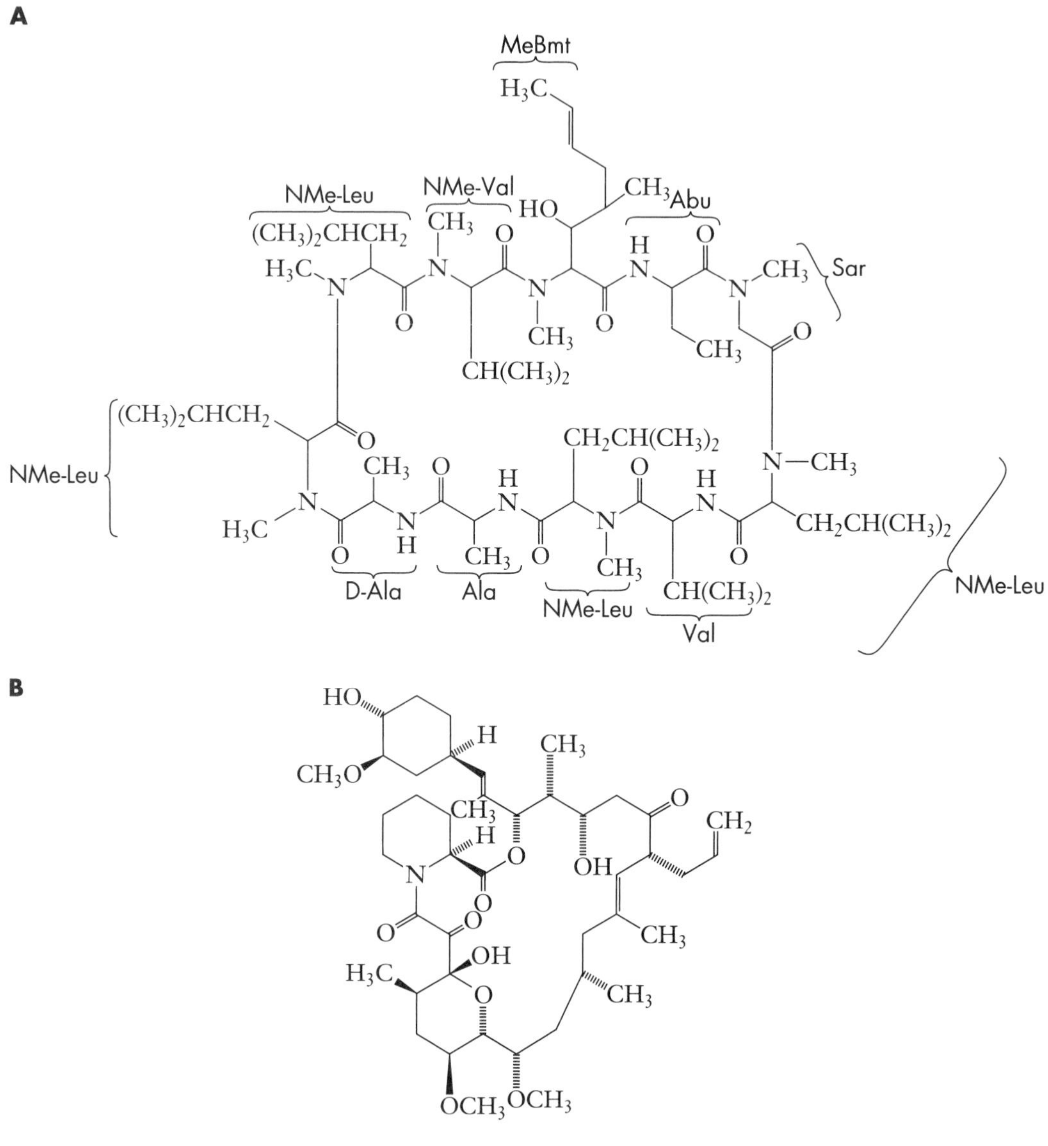

A

B

Figure 13.24
A. Cyclosporin A. NMe, *N*-methyl; Sar, sarcosine; Abu, 2-aminobutyric acid; MeBmt, 4-(2-butenyl)-4,*N*-dimethyl-L-threonine. The new amino acid, MeBmt, is apparently essential for the biological activity, as is the presence of numerous *N*-methyl groups, which inhibit hydrogen bonding between backbone residues and presumably help cyclosporin to assume a particular conformation. **B.** FK-506.

Compounds Binding to Specific Receptors. Merck scientists have identified a compound they call asperlicin (Figure 13.25), which is produced by *Aspergillus* sp. and binds strongly to the cholecystokinin receptor. Cholecystokinin is a peptide hormone that was initially isolated from

A

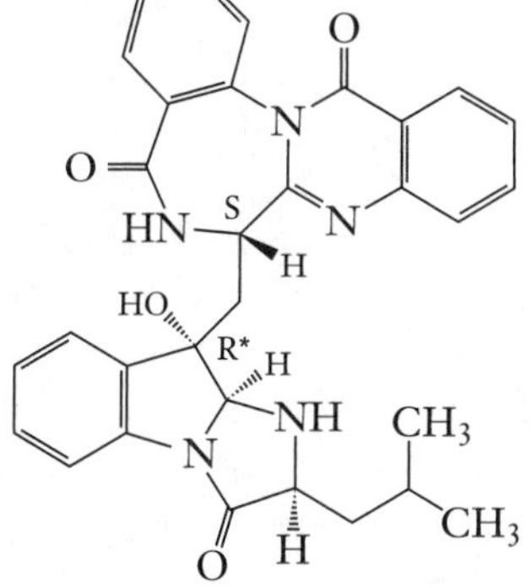

B

Lys—Ala—Pro—Ser—Gly—Arg—Met—Ser—Ile—Val—
Lys—Asn—Leu—Gln—Asn—Leu—Asp—Pro—Ser—His—
Arg—Ile—Ser—Asp—Arg—Asp—Tyr—Met—Gly—Trp—
                                   |
                                 SO_3H

Met—Asp—Phe—NH_2

Figure 13.25
A. Asperlicin. **B.** Human cholecystokinin.

intestinal tissue and is known to stimulate gut motility and contraction of the gallbladder. The relationship between cholecystokinin and aspercillin could not have been discovered without screening. No one would have imagined from the structure of cholecystokinin alone that aspercillin was its analog, especially because little was known about the structure of the cholecystokinin receptor and the manner in which cholecystokinin binds to it. Another example of a microbial product that binds to a specific receptor in higher animals is MY 336-a (Figure 13.26), which associates tightly with β-adrenergic receptors and is made by a *Streptomyces* species.

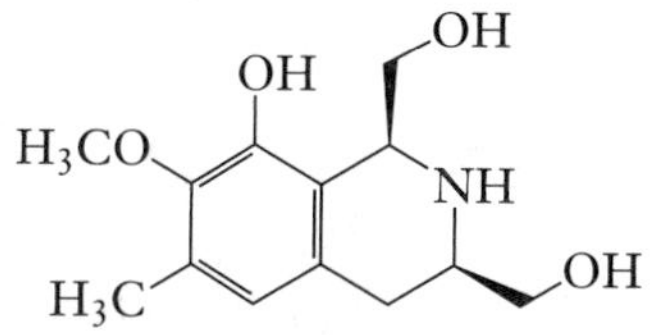

Figure 13.26
MY 336-a, a *Streptomyces* product that binds to β-adrenergic receptors.

GOALS OF ANTIBIOTIC RESEARCH

The study and development of antibiotics certainly share some of the same aims as other areas of biotechnology. For example, it is always desirable to try to improve the yield of an antibiotic during fermentation and subsequent processing steps. Indeed, in the early stage of the development of penicillin, the crucial achievement was to obtain the compound in amounts sufficient for therapeutic use. Also, for antibiotics that are put mainly to veterinary or agricultural uses, reducing cost through improved yields is of paramount importance. However, much of the research on the antibiotics for human use has a very different focus. When we are trying to develop a cure for a life-threatening infection, the cost of the treatment is not the most important factor.

A very large fraction of antibiotic research is directed toward the development of new agents. They are needed because of the many microorganisms, including most fungi and viruses, for which we do not yet possess truly effective and safe antibiotic agents. Some bacteria, such as *Pseudomonas aeruginosa*, also are intrinsically resistant to most antibiotics.

An even more vexing problem is the emergence of resistant strains among the organisms that were sensitive to antibiotics before the drugs became widely used. We will encounter numerous examples later in this chapter. This phenomenon tends to limit severely the useful life of any new antibiotic, requiring the pharmaceutical industry to come up with new compounds continually. The need is especially acute because of the following unfortunate situation. In any modern hospital, huge amounts of antibiotics are used in the treatment as well as the prevention of infectious disease. As a result, the hospital environment becomes highly enriched for bacteria that are resistant to those antibiotics. At the same time, the immune and other defense mechanisms of the body are not functioning well in many hospitalized patients, who are thus especially vulnerable to hospital-acquired ("nosocomial") infection by these resistant bacteria.

Although many different kinds of antibiotics have been developed, the same basic concepts and principles have governed the research in almost every case. For this reason, we will describe the steps in the development of only two classes of antibiotics: aminoglycosides and β-lactams. Readers who wish to learn the particulars of how other classes of antibiotics were developed should consult the references listed at the end of this chapter.

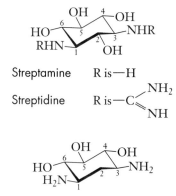

Streptamine   R is —H

Streptidine   R is —C
                      NH₂
                      NH

2-Deoxystreptamine

Figure 13.27
Streptamine and 2-deoxystreptamine. In streptomycin, the two amino groups of streptamine are substituted by amidino groups, producing a cyclitol often called streptidine.

DEVELOPMENT OF THE AMINOGLYCOSIDES

More than 150 naturally occurring aminoglycoside antibiotics have been isolated from culture filtrates of *Streptomyces* species and other members of the actinomycete line. In most of these compounds, the aminocyclitol moiety is either streptamine (as in streptomycin) or 2-deoxystreptamine (as in most other compounds; Figure 13.27). Usually the aminocyclitol is further substituted by sugars, very often amino sugars. It appears that some of the enzymes involved in aminoglycoside biosynthesis do not have a very stringent substrate specificity, and a single strain often produces a number of structurally related compounds. For example, one gentamicin-producing strain was found to produce more than 20 compounds, all related in structure, with only minor variations.

The history of the development of various aminoglycoside antibiotics is very interesting, because it was influenced strongly by rational, synthetic approaches yet at several points was affected also by the discovery of novel types of natural compounds that could not have been imagined even by the most capable medicinal chemists. In order to follow this history, one must first know something about the mechanism of aminoglycoside action and that of aminoglycoside resistance.

The aminoglycosides inhibit protein synthesis. Their main target of action is the eubacterial—that is, the 70S—ribosome. Aminoglycoside molecules contain two or more amino groups, and their resulting polycationic nature is probably important in the binding to the ribosome. Many of the antibiotics that inhibit protein synthesis (for example, chloramphenicol, tetracycline, and erythromycin) are merely bacteriostatic; they stop the growth of bacteria but do not kill them. However, as Selman Waksman and co-workers noted in their first report on streptomycin in 1944, aminoglycosides are unusual in that they are clearly bactericidal. After sensitive bacteria are exposed to aminoglycosides for several minutes, even extensive washing cannot revive them.

Apparently the irreversible nature of the aminoglycosides' action is related to their irreversible entry into the cells. A plausible scenario, proposed by Bernard D. Davis, is as follows: The small number of aminoglycoside molecules that enter the cell shortly after exposure to the antibiotic bind to the ribosomes, not only inhibiting protein synthesis but also producing the misreading and truncation of polypeptides. These abnormal polypeptides become inserted into membranes and make them leaky. This is the irreversible step in aminoglycoside action. Because aminoglycosides are polycations, large amounts of the drug are now "sucked in" through the leaky membrane by the interior-negative membrane potential. This massive influx finally produces a complete cessation of protein synthesis. Regardless of the validity of the details of this hypothesis, the polycationic nature of aminoglycosides is clearly an absolute prerequisite for their rapid entry into the bacterial cell.

In the laboratory, it is easy to isolate streptomycin-resistant mutants of *E. coli* by plating 10^9 cells or more on streptomycin-containing media. Such mutants are altered in one of the proteins in the small subunit of the ribosome, which, as we have noted, is the target of streptomycin action. However, *E. coli* mutants of this type are practically never found in patients. Instead, among enteric bacteria such as *E. coli*, naturally occurring aminoglycoside-resistant bacteria almost always carry resistance plasmids (R plasmids), which contain genes coding for the inactivation of aminoglycosides. The reason seems to be that in the intestine, so many bacteria coexist at such high population densities that there is a high probability of genetic exchange between them. (In contrast, the streptomycin-resistant strains of *Mycobacterium tuberculosis* are apparently ribosomal mutants.) The resistant *E. coli* inactivate the drug by enzymatically attaching various groups onto it and thus diminishing its polycationic character. A group of enzymes called AACs (aminoglycoside acetyltransferases) transfer acetyl groups to the amino groups, thereby decreasing the number of positive charges on the drug molecule. Another enzyme group, the APHs (aminoglycoside phosphoryltransferases), transfer phosphoryl groups to hydroxyl groups on the drug, thus adding negative charges and decreasing the molecule's net positive charge. ANTs (aminoglycoside nucleotidyltransferases, some of which were once also called AADs, or aminoglycoside adenylyltransferases) work in the same way, by adding nucleotide groups—most often adenylyl groups—to hydroxyl groups on the drug. The way the kanamycin B molecule can be inactivated by various enzymes is illustrated in Figure 13.28 (see Table 13.1 as well).

The toxicity of aminoglycosides is probably associated with their polycationic nature. All aminoglycosides are somewhat toxic to humans and often damage the inner ear and kidney. Thus the use of aminoglycosides has become limited to certain applications.

1. Aminoglycosides, especially streptomycin, were the first drugs to prove effective against tuberculosis, caused by *M. tuberculosis*. Tuberculosis is a serious disease that claimed the lives of many young people in the first half of the twentieth century, and it still is the leading agent of death in the developing parts of the world, causing nearly 3 million deaths every year. Although the incidence of tuberculosis in the United States was declining

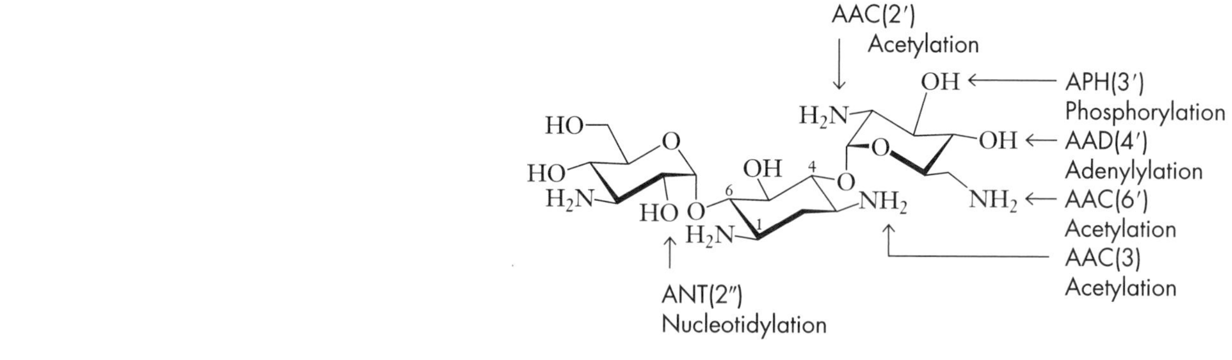

Figure 13.28
Kanamycin B can be inactivated in many different ways. The name of the enzyme and the nature of the modification reaction are shown. Note that the atoms in the sugar linked to the 4-position of the aminocyclitol (at the right end of the molecule) are numbered with primes and that the atoms on the 6-substituent sugar (at the left end of the molecule) are numbered with double primes. The enzymes include all those listed in Table 13.1 as being capable of inactivating kanamycin A, plus AAC(2′), which inactivates kanamycin B but not kanamycin A (because the latter lacks the 2′-amino group).

Susceptibility of Aminoglycosides to Modifying Enzymes

T
A
B
L
E

13.1

Enzyme	Occurrence[a]	Aminoglycoside						
		Streptomycin	Kanamycin	Dibekacin	Tobramycin	Amikacin	Gentamicin	Netilmicin
AAC(3)-I	11%	−	(−)	[−]	[−]	(−)	+	(−)
AAC(3)-II	60%	−	(−)	+	+	(−)	+	+
AAC(3)-III	4%	−	+	+	+	(−)	+	(−)
APH(3′)-I	46%	−	+	−	−	[−]	−	−
ANT(4′)-I	b	−	+	+[c]	+	+	−	−
AAC(6′)-I	60%	−	+	+	+	+	[−]	+
AAC(6′)-II	d	−	+	+	+	(−)	+	+
AAC(6′)-APH(2″)	e	−	+	+	+	+	+	+
ANT(2″)-I	15%	−	+	+	+	(−)	+	(−)
ANT(3″)-I	f	+	−	−	−	−	−	−
APH(3″)-I		+	−	−	−	−	−	−

Antibiotics: kanamycin and gentamicin denote, respectively, kanamycin A and gentamicin C_{1a}.

Symbols: +, modification resulting in resistance; −, the target functional group is absent and the drug is not modified; (−) the target functional group is present but is not modified; [−], the drug is modified *in vitro*, but the rate (or affinity) is insufficient to produce significant resistance.

[a] This column shows the percentage of general Gram-negative bacterial strains (excluding *Serratia*, *Acinetobacter*, and *Pseudomonas*) that show the particular phenotype expected of the enzyme-carrying strain, among recent clinical isolates showing resistance to at least one aminoglycoside (excluding streptomycin). Based on Shaw, K.J., et al. (1993), *Microbiol. Rev.* 57:138–163.

[b] Present in 30% of resistant *S. aureus* strains.

[c] Although dibekacin lacks the 4′-OH group, the enzyme modifies the 4″-OH group instead.

[d] Present in 48% of resistant *P. aeruginosa* strains.

[e] This hybrid enzyme, made by fusing AAC(6′) to ANT(2″), is present in 99% of the resistant *S. aureus* strains.

[f] In one study, 59% of the strains that were resistant to other aminoglycosides were also resistant to streptomycin, and of those, 56% carried this gene.

steadily in the second half of the century, the trend was reversed in 1985, and the reemergence of tuberculosis is causing much concern. A treatment regimen including streptomycin made a strong contribution to our fight against tuberculosis and was standard until the early 1980s. More recently rifampicin (a rifamycin) has replaced streptomycin, because rifampicin can kill even those bacteria found in closed caseous lesions (streptomycin cannot) and because it can be taken orally.

2. Aminoglycosides are poorly absorbed from the intestinal tract when taken orally. This means, however, that the aminoglycoside concentration *within* the intestinal tract remains high and that enteric bacteria are killed efficiently. Poor absorption also means that patients are less likely to suffer from the inherent toxicity of these agents. The drugs were thus used widely in

past decades to treat gastrointestinal infections, especially in the developing countries where such infections are frequent and where the low cost of some of the aminoglycosides was an attractive feature. Unfortunately, this usage may have contributed heavily to the selection and dissemination of R plasmids coding for inactivating enzymes of different types. Such use of aminoglycosides has now been discontinued in most of the industrialized countries, because even slow absorption can produce side effects in patients with impaired kidney function, and because safer alternatives are available.

3. Currently, the most important use of aminoglycosides is in the treatment of systemic infections caused by species of Gram-negative bacteria that are intrinsically resistant to most other antibiotics. Under these circumstances, we tolerate the inherent toxicity of these antibiotics because we do not have many other choices and because we are treating life-threatening infections. Aminoglycosides can cross even outer membranes of very low permeability, possibly because these polycations can bind to the anionic surface of the outer membrane, disorganize it, and thereby destroy the permeability barrier in a manner reminiscent of the action of polymyxin (page 442). The presence of inactivating enzymes, however, can make the drug totally ineffective, so one major thrust of aminoglycoside research has been the search for compounds that are not significantly inactivated by commonly encountered enzymes.

Streptomycin

This agent (Figure 13.29) was discovered by Selman A. Waksman, a soil microbiologist interested in the antagonism between different groups of soil organisms. Waksman and co-workers were examining the products of *Streptomyces*, a typical inhabitant of soil (Chapter 2), looking for

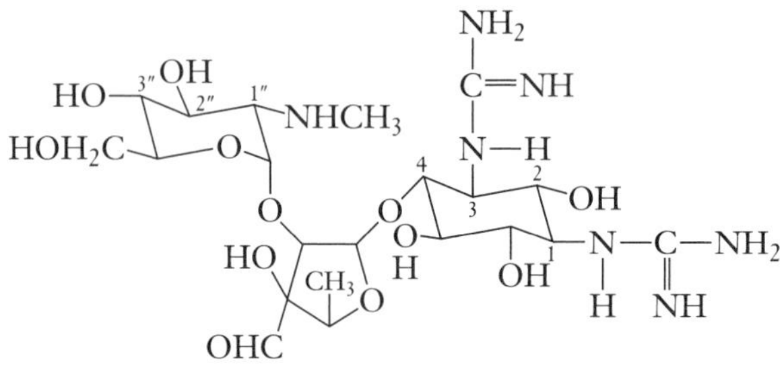

Figure 13.29
Streptomycin.

agents that were active against Gram-negative bacteria, for most of which penicillin G, the only commercially produced antibiotic at that time, was totally ineffective. (*Neisseria* spp. are susceptible to penicillin G, possibly because their porin channel favors anionic solutes, and thus constitute an exception among Gram-negative bacteria.) Their short note published in 1944, announcing the discovery of streptomycin, is impressive because it describes not only the activity of streptomycin against both Gram-positive and Gram-negative bacteria but also its bactericidal property and the fact that its production is greatly affected by the nature of the growth medium and the growth phase.

Although its toxicity prevented streptomycin from becoming the "penicillin for Gram-negative infections," the discovery of its activity against *M. tuberculosis* made this antibiotic extremely important. Streptomycin is no longer a first-line drug for the treatment of tuberculosis. However, because *M. tuberculosis* can develop resistance to rifampicin and the other first-line drugs, streptomycin still remains an important agent for the treatment of cases caused by such resistant organisms.

Newer Aminoglycosides

The wide use of streptomycin to treat gastrointestinal infections by Gram-negative bacteria in the 1950s and 1960s resulted in the selection of enteric bacteria carrying R plasmids that bore the streptomycin-resistance gene. Since then, the major focus of aminoglycoside research has been the discovery and development of new agents that are not inactivated by commonly encountered aminoglycoside-modifying enzymes.

Kanamycin

Kanamycin (Figure 13.30), the second commercially produced aminoglycoside, was discovered by Hamao Umezawa and co-workers in 1957 from the culture filtrate of another *Streptomyces* species. This was a timely discovery, because kanamycin proved to be active on enteric bacteria that produce streptomycin-inactivating enzymes. When studies of the plasmid-coded resistance to streptomycin showed that the antibiotic is inactivated through adenylylation or phosphorylation of its 3″-hydroxyl group, it became clear why the kanamycins remain fully active against strains that contain the streptomycin resistance plasmids. In streptomycin the amino sugar is connected via another sugar to the 4-position of the aminocyclitol streptamine, but in kanamycins the amino sugar is connected directly to the corresponding position of the aminocyclitol. Thus in kanamycins, there is no group that corresponds to the 3″-hydroxyl group of streptomycin.

Antibiotic	R	R_1	R_2	R_3
Kanamycin A	H	OH	OH	OH
Kanamycin B	H	NH_2	OH	OH
Tobramycin	H	NH_2	H	OH
Dibekacin	H	NH_2	H	H
Amikacin	HABA	OH	OH	OH

Figure 13.30
Kanamycin A and its relatives. The molecule is drawn in such a way as to represent roughly its three-dimensional structure as determined by X-ray crystallography and NMR studies. HABA: 2-hydroxy-4-aminobutyryl. (See the chapter by Nagabhushan et al. in Whelton, A., and Neu, H.C. (1982), *The Aminoglycosides: Microbiology, Clinical Use, and Toxicology* (Marcel Dekker).

The aminocyclitol in streptomycin is streptamine. In kanamycin and in most of the other aminoglycosides either isolated or synthesized since the discovery of kanamycin, the aminocyclitol is 2-deoxystreptamine (see Figure 13.27). In most compounds, 2-deoxystreptamine is substituted by amino sugars at both the 4- and 6-positions (Figure 13.13), although exceptions such as neomycin, ribostamycin, and butirosin (Figure 13.31), which are substituted at the 4- and 5-positions, exist.

Kanamycin proved to be significantly more active than streptomycin against most Gram-negative bacteria. However, it had poor activity

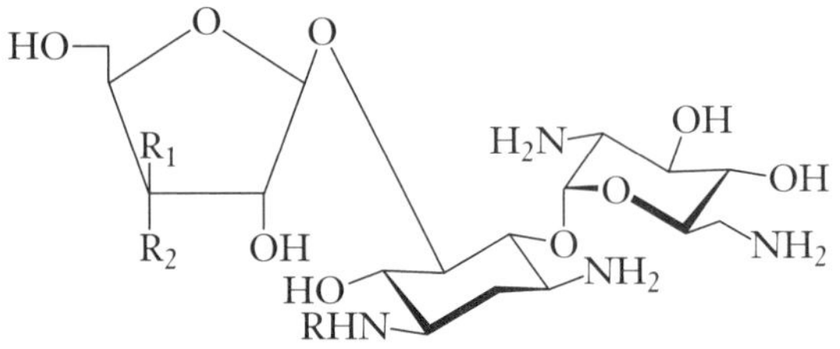

	R	R_1	R_2
Ribostamycin	H	OH	H
Butirosin A	NH_2—$(CH_2)_2$—CHOH—CO	H	OH

Figure 13.31
Ribostamycin and butirosin A.

against *P. aeruginosa*, partly because many strains of this organism contain enzymes, sometimes coded for by a chromosomal gene, that inactivate kanamycin either through phosphorylation of the 3'-hydroxyl group by APH(3') or through acetylation of the 6'-amino group by AAC(6').

Semisynthetic Aminoglycosides: Dibekacin, Amikacin, and Netilmicin

In the 1960s, kanamycin was used extensively in Japan and elsewhere to treat bacterial infections of the intestinal tract. This led to the spread of R plasmids with genes coding for kanamycin resistance. H. Umezawa and co-workers found that the most common mechanism of resistance was phosphorylation of the drug at 3'-position by APH(3'), although among kanamycin-resistant strains of Gram-positive bacteria, adenylylation at the 4'-position by ANT(4') was often seen. Sumio Umezawa, a brother of H. Umezawa, then took the unprecedented "rational" step of removing the targets of these modification reactions from the kanamycin B molecule—that is, the hydroxyl groups at the 3'- and 4'-positions. This was first achieved by a total chemical synthesis. The resulting compound, dibekacin, retained the efficacy of kanamycin and was resistant to the kanamycin-modifying enzymes mentioned above, as expected. Since then, ways of producing dibekacin semisynthetically, by chemical modifications of kanamycin B, have been found, and dibekacin has been used extensively in Japan and neighboring South Korea and Taiwan.

Searches were carried out in many laboratories for other new aminoglycosides that would not be modified by existing enzymes. One notable success was the development of amikacin (Figure 13.32). The scientists involved in this endeavor had noticed that a *Bacillus* species produced an aminoglycoside, butirosin A, that was identical to the *Streptomyces* product ribostamycin except that it carried a 4-amino-2-hydroxybutyryl substituent on the 1-amino group of the 2-deoxystreptamine moiety (see Figure 13.31). Butirosin seemed to be unaffected by the APH(3') enzyme that readily inactivated the ribostamycin, so scientists at Bristol-Myers added the 4-amino-2-hydroxybutyryl substituent to the 1-amino group of kanamycin A, producing a new compound that they named amikacin. The addition of this substituent had a remarkable effect. First, the modification of the 2″-hydroxyl group of kanamycin by the ANT(2″) enzyme was abolished. This could not have been predicted from the behavior of butirosin, because butirosin does not have a substituent sugar at the 6-position. However, in aqueous solution, the position of amikacin's 2″-hydroxyl group is very close to that of the 1-amino group (see Figure 13.32); thus the bulky substituent on the latter group presents a steric hindrance to the approach of the modifying enzyme.

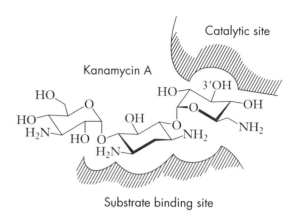

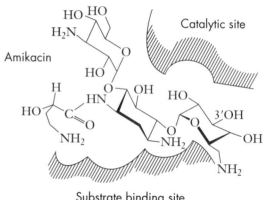

Figure 13.32

A possible mechanism for the stability of amikacin against enzymes that
would modify the 3′, 4′, or 6′-position of the aminoglycoside. Kanamycin A
binds to the APH(3′) enzyme by using the amino groups at the 1- and 3-
positions of the 2-deoxystreptamine; thus the catalytic site of APH(3′) comes
to be positioned just above the 3′-OH group (above). In comparison, the
binding of amikacin to the enzyme is skewed, because the amino group of
the 4-amino-2-hydroxybutyryl substituent is displaced in relation to the 1-
amino group of the 2-deoxystreptamine. This skewed binding holds the 3′-
OH group away from the catalytic site, thus making amikacin resistant to
the APH(3′) enzyme. Similar mechanisms can be proposed for ANT(4′) and
AAC(6′).

Second, amikacin was resistant to enzymes that modify the 3-amino
group of the 2-deoxystreptamine residue, again presumably because of
the proximity of the modified 1-amino group. Third, amikacin was resistant to some of the enzymes that modify the groups at the 3′- and
6′-positions. This was expected from the resistance of butirosin to
APH(3′), but it is somewhat difficult to explain in terms of theoretical

considerations because these groups are located very far away from the 1-amino substituent. One proposed explanation is that the 4-amino group of the aminohydroxybutyryl substituent and the 3-amino group of 2-deoxystreptamine may fit into the binding sites of the enzyme that normally bind the 1-amino and 3-amino groups of kanamycin, and this may move the potential modification sites far from the active site of the enzyme (see Figure 13.32). In any case, amikacin is currently one of the compounds that are resistant to the widest range of aminoglycoside-modifying enzymes (Table 13.1).

It is interesting (parenthetically) that two genera as distant from each other phylogenetically as *Bacillus* and *Streptomyces* produce nearly identical versions of very complex metabolites like butirosin A and ribostamycin. Yet the search for antibiotics has repeatedly turned up similar situations with many other species. It is possible that the ability to synthesize these secondary metabolites has been transferred between distantly related organisms by the "horizontal gene transfer" process. On the other hand, the presence of the unusual bulky substituent at the 1-position of butirosin A is a feature that has not been seen in the products of *Streptomyces;* it may illustrate the benefit of going to phylogenetically distant organisms in search for compounds of novel structure.

Because the ongoing use of antibiotics causes the proliferation of resistance factors able to inactivate an increasingly wide range of compounds, finding new compounds that are immune to all the resistance enzymes is getting more and more difficult. It is thus unrealistic to expect to find a natural compound that, without further chemical modification, already has all the desirable properties (resistance to enzymatic modification, low toxicity, high efficacy, and so on). Accordingly, a trend in recent years has been to use natural antibiotic products either as a model (as in the case of butirosin) or as the starting material for further chemical modification (as with kanamycin for the production of amikacin and dibekacin).

Another example of this phenomenon is seen in the development of netilmicin (Figure 13.33). The starting material in this case was sisomicin, which is produced not by *Streptomyces* but by another actinomycete,

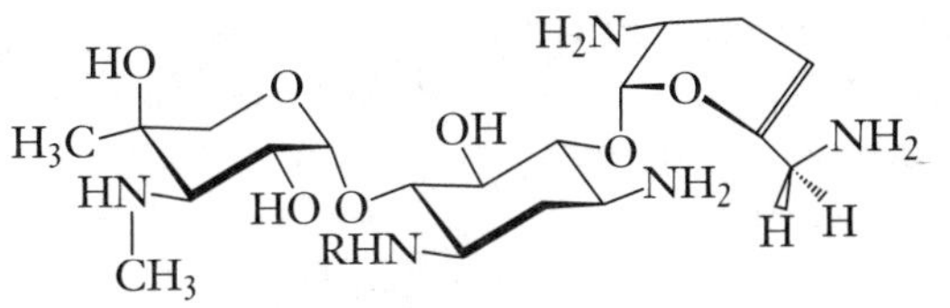

Sisomicin	R is —H
Netilmicin	R is —CH_2CH_3

Figure 13.33
Sisomicin and netilmicin.

Micromonospora. Sisomicin is unusual in that the amino sugar linked to the 4-position of the 2-deoxystreptamine is an unsaturated sugar and lacks the major targets of enzymatic modification, such as the hydroxyl groups at the 3′- and 4′-positions. Thus it is an excellent antibiotic and is produced commercially. However, the addition of a bulky substituent on the 1-amino group resulted in the blocking of another potential modification site, the 2″-hydroxyl group, just as we saw with amikacin. The new product, netilmicin, retains all the good properties of sisomicin *and* is not inactivated by ANT(2″) or APH(2″).

Tobramycin and Gentamicin C: Nature's Work as a Medicinal Chemist

The "rational" approach of eliminating the target sites of modification and introducing bulky substituents is not the only route for the production of drugs that would withstand the inactivating enzymes. At about the same time when medicinal chemists were producing dibekacin and amikacin by chemical modification, screening for natural microbial products led to the isolation of a compound that showed good activity against many kanamycin-resistant Gram-negative bacteria. This compound, tobramycin, already lacks the 3′-hydroxyl group, one of the prime targets of modification of kanamycin (see Figure 13.30). Nature had long ago accomplished something that the best medicinal chemists were trying to do! As Table 13.1 shows, tobramycin is as inert to enzymatic modification as dibekacin. It is also active, unlike kanamycins, against many strains of *P. aeruginosa*, because it is not inactivated by APH(3′) and possibly because it penetrates the outer membrane somewhat better. Being a natural product, it is less expensive than some of the semisynthetic compounds, and it is still used widely, especially in the United States.

Another natural product is gentamicin C (Figure 13.34A). If we did not know that it is synthesized by *Micromonospora* species, we would surely believe it to be a semisynthetic compound developed by medicinal chemists, because it lacks both the 3′- and the 4′-hydroxyl groups that Umezawa's team removed chemically from kanamycin. Gentamicin C is not inactivated by several enzymes that modify the amino sugar linked to the 4-position (and therefore is also active against many *P. aeruginosa* strains), although it is still susceptible to the enzymes that modify the aminocyclitol [AAC(3)] or the amino sugar linked to the 6-position [ANT(2″)].

Gentamicin C is produced together with gentamicin B, which does contain both the 3′- and the 4′-hydroxyl groups and is thus susceptible to inactivation by common enzymes. However, a semisynthetic derivative of gentamicin B, isepamicin (Figure 13.34B), has on its 1-amino group a substituent similar to that present on amikacin (the 3-amino-2-hydroxypropyl group) and is not modified by most commonly occurring enzymes.

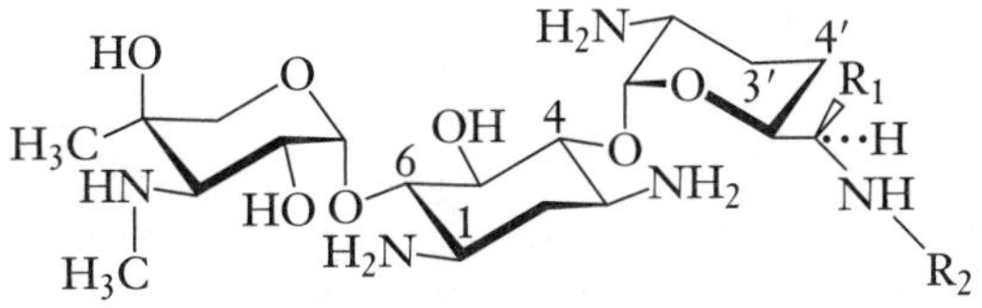

Gentamicin C

	R_1	R_2
C_{1a}	H	H
C_1	CH_3	CH_3
C_2	CH_3	H

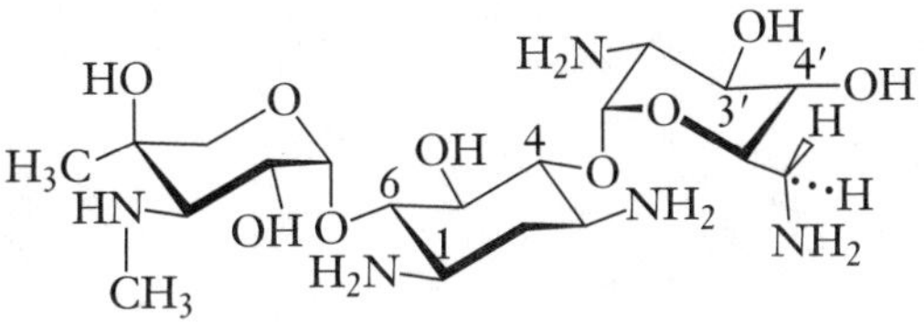

Gentamicin B

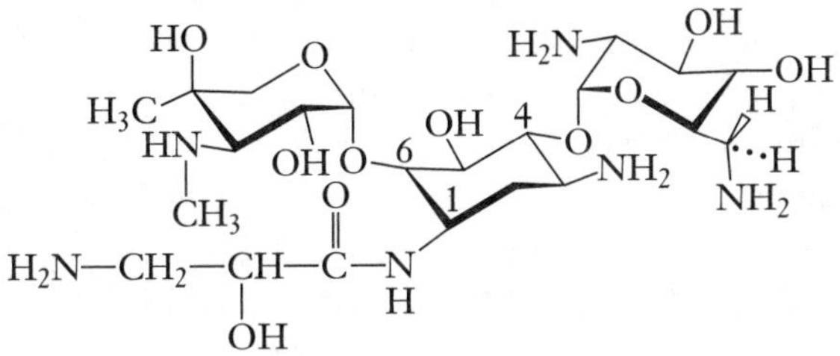

Isepamicin

Figure 13.34
Gentamicins and isepamicin.

Fortimicin A

This antibiotic (Figure 13.35), introduced into clinical use in Japan, has an aminocyclitol moiety different from 2-deoxystreptamine and based instead on 1,4-diaminocyclitol. Its 4-amino group is substituted by a glycyl group and a methyl group, and it has only one amino sugar substituent at the 6-position. This compound was isolated from *Micromonospora* in 1977. Because its structure is radically different from those of all other aminoglycosides currently in use, it is not inactivated by many aminoglycoside-modifying enzymes coded by commonly occurring R plasmids. Surprisingly, AAC(3)-I modifies this antibiotic despite the absence of its normal target, the 3-amino group, by acetylating the 1-amino group. Thus the aminocyclitol must fit into the substrate-binding site of this enzyme in a manner quite different from that of the 2-deoxystreptamine-containing aminoglycosides. Fortimicin A, however, does not have a good activity against *P. aeruginosa*, and this limits its application somewhat.

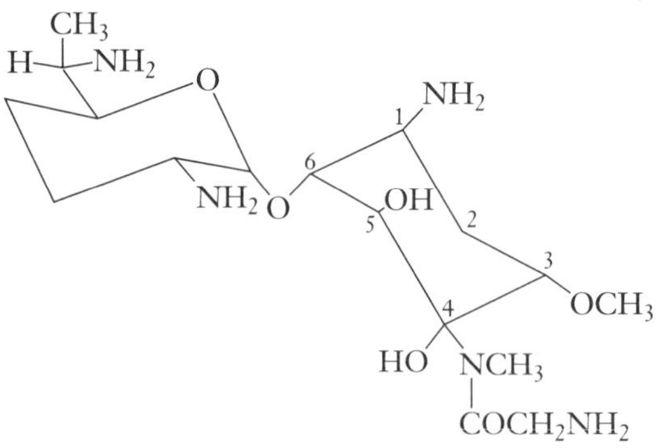

Figure 13.35
Fortimicin A.

Use of Mutasynthesis for Developing New Aminoglycosides

Mutasynthesis, which was proposed in the 1960s, is a method for developing new aminoglycosides that takes advantage of the fact that some of the enzymes involved in antibiotic biosynthesis may not have an absolutely strict substrate specificity and may incorporate structural analogs of an intermediate into the final product. If a wild-type organism is used to make the antibiotic, the normal intermediate is present in its cells and competes successfully against the analog supplied. Thus one must use mutant strains that are unable to synthesize the natural substrate. In one example, chemically synthesized 3′,4′-dideoxy-6-*N*-methylneamine was added to a *Bacillus* strain that was defective in the biosynthesis of neamine and therefore required neamine in the medium in order to synthesize the aminoglycoside butirosin. The resulting compound (Figure 13.36) was missing the 3′-hydroxyl group and had a bulky substituent on the amino nitrogen at the 6′-position. Thus it was resistant to two modifying enzymes to which butirosin is susceptible, APH(3′) and AAC(6′).

Origin and Pattern of Aminoglycoside Resistance

Table 13.1 lists only the most common aminoglycoside-modifying enzymes. Still, the number and variety of these enzymes are almost bewildering. Most of them are coded for by genes on resistance plasmids, although a few bacterial species, such as *Serratia marcescens* and *P. aeruginosa*, are known to contain many strains that possess chromosomal genes coding for some of them. It seems possible that the chromosomal genes are the result of rather recent transposon insertion events from the plasmids.

Even the strong selective pressure imposed by the extensive clinical use of aminoglycosides is unlikely to account for the evolution of so

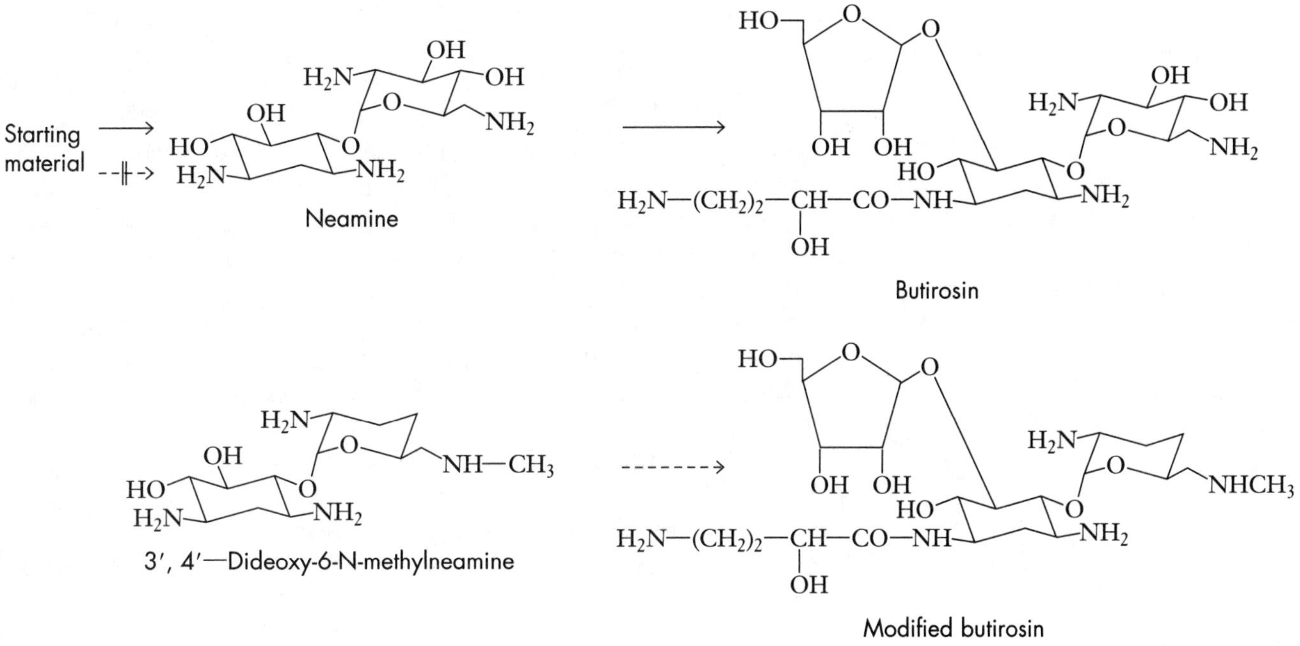

Figure 13.36

A new aminoglycoside produced by a mutasynthesis approach. In the wild-type organism (above), an intermediate, neamine, is converted into butirosin. Scientists used a mutant *Bacillus* strain blocked in the biosynthesis of neamine (broken arrow) and fed it a synthetic analog of neamine (below). As a result, the bacteria synthesized an analog of butirosin that lacked several potential inactivation sites.

many enzymes so quickly. Thus Julian Davies proposed that these enzymes have existed in nature for a long time and that the effect of the recent selective pressure was simply to recruit them. Davies also suggested that the original sources of these genes were probably the organisms producing the target antibiotics; he reasoned that the antibiotic-producing organisms had to protect themselves against their own products. The cloning and sequencing of genes for aminoglycoside-modifying enzymes from resistance plasmids and from the producing strains have in fact revealed extensive homology between the DNA sequences of many genes from these two sources.

The cloning of various genes for aminoglycoside-modifying enzymes now enables researchers to classify the enzyme, at the genotype level, by using DNA probe hybridization assay (see Figure 3.18). Such a study using bacterial strains collected between 1987 and 1991 showed that the most prevalent enzymes found in Gram-negative bacteria were AAC(3), AAC(6′), and APH(3′); see Table 13.1. Because assaying for enzyme activity and determining the specificity of an enzyme are not easily carried out for a large number of strains, there was little data on the occurrence of various enzymes before the DNA probes became available. One source is a Bristol-Myers study, conducted in the 1970s, that analyzed the enzyme patterns for aminoglycoside-resistant isolates as a service to hospitals in the United States. The most prevalent enzymes among Gram-negative bacteria were APH(3′), 58%; AAC(6′), 38%; and AAC(3), 18%. Thus during the last 20 years there has been no drastic alteration in the relative abundance of various enzymes. However, researchers have noted that more recently isolated bacteria usually contain more than one aminoglycoside-inactivating enzyme, a situation that did not occur frequently in the past.

In a study published in 1985, aminoglycoside-resistant bacterial strains obtained in different countries were tested, and the identities of various inactivating enzymes were deduced from the pattern of resistance to a series of antibiotics. In strains from Japan, Taiwan, and South Korea, where kanamycin and its derivatives, especially dibekacin and amikacin, were used extensively, 78% of the strains contained AAC(6′), which inactivates these kanamycin derivatives. In contrast, in strains from sources in the United States, where gentamicin was quite popular, the most commonly found (42%) enzyme was ANT(2″), which inactivates gentamicin very efficiently. Furthermore, AAC(3), which is another efficient inactivator of gentamicin but is often ineffective against kanamycin derivatives, was present in 17% of the U.S. isolates but was virtually absent from the isolates from the Far East. Analyzed in terms of the resistance pattern, 99% of the Far Eastern isolates were resistant to dibekacin, probably the most popular aminoglycoside in that region but not licensed for use in the United States, whereas 92% of the U.S. isolates were resistant to gentamicin. These data show beyond any doubt that the usage pattern of antibiotics contributes significantly to the prevalence of various types of drug resistance genes in the local microbial population.

This proof of the relationship between antibiotic use and the prevalence of resistance plasmids is sobering. But unfortunately, it is not easy to agree on hard-and-fast rules for the selection of antibiotics. Some in the United States argue that because amikacin is still very effective against the majority of Gram-negative bacteria, the people there should try not to lose this precious advantage. They discourage the use of amikacin as a first-line drug, even in life-threatening infections by Gram-negative bacteria, until the causative organism is proved to be resistant to all other aminoglycosides. This policy makes sense from one point of view, but its merits must be weighed against the fact that implementing it would delay the use of amikacin in those cases where it was absolutely needed.

Aminoglycosides Used for Nonmedical Purposes

Not all aminoglycosides are appropriate for human use. Some have important agricultural and veterinary applications.

Hygromycin B

Hygromycin B (Figure 13.37), like many of the compounds we have discussed, is based on 2-deoxystreptamine. However, unlike many other aminoglycosides, the sugar substituent occurs not on the 4- or 6-position but on the 5-position of the aminocyclitol. Furthermore, the substituent is a 6-hydroxy-6-amino sugar linked to a neutral sugar (see Figure 13.37). Interestingly, hygromycin B inhibits both eubacterial and eukaryotic ribosomes. This implies strong toxicity, so the medicinal role of the antibiotic has been limited to occasional use as an anthelmintic (worms are eukaryotic) in poultry and swine.

In the laboratory, hygromycin has been very useful. If a plasmid contains a hygromycin resistance gene, then the plasmid-containing cell, regardless of whether it is eubacterial or eukaryotic, can be selected for by the use of this drug. This selection method has been used widely in shuttle cloning between bacteria and plant cells (Chapter 9), for example.

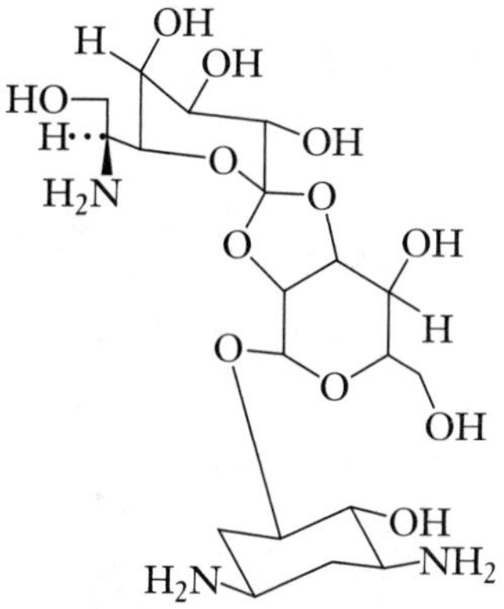

Figure 13.37
Hygromycin B.

Kasugamycin

Kasugamycin does not contain an aminocyclitol, and the cyclitol in it is an uncharged D-inositol. This is substituted by a diamino sugar, but the positive charge on one of the amino groups is neutralized by the presence of a carboxyl group (Figure 13.38). Thus the net charge of this antibiotic is only $+1$, in contrast to all the other aminocyclitol antibiotics, which are polycationic. Kasugamycin, however, is known to inhibit bacterial protein synthesis.

An important use of kasugamycin is to protect rice plants from infection by the fungus *Piricularia oryzae*. The basis for the antibiotic's efficacy against this fungus has not yet been investigated in depth. With its highly unusual structure, kasugamycin may exhibit a mechanism quite unlike that of the traditional aminoglycosides, perhaps by acting as an analog of the polysaccharide building blocks in the fungal cell wall.

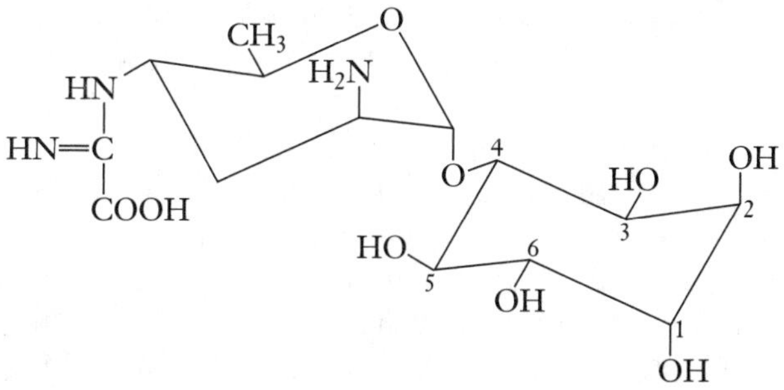

Figure 13.38
Kasugamycin.

DEVELOPMENT OF THE β-LACTAMS

β-Lactams are probably the most important class of antibiotics, for several reasons. For one thing, penicillin, the first antibiotic isolated and characterized, is a β-lactam. Since its discovery, an enormous amount of work has been done with the compounds of this class (which therefore provide ideal examples of many of the principles involved in antibiotic development). Second, the inherent toxicity of β-lactams against higher animals is extremely low, yet they are often very effective at killing infecting bacteria. This is because they inhibit peptidoglycan synthesis, a reaction that does not exist in the eukaryotic world. Consequently, they are used widely and represent much more than half the monetary value of all the antibiotics used for the treatment of human diseases worldwide. Penicillins and cephalosporins are the classic members of the β-lactam group. In recent years, however, compounds with novel nuclei have been added, such as penems, and carbapenems. Some γ-lactams are also being studied for antibiotic activity.

Penicillin G

The penicillin-producing organism in Alexander Fleming's laboratory was a true fungus, *Penicillium notatum*. When Fleming discovered the antibacterial properties of this mold in 1928, he showed remarkable foresight in applying his culture filtrate, "penicillin," to animal infection models. There was no way he could have predicted how important penicillin would become in the treatment of infectious diseases, for even sulfonamides were not known at that time. Thus the further development of penicillin as a revolutionary drug began a decade later and required the efforts of a team headed by two remarkable scientists, Howard Florey and Ernst Chain.

Florey was a medical scientist, a pathologist. However, unlike most pathologists, who deal essentially with tissue specimens from cadavers and thus have a rather static view of disease, he was a physiologist and biochemist as well, and he tried to look at disease processes as dynamic events. Chain was a chemist with unusually broad interests. These men decided in 1938 to study natural products with antimicrobial properties, and thus they continued the work of Fleming and his group on penicillin. As early as 1940 they showed that a crude preparation of penicillin, given by subcutaneous injection, was effective in treating infections of Gram-positive bacteria in mice. The first clinical trial was conducted a year later and showed promising results.

These developments occurred in the middle of World War II, when the need to treat wound infections in civilians and soldiers gave added impetus to the research. In 1941 Florey's team went to the United States to solicit advice and collaboration, and their activities spurred an unprecedented effort on the part of both the British and the U.S. governments, several pharmaceutical companies, and various laboratories. Mycologists at the Northern Regional Research Laboratory of the U.S. Department of Agriculture examined their stock collections and found that a strain of *P. chrysogenum* there produced far more penicillin than Fleming's original isolate of *P. notatum*. These people were experts on the production of secondary metabolites by fungi, and they suggested the use of submerged deep fermentation instead of the surface culture for growing *Penicillium;* this led to further improvement in yields. Subsequent improvements in penicillin-producing strains (described below) and in purification protocols finally resulted in the production of enough penicillin to treat all of the injured allied soldiers at the time of the Normandy landing in 1944.

Because penicillin and other antibiotics have so dramatically changed the nature of the infectious diseases we suffer from, it is not easy, in the 1990s, to imagine the impact penicillin had at the time of its introduction. Before the introduction of antibiotics, the major killer diseases were those caused by Gram-positive bacteria such as *Streptococcus pneumoniae* (pneumonia), *S. pyogenes* (scarlet fever, nephritis), and *Staphylococcus aureus* (various purulent infections, sepsis). Today the most troublesome bacterial infections are those caused by Gram-negative bacteria, which are resistant to antibiotic treatment mainly because of their rather impermeable outer membrane. This situation is a direct result of the near-complete success of antibiotic therapy in curing most of the "classic" bacterial infectious diseases. Perhaps this trend can best be appreciated by comparing the principal causative organisms of hospital-acquired serious infections (resulting in bacteremia—that is, multiplication of bacteria in the bloodstream) in the mid-1930s and the late 1960s (Table 13.2). Clearly, the major causative organisms of such infectious diseases in the pre-antibiotic era were *Streptococcus* and *Staphylococcus* species, whose role has now been taken over by Gram-negative rods such as *Enterobacter, Klebsiella, Serratia, Proteus, Providencia,* and *Pseudomonas aeruginosa.* Penicillin's fame as a "miracle drug" was well deserved, because the drug was extremely active against those Gram-positive bacteria that were most important at the time of its introduction.

Together with the efforts to produce more penicillin, chemical studies of its structure were actively pursued. However, because it was thought that chemists in Germany and Japan would synthesize this great drug once the structure became known, a ban was imposed, during World War II, on publication of the results of any chemical studies on penicillin. This is amusing in hindsight because penicillin has such an unusual structure (a β-lactam ring with highly distorted bond angles), which had to be deduced from the X-ray crystallographic results of Dorothy Hodgkin and Barbara Low. Ten more years passed before

Microorganisms Responsible for Hospital-Acquired Infections Leading to Bacteremia in Boston City Hospital

	Percent of All Bacteremia Cases in Year	
Causative Organism	1935	1969
Streptococci	62	12
Staphylococci	19	19
Enterococci	0	7
E. coli	11	11
Enterobacter/Klebsiella/Serratia	0	22
Proteus/Providencia	4	10
P. aeruginosa	1	8
Other Gram-negatives	3	3
Fungi	0	8

TABLE 13.2

SOURCE: J.E. McGowan, Jr. (1985), Changing etiology of nosocomial bacteremia and fungemia and other hospital-acquired infections at Boston City Hospital, *Rev. Inf. Dis.* 7:S357–370.

successful chemical synthesis finally confirmed the accuracy of this structure.

One structural feature that became apparent early was the influence of the growth medium on the substituent at position 6 (Figure 13.39). Thus in the United States, where corn-steep liquor containing phenylacetic acid was used as a carbon source in fermentation, the main product was penicillin G, with its benzyl side chain. In contrast, the product that British workers obtained during their early efforts was penicillin F, with a 2-pentenyl side chain. Various compounds were then tested for their possible incorporation into the penicillin molecule, and penicillin V, with a phenoxymethyl side chain, was produced by the addition of phenoxyacetic acid, rather than phenylacetic acid, to the medium. Penicillin V better withstands the acidic pH in the stomach, and it can be administered orally.

Currently, fermentation runs are performed with the side-chain precursors intentionally added to the medium—phenylacetic acid to produce penicillin G and phenoxyacetic acid to produce penicillin V. These compounds are introduced slowly by a continuous-feed process, because they are toxic to fungi at high concentrations and because prolonged contact with the fungal cells tends to convert these compounds into products with a hydroxyl group on the aromatic ring.

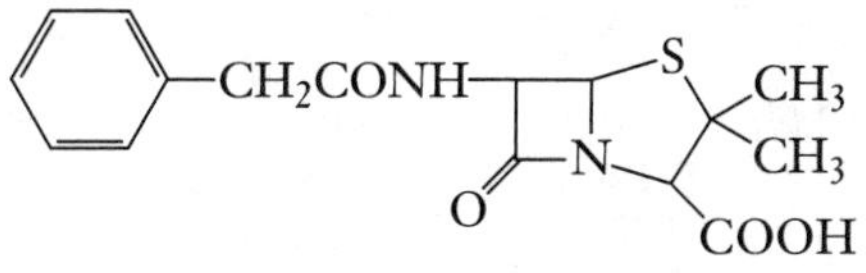

Penicillin G

Penicillin F

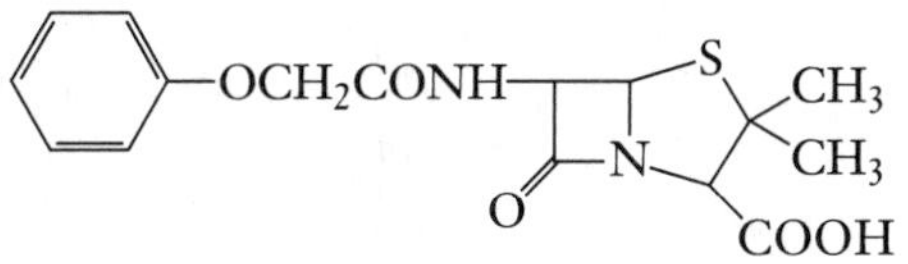

Penicillin V

Figure 13.39
Penicillins found in different culture filtrates.

Statistics from the mid-1970s show that about 20% of the penicillin produced by fermentation was penicillin V and 80% was penicillin G. Only about half the penicillin G was administered without further modification, the rest being used as the starting material for production of the semisynthetic penicillins described below.

Semisynthetic Penicillins

Although penicillin was indeed an amazingly effective drug, improvement was still desired in two areas. First, after just a few years of its widespread use, some of the *S. aureus* strains had become penicillin-resistant. They produced a penicillinase, a penicillin-hydrolyzing enzyme, that was coded by a gene on a plasmid (Box 13.2). Although the plasmid was not easily transferred from one strain to another, the resistant staphylococci were perfectly suited to cause hospital-acquired infections. Staphylococci are resistant to drying and thus remain viable in dust particles: In a hospital environment, where all staphylococcal infections are treated with penicillin G, the plasmid-containing strain had a strong selective advantage and spread easily from patient to patient through the air. There was thus an urgent need to produce penicillins that could withstand the staphylococcal penicillinase. Second, penicillin

Most bacteria that are resistant to penicillins and cephalosporins owe this pheno-type to the production of β-lactamases. They have been classified in detail (for example, see Bush, K., 1989, *Antimicrob. Agents Chemother.* 33:264–276). For our purposes, however, the most useful classification is the one based on sequence homology. This system divides commonly occurring β-lactamases into three homol-ogous groups. Class A includes the staphylococcal penicillinase and most β-lac-tamases found in Gram-positive bacteria, as well as the TEM enzyme found com-monly in Gram-negative bacteria and coded by a gene on R plasmids. Class C includes many of the chromosomally coded β-lactamases of Gram-negative bacte-ria. Both class A and class C enzymes have serine at the active site, but class B enzymes have zinc at the active center. The three-dimensional structures of some of these enzymes are shown in Figure 13.58.

R-CONH
S
CH₃
CH₃
O
N
COOH

Penicillin G or V

↓ Penicillin acylase

H₂N
S
CH₃
CH₃
O
N
COOH

6-Aminopenicillanic acid

R′COCl ↓

R′CONH
S
CH₃
CH₃
O
N
COOH

Semisynthetic penicillin

Figure 13.40
Production of semisynthetic penicil-lins. The starting material, usually penicillin G, is cleaved enzymati-cally by penicillin acylase to yield 6-aminopenicillanic acid. This is then chemically acylated with acyl chlo-rides to produce various semisyn-thetic compounds. Acid anhydrides can also be used as the acyl donor. When the R′ group contains an amino function, as in the synthesis of ampicillin, this amino group must be blocked first and then deblocked after the acylation reaction.

G (like penicillin V) is essentially inactive against Gram-negative bacte-ria (with the exception of *Neisseria gonorrhoeae*, the causative organism of the sexually transmitted disease gonorrhea). Thus it was desirable to expand the spectrum of penicillin's targets to include Gram-negative bacteria as well.

Because penicillin V was much more acid-resistant than penicillin G, scientists hoped that further modification of the acyl substituent at the 6-position would produce compounds with the properties they de-sired. However, adding potential precursors of acyl substituents to the growth medium produced no other useful antibiotics. Researchers thus turned to the strategy of chemically acylating 6-aminopenicillanic acid, the deacylated derivative of penicillin G, instead (Figure 13.40).

The first issue that arose in implementing this approach was how to produce the starting material, 6-aminopenicillanic acid. Apparently, this compound had been noticed in the fermentation broth during penicillin production in Japan, but the people involved presumably failed to rec-ognize its potential significance. In the late 1950s, however, Beecham scientists found that 6-aminopenicillanic acid was most abundant in fer-mentation broths that lacked potential precursors of the side chain, and they succeeded in obtaining quantities sufficient for the synthesis of semisynthetic penicillins. Later, in the 1960s and 1970s, a large number of organisms were found to produce penicillin acylase, which cleaves off the 6-acyl side chain of penicillins G and V without touching the rest of the molecule. The sources of this enzyme include bacteria (more than two dozen species are known to produce it), fungi, and higher plants. Furthermore, chemical procedures for splitting the side chain were de-veloped in the early 1970s.

Penicillins Not Hydrolyzed by the Staphylococcal Penicillinase

By systematically going through various substituents, scientists found that the presence of bulky substituents on the α-carbon of the 6-sub-stituents produced compounds that were hydrolyzed more slowly by the

Methicillin

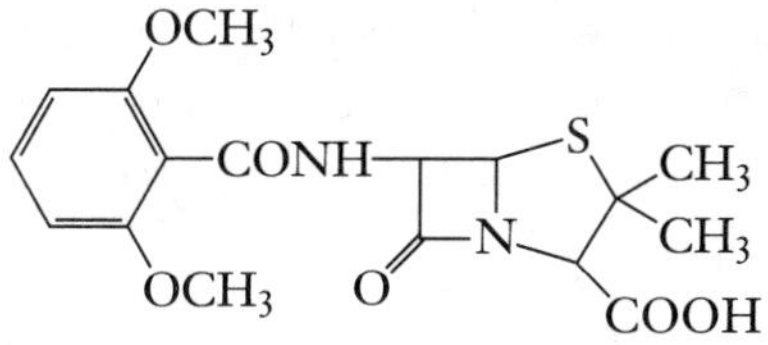

Nafcillin

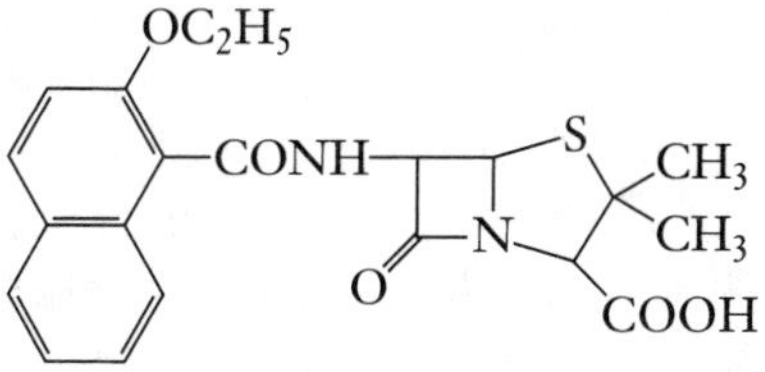

Isoxazolyl penicillins

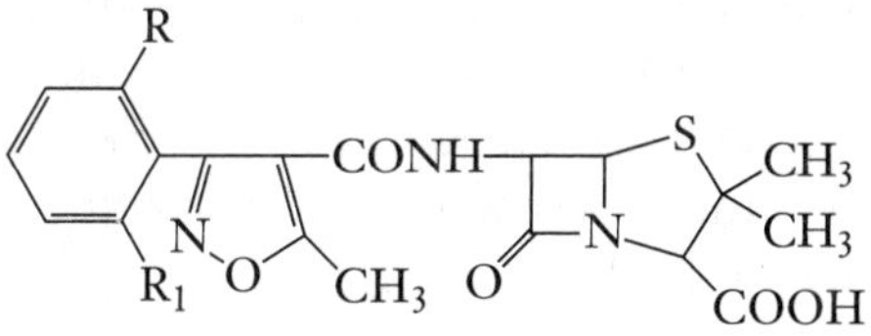

	R	R_1
Oxacillin	H	H
Cloxacillin	Cl	H
Dicloxacillin	Cl	Cl
Flucloxacillin	Cl	F

Figure 13.41
Penicillins impervious to the action of staphylococcal penicillinase.

staphylococcal enzyme. Methicillin and naphcillin (Figure 13.41) are examples of compounds that were produced by refining these structure–function studies, and they are almost absolutely resistant to the enzymatic hydrolysis. In these compounds, the flexible α-methylene group of benzylpenicillin is absent, and the α-carbon is now a part of a bulky and rigid aromatic ring system. In addition, an O—CH₃ or O—C₂H₅ group at the ortho position of the ring further contributes to the steric hindrance of the interaction between the penicillinase and these penicillins. Similar stability against the staphylococcal penicillinase enzyme was obtained by the use of 3-phenyl-5-methylisooxazolyl substituents, as in oxacillin, cloxacillin, and the other isoxazolyl penicillins (see Figure 13.41), which have the added advantage that they can be administered orally.

These drugs remained effective against penicillinase-producing staphylococci for a remarkably long time — from the early 1960s to the late 1980s — but very recently, methicillin-resistant staphylococci with altered targets (that is, altered penicillin-binding proteins) have appeared. Resistant strains with altered penicillin-binding proteins have also emerged in recent years among species that used to be exquisitely susceptible to penicillin G, such as pneumococci (*Streptococcus pneumoniae*) and gonococci (*Neisseria gonorrhoeae*).

Penicillins with Gram-negative Activity

Penicillin N, a natural fermentation product that carries an aminoadipyl side chain, has been shown to have lower Gram-positive activity but significantly higher Gram-negative activity than penicillin G. Similarly, introduction of an amino group on the α-carbon of the benzyl substituent of penicillin G produces ampicillin (Figure 13.42), which is about ten times more active against Gram-negative rods, such as *E. coli*, and retains half the activity of penicillin G against Gram-positive bacteria. Ampicillin is very widely used as a safe, inexpensive, broad-spectrum antibiotic. A close relative of ampicillin is amoxycillin (see Figure 13.42), which is absorbed more efficiently than ampicillin upon oral administration and which also shows slightly increased activity against some Gram-positive bacteria.

A major reason why ampicillin is active against Gram-negative bacteria is its rate of diffusion across the outer membrane. As we noted in Chapter 2, the main difference between Gram-positive and Gram-negative bacteria is that the latter are protected by an additional permeability barrier, the outer membrane. Most antibiotics must cross the outer membrane barrier by diffusing through the narrow, water-filled porin channels (page 436). Diffusion through the porin channels of enteric bacteria for zwitterionic compounds is faster than for anionic compounds, and the diffusion of hydrophilic compounds is faster than that of hydrophobic ones. Ampicillin is both zwitterionic and more hydrophilic, in contrast to benzylpenicillin, which is anionic and quite hydrophobic. Thus ampicillin can cross the outer membrane barrier much faster. However, this is only a superficial description of a more complex situation.

Measurements of the actual rates of penetration of various β-lactam antibiotics across the *E. coli* outer membrane have shown them to be quite fast: Most compounds achieve half-equilibrium between both

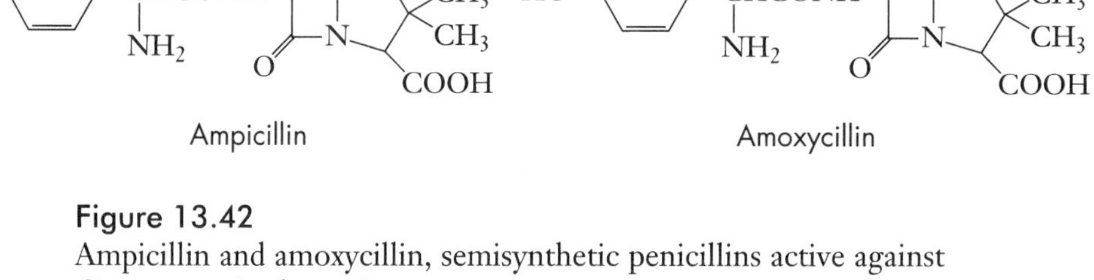

Ampicillin   Amoxycillin

Figure 13.42
Ampicillin and amoxycillin, semisynthetic penicillins active against Gram-negative bacteria.

sides of the membrane in less than 10 seconds. Thus differences in permeation rate alone cannot explain the differences in bacterial susceptibility to various penicillins. Most Gram-negative bacteria, however (*N. gonorrhoeae* is a possible exception), produce β-lactamase, an enzyme that hydrolyzes penicillins and cephalosporins even when they do not contain any resistance plasmids. The β-lactamase is located in the periplasm, the narrow space between the outer and inner (cytoplasmic) membranes. Thus the small number of β-lactam molecules that succeed in traversing the outer membrane immediately encounter these enzyme molecules and are likely to be hydrolyzed by them. What is important in Gram-negative bacteria is the synergistic interaction between the outer membrane barrier and the periplasmic, β-lactamase "barrier." Compared to β-lactamases in the medium or even on the cell surface, the β-lactamase molecules in the periplasm are many orders of magnitude more effective, because they have to deal with only a very small number of substrate molecules trickling through the porin channels into the periplasm.

Thus the stability of β-lactam antibiotics in the presence of these chromosomally determined β-lactamases is as important as their ability to permeate the outer membrane in determining their efficacy against Gram-negative bacteria. In the case of ampicillin, not only does that compound cross the membrane faster, but it is also more resistant to enzymatic hydrolysis than penicillin G.

Penicillins with Negatively Charged Substituents

Ampicillin is active against *E. coli* and some other Gram-negative bacteria, but it shows little activity against the Gram-negative, rod-shaped bacteria that are more and more frequently the causative organisms of hospital-acquired infections. These include *P. aeruginosa*, *Enterobacter cloacae*, *Serratia marcescens*, and some *Proteus* species. In *E. coli*, the level of β-lactamase remains low even after exposure to drugs. These other bacteria, however, produce a very high level of the enzyme whenever they sense the presence of β-lactam in the medium. In other words, the enzyme is inducible in these species. Because of this, and because some of these bacteria produce outer membranes whose permeability is quite low, only the β-lactams with exceptional stability against β-lactamase can kill these bacteria. It would be even more desirable if those exceptionally stable β-lactams were also poor inducers of the enzyme.

Penicillins with negatively charged groups in the alpha position of the 6-substituent fulfill these conditions. Carbenicillin, ticarcillin, and sulbenicillin (Figure 13.43) show activity against such organisms as *P. aeruginosa* and *E. cloacae* (see carbenicillin in Table 13.3). The activity levels are marginal. Nevertheless, at the time of their introduction, these antibiotics were valuable weapons against these "intrinsically resistant" bacteria.

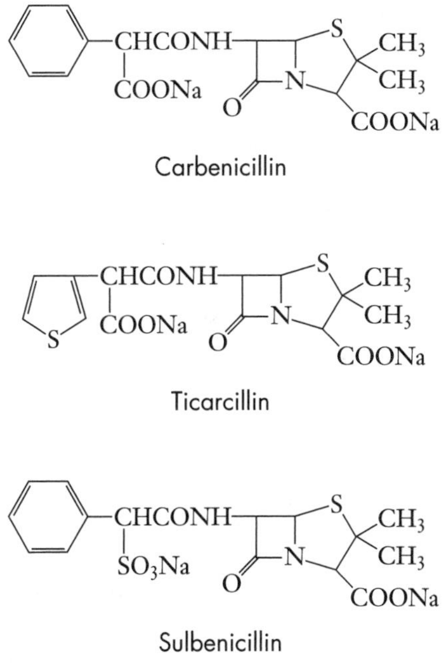

Figure 13.43
Carbenicillin, ticarcillin, and sulbenicillin, semisynthetic penicillins with negatively charged substituents.

Typical MIC (Minimal Inhibitory Concentration) Values for Some β-Lactams (μg/ml)

β-Lactam	S. aureus (S)	S. aureus (R)	E. coli (S)	E. coli (R plasmid)	E. cloacae (inducible)	E. cloacae (constitutive)	P. aeruginosa
Penicillin G	0.02	>128	64	>128	>128		>128
Methicillin	1	2	>128	>128	>128		>128
Ampicillin	0.05	>128	2	>128	>128		>128
Carbenicillin	1	16	4	>128	16	>128	64
Azlocillin	1	>128	16	>128	32	>128	4
Cephalothin	0.2	0.4	4	>128	>128	>128	>128
Cephaloridine	0.05	0.8	2	16	>128	>128	>128
Cefoxitin	4	4	4	4	>128	>128	>128
Cefotaxime	2		0.05	0.05	0.2	>128	16
Ceftazidime	8		0.05	0.05	0.1	32	2
Cefepime	1	1	0.05	0.05	0.1	1	2
Aztreonam	>128	>128	0.2	0.2	0.2	16	4
Imipenem	0.01	0.02	0.1	0.2	1	1	2

"S" and "R" indicate susceptible and resistant strains, respectively. "R plasmid" denotes strains containing the common R plasmid, producing TEM-type β-lactamase. Strains producing the chromosomally coded β-lactamase in the inducible (as in the wild-type strains) and constitutive manner are shown as "inducible" and "constitutive."

SOURCES: G.N. Rolinson (1983), *J. Antimicrob. Chemother.* 17:5–36; H. Nikaido, unpublished data.

Acylampicillins

In a number of recently introduced semisynthetic penicillins, the amino group of ampicillin is substituted with a carboxyl group connected to a heterocyclic ring (Figure 13.44). These compounds are similar to carbenicillin in their stability in the presence of the chromosomally coded

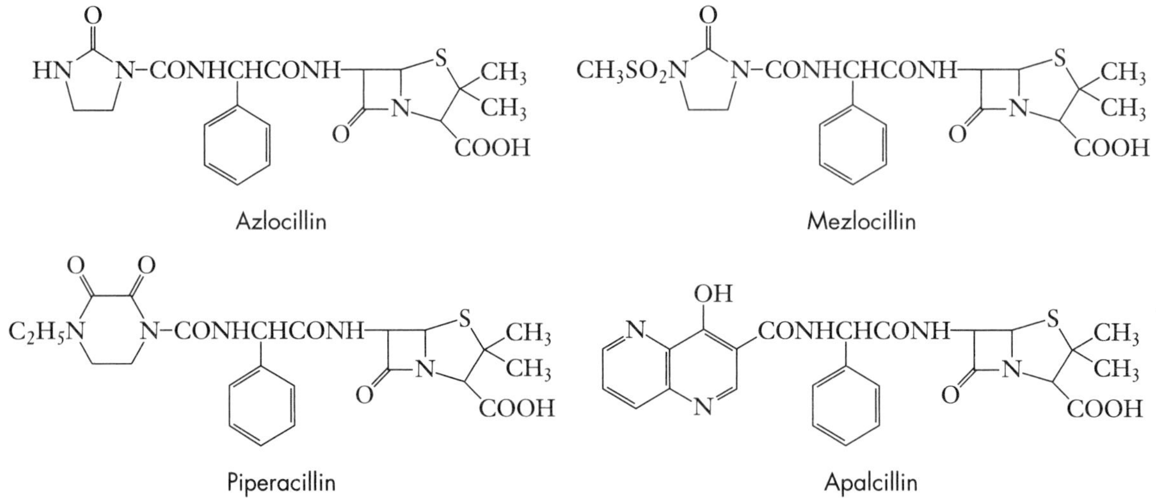

Figure 13.44
Acylampicillins.

β-lactamases of Gram-negative bacteria, and they are even weaker inducers of these enzymes than is carbenicillin or sulbenicillin. Thus they have a similar spectrum, but at least some of the acylampicillins are significantly more active against *P. aeruginosa* (see azlocillin in Table 13.3).

Cephalosporins

In cephalosporins, the β-lactam ring is fused to a six-membered dihydrothiazine ring rather than to the five-membered thiazolidine ring found in penicillins (Figure 13.45). The natural product, cephalosporin C, was discovered accidentally in 1955. Edward Abraham's group was studying the products secreted by *Cephalosporium acremonium*, a very different kind of mold from the classical penicillin producer, *Penicillium*. The major product, penicillin N, with its hydrophilic aminoadipyl side chain, was of interest at the time because it had a low but significant activity against Gram-negative bacteria (see page 472). When these researchers tried to purify penicillin N, they found a smaller amount of a second product that turned out to be cephalosporin C. It too had activity against Gram-negative bacteria. At that time, the penicillinase-producing staphylococci were already becoming a major problem. Furthermore, Abraham was especially interested in this enzyme, having published in 1940 the first report on the presence of β-lactamase in some bacteria. For these reasons, he tested the stability of the newly isolated compounds to staphylococcal penicillinase. Although penicillin N was as susceptible to enzymatic hydrolysis as penicillin G, cephalosporin C was much more stable.

This finding immediately suggested that cephalosporins might have significant advantages over penicillin G and might also be active on Gram-negative bacteria and on penicillinase-producing *S. aureus*. Also promising was the fact that cephalosporin C had an extremely low toxicity. Unfortunately, cephalosporin C itself had only a low antibacterial activity, but its chemical structure, when elucidated, showed that a range

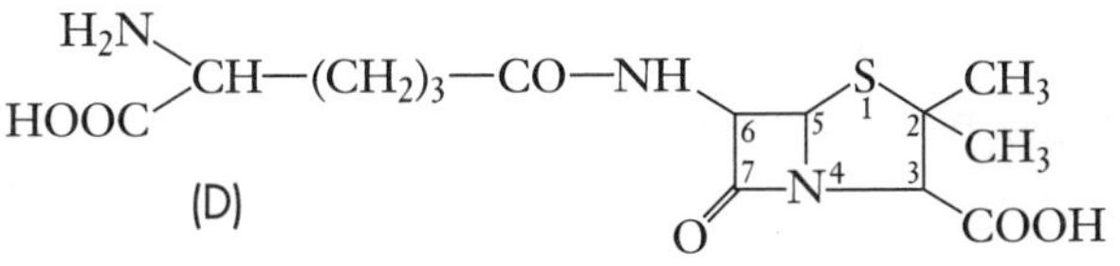

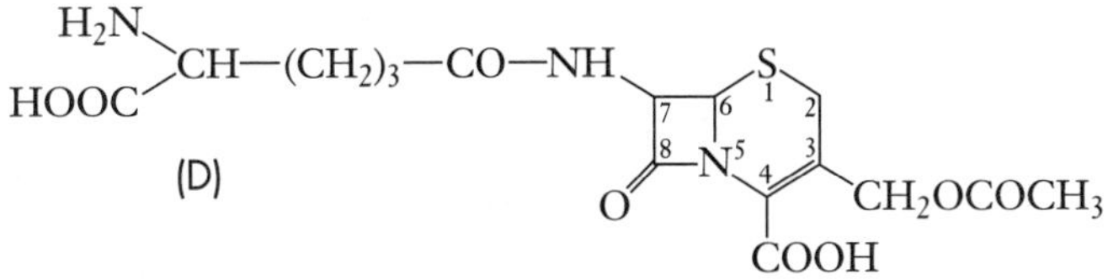

Figure 13.45
Penicillin N and cephalosporin C. In both compounds, the α-aminoadipic acid moiety in the side chain has a D-configuration.

of chemical modifications—not only at the 7-position (which corresponds to the 6-substituent in penicillins) but also at the 3-position—should be possible.

The first step toward synthesizing new cephalosporin compounds was to remove the 7-substituent from cephalosporin C. The resulting 7-aminocephalosporanic acid would then serve as the starting material. Because enzymes that remove the 6-substituent of penicillin G occur in a very large number of organisms, a great deal of effort was devoted to searching for a cephalosporin C acylase. Surprisingly, such an enzyme was not found. Thus it was Robert B. Morin's 1962 discovery of a chemical method for cleaving off the 7-substituent that opened a way to the development of a very large number of semisynthetic cephalosporins (Figure 13.46). Some of the penicillins sold as commercial antibiotics are natural fermentation products (penicillins G and V, for example); in contrast, every cephalosporin marketed is a semisynthetic compound.

The biosynthesis of cephalosporin C starts with penicillin N, so it is not surprising that the culture filtrate examined by Abraham contained

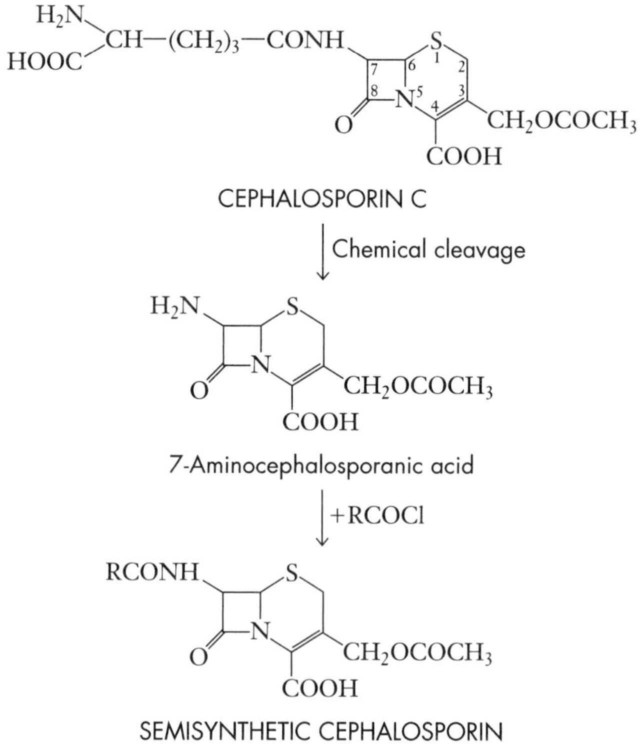

Figure 13.46
Conversion of cephalosporin C into semisynthetic cephalosporins. In addition to substitutions at the 7-position, the acetoxy substituent at the 3-position can easily be replaced by a nucleophile, opening up additional possibilities for chemical modification of the cephalosporin structure.

both of these compounds. The penicillin nucleus undergoes a ring expansion reaction that converts it into the cephalosporin nucleus. Interestingly, a similar conversion can be achieved chemically. And this is fortunate, because producing cephalosporin C by fermentation is still much more costly than the production of penicillins. The chemical conversion of penicillin V into a cephalosporin was devised in 1962, and reactions of this type have been valuable in the production of certain classes of cephalosporins.

"First-Generation" Cephalosporins

The first semisynthetic cephalosporin antibiotics, introduced between 1962 and 1965, include cephalothin, cephaloridine, and cefazolin (Figure 13.47). At that time, there were two major groups of bacteria for which penicillin G had little activity: penicillinase-producing staphylococci and Gram-negative rods. Although methicillin and its relatives were active against the former, and although ampicillin was active against the latter, no penicillin derivative showed activity against both of these groups. The cephalosporins had the advantage of being quite active against both groups, thus fulfilling the promise that Abraham saw in them.

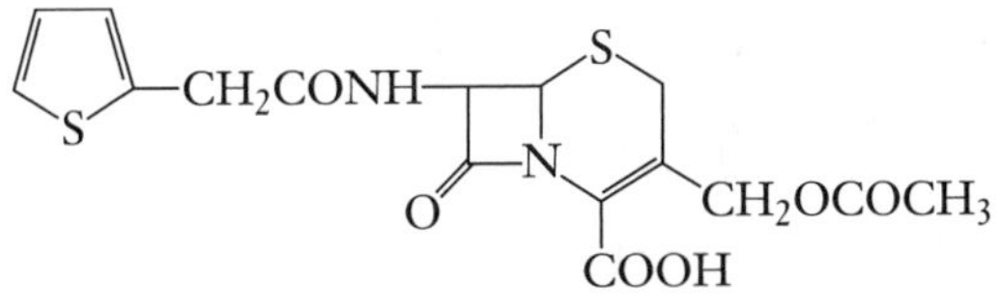

Cephalothin

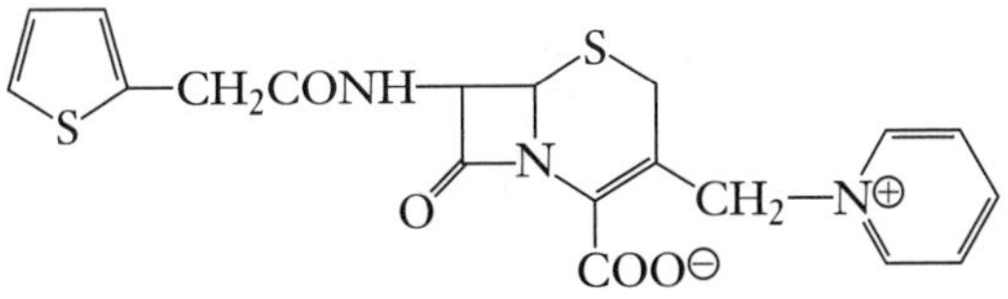

Cephaloridine

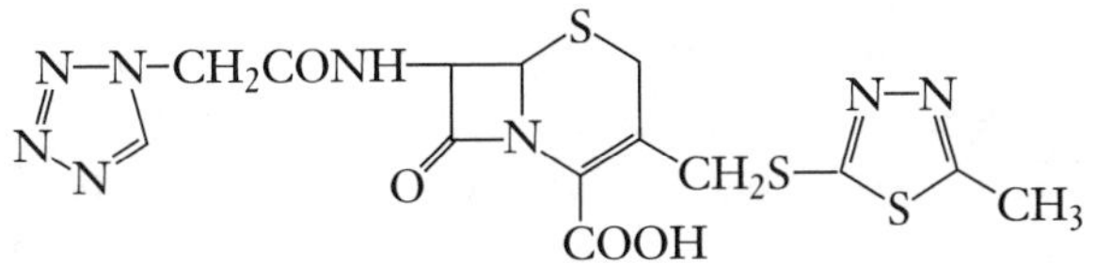

Cefazolin

Figure 13.47
The "first-generation" cephalosporins.

The action of cephalosporins against Gram-negative rods merits some comment. As we have said, the major permeability barrier in these bacteria is the outer membrane, and cephalosporins as a class have a slight advantage over penicillins in passing through the porin channel, because the cephalosporin nucleus is significantly less hydrophobic than the penicillin nucleus. Cephaloridine, especially, has a very high rate of diffusion through the outer membrane, thanks to its zwitterionic nature. We also saw that, in order to reach the targets (that is, penicillin-binding proteins; see page 437), these compounds have to overcome the second "barrier": periplasmic β-lactamases. These chromosomally coded enzymes, which are present in most Gram-negative rods, have been called "cephalosporinases" because they were believed to hydrolyze cephalosporins so much better than they do penicillins. Herein lies a paradox: If cephalosporins are hydrolyzed more readily by these ubiquitous enzymes, how could they be more effective than penicillins against Gram-negative bacteria?

To answer this question, we must begin with the observation that the penicillin-binding proteins are irreversibly inactivated at very low drug concentrations, usually in the range of $0.1 - 1$ μM. Whether the enzymatic hydrolysis can protect the bacterial cells against attack by β-lactams is therefore decided by the behavior of the β-lactamases at micromolar concentrations of the substrate. However, because of the poor sensitivity of the laboratory method for measuring β-lactamase activity, the susceptibility of β-lactams to enzymatic hydrolysis is customarily measured by using substrate concentrations in the range of $0.1 - 5$ mM, probably a 1000-fold higher concentration than those that occur in the periplasm of bacteria in the tissues of patients who are being treated with these drugs. Consequently, situations sometimes arise wherein the "β-lactamase stability" data published in respected journals mean absolutely nothing, or are even positively misleading, in terms of the behavior of these enzymes in the bacteria being exposed to the β-lactam. This is precisely the case with the real-life behavior of the chromosomally determined "cephalosporinases" in Gram-negative rods toward the first-generation cephalosporins. As seen in the left half of Table 13.4, the *E. coli* chromosomal enzyme, a "cephalosporinase," hydrolyzes cefazolin much more rapidly than it hydrolyzes penicillin G, when assayed at 5 mM substrate concentrations, thanks to the much higher V_{max} value for the former compound. However, at a real-life concentration of 0.1 μM, cefazolin is about 70 times more stable than penicillin G, because the enzyme has a much higher affinity (or a lower value of K_m) for penicillin G than for cefazolin. Apparently the Gram-negative activity of cephalosporins owes as much to their lower affinity for the chromosomally coded β-lactamases (which, to avoid misunderstanding, should *not* be called cephalosporinases) as to their higher rate of diffusion across the outer membrane.

Hydrolysis of β-Lactams by *E. coli* Enzymes

Antibiotic	Chromosomally Coded Enzyme				Plasmid-Coded TEM Enzyme			
	K_m (μM)	V_{max}	V(5 mM) *(relative rates)*	V(0.1 μM)	K_m (μM)	V_{max}	V(5 mM) *(relative rates)*	V(0.1 μM)
Cefazolin	1900	100	100	100	320	100	100	100
Penicillin G	1.9	7.6	10.5	7200	18	880	930	15500
Cefoxitin	0.22	0.02	0.03	120	3600	0.03	0.02	<0.001
Cefotaxime	0.16	0.007	0.01	51	9500	15	5.5	0.5
Cefepime	80	0.001	0.01	0.2	5000	15	8	0.5

V(5 mM) and V(0.1 μM) represent rates of hydrolysis at 5 mM and 0.1 μM substrate concentration, respectively. Values of V_{max}, V(5 mM), and V(0.1 μM) were all expressed as relative values by setting the cefazolin rate as 100. The chromosomal enzyme data on cefepime were obtained by using *E. cloacae* enzyme rather than the *E. coli* enzyme.

SOURCES: H. Nikaido and S. Normark (1987), *Mol. Microbiol.* 1:29–36; H. Nikaido et al. (1990), *Antimicrob. Agents Chemother.* 34:337–342.

Discovery of Cephamycins

Because of their wide spectrum, the first-generation cephalosporins were used extensively, a situation that unfortunately resulted in the selection of Gram-negative bacteria containing R plasmids. The majority of these plasmids, isolated from bacteria all over the world, carried a gene for a β-lactamase called TEM β-lactamase (these are the initials of the patient from whom it was first isolated). The TEM enzyme has a wide substrate specificity that includes most penicillins and all of the first-generation cephalosporins. Although its affinity toward cephalosporins is not particularly high (see Table 13.4), it can hydrolyze these compounds effectively even at low concentrations, because the V_{max} values are high and because the presence of multiple copies of the plasmid, and therefore of the gene, in a cell leads to very high levels of the enzyme.

Compounds that are stable in the presence of TEM β-lactamase were not found among semisynthetic derivatives of cephalosporin C at that time, so, as happens often in the development of antibiotics, scientists turned to nature for inspiration. Specifically, and significantly, they turned to organisms that are *not* classic producers of β-lactams. A search through the culture filtrates of *Streptomyces* species, which are eubacteria and thus very far from fungi in phylogenetic terms, yielded cephamycins, with a methoxy group at the 7-α-position of the nucleus (Figure 13.48). These compounds were found to be absolutely resistant to the TEM enzyme (see the behavior of the plasmid-coded TEM enzyme against cefoxitin, in Table 13.4).

The cephamycin cefoxitin was introduced commercially in the United States in 1978. In Japan, another cephamycin derivative, cefmetazole, was introduced soon after. These compounds, sometimes called

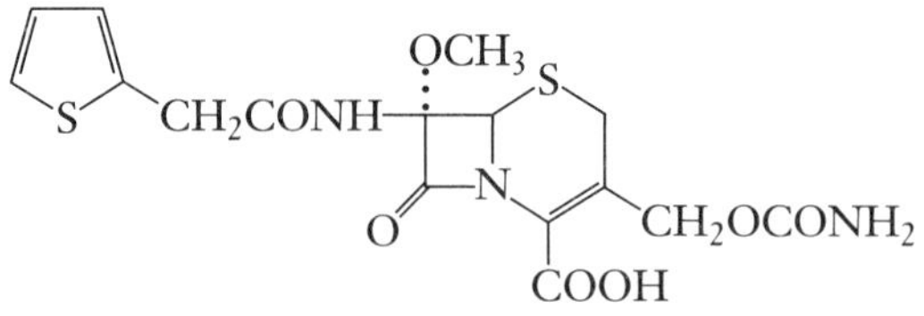

Figure 13.48
Cephamycins.

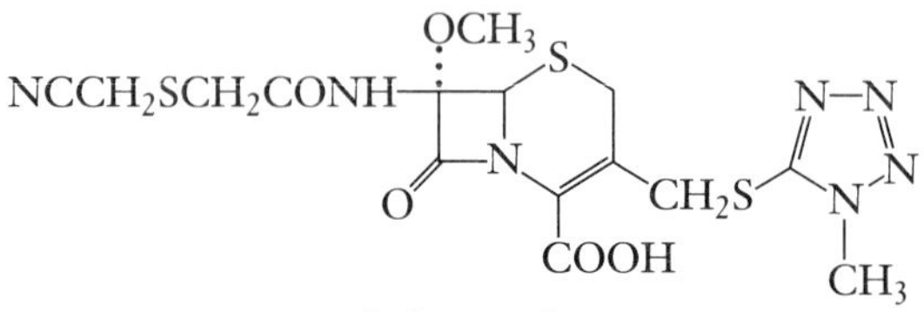

second-generation cephalosporins, were hailed as the last word in the
β-lactams, because they were effective against all the target groups
thought to be important at that time, including the penicillinase-producing staphylococci and the Gram-negative rods, even when the latter
contained R plasmids (see Table 13.3).

"Third-Generation" Cephalosporins

Within a few years after the introduction of cephamycins, physicians
were noticing the presence of pathogens such as *Enterobacter cloacae*,
Serratia marcescens, and *Pseudomonas aeruginosa*, especially as the causes
of hospital-acquired infections. These bacteria are intrinsically resistant
to cephamycins (that is, even without the acquisition of R plasmids or
new mutations), but it is doubtful whether the use of cephamycins actually resulted in a strong selection for these organisms in the environment. Unlike the drastic change that the widespread use of β-lactams
has brought about in the prevalence of various bacterial pathogens (see
Table 13.2), the limited statistical data available do not reflect a large
increase in the incidence of infections caused by these cephamycin-re-
sistant species. In fact, Table 13.2 shows that such an increase had already occurred in 1969, long before the introduction of cephamycins.
What probably happened was that once infections caused by most other
classes of bacteria could be treated effectively with cephamycin, physicians became especially aware of the infections caused by cephamycin-
resistant bacteria.

In any case, a search for cephalosporins active against these resistant
bacteria produced fruitful results, this time from synthetic chemistry. It

was discovered in the early 1980s that the addition of a substituted oxime group on the α-carbon of the side chain, as well as the use of a hydrophilic aminothiazole group in the side chain, produced cephalosporins with revolutionary efficacy against Gram-negative bacteria, including many of the "intrinsically resistant" bacteria mentioned above. These compounds, often called third-generation cephalosporins, include cefotaxime, ceftizoxime, ceftazidime (Figure 13.49), and ceftriaxone. As illustrated in Table 13.3, cefotaxime and ceftazidime show strong activity against the wild-type strains of *E. cloacae* (strains that have inducible β-lactamase and are therefore labeled "inducible"), and ceftazidime shows excellent activity against *P. aeruginosa;* both of these organisms were totally refractory to cefoxitin therapy.

"Fourth-Generation" Cephalosporins

The third-generation cephalosporins appeared to have solved practically all the problems remaining in the development of the semisynthetic β-lactams. These cephalosporins were said to be absolutely

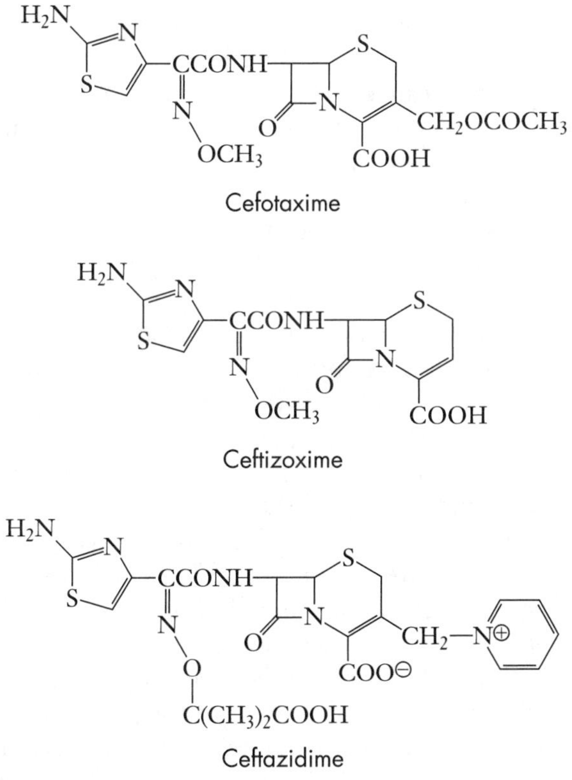

Figure 13.49
Some examples of "third-generation" cephalosporins.

impervious both to plasmid-coded TEM β-lactamase and to chromosomally coded β-lactamases of Gram-negative bacteria. However, the clinical use of these drugs led to the emergence of highly resistant mutants of *E. cloacae*, *S. marcescens*, and *P. aeruginosa*; an example is shown in Table 13.3 as "*E. cloacae* (constitutive)." These mutants produced the normally inducible β-lactamase in a constitutive manner and at a very high level. How could a bacterium become resistant to a drug by producing an enzyme that is supposed to be absolutely inactive against that drug? This dilemma led to some creative hypotheses. In retrospect, we can see once again how the practice of using very high, arbitrarily chosen substrate concentrations in laboratory tests of enzyme activity has misled many scientists.

The chromosomally coded β-lactamases of Gram-negative rods have a very low maximum velocity of hydrolysis with the third-generation compounds (see Table 13.4, which shows that the V_{max} value for cefotaxime hydrolysis of the *E. coli* chromosomal enzyme is less than 0.01% of that for cefazolin hydrolysis). However, these enzymes have very strong affinity, or low K_m, for the third-generation compounds (Table 13.4 again shows that the K_m value for cefotaxime is more than 10,000-fold lower than that for cefazolin). Thus with a traditional assay using 5 mM substrate, for example, one would get the impression that a third-generation compound, cefotaxime, is completely unaffected by the chromosomal enzyme, its measured hydrolysis rate being 0.01% of that of the first-generation compound, cefazolin. However, we can predict from the kinetic constants that at 0.1 μM, cefotaxime will be hydrolyzed about half as rapidly as cefazolin (see Table 13.4), and it is the hydrolysis rate at these low concentration ranges that determines the sensitivity or resistance of the organism.

We thus understand why the β-lactamase-constitutive mutants can develop high levels of resistance to third-generation compounds. But why were the inducible parent strains so exquisitely sensitive to the same compounds? In answering this question, we should compare the cephamycins with the third-generation compounds. We now know that cephamycins also are rapidly hydrolyzed at low concentrations by chromosomal enzymes (see Table 13.4). Thus both cephamycins and third-generation compounds are hydrolyzed rapidly by the chromosomal enzyme, yet the wild-type strains of *E. cloacae* are very resistant to the former and very sensitive to the latter. The major difference in the behavior of these two classes of compounds is that cephamycins are very strong inducers of the chromosomal enzyme, whereas the third-generation compounds have almost no inducer activity. Clearly, cephamycins become hydrolyzed after the induction of these efficient enzymes, but the third-generation compounds remain active because in the absence of induction, the enzyme levels remain extremely low. The constitutive mutants can of course hydrolyze efficiently both cephamycins and the third-generation cephalosporins.

The "fourth-generation" cephalosporins, β-lactam antibiotics that can truly withstand the chromosomal β-lactamases, have recently been developed and are being introduced at the time of this writing. These new compounds retain the oxyimino substituents of the third-generation compounds but also contain a 3-substituent with a quaternary nitrogen atom (Figure 13.50). As Table 13.4 shows, cefepime, an example of these compounds, has a much lower affinity for chromosomally coded enzymes than do the third-generation compounds and is therefore quite stable in these enzymes' presence. It shows good activity against constitutive mutants of *E. cloacae, S. marcescens,* and *P. aeruginosa* (see Table 13.3).

Compounds with Nontraditional Nuclei

As we have seen already, the spectrum of β-lactam antibiotics has been expanding steadily. At the same time, the introduction of each new "generation" has resulted in the emergence, either real or conceptual, of strains that show resistance to these drugs. In fact, there is already a

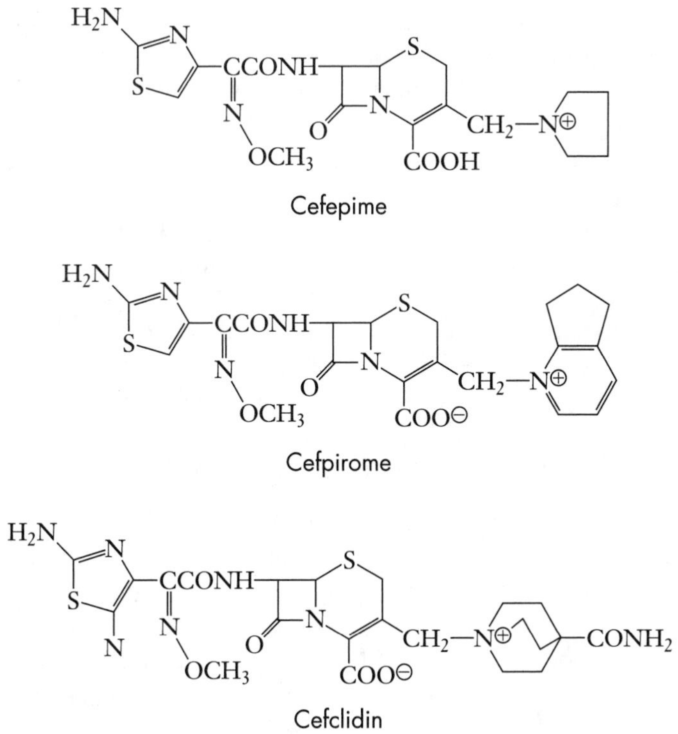

Cefepime

Cefpirome

Cefclidin

Figure 13.50
Some examples of "fourth-generation" cephalosporins.

significant threat to the continued efficacy of third- and fourth-generation cephalosporins, because mutants of the TEM enzyme that show drastically increased activity against these compounds are appearing in different parts of the world.

In response to this situation, researchers are developing new compounds on a continuous basis. It is not possible to get a precise number, but one review estimates that at least 50,000 semisynthetic antibiotics have been made and tested already, and an overwhelming majority among them must have been β-lactams. Thus some workers feel that at this point, only compounds with radically new structures are worth investigating. They may also hope that such compounds will be less likely to lead to a rapid emergence of resistant organisms. Compounds that are not based on the traditional penicillin or cephalosporin nucleus certainly fall into the category of radically new compounds, and in recent years there have been many efforts to develop such compounds.

In some cases the nucleus has been totally synthetic. On the other hand, many of the newest antibiotics were discovered, as in the past, by the screening of natural products. Of course, the traditional screening methods using growth-inhibition assays are unlikely to turn up anything very new, because most of the compounds that are produced by easily cultivated microorganisms and show strong antibacterial activity have already been discovered. Thus much more sensitive or narrowly focused assays are now being used—for example, assays that test for the inhibition of cell wall biosynthetic reactions in cell-free systems, the inhibition of β-lactamase, or the inhibition of hypersensitive mutants of bacteria. When a novel compound shows even traces of antimicrobial activity, then semisynthetic approaches can be utilized to produce more active derivatives.

It is also important that in recent years, nontraditional source material has been used for screening. Microorganisms that are only remotely related to the fungi or to *Streptomyces* (the traditional sources of antibiotics) have been examined—even unicellular bacteria that show no signs of cellular differentiation—and this has resulted in major discoveries. Although the inhabitants of unusual environments have not yet been examined very extensively, the use of acidic soil alone has already led to the discovery of monobactams (see below).

Oxacephems. One totally synthetic compound developed by Shionogi & Co. Ltd. is moxalactam (also called latamoxef), which contains oxygen instead of sulfur in the cephalosporin nucleus (cephem; Figure 13.51). The presence of the carboxyl group on the α-carbon of the 7-substituent, as well as the presence of a 6-α-methoxy group, presumably contributes to moxalactam's exceptional stability in the presence of various kinds of β-lactamases.

Carbapenems. Carbapenems, which include thienamycin, olivanic acid, and carpetimycin, are compounds isolated from *Streptomyces* species.

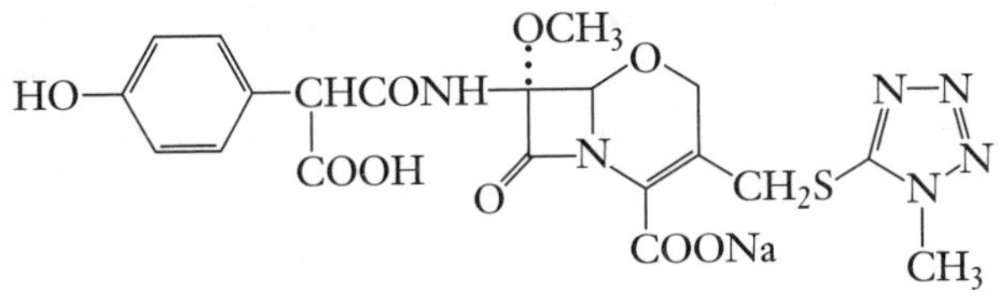

Figure 13.51
Latamoxef (moxalactam).

They have a carbon atom in place of the sulfur atom of the penicillin nucleus, and they contain a double bond in the five-membered ring of the nucleus. (The nucleus of the penicillin is called a *penam*, but when a double bond exists, the nucleus becomes a *penem*). Some of these carbapenems were detected by screening for inhibitors of bacterial cell wall synthesis, others by looking for inhibitors of β-lactamases.

Thienamycin is a potent antibiotic with a broad spectrum. However, thienamycin is chemically unstable, and some ingenious chemical modification was necessary to convert it to a clinically useful compound, imipenem (Figure 13.52). Becaues imipenem is hydrolyzed by a peptidase present in human tissues, it is administered together with the peptidase inhibitor cilastatin.

Although the starting material for imipenem synthesis, thienamycin, was a natural product of *Streptomyces*, its mass production by fermentation was hampered by low yield, the instability of the compound, and the production of several related compounds that complicated the isolation process. Chemical synthesis was also difficult because of the presence of three contiguous chiral centers and because of the chemical instability of intermediates; it has been successfully achieved, however, and imipenem is now produced commercially by a totally synthetic process.

Imipenem is stable in the presence of many β-lactamases. With others it acts as a suicide inhibitor; these include the chromosomally coded enzymes of Gram-negative rods that are responsible for the resistance of *E. cloacae*, *S. marcescens*, and *P. aeruginosa* mutants to the third-generation cephalosporins. Thus imipenem has an extremely wide spectrum (see Table 13.3). More recently, however, it was found that the exceptional activity of imipenem against *P. aeruginosa* was largely due to its rapid penetration through the specific outer membrane channel, which normally functions in the uptake of basic amino acids. Because this channel can be lost by mutation, imipenem resistance can arise very rapidly in *P. aeruginosa* populations, limiting the drug's usefulness against this important pathogen.

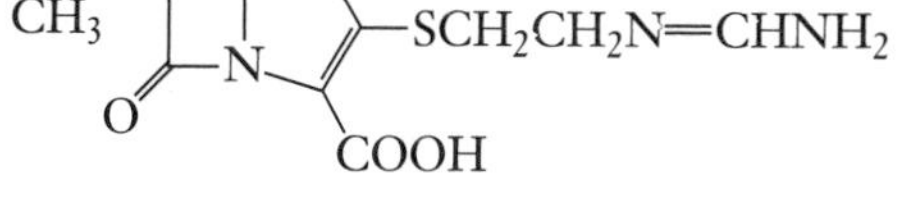

Clavams. Clavams have no side chain corresponding to the 6-substituent of penicillins. They have oxygen instead of sulfur in the penicillin (penam) nucleus and are reminiscent of the oxacephems in this regard.

Figure 13.52
Imipenem.

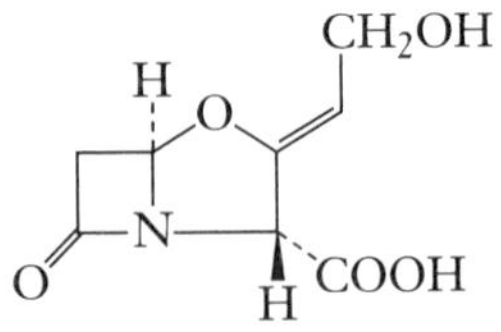

Figure 13.53
Clavulanic acid.

In contrast to oxacephems, clavams were isolated from *Streptomyces* through screening to detect β-lactamase inhibitors. Because they usually have only very low antibiotic activity, they would not have been detected in the conventional screening for antibiotics. One clavam, called clavulanic acid (Figure 13.53), is an efficient and irreversible inhibitor of class A β-lactamases, including the staphylococcal enzyme and the TEM enzyme. It is sold in combination with amoxycillin under the trade name Augmentin.

Nocardicins and Monobactams. Nocardicins and monobactams are characterized by an isolated β-lactam ring, without the fused thiazolidine or dihydrothiazine ring (Figure 13.54). In nocardicins the lactam nitrogen is connected to a carbon atom, whereas in monobactams it is connected to the sulfur atom of a sulfate group. It is remarkable that these compounds occur as metabolites of bacteria that are located far from *Streptomyces* in the phylogenetic tree. Nocardicins are produced by *Nocardia*, and monobactams come from groups that are even more distant; Gram-negative rods that do not go through any developmental cycle. Many of these, such as *Gluconobacter*, *Acetobacter*, and *Agrobacterium*, belong to the α division of the purple bacteria branch (Chapter 2). Monobactam-producing organisms reported as new species of *Pseudomonas* have properties very different from those of the fluorescent pseudomonads and may also belong to the α division instead. Other monobactam producers include members of *Chromobacterium* (the β division of the purple bacteria branch) and even *Flexibacter* species, which, if

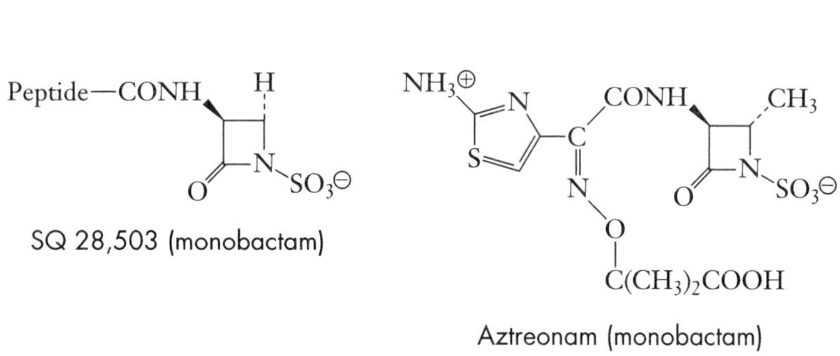

Nocardicin A

SQ 28,503 (monobactam)

Aztreonam (monobactam)

Figure 13.54
Nocardicin and the monobactams SQ 28,503 and aztreonam.

their identification is correct, may belong to the unusual group that constitutes the *Bacteroides-Flavobacterium* branch.

Nocardicins have low antibacterial activity and have not so far led to any commercially useful product. Monobactams isolated from culture filtrates also have very low antibacterial activity. For example, SQ 28,503 (see Figure 13.54), isolated from a *Flexibacter* strain, showed MIC values of 50–100 µg/ml against many pathogenic bacteria. Nevertheless, Squibb chemists used the basic monobactam structure as a starting point and, after extensive studies of structure-activity relationships, produced a totally synthetic compound, aztreonam, that has good activity against a wide spectrum of Gram-negative bacteria (see Table 13.3).

γ-Lactams. A compound with a five-membered lactam ring (LY173013) was synthesized at Eli Lilly & Co. on the basis of theoretical predictions and was shown to have significant antibacterial activity (Figure 13.55). At about the same time, Takeda Chemical Industries scientists isolated a compound called lactivicin (see Figure 13.55), also containing a 5-membered lactam ring, from a mold by screening for antibacterial activity. Thus a purely synthetic approach and screening for a natural product yielded two compounds of the same new class!

Enzymatic Synthesis with Substrate Analogs. Some of the enzymes that catalyze key steps in antibiotic biosynthesis (Figure 13.56) appear to have a rather broad substrate specificity. Products with novel structures can be obtained by supplying these enzymes with synthetic compounds that are similar to their natural substrates. If a process of this type is going to be used in commercial production, or even in large-scale experiments, it must be possible to produce the enzyme in large amounts and in reasonable purity by gene amplification via cloning.

An excellent example of this approach is the use of isopenicillin N synthetase (see Figure 13.56) for the production of totally new kind of

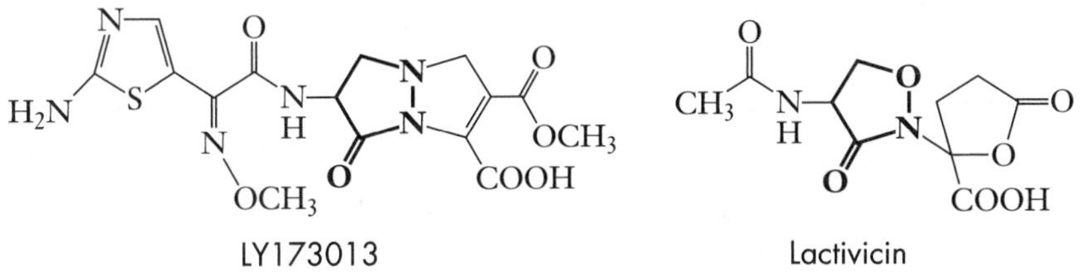

Figure 13.55
γ-Lactams. LY173013 is a synthetic compound containing a five-membered ring with a lactam structure (in boldface). Lactivicin is a natural compound that also has a five-membered ring and a lactam structure (in boldface). The lactam ring in LY173013 contains an extra nitrogen, whereas that in lactivicin contains an oxygen atom.

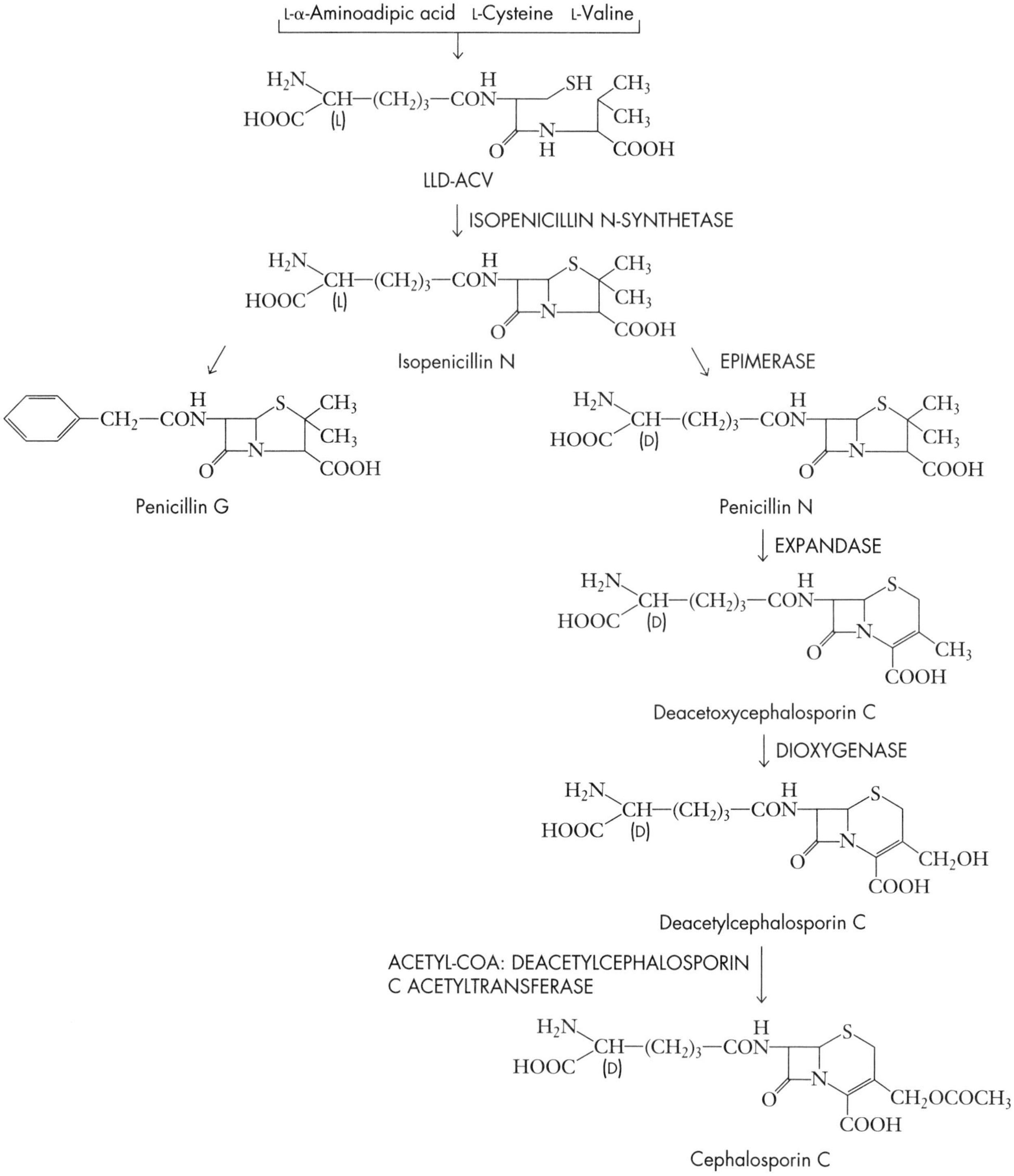

Figure 13.56

Synthetic pathway of penicillins and cephalosporins. The tripeptide L-δ-aminoadipyl-L-cysteinyl-D-valine (LLD-ACV) is first cyclized by the enzyme isopenicillin N synthetase, or cyclase, to produce a key intermediate isopenicillin N. From this point, exchange of the side chain will produce penicillin G, and epimerization of the side chain to change the chiral center of the α-ami-

noadipic acid moiety from the L- to the D-configuration will produce penicillin N. Penicillin N will be converted by ring expansion to a cephalosporin, which will eventually be converted to cephalosporin C. Modified from Martin, J.F., and Liras, P. (1985), Biosynthesis of β-lactam antibiotics: Design and construction of overproducing strains, *Tr. Biotechnol.* 3:39–44.

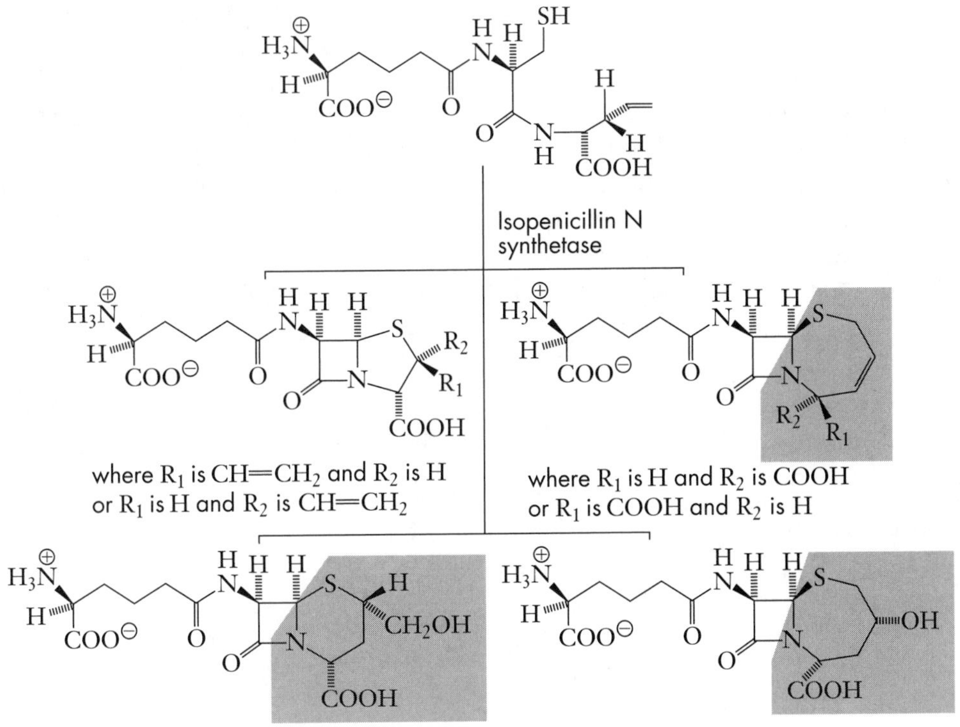

Figure 13.57
Examples of compounds synthesized by feeding an unnatural substrate to isopenicillin N synthase. The analog introduced here, instead of the natural substrate α-aminoadipyl–cysteinyl–valine, contained an allyl group in the valine residue. Based on Floss, H.G. (1987), *Tr. Biotechnol.* 5:11–115.

β-lactam. This enzyme, which catalyzes the crucial ring-closure step in the synthesis of penicillins, cephalosporins, and cephamycins, has been cloned from various sources. From the cephalosporin-producing fungus *Cephalosporium acremonium*, for example, it was cloned by using synthetic oligonucleotides based on the amino acid sequence of the purified enzyme. Figure 13.57 shows that by adding to this enzyme an allyl analog of the tripeptide α-aminoadipyl–cysteinyl–valine, one can generate a whole new set of β-lactams, most of which contain nuclei of novel structure (shaded in Figure 13.57).

Rational Design of New β-Lactams

The three-dimensional structures of several β-lactamases and that of a *Streptomyces* DD-peptidase, which is presumed to be very similar to the cell-wall-synthesizing enzymes, have been determined by X-ray crystallography during the last several years. Remarkably, these proteins, which do not show extensive homology in amino acid sequences, exhibit overall folding patterns that are very much alike (Figure 13.58), and the substrate-binding cavities are lined by similar amino acid residues. Thus the right half of the molecule as depicted is composed of a β-sheet containing five strands, protected by three α-helices in front and two α-helices in the back. The left half of the molecule consists of a cluster of α-helices. We can already explain, from the structure of β-lactamases, why certain β-lactams withstand the attack of these enzymes (see Figure 13.58). The logical extension of these studies is to produce

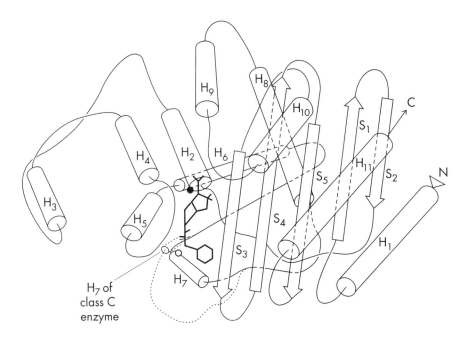

Figure 13.58

The folding patterns of β-lactamases and an enzyme believed to resemble the peptidoglycan transpeptidase, the DD-peptidase from *Streptomyces* R61. "H" indicates α-helical segments, and "S" indicates β-strands. The structure of class A β-lactamases, which include the *S. aureus* penicillinase and the TEM-β-lactamase, is shown in solid lines. Class C β-lactamases, which are the chromosomally coded β-lactamases of Gram-negative bacteria such as *Escherichia*, *Enterobacter*, and *Pseudomonas*, as well as the *Streptomyces* R61 DD-peptidase, show very similar folding patterns except that H_7 is truncated and displaced (dotted line), and small extra segments are present on the surface of the enzyme (not shown). Closed and open circles show the positions, respectively, of the active site serine and the glutamate$_{166}$ of the class A enzyme, which is thought to be involved in hydrolysis of the acyl-enzyme. The class C enzyme does not have glutamate$_{166}$, and the mechanism of hydrolysis of the acyl-enzyme is still controversial in this case. β-Lactam substrates (we show benzylpenicillin as an example) are bound to the backbone of S_3 by hydrogen bonding in the antiparallel manner. Methicillin is inert to staphylococcal enzyme (class A), because bulky side chains near the bottom end of S_3 produce steric hindrance. Cephamycins are not hydrolyzed by class A enzymes such as the TEM enzyme (see Table 13.4), because the water molecule used for the hydrolysis of the acyl-enzyme, held by glutamate$_{166}$, becomes displaced by the α-methoxy group. In contrast, these compounds are hydrolyzed by class C enzymes (see Table 13.4), because water for hydrolysis comes in by a different route. Modified, on the basis of more recent data, from Joris, B., et al. (1988), *Biochem. J.* 250:313–324.

β-lactam compounds that more effectively withstand degradation by the β-lactamases prevalent among the hospital isolates.

Knowledge about the structure of a cell-wall-synthesizing enzyme (see Figure 13.58) may one day lead to the design of a drug that will bind to the target transpeptidase yet avoid being captured by β-lactamases. At present, however, the data on the DD-peptidase are not of high enough resolution for such an effort.

The similarity between the folding patterns of β-lactamases and peptidoglycan transpeptidases suggests that these enzymes have a common evolutionary origin. One theory is that the excretion of transpeptidases in a soluble form, like the *Streptomyces* DD-peptidase described above, may represent, for bacterial cells, the first primitive mechanism of defense against β-lactams. Such soluble enzymes can at least bind and inactivate β-lactams. From there it is a short step, in evolutionary terms, to modifications that allow the entry of water into the active site of the enzyme, making it possible for the penicilloyl-enzyme to be hydrolyzed rapidly and the active enzyme regenerated. Thus the very fact that β-lactams act by producing a covalent acyl-enzyme probably led to the evolution of modified targets that could hydrolyze this acyl-enzyme complex.

PRODUCTION OF ANTIBIOTICS

As we mentioned at the beginning of this chapter, antibiotic research is unusual among the fields of microbial biotechnology in that a large part of the scientific effort has always gone into the development of new compounds. The foregoing examination of aminoglycosides and β-lactams provided numerous examples. It also showed, however, that most antibiotics are still produced by fermentation or by the chemical modification of fermentation products. It follows, therefore, that research devoted to improving the fermentation processes can have a significant beneficial impact.

Physiology of Antibiotic Production

Antibiotics are small molecules whose synthesis often requires dozens of enzymes. Enzyme activities are of necessity closely regulated in such complex pathways. It is therefore important to understand the physiology of the producing organisms in order to maximize the fermentative production of antibiotics.

Secondary Metabolism and Its Control

What sets the fermentative production of antibiotics apart from other types of fermentation is that antibiotics are typical *secondary metabolites*. In other words, they do not seem to play a central role in the growth and catabolism of the organism; they tend to have complex, unusual structures rather than the simple structures seen in primary metabolites; and they tend to be produced after a population has ceased to grow.

Primary metabolites—ethanol and lysine, for example—are produced during the entire growth phase of a culture. In fact, the production of these compounds is tightly coupled to growth. Thus all an engineer has to do to achieve the efficient production of these metabolites, at least in principle, is to optimize the conditions under which the culture is grown. Continuous processes are often feasible in such cases, and as we have seen, these are preferable. In contrast, the production of secondary metabolites is regulated in a complex and poorly understood manner, and designing a fermentation procedure is often difficult. In most cases, continuous fermentation (Box 13.3) is not an option, because these cultures do not produce antibiotics unless they are entering, or are already in, a stationary phase.

Molecular genetic studies of antibiotic-producing prokaryotes, such as *Streptomyces*, are finally shedding some light on the complex regulatory mechanisms that operate in these organisms. One striking fact is that the classic producers of antibiotics are soil microorganisms that go through sporulation processes, such as *Streptomyces*, *Bacillus*, fruiting myxobacteria, and eukaryotic fungi. It has been established in *Bacillus* that the sporulation process, at the end of the growth phase, requires the differential transcription of various types of promoters by several different sigma subunits of RNA polymerase. (The sigma subunit is one of the subunits of this giant enzyme; it plays a decisive role in recognizing a specific DNA sequence as a promoter.) A similar situation apparently

---

> **BOX** 13.3   Continuous Fermentation

In the usual fermentation process, called *batch fermentation,* a given amount of medium in a tank is sterilized and inoculated with the microorganism. The culture goes through lag phase and exponential phases of growth and finally reaches the stationary phase at which there is little or no net increase in the density of the organisms. In contrast with this closed system, *continuous fermentation* is an open system: Fresh, sterile medium is constantly added, and the same amount of medium containing microorganisms and products is constantly taken out. Continuous fermentation is advantageous because the culture can always be at a highly productive concentration; in batch fermentation, much time is wasted waiting for the culture to attain productive concentrations. In spite of this advantage, only a few products are currently made by continuous fermentation on an industrial scale, partly because avoiding contamination in such an open system is difficult.

exists in *Streptomyces*. Recently, a sigma subunit called *whiG* was shown to be necessary for the formation of spores in this organism. Inactivation of this subunit results in an inability to form spores, whereas an increase in the number of copies of the normal *whiG* gene causes premature sporulation. Interestingly, *whiG* protein is highly homologous to a *Bacillus* sigma subunit that is known to recognize a special set of genes. *Streptomyces* is thought to contain at least seven species of sigma subunits, each of which presumably recognizes a different promoter sequence. Thus it is not surprising that the sequences of DNA upstream from coding regions appear to be very complex in *Streptomyces*. It has not been proved that the genes of antibiotic biosynthesis are regulated in exactly the same way as the genes involved in sporulation, and some differences are likely to exist. Nevertheless, similar principles probably operate in both cases.

There are certain to be other mechanisms that also regulate sporulation and antibiotic production in *Streptomyces*. For example, one locus, *bldA*, which is essential for sporulation, was found to code for a tRNA-like molecule. Apparently, some of the regulatory proteins in these processes use a rare codon, UUA, for leucine, and production of the tRNA that recognizes this unusual codon may serve as a mechanism to start the synthesis of these regulatory proteins.

Catabolite Repression

Many antibiotic-producing organisms are less productive in the presence of excess carbon source, especially of compounds that are rapidly degraded, such as glucose. Some studies have found that the levels of the enzymes involved in antibiotic synthesis become much lower under these conditions, suggesting that regulation occurs at the level of enzyme synthesis, presumably transcription, rather than affecting the activity of enzymes already present. This finding is reminiscent of the catabolite repression phenomenon that is well known in *E. coli*. Thus in the presence of good carbon sources such as glucose, *E. coli* cells stop synthesizing the enzymes for the degradation of less preferable carbon sources, such as lactose (Box 13.4). Similarly, we can imagine that antibiotic-producing soil organisms need to produce antibiotics and kill off their competitors only in a nutritionally poor environment, and catabolite repression is useful for this arrangement. In order to overcome catabolite repression, carbon sources must be added to the culture medium in carefully adjusted, small increments so that no large build-up of these compounds occurs at any given time. Alternatively, scientists may try to isolate mutants in which the antibiotic production is no longer repressed by the presence of excess carbon source.

> ▶ **BOX** 13.4 **Mechanism of Catabolite Repression**
>
> Most cellular regulatory mechanisms respond to *specific* signals. For example, the biosynthesis of the amino acid histidine is regulated specifically by the availability of histidine (Chapter 12). Similarly, the transcription of *lac* genes that code for enzymes of lactose utilization is regulated specifically by the availability of lactose in the medium in *E. coli*. However, it is more advantageous for *E. coli* to have a regulatory mechanism that takes into account its *general* nutritional state as well. Catabolite repression is an example of a regulatory process that responds to such a general or global signal—in this case, the abundance of energy supply. Thus even in the presence of lactose, *E. coli* does not synthesize the enzymes of the *lac* operon if a preferable carbon and energy source, glucose, is also present in the medium. In *E. coli*, catabolite repression works through the alteration of an intracellular signal compound, cyclic AMP, but other bacteria are known to use different mechanisms.

Nitrogen and Phosphate Repression

In many cases, the presence of excess nitrogen compounds or phosphate in the fermentation medium decreases antibiotic production severely; the ecological advantage of such regulation is probably similar to that of catabolite repression. Phosphate has been shown to inhibit the transcription of some of the genes of antibiotic synthesis. In organisms that are sensitive to nitrogen or phosphate regulation, it is necessary either to regulate the concentrations of these compounds very carefully during fermentation runs or to try to obtain mutants that are less sensitive to these regulatory processes. It must be noted, however, that such approaches are not feasible when one is screening many strains for potential producers of antibiotics. Satoshi Omura devised an interesting approach to circumvent the problem. In Japan, soil from a special area, called Kanuma-tsuchi, is much prized by people who grow bonsai (miniature trees in pots). Apparently Kanuma-tsuchi contains a mineral that binds phosphate efficiently and releases it slowly; in this soil, trees remain alive but do not grow rapidly. Omura added Kanuma-tsuchi to the medium used in antibiotic screening and very often found that the microorganisms produced much higher concentrations of antibiotics, thus improving the sensitivity of the screening process.

Feedback Regulation

Some scientists suspect that the antibiotics themselves, as end products, may exert negative-feedback regulation on their synthesis. Supporting data come from experiments in which penicillin added to the culture of penicillin-producing fungi apparently inhibited the synthesis of the antibiotic. Furthermore, the level of exogenous penicillin required for this inhibition was much higher with overproducers of penicillin, suggesting that resistance to this feedback inhibition was a major cause of overproduction in these strains. (This does not seem to be the whole explanation, however, because overproducers also contained much higher levels of penicillin-biosynthetic enzymes.)

If feedback inhibition plays an important role in regulating the production of an antibiotic, it should be possible to increase its production by promoting excretion of the antibiotic and thus lowering its intracellular concentration. Encouraging results have been obtained by promoting antibiotic leakage with polyene antibiotics (see page 443), but then one is faced with the problem of separating the produced antibiotic from the polyenes. Perhaps the method of making the cell membrane leaky without adding extraneous drugs, developed with amino acid producers (Chapter 12), may be useful in this regard.

Effects of Precursors

Secondary metabolites have to be synthesized from primary metabolites. Thus the efficient production of antibiotics requires a steady flow of their precursors. In many cases the production of these precursors is regulated by known mechanisms.

An interesting example of how the supply of precursors is regulated and how it affects production of the antibiotic is the effect of culture conditions on the production of α-aminoadipic acid, a precursor for penicillin biosynthesis. In fungi, α-aminoadipic acid is an intermediate in the pathway of lysine biosynthesis. Because lysine is the end product of a biosynthetic pathway, a high level of lysine in the medium shuts off that biosynthesis by inhibiting the first enzyme of the pathway (feedback inhibition). This results in a shortage of all the intermediates on the pathway, including α-aminoadipic acid. Thus the presence of excess lysine strongly inhibits penicillin production in *P. chrysogenum* fermentations. In striking contrast, the addition of excess lysine *stimulates* the production of cephamycin C in *Streptomyces*. This is because α-aminoadipic acid is synthesized by a totally different route in eubacteria (Figure 13.59), lysine being the precursor. As these cases illustrate, similar secondary products are often synthesized by similar pathways in very different organisms, but the primary metabolites used as the starting material may be made in different pathways, which are governed by different regulatory mechanisms.

In addition to α-aminoadipic acid, the biosynthesis of penicillin or cephalosporin requires the presence of cysteine and valine (see Figure 13.56). The way cysteine is made is different in different species and even in different strains. In *P. chrysogenum*, much of the sulfur atom of cysteine is derived from inorganic sulfate in the medium. In contrast, *C. acremonium*, which produces cephalosporin, derives much of its cysteine from methionine via a transsulfuration reaction (Figure 13.60). In this case, the addition of methionine to the medium strongly stimulates

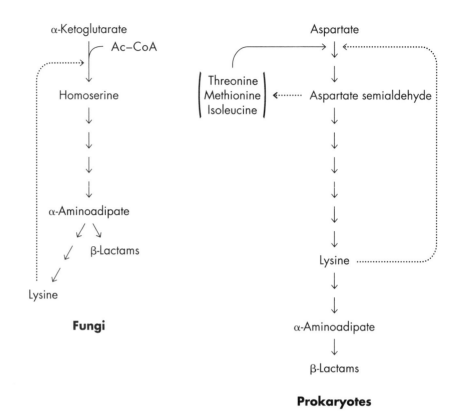

Figure 13.59
Pathways for the biosynthesis of α-aminoadipic acid. In fungi, α-aminoadipic acid is a precursor of lysine. Because lysine regulates by negative feedback the first step of the pathway (dotted line), the addition of lysine inhibits β-lactam synthesis by decreasing the supply of α-aminoadipic acid. In contrast, bacteria such as *Streptomyces* make α-aminoadipic acid from lysine. Although the addition of lysine inhibits its own synthesis (dotted line), exogenous lysine is converted efficiently into α-aminoadipic acid and stimulates the synthesis of β-lactam.

cephalosporin production, at least partly by increasing the supply of cysteine. Furthermore, when several strains producing higher amounts of cephalosporin C were examined, a proportional relationship emerged between the level of the cystathionine γ-lyase, an enzyme involved in cysteine production, and the yield of the drug. The mechanism of action of methionine is more complex than this suggests, however. The addition of norleucine, an analog of methionine, also stimulates cephalosporin C synthesis in *C. acremonium*, although norleucine obviously cannot serve as a precursor of cysteine. Some studies suggest that the levels of the enzymes of β-lactam synthesis go up when the cells are grown in the presence of methionine or norleucine and thus that these compounds might act as general regulatory molecules for cephalosporin synthesis.

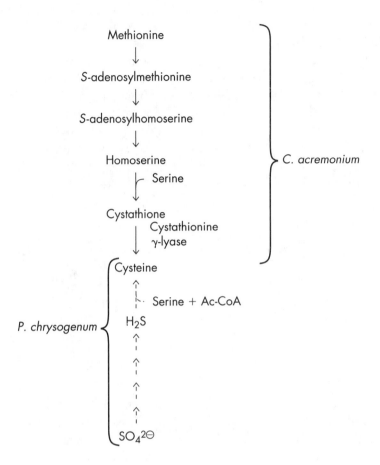

Figure 13.60

Major pathways for the biosynthesis of cysteine, which in turn is used in
β-lactam synthesis. Available data suggest that in *C. acremonium*, much of the
cysteine is made by transsulfuration via cystathionine (solid arrows). In contrast, in *P. chrysogenum*, the conversion of sulfate into cysteine (broken arrows) appears to play a much more important role.

Conditions for Fermentation

Numerous empirical improvements have doubtless been made in the
conditions for fermentative production of antibiotics, but most of them
constitute proprietary knowledge. Apparently, most fermentation processes are run in two stages. The first stage starts from spores, usually of
established strains, because antibiotic production is often an unstable
trait that can be lost if the stock is kept growing constantly through
serial culture. The organisms are cultured under submerged conditions,
with sufficient aeration and a generous supply of nutrients so that they
will attain near-maximal density in a short time. In the second stage,
when the culture reaches the stationary phase, or stops growing, and
begins to produce antibiotics, the concentration of the key nutrients,

such as carbon source, phosphate, and nitrogen source must be controlled carefully by continuous-feed processes.

The fermentative production of penicillin is much better described in the literature than that of other antibiotics. Published data indicate that in the best currently available strain of *P. chrysogenum*, about 10% of the carbon in the glucose added finds its way into penicillin G, whose final concentration may reach almost 30 g/liter. They note that particular attention must be paid to supplying just-sufficient amounts of the side-chain precursor phenylacetic acid, which is toxic and therefore must be added slowly by a controlled-feed process.

Genetics of Antibiotic Production

The efficiency of production of antibiotics is determined largely by the genetic make-up of the producing strains, and much effort has been spent on improving these strains.

Traditional Method of Strain Improvement

The traditional genetic approach to improving the yield of an antibiotic-producing organism depends entirely on random mutagenesis and screening of high producers. Until recently, the screening had to be done by growing each progeny clone in liquid media and assaying for the antibiotic in the culture filtrate. Because such screening was laborious and slow, only a small number of progeny could be tested in one experiment. Induction of a large number of mutations—many of them with deleterious effects on the growth of the organism—through heavy mutagenesis was necessary in order to boost the probability that the small number of strains tested would contain interesting mutants. In spite of these disadvantages, practically all the improvements achieved to date in the strains that produce commercially important antibiotics have come about in this manner, because there was no better method. A part of the genealogy of penicillin-producing strains is shown in Figure 13.61. Interestingly, the magnitude of the improvement per step was greater during the early stages. Thus a nearly 10-fold increase was achieved in just three steps between NRRL-1951 and Wis Q-176, whereas another 10-fold increase from that point required 18 steps—a situation reminiscent of what economists call the law of diminishing returns.

Not a great deal is known about the biochemical and genetic nature of the improvements that have been made. We noted earlier that many overproducing strains have a higher intracellular concentration of precursors and also higher levels of enzymes involved in precursor

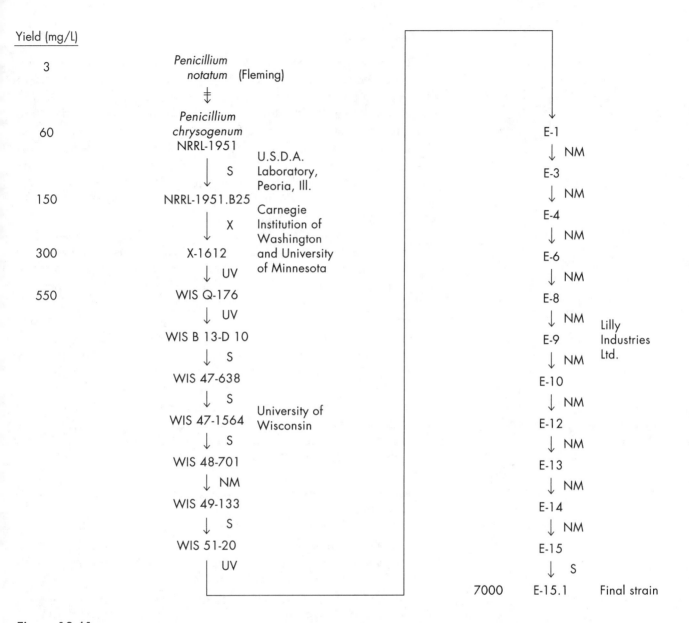

Figure 13.61
The improvement of penicillin-producing strains. The mutagenesis procedures used for each step: S, spontaneous; X, X-ray; UV, ultraviolet light;
and NM, nitrogen mustard. Modified from Aharonowitz, Y., and Cohen, G.
(1981), The microbial production of pharmaceuticals, *Sci. Amer.*, September,
141–152.

synthesis. In addition, overproducing strains show a striking increase in
the activity of enzymes involved in penicillin synthesis. In some cases,
gene duplication has been documented and the gene-dosage effect explains the higher levels of enzymes present.

Methods of Classical Genetics

In the ideal scenario, a scientist would identify desirable mutant alleles of various genes that affect antibiotic production and recombine them into a single organism, just as was done with amino-acid–overproducing strains. Unfortunately, this goal has been difficult to attain; the pathways of biosynthesis were too complex, and our knowledge about the genetics of antibiotic-producing organisms was too limited. (Today, however, we are at the start of new stage. At least with the prokaryotic *Streptomyces* species, many genes of antibiotic biosynthesis have been identified through cloning. Researchers will soon begin to recombine these genes at will, as we shall see below.)

The most effective use of classical genetics in the past was the backcrossing of overproducing strains with parent strains to improve the vigor of the mutant strains. Because the traditional approach to strain improvement involves many steps of heavy mutagenesis, and because each step introduces many unwanted mutations into the organism, the overproducing strain that results is invariably a weakened strain, one that grows poorly and is hypersensitive to various stresses. After backcrossing such a strain with the wild type, one is likely to find progeny cells that have inherited the overproducing traits from the mutant parent and the wild-type hardiness and vigor from the wild-type parent.

At first this did not seem possible with *Penicillium*, which does not have a true sexual cycle although it is a eukaryotic fungus. However, a parasexual cycle resulting in the production of heterocaryons was discovered in *Penicillium* in 1958 and was used to improve the strains.

Some strains of *Streptomyces*, a eubacterium, carry out conjugational transfer of DNA. David Hopwood and co-workers, who have made extensive studies of *Streptomyces* genetics, found that in many cases, genes of antibiotic synthesis are clustered together, as are the *act* genes and *red* genes in Figure 13.62. Recombinant DNA methods have shown that this is also true of the genes involved in the biosynthesis of commercially important antibiotics (see below).

Most of the antibiotic biosynthesis genes of *Streptomyces* are believed to be located on the chromosome, although an interesting phenomenon gave earlier workers the impression that these genes are on plasmids. Thus in the genome of *Streptomyces*, very large-scale deletions involving several hundred kilobases occur at a high frequency, and these deletions frequently involve the antibiotic production genes. The deletions are often accompanied by an extensive (up to 500-fold) amplification of a small piece of DNA. This frequent loss of antibiotic biosynthesis genes resembled the loss of a plasmid, which causes the simultaneous loss of all the genes located on it. Another recently discovered peculiarity of the *Streptomyces* genome is that the chromosome may sometimes exist as linear DNA with telomeric sequences (for telomeres see Box 4.4).

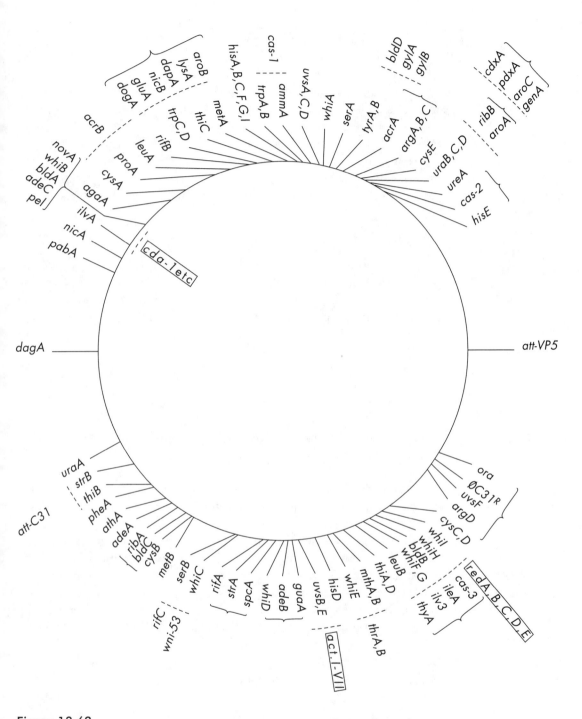

Figure 13.62
The chromosomal map of *S. coelicolor*. The boxes show the clustering of genes for the biosynthesis of antibiotics; *act*, for example, indicates the position of genes for actinorhodin synthesis. Based on Chater, K.F., and Hopwood, D.A. (1984), *Streptomyces* genetics, in Goodfellow, M., Modarski, M., and Williams, S.T. (eds), *The Biology of Actinomycetes*, pp. 229–286 (Academic Press).

The antibiotic producers often have to protect themselves against the toxic effects of their own products. This need is especially acute for eubacterial *Streptomyces* that produce antibiotics with anti-eubacterial activity. Many antibiotic-producing species thus have antibiotic resistance mechanisms, which inactivate or pump out the antibiotics, or produce resistance by modifying the target. In most cases studied, the genes responsible for resistance are part of the cluster of antibiotic synthesis genes; presumably this ensures the expression of resistance genes at the time of active antibiotic synthesis. As we will see later, this arrangement has been very useful in the cloning of antibiotic synthesis genes.

Targeted Mutagenesis

As we saw above, the major problems plaguing the classic strain improvement procedure, which is based on random mutagenesis, were the very low probability of introducing mutations into relevant genes and the high rate of unwanted mutations in other, unrelated genes. These problems can be avoided by mutagenizing only the relevant pieces of DNA.

Step 1 of a now classic approach developed by Jen-Hsiang Hong and Bruce Ames is to mutagenize a population of transducing phages (see Chapter 3), introducing mutations indiscriminately into all the genes. Later, transductants that received the donor copy of a gene located next to the antibiotic synthesis gene complex are selected, and in this way it is possible to recover only those pieces of mutagenized DNA that contained the particular gene used for selection (Figure 13.63). All the rest of the DNA, containing mutations in unrelated genes, is discarded during the selection for transductants. To use this approach for mutagenizing the antibiotic synthesis genes, one must first obtain a mutant in a gene closely linked to the antibiotic gene cluster. One can then transduce with a mutagenized phage population and select for the donor allele of this neighboring gene.

If the antibiotic production genes have been cloned, targeted mutagenesis of the cloned DNA can be performed *in vitro*, followed by transformation of the recipient bacterium. Although intact cells of *Streptomyces* are difficult to transform, their protoplasts, without the cell wall barrier, are easily transformed. This is one reason why the cloning of these genes is potentially so important.

Rational Selection

In traditional strain improvement methods, the progeny produced after the mutagenesis have to be *screened*—looked at one by one—to find a line that produces more of the antibiotic. If methods were available for *selecting* an improved producer out of a very large population of progeny, say 100 million cells, then the strain improvement process would

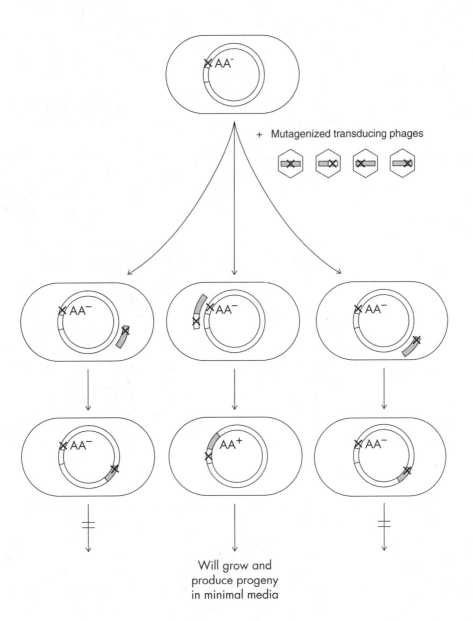

+ Mutagenized transducing phages

Will grow and
produce progeny
in minimal media

Figure 13.63

The principle of the targeted mutagenesis method, according to Hong and Ames. In this example, let us assume that the target gene we want to mutagenize (empty box) is linked closely to a gene for amino acid biosynthesis (AA). We infect a mutant that is defective in the AA gene (AA⁻) with a heavily mutagenized population of transducing phages (the mutations are marked by crosses). When the transducing DNA enters the cell, it aligns itself with the homologous segment of the chromosome and then replaces that part of the chromosome by the process of recombination. When the population of the progeny cells is transferred to minimal media, which do not contain the amino acid, the only transductant cells (see Chapter 3) to grow are cells that received the gene for amino acid synthesis (AA) and that therefore are likely to have also received a mutagenized target gene. These strains are unlikely to have any other mutations in other areas of the chromosome.

be infinitely more efficient (for the distinction between screening and selection, see Box 3.3). Although nobody has yet been able to devise a general protocol for such a selection, some indirect selection methods have been used in certain cases.

For example, some antibiotics, notably penicillins and tetracyclines, are chelators of heavy metal ions. The more of these antibiotics an organism produces, the more resistant it will be to heavy metals in the medium. Thus selection for mutants resistant to heavy metals was used in the improvement of the penicillin producers.

As described earlier, the biosynthesis of amino acids, which are the starting material for the biosynthesis of many antibiotics, including the β-lactams, is regulated by the usual mechanisms that control primary metabolism. For example, lysine biosynthesis is regulated by feedback inhibition by the end product of the pathway, lysine. In the eubacterium *Streptomyces*, as we saw, lysine is the precursor of α-aminoadipic acid, which is needed for cephamycin biosynthesis. Thus it is logical to assume that prevention of this feedback inhibition would improve the flow of carbon into cephamycin. Feedback-resistant mutants of the first enzyme of the lysine biosynthetic pathway can be selected positively by growing the population in a minimal medium in the presence of a lysine analog, S-(2-aminoethyl)-L-cysteine (see Chapter 12). Such mutants do overproduce lysine and, as expected, produce several times more cephamycin C in minimal medium, as well.

Protoplast Fusion

Protoplast fusion is a technique that achieves the efficient intercellular transfer of genetic material in species that, like many antibiotic-producing organisms, have no natural mechanism for conjugation. Its impact on the field of antibiotic production has been quite significant. Researchers finally have a way to backcross their overproducing mutant strains with the wild-type parent strain and restore some "vigor" to the line of the producing strains.

In this technique, bacterial or fungal cells are converted to protoplasts by dissolving the cell wall with lytic enzymes. The membranes of two protoplasts are then encouraged to fuse together by the addition of high concentrations (typically 20–35%) of polyethyleneglycol or other agents. With some strains, up to 20% fusion has been reported under optimal conditions. Then cell walls are regenerated in the progeny in suitable protective media. Usually the chromosomes from the two parental protoplasts undergo recombination, and the redundant material is eventually discarded in the process of successive cell division. Thus many of the progeny cells recovered at the end of this process have only one equivalent of genetic material, which consists of a mixture of genes from both parents.

This method can also be used to develop new antibiotics. Fusing the cells of two species that synthesize different antibiotics may produce progeny cells that contain a mixture of the two sets of antibiotic-synthesizing genes, and these in turn may produce a new, hybrid antibiotic. Among the aminoglycoside and macrolide antibiotics, for example, are a great many products that are synthesized by minor variations on a central theme. With the organisms that produce these compounds, one might expect that some progeny from interspecies fusion would produce a new, hybrid antibiotic. This "shotgun" approach is less sophisticated and less efficient than the cloning approaches described below. It has the advantage, however, of sometimes producing a fortuitous combination of genes that could not easily be predicted by rational means.

Cloning

Because genes involved in the production of a single antibiotic tend to be grouped in closely linked clusters, several complete gene clusters have successfully been cloned into a single plasmid. Francisco Malpartida and David Hopwood achieved this in *Streptomyces* in 1984, cloning genes for the entire biosynthetic pathway of actinorhodin. They did this by screening recombinant plasmids in a non-actinorhodin-producing mutant host cell and looking for actinorhodin's characteristic blue color. A plasmid that had acquired an insert of about 30 kb apparently contained the entire set of genes: It complemented all known classes of non-actinorhodin-producing mutants, and it directed the synthesis of actinorhodin in an unrelated strain. The complete gene clusters coding for the production of numerous commercially important antibiotics have also been cloned. The screening methods included complementation of defective mutants, selection for an antibiotic resistance gene that is usually a part of a particular biosynthetic gene cluster, and use of oligonucleotide probes made to match the amino acid sequence of isolated enzymes.

Theoretical Insights Obtained via Cloning

The cloning of antibiotic production genes has had a profound impact on the field. First, the cloning and sequencing of these genes gave us new and important insights into the nature of antibiotics. For example, the genes coding for the first two enzymes of β-lactam biosynthesis, ACV synthetase (*pcbAB*) and isopenicillin N synthase (*pcbC*), were found to exist next to each other in both the prokaryotic (eubacterial) and the eukaryotic producers of β-lactams. In addition, strong sequence similarity was found among homologous genes of these clusters from the diverse organisms. These observations suggest that the capacity to produce β-lactams was spread from a prokaryote into eukaryotic fungi by a horizontal gene-transfer process fairly recently on an evolutionary time scale (370 million years ago, according to one estimate), which would explain why β-lactam production is found not only both in eukaryotes and in eubacteria but also in unrelated and scattered groups among the eubacteria (for example, *Pseudomonas, Erwinia, Nocardia, Streptomyces*).

A second area in which our understanding of antibiotics has been enhanced by molecular cloning is the study of polyketide antibiotics. As we mentioned at the beginning of this chapter, the structures of the known antibiotics are bewilderingly diverse. However, many of them, including macrolides (Figure 13.8), rifamycins (Figure 13.9), tetracyclines (Figure 13.10), polyenes (Figure 13.16), anthracyclines (Figure 13.17), and FK-506 (Figure 13.24B), are thought to be produced through a pathway very similar to the one that leads to fatty acids. In fatty acid synthesis, an acetyl group is added at each round of synthesis, to produce a long chain without branches, and the carbonyl group introduced at the condensation step in each round is usually reduced to the level of $-CH_2-$. In the biosynthesis of polyketide antibiotics, the unit added is often larger than an acetyl, yet each condensation step adds two carbon atoms to the elongating main chain, so the remaining part of the unit protrudes from the main chain as a branch. This is seen in the structure of macrolides, for example, where a propionate unit is added at each step such that every second carbon in the main chain bears a methyl branch (see Figure 13.8). Some of the carbonyl groups are not reduced at all, and others are reduced only to the level of CHOH (see Figure 13.8, for example). In spite of these differences, cloning the macrolide biosynthesis genes and sequencing them revealed a remarkable homology between these polyketide synthesis genes and the fatty acid synthetase genes.

The production of the erythromycin macrocyclic ring by *Saccharopolyspora erythrea* (*Streptomyces erythreus*, in older literature) involves only three giant genes, each of which is predicted to code for a protein of more than 300,000 daltons. Each protein consists of two inexact repeats, so the entire enzyme complex contains six "modules," each of which in turn contains acyl-carrier protein, acyl transferase, ketoacyl-ACP synthase, and other domains necessary for the biosynthetic process (Figure 13.64). In erythromycin A, only C_7 is reduced to the level of CH_2 (seen in Figure 13.8), and only module 4 contains the domains needed for this reduction: sequences homologous to the β-hydroxyacyl-ACP dehydratase and enoyl-ACP reductase. Thus we can deduce that each of modules 1 through 6 catalyzes, in precise succession, the addition of C_3 units and that the fate of the keto group depends on the enzyme content of the particular module. These findings have far-reaching implications. Although a large number of polyketide antibiotics exist, it appears that the specific structure of each compound is determined by the substrate specificity and composition of each enzyme

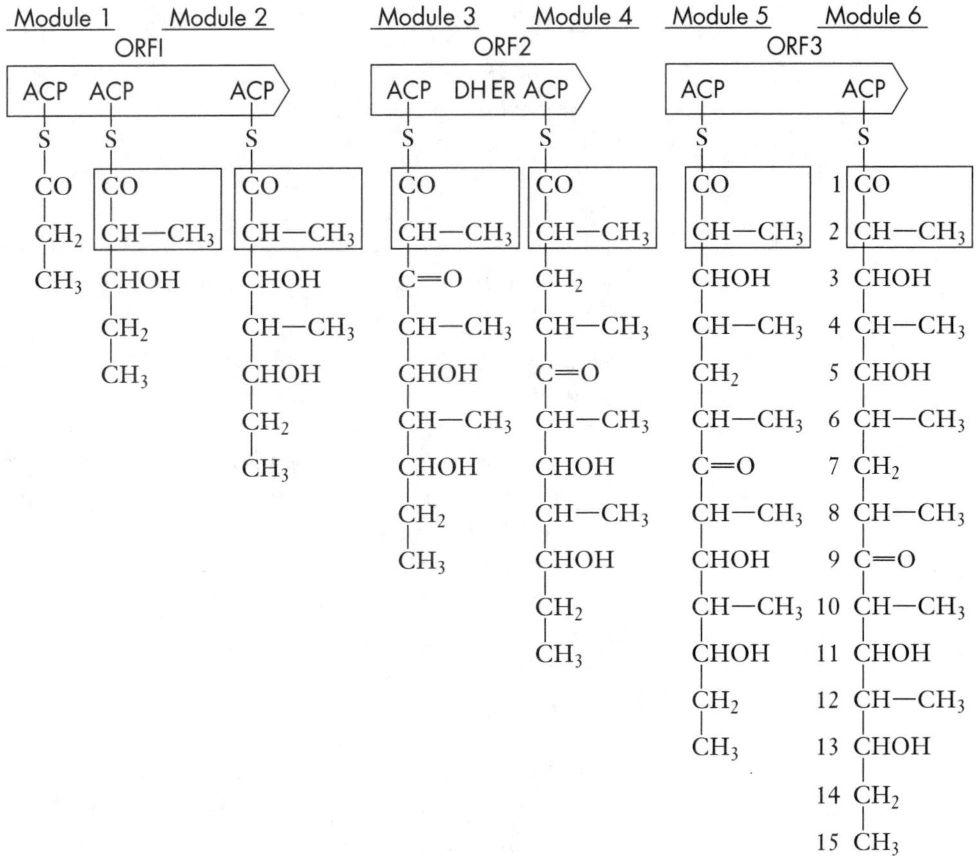

Module 1   Module 2   Module 3   Module 4   Module 5   Module 6

ORFI   ORF2   ORF3

ACP ACP ACP⟩	ACP DH ER ACP⟩	ACP ACP⟩

S	S	S	S	S	S	S
CO	CO	CO	CO	CO	CO	1 CO
CH₂	CH—CH₃	CH—CH₃	CH—CH₃	CH—CH₃	CH—CH₃	2 CH—CH₃
CH₃	CHOH	CHOH	C=O	CH₂	CHOH	3 CHOH
	CH₂	CH—CH₃	CH—CH₃	CH—CH₃	CH—CH₃	4 CH—CH₃
	CH₃	CHOH	CHOH	C=O	CH₂	5 CHOH
		CH₂	CH—CH₃	CH—CH₃	CH—CH₃	6 CH—CH₃
		CH₃	CHOH	CHOH	C=O	7 CH₂
			CH₂	CH—CH₃	CH—CH₃	8 CH—CH₃
			CH₃	CHOH	CHOH	9 C=O
				CH₂	CH—CH₃	10 CH—CH₃
				CH₃	CHOH	11 CHOH
					CH₂	12 CH—CH₃
					CH₃	13 CHOH
						14 CH₂
						15 CH₃

Figure 13.64

The organization of genes for erythromycin A biosynthesis in *Saccharopolyspora erythrea*. The 30-kb region of DNA is divided into three open-reading frames (ORFs), each of which codes for a large, complex enzyme molecule. Each enzyme is divided into two modules. Each module contains an acyl-carrier protein domain (ACP), an acyl transferase domain, and a β-ketoacyl-ACP synthase domain (for clarity, only the ACP domain is shown). Each module adds a new propionic acid unit (shown in boxes) to the growing chain. All of the modules except module 3 also contain a β-ketoacyl-ACP reductase domain. Thus the keto group of the newly added unit is reduced (usually to -CHOH) except in the unit added by module 3 (C9 in the completed open-chain structure, far right, where the numbering scheme for the carbon atoms is shown.) Module 4 contains dehydratase (DH) and enoyl-ACP reductase (ER) domains in addition to all four domains mentioned already; these two additional domains convert the original keto group in the propionic acid moiety added in the preceding step (by module 3) into a methylene group (C7 in the completed chain, far right). Based on the results of Donadio, S., et al. (1991), *Science* 252:675–679.

module. This means that the apparent diversity of the structure of polyketide antibiotics is rather deceptive and that the producing organisms can create an impressive variety of antibiotics by introducing minor variations in the biosynthetic apparatus. It also shows that the biosynthesis of exotic structures—polyketide antibiotics—can be carried out by not-too-extensive modifications of common enzymes such as fatty acid synthetase.

Practical Benefits Brought About by Cloning

Molecular cloning of the genes obviously opened up the way for many practical improvements in the commercial production of antibiotics. One result has been the creation of new compounds.

1. We have already mentioned (page 489) that feeding chemically synthesized analogs to isopenicillin N synthase, produced by overexpression of the cloned gene, led to the production of *novel β-lactam compounds*.

2. A similar approach can be used in intact cells to create *hybrid antibiotics*. The principle of this method was first demonstrated in a collaborative project involving groups headed by David Hopwood in England, Heinz Floss in the United States, and Satoshi Omura in Japan. A plasmid containing a gene or genes involved in actinorhodin biosynthesis was transferred to a *Streptomyces* species that produces a structurally related antibiotic, medermycin. As shown in Figure 13.65, actinorhodin contains a hydroxyl group on one of the rings. A hybrid antibiotic, mederrhodin A, with essentially the same structure as medermycin but with the hydroxyl group characteristic of actinorhodin, was produced by the plasmid-containing strain. One can hypothesize in this case that the plasmid contained the gene for a hydroxylase. Similar approaches may be useful in the development of new agents in commercially important classes of antibiotics, such as macrolides and aminoglycosides.

3. A *direct manipulation* of biosynthetic pathways can be accomplished. The polyketide biosynthetic pathway discussed above can be manipulated to alter the structure of the resulting antibiotic in a predictable manner. For example, by deleting the β-ketoacyl reductase domain in one of the modules, it was possible to prevent the reduction of a particular keto group and to produce an erythromycin analog in which that position of the ring was deliberately changed from CHOH to C=O.

A second practical result has been the construction of strains for the production of what are now semisynthetic compounds, a goal accomplished by bringing in the desirable genes and inactivating unwanted ones. For example, a strain producing 7-aminocephalosporanic acid fermentatively was constructed by bringing into *C. acremonium* a cephalosporin acylase gene from *Pseudomonas diminuta*. Because this enzyme does not deacylate the D-aminoadipic acid side chain of cephalosporin C, a D-amino acid oxidase gene from *Fusarium solani* was also brought in to convert the side chain to a keto acid. The chemical deacylation of cephalosporin C currently requires complex chemical steps, so this is a potentially useful way of producing 7-aminocephalosporanic acid, the starting material for many semisynthetic cephalosporin drugs.

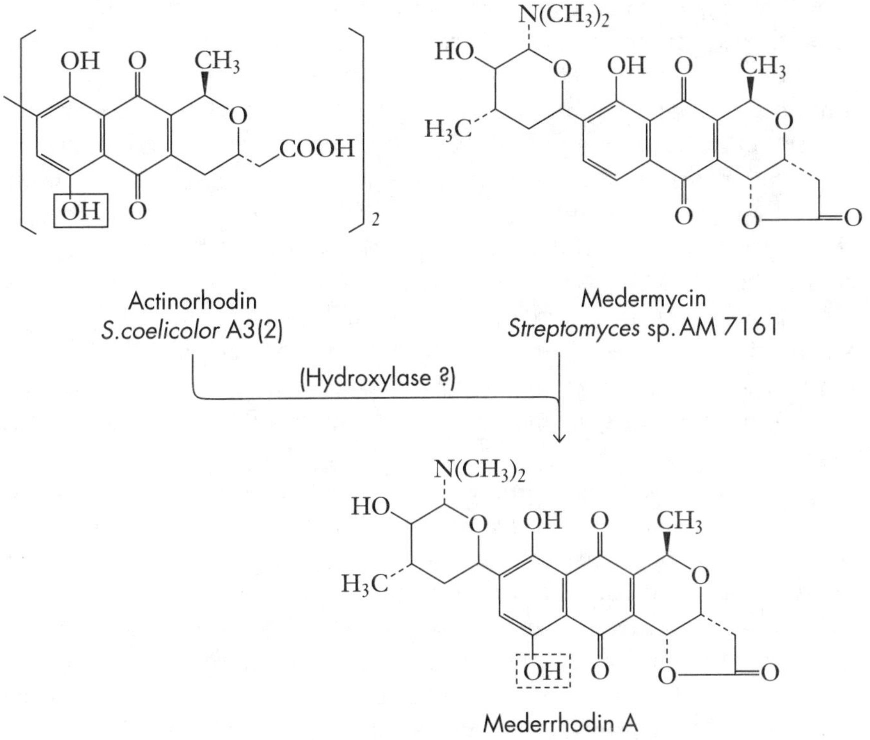

Figure 13.65
Production of a hybrid antibiotic by the use of recombinant DNA. The gene or genes coding for some steps (probably hydroxylation) in the synthesis of the antibiotic actinorhodin were cloned from the DNA of *S. coelicolor* A3(2), and the recombinant plasmid was transformed into another *Streptomyces* strain that produces medermycin. The plasmid-containing strain then produced a new antibiotic, mederrhodin A, presumably as a consequence of collaboration between the medermycin-biosynthetic enzymes of the host and the hydroxylase brought in from *S. coelicolor*. Based on Omura, S. (1986), *Microbiol. Rev.* 50:259–279.

Gene-cloning studies also reveal possible ways to increase the yield of fermentation. For example, it may be possible to enhance antibiotic production by increasing the copy number of strategic genes, such as a gene whose low level of expression creates a "bottleneck" in the pathway. An approach of this kind was successfully used with the cephalosporin C-producing organism, *Cephalosporium acremonium*. Cephalosporin is produced by the pathway shown in Figure 13.56. Because this organism produced about one-third as much penicillin N as cephalosporin C, the final product, investigators hypothesized that a bottleneck was occurring at the conversion of penicillin N to a cephalosporin by "expandase" (deacetoxycephalosporin C synthetase/deacetylcephalosporin synthetase), an oxygenase catalyzing conversion of the penicillin nucleus to the cephalosporin nucleus (see Figure 13.56). Researchers introduced the cloned expandase gene into the organism, where it apparently was integrated into a chromosome, thereby doubling the copy number. The engineered strain produced slightly larger (by 20–40%) amounts of cephalosporin C. This was accompanied by a drastic decrease in the secretion of penicillin N by the engineered strain, perhaps a clearer indicator of the experiment's success. Although this example raises the hope that similar approaches will one day result in a substantial improvement in production yield, such approaches have not always been successful. Even in these cases, however, molecular-cloning research suggests plausible reasons for the failure. For example, it is now known that some penicillin-producing strains of *P. chrysogenum* already contain up to a dozen copies of *pcb* genes; here one would not expect a further increase in gene dosage to enhance production to any great extent.

In the area of strain improvement, the biggest obstacle right now is our ignorance of how the transcription of these genes of secondary metabolism is regulated. Because antibiotic production often involves complex pathways, and because it requires a proper supply of primary metabolites and mechanisms to prevent self-intoxication, simply cloning the genes and overexpressing them in totally unrelated microorganisms is not likely to work (although that approach *is* effective in the production of the primary gene products such as hormones, discussed in Chapters 3 and 4). This is why recombinant DNA technology has not yet made so great a contribution to the effort to improve antibiotic-producing strains as might have been expected. Perhaps as our knowledge increases, we will see accelerated development.

A fourth practical result of cloning research has been that, when an antibiotic is synthesized by a relatively simple pathway, the enzymes of that pathway can be overproduced from cloned genes and then placed in reactor columns to make antibiotics, or at least intermediates of antibiotic biosynthesis, *in vitro*. Investigators have done this successfully on an experimental scale, producing penicillins and cephalosporins in immobilized enzyme reactors by using crude extracts of producing organisms. The starting materials for β-lactam biosynthesis are common, inexpensive amino acids, so the overproduction of critical enzymes such as isopenicillin N synthetase and "expandase" might one day make this process a commercially viable one.

PROBLEM OF ANTIBIOTIC RESISTANCE

A unique feature of the antibiotic field is the constant need to develop new agents, to keep pace with the constant appearance of new, resistant organisms. As we mentioned earlier, in most European countries it is illegal to use an antibiotic to promote growth and prevent disease in farm animals if that antibiotic is also used to treat humans. This is a sensible policy, in view of the known instances in which R-factor-carrying pathogenic bacteria spread from farm animals to humans, and in view of the correlation between widespread use of an antibiotic and the prevalence of R factors coding for resistance against that antibiotic (see the section on aminoglycosides). Nevertheless, the governments of Europe are under constant pressure to rescind this restriction. The "scientific" argument being advanced in favor of changing the law is that the frequency of isolation of tetracycline-resistant *E. coli* strains from farm animals in the United Kingdom has not declined significantly during the ten years that have passed since the use of tetracycline in animal feed was banned there (it hovers around 40–50% of all isolated strains). However, studies have also shown that the total amount of tetracycline used for farm animals in the United Kingdom per year has *not decreased at all* in spite of the legal restrictions, there apparently being enough sympathetic veterinarians who will write prescriptions for spurious "health problems" in the animal herds. Thus it is hardly surprising that the frequency of tetracycline-resistant bacteria has not decreased. The government restriction, however, *has* had the effect of encouraging the pharmaceutical industry to develop a number of antibiotics specifically for addition to animal feeds. These include moenomycin (a phosphoglycolipid antibiotic), tylosin (a macrolide), and thiostrepton (a peptide antibiotic). Although it is clearly reasonable to use a special group of antibiotics as feed additives, it is also possible that some of these antibiotics may select for strains that show cross resistance to antibiotics used for humans.

In the United States, the FDA has to have a proof of a "clear and imminent danger" in order to ban the use of common antibiotics in farm animals, and such proof is difficult to produce, especially in a legally watertight form. Thus enormous amounts of penicillin, streptomycin, and especially tetracyclines (the common recipe for swine feed specifies 110–220 μg/ml final concentration of tetracycline) are used as feed additives, which undoubtedly contributes to the selection of R-factor-containing strains in this country.

It is obvious that various resistance genes are occurring more and more frequently among strains isolated from clinical sources throughout the world and that new mechanisms of resistance are emerging. For example, 99% of the *Shigella* strains (which cause bacillary dysentery) isolated in Vietnam during the 1980s were resistant to at least one antibiotic; about 70% of them were resistant to at least four (ampicillin, tetracycline, chloramphenicol, and sulfonamides). As another example, each of the R plasmids in aminoglycoside-resistant Gram-negative bacteria isolated before 1983 usually contained only one gene coding for an aminoglycoside-modifying enzyme; in contrast, each of the similar plasmids isolated during the last several years usually contains several different genes. As we saw in the methicillin resistance of staphylococci and the penicillin resistance of gonococci, in some species the targets of β-lactam action, penicillin-binding proteins, have been drastically altered in structure, apparently because of the introduction of genetic material from other bacteria. Another resistance mechanism whose importance is becoming more widely recognized is a drug efflux pump with very broad specificity. This has been shown to confer multidrug resistance on Gram-positive cocci and to make a major contribution to the well-known intrinsic antibiotic resistance of *P. aeruginosa*.

Clearly, the need for new antibiotic agents is more acute than ever. However, when new compounds are derived from natural products, there is always a possibility that enzymes capable of inactivating these compounds already exist in the producing microorganisms and that the genes coding for these enzymes could quickly become widespread among pathogenic microorganisms through the lateral transfer of plasmids and phages. In contrast, resistance mechanisms based on enzymatic hydrolysis or inactivation have not appeared for the totally synthetic agents, such as fluoroquinolones, sulfonamides, and trimethoprim (although bacteria can develop resistance to these agents by altering the drugs' targets, sometimes quite easily). For this reason, it may be very fortunate that methods now exist to produce inhibitors that may not exist in nature.

Several laboratories are now advocating development of the following *combinatorial* approach to creating new antimicrobial agents.

1. A potential target protein in a pathogenic microorganism is chosen. The gene for it is then cloned and overexpressed, and the protein is purified and immobilized on a matrix.

2. Peptides with random sequences are expressed as fusion proteins on the surface of a filamentous bacteriophage vector. (The random sequences are generated by synthesizing oligonucleotides using a mixture of the four deoxynucleotides.)

3. These phages are then added to affinity matrix holding the target protein. Any phage expressing a peptide that binds tightly to the target protein will become adsorbed to this affinity matrix.

4. Such phages are then amplified, and the sequence of the peptide is deduced by sequencing the DNA.

A similar process can also be carried out by using totally synthetic peptides. In principle, this procedure should produce peptides with an affinity to the critical target protein—and hence with some antimicrobial activity. It should also be possible to go one step further and synthesize a chemical with a rigid structure that would mimic the active conformation of the peptide. Although no commercial agent has yet been produced in this manner, it is encouraging that innovative methods such as this are being developed in response to the challenge of countering drug resistance among pathogenic microorganisms.

SUMMARY

Antibiotics are secondary metabolites of microorganisms and are usually produced in the stationary phase of growth. The majority of antibiotics come from eukaryotic fungi or from streptomycetes, a prokaryotic group with a complex life cycle reminiscent of that of fungi; a few are produced by unicellular bacteria not known to have developmental cycles. Many of the genes of antibiotic biosynthesis have been cloned and sequenced. And in many cases they occur as a cluster, which often also contains genes that confer resistance to the particular antibiotic produced. This arrangement explains the apparent occurrence of the horizontal transfer of these genes, during evolution, between distantly related organisms. The gene sequence also explains (for example, with the producers of polyketide antibiotics) why a single antibiotic producer often secretes many related antibiotics that are different from each other in minor details. Cloning of production genes is opening up ways to produce novel antibiotics by the combination of defined genes. However, it has had only limited success in increasing the yields of conventional antibiotics. Possible reasons for this include the complex nature of the regulation of secondary metabolism, which is still poorly understood, and the fact that the supply of primary metabolites often becomes limiting for the overproduction of antibiotics.

Antibiotics have been extremely effective in our battle against bacterial infections, and they have changed drastically the ecological balance between humans and various pathogenic microorganisms. Yet every introduction of a new antibiotic into clinical practice has been followed by the emergence of resistant organisms. Very often these organisms harbored R plasmids that contained resistance genes specifying the degradation, inactivation, or pumping out of the antibiotic molecules. Sometimes the resistant strains were opportunistic wild-type organisms that are intrinsically resistant to antibiotics but had not been thought of as significant pathogens. Such wild-type strains resistant to β-lactams unexpectedly contain β-lactamase genes on their chromosome; this situation presumably arose because these species, normally inhabitants of soil, have been in contact with natural β-lactam compounds in their environment throughout their evolutionary history.

Production of new antibiotics that are effective against resistant organisms has therefore been one of the most important aspects of antibiotic research. Two approaches have played prominent roles in this endeavor. First, discovery of new antibiotics through the screening of natural products led to many new classes of compounds, including cephalosporins, cephamycins, carbapenems, and monobactams among β-lactam antibiotics, and gentamicin, tobramycin, and fortimicin A among aminoglycosides. Second, chemical modification of natural products led to a number of compounds with increased stability to enzymatic degradation or inactivation mechanisms. Examples include methicillin and the third-generation cephalosporins among β-lactams, and dibekacin among aminoglycosides. In spite of these successful efforts, the frequency of occurrence of resistance genes among bacterial populations is increasing, and new mechanisms of resistance, such as target modification and the multiple drug efflux pump, are becoming clinically important. New approaches are vital in our fight against the emergence of antibiotic resistance. The combination of cloned genes for the production of new antibiotics and the screening of combinatorial libraries for affinity to defined target structures are examples of these novel methods.

SELECTED REFERENCES

General

Rehm, H.-J., and Reed, G. (eds.), 1986. *Biotechnology*, vol. 4. Verlag Chemie.

Microbial Products with Biological Activity

Hall, M.J., 1989. Microbial product discovery in the biotech age. *Bio/Technology* 7:427–430.

Franco, C.M.M., and Coutinho, L.E.L., 1991. Detection of novel secondary metabolites. *Crit. Rev. Biotechnol.* 11:193–276.

Davies, J., 1990. What are antibiotics? Archaic functions for modern activities. *Mol. Microbiol.* 4:1227–1232.

Aminoglycosides

Umezawa, H., and Hooper, I.R. (eds.), 1982. *Aminoglycoside Antibiotics*. Springer-Verlag.

Whelton, A., and Neu, H.C. (eds.), 1982. *The Aminoglycosides: Microbiology, Clinical Use, and Toxicology*. Marcel Dekker.

Davis, B.D., 1987. Mechanism of bactericidal action of aminoglycosides. *Microbiol. Rev.* 51:341–350.

Price, K.E., 1986. Aminoglycoside research 1975-1985: Prospects for development of improved agents. *Antimicrob. Agents Chemother.* 29:543–548.

Shaw, K.J., Rather, P.N., Hare, R.S., and Miller, G.H., 1993. Molecular genetics of aminoglycoside resistance genes and familial relationships of the aminoglycoside-modifying enzymes. *Microbiol. Rev.* 57:138–163.

β-Lactams

Demain, A.L., and Solomon, N.A. (eds.), 1983. *Antibiotics Containing the β-Lactam Structure*. Vols. I and II. Springer-Verlag.

Morin, R.B., and Gorman, M. (eds.), 1982. *Chemistry and Biology of β-Lactam Antibiotics*. Vols. 1–3. Academic Press.

Page, M.I. (ed.), 1992. *The Chemistry of β-Lactams*, Blackie Academic and Professional.

Gadebusch, H.H., Stapley, E.O., and Zimmerman, S.B., 1992. The discovery of cell wall active antibacterial antibiotics. *Crit. Rev. Biotechnol.* 12:225–243.

Strynadka, N.C.J., et al., 1992. Molecular structure of the

acyl-enzyme intermediate in β-lactam hydrolysis at 1.7 Å resolution. *Nature* 359:700–705.

Matagne, A., Lamotte-Brasseur, J., Dive, G., Knox, J.R., and Frere, J.-M., 1993. Interactions between active-site-serine β-lactamases and compounds bearing a methoxy side chain on the α-face of the β-lactam: Kinetic and molecular modelling studies. *Biochem. J.* 293:607–611.

Other Classes of Antibiotics

Kirst, H.A., and Sides, G.D., 1989. New directions for macrolide antibiotics: Structural modifications and *in vitro* activity. *Antimicrob. Agents Chemother.* 33:1413–1418.

Hlavka, J.J., and Boothe, J.H. (eds.), 1985. *The Tetracyclines*. Springer-Verlag.

Novel Aspects of Antibiotic Resistance

Nikaido, H., 1994. Prevention of drug access to target: Resistance mechanisms in bacteria based on permeability barriers and active efflux. *Science* 264:382–388.

Spratt, B.G., 1994. Resistance to antibiotics mediated by target alterations. *Science* 264:388–393.

Antibiotic Production

Goodfellow, M., Modarski, M., and Williams, S.T. (eds.), 1984. *The Biology of Actinomycetes*. Academic Press.

Embley, T.M., and Stackebrandt, E., 1994. The molecular phylogeny and systematics of the actinomycetes. *Ann. Rev. Microbiol.* 48:257–289.

Omura, S., 1986. Philosophy of new drug discovery. *Microbiol. Rev.* 50:259–279.

Chater, K.F., 1989. Multilevel regulation of *Streptomyces* differentiation. *Tr. Genet.* 5:372–377.

Birch, A., Hausler, A., and Hutter, R., 1990. Genome rearrangement and genetic instability in *Streptomyces* spp. *J. Bacteriol.* 172:4138–4142.

Chater, K.F., 1990. The improving prospects for yield increase by genetic engineering in antibiotic-producing streptomycetes. *Bio/Technology* 8:115–121.

Kirby, R., 1992. The isolation and characterization of antibiotic biosynthesis genes. *Biotechnol. Adv.* 10:561–576.

Donadio, S., Staver, M.J., McAlpine, J.B., Swanson, S.J., and Katz, L., 1991. Modular organization of genes required for complex polyketide biosynthesis. *Science* 252:675–679.

McDaniel, R., Ebert-Khosla, S., Hopwood, D.A., and Khosla, C., 1993. Engineered biosynthesis of novel polyketides. *Science* 262:1546–1550.

Hutchinson, C.R., 1994. Drug synthesis by genetically engineered microorganisms. *Bio/Technology* 12:375–380.

Aharonowitz, Y., Cohen, G., and Martin, J.F., 1992. Penicillin and cephalosporin biosynthetic genes: Structure, organization, regulation, and evolution. *Ann. Rev. Microbiol.* 46:461–495.

Skatrud, P.L., 1992. Genetic engineering of β-lactam antibiotic biosynthetic pathways in filamentous fungi. *Tr. Biotechnol.* 10:324–329.

Combinatorial Peptide Library

Scott, J.K., and Smith, G.P., 1990. Searching for peptide ligands with an epitope library. *Science* 249:386–390.

Houghten, R.A., Pinilla, C., Blondelle, S.E., Appel, J.R., Dooley, C.T., and Cuervo, J.H., 1991. Generation and use of synthetic peptide combinatorial libraries for basic research and drug discovery. *Nature* 354:84–86.

ORGANIC SYNTHESIS AND DEGRADATION

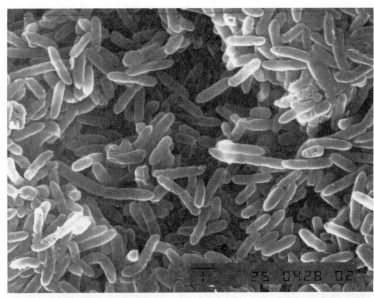

The top panel shows a scanning electron photomicrograph of *Thiobacillus ferrooxidans*, a bacterium important in the heap leaching of low-grade metal sulfide ores. The bacteria are magnified about 7500 times. The bottom left panel shows a gold-containing pyrite particle before biooxidation with bacteria. The bottom right panel shows the same pyrite particle after about 30 days of exposure to bacteria (primarily *Thiobacillus ferrooxidans* and *Leptospirillum ferrooxidans*). Courtesy of Dr. James A. Brierley, Newmont Metallurgical Services.

14

Organic Synthesis

In carrying out their metabolic processes, microorganisms inter-convert diverse organic compounds. These "biotransforma-tions" occur with high specificity and efficiency because they are catalyzed by enzymes. The active site of an enzyme, where substrate binding and catalysis are carried out, is an asymmetric surface whose special geometry frequently guarantees that the enzyme-catalyzed reaction will yield a particular stereoisomer as the sole product. Such stereospecific, or enantioselective, reactions can be difficult or impossible to achieve by purely chemical means. The terms used to describe the stereochemistry of organic compounds are defined in Box 14.1.

Even when an organic compound *can* be synthesized chemically, the process may require many steps, whereas a single enzyme-catalyzed reaction can often achieve the same end. Also, enzymes can catalyze reactions under mild conditions of temperature and pH, a particular advantage when the desired product is rather labile. Finally, enzymes can accelerate the rates of chemical reactions by factors of 10^8 to 10^{12}. For all of these reasons, biotransformations by microorganisms, or by enzymes purified from microorganisms, are highly useful in preparative organic chemistry.

Certain disadvantages do limit the use of enzymes in organic chemical processes, but these limitations have frequently proved to be surmountable obstacles rather than impenetrable barriers. One problem is that enzymes exhibit low or no activity in most or-ganic solvents and are denatured both at high temperatures and under strongly acidic or basic conditions. These difficulties can be

BOX 14.1 Stereochemistry of Organic Compounds—
A Glossary of Terms

Enantiomers Compounds that are mirror images of each other.
Optically active compound Compound that is not superimposable with its enantiomer, or mirror image, and rotates the plane of polarization of plane-polarized light. A simple diagnostic test for superimposability is the presence of a plane or a center of symmetry. The presence of such symmetry indicates lack of optical activity; its absence indicates optical activity.
Chiral center Asymmetric carbon atom with four different groups around it that causes the optical activity of many organic molecules of biological interest.
Prochiral center The carbon atom in $CR_2R'R''$. Although it is not optically active because it is bound to two identical groups (and thus has a plane of symmetry), it is potentially chiral because one of the R groups could be chemically replaced by another group (not R' or R'').
Diastereoisomers Isomeric molecules with two or more chiral centers that are not mirror images of one another.
Meso isomer Optical isomer whose lack of optical activity is due to internal compensation.
Absolute molecular asymmetry According to the RS convention, the groups around the chiral carbon atom are assigned an order of priority according to three basic rules.

1. Assign priority to functional groups in order of decreasing atomic number. For isotopes, the higher mass number has priority. For example,

$$O > N > C > H$$
$$^3H > {}^2H > {}^1H$$
$$C{-}OH > C{-}CH_2Cl > C{-}CH_2OH > C{-}CH_3 > C{-}H$$

   Unsaturated centers should be treated as though carbon atoms were attached.

2. Orient the center under examination so that either (a) the viewer is farthest away from the lowest-priority substituent in a tetrahedral projection or (b) the lowest-priority substituent occupies the bottom position in a Fischer projection.

Tetrahedral projection   Fischer projection
(H behind plane of paper)   (H at bottom)

3. With the center thus oriented, count around the remaining three substituents in order of decreasing priority. If these three substituents thereby describe a clockwise turn, the center is designated **R** (for *rectus*, which is Latin for "right[handed]"). If these three substituents thereby describe a counterclockwise turn, the center is designated **S** (for *sinister*, which is Latin for "left[handed]").

overcome by immobilizing the enzymes on a solid support or carrying out the reaction in a two-phase solvent system. Another limitation is that enzymes are subject to product and substrate inhibition, so the reaction may have to be performed at low substrate and/or product concentrations to achieve optimal reaction rates. Sophisticated systems have been developed that continuously feed substrate to a reaction mixture at an appropriately low concentration and continuously remove the product. Ingenious approaches have also been developed to regenerate cofactors and replenish cosubstrates when these are costly or must be constantly replaced.

Bacteria and fungi colonize virtually every ecological niche. Each of these organisms produces the enzymes it needs to survive in its particular environment and to utilize whatever nutrients are available there. Collectively, therefore, microbial enzymes catalyze an enormous variety of chemical reactions that transform both naturally occurring and human-made organic compounds. This immense source of novel enzymes remains largely untapped, but recombinant DNA technology has made it possible to acquire virtually any enzyme in the living world and produce it on a large scale in microbial cells. Consequently, the number of microbial and microbially produced enzymes used for the industrial generation of chemicals is growing rapidly. The examples described in this chapter highlight some of the most important applications of biocatalysis and illustrate how particular enzymes have been successfully exploited in preparative organic chemistry.

CLASSIFICATION OF ENZYMES

Enzymes are named according to the type of reaction that they catalyze.

- *Oxidoreductases* remove hydrogen atoms from (oxidize) a donor molecule and concomitantly add hydrogen atoms to (reduce) an acceptor molecule. Such enzymes catalyze reactions like $-CH_2OH \leftrightarrow >CH=O$, $>CHOH \leftrightarrow >C=O$, and $>CH-CH< \leftrightarrow >C=C<$.

- *Transferases* catalyze the transfer of groups (such as acyl or phosphoryl groups) from one molecule to another.

- *Hydrolases* cleave a bond by adding the atoms from a water molecule across it. These enzymes hydrolyze the anhydride bonds in biopolymers, such as proteins (amide bonds), polysaccharides (glycoside bonds), and lipids (ester bonds).

- *Lyases* remove a group of atoms (such as CO or HOH) from a substrate, leaving a double bond in its place; conversely, they can also add a group of atoms to a double bond. These enzymes can add (or remove) a group HX (where X is a substituent other than OH) across alkene, imine, and carbonyl bonds.

- *Isomerases* change the configuration of atoms within a molecule. Such enzymes can carry out epimerization, racemization, and other interconversions of stereoisomers.

- *Ligases (synthetases)* catalyze the joining of two molecules at the expense of the cleavage of a pyrophosphate bond in ATP or a similar triphosphate. Such enzymes catalyze the formation of C—O, C—N, C—S, and C—C bonds.

Enzymes are currently used as biocatalysts (catalysts of biological origin) in the manufacture of a wide variety of substances, including steroids, semisynthetic antibiotics, carbohydrate derivatives, amino acids, and alkaloids. Hydrolases are in particularly common use. For example, amidases and hydantoinases, which catalyze the cleavage of amide bonds, are used in the production of L-amino acids from racemic N-acetyl precursors. Esterases and lipases, which act on ester linkages, are widely used as catalysts both in esterification and transesterification and in syntheses of optically active acids and alcohols. Oxidoreductases, which rival the hydrolases in the extent of their commercial use, are important in the production of steroids, alkaloids, and terpenes. Enzyme-catalyzed reactions involving asymmetric C—C bond formation (or cleavage) based on aldol condensations are important in many biological biosynthetic (and degradative) pathways. The condensation of two molecules of acetaldehyde to form an *aldol* provides the generic name (aldol condensation) for this class of reactions.

$$2H_3C-\overset{\overset{\displaystyle O}{\|}}{C}H \longrightarrow H_3C-\overset{\overset{\displaystyle OH}{|}}{\underset{\underset{\displaystyle H}{|}}{C}}-\overset{\overset{\displaystyle H}{|}}{\underset{\underset{\displaystyle H}{|}}{C}}-\overset{\overset{\displaystyle O}{\|}}{C}H$$

Acetaldehyde   Aldol

Enzyme-catalyzed aldol condensations are valuable in the commercial synthesis of sugar derivatives.

MICROBIAL TRANSFORMATION OF STEROIDS AND STEROLS

The oxidation and reduction reactions that microorganisms perform on steroid and sterol substrates provide particularly impressive examples of regioselective and stereospecific biotransformations and also demonstrate the ability of enzymes to promote reactions at unactivated centers in hydrocarbons.

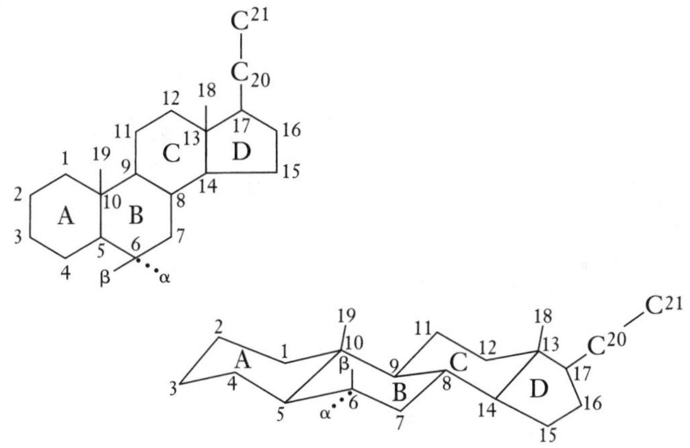

Figure 14.1
Structure, stereochemistry, and numbering of the nucleus of adrenocortico-steroids. The four rings, A through D, do not lie in a flat plane as conventionally represented (above at the left) but rather have the configuration shown at the right. The biological activity of steroids depends on the orientation of the groups attached to the ring system. As shown at C6, groups that project above the plane of the steroid are designated β. Their connection to the ring system is shown by solid-line bonds; those that project below the plane are designated α, and their connection to the ring is shown by a dotted-line bond.

Virtually any position in the carbon skeleton of a steroid nucleus (Figure 14.1) can be hydroxylated stereospecifically by enzymes present in some microorganism. Steroid hydroxylases are named according to the position they attack on the rings or the side chain of the steroid nucleus. There are three primary carbon atoms (C18, C19, and C21). An enzyme that catalyzes hydroxylation at C21, for example, is designated 21-hydroxylase. There are 18 secondary carbon atoms. At the secondary carbon atoms within the ring system, there are two alternative ways, designated α and β, to attach the —OH group. The α position lies below the plane of the steroid ring, and the β position lies above the plane (see Figure 14.1). Every one of the 18 secondary carbon atoms can be hydroxylated, in either the α or β configuration, each by a different known microbial hydroxylase (Table 14.1). In addition to hydroxylations (Table 14.2), certain microbial enzymes can aromatize ring A, reduce double bonds in the rings, and reduce specific ketone substituents. The microbial transformations of steroids and sterols have dramatically lowered the cost of manufacturing steroid hormones (Figure 14.2).

In the early 1930s, Edward C. Kendall of the Mayo Foundation and Tadeus Reichstein of the University of Basel isolated cortisone, a steroid secreted by the adrenal gland. In 1949, Philip S. Hench of the

Selective Microbial Hydroxylation of Steroids

T
A
B
L
E

14.1

Position of Hydroxylation	Stereochemistry of Incoming Hydroxyl Group	Position of Hydroxylation	Stereochemistry of Incoming Hydroxyl Group
1	α	10	β
1	β	11	α
2	α	12	β
2	β	13	α
3	α	14	α
3	β	15	α
4	α	15	β
4	β	16	α
5	α	16	β
6	α	17	α
6	β	17	β
7	α		
7	β		
9	α		

SOURCE: Davies, H.G., et al. (1989), *Biotransformations in Preparative Organic Chemistry: The Use of Isolated Enzymes and Whole Cell Systems in Synthesis*, pp. 175–176 (Academic Press).

Examples of Steroid Hydroxylations by Different Fungi

T
A
B
L
E

14.2

Hydroxylation Position	Substrate	Product	Microorganism
1α	Androst-4-ene-3,17-dione	1α-Hydroxyandrost-4-ene-3,17-dione	*Penicillium* sp.
1β	Androst-4-ene-3,17-dione	1β-Hydroxyandrost-4-ene-3,17-dione	*Xylaria* sp.
3α	Androstane-7,17-dione	3α-Hydroxyandrostane-7,17-dione	*Diaporthe celastrinia*
3β	17β-Hydroxyandrostan-11-one	3β,17β-Dihydroxy-androstan-11-one	*Wojnowicia graminis*
11α	Progesterone	11α-Hydroxyprogesterone	*Rhizopus* sp.
11β	11-Deoxycortisone	Hydrocortisone	*Curvularia lunata*
12β	17β-Hydroxy-estr-4-ene-3-one	12β,17β-Dihydroxy-estr-4-ene-3-one	*Collectotrichum derridis*

SOURCE: Neidleman, S.L. (1991), Industrial chemicals: Fermentation and immobilized cells, in Moses, V., and Cape, R.E. (eds.), *Biotechnology: The Science and the Business*, pp. 306–307 (Harwood Academic).

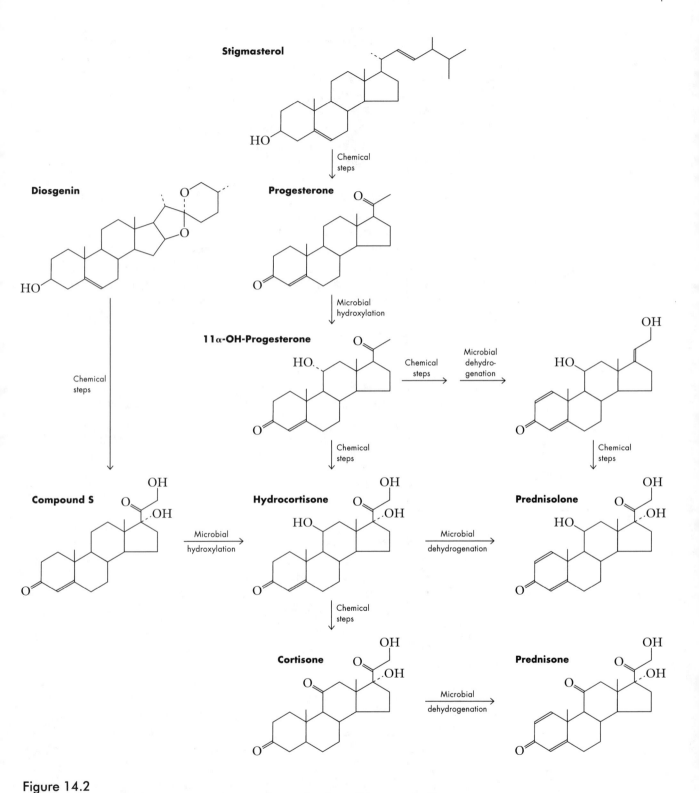

Figure 14.2

Chemical and microbial transformations in the production of therapeutically useful steroids. Based on Primrose, S.B., ed., (1987), *Modern Biotechnology*, p. 76, Fig. 6.11 (Blackwell Scientific Publications).

Mayo Foundation found that administration of cortisone led to remission in patients with acute rheumatoid arthritis. Discovery of the anti-inflammatory effects of cortisone had a profound impact on the medical world and earned Kendall, Reichstein, and Hench the Nobel Prize in 1950.

The demand for large amounts of cortisone spurred the development of a chemical synthesis for the hormone. It was an elaborate synthesis, requiring 31 steps, and its final yield was extremely low. A starting batch of 615 kg of deoxycholic acid (purified from beef bile) was converted to 1 kg of cortisone acetate. The market price for the synthetic hormone was $200 per gram.

A major complication in the synthetic route from deoxycholic acid to cortisone is the need to shift the $C12\beta$ hydroxyl in deoxycholic acid to C11. In the chemical synthesis this required nine steps. In 1952, however, researchers at Upjohn Company discovered that an aerobically grown bread mold, *Rhizopus arrhizus*, could hydroxylate progesterone (another steroid and an early intermediate in cortisone synthesis) at $C11\alpha$, and workers at the Squibb Institute found that another common mold, *Aspergillus niger*, carried out the same reaction. Exploiting microbial hydroxylation at C11 shortened industrial cortisone synthesis from 31 to 11 steps. Moreover, the microbial hydroxylation of progesterone had economic benefits beyond those resulting from abbreviation of the chemical synthesis. This biotransformation takes place at 37°C in aqueous solution at atmospheric pressure—conditions that are much less expensive than the high temperature and pressure and nonaqueous solvents required for the equivalent steps in the chemical synthesis. The commercial price of cortisone dropped to $6 per gram shortly after these discoveries.

Further reductions in the cost of cortisone came from using inexpensive sterols, instead of deoxycholate, as the starting material. Two such sterols, stigmasterol and sitosterol, are generated in large amounts as by-products in the production of soybean oil; a third, diosgenin, comes from the roots of the Mexican barbasco plant. To make steroids from these plant sterols, the side chain beyond C21 must be removed. Although chemical degradation can accomplish this step, it is achieved much more economically by mycobacteria, aerobic Gram-positive eubacteria that can utilize sterols as a carbon and energy source. To prevent mycobacteria from breaking the sterols down totally, mutant strains have been developed that are unable to degrade the sterols beyond the desired stage. The introduction of these process changes brought the price of cortisone in the United States down to 46 cents per gram by 1980, a 400-fold reduction from the original price without even adjusting for inflation!

In addition to being used to treat rheumatoid arthritis, steroids are prescribed for allergies and other inflammatory diseases (especially of the skin), for contraception, and for hormonal insufficiencies. Various steroids useful for these purposes are produced with the aid of microbes capable of modifying the steroid nucleus in specific ways. Bulk sales worldwide of the four major steroids—cortisone, aldosterone, prednisone, and prednisolone—amount to over 700,000 kg per year.

Biocatalysis by Amidases, Peptidases, and Lipases

In searching for enzymes that can be used in organic synthesis, scientists are frequently hampered by the problem that the substrate they wish to alter is not a compound found in nature. Fortunately, they often succeed in finding an enzyme that catalyzes the desired reaction by using the "unnatural" substrate. Some classes of enzymes have, as a group, a more relaxed substrate specificity than others, and with these enzymes there is a higher probability of screening successfully for such a biocatalyst. Many lipases, peptidases, and amidases, for example, will catalyze reactions with substrates very different from their natural ones. This general observation is illustrated by specific examples of major importance in industry: the use of amidases and peptidases in the production of semisynthetic penicillins and the use of lipases in the production of fats of defined composition.

Semisynthetic Penicillins

Penicillins and cephalosporins are antibacterial agents of major importance. Worldwide sales of these compounds (formulated and ready to use) amounted to $13 billion in 1988. As we discussed in Chapter 13, the continued emergence of bacterial strains that are resistant to commonly used penicillin and cephalosporin derivatives necessitates ongoing efforts to synthesize new antibiotic derivatives. Such "semisynthetic" antibiotics can be produced by enzymatically cleaving the amide linkage between the side chain of penicillin G and the 6-aminopenicillanic acid portion (Figure 14.3) and then attaching a new side chain by acylating 6-aminopenicillanic acid on the amine group released by the cleavage. The first semisynthetic antibiotic commercially produced in this manner was ampicillin, which was introduced in 1961.

The production of ampicillin illustrates the importance of microbial amidases in the commercial development of semisynthetic penicillins (see Figure 14.3). The breakthrough that made this development possible was the finding that under certain conditions, the fungus *Penicillium chrysogenum* excretes 6-aminopenicillanic acid, which can then be converted into new penicillin derivatives. Although a chemical synthetic

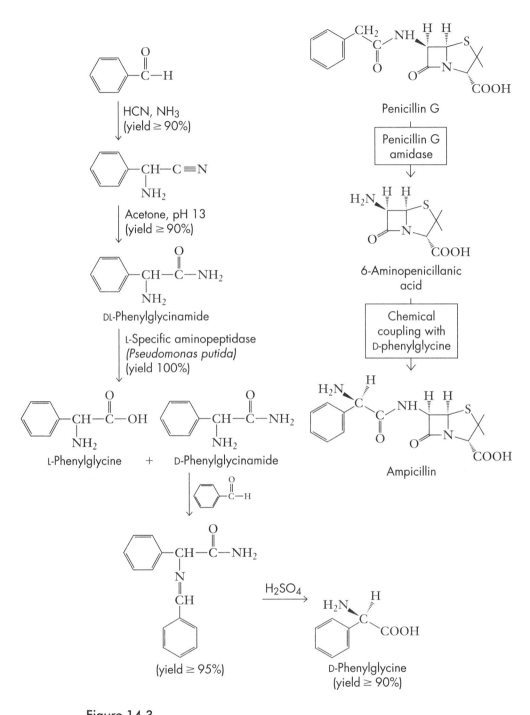

Figure 14.3
A combined biocatalytic and chemical route for producing the semisynthetic penicillin ampicillin.

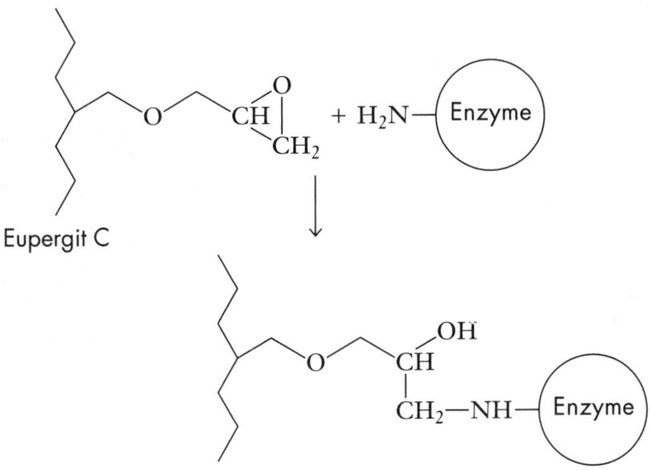

Figure 14.4
Covalent immobilization of penicillin G amidase on Eupergit C.

route to 6-aminopenicillanic acid had already been discovered, the enzyme-based conversions were more economical than both the chemical process and the fermentative production of 6-aminopenicillanic acid.

The enzyme-dependent production of ampicillin consists of three steps. In step one, penicillin G amidase, which is used to convert penicillin to 6-aminopenicillanic acid, is produced in high yield by appropriately engineered fungal strains. The enzyme is then attached covalently to an epoxy acrylic bead polymer, Eupergit C (Figure 14.4). The enzyme immobilized in this manner is a highly active and very stable catalyst that after 500 cycles still retains over 90% of its enzymatic activity.

In step two, the "side chains" of the semisynthetic penicillins are produced (examples are D-phenylglycine and D-[p-hydroxyphenyl]glycine). This step also depends on enzyme-catalyzed reactions: DL-Amino acid amides can be made inexpensively by conventional chemical processes, but the most selective and inexpensive way to resolve such racemic mixtures is with the help of enzymes. First, the microbial aminopeptidase selectively catalyzes the cleavage of the amide bond in L-phenylglycinamide to yield L-phenylglycine and D-phenylglycinamide. D-Phenylglycinamide is then quantitatively separated from the reaction mixture by the formation of an insoluble Schiff's base with benzaldehyde. After acid hydrolysis of the latter compound, D-phenylglycine is obtained without racemization (see Figure 14.3).

Step three is the coupling of the side chain to the penicillin nucleus (6-aminopenicillanic acid). On an industrial scale, this has traditionally been performed chemically; however, an enzyme-catalyzed route is available (Figure 14.5). At acidic pH (pH 4.5–5.5), penicillin G acylases from *E. coli, Bacillus circulans*, and *B. megaterium* have been successfully used to synthesize ampicillin from D-phenylglycine methyl ester and

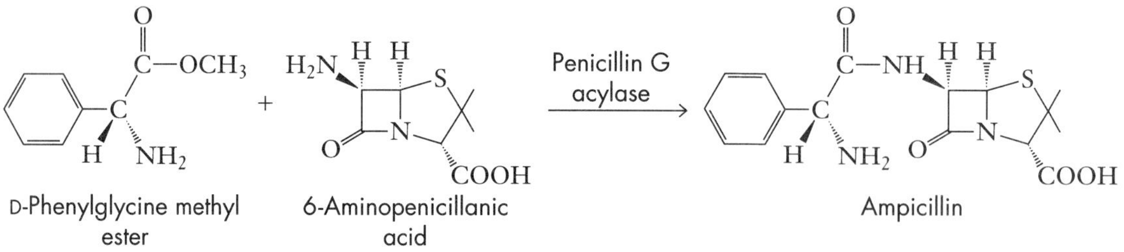

Figure 14.5
Enzyme-catalyzed formation of ampicillin from 6-aminopenicillanic acid and D-phenylglycine methyl ester.

6-aminopenicillanic acid. The widely used semisynthetic penicillin amoxycillin is produced in the same manner by using D-(p-hydroxyphenyl)glycine methyl ester.

The production of amoxycillin, unlike that of ampicillin, incorporates an enzymatic process that totally converts a racemic substrate to the desired D-amino acid, in this case D-(p-hydroxyphenyl)glycine (Figure 14.6). Many organisms possess dihydropyrimidinases that cleave a

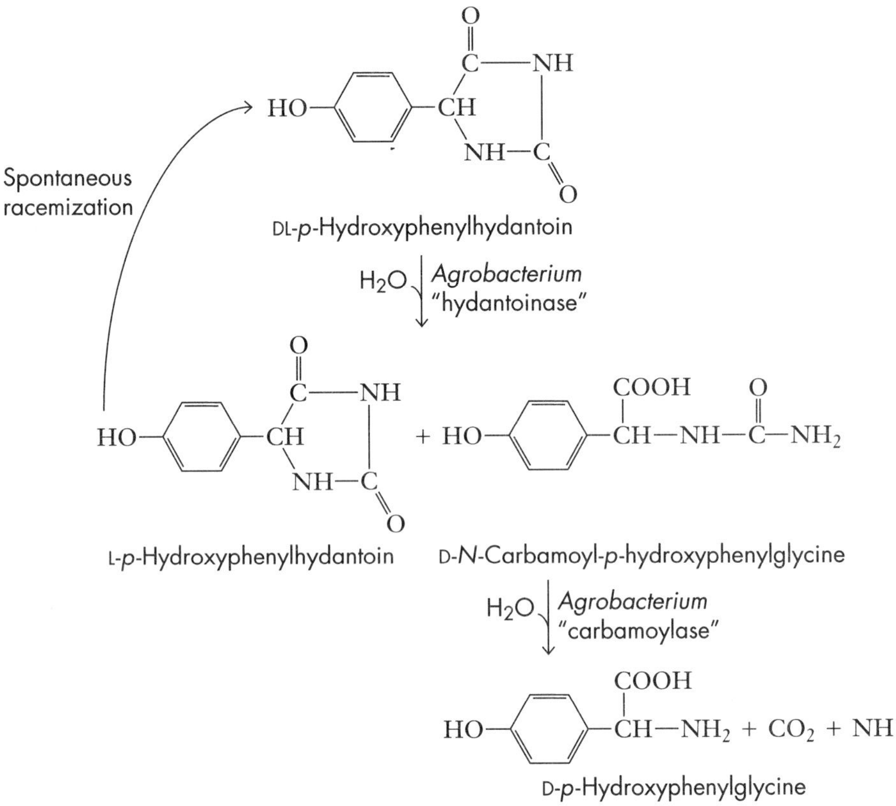

Figure 14.6
Preparation of D-p-hydroxyphenylglycine from DL-p-hydroxyphenylhydantoin with hydantoinase and carbamoylase from *Agrobacterium* sp.

CO—NH bond in pyrimidines. These enzymes, also called "hydantoinases," stereospecifically hydrolyze D-*p*-hydroxyphenylhydantoin to form a D-carbamoyl derivative. Certain strains (such as *Agrobacterium radiobacter* NRRL-B11291) possess both a hydantoinase activity and a D-carbamoylase activity. Because alkaline conditions cause L-*p*-hydroxyphenylhydantoin to racemize spontaneously to the D isomer, such strains can convert DL-*p*-hydroxyphenylhydantoin completely to D-(*p*-hydroxyphenyl)glycine in a single reactor (see Figure 14.6). It has been reported that a *Flavobacterium hydantoinophilum* strain can carry out the foregoing reactions and also couple the resulting D-(*p*-hydroxyphenyl)glycine to 6-aminopenicillanic acid to form amoxycillin.

Properties and Applications of Lipases

Lipases (triacylglycerol acylhydrolases) are present in all organisms, where they catalyze the synthesis or hydrolysis of fats. Depending on the source, these enzymes vary widely in pH optima and thermostability, in positional specificity, and in selectivity with regard to the structure or length of the fatty acid chains they either hydrolyze off or utilize for esterification (Table 14.3). Some lipases show high regiospecificity for the 1 and 3 positions of a triglyceride (Figure 14.7), whereas others show no positional preference. Other lipases show an intermediate level of specificity.

Lipases catalyze the following types of reactions:

1. Ester hydrolysis and synthesis

    Ester hydrolysis
    $$R\text{-}COOR' + H_2O \rightarrow R\text{-}COOH + R'\text{-}OH$$

    Ester synthesis
    $$R\text{-}COOH + R'\text{-}OH \rightarrow R\text{-}COOR' + H_2O$$

2. Transesterification

    Transesterification by acidolysis
    $$R_a\text{-}COOR' + R_b\text{-}COOH \rightarrow R_b\text{-}COOR' + R_a\text{-}COOH$$

    Transesterification by alcoholysis
    $$R\text{-}COOR'_a + R'_b\text{-}OH \rightarrow R\text{-}COOR'_b + R'_a\text{-}OH$$

    Ester exchange (interesterification)
    $$R_a\text{-}COOR'_a + R_b\text{-}COOR'_b \rightarrow R_a\text{-}COOR'_b + R_b\text{-}COOR'_a$$

3. Aminolysis

    $$RCOOR_a + R_b\text{-}NH_2 \rightarrow R\text{-}CONH\text{-}R_b + R_a\text{-}OH$$

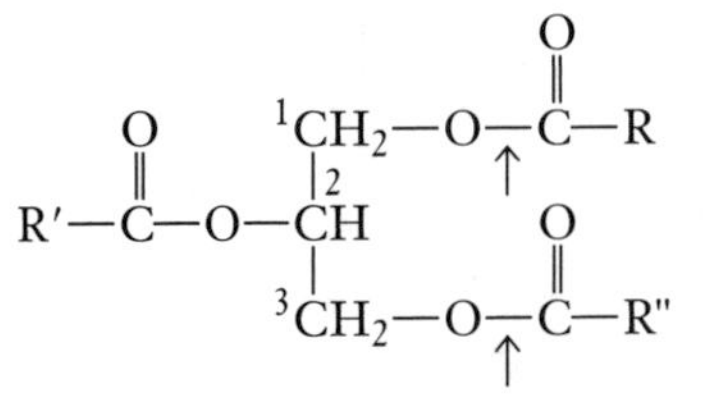

1, 2, 3-Triacyl-sn-glycerol
(triglyceride)

Figure 14.7
Structure and conventional numbering of a triacylglycerol. Arrows indicate bonds whose cleavage is catalyzed by a 1,3-regiospecific lipase.

TABLE 14.3

Properties of Some Microbial Lipases Produced Industrially on a Large Scale

			Thermostability				
Microorganism	Optimum pH	Temperature (°C)	Time (min)	Activity Remaining (%)	Positional Specificity	Molecular Weight
Candida cylindracea	7.0	50	10	40	nonspecific	55,000
Aspergillus niger	5.6	60	15	50	1,3	38,000
Rhizopus japonicus	5.0	55	30	50	1,3	30,000
Pseudomonas fluorescens	7.0	60	30	60	1,3	31,000

SOURCE: Yamane, T. (1987), Enzyme technology for the lipids industry: An engineering overview, *J. Am. Oil. Chem. Assoc.* 64:1657–1662.

The food industry exploits lipase-catalyzed reactions to manufacture fats of defined composition and to improve the flavor of food. Lipases are also used in many organic syntheses that require the resolution of racemic mixtures.

Interesterification of Fats and Oils

The reaction enthalpy of triglyceride hydrolysis is exceptionally small, and the net free-energy change in transesterification reactions is zero. Consequently, both hydrolysis and resynthesis of triacylglycerols occur when lipases are incubated with fats and oils. The resulting interchange of fatty acyl groups between triacylglycerol molecules gives rise to interesterified products.

The regiospecificity of lipases makes it possible to produce triacylglycerol mixtures that cannot be obtained by conventional chemical methods (Figure 14.8). Numerous microorganisms, including the bacteria *Pseudomonas fluorescens* and *Chromobacterium vinosum* and the fungi *Aspergillus niger* and *Humicola lanuginosa*, secrete 1,3-regiospecific lipases into their growth medium to catalyze the degradation of lipids. These enzymes can thus be produced on a large scale by fermentation and have come to be widely used.

An industrial process for the production of cocoa butter substitutes of high commercial value uses a fungal lipase to catalyze the transesterification of readily available inexpensive oils (for example, 1,3-dipalmitoyl-2-oleyl glycerol from palm oil) with stearic acid to produce 1,3-distearoyl-2-oleyl glycerol, the major component of cocoa butter (Table 14.4). The industrial transesterification process has two features of general interest to the field of biocatalysis. *First, the reaction takes place in a*

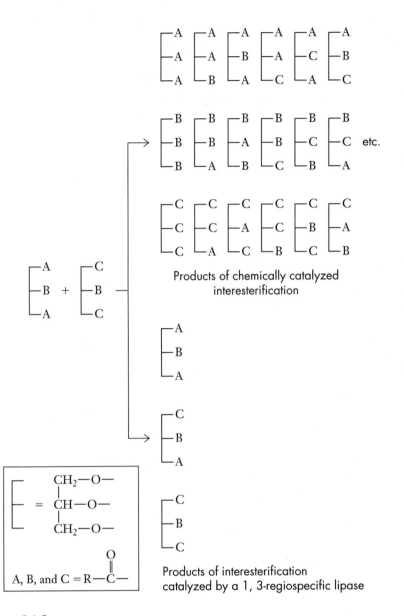

Figure 14.8
Comparison of the products of chemical and enzymatic catalysis of interesterification, derived from a mixture of two triacylglycerols. Because the 1 and 3 positions of triacylglycerols are not equivalent, compound ABC is different from CBA. This complication, however, was ignored in this figure to simplify the scheme.

two-phase system: The reactants and products are present in a water-immiscible liquid organic phase, and the hydrated enzyme protein is present in the small volume of aqueous phase. Second, in this process a hydrolytic enzyme is used to catalyze the reverse of its natural reaction.

Composition of a Cocoa Butter Substitute Obtained from the Interesterification of Palm Oil Fraction (1 part) and Stearic Acid (0.5 parts), Catalyzed by *Mucor miehei* 1,3-Regiospecific Lipase

Triacylglycerol[a]	Amount in Cocoa Butter (%)	Amount in Interesterification Product (%)
SSS	1.0	3.0
POP	16.3	16.2
POS	40.8	38.5
SOS	27.4	28.5
SLnS	7.5	8.0
SOO	6.0	4.0
Others	1.0	1.5

[a] The abbreviations are S, stearoyl; P, palmitoyl; O, oleyl: Ln, linoleyl. POP, etc., triacyl glycerol with the specified acyl groups in the 1, 2, and 3 positions.

A Biokinetic Transformation: Enantioselective Esterification by a Lipase

In the examples presented above, biocatalysis exploited the *absolute* stereochemical preference of the enzymes for the different isomers of the substrate. *The enzyme-catalyzed transesterification reaction described below is an example of a* biokinetic *transformation — one that takes advantage of the kinetic stereochemical preference of the enzyme for the different isomers of the substrate.* We discuss the key steps involved in finding a suitable enzyme for a particular transformation and explain how to define reaction conditions that achieve a compromise between the need to maintain an aqueous environment for the enzyme and the need to dissolve the substrate in an organic solvent.

4-Methyl-1-heptyn-4-en-3-ol is a building block of S-2852, a synthetic pyrethroid insecticide used against household pests (Figure 14.9). Because of its volatility at room temperature, this insecticide is used primarily as a fumigant to eliminate pests from enclosed storage areas for fabrics. The insecticidal activity of S-2852 requires that its 4-methyl-1-heptyn-4-en-3-ol precursor be the S enantiomer. The optical resolution of 4-methyl-1-heptyn-4-en-3-ol exploits enantioselective esterification by a lipase.

Choice of Lipase. Several commercially available lipases were tested to determine the rate and enantioselectivity of their transesterification of (R,S)-4-methyl-1-heptyn-4-en-3-ol, with vinyl acetate as acyl donor and *n*-hexane as solvent. The results presented in Table 14.5 show that *Pseudomonas* lipase had the highest reaction rate and enantioselectivity (the highest E value) of the enzymes tested.

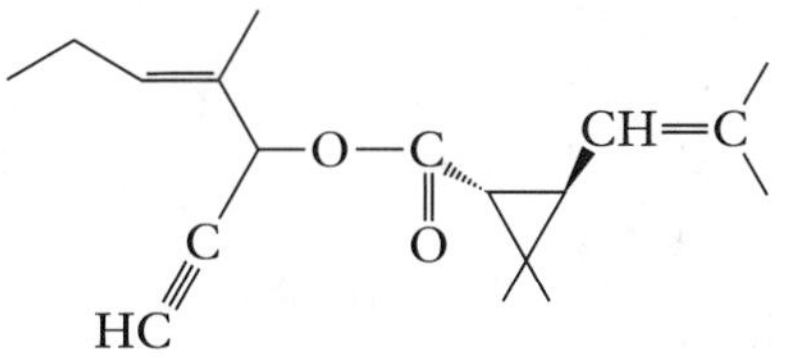

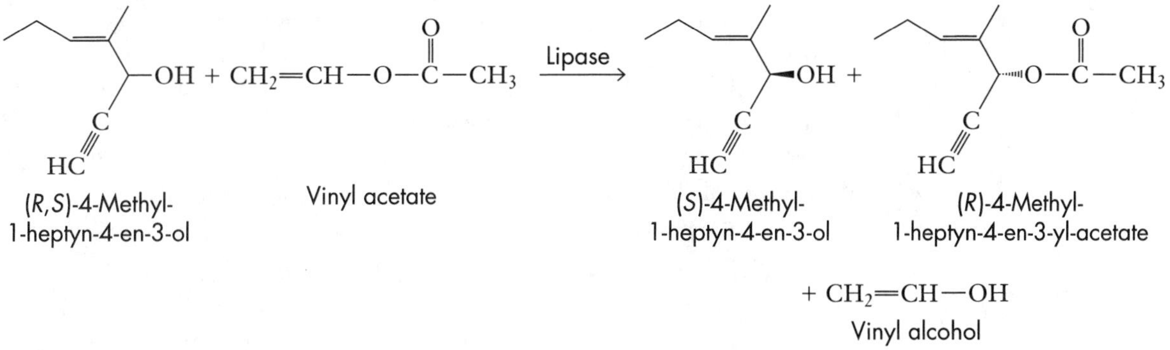

Figure 14.9
Chemical structure of the pyrethroid insecticticide S-2852. The lipase-catalyzed enantioselective esterification of (R,S)-4-methyl-1-heptyn-4-en-3-ol
leaves the S-2852 precursor (S)-4-methyl-1-heptyn-4-en-3-ol unmodified.
From Mitsuda, S., and Nabeshima, S. (1991), Enzymatic optical resolution of
a synthetic pyrethroid alcohol: Enantioselective transesterification by lipase
in organic solvent, *Rec. Trav. Chim. Pays Bas* 110:151–154.

Enantioselective Esterification of (R,S)-4-Methyl-1-heptyn-4-en-3-ol by Various Microbial Lipases

TABLE 14.5

Source of Lipase[a]	Time (hours)	Conversion (%)	E[b]
Alcaligenes sp.	52	51.4	5
Arthrobacter sp.	52	23.9	<1
Candida cylindracea	52	20.9	<1
Pseudomonas sp.	10	48.3	16

[a] The reaction was performed in *n*-hexane at 30°C with vinyl acetate as acyl donor.

[b] The enantioselectivity is expressed by the parameter E, where $E = \ln[(1 - c) \cdot [1 - ee(S)]]/\ln[(1 - c) \cdot [1 + ee(S)]]$; c is the fraction of (R,S)-4-methyl-1-heptyn-4-en-3-ol esterified, and $ee(S)$ is $([S] - [R])/([S] + [R])$, where $[S]$ and $[R]$ correspond to the concentrations of S and R isomers remaining after enzyme reaction.

Choice of Acyl Donor. Several acyl donors were also tested to determine their influence on the rate and enantioselectivity of the *Pseudomonas* lipase-catalyzed transesterification (Table 14.6). Two enol esters, vinyl acetate and isopropenyl acetate, gave high enantioselectivity. The reaction rate was significantly higher with vinyl acetate, so this compound was chosen as the acyl donor.

Choice of Solvent. Eight organic solvents were tested as reaction media for the *Pseudomonas* lipase-catalyzed transesterification of (R,S)-4-methyl-1-heptyn-4-en-3-ol with vinyl acetate. There was a reasonably linear relationship between the logarithm of the initial rate of the enzymatic reaction and the hydrophobicity (log P; Figure 14.10A) of the solvent. The highest initial reaction rate was obtained in the most hydrophobic of the solvents, *n*-hexane. The solvent had little influence on the enantioselectivity of the reaction.

Effect of Water Content of the Solvent. The amount of water in the organic solvent had a strong influence on the success of the biocatalysis. The enzyme protein requires water for hydration, but excess water allows it to catalyze hydrolysis of the acyl donor instead of transesterification. The data plotted in Figure 14.10B provide a dramatic illustration: The highest activity occurred at the very low water content of 0.017 weight/volume %; above 0.64 water weight/volume %, transesterification ceased and the dominant reaction was the enzymatic hydrolysis of vinyl acetate.

Effect of temperature. The Arrhenius plot relating the initial transesterification rate to the absolute temperature was linear (Figure 14.11), indicating that the enzyme-catalyzed reaction behaved similarly to an ideal chemical reaction. The remarkable finding was that the reaction rate increased with temperature up to the boiling point of *n*-hexane

TABLE 14.6 Enantioselective Esterification of (R,S)-4-Methyl-1-heptyn-4-en-3-ol by *Pseudomonas* Lipase with Various Acyl Donors

Acyl donor[a]	Time (hours)	Conversion (%)	E[b]
Vinyl acetate	10	48.3	16
Isopropenyl acetate	50	45.4	18
Ethyl acetate	29	2.1	–
Acetic anhydride	29	68.5	<5

[a] Reaction was performed in *n*-hexane at 30°C.

[b] See footnote *b* to Table 14.5.

SOURCE: Mitsuda, S., and Nabeshima, S. (1991), Enzymatic optical resolution of synthetic pyrethroid alcohol: Enantioselective transesterification by lipase in organic solvent, *Rec. Trav. Chim. Pays Bas* 110:151–154.

Figure 14.10
Relative transesterification activity of *Pseudomonas* lipase in various organic solvents at 30°C, with (*R,S*)-4-methyl-1-heptyn-4-en-3-ol as substrate and vinyl acetate as acyl donor. **A.** The logarithm of the partition coefficient (log *P*) of an organic solvent in an octanol-water two-phase system is used as a measure of its hydrophobicity. Solvents that have log *P* values between about -2.5 and 0 are water-miscible; those that have positive log *P* values are water-immiscible. Key to organic solvents (water content, weight/volume): **1**, *n*-hexane (0.009%); **2**, cyclohexane (0.005%); **3**, toluene (0.02%); **4**, benzene (0.007%); **5**, pyridine (0.04%); **6**, tetrahydrofuran (0.04%); **7**, acetonitrile (0.1%); **8**, dioxane (0.1%). **B.** The effect of water content on the relative transesterification activity of lyophilized (freeze-dried) *Pseudomonas* lipase in *n*-hexane. At the lowest water content of the reaction mixture (0.017 weight/volume %), the water content of the lipase was 0.8 weight/weight % (the lipase was separated from the reaction mixture by filtration), and that of the liquid phase was 0.009 weight/volume %. Other water contents of reaction mixtures were obtained by increasing the moisture of the lipase or by adding water to the reaction mixture. From Mitsuda, S., and Nabeshima, S. (1991), *Rec. Trav. Chim. Pays Bas* 110:151–154.

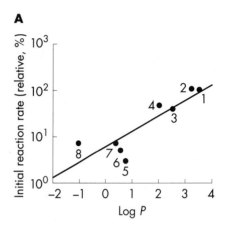

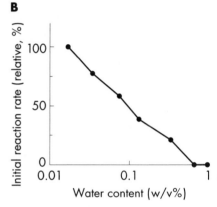

(69°C). Gradual inactivation of the enzyme was observed above 40°C. Approximately 10–15% of the activity was lost at 60°C in 50 h.

Yield of (S)-4-methyl-1-heptyn-4-en-3-ol Under Optimized Conditions. When the transesterification reaction was performed in *n*-hexane at 30°C for 21 h with a molar excess of the acyl donor (vinyl acetate) to (*R,S*)-4-methyl-1-heptyn-4-en-3-ol, the *S* enantiomer was obtained with 99% enantiomeric excess (see Table 14.5 for a definition) at a conversion of 68% — that is, *at the point when all of the (R)-4-methyl-1-heptyn-4-en-3-ol had been converted to the acetate ester, but only 18% of the (S) enantiomer had been esterified.*

To reemphasize, the success of this *biokinetic* transformation depends on the much more rapid rate at which the enzyme catalyzes the formation of the ester of the *R* isomer than that of the *S* isomer of 4-methyl-1-heptyn-4-en-3-ol.

EXAMPLES OF THE SYNTHESIS OF OPTICALLY PURE DRUGS

Biocatalysis represents an important approach to the synthesis of *chiral synthons* (optically active building blocks) for a wide variety of pharmaceuticals and other high-quality chemicals. As we saw with the pyrethroid insecticide S-2852, biological activity and stereochemistry go hand in hand. In addition to answering the need for stereoselectivity,

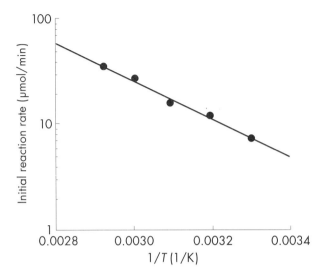

Figure 14.11

Effect of temperature on the *Pseudomonas* lipase-catalyzed transesterification of (R,S)-4-methyl-1-heptyn-4-en-3-ol with vinyl acetate in *n*-hexane containing 0.017% (weight/volume) water. The results are presented in the form of an Arrhenius plot (log[initial reaction rate] *versus* 1/T, where T is the absolute temperature in kelvins). The linearity of this plot from 30°C (303 K) to 69°C (342 K) indicates that the enzyme is stable over this temperature range. From Mitsuda, S., and Nabeshima, S. (1991), *Rec. Trav. Chim. Pays Bas* 110:151–154.

BOX 14.2 **Importance of Chirality in the Action of Synthetic Drugs**

"Perhaps, looking-glass milk isn't good to drink?"
—Lewis Carroll

Drugs, herbicides, and pesticides frequently act by interacting with receptors, enzymes, carrier macromolecules, and the like. All such interactions are highly stereospecific. About 25% of the drugs now in use are chiral. The tacit assumption that only one of the stereoisomers is active and the other inactive is dangerous. One stereoisomer of a drug may interact tightly with a particular receptor, whereas the other stereoisomer may have a different target altogether, as indicated in the following examples.

- The dextrorotatory isomer of the antitubercular drug ethambutol, 2,2'-(ethylenediimino)-di-1-butanol dihydrochloride, has potent antitubercular activity, whereas the levorotatory isomer causes degeneration of the optic nerve, leading to blindness.
- The dextrorotatory isomer of propoxyphene, α-(+)-4-(dimethylamino)-3-methyl-1,2-diphenyl-2-butanol propionate, is an analgesic, whereas the levorotatory isomer is a cough suppressant.

biocatalysis also answers the need to minimize the environmental impact of chemical processes, many of which employ a variety of chemical intermediates and organic solvents. There are other important benefits as well. For example, *optically pure drugs frequently cause fewer side effects than racemates* (Box 14.2).

Selective Oxidation of (S)-1,2-O-Isopropylideneglycerol

Atenolol (Figure 14.12) is a drug used for the treatment of hypertension. It acts by selectively blocking the β_1-adrenergic receptors in the heart,

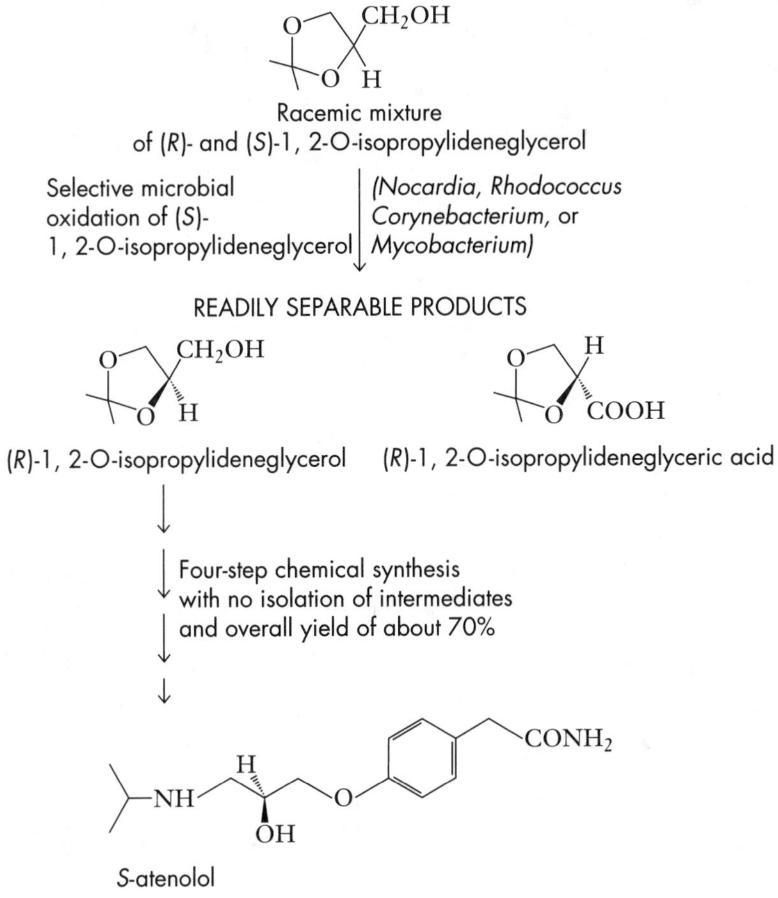

Racemic mixture
of (R)- and (S)-1, 2-O-isopropylideneglycerol

Selective microbial oxidation of (S)-1, 2-O-isopropylideneglycerol

(Nocardia, Rhodococcus Corynebacterium, or Mycobacterium)

READILY SEPARABLE PRODUCTS

(R)-1, 2-O-isopropylideneglycerol   (R)-1, 2-O-isopropylideneglyceric acid

Four-step chemical synthesis with no isolation of intermediates and overall yield of about 70%

S-atenolol

Figure 14.12
Resolution of racemic 1,2-O-isopropylideneglycerol, a precursor of the β-blocker (S)-Atenolol, by selective oxidation of the (S) isomer by cultures of various bacteria. Phillips, G.T. (1990), Biotransformations and their role in industrial synthesis, in *Proceedings of the Chiral 90 Symposium*, pp. 17–22 (Manchester, England).

with a resulting reduction in heart rate, cardiac output, and blood pressure. *The use of bacteria to resolve the racemic mixture of a key precursor has enabled manufacturers to synthesize Atenolol in ton amounts.*

The physiological activity of this compound is primarily associated with the *S* enantiomer, (*S*)-Atenolol, and its synthesis, via a straightforward chemical route requires the optically active precursor (*R*)-1,2-*O*-isopropylideneglycerol. Racemic (*RS*)-1,2-*O*-isopropylideneglycerol is readily synthesized chemically. Cultures of a variety of bacteria are able to oxidize all the *S* isomer of this compound to (*R*)-1,2-*O*-isopropylideneglyceric acid while leaving the *R* isomer unchanged. As shown in Figure 14.12, this transformation produces an equal mixture of (*R*)-1,2-*O*-isopropylideneglycerol and (*R*)-1,2-*O*-isopropylideneglyceric acid. The acid and alcohol are easily separated, after which the (*R*)-1,2-*O*-isopropylideneglycerol is converted to (*S*)-Atenolol by a sequence of chemical reactions. No intermediates need be isolated and the overall yield is high. The low cost of the racemic substrate, the good selectivity of the oxidation, the high rate of oxidation of the *S* isomer by the culture, and the lack of product inhibition (allowing the formation of high concentrations of product) all contribute to the technical success of this process.

Isomerization of (*R*)-Naproxen to (*S*)-Naproxen

(*S*)-Naproxen is a nonsteroidal anti-inflammatory drug with analgesic and antipyretic properties. Straightforward chemical synthesis yields a racemic mixture of (*R*)- and (*S*)-Naproxen (Figure 14.13).

Certain species of fungi are able to isomerize (*R*)-Naproxen to (*S*)-Naproxen. The experimental results in Table 14.7 show that the fungi *Cordyceps militaris*, *Beauvaria bassiana*, and *Exophiala wilhansii* all convert the *R* isomer entirely to the *S* form. However, the recovery of product ranges from 6% to 92%. For each fungus, the recovery of product is dependent on the relative rates of isomerization and degradation of the substrate and/or the product. In other words, these fungi not only catalyze the conversion of the *R* to the *S* but also convert one or the other of these isomers to other products. A particularly interesting aspect of this biotransformation is the absence of the reverse reaction, the conversion of (*S*)-Naproxen to the (*R*) isomer. Apparently, the equilibrium of the isomerization reaction lies far to the side of the (*S*) isomer.

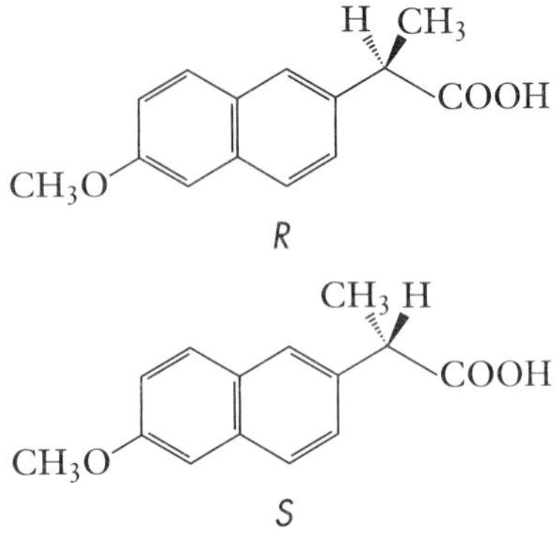

Figure 14.13
Naproxen (6-methoxy-α-methyl-2-naphthaleneacetic acid).

MICROBIAL SYNTHESIS OF THE COMMODITY CHEMICAL ACRYLAMIDE

The examples we have examined up to this point illustrate the major, long-standing contributions of biotransformations to the commercial

Isomerization of Naproxen by Different Fungi

Microorganism	Enantiomer Added	Enantiomeric Ratio in Product, S:R	Recovery (%)
Cordyceps militaris	R	100:0	39
	S	100:0	
Cladosporium resinae	R	74:26	15
	S	100:0	
Beauvaria bassiana	R	100:0	6
	S	100:0	
Exophiala wilhansii	RS	100:0	92[a]
	R	100:0	80

TABLE 14.7

[a] Yield: 450 mg per liter.

SOURCE: Phillips, G.T. (1990), Biotransformations and their role in industrial synthesis. *Proceedings of the Chiral 90 Symposium*, pp. 17–22, September (Manchester, England).

production of high-quality ("fine") chemicals, with particular focus on chiral syntheses. Not all biocatalytic transformations involve stereospecificity, however. One important commercial microbial process, developed in the 1980s for the conversion of acrylonitrile to acrylamide on a multi-ton scale, owes its value instead to its low cost and low environmental impact, suggesting that the mass production of many other commodity chemicals can be achieved economically by exploiting biocatalysis and that biocatalytic processes may do less harm to the environment than conventional large-scale chemical syntheses.

Acrylamide is a building block of various polymers used in petroleum recovery or as flocculants or additives in various products. About 200,000 tons of this chemical are produced annually worldwide. Unfortunately, the conventional chemical synthesis of acrylamide (Figure 14.14), in which copper salts catalyze the hydration of acrylonitrile, has several drawbacks. The procedure for preparation of the catalyst is complex, and used catalyst is difficult to regenerate. Purification of the acrylamide is necessary to remove unreacted acrylonitrile. Undesirable by-products, such as hydrocyanic acid and large amounts of salts, are formed. And the process needs to be carefully controlled because of the ease with which acrylamide polymerizes. The new biocatalytic process, in which a bacterial nitrile hydratase converts acrylonitrile to acrylamide, has none of these disadvantages and is significantly more economical.

Structurally dissimilar nitriles (organic compounds containing a -CN group) are formed by various plants, bacteria, and fungi (Figure 14.15), and many microorganisms can use nitriles as a source of carbon or nitrogen for growth. Saturated aliphatic nitriles are catabolized to the corresponding acids in two steps:

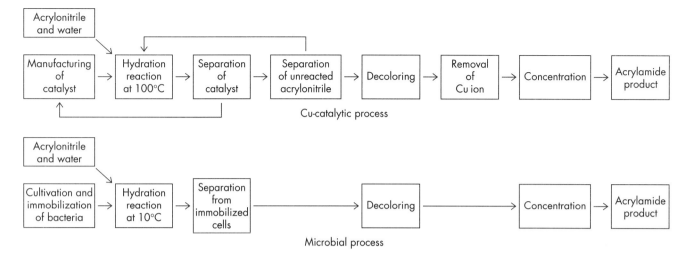

Figure 14.14
Comparison of the flowsheets for the conversion of acrylonitrile to acrylamide by a process employing a copper salt as catalyst and by a process employing immobilized microbial cells containing high levels of nitrile hydratase. Based on Nagasawa, T., and Yamada, H. (1990), Large-scale bioconversion of nitriles into useful amides and acids, in Abramowicz, D.A. (ed.), *Biocatalysis*, pp. 277–318 (Van Nostrand Reinhold).

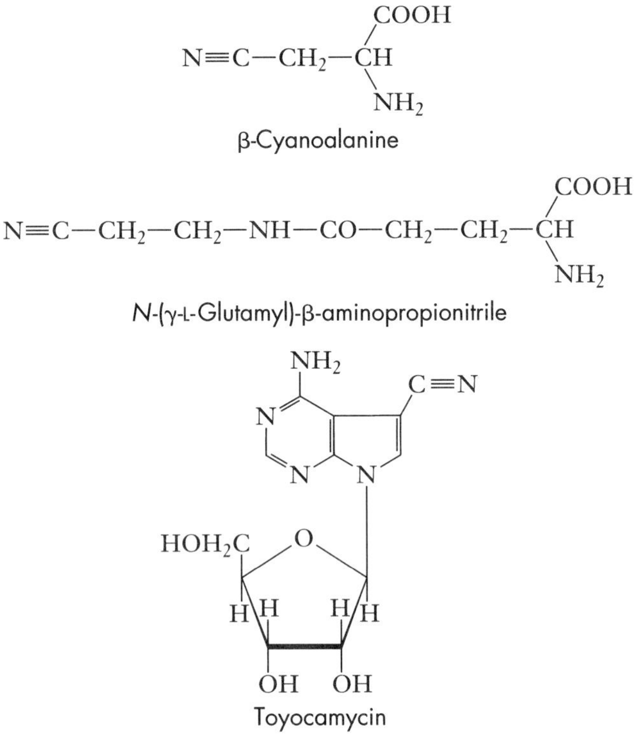

$$N\equiv C-CH_2-\underset{NH_2}{\overset{COOH}{\underset{|}{\overset{|}{CH}}}}$$

β-Cyanoalanine

$$N\equiv C-CH_2-CH_2-NH-CO-CH_2-CH_2-\underset{NH_2}{\overset{COOH}{\underset{|}{\overset{|}{CH}}}}$$

N-(γ-L-Glutamyl)-β-aminopropionitrile

Toyocamycin

Figure 14.15
Examples of naturally occurring nitriles. β-Cyanoalanine is produced by leguminous plants, and *N*-(γ-L-glutamyl)-β-aminopropionitrile by *Lathyrus* spp. Toyocamycin is an antibiotic substance produced by *Streptomyces toyocaensis*.

$$R\text{-}CN + H_2O \rightarrow R\text{-}CONH_2$$
$$R\text{-}CONH_2 + H_2O \rightarrow R\text{-}COOH + NH_3$$

where R is an alkyl group. The first step is catalyzed by a nitrile hydratase and the second by an amidase.

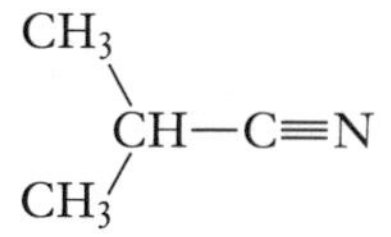

Catabolism of Acrylonitrile by *Pseudomonas chlororaphis* Strain B23

Pseudomonas chlororaphis strain B23, the organism chosen for the commercial bioconversion of acrylonitrile to acrylamide, was isolated from soil as an isobutyronitrile-utilizing bacterium (Figure 14.16). Catabolism of acrylonitrile by stationary-phase cells of *P. chlororaphis* B23 led to the accumulation of acrylamide in the medium up to 40% by weight. Three factors account for this accumulation of acrylamide. First, in *P. chlororaphis* B23 cells, the nitrile-hydratase–catalyzed conversion of acrylonitrile to acrylamide is at least 4000 times faster than the subsequent reaction, the amidase-catalyzed conversion of acrylamide to acrylic acid and ammonia. Second, the nitrile hydratase is not inhibited by high concentrations of the reaction product—that is, by acrylamide. Third, acrylonitrile, a powerful nucleophile, inactivates the amidase by reacting with the sulfhydryl group at that enzyme's active site. Nitrile hydratase is much less sensitive to acrylonitrile and remains active so long as the concentration of the substrate is kept at about 2% by weight or lower. This condition is readily met by introducing the acrylonitrile in small portions (Figure 14.17).

Modulation of the Level of Nitrile Hydratase

The conditions for obtaining very high levels of nitrile hydratase per cell include the use of an appropriate inducer, a properly formulated growth medium, and a strain improved by chemical mutagenesis.

Inducer. Nitrile hydratase is a catabolic enzyme, and like many other such enzymes in bacteria, its synthesis is induced by the presence of a substrate or related compound. A barely detectable level of nitrile hydratase is produced in the absence of substrate. A study of the ability of different nitriles, amides, and acids to induce the enzyme showed that methacrylonitrile (see Figure 14.16) was the most effective inducer.

Growth Medium. Nitrile hydratase is a nonheme iron enzyme and contains tightly bound Fe(III). The iron is required for catalytic activity. Adding either ferrous or ferric ions (up to 1 mg/100 ml) to the medium greatly increased the amount of nitrile hydratase in the cells. At optimal levels of iron, in a medium containing methacrylamide as inducer and

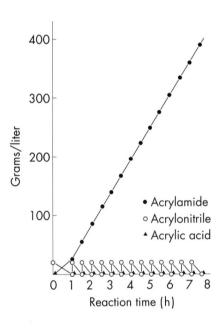

Figure 14.16
Isobutyronitrile and methacrylonitrile.

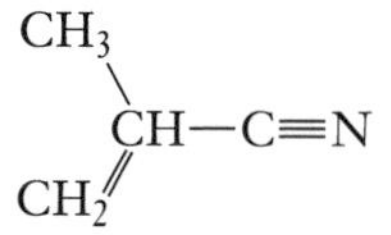

Figure 14.17
The production over time of acrylamide by *Pseudomonas chlororaphis* B23 cells high in nitrile hydratase in phosphate buffer at pH 7.0. Acrylonitrile was added in small portions at 30-minute intervals, and the reaction was carried out at 4°C. Based on Nagasawa, T., and Yamada, H. (1990), Large-scale bioconversion of nitriles into useful amides and acids, in Abramowicz, D.A. (ed.), *Biocatalysis*, pp. 277–318, Fig. 14.1 (Van Nostrand Reinhold).

inexpensive soybean hydrolysate as a source of nutrients, *P. chlororaphis* B23 produced nitrile hydratase in amounts exceeding 20% of total soluble protein.

Chemical Mutagenesis. Cultures of *P. chlororaphis* B23 are viscous because the cells produce extracellular polysaccharide, and this viscosity interferes with rapid large-scale harvesting of cells by centrifugation, an essential step in bioreactor preparation. Mutagenesis with *N*-methyl-*N'*-nitro-*N*-nitrosoguanidine was undertaken to solve this problem. On agar plates, wild-type *P. chlororaphis* B23 cells formed large, translucent colonies with irregular morphology. In mutagenized cultures, smaller, well-defined, and less sticky colonies were seen at a frequency of about 1%. Cultures of cells isolated from such colonies could be centrifuged down very readily. These cells either no longer produce extracellular polysaccharide or produce it in very small amounts. In addition, certain of these polysaccharide-minus mutants exhibited higher levels of nitrile hydratase activity than the wild-type cells. Under the optimal conditions described above, the best of these mutants produced 3000 times more nitrile hydratase than did uninduced wild-type cells.

Bioreactor. P. chlororaphis B23 cells containing very high levels of nitrile hydratase were immobilized on a cationic acrylamide-based polymer gel within a bioreactor. Large amounts of acrylonitrile were passed through the bioreactor and converted in their entirety to acrylamide. As Figure 14.14 illustrates, the microbial process is simpler than the chemical one.

BIOCATALYTIC SYNTHESES REQUIRING MULTIPLE STEPS

Genetic engineering makes possible the assembly of high-throughput biosynthetic pathways for the synthesis of complex and/or labile molecules from plentiful, inexpensive precursors. We provide two examples of such pathway engineering: the production of glutathione and that of ascorbic acid.

Overproduction of Glutathione by Genetically Engineered Microbial Cells

Glutathione has various biochemical functions. Perhaps its most important role is as a protective intracellular antioxidant. It also functions as a cofactor—either as a reductant (when in its sulfhydryl form) or as an oxidant (when in its disulfide form)—for many enzymes, such as glyoxalase, formaldehyde dehydrogenase, and so on.

$$
\begin{array}{l}
\gamma\text{-Glu-Cys-Gly} \\
\quad | \\
\quad S + 2H^+ + 2e^- \rightleftharpoons \quad 2\,\gamma\text{-Glu-Cys-Gly} \\
\quad | \\
\quad S \\
\quad | \\
\gamma\text{-Glu-Cys-Gly}
\end{array}
$$

Oxidized glutathione   Reduced glutathione

Some 20 chemical syntheses of glutathione have been described, but the most economical procedure for obtaining it has been purification of the peptide from yeast cells. When glutathione was isolated from yeast by F.G. Hopkins in 1921, it was the first peptide containing a non-α-peptide bond to be characterized.

The biosynthesis of this tripeptide, which consists of L-glutamate, L-cysteine, and glycine, proceeds in two steps.

$$
\text{Glu} + \text{Cys} + \text{ATP} \xrightarrow[\text{Enzyme GshA}]{\text{(γ-glutamylcysteine synthetase)}} \gamma\text{-Glu-Cys} + \text{ADP} + \text{HPO}_4^{2-}
$$

$$
\text{Gly} + \gamma\text{-Glu-Cys} + \text{ATP} \xrightarrow[\text{Enzyme GshB}]{\text{(glutathione synthetase)}} \gamma\text{-Glu-Cys-Gly} + \text{ADP} + \text{HPO}_4^{2-}
$$

The development of a process now used for the large-scale biocatalytic production of glutathione exemplifies the coordinated use of molecular biology, biochemistry, enzyme engineering, and bioreactor design in the achievement of a commercial goal. The account that follows describes how the process was put together and the rationale for the various steps.

Properties of E. coli B GshA and GshB

γ-L-Glutamyl-L-cysteine synthetase (GshA) was purified from *E. coli* B. The purification of the enzyme was monitored by assaying for the ATP-dependent formation of γ-Glu-Cys from glutamic acid and cysteine. Native GshA was found to be a single polypeptide chain of 55,000 daltons. The sequence of the amino-terminal 10 residues was determined by automated Edman degradation.

The enzyme is most active at pH 8.5 and 45°C, with apparent K_m values for L-glutamate, L-cysteine, and ATP of 0.50, 0.09, and 0.01 mM, respectively. It requires Mg^{2+} for activity at a Mg^{2+}/ATP molar ratio of 2:1. Inhibitor studies indicated that the enzyme did not have a sulfhydryl group at the active site. The enzymatic activity of GshA is specifically inhibited by reduced (but not oxidized) glutathione (Figure 14.18). This inhibition appears to be physiologically significant, because the K_i value of 2.5 mM (the concentration of reduced glutathione required for 50% inhibition) is comparable to the intracellular level of glutathione in stationary-phase cells.

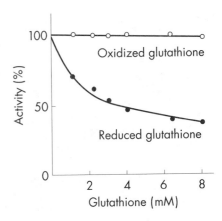

Figure 14.18
Effect of oxidized or reduced glutathione on the relative activity of GshA. From Murata, K., and Kimura, A. (1990), Overproduction of glutathione and its derivatives by genetically engineered microbial cells, *Biotechnol. Adv.* 8:59–96.

Glutathione synthetase (GshB), purified from *E. coli* B, had a molecular weight, obtained by gel filtration, of 152,000. Polyacrylamide gel electrophoresis in the presence of sodium dodecyl sulfate showed a single polypeptide of 38,000 daltons, indicating that the native enzyme is a tetramer. The sequence of the amino-terminal 23 residues was determined.

This enzyme is most active at pH 8.0 and 40°C, with apparent K_m values for γ-Glu-Cys, glycine, and ATP of 2.6, 2.0, and 1.9 mM. It too requires Mg^{2+} for activity, also at a Mg^{2+}/ATP molar ratio of 2:1. In contrast to GshA, studies with specific sulfhydryl reagents indicated that GshB does have a sulfhydryl group at the active site. At physiological concentrations, neither reduced nor oxidized glutathione inhibited the enzyme.

Elimination of Feedback Inhibition: Isolation of a Mutant with a "Desensitized" GshA

Experiments with purified GshA and GshB indicated that the simple two-step biosynthetic pathway leading to glutathione is controlled by feedback inhibition. The first enzyme in the pathway, GshA, is inhibited when large amounts of glutathione accumulate in the cell. *In general, feedback inhibition operates by an allosteric mechanism*, with the inhibitor binding to a site that does not overlap the substrate binding site (see Chapter 12). Consequently, it is possible to isolate mutant enzymes in which the binding site for the feedback inhibitor is altered but which retain the enzyme's catalytic activity. Mutant enzymes of this type are said to be "desensitized."

Strains with desensitized GshA would be expected to accumulate higher levels of glutathione than the wild type and hence to show enhanced reactivity with the nitroprusside–ammonia sulfhydryl reagent (as assessed by intensity of purple color) when the reagent is added to the agar plates. Exploitation of such a phenotype led to the isolation of a mutant strain of *E. coli* B, strain RC912, with a desensitized GshA.

E. coli *B Mutants Defective in GshA or GshB*

Although glutathione represents some 40% of the total sulfhydryl group content of *E. coli* B cells, the growth of glutathione-synthesis–deficient mutants on a minimal medium is equal to that of wild-type cells. This indicates that glutathione is not required for any metabolic processes critical to the growth and multiplication of *E. coli*. However, glutathione-deficient mutants are much more sensitive to a variety of toxic compounds, including heavy metals and other reagents that react with sulfhydryl groups. Such enhanced sensitivity can be exploited for the isolation of mutants deficient in GshA or GshB.

Mutants deficient in GshA or GshB were detected by plating cells on agar containing minimal medium supplemented with different concentrations of the sulfhydryl reagent tetramethylthiuram disulfide (TMTD; Figure 14.19). Two kinds of mutants strains, exemplified by C912 and C1001, were derived from *E. coli* B. Mutant C912 lacked GshA activity, and mutant C1001 lacked GshB activity. The minimal inhibitory concentrations of TMTD were lower for strain C912 (40 μg/ml) than for C1001 (100 μg/ml). The wild type is unaffected by these levels of TMTD. *The higher resistance of strain C1001 indicates that γ-glutamylcysteine, a product of the reaction catalyzed by GshA, can partially substitute for glutathione as a protective agent.*

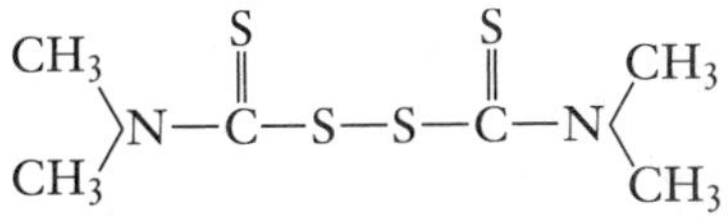

Figure 14.19
Tetramethylthiuram disulfide.

Cloning of the Gene Encoding GshA

Mutant C912, which lacks GshA activity, was utilized to clone the *gshA* gene. The chromosomal DNA from strain RC912 (in which the GshA is desensitized to feedback inhibition by reduced glutathione) was partially digested with the restriction enzyme *Pst*I. DNA of the plasmid vector pBR322 was also digested with *Pst*I. The two digests were then mixed, ligated, and used for the transformation of strain C912. Transformed cells were plated on agar containing TMTD (to select for the production of glutathione) and tetracycline (to select for the presence of the plasmid pBR322). The colonies that formed on such plates were tested for the development of purple color in the presence of the nitroprusside–ammonia reagent. The transformant that gave the strongest purple color was isolated, and the hybrid plasmid it contained was designated pGS100-2.

Plasmid pGS100-2 (Figure 14.20) contained a chromosomal DNA fragment of 3.6 kb. Detailed restriction mapping indicated that *gshA* was located within a 2.098-kb *Hinc*II-*Pst*I fragment, the sequence of which was determined next. Only one open reading frame was present on this fragment, and this reading frame encoded *gshA*. The initiation codon was identified by comparison of the gene sequence and the amino-terminal sequence of GshA, and the sequence 5′ to the reading frame was examined for features important to the binding of RNA polymerase. The "−35 sequence" upstream of the initiation codon was TGCACA, and the "−10 sequence" (Pribnow box) was TATAAT. These sequences were separated by 18 bp. The sequences preceding *gshA* and their spacing are typical of the *E. coli* promoter consensus −35 (TTGACA) and −10 (TATAAT) sequences and of the consensus 16–18-bp separation between these sequences. A potential Shine–Dalgarno sequence (encoding a ribosome-binding site on the mRNA) was identified about 10 nucleotides 5′ of the initiation codon of *gshA*. *These results indicated that all of the information required for the expression of gshA in* E. coli *was present on the fragment of chromosomal DNA in pGS100-2.*

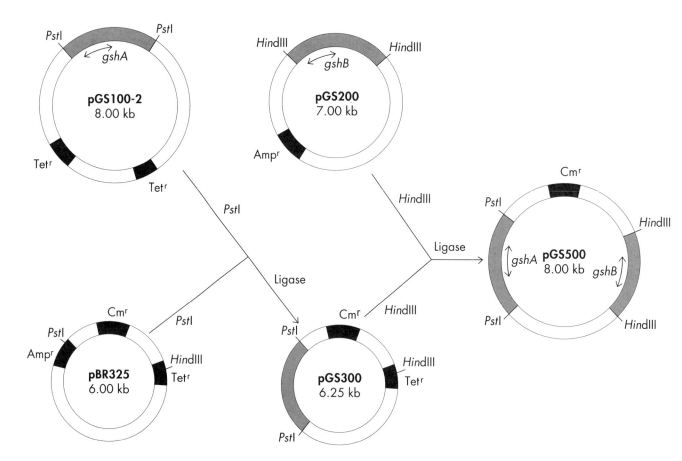

Figure 14.20

Steps leading to the construction of the hybrid plasmid pGS500, containing the genes *gshA* and *gshB*. pGS100-2 and pGS200 contain fragments of *E. coli* B strain RC912 chromosomal DNA that include the *gshA* and *gshB* genes, respectively. The isolation of these hybrid plasmids is discussed in the text. The *Pst* I restriction fragment with the *gshA* gene from pGS100-2 was inserted into the *Pst* I site of pBR325 to give plasmid pGS300. pGS500 was constructed by inserting the *Hin*d III fragment of pGS200, which contains the *gshB* gene, into the *Hin*d III site of pGS300. Based on Murata, K., and Kimura, A. (1990), Overproduction of glutathione and its derivatives by genetically engineered microbial cells, *Biotechnol. Adv.* 8:59–96, Fig. 8.

Cloning of the Gene Encoding GshB

The strategy for cloning *gshB* paralleled that described above for the cloning of *gshA*, but there were some important technical differences. Strain C1001, lacking GshB activity, was utilized as the recipient strain. With an eye to the later construction of a plasmid carrying both *gshA* and *gshB*, the chromosomal DNA isolated from strain RC912 was digested with *Hin*dIII. The ligation of the *Hin*dIII fragments into pBR322 interrupted the *tet* gene, whereas in the *gshA* construction, the insertion of *Pst*I fragments interrupted the *amp* gene. Because strain C1001 produces γ-glutamylcysteine, the agar plates used for the selection of

TMTD-resistant colonies contained a much higher level of TMTD than those used for the isolation of RC912 transformants with plasmids containing *gshA*.

Hybrid plasmid pGS200, isolated from a transformed, TMTD-resistant colony of strain C1001, contained the gene for *gshB*. As seen for *gshA*, the region of chromosomal DNA upstream of *gshB* in the hybrid plasmid contained "−35" and "−10" sequences typical of *E. coli* consensus sequences and a Shine–Dalgarno sequence. Moreover, the codon usage for both *gshA* and *gshB* was close to that optimal for *E. coli*, suggesting that these genes are expressed at a reasonable level in wild-type cells.

Construction of a Hybrid Plasmid Carrying gshA and gshB

The construction of hybrid plasmid pGS500 containing genes *gshA* and *gshB* is illustrated in Figure 14.20. Plasmid pGS500 was then transformed into strain RC912 (described above), and transformant colonies were selected for resistance to both TMTD and chloramphenicol. The transformed cells showed GshA and GshB activities 15-fold higher and 18-fold higher, respectively, than the untransformed RC912 strain.

Production of Glutathione by E. coli B Strain RC912/pGS500

Cells of strain RC912/pGS500 were immobilized by a gentle procedure. The cells were mixed with a solution of a low-melting red algal polysaccharide κ-carrageenan and trapped in the gel formed upon cooling. The trapped cells were then treated with glutaraldehyde in the presence of hexamethylenediamine to form extended crosslinked networks of proteins within the cells (Figure 14.21). Even though GshA and GshB are soluble proteins, they are trapped within the cells by this crosslinking procedure and are not removed by extended washing. Moreover, the crosslinking leads to stabilization rather than to loss of enzymatic activities. After crosslinking, the immobilized cells were permeabilized by treatment with a low concentration of toluene to allow free passage of substrates and product.

Two moles of the expensive cofactor ATP are required in the synthesis of one mole of glutathione from its amino acid precursors. However, the ATP need only be added in catalytic amount. The *E. coli* cells contain the enzyme acetate kinase, which catalyzes the reaction

$$CH_3CO-PO_3^{2-} + ADP \longrightarrow CH_3COO^- + ATP$$

Acetyl phosphate   Acetate

Therefore, the addition of substantial amounts of the inexpensive phosphoryl donor acetylphosphate leads to continuous regeneration of ATP.

$$H-\overset{\displaystyle O}{\overset{\|}{C}}-CH_2-CH_2-CH_2-\overset{\displaystyle O}{\overset{\|}{C}}-H$$

Glutaraldehyde

$$H_2N-CH_2-CH_2-CH_2-CH_2-CH_2-CH_2-NH_2$$

Hexamethylenediamine

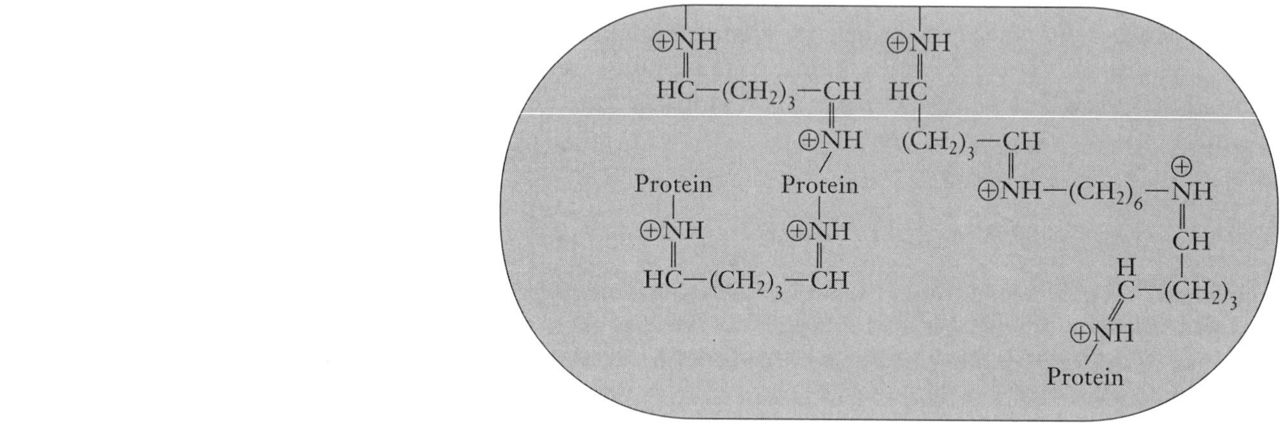

Figure 14.21
Examples of crosslinks formed in a bacterial cell treated with glutaraldehyde and hexamethylenediamine.

The efficiency of glutathione production by the immobilized cells was optimized by varying the concentration of the three precursor amino acids. The highest yield was obtained at 80 m*M* L-glutamate, 20 m*M* L-cysteine, and 20 m*M* glycine. The high nonstoichiometric requirement for glutamate appears to be a consequence of the utilization of this amino acid by other reactions catalyzed by the *E. coli* cells. Under optimal conditions (Figure 14.22), a 10-ml column of immobilized *E. coli* cells produced 5 g/liter of glutathione continuously, with a column activity half-life of 28 days. Glutathione was easily isolated from the effluent by ion exchange chromatography.

Overview of the Glutathione Process

The development of a commercial synthesis of glutathione using genetically engineered microbial cells encapsulates the essence of research in many areas of biocatalysis. The problems that were encountered in this effort are typical, as are many of the solutions designed to bridge the gap between the metabolic potential of microorganisms and the production of economically viable products.

Classical biochemistry contributed the information about glutathione's structure, functions, and biosynthetic pathway. Armed with that

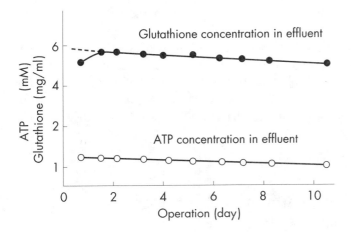

Figure 14.22

Continuous production of glutathione in a flow reactor packed with immobilized cells of *E. coli* B strain RC912 transformed with plasmid pGS500. From Murata, K., and Kimura, A. (1990), Overproduction of glutathione and its derivatives by genetically engineered microbial cells, *Biotechnol. Adv.* 8:59–96.

knowledge, it was possible to develop assays for the purification of GshA and GshB. Biochemical characterization of the purified enzymes led to recognition of the feedback inhibition of GshA, provided amino acid sequence information so that putative candidates for the structural genes encoding GshA and GshB could be recognized, and yielded kinetic parameters essential to the later development of the continuous-flow reactor. Understanding of the mechanism of inhibition by tetramethylthiuram disulfide provided a screen for mutants deficient in GshA or GshB activity. Finally, the biochemistry of the acetate kinase reaction made it possible to avoid the need for large amounts of an expensive cofactor.

Molecular biology was central to the cloning of *gshA* and *gshB*. DNA sequence analysis provided both the primary structure of the enzymes and information about the sequences that govern the transcription and translation of the genes in *E. coli*. Cloning, genetic engineering, transformation, and antibiotic selection all played crucial roles in constructing the production strain.

Bioprocess engineering made the practical use of the production strain possible. The gentle immobilization technique allowed biocatalysis by intact cells, and, aside from the obvious advantage that the GshA and GshB need not be purified, the use of whole cells offered the important bonus that the cells' endogenous acetate kinase catalyzes the continuous production of ATP. Finally, the column format and optimization of the concentrations of substrates and of other parameters

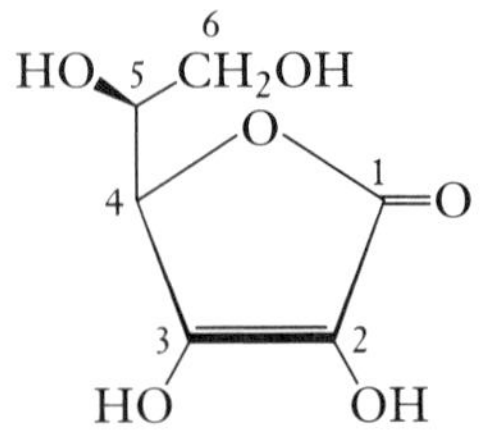

Figure 14.23
Structure of vitamin C (L-ascorbic acid).

(Mg^{2+}, buffer concentration, pH, temperature) were all essential to the success of the process.

The engineered glutathione production strain would probably not be appropriate for use in a fermentative glutathione production process: The plasmid is likely to be unstable because its *gshA* and *gshB* sequences are homologous to those on the chromosome. This situation does not cause problems in the bioprocess, however, because it employs non-growing cells.

Metabolic Pathway Engineering for Vitamin C Production

Two key requirements in the production of vitamin C (Figure 14.23) are that the synthesis be chiral, because only the L-enantiomer of ascorbic acid is biologically active, and that the final step in the process be non-oxidative, because ascorbate is very easily oxidized. In 1934, T. Reichstein and A. Grussner reported in *Helvetica Chimica Acta* on "A good synthesis of L-ascorbic acid (vitamin C)." Their somewhat lengthy synthesis, in modified form, is still used commercially today (Figure 14.24). In it, D-glucose is chemically reduced to D-sorbitol, which is then oxidized to L-sorbose by an *Acetobacter suboxydans* fermentation. L-Sorbose is converted to the di-isopropylidene derivative, the derivative is chemically oxidized at C1, and then the isopropylidene groups are removed under acidic conditions to give 2-keto-L-gulonic acid (2-KLG). This stable compound is the last intermediate in the synthesis. It can be readily converted to L-ascorbic acid by either an acid- or a base-catalyzed cyclization. The overall yield of L-ascorbic acid from D-glucose is greater than 50%.

Recently, metabolic pathway engineering of *Erwinia herbicola* bacteria created a genetically modified strain capable of producing 2-KLG directly from D-glucose at a rate of 1 g/liter/h, to give a final concentration of 120 g/liter, a greater than 60% yield (Figure 14.25). This feat of metabolic engineering was by no means simple, and the work of translating it into a commercially viable process is not yet complete. Superficially, the task appeared simple: extension of a metabolic pathway existing in one bacterium by one step catalyzed by an enzyme obtained from another bacterium. As revealed by the following account, however, many obstacles stood in the way of this accomplishment.

Microbial Conversion of D-Glucose to 2-KLG

Bacteria belonging to the genera *Erwinia*, *Acetobacter*, and *Gluconobacter* oxidize D-glucose efficiently to 2,5-diketo-D-gluconic acid (2,5-DKG).

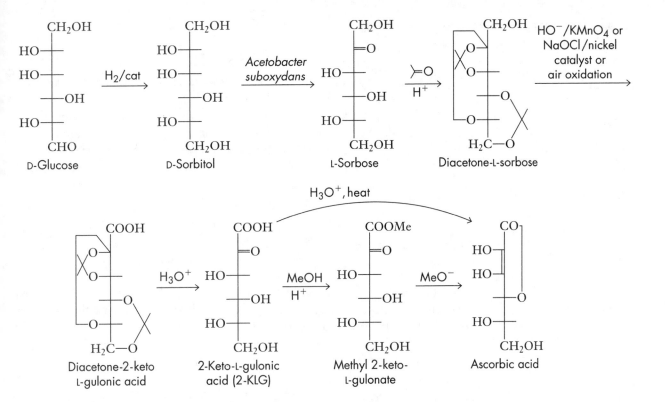

Figure 14.24
Reichstein–Grussner synthesis of L-ascorbic acid. The only biotransformation step is the oxidation of D-sorbitol to L-sorbose by an *Acetobacter suboxydans* fermentation. From Anderson, S., et al. (1985), Production of 2-keto-L-gulonate, an intermediate in L-ascorbate synthesis, by a genetically modified *Erwinia herbicola*, *Science* 230:144–149.

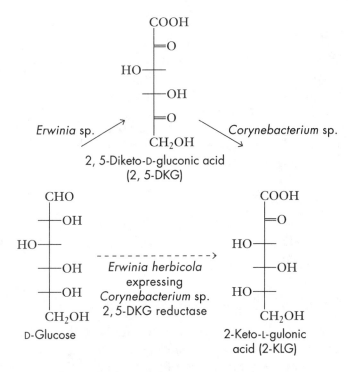

Figure 14.25
Synthesis of 2-keto-L-gulonic acid by fermentation. The upper pathway shows the conversion of glucose to 2-keto-L-gulonic acid by *Erwinia* sp. and *Corynebacterium* sp. fermenting in tandem. The lower pathway illustrates the fermentation by a recombinant strain of *Erwinia herbicola*, engineered to express *Corynebacterium* 2,5-diketo-D-gluconate reductase. From Anderson, S., et al. (1985), Production of 2-keto-L-gulonate, an intermediate in L-ascorbate synthesis, by a genetically modified *Erwinia herbicola*, *Science* 230:144–149.

Organisms from the coryneform group of bacteria, such as *Corynebacterium*, *Brevibacterium*, and *Arthrobacter*, can convert 2,5-DKG to 2-KLG. When a cofermentation process using appropriate strains of *Erwinia* and *Corynebacterium* coupled these capabilities, D-glucose was efficiently converted to 2-KLG (see Figure 14.25). In a quest to simplify the process, and thus to minimize its cost, investigators have combined the relevant metabolic traits of these two microorganisms in one, a genetically modified strain of *E. herbicola*.

Identification and Characterization of Corynebacterium *sp. ATCC31090 2,5-DKG Reductase*

The ability of *Corynebacterium* to convert 2,5-DKG to 2-KLG implied that it possessed a 2,5-DKG reductase. Because many of the enzymes that catalyze the reduction of ketones to alcohols use NADH or NADPH as cofactors, these cofactors were added to crude lysates of *Corynebacterium* sp. cells, which were then assayed under a variety of conditions for 2,5-DKG reductase activity. Activity was in fact detected in the presence of either NADH or NADPH.

As far as researchers knew, any or all of several different enzymes might have been responsible for the observed 2,5-DKG reductase activity. For example, reduction of 2,5-DKG at C5 would give either 2-keto-D-gluconate or 2-KLG; reduction at C2 would give either 5-keto-D-gulonate or 5-keto-D-mannonate; and reduction at C1 would give 2,5-diketogluconaldehyde (Figure 14.26). Indeed, several 2,5-DKG reductases were found in the crude lysate.

An assay methodology was developed to separate and quantitate the various ketoaldonic and aldonic acids that can result from the reduction of 2,5-DKG. Purification by a variety of chromatographic procedures in conjunction with this assay methodology has made possible the identification and purification to homogeneity of a 30,000-dalton NADPH-dependent 2,5-DKG reductase that produced 2-KLG. At pH 9.0–9.5, the measured equilibrium constant K_{eq} (see below) for the reaction catalyzed by this reductase was 5×10^{-13}, indicating that the formation of 2-KLG was highly favored.

$$K_{eq} = [\text{2,5-DKG}][\text{NADPH}][\text{H}^+]/[\text{2-KLG}][\text{NADP}^+] = 5 \times 10^{-13}$$

However, because *Corynebacterium* sp. does not normally produce 2,5-DKG as a metabolite, the true physiological substrate for the NADPH-dependent 2,5-DKG reductase is not yet known.

Cloning of the 2,5-Diketogluconate Reductase Gene

The N-terminal amino acid sequence of the NADPH-dependent 2,5-DKG reductase (DKGR) provided essential, but incomplete, information for the design of the synthetic oligonucleotide DNA probes needed

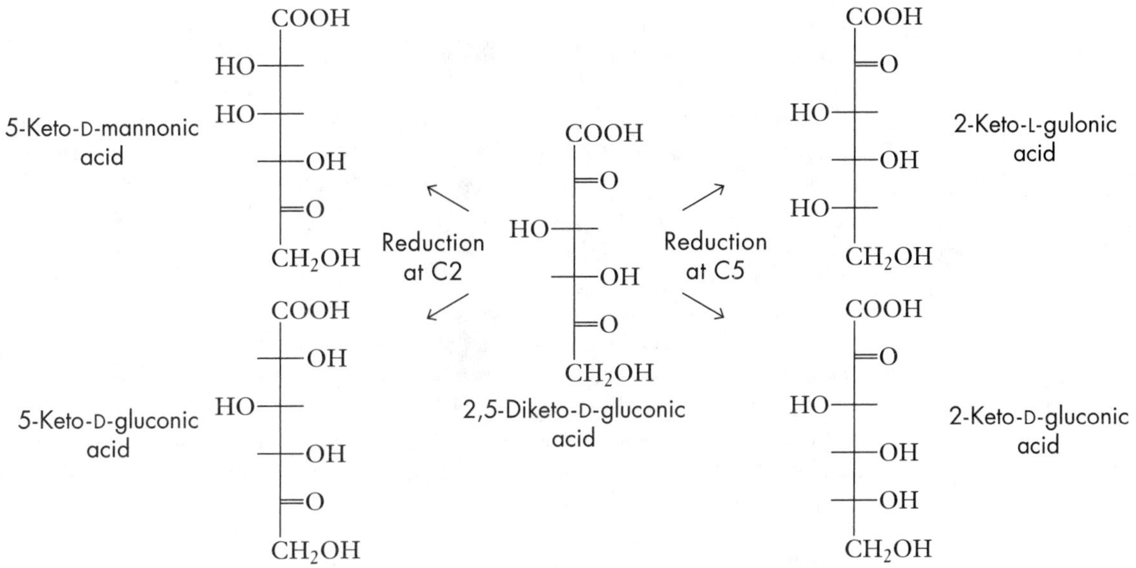

Figure 14.26
Products that may be generated by the reduction of 2,5-diketo-D-gluconic acid at the C2 and C5 positions.

to detect clones carrying the reductase gene. The information was inadequate because the amino-terminal sequence contained no regions of minimal codon degeneracy. Other information was used to decrease the complexity of the probes: For one thing, the thermal melting curves of *Corynebacterium* DNA indicated that it had a mole % G+C content of 71; for another, bacterial DNAs with extreme base compositions are known to have highly biased codon usage patterns. Thus two unique 43-mer nucleotide probes were synthesized based on regions of the amino-terminal sequence with codons known to be prevalent in bacterial DNAs of high G+C content. The two probes both hybridized to a 2.2-kbp *Bam*H1 fragment of *Corynebacterium* DNA. They were then successfully used to screen a partial genomic library of *Corynebacterium* sp. DNA cloned into the *Bam*H1 site of pBR322. The DNA sequence of the 2.2-kb *Bam*H1 fragment proved to contain only a portion of the NADPH-dependent 2,5-DKGR gene, but isolation of an overlapping 0.9-kb *Pst*I-*Bam*H1 fragment allowed completion of the gene sequence.

Expression of the 2,5-DKGR Gene in E. herbicola and Attendant Problems

A vector for the expression of the 2,5-DKG reductase in *E. herbicola* was constructed as follows: The *Corynebacterium* sp. DNA upstream of the start codon of the 2,5-DKGR gene was deleted, and the gene was inserted, immediately downstream from the *E. coli trp* promoter and a

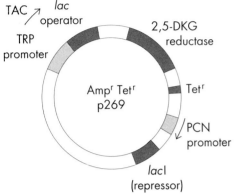

Figure 14.27
A regulable expression plasmid, derived from pBR322, for 2,5-diketo-D-glu-
conate reductase (2,5-DKGR) in *E. herbicola*. The 2,5-DKGR gene is under
the control of the TAC promoter. This promoter is made up of the *E. coli trp*
promoter and the *lac* operator. The *lac*I repressor gene is under the control
of a *Bacillus licheniformis* (PCN) promoter. The plasmid also carries genes for
ampicillin (Amp^r) and tetracycline (Tet^r) resistance. All of the regulatory
elements on the plasmid function in *E. herbicola*; no 2,5-DKGR activity was
observed in the absence of inducers of the *lac* operon. Addition of lactose,
or of its analog isopropyl-β-D-thiogalactopyranoside (IPTG), led to the appearance of 2,5-DKGR activity. There was a direct correlation between the
amount of IPTG added and the rate of 2,5-diketo-D-gluconate reduction
by whole cells. From Lazarus, R.A., et al. (1990), A biocatalytic approach to
vitamin C production: Metabolic pathway engineering of *Erwinia herbicola*,
in Abramowicz, D.A. (ed.), *Biocatalysis*, pp. 135–155 (Van Nostrand Reinhold).

synthetic ribosome-binding site, into a plasmid derived from pBR322
(and carrying a tetracycline-resistance gene). The resulting plasmid,
designated ptrp1-35, was used to transform *E. herbicola*.

The tetracycline-resistant transformants expressed a high level of
active 2,5-DKGR. Compared to a control strain transformed with
pBR322, the level of 2,5-DKG reductase activity in a strain transformed
with ptrp1-35 was 46-fold higher. However, the high level of expression
of 2,5-DKGR was accompanied by a significant parallel decrease in
both 2-KLG production and cell viability, the latter attributable to the
heavy drain on energy in the form of NADPH to reduce 2,5-DKG. Use
of a different expression plasmid, p269 (Figure 14.27), one that allowed
the level and timing of 2,5-DKGR expression to be regulated by lactose
or one of its analogs, such as isopropyl-β-D-thiogalactopyranoside
(IPTG), greatly improved cell viability. With the gene fully induced, the
E. herbicola strain carrying p269 expressed 2,5-DKGR at a level about
20% of that of the strain that had been transformed with ptrp1-35.

Surprisingly, in the *E. herbicola* p269 fermentation, the major product was L-idonate (IA), formed by reduction of 2-KLG, rather than the
desired 2-KLG (Figure 14.28). The solution of this second-level problem required a thorough investigation of the reductive and oxidative
pathways of carbohydrate metabolism in *E. herbicola*.

Identification of a 2-Ketoreductase Catalyzing the Conversion of 2-KLG to IA

In "ketogenic" bacteria such as *E. herbicola*, glucose is oxidized to ke-
togluconates through the activity of membrane-bound dehydrogenases
linked to an electron transport chain. Cytoplasmic NAD(P)H-depen-
dent ketogluconate reductases then partially re-reduce the oxidized carbohydrates to gluconate (Figure 14.29), which can enter the central

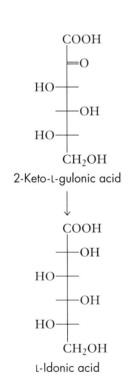

Figure 14.28
Formation of L-idonate from 2-
keto-L-gluconate.

metabolic pathways of the cell. This reaction also regenerates NADP⁺. Presuming that the conversion of 2-KLG to IA was being catalyzed in this way by a cytoplasmic NAD(P)H-dependent ketogluconate reductase, researchers launched a search for the enzyme. They soon discovered and purified from *E. herbicola* a cytosolic 2-ketoreductase (designated 2-ketoreductase A) that catalyzed the conversion of 2-KLG to IA. This enzyme was relatively nonspecific, catalyzing the NADH- or NADPH-dependent reduction of 2-KLG, 2-keto-D-gulonate, and 2,5-KDG to IA, gluconic acid, and 5-keto-D-gluconic acid, respectively.

Because reactions catalyzed by 2-ketoreductase A convert both 2,5-DKG and 2-KLG to other products and thereby decrease the yield of 2-KLG, the researchers wished to delete this enzyme from the producing strain.

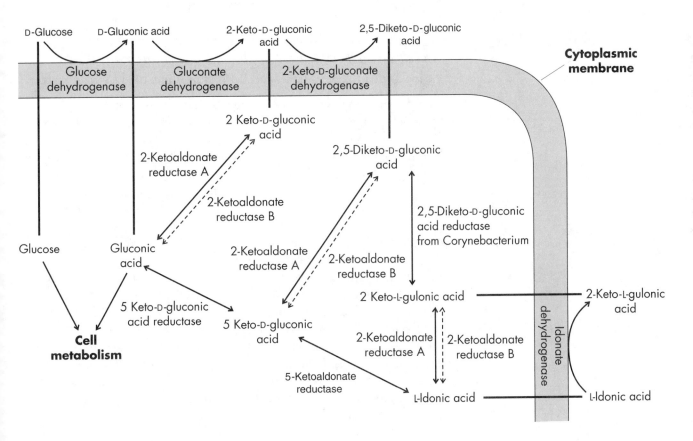

Figure 14.29

Features of carbohydrate metabolism in the recombinant *E. herbicola* strain producing 2-keto-L-gulonate. Based on Lazarus, R.A., et al. (1990), Metabolic and genetic aspects of a recombinant bioconversion leading to ascorbic acid, in Heslot, H., et al. (eds.), *6th International Symposium on Genetics of Industrial Microorganisms*, vol. II, pp. 1073–1082, Fig. 2 (Proceedings of Société Française de Microbiologie).

Deletion Mutagenesis of the Chromosomal Gene Encoding
2-KLG Reductase A and Discovery of 2-KLG Reductase B

The gene *trk*A, encoding 2-ketoreductase A, was cloned, and the sequence of the gene was modified in such a way as to cause a translation termination within the coding region. The *trk*AΔ3 deletion was then transferred to the chromosomal copy of *trk*A by a double-reciprocal crossover event to generate an *E. herbicola trk*A⁻ strain. Strains carrying the modified gene (which was named *trk*AΔ3) did not produce active 2-ketoreductase A.

Unfortunately, the fermentation of D-glucose to 2-KLG by *E. herbicola trk*A⁻ p269 was still accompanied by the formation of a significant amount of L-idonate. This observation led researchers to look for and isolate a 2-ketoreductase B, an enzyme distinct from 2-ketoreductase A and also capable of converting 2-KLG to IA.

Aspects of Carbohydrate Metabolism in Recombinant
E. herbicola *Relevant to the Conversion of 2,5-DKG to 2-KLG*

The diagram in Figure 14.25 suggests that engineering an *E. herbicola* strain to produce high yields of 2-KLG might be a simple matter. However, Figure 14.29, summarizing aspects of carbohydrate metabolism in the recombinant *E. herbicola* strain that contains the *Corynebacterium* sp. 2,5-diketoreductase, puts the matter in a more realistic perspective. The creation of such a metabolically engineered strain has required considerable biochemical research. Along the way, researchers have confronted unexpected obstacles resulting from their initially incomplete knowledge of microbial metabolism and of the versatility of the many enzymes that catalyze the numerous reactions within the cell. For example, the *E. herbicola* strain chosen for metabolic engineering can utilize neither 2-KLG nor IA as a substrate for growth; yet, as we have seen, two relatively nonspecific cytoplasmic NAD(P)H reductases in this strain convert 2-KLG to IA. Moreover, one of the membrane-bound dehydrogenases in this strain can catalyze the oxidation of IA to 2-KLG. In the end, however, the knowledge and experience gained from studying these unexpected complexities have brought a bioconversion process for the production of vitamin C close to scientific and commercial realization

SUMMARY

Microorganisms are an immense source of novel enzymes capable of catalyzing chemical reactions that transform both naturally occurring

SELECTED REFERENCES **559**

and human-made organic compounds. Enzymes show high regio- and enantioselectivity and specificity and work under mild conditions of pH, temperature, and pressure. Moreover, biotransformations minimize the use of toxic organic reagents and solvents and the attendant safety and pollution problems. Enzymes function effectively in two-phase, water–organic-solvent systems with very low water content. In such systems, they catalyze the transformation of water-insoluble compounds. Immobilization of enzymes on solid supports frequently stabilizes them and permits multiple reuse of the biocatalyst in reactor and batch procedures.

Enzymes find large-scale use in the stereoselective synthesis of steroids, semisynthetic antibiotics, and many other drugs. They are extensively used in the enantioselective resolution of racemic mixtures of derivatives of amino acids and other compounds. Hydantoinases, β-lactamases, acylases, lipases, and steroid hydroxylases are the enzymes most extensively used in commercial biotransformations.

The microbial synthesis of acrylamide is an example of a synthesis of a commodity chemical, performed on a multi-ton scale, that does not require stereospecificity yet offers major advantages over conventional chemical routes.

The processes used in the synthesis of glutathione by immobilized, engineered *E. coli* cells, and for the synthesis of ascorbic acid (vitamin C) by fermentation employing a recombinant strain of *Erwinia herbicola*, illustrate the potential of metabolic pathway engineering.

SELECTED REFERENCES

General

Eliel, E.L., and Otsuka, S. (eds.), 1982. *Asymmetric Reactions and Processes in Chemistry*. ACS Symposium Series 185. American Chemical Society.

Sih, C.J., and Chen, C-S., 1984. Microbial asymmetric catalysis—Enantioselective reduction of ketones. *Angew. Chem. Int. Ed. Engl.* 23:570–578.

Simon, H., Bader, J., Günther, H., Neumann, S., and Thanos, J., 1985. Chiral compounds synthesized by biocatalytic reductions. *Angew. Chem. Int. Ed. Engl.* 24:539–553.

Rehm, H.-J., and Reed, G. (eds.), 1987. *Biotechnology*. Vol. 7a. *Enzyme Technology*. VCH Publishers.

Davies, H.G., Green, R.H., Kelly, D.R., and Roberts, S.M., 1989. *Biotransformations in Preparative Organic Chemistry. The Use of Isolated Enzymes and Whole Cell Systems in Synthesis*. Academic Press.

Abramowicz, D.A. (ed.), 1990. *Biocatalysis*. Van Nostrand Reinhold.

Poppe, L., and Novák, L., 1992. *Selective Biocatalysis. A Synthetic Approach*. VCH Publishers.

Faber, K., 1992. *Biotransformation in Organic Chemistry*. Springer-Verlag.

Holland, H.L., 1992. *Organic Synthesis with Oxidative Enzymes*. VCH Publishers.

Laane, C., and Tramper, J., 1990. Tailoring the medium and reactor for biocatalysis. *Chemtech* 20:502–506.

Bailey, J.E., 1991. Toward a science of metabolic engineering. *Science* 252:1668–1675.

Elferink, V.H.M., Breitgoff, D., Kloosterman, M., Kamphuis, J., van den Tweel, W.J.J., and Meijer, E.M., 1991. Industrial developments in biocatalysis. *Recl. Trav. Chim. Pays-Bas* 110:63–74.

Cofactor Regeneration

Wong, C-H., and Whitesides, G.M., 1982. Enzyme-catalyzed organic synthesis: NAD(P)H cofactor regeneration using ethanol/alcohol dehydrogenase/aldehyde dehydrogenase and methanol/alcohol dehydrogenase/aldehyde dehydrogenase/formate dehydrogenase. *J. Org. Chem.* 47:2816–2818.

Kazlauskas, R.J., and Whitesides, G.M., 1985. Synthesis of methoxycarbonylphosphate, a new reagent having high phosphoryl donor group potential for use in ATP cofactor regeneration. *J. Org. Chem.* 50:1069–1076.

Importance of Chirality in the Action of Synthetic Drugs

Crossley, R., 1992. The relevance of chirality to the study of biological activity. *Tetrahedron* 38:8155–8178.

Wilson, K., and Walker, J. (eds.), 1991. Chirality and its importance in drug development. *Biochem. Soc. Trans.* 19:443–475.

Ariëns, E.J., Wuis, E.W., and Veringa, E.J., 1988. Stereoselectivity of bioactive xenobiotics. *Biochem. Pharmacol.* 37:9–18.

Antibiotics

Vandamme, E.J., 1983. Peptide antibiotic production through immobilized biocatalyst technology. *Enzyme Microb. Technol.* 5:403–416.

Lipids

Ratledge, C., 1987. Lipid biotechnology: A wonderland for the microbial physiologist. *J. Am. Oil Chem. Soc.* 64:1647–1656.

Yamane, T., 1987. Enzyme technology for the lipid industry: An engineering overview. *J. Am. Oil Chem. Soc.* 64:1657–1662.

Transformations of Nitriles

Nagasawa, T., and Yamada, H., 1989. Microbial transformations of nitriles. *Tr. Biotechnol.* 7:153–158.

Nagasawa, T., and Yamada, H., 1990. Large-scale bioconversion of nitriles into useful amides and acids. In Abramowicz, D.A. (ed.), *Biocatalysis*, pp. 277-318. Van Nostrand Reinhold.

Glutathione

Kimura, A., 1986. Application of recDNA techniques to the production of ATP and glutathione by the "syntechno system." *Adv. Biochem. Eng./Biotechnol.* 33:29–51.

Murata, K., and Kimura, A., 1990. Overproduction of glutathione and its derivatives by genetically engineered microbial cells. *Biotech. Adv.* 8:59–96.

Vitamin C (L-Ascorbate)

Anderson, S., et al., 1985. Production of 2-keto-L-gulonate, an intermediate in L-ascorbate synthesis, by a genetically modified *Erwinia herbicola*. *Science* 230:144–149.

Lazarus, R.A., et al., 1990. Metabolic and genetic aspects of a recombinant bioconversion leading to ascorbic acid. In Heslot, H., et al. (eds.), *6th International Symposium on Genetics of Industrial Microorganisms*. Vol. II, pp. 1073–1082. Proceedings of Société Française de Microbiologie.

Lazarus, R.A., et al., 1990. A biocatalytic approach to vitamin C production. Metabolic pathway engineering of *Erwinia herbicola*. In Abramowicz, D.A. (ed.), *Biocatalysis*, pp. 135–155. Van Nostrand Reinhold.

15

Environmental Applications

By releasing carbon, nitrogen, phosphorus, and sulfur from an immense variety of complex organic compounds, thereby enabling these elements to be reused by living organisms, microorganisms play a fundamental role in the global recycling of matter. Bacteria and fungi, in particular, display spectacular metabolic versatility in the three areas of environmental microbiology examined in this chapter: sewage and wastewater treatment, degradation of xenobiotics, and mineral recovery.

In an effort to describe their behavior in the environment, organic compounds are often classified as biodegradable, persistent, or recalcitrant. A *biodegradable* organic compound is one that undergoes a biological transformation. A *persistent* organic compound does not undergo biodegradation in certain environments. A *recalcitrant* compound resists biodegradation in a wide variety of environments. Note that biodegradation is not always desirable. In some instances, degradation may convert an innocuous compound into a toxic one or may alter an already toxic compound to generate a product toxic to even more organisms.

By itself, the term *biodegradable* does not imply any particular extent of degradation. The transformation may involve one or several reactions, and the effect may be slight or significant. *Primary biodegradation* usually refers to alteration by a single reaction, whereas *partial biodegradation* involves a more extensive chemical change. In common parlance, however, when people say that a compound is biodegradable, they mean it can be mineralized. *Mineralization* is complete degradation to the end products of CO_2, water, and other inorganic compounds.

DEGRADATIVE CAPABILITIES OF MICROORGANISMS AND ORIGINS OF ORGANIC COMPOUNDS

Microorganisms excel at using organic substances, natural or synthetic, as sources of nutrients and energy. These include certain human-made compounds—detergents, transformer fluids (polychlorobiphenyls), and solvents (trichloroethane, toluene, xylenes)—that seem very different from any natural compounds such an organism would be likely to encounter. The explanation for this remarkable range of degradative abilities is that by the time humans came on the scene, microorganisms had already coexisted for billions of years with an immense variety of organic compounds. The vast diversity of potential substrates for growth led to the evolution of enzymes capable of transforming many unrelated natural organic compounds by many different catalytic mechanisms. The resulting giant "library" of microbial enzymes serves as raw material for further evolution whenever a new synthetic organic chemical becomes available: Mutations in existing enzymes generate catalysts capable of utilizing new substrates; the possessor of such an enzyme, which is capable of degrading a substrate that other organisms cannot, gains a growth advantage or the ability to exploit a new ecological niche.

How many natural organic chemicals are there, and how are these compounds generated? Scientists have already identified many thousands of them, but the list is far from complete. In addition to the organic constituents typical of all living organisms, the natural world contains an abundance of chemicals produced either by metabolic pathways unique to specific groups of organisms or by biogeochemical processes. For example, mixtures of highly reduced carbon compounds—coal, petroleum, and natural gas—are produced by the combined effects of high pressure and heat on buried remains of plants and phytoplankton.

Petroleum originates primarily from phytoplankton residues that accumulated in the depressions of shallow seas. It consists of a complex array of gaseous, liquid, and solid *n*-alkanes, branched paraffins, cyclic paraffins and substituted cycloparaffins, aromatic compounds, sulfur compounds including benzo(*b*)thiophene and dibenzothiophene, and many other organic compounds (Box 15.1). These compounds re-enter the biosphere as a result of both upward seepage through porous rocks and sediments and various kinds of geological upheavals. About a million metric tons of petrochemicals annually are transported into the oceans and the surface soil of the continents in these ways, there to be fed on by microorganisms. Microorganisms even grow in certain underground petroleum reservoirs when a steady seepage of ground water brings them sufficient supplies of mineral nutrients and oxygen. In fact, it is estimated that some 10% of global oil deposits have been destroyed in this manner.

► BOX 15.1 **Petroleum Constituents**

The major constituents of crude oils (petroleum) are paraffins (30–50%), cycloparaffins (20–65%), and aromatics (6–14%). The actual amounts depend on the origin of the crude oil.

Paraffins include compounds such as *n*-heptane, *n*-hexane, 2-methylpentane, and 2,3-dimethylhexane. Cycloparaffins include cyclohexane, methylcyclopentane, methylcyclohexane, and trimethylcyclopentane. Toluene and benzene are major aromatic components. Petroleum contains many other organic compounds.

Coal, which is derived chiefly from terrestrial plant matter, is another common reservoir of highly reduced carbon compounds. It is made up of aromatic rings fused into different small polycyclic clusters that are linked by aliphatic structures. The aromatic rings carry phenolic, hydroxyl, quinone and methyl substituents (Figure 15.1). Coal, too, contains a variety of sulfur compounds. Many microorganisms can utilize the carbon compounds in coal as substrates and can metabolize the sulfur compounds also.

Organohalogen compounds such as polychlorobiphenyls (PCBs), dichlorodiphenyltrichloroethane (DDT), and trichloroethane are widespread and long-lived human-made contaminants. In recent years, researchers have discovered that organohalogen compounds also occur naturally in many parts of the world. Soil samples collected from far-flung sites that are unpolluted, or at least are only minimally contaminated by long-range atmospheric transport of industrially produced compounds, contain organohalogens in significant amounts. Indeed, the ratio of carbon-bound halogen to total organic carbon ranges from 0.2 to 2.8 mg of chloride per gram of carbon in soil worldwide, and the global contribution of human-made organohalogen compounds can constitute only a small fraction of this amount. Various marine organisms and many plants produce halogenated organic compounds. So do fungal haloperoxidases, through action on soil compounds derived from lignin degradation. Thus microorganisms have long faced the challenge of breaking down many natural organohalogen compounds. Not surprisingly, the discovery of the abundance of natural organohalogens has been accompanied by the discovery of microorganisms able to degrade all but the most highly halogenated synthetic organic compounds.

SEWAGE AND WASTEWATER MICROBIOLOGY

Arguably the most important, if the least glamorous, practical application of microbiology is in the treatment of sewage and wastewater. The amounts of waste materials discharged into bodies of water around the world are rising steadily. Increases in population, the dependence of agriculture on massive amounts of fertilizers and pesticides, expansion

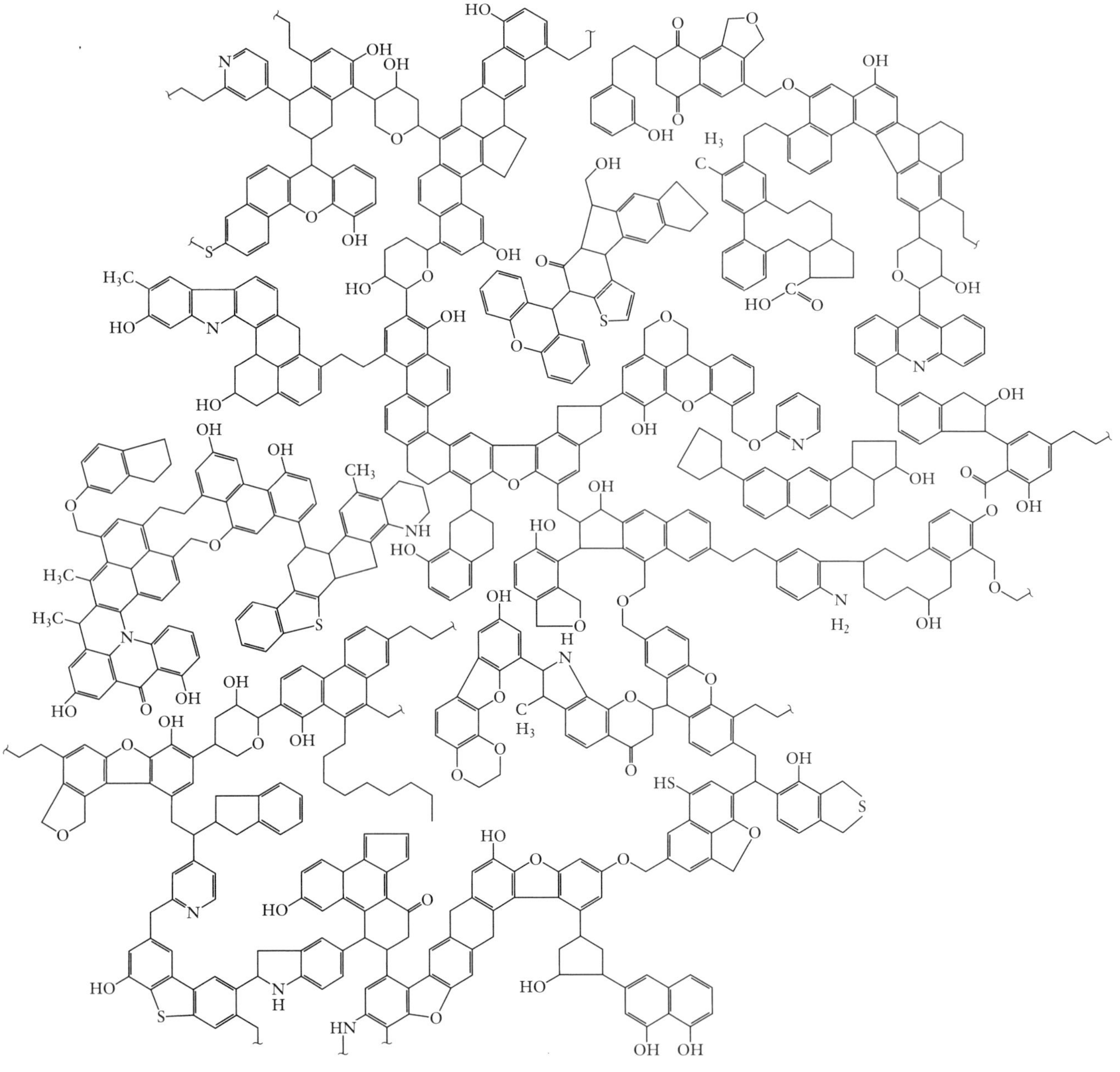

Figure 15.1
Model of the structure of bituminous coal. Reference: Crouch, G.R. (1990), Biotechnology and coal: A European perspective, in Wise, D.L. (ed.), *Bioprocessing and Biotreatment of Coal*, pp. 29–55 (Marcel Dekker).

of the food industry, and growth of other industrial processes all contribute to the volume of sewage and wastewater and to their content of undesirable substances. The volume of water affected by human activities is huge. Water use worldwide appropriates about 10% of the average runoff in all continental river basins. Of this usage, irrigation consumes some 70%, industry about 20%, and livestock, domestic, and other uses the rest.

The challenge of wastewater treatment is to remove (1) compounds with a high biochemical oxygen demand, (2) pathogenic organisms and viruses, and (3) a multitude of human-made chemicals. The price of failure to protect the purity of groundwater and that of lakes and rivers is very high. In developing countries that have insufficient sewage treatment facilities (particularly in rural areas), and thus have water supplies contaminated with fecal matter, diarrheal diseases are estimated to cause about 10% of all deaths. Other common, communicable water-borne diseases includes cholera, hepatitis, paratyphoid, and typhoid.

Treatment Plant Processing of Sewage and Wastewater

The wastewater stream entering a treatment plant typically goes through the operations that are diagrammed in Figure 15.2.

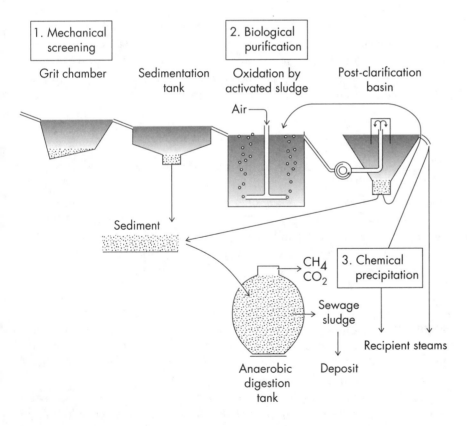

Figure 15.2
Flow of material through a sewage treatment plant employing both biological and chemical purification. Based on Schlegel, H.G. (1986), *General Microbiology*, 6th ed., Fig. 17.3 (Cambridge University Press).

1. The *primary treatment* is the removal of large objects (such as rags and pieces of wood) by passing the wastewater through a series of screens. Rapidly sedimentable solid particles ("grit") next settle out during slow flow through a grit chamber. The effluent then settles in a sedimentation tank.

2. The *secondary treatment* is microbiological and exploits the degradative capabilities of various bacteria and fungi. The supernatant from the sedimentation tank, containing dissolved organic materials, undergoes microbial oxidation in an aerated tank. The sediment from the tank is transferred to an anaerobic digestion tank for degradation by a different population of microorganisms.

   After a period of microbial oxidation, effluent from the aerated tank goes to a second sedimentation tank. The sediment from this tank is also transferred to the anaerobic digestion tank. The supernatant from the second clarification tank can be released to a natural body of water. However, it contains a high level of the inorganic nutrients phosphate and nitrate.

3. *Tertiary treatment* is necessary to prevent eutrophication (reduction in the dissolved oxygen in the water, resulting from bacterial and algal growth promoted by the discharge of phosphate and nitrate) and to render the treated water potable. Chemical precipitation steps remove phosphate and nitrate, and disinfection is achieved by chlorination.

Microbes in Secondary Treatment

The purpose of secondary treatment is to lower the biochemical oxygen demand (BOD; see Box 1.2) of liquid waste. *BOD is a measure of the amount of dissolved oxygen taken up by aerobic microorganisms as they metabolize the degradable organic material in the waste.* BOD values for different types of wastes are given in Table 15.1. Release of excessive amounts of organic waste into natural waters leads to exhaustion of dissolved oxygen, with the concomitant death of obligately aerobic microorganisms, invertebrates, and fish.

Aerobic Secondary Treatment Process. In the aerobic phase of the treatment process, a mixed microbial population degrades the complex mixture of organic compounds in liquid waste to carbonate, nitrate, ammonia, phosphate, and sulfate. Either trickling (percolating) filter tanks or activated sludge tanks are used for this purpose.

A *trickling filter tank* is 3–10 feet deep, packed with either crushed rock or some other inert support material on which the microbial population forms a thin film. The liquid waste is applied evenly to the top of the filter and percolates downwards, depositing organic matter on the support and the microbial film. Aerobic conditions are maintained in

Representative BOD Values for Sewage and Wastes from Agriculture and the Food Industry

T
A
B
L
E

15.1

Source	BOD (mg/liter)[a]
Domestic sewage	200–600
Cattle shed and piggery effluents	10,000–25,000
Dairies	500–2000
Whey from cheese making	40,000–50,000
Meat packing and processing	100–3000
Fruit and vegetable canning	200–5000
Sugar refining	200–2000
Breweries	500–2000
Distilleries	>5000
Palm oil processing	15,000–25,000

[a] Milligrams of dissolved oxygen consumed per liter on incubation for 5 days at 20°C.

SOURCE: Grainger, J.M. (1987), Microbiology of waste disposal systems, in Norris, J.R., and Pettipher, G.L. (eds.), *Essays in Agricultural and Food Microbiology*, pp. 105–134 (Wiley).

the filter by an upward flow of air through the spaces between the particles of the packed solid support. Heterotrophic organisms in the upper part of the filter obtain energy and nutrients by oxidizing the organic matter. Thus their numbers multiply and new film is formed.

The mixed microbial population in the upper layers of trickling filters consists mainly of aerobic chemoheterotrophic bacteria and fungi. This population arises by selection, from the microorganisms initially present in the waste, of the types best adapted to the nutritional and physical conditions in the treatment system. Characteristic Gram-negative bacterial genera include *Zoogloea*, *Pseudomonas*, *Alcaligenes*, *Achromobacter*, and *Flavobacterium*. Gram-positive bacteria, mainly coryneform, are also present. Frequently represented fungal genera include *Sepedonium*, *Subbaromyces*, *Ascoidea*, *Fusarium*, *Geotrichium*, and *Trisporon*. Certain of the chemoheterotrophs degrade many organic nitrogen-containing substrates, with concomitant release of ammonia. In the lower layers of the filter, the autotrophic nitrifying bacteria oxidize ammonia to nitrite (*Nitrosomonas*) and then nitrite to nitrate (*Nitrobacter*). Protozoa are present in large numbers in sewage treatment plants; they consume bacteria freely suspended in the film, thus preventing the build-up of excess film and clarifying the filter effluent. A grazing fauna that consumes protozoa and bacteria is also present. It includes nematodes, rotifers, worms, and larval flies.

The effluent from the filter tank is transferred to a sedimentation tank where suspended solids (sloughed microbial film, solid waste from the grazer community, and particles of undegraded organic matter) are allowed to settle out. The BOD has now been reduced by up to 95%. The efficiency of a trickling filter, as measured by removal of BOD, depends primarily on contact area and contact time. Contact area varies with the size and shape of the particles in the solid support material. Contact time is controlled by the rate of sewage loading.

In the *activated sludge process*, sewage that has received primary treatment is mixed with "activated" sludge (an inoculum of microorganisms) and continually aerated, occasionally with pure oxygen, for periods of up to 15 hours. The organisms that degrade organic matter in an activated sludge tank are the same as those in the trickling filter tank, but slime-forming bacteria, primarily *Zoogloea ramigera*, play a particularly important role. As these organisms grow, they form clumps called flocs to which soluble organic matter, as well as protozoa and other organisms, become attached. When the effluent from an activated sludge tank passes into a sedimentation tank, these flocs, and all they are carrying, sediment out and are transferred to an anaerobic sludge digestor. A small proportion of the sediment is returned to the aeration tank to act as a microbial inoculum for fresh incoming sewage. In the activated sludge process, much of the reduction in BOD occurs as a result of this floc-dependent removal of organic matter.

Anaerobic Secondary Treatment Process. After a period of aerobic digestion, sewage sludge is generally digested under anaerobic conditions as well. Anaerobic digestion is also commonly used to treat materials with a high content of insoluble organic matter, such as cellulose, and to degrade concentrated industrial wastes such as those from the food-processing industry.

The degradative and fermentative reactions in the anaerobic secondary treatment process can be divided into two stages: acid-forming and methane-forming (Figure 15.3). Numerous different bacterial species participate in the acid-forming stage. These species include some that are known to be obligate aerobes but that may grow by utilizing alternative electron acceptors, such as nitrate (see Chapter 2, page 58). In the methane-forming stage, strict anaerobes of the genera *Methanobacterium*, *Methanobacillus*, *Methanococcus*, and *Methanosarcina* convert the acetate, hydrogen, and carbon dioxide produced by the fermenters to methane (Box 15.2).

In the acid-forming stage, complex organic polymers, including carbohydrates, fats, and proteins, are hydrolyzed by extracellular hydrolytic enzymes (polysaccharidases, lipases, and proteases) and converted to volatile (short-chain) fatty acids, alcohols, and ketones. The fatty acids are fermented to acetate, carbon dioxide, and hydrogen (Table 15.2). Interestingly, some of the reactions listed in Table 15.2 (for example, the oxidation of butyrate by H_2O) have a large positive $\Delta G^{\circ\prime}$ and would not have been predicted to occur. These unusual fermentation reactions occur in the anaerobic sludge because the partial

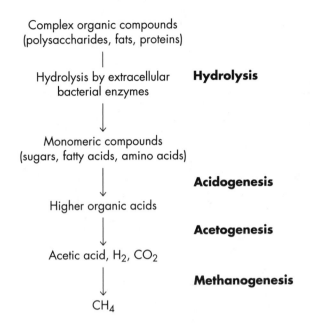

Complex organic compounds
(polysaccharides, fats, proteins)

|

Hydrolysis by extracellular   **Hydrolysis**
bacterial enzymes

↓

Monomeric compounds
(sugars, fatty acids, amino acids)

↓   **Acidogenesis**

Higher organic acids

↓   **Acetogenesis**

Acetic acid, H_2, CO_2

↓   **Methanogenesis**

CH_4

Figure 15.3
The major metabolic stages in the decomposition of organic wastes into methane and carbon dioxide.

pressure of H_2, one of the products in these fermentation reactions, is kept exceptionally low, at 10^{-4} to 10^{-5} atmospheres, by the consumption of H_2 by CO_2-reducing methanogens and hydrogen-consuming acetogenic bacteria (see below).

BOX 15.2   Methanogens

Methane (CH_4) is produced in many natural anaerobic habitats where organic matter is degraded by microorganisms and carbon dioxide is the only available electron acceptor—in other words, in the absence of oxygen, sulfate, or nitrate. Alessandro Volta (1745–1827), the Italian physicist famous for his invention of the electric battery, was the first to discover, in 1776, that "combustible air" (CH_4) was generated in sediments rich in organic matter at the bottom of marshes, lakes, and rivers. All methanogens—microorganisms responsible for CH_4 production— are archaebacteria that derive their energy by reduction of one of several compounds (such as CO_2, acetate, or methanol) by H_2 with the formation of CH_4.

Methanogens are found in abundance in diverse natural anaerobic habitats, such as the rumen and intestinal tract of animals, freshwater and marine sediments, and water-logged soils, and near volcanic hot springs and deep-sea hydrothermal vents. Many are thermophiles with growth temperature optima between 60 and 95°C.

Between 500 and 800 million tons of biologically generated CH_4 are released into the atmosphere per year. This is equivalent to about 2% of the carbon dioxide fixed annually by photosynthesis. A cow produces about 200 liters of CH_4 per day, which it releases by belching.

⤳ **Examples of Reactions Occurring in an Anaerobic Digestion Tank** ⋯⋯
and Their Free Energies under Standard and "Typical" Conditions

			Free energy[a] (kcal/reaction)	
Reaction	*Reactants*	*Products*	$\Delta G^{0\prime}$	$\Delta G'$
Conversion of glucose to CH_4 and CO_2	Glucose + $3H_2O$	$3CH_4 + 3HCO_3^- + 3H^+$	−96.5	−95.3
Conversion of glucose to acetate and H_2	Glucose + $4H_2O$	$2CH_3COO^- + 2HCO_3^- + 4H^+ + 4H_2$	−49.3	−76.1
Methanogenesis from acetate	$CH_3COO^- + H_2O$	$CH_4 + HCO_3^-$	−7.4	−5.9
Methanogenesis from H_2 and CO_2	$4H_2 + HCO_3^- + H^+$	$CH_4 + 3H_2O$	−32.4	−7.6
Acetogenesis from H_2 and CO_2	$4H_2 + 2HCO_3^- + H^+$	$CH_3COO^- + 2H_2O$	−25.0	−1.7
Amino acid oxidation	Leucine + $3H_2O$	Isovalerate + $HCO_3^- + NH_4^+ + 2H_2$	+1.0	−14.2
Butyrate oxidation to acetate	Butyrate + $2H_2O$	$2CH_3COO^- + H^+ + 2H_2$	+11.5	−4.2
Propionate oxidation to acetate	Propionate + $3H_2O$	$CH_3COO^- + HCO_3^- + H^+ + 3H_2$	+18.2	−1.3
Benzoate oxidation to acetate	Benzoate + $7H_2O$	$3CH_3COO^- + HCO_3^- + 3H^+ + 3H_2$	+21.4	−3.8
Reductive dechlorination	$H_2 + CH_3Cl$	$CH_4 + H^+ + Cl^-$	−39.1	−29.0

TABLE 15.2

[a] For the calculation of $\Delta G^{0\prime}$ the standard conditions are as follows: solutes, 1 molar; gases, 1 atmosphere; 25°C; pH 7. For the calculation of $\Delta G'$, the "typical" conditions for an anaerobic digestion tank were estimated to be 37°C and pH 7, and the concentrations of products and reactants were as follows: glucose, leucine, benzoate, and methyl chloride (CH_3Cl), 10 micromolar; acetate, butyrate, propionate, and isovalerate, 1 millimolar; HCO_3^- and Cl^-, 20 millimolar; CH_4, 20 millimolar; CH_4, 0.6 atmospheres; and H_2, 10^{-4} atmospheres.

SOURCE: Zinder, S.H. (1984), Microbiology of anaerobic conversion of organic wastes to methane: Recent developments, *ASM News* 50:294–298.

In the methane-forming stage, under the conditions prevailing in an anaerobic digestion tank (see Table 15.2), methanogenic bacteria derive energy by converting hydrogen plus carbon dioxide and acetate to methane by the following reactions:

$$4H_2 + HCO_3^- + H^+ \longrightarrow CH_4 + 3H_2O \qquad \begin{aligned} \Delta G^{\circ\prime} &= -32.4 \text{ kcal/reaction} \\ \Delta G' &= -7.6 \text{ kcal/reaction} \end{aligned}$$

$$CH_3COO^- + H_2O \longrightarrow CH_4 + HCO_3^- \qquad \begin{aligned} \Delta G^{\circ\prime} &= -7.4 \text{ kcal/reaction} \\ \Delta G' &= -5.9 \text{ kcal/reaction} \end{aligned}$$

where $\Delta G'$ refers to the reactant and product concentrations in an anaerobic digestion tank at 37°C and pH 7 (see Table 15.2). Note that the first of these reactions is so strongly favored that $\Delta G'$ remains negative even at 10^{-4} atmospheres of H_2.

Anaerobic organisms in general show a high degree of metabolic specialization. The success of the anaerobic digestion process therefore depends on the cooperative interactions between microorganisms with different metabolic capabilities. This is illustrated by the data in Table 15.2. These data show that the negative-free energy change associated with the anaerobic oxidation of amino acids, butyrate, propionate, and benzoate by fermentative and hydrogen-producing acetogenic bacteria results from the very low partial pressure of hydrogen in the anaerobic digestion tank. An important practical feature of anaerobic digestion is that most of the free energy present in the substrate is conserved in the methane that is produced. Using the methane for energy generation offsets the cost of the sewage treatment process.

Degradation of Synthetic Organic Compounds During Sewage Treatment

Many human-made organic compounds are degraded during sewage treatment. The extent of degradation depends critically on the rate of biodegradation of the compound in question. If the rate is slow, the duration of the compound's residence in the sewage treatment plant may be too short for complete degradation. The fate of alkylbenzene sulfonate detergents illustrates this point.

A highly branched alkylbenzene sulfonate (BAS) was first introduced as a surfactant for synthetic household detergents in the 1940s and has been used worldwide since the 1950s. Current usage of BAS is approximately 1.2 billion pounds per year. Because of the environmental concerns described below, a linear alkylbenzene sulfonate (LAS) was developed in the 1960s. Worldwide usage of LAS is about 3.2 billion pounds per year.

BAS

LAS

Alkylbenzene sulfonates enter lakes, rivers, and the ocean as components of household sewage (at a concentration of $1–20$ ppm in the sewage). Aqueous solutions of alkylbenzene sulfonates foam at these concentrations when agitated, causing problems in sewage treatment plants and at their outfalls. Moreover, at levels of $1–5$ ppm in clean water, surfactants such as alkylbenzene sulfonates are toxic to some fish. These problems were solved by the substitution of LAS for BAS.

In spite of the fact that BAS and LAS are compounds foreign to the natural world, they are biodegradable. The rate-limiting step in the microbial decomposition of these detergents is cleavage of the alkyl chain from the benzene sulfonate head group. Thereafter, biodegradation of the resulting fatty acid proceeds by the pathways used for the β-oxidation of naturally occurring fatty acids to CO_2 and water. The benzene sulfonate is degraded to CO_2, water, and sulfate. Because of the branching of its alkyl group, the initial step in the degradation of BAS is more than one order of magnitude slower than that in the degradation of LAS. Consequently, a secondary aerobic treatment time that suffices to decrease the BOD of sewage by over 90% is grossly insufficient for the satisfactory degradation of BAS, but it allows for the virtually complete degradation of LAS.

> "My first awareness of this subject came when, in the mid-1950's, I was released from the armed services and returned to a woodland near my home to observe the spring bird migration. A small river wound its way through the woodland, and to my surprise I observed suds developing wherever there was a small waterfall in the river. The suds floated downstream in swan-like masses, sometimes forming a blanket which covered many square metres of the stream. Several miles upstream I found the source of the suds: a small sewage treatment plant with an outfall in the river. As I later learned, the suds were a consequence of the use of branched-chain alkylbenzene surfactants in the manufacture of synthetic detergents. It was all too evident that the detergents were passing through the treatment plant without being broken down." Wright, R.T. (1987), Microbial degradation of organic compounds in soil and water, in Norris, J.R., and Pettipher, G.L. (eds.), *Essays in Agricultural and Food Microbiology*, pp. 75–103 (Wiley).

The case of the alkylbenzene sulfonates highlights two very important points. The first is that the complex microbial populations present in the aerobic and anaerobic sewage treatment tanks can degrade both naturally occurring compounds and synthetic ones. In fact, the degradation of a number of other industrial synthetic waste products (such as phenols and chlorobenzenes) has been shown to be particularly efficient in the presence of large amounts of sewage sludge. The second point is that it is important to conduct studies of biodegradability under real-world conditions. It is not sufficient to establish that a compound is biodegradable. It must also be shown that the compound is degraded rapidly enough in the treatment facility to ensure its removal from the environment. In many products and processes, it may be possible to substitute a compound that is more readily decomposed for one that biodegrades slowly. This point, which now seems self-evident, was first appreciated when abundant foam was recognized as a sign of contamination by detergent.

Biodegradation of Organic Compounds in Groundwater Under Anaerobic Conditions

In many parts of the world, aquifers (water-bearing strata of earth, gravel, or rock) containing essential groundwater supplies are contaminated with toxic chemicals that leach from terrestrial dump sites or enter the ground from other sources because of improper handling or storage. Contamination of groundwater also results from massive treatment of land with fertilizers and pesticides.

Concern about the presence of undesirable organic contaminants (nonhalogenated and halogenated aromatic hydrocarbons, haloalkanes, and the like) in groundwater has led to extensive studies of their biodegradation. One important finding has been that some of these compounds are degraded under both aerobic and anaerobic conditions, and some only under aerobic conditions. Still others may be recalcitrant under aerobic conditions but readily degradable under anaerobic conditions. The fate of a particular compound is largely decided by its intrinsic chemical properties and by the metabolic capabilities of microorganisms that have access to it.

A list of the possible reactions that chlorinated aliphatic hydrocarbons can undergo illustrates the influence of chemical properties. Increased chlorination increases the electrophilicity and oxidation state of an aliphatic hydrocarbon, making it more susceptible to dehydrohalogenation and reduction and less susceptible to substitution and oxidation. An example of each of these types of reactions follows.

1. Dehydrohalogenation $CH_3CCl_3 \rightarrow CH_2{=}CCl_2 + HCl$
2. Reduction $CCl_4 + H^+ + 2e^- \rightarrow CHCl_3 + Cl^-$

3. Substitution

$$CH_3CH_2CH_2Cl + H_2O \rightarrow$$
$$CH_3CH_2CH_2OH + HCl$$

4. Oxidation

$$CH_3CHCl_2 + H_2O \rightarrow$$
$$CH_3CCl_2OH + 2H^+ + 2e^-$$

The fate of the dry cleaning solvent perchloroethylene (tetrachloroethylene), a common groundwater contaminant, serves as an example. Because of the highly oxidized nature of this compound, it is very stable in the environment under aerobic conditions. However, an anaerobic methanol–perchloroethylene enrichment culture, which uses methanol as the electron donor for the reductive dehalogenation, completely converts perchloroethylene to ethylene. The details of the biochemistry of this system are not yet known. Mixed cultures of the strictly anaerobic acetate-utilizing methanogens catalyze the stepwise reductive dehalogenation of perchloroethylene to vinyl chloride.

$$CCl_2\!\!=\!\!CCl_2 \xrightarrow{\;H_2\;\; HCl\;} CHCl\!\!=\!\!CCl_2 \xrightarrow{\;H_2\;\; HCl\;} CH_2\!\!=\!\!CCl_2 \xrightarrow{\;H_2\;\; HCl\;} CH_2\!\!=\!\!CHCl$$
$$\text{or}$$
$$CHCl\!\!=\!\!CHCl$$

In summary, because highly chlorinated compounds are highly oxidized, on thermodynamic grounds they could serve as excellent electron acceptors for bacteria where alternative acceptors (such as oxygen) are absent. Moreover, where suitable enzyme systems exist, heavily chlorinated compounds would be preferentially dechlorinated, because a bacterium could transfer more electrons by reductively dehalogenating such compounds rather than those with lower degrees of chlorination. These considerations explain how the reductive dehalogenation of a compound such as perchloroethylene may provide a microorganism with sufficient energy to support growth.

Oxidative Dehalogenation in Groundwater Under Anaerobic Conditions

In Chapter 2 we emphasized that a much wider range of modes for energy generation is seen among bacteria than among eukaryotes. This facet of microbial metabolism plays a decisive role in enabling microorganisms to carry out oxidative dehalogenation reactions in the absence of oxygen.

Most uncontaminated aquifers contain significant amounts of oxygen. However, continuous contamination of the aquifer with excess dissolved organic carbon compounds leads to oxygen depletion. The resulting anoxic conditions favor the growth of microorganisms that utilize electron acceptors other than oxygen, such as nitrate, ferric ion, sulfate, and bicarbonate.

Sulfate and bicarbonate are present in abundance in shallow subsurface zones. It is therefore not surprising that sulfate-reducing bacteria and bicarbonate-reducing (methanogenic) bacteria are particularly numerous in anaerobic aquifers. In fact, it has long been recognized that sulfate-reducing microorganisms make an important contribution to the oxidation of carbon compounds in anaerobic marine sediments and coastal salt marshes: At least half of the organic matter in marine sediments is mineralized by sulfate-reducing bacteria.

The sulfate reducers, the methanogens, and the nitrate reducers (denitrifying bacteria) can degrade a variety of nonhalogenated and halogenated aromatic compounds as well as various halogenated aliphatic compounds, and individual strains differ greatly in the range of their catabolic capabilities. Working together, these microorganisms are able to degrade many of the human-made compounds that enter terrestrial subsurface anaerobic zones. However, the subsurface microbial population may be limited in some locations by extremely low levels of ammonium ion, nitrate, and phosphate, and there the degradation may be slow.

MICROBIOLOGICAL DEGRADATION OF XENOBIOTICS

As we noted at the beginning of this chapter, biological and geochemical processes produce enormous quantities of organic compounds with a great diversity of structures. Nearly every one of these compounds can be utilized by some microorganism as a source of energy and/or cell building blocks. However, many of the tens of thousands of organic compounds produced artificially by chemical synthesis for industrial or agricultural purposes have no obvious counterparts in the natural world. Such synthetic novel compounds are called *xenobiotics* (*xenos* means "foreign" in Greek), and many of these compounds are stable in the environment under both aerobic and anaerobic conditions.

The ever-growing list of xenobiotics released into the environment on a large scale includes numerous halogenated aliphatic and aromatic compounds, nitroaromatics, phthalate esters, and polycyclic aromatic hydrocarbons. These compounds enter the environment through many different paths. Some, as components of fertilizers, pesticides, and herbicides, are distributed by direct application. Others, such as the polycyclic aromatic hydrocarbons, dibenzo-*p*-dioxins, and dibenzofurans, are released by combustion processes. And of course many kinds of xenobiotics are found in the waste effluents produced by the manufacture and consumption of all the commonly used synthetic products.

Various xenobiotics are found in particular environments in concentrations ranging from parts per thousand (ppt) to parts per billion (ppb). The local concentration depends on the amount of the compound released, the rate at which it is released, the extent of its dilution in the environment, the mobility of the compound in a particular environment (for example, in soil), and its rate of degradation, both biological and nonbiological. Many toxic xenobiotics present in the environment in parts per billion (levels at which toxicity cannot be demonstrated) are nevertheless strictly regulated. Such regulation is necessary when a compound becomes progressively more concentrated in each link of a food chain—a process called *biomagnification.*

The first study to measure biomagnification was carried out in Clear Lake in northern California. In 1949 Clear Lake had been treated with the persistent pesticide dichlorodiphenyldichloroethane (DDD, a close relative of DDT) at 0.01–0.02 parts per million (ppm) of water to control the gnat *Chaoborus astictopus.* By 1954 western grebes, ducklike birds, began dying around the lake. The levels of DDD in the grebes' body fat were found to be 1600 ppm, some 100,000 times higher than the DDD concentration in lake water. The DDD was accumulating to progressively higher concentrations first in the plankton in the water, then in the fish that ate the plankton, and finally in the grebes that ate the fish.

Many other persistent fat-soluble organic compounds become increasingly concentrated as they travel up the food chain. Prominent examples are the phthalate esters and the PCBs discussed in detail below.

Both actual measurements and theoretical models indicate that biomagnification of persistent chemicals within food chains is not universal or inevitable. Extensive field data on Cd^{2+} and Cu^{2+}, for example, indicate that it is the effectiveness of active excretion of heavy metal ions by a given organism, rather then the position of the organism in a food chain, that determines the ions' concentration in terrestrial animals. For a particular persistent chemical, biomagnification cannot be assumed *a priori.*

Priority Pollutants and Their Health Effects

The U.S. Environmental Protection Agency's list of "priority pollutants" (Table 15.3) includes widely used industrial solvents, building blocks of plastics, PCBs, pesticides, and certain potent carcinogens. Some of these compounds are or have been produced in massive amounts. For example, billions of pounds of *o*-phthalic and terephthalic acids have been used in the plastics and textile industries. Phthalic acid esters are the most important class of plasticizers for cellulose and vinyl

Environmental Protection Agency (EPA) Priority Pollutant List[a,b]

Purgeable (Volatilizable) Organic Compounds

Acrolein	1,1,2-Trichloroethane	Bromoform
Acrylonitrile	1,1,2,2-Tetrachloroethane	Dichlorobromomethane
Benzene[c]	Chloroethane	Trichlorofluoromethane
Toluene	2-Chloroethyl vinyl ether	Dichlorodifluoromethane
Ethylbenzene	Chloroform	Chlorodibromomethane
Carbon tetrachloride	1,2-Dichloropropane	**Tetrachloroethylene**
Chlorobenzene	1,3-Dichloropropene	**Trichloroethylene**
1,2-Dichloroethane	**Methylene chloride**	Vinyl chloride
1,1,1-Trichloroethane	Methyl chloride	1,2-*trans*-Dichloroethylene
1,1-Dichloroethane	Methyl bromide	bis(Chloromethyl) ether
1,1-Dichloroethylene		

Compounds Extractable Into Organic Solvent Under Alkaline Or Neutral Conditions

1,2-Dichlorobenzene	Di-*n*-octyl phthalate	Benzo(k)fluoranthene
1,3-Dichlorobenzene	Dimethyl phthalate	Benzo(a)pyrene
1,4-Dichlorobenzene	Diethyl phthalate	Indeno(1,2,3-c,d)pyrene
Hexachloroethane	**Di-*n*-butyl phthalate**	Dibenzo(a,h)anthracene
Hexachlorobutadiene	Acenaphthylene	Benzo(g,h,i)perylene
Hexachlorobenzene	Acenaphthene	4-Chlorophenyl phenyl ether
1,2,4-Trichlorobenzene	Butyl benzyl phthalate	3,3′-Dichlorobenzidine
bis(2-Chloroethoxy)methane	Fluorene	Benzidine
Naphthalene	Fluoranthene	bis(2-Chloroethyl) ether
2-Chloronaphthalene	Chrysene	1,2-Diphenylhydrazine
Isophorone	Pyrene	Hexachlorocyclopentadiene
Nitrobenzene	**Phenanthrene**	N-Nitrosodiphenylamine
2,4-Dinitrotoluene	**Anthracene**	N-Nitrosodimethylamine
2,6-Dinitrotoluene	Benzo(a)anthracene	N-Nitrosodi-*n*-propylamine
4-Bromophenyl phenyl ether	Benzo(b)fluoranthene	bis(2-Chloroisopropyl) ether
bis(2-Ethylhexyl) phthalate		

Compounds Extractable Into Organic Solvent Under Acid Conditions

Phenol	4,6-Dinitro-*o*-cresol	2,4-Dichlorophenol
2-Nitrophenol	Pentachlorophenol	2,4,6-Trichlorophenol
4-Nitrophenol	*p*-Chloro-*m*-cresol	2,4-Dimethylphenol
2,4-Dinitrophenol	2-Chlorophenol	

(continued)

plastics and are also used in insect repellents, munitions, and cosmetics and as pesticide carriers. It is estimated that over 50 million pounds of phthalate esters enter the environment yearly in the United States by leaching out of solid plastic wastes and as a result of direct application (for example, as pesticide carriers). Phthalate contamination goes hand in hand with the pervasive use of plastics. In the early 1970s, phthalate esters were discovered in blood that had been collected for transfusion and stored in plastic bottles.

Table 15.3 *continued*

Pesticides, Polychlorobiphenyl (PCB) And Related Compounds

α-Endosulfan	4,4'-DDE	Toxaphene
β-Endosulfan	4,4'-DDD	Aroclor 1016[d]
Endosulfan sulfate	4,4'-DDT	Aroclor 1221
α-BHC	Endrin	Aroclor 1232
β-BHC	Endrin aldehyde	Aroclor 1242
γ-BHC	Heptachlor	Aroclor 1258
Aldrin	Heptachlor epoxide	Aroclor 1254
Dieldrin	Chlordane	Aroclor 1260
		2,3,7,8-Tetrachlorodibenzo-*p*-dioxin (TCDD)

Metals

Antimony	Copper	Selenium
Arsenic	Lead	Silver
Beryllium	Mercury	Thallium
Cadmium	Nickel	Zinc
Chromium		

Miscellaneous

Cyanides	Asbestos (fibrous)

[a] The EPA list of priority pollutants was developed as a consequence of a court Consent Decree on June 7, 1978, in the settlement of a suit brought against the EPA by several plaintiffs (Natural Defense Council, Inc.; Environmental Defense Fund, Inc., Businessmen for the Public Interest, Inc.; National Audubon Society, Inc.; and Citizens for a Better Environment) for failing to implement portions of the Federal Water Pollution Control Act (P.L. 92-500).

[b] The priority pollutants are divided into groups on the basis of properties that are relevant to the analysis for these compounds in industrial wastewaters.

[c] Compounds shown in boldface type were found in 10% or more of over 2600 samples of wastewater from 32 different industrial categories analyzed in August 1978.

[d] Aroclor designations are explained in Box 15.3.

SOURCE: Keith, L.H. and Telliard, W.A. (1979), Priority pollutants I—A perspective view, *Env. Sci. Technol.* 13:416–423.

Commercial PCBs are mixtures prepared by partial chlorination of biphenyl (Box 15.3). Their notable differences in physical properties, chemical stability, and miscibility with organic solvents are all due to the degree of biphenyl chlorination in a given mixture. Thus PCB formulations have served a wide range of purposes, including use as hydraulic fluids, plasticizers, adhesives, lubricants, flame retardants, and dielectric fluids in capacitors and transformers. Widespread PCB pollution was first detected in 1966 during analysis of environmental samples for DDT, and the manufacture of PCB was halted by 1977. Between 1929 and 1977, however, about 1.2 billion pounds of PCBs were produced in the United States, and it is estimated that several hundred million

BOX 15.3 Polychlorobiphenyls

PCBs are a class of 209 distinct synthetic chemical compounds, in which one to ten chlorine atoms are attached to biphenyl.

Biphenyl

The empirical formula for PCBs is $C_{12}H_{12-n}Cl_n$, where $n = 1-10$. Closely related compounds such as these are called *congeners*. PCB *isomers* are compounds with the same number of chlorine atoms—for example, 2′,3,4-trichlorobiphenyl and 2′,4,4′-trichlorobiphenyl.

PCBs were manufactured and sold as complex mixtures differing in their average chlorination level. The manufacturers attached, to the trade name of their products, numbers that conveyed information about the weight percent chlorine in the mixture. For example, Aroclor 1242 (Monsanto; see Table 15.3) indicates 12 carbon atoms and 42% chlorine by weight. Because of steric hindrance, only about half of the 209 possible congeners were actually produced in the synthesis of PCBs.

pounds have been released into the environment. PCBs biodegrade very slowly and will persist in the environment for decades.

Thousands of toxic waste sites in the United States release compounds on the priority pollutant list, as well as many others, to the environment. Substantial contamination of land, surface water, ground water, and air in virtually every part of the country has been exhaustively documented. Where the population has been exposed to high levels of such contaminants, there is convincing evidence of adverse health effects. Certain pollutants have achieved particular notoriety because of acute ill effects of accidental high-level exposures.

- *The Seveso Affair.* In 1976 an explosion at a chemical factory in the Italian town of Seveso released a cloud of vapor over the surrounding area. The vapor consisted primarily of trichlorophenol, but it also contained a total of 4–5 pounds of 2,3,7,8-tetrachlorodibenzo-*p*-dioxin (dioxin, TCDD). Dioxin is a by-product in the manufacture of 2,4,5-trichlorophenoxyacetic acid butyl ester, a defoliant extensively used in the Vietnam War. It is very toxic to some mammals; a lethal dose in guinea pigs is 0.6 μg/kg. In Seveso the most obvious effect on people exposed to the vapor cloud was chloracne, a persistent skin disease with disfiguring sores.

- *The Yusho and Yu Cheng Poisonings.* The Yusho poisoning, caused by the ingestion of rice oil that had been accidentally contaminated with PCBs, occurred in the summer of 1968 among residents of the western part of the Japanese archipelago. Over a thousand people were affected, and their illnesses—chloracne, headaches, nausea, and diarrhea—were traced to cooking oil that had been contaminated with a heat-transfer agent during manufacture. The oil contained 2000–3000 ppm of PCBs, polychlorinated dibenzofurans, and quaterphenyls (Figure 15.4). The latter two types of compounds are considerably more toxic than PCBs in animal tests. The average total level of ingestion of PCBs was estimated at 0.5–2.0 grams. A second chloracne epidemic caused by PCB-contaminated rice oil occurred in 1979 in Yu Cheng, a town in central Taiwan. Clinical manifestations were similar to those observed in the episode in Japan.

- *Love Canal, New York.* During the 1940s, thousands of 55-gal drums containing various toxic wastes were dumped into Love Canal, a partially excavated canal on the edge of Niagara Falls City. In 1953 the site was filled in and capped with clay. Later the site was purchased by the city, which built a primary school and playground on top of the site and a housing development nearby. Perhaps the construction work damaged the cap. In any case, rain water gained access to the waste. In time, leaking toxic waste contaminated water, soil, and air. Residents reported increases in health problems, including central nervous system disorders, miscarriages, stillbirths, and birth defects. In 1978 New York State declared a health emergency, closed the school, and moved the 235 families living nearest the site to new homes.

There are also numerous instances of population exposures to *low* levels of contaminants from toxic waste sites. For example, in San Jose, California, a municipal well was contaminated by waste leaking from an underground storage tank. 1,1,1-Trichloroethane and 1,1-dichloroethylene were found in the well, which was subsequently closed. Studies of health effects compared the rates of spontaneous abortions and birth

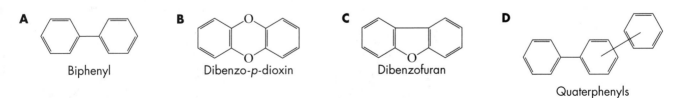

A Biphenyl   B Dibenzo-*p*-dioxin   C Dibenzofuran   D Quaterphenyls

Figure 15.4
Parent compounds (**A**) of chlorobiphenyls, (**B**) of chlorodibenzodioxins, (**C**) of chlorodibenzofurans, and (**D**) of quaterphenyls.

defects in the residents of the exposed area to rates in the rest of the country, with inconclusive and conflicting results.

There is no argument about the toxic effects of large-scale exposure to chemicals, such as that in Seveso. However, the health effects of low levels of chemical pollutants have been very difficult to document, and virtually all epidemiological studies indicating a link between such pollutants and increases in relative risks for a wide variety of illnesses are controversial. Nevertheless, there are strong reasons for minimizing the future release of chemicals into the environment and for eliminating those already present. It is well documented, for example, that chemicals that are not known to be harmful to humans can be highly toxic to other organisms. And as we have seen, many toxic chemicals enter food chains at a low level and, through biomagnification, reach concentrations sufficiently high to cause health problems for humans and other living organisms. Furthermore, health effects may surface a long time after the exposure, when the cause-and-effect relationship will be difficult to prove. *In summary, the incentive to free the environment of chemical pollutants is in large part a concern about future repercussions.*

Microbiological Basis of Biodegradation

Natural microbial communities are complex assemblages in which the various microorganisms (such as those listed in Table 15.4) are highly interdependent. This interdependence is evident in the high frequencies of commensalism and mutualism that are observed. *Commensalism* is an interactive association between two populations of different species that live together in which one population benefits from the association and

Population Densities of Soil Microorganisms in Samples Obtained from Common Soils in a Temperate Region

TABLE 15.4

Organism	Cells/g
Bacteria[a]	$10^6 - 10^9$
Actinomycetes	$10^5 - 10^8$ (based on spore number)
Filamentous hyphae	$10^1 - 10^2$ meters (hyphal length)
Yeasts	10^3
Algae and cyanobacteria	$10^2 - 10^4$
Protozoans	$10^4 - 10^6$

[a] Bacteria other than actinomycetes and cyanobacteria.

SOURCE: Based on data in Yanagita, T. (1990), *Natural Microbial Communities. Ecological and Physiological Features*, Table 11 (Japan Scientific Societies Press).

the other is not affected. *Mutualism* is a symbiosis, an interaction in which two organisms of different species live in close physical association to their mutual benefit.

Commensalism takes different forms. Many microorganisms that are isolated from soils require amino acids or vitamins for growth, and many other such organisms produce those compounds. For example, some 19% of these isolates require thiamine, whereas about 36% of the isolates secrete it. The corresponding numbers for vitamin B_{12} are about 7% and about 20%, respectively. Thus *crossfeeding* is a general feature of natural microbial communities. The interaction between organisms through the production and consumption of oxygen is also of particular importance; the ability of anaerobes to survive in surface layers of soil depends on the efficient consumption of oxygen by aerobes. Organic acids produced by fungal decomposition of cellulose are utilized as nutrients by bacteria. Ethanol that yeasts produce from sugars in fruit is oxidized to acetic acid by *Acetobacter* species. Methane (CH_4), the product of anaerobic breakdown of various organic compounds by methane-producing bacteria (*methanogens*), is oxidized aerobically by methane-oxidizing bacteria (*methylotrophs*).

Mutualistic interactions are likewise diverse. Where nitrogen fixers and cellulose decomposers coexist, for example, each organism utilizes compounds produced by the other. The anaerobic *Desulfovibrio* uses SO_4^{2-} as a terminal electron acceptor in its energy-producing respiratory pathway and converts it to H_2S; purple sulfur bacteria, which use sunlight for photosynthetic production of ATP, utilize the H_2S as an electron donor and oxidize it to SO_4^{2-}. *Lactobacillus arabinosus* and *Streptococcus faecalis* depend on each other to satisfy nutritional requirements; *L. arabinosus* makes the folic acid required by *S. faecalis*, and *S. faecalis* makes the phenylalanine that *L. arabinosus* needs.

As a general rule, organic compounds are more effectively degraded in environments containing many microorganisms than in a pure culture of a single organism. This is due to several factors. The range of degradative capabilities represented in a complex community of many bacteria and fungi is far greater than the capabilities of any single organism alone. Further, the product of partial biodegradation of a xenobiotic by one organism may serve as a substrate for another organism. The concerted action of several different organisms may lead to complete mineralization of the xenobiotic. A microbial community is also likely to be more resistant than a solitary species to a toxic product of biodegradation, because one of its members may be able to detoxify it.

A microbial community is dynamic: Its composition responds to environmental conditions, coming over time to exploit available nutrients in the most effective manner. Thus when a new biotransformable organic compound is presented at a constant level to such a community, a period of adaptation ensues, after which the rate of biotransformation of the organic compound is generally seen to be much increased. This reflects a selective enrichment within the community for organisms that are resistant to the new organic compound and are able to utilize it or transform it.

The fate of organic compounds introduced into the soil is determined by a combination of physical, chemical and biological factors. A particular molecule may be removed by volatilization or leaching, or it may be strongly adsorbed and remain near the site of entry for a long time. The molecule may be degraded photochemically, or it may undergo abiotic oxidation or hydrolysis. Finally, the molecule may undergo biodegradation through the action of bacteria and fungi. In some instances, the products of nonbiological and biological degradation are identical. In other instances, nonbiological degradation is very slow compared to biotransformation and gives rise to different products.

Laboratory studies of the fate of single organic compounds do not provide clear, accurate forecasts of the persistence of xenobiotics in the environment, because in the real world, microbial communities are likely to be exposed to mixtures of organic compounds and heavy metals. Many strains are unable to grow in the presence of heavy metals, the general order of resistance being fungi > actinomycetes > Gram-negative bacteria > Gram-positive bacteria. Thus the bacteria that are able to degrade certain persistent organic chemicals, such as polycyclic aromatic hydrocarbons or chlorinated organic compounds, are likely to disappear from the environment when heavy metals are also present. As a consequence, those organic chemicals will persist much longer in such an environment than simple laboratory experiments might suggest.

Two other key factors influencing the degradation of xenobiotics are the phenomena of gratuitous biodegradation and co-metabolism. *Gratuitous biodegradation* occurs when an enzyme is able to transform a compound other than its natural substrate. The prerequisites are that the unnatural substrate be able to bind to the active site of the enzyme and that it do so in such a manner that the enzyme can exert its catalytic activity. And as we have seen, bacteria and fungi are so diverse in their metabolic capabilities that they produce enzymes able to act on a wide range of organic molecules. *Co-metabolism*, on the other hand, is the ability of an organism to transform a *nongrowth substrate* as long as a growth substrate or other transformable compound is also present. This requirement distinguishes co-metabolism from gratuitous metabolism. A nongrowth substrate is one that cannot serve as the sole source of carbon and energy for a pure culture of a bacterium and hence cannot support cell division.

Types of Bioremediation

Bioremediation is defined as a spontaneous or managed process in which biological (especially microbiological) catalysis acts on pollutants and thereby remedies or eliminates environmental contamination. Bioremediation is currently being used to decrease the organic chemical waste content of soils, ground water, effluent from food processing and chemical plants, and oily sludge from petroleum refineries. Bioremediation techniques fall into four categories: *in situ* treatment, composting, landfarming, and above-ground reactors.

In situ Bioremediation

In Latin, *in situ* means "in the original place." Thus *in situ* bioremediation relies on the indigenous microbial flora of subsurface soils and groundwater. It rests on the premise that the microorganisms already present in a contaminated site have adapted to the organic chemical wastes there and are able to degrade some or all of the components of these wastes.

The degradation by these adapted organisms proceeds until some nutrient or electron acceptor reaches a limiting concentration. Oxygen level is most often the limiting factor, but nitrate and phosphate limitation frequently play a role too. The stimulation of natural biotransformations by adding such nutrients to the environment is called *enhanced* in situ *bioremediation*.

The clean-up of the *Exxon Valdez* oil spill in Alaska provided a large-scale field test of the effectiveness of enhanced *in situ* bioremediation. In March 1989, the supertanker *Exxon Valdez* ran aground on Bligh Reef in Prince William Sound. The resulting spill of about 11 million gallons of crude oil severely affected 350 miles of shoreline in the sound. In this case, fertilizers were used to accelerate the removal of oil from the beaches, supplying extra nutrients that would otherwise have dwindled to limiting concentrations. A single application of inorganic fertilizer was shown to speed the disappearance of oil by a factor of two to three over its rate of disappearance on untreated shoreline. Moreover, the accelerated rate was maintained for several weeks, even after nutrient concentrations returned to background level. Samples of oil taken at the end of that time from surfaces of treated beaches showed changes in composition consistent with extensive biodegradation. Enhanced *in situ* bioremediation offers several potential advantages in the elimination of hazardous wastes: It is cheaper than incineration, and workers are not exposed to the risks associated with excavation and removal of contaminated soils. It is well suited to treating large areas contaminated with low levels of wastes.

Composting

Compost, a mixture of soil, partially decayed plants, and sometimes manure and commercial fertilizer, is very rich in microorganisms. It has long been used by farmers and gardeners to make soils more fertile and

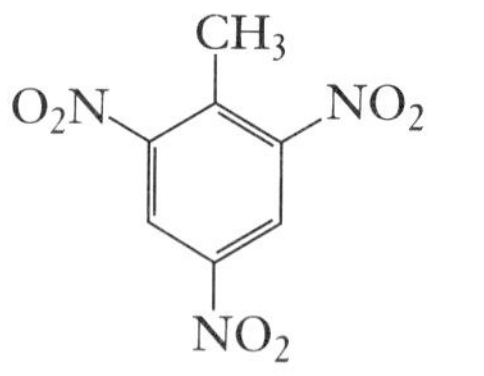

TNT

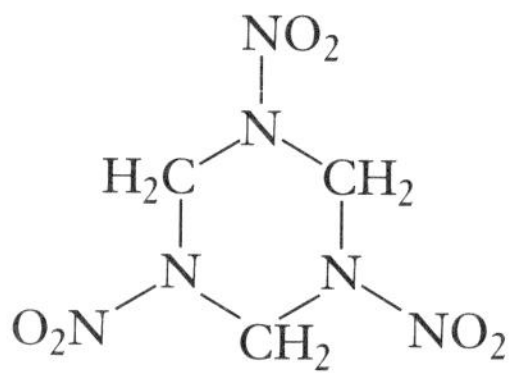

RDX

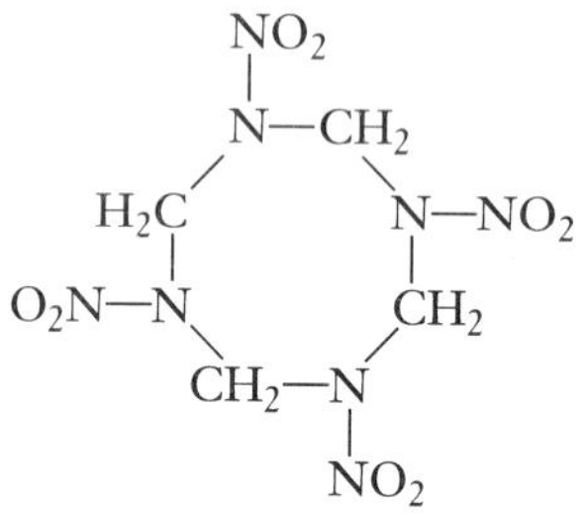

HMX

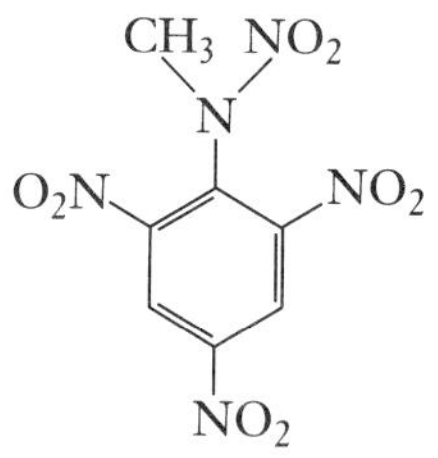

Tetryl

Figure 15.5
Chemicals in waste streams from the explosives industry. TNT is an explosive that can be molded. RDX (Research Division X, Formula X, Cyclonite, or Hexogen) and HMX (Octogen) are primary ingredients of certain military plastic explosives. Tetryl (Tetralite) is the most common military booster chemical.

to improve crop yields. Composting shows promise in the treatment of high concentrations of resistant chemical wastes, as illustrated by a recent application to the degradation of explosives.

In the past, waste streams from the explosives industry were often discharged to settling basins or lagoons. The waste streams contained 2,4,6-trinitrotoluene (TNT), hexahydro-1,3,5-trinitro-1,3,5-triazine (RDX), octahydro-1,3,5,7-tetranitro-1,3,5,7-tetraazocine (HMX), and N-methyl-1,2,4,6-tetranitroaniline (Tetryl) (Figure 15.5). At ambient temperature, these compounds are solids sparingly soluble in water, and as a result, they have largely remained in the soil at the discharge sites. The strategy that is currently most often applied for decontaminating soils that contain explosives is to incinerate them, a process that costs approximately $300 per ton of soil to be treated. For the estimated 5,200,000 tons of soil requiring decontamination, this treatment would cost in excess of $1.5 billion. Thus there is considerable incentive to develop cheaper methods, and composting is emerging as a potential low-cost alternative. Studies in the 1970s showed that RDX and HMX can be degraded by anaerobic bacteria and that TNT can be biodegraded under both aerobic and anaerobic conditions. Biodegradation or biotransformation of over 90% of these explosives was achieved within 80 days in a compost pile maintained at 55°C. After 150 days, a starting concentration of 18,000 mg of explosives per kilogram of soil was reduced to 74 mg/kg.

Landfarming

Landfarming is used to dispose of oily sludges from petroleum refinery operations. In this process, oily sludge from refinery wastes is mixed with soil and subjected to enhanced *in situ* bioremediation. The sludges may be pretreated or not. Biological pretreatment of refinery effluents partially mineralizes the organic waste components; the residual solid waste (sludge) then has a high content of aromatic hydrocarbon compounds and is low in aliphatic hydrocarbon compounds (5–10% by weight). In contrast, untreated settled solids (such as the solids from tank bottoms) contain high amounts of aliphatic hydrocarbons (30–50%) and inorganic solids (silt).

The terrain of a landfarm must be flat to minimize runoff, the soil should be light and loamy for adequate aeration, and a clay layer should underlie the porous surface soil to reduce the possibility of groundwater contamination through seepage. The landfarm is graded to a very gentle slope to prevent standing water from collecting after rain, and it is surrounded by a moat to contain runoff. The geographic location of the landfarm is also chosen with precipitation and temperature in mind. The air-filled pores in the light landfarm soil ensure rapid access of oxygen to the organisms there, but excessive rain would waterlog the soil and eliminate the air-filled pores. About 20% water saturation of the soil is sufficient for maximal oil degradation. The optimal temperature range for biodegradation is 20–30°C, and most activity ceases below 5°C.

Inorganic fertilizer is applied to the site to provide fixed nitrogen, and phosphate and pulverized limestone ($CaCO_3$) are added to raise the pH of the soil-waste mixture to about 7.8. For untreated sludge, maximal oil biodegradation rates in soil are achieved at a hydrocarbon load of 5–10% by weight—that is, about 100–200 metric tons of hydrocarbon per hectare (an area of about 12,000 yd^2). Under favorable conditions, a 5% application can be repeated at intervals of approximately 4 months. In such landfarms, approximately 50–70% of the applied organic waste is degraded before the next batch of sludge is applied.

A disadvantage of landfarming is that the process is slow and incomplete. Moreover, the heavy metal constituents of the sludge gradually accumulate in the landfarm soil. Consequently, a plot of land used intensively as a landfarm cannot later be used for growing crops or grazing livestock.

Above-Ground Bioreactors

Above-ground bioreactors are based on the same technology as fermenters (see Box 7.1). They are used for the treatment of either excavated soil or groundwater containing high levels of contaminants (such as chemical landfill leachates). Contaminated soil is mixed with water and introduced into the reactor as a slurry. Granulated charcoal, plastic spheres, glass beads, or diatomaceous earth provide a large surface area for microbial growth in such bioreactors. The large surface area of the microbial biofilm that forms on such supports leads to a rapid rate of biodegradation. The microbial inoculum may come from an indigenous population at the contaminated site, from activated sludge from a sewage treatment plant, or from a pure culture of an appropriate organism. Because the reactors are enclosed, the use of genetically engineered organisms as inocula is also feasible. Bioreactors can be used in series to accomplish different kinds of degradation. For example, the first reactor can be operated in an anaerobic mode and its effluent transferred to a second reactor operated in an aerobic mode. Some biotransformations, such as dehalogenations of certain compounds, proceed optimally under anaerobic conditions, whereas mineralization requires aerobic conditions.

Challenges of Evaluating in situ Biodegradation

As we have seen, the fate of an organic compound introduced into the environment depends on the properties of both the compound and the site. The compound may be tightly adsorbed to the soil at the site, it may be loosely bound and leach away from the application site into deeper layers of the soil and into ground water, it may be photooxidized or decomposed by other abiotic processes, or it may be taken up by plants or transformed by microorganisms. Any or all of these events may contribute to the real or apparent time-dependent disappearance of the compound from the site.

Attempts to decontaminate the IBM Dayton hazardous waste site in New Jersey illustrate the strength with which organic chemicals can adhere to soil particles and the potential difficulty of extracting even highly soluble contaminants by running water through the soil. Water flows preferentially through high-permeability zones in the soil and equilibrates slowly with any water present in the low-permeability zones. The groundwater at the IBM Dayton site had been contaminated with about 400 gal of 1,1,1-trichloroethane and tetrachloroethylene. The maximal concentrations recorded were 9600 ppb of 1,1,1-trichloroethane and 6130 ppb of tetrachloroethylene. Between 1978 and 1984, pumping water through the site at an average on-site extraction rate of 300 gallons per minute lowered the concentrations of these compounds in the water to below 100 ppb. The pumping was suspended in 1984, and by 1988 the tetrachloroethylene concentrations in the groundwater had risen to 12,560 ppb. In effect, the chlorohydrocarbons adsorbed to the soil, or sequestered in fine pores, represented a practically inexhaustible slow-release reservoir of pollutant.

Laboratory simulations of on-site conditions are rarely authentic enough to provide trustworthy insights into a field situation. It would take an extraordinary effort at simulating real-life environmental conditions to answer crucial questions: What fraction of a pollutant is strongly adsorbed to soil or sediment? How much of the pollutant is destroyed by abiotic processes and how much is biodegraded? What are the rates of these processes at different pollutant concentrations? What is the response of the microbial community to introduction of the pollutant? What is the habitat of the organisms that contribute most decisively to degradation of the pollutant? Is the transformation or degradation favored by oxygen-rich or oxygen-poor conditions? What is the impact of natural variation in other conditions, such as temperature and pH? Very few studies can answer even a few of these questions about the *in situ* fate of an important pollutant. A well-designed quantitative study of pentachlorophenol degradation in an experimental channel fed by

Mississippi River water, which represents an attempt at a comprehensive analysis, is described below.

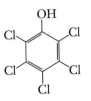

Figure 15.6
Pentachlorophenol.

Degradation of pentachlorophenol in an artificial freshwater stream. The compound pentachlorophenol (PCP; Figure 15.6) is generally toxic to living organisms. First introduced during the 1930s as a wood preservative, it has proved effective as a general-purpose killer of algae, bacteria, fungi, weeds, mollusks, and insects in a variety of agricultural and industrial settings (though commercial wood treatment remains the major application). Worldwide production of PCP is about 50 million kg/yr. Streams and groundwater commonly contain PCP in concentrations of micrograms per liter. The purpose of the study was to predict the fate of PCP in these natural aquatic systems.

This field study was carried out during the summers of 1982 and 1983 at the Monticello Ecological Research Station of the U.S. Environmental Protection Agency at Duluth, Minnesota. The station has outdoor experimental channels, each 488 meters in length, fed on a year-round basis by water pumped from the Mississippi River. The stretch of each channel that is utilized for experiments consists of eight ponds with mud at the bottom alternating with eight coarse gravel riffles (Figure 15.7). A riffle is a sandbar lying just below the surface of a waterway. Of the various water plants (macrophytes) colonizing the ponds, the predominant species included *Potamogeton crispus*, a rooted pond weed, and *Lemna minor*, a floating (nonrooted) aquatic plant.

These channels are well suited for the study of factors that contribute to the degradation of xenobiotics. They are fed from a natural source of water and contain diverse microbial habitats: the water column, the microaerophilic sediment surface of the mud at the bottom of the ponds, the anoxic deeper layers of sediment, the surfaces of water plants, and the rock surfaces in the riffles. The channels were treated continuously for a period of 88 days with PCP, which was introduced as a concentrated solution of its sodium salt. The effects discussed below were observed in a channel treated with 144 μg of PCP per liter of water. Rates of photodegradation of PCP by sunlight were determined by suspending, at known depths in the channel pool, glass vials containing the appropriate PCP solution. Analyses of the rates of PCP degradation by microorganisms in different habitats were performed using the following samples:

- *Rock surfaces.* Rocks collected individually from the riffles were placed in beakers in a known volume of water containing a known concentration of PCP from the same riffle.

- *Sediment cores.* Cores 3 cm in diameter were removed from the pool bottom. The degradation of PCP was then measured under different conditions: aerobic (bubbling air above the sediment surface), microaerophilic (cores left open to air but otherwise

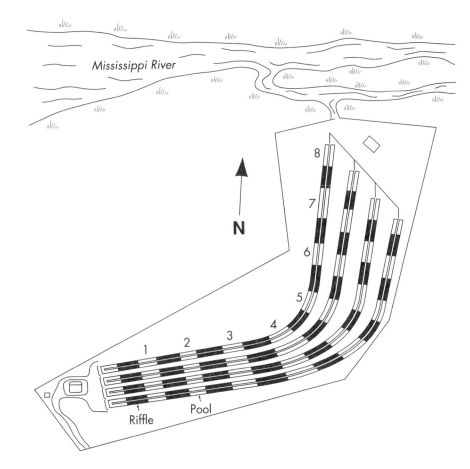

Figure 15.7
Location and configuration of outdoor artificial streams at the Monticello
Ecological Research Station of the U.S. Environmental Protection Agency at
Duluth, Minnesota. Adapted from Arthur, J.W., Zischke, J.A., and Erickson,
G.L. (1982), Effect of elevated water temperature on macroinvertebrate
communities in outdoor experimental channels, *Water Res.* 16:1465–1474,
Fig. 1.

undisturbed), and anaerobic (bubbling high-purity nitrogen
through the cores).

- *Macrophyte surfaces.* The top portions of *Potamogeton* plants were
  collected from a pool, and *Lemna* were scooped up from the
  surface. These plants were then carefully submerged in beakers
  of PCP-containing water from the same site.

- *Microorganisms floating free in the water column and those attached
  to particles.* Samples of water from various locations were divided
  into two equal portions. One portion was passed through a 1.0-
  μm filter to remove suspended particles but leave free-floating
  bacteria in the filtrate; the other portion was unfiltered.

The following observations were made:

- Microbial degradation of PCP in the treated channel became significant about 3 weeks after the PCP was first introduced, as indicated by (1) a sharp decline in PCP concentration down the length of the dosed channel; (2) rapid degradation of PCP by microorganisms in samples removed after week 4 from a PCP-dosed channel but not from a control channel; (3) the appearance in the dosed channel of bacteria capable of mineralizing uniformly labeled [^{14}C]PCP with release of $^{14}CO_2$; and (4) a large decline in the PCP concentrations in the sediment between weeks 3 and 5. Laboratory studies showed that the timing of the appearance of PCP-degrading activity conformed to the time that would have been necessary for the selective enrichment of an initially low population of PCP-degrading microorganisms in the channel. Many of the pure cultures of bacteria isolated from dosed channels were able to use PCP as a sole source of carbon and energy for growth. One such organism grew on PCP at concentrations as high as 100 mg/liter, releasing all of the organically bound chlorine as Cl^-.

- After the microbial community had fully adapted, water passing through the channel was cleansed of 50–60% of its PCP. Microbes, especially those attached to rock and plant surfaces, were responsible for most of the observed degradation.

- The rate of PCP disappearance in the water column above the sediment cores was more rapid under aerobic than under anaerobic conditions.

- The rate of PCP degradation was virtually temperature-independent during the summer, when water temperatures ranged from 19 to 30°C. However, degradation gradually slowed at lower temperatures and ceased at 4°C.

- Photodegradation of PCP was rapid at the water's surface but decreased rapidly with depth, owing to the attenuation of light by suspended particles and dissolved materials in the channel water. Depending on sunlight, photodegradation accounted for a 5–28% decline from the initial PCP concentration during the water's passage down the channel. Adsorption, sedimentation, or volatilization of PCP and its uptake by living organisms accounted for less than a 5% decrease in unacclimated water (that is, in water immediately after the initial addition of PCP).

The study concluded that PCP is degraded in the aquatic environment and that microorganisms attached to surfaces are responsible for most of this degradation. The biodegradation of PCP requires an adaptation period on the order of weeks, but once the microbial populations have adapted, the degradation process is quite rapid, with a PCP half-life of less than 12 hours. PCP mineralizing activity is greatly reduced at low temperatures, and stream temperatures in northern climates may be too low for biodegradation to occur during much of the year. Note too that other investigations had shown that contamination with PCP is ubiquitous, and low background levels of PCP were found in Mississippi River sediments and macrophytes and also in the sediments of the control channels. Thus it is possible that the channels were "primed" with PCP before dosing began and that some enrichment for PCP degraders preceded the start of the study.

Genetic and Metabolic Aspects of Biodegradation

As we saw earlier, much of our understanding of microbial genetics and metabolic regulation has come from extensive studies of the enteric bacterium *Escherichia coli*. This organism utilizes a wide range of substrates for growth. In general, the transcription of an operon encoding a particular catabolic pathway in *E. coli* is induced only when a relevant substrate is present. This is a common control mechanism in microorganisms that occupy ecological niches where the type and availability of substrates vary in space and time (see Chapter 7, page 246).

Soil is an example of an important and extensive habitat that varies from one spot to the next and from one time to another in the nature of the organic compounds present. The territory that a particular microorganism can colonize may be limited to a patch of a few square feet and to a depth of an inch or less. An organism capable of using many different organic compounds as a sole source of carbon and energy for growth is likely to flourish in more patches than one that is fastidious about its diet. Pseudomonads (Gram-negative, rod-shaped, polarly flagellated, aerobic bacteria) exemplify such versatile organisms common in soil. Taxonomically, the pseudomonads form a very large and heterogeneous group (see Chapter 2). One of their hallmarks is the ability to grow on any one of a large number of organic compounds.

At about the same time that the phenomenon of enzyme induction was being established in *E. coli* by Jacques Monod and François Jacob at the Pasteur Institute in Paris, a similar phenomenon was discovered in one of the pseudomonads by Roger Stanier at the University of California at Berkeley. Stanier noted in 1947 that many enzymes of the pathway involved in the degradation of benzoate are induced when nongrowing *Pseudomonas* cells "adapt" to the presence of benzoate in the medium, and he coined the term "simultaneous adaptation" to describe the phenomenon.

Many years later, it was found that some of the versatility and adaptability of soil microorganisms stems from their possession of *catabolic plasmids*, plasmids that specify a degradative pathway. Bacteria carrying catabolic plasmids have been isolated principally from soil, and the majority of these strains have been classified as pseudomonads, although catabolic plasmids have also been found in organisms belonging to many other genera isolated from a wide variety of environments. The majority of catabolic plasmids are *self-transmissible*, and many have a broad host range. That is, transmissible catabolic plasmids represent a pool of metabolic potential available to many strains in a microbial community through interspecies transfer of genetic information.

The proliferation of catabolic plasmids is analogous to that of R plasmids, which carry genes conferring antibiotic resistance (Chapter 13). R plasmids spread through bacterial populations under antibiotic selection pressure. Similarly, the ability to utilize a novel source of nutrient or to eliminate a potentially toxic compound promotes the spread of a catabolic plasmid. For example, unrestricted use of a pesticide frequently results in the development of microbial populations capable of degrading it. The ability to degrade the pesticide spreads among the different bacterial strains in the natural population by interspecies transfer of a catabolic plasmid that carries the genes for the degradative enzymes.

Catabolic plasmids have been found that encode enzymes for degrading such naturally occurring compounds as camphor, octane, naphthalene, salicylate, and toluene. Other plasmids make possible the degradation of various synthetic compounds, including certain widely used herbicides and insecticides (Figure 15.8; Table 15.5). The degradative capabilities of microorganisms carrying catabolic plasmids result from a cooperative interaction between the genes carried on the plasmid and those on the chromosome of the host cell. *Such interactions are particularly important when a single compound is to be used as a sole source of carbon and energy.* Many plasmids encode only part of the catabolic pathway for a given compound. The products of the transformations mediated by the plasmid-encoded enzymes must be those that can then be utilized by chromosomally encoded enzymes functioning in the central energy-producing metabolic pathways of the cell.

It is not surprising that many catabolic plasmids carry pathways for the degradation of aromatic compounds. The benzene ring is second only to glucose as a building block in nature. Whereas glucosyl residues are the monomer units of cellulose, the most abundant organic compound in nature, benzene rings form part of the precursors of lignin, the second most abundant constituent of biomass (Chapter 10).

Aerobic Biodegradation of Benzene and Other Aromatic Hydrocarbons

The first step in an oxidative microbial attack on benzene is hydroxylation. Bacteria employ a dioxygenase to catalyze the simultaneous incorporation of two atoms of oxygen from an oxygen molecule into the ring.

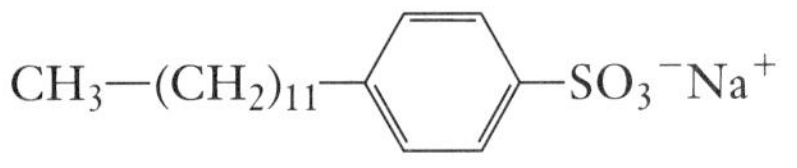

$CH_3-(CH_2)_{11}-$⟨benzene ring⟩$-SO_3^-Na^+$

Dodecylbenzenesulfonate
sodium salt
(an alkylbenzene sulfonate)

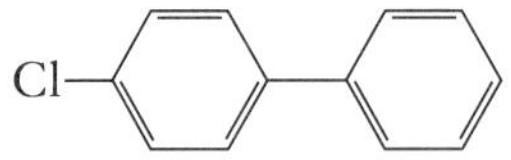

$Cl-$⟨biphenyl⟩

4-Chlorobiphenyl

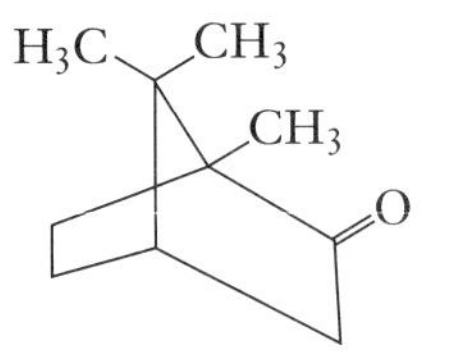

Camphor

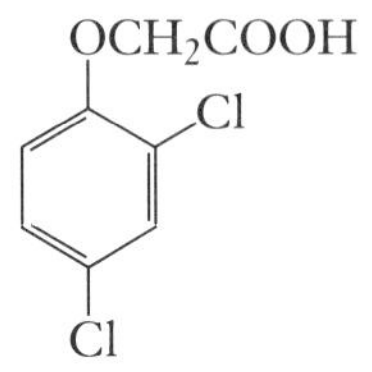

OCH_2COOH
Cl
Cl

2, 4-Dichlorophenoxyacetic
acid (herbicide)

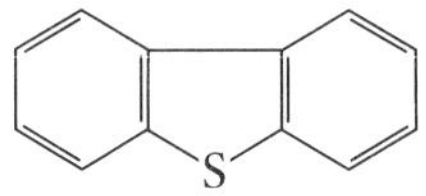

Dibenzothiophene

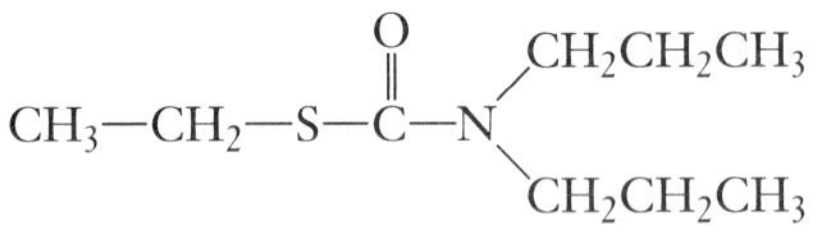

$CH_3-CH_2-S-\overset{\overset{O}{\|}}{C}-N\overset{CH_2CH_2CH_3}{\underset{CH_2CH_2CH_3}{}}$

S-Ethyl-N, N-dipropyl
thiocarbamate (herbicide)

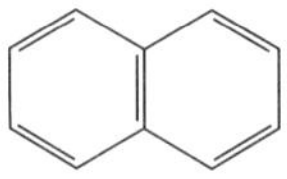

Naphthalene

$CH_3-(CH_2)_6-CH_3$
Octane

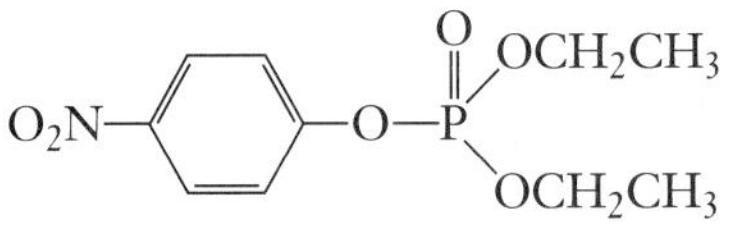

O_2N-⟨benzene⟩$-O-\overset{\overset{O}{\|}}{P}\overset{OCH_2CH_3}{\underset{OCH_2CH_3}{}}$

Parathion
(O,O-diethyl-o-nitrophenyl-
phosphorothioate; insecticide)

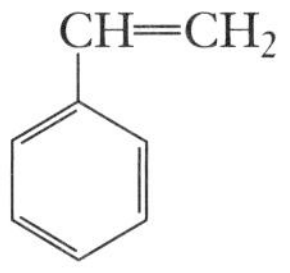

$CH=CH_2$

Styrene

Figure 15.8
Examples of herbicides and insecticides subject to degradation by enzymes
encoded on catabolic plasmids.

The product, *cis*-1,2-dihydroxy-1,2-dihydrobenzene, is then converted
to catechol (1,2-dihydroxybenzene; Figure 15.9) in a reaction catalyzed
by the enzyme *cis*-benzene glycol dehydrogenase. These initial two
steps in benzene biodegradation—dioxygenase-mediated hydroxylation followed by dehydrogenation—are common to the pathways of
bacterial degradation of numerous other aromatic hydrocarbons (Figure 15.10).

The subsequent metabolism of catechol follows one of two divergent pathways (see Figure 15.9). At the branch point, catechol either is
oxidized in a reaction catalyzed by catechol-1,2-dioxygenase, the so-
called *ortho* (or intradiol) cleavage, to *cis,cis*-muconate or is oxidized in a
reaction catalyzed by catechol-2,3-dioxygenase, the so-called *meta* (or

Examples of Naturally Occurring Transmissible Catabolic Plasmids

Primary substrate[a]	Plasmid	Size(kb)	Bacterial host strain
Alkylbenzene sulfonate	ASL	91.5	*Pseudomonas testosteroni*
Benzoate	pCB1	17.4	*Alcaligenes xylosoxidans* subsp. *denitrificans* PN-1
Biphenyl	pBS241	195	*P. putida* BS893
Camphor	PpG1(CAM)	~500	*Pseudomonas* sp.
4-Chlorobiphenyl	pSS50	53.2	*Alcaligenes* spp.
2,4-Dichlorophenoxyacetate	pJP1	87	*A. paradoxus* Jmp116
Dibenzothiophene	—[b]	82.5	*Pseudomonas* sp. DBT2
S-Ethyl-N,N-dipropylthiocarbamate	—[b]	75.7	*Arthrobacter* sp. TE1
Naphthalene	Nah7	83	*P. putida* PpG7
Octane	OCT	~500	*P. oleovorans*
Parathion	pPDL2	43	*Flavobacterium* ATCC27551
Styrene	pEG	37	*P. fluorescens* PAW340
Toluene	pWW0 (TOL)	117	*P. putida* mt-2

[a] The structures of some of the substrates are shown in Fig. 15.8.

[b] These plasmids have not been named.

SOURCE: Sayler, G.S., Hooper, S.W., Layton, A.C., and King, J.M.H. (1990), Catabolic plasmids of environmental and ecological significance, *Microb. Ecol.* 19:1–20.

extradiol) cleavage, to 2-hydroxymuconic semialdehyde. The final products of both pathways are molecules that can enter the tricarboxylic acid cycle.

In addition to benzene itself, the *ortho* and *meta* pathways can catabolize a number of benzene derivatives. As illustrated in the following discussion of the TOL catabolic plasmid, different benzene derivatives induce either one pathway or the other, because the enzymes in the *ortho* and *meta* pathways differ in their ability to utilize particular catechol derivatives as substrates. Possessing two pathways increases the range of benzene derivatives that an organism can utilize as substrates.

TOL (pWW0) Catabolic Plasmid

Examination of *Pseudomonas putida* mt-2 and its associated transmissible catabolic plasmid TOL (pWW0; Figure 15.11) gives a glimpse of the complexity of the pathways whereby bacteria utilize aromatic hydrocarbons and illustrates the interplay between chromosomal and plasmid genes. In *Pseudomonas putida* mt-2, chromosomal genes encode the *ortho* pathway, and the TOL plasmid encodes the *meta* pathway (see Figure 15.9). Benzoate, a product of toluene degradation, induces the expression of the genes of the *meta* pathway, and catechol, like benzoate a

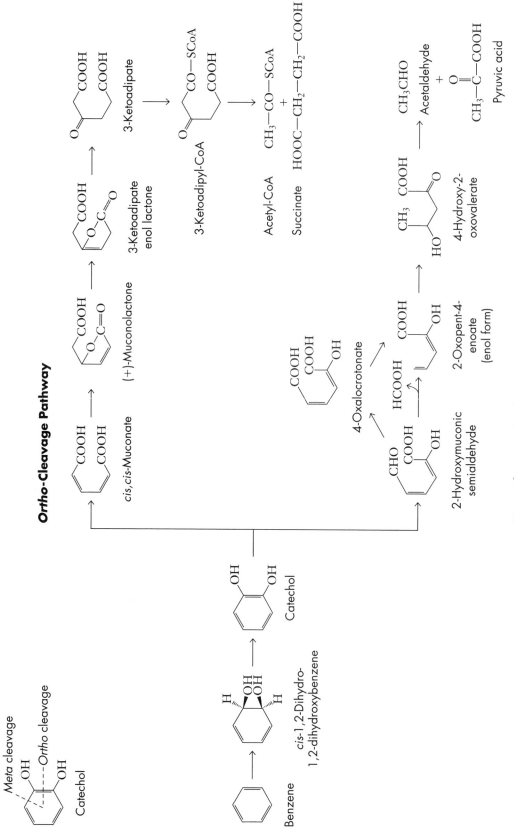

Figure 15.9
Pathways of benzene degradation by pseudomonads.

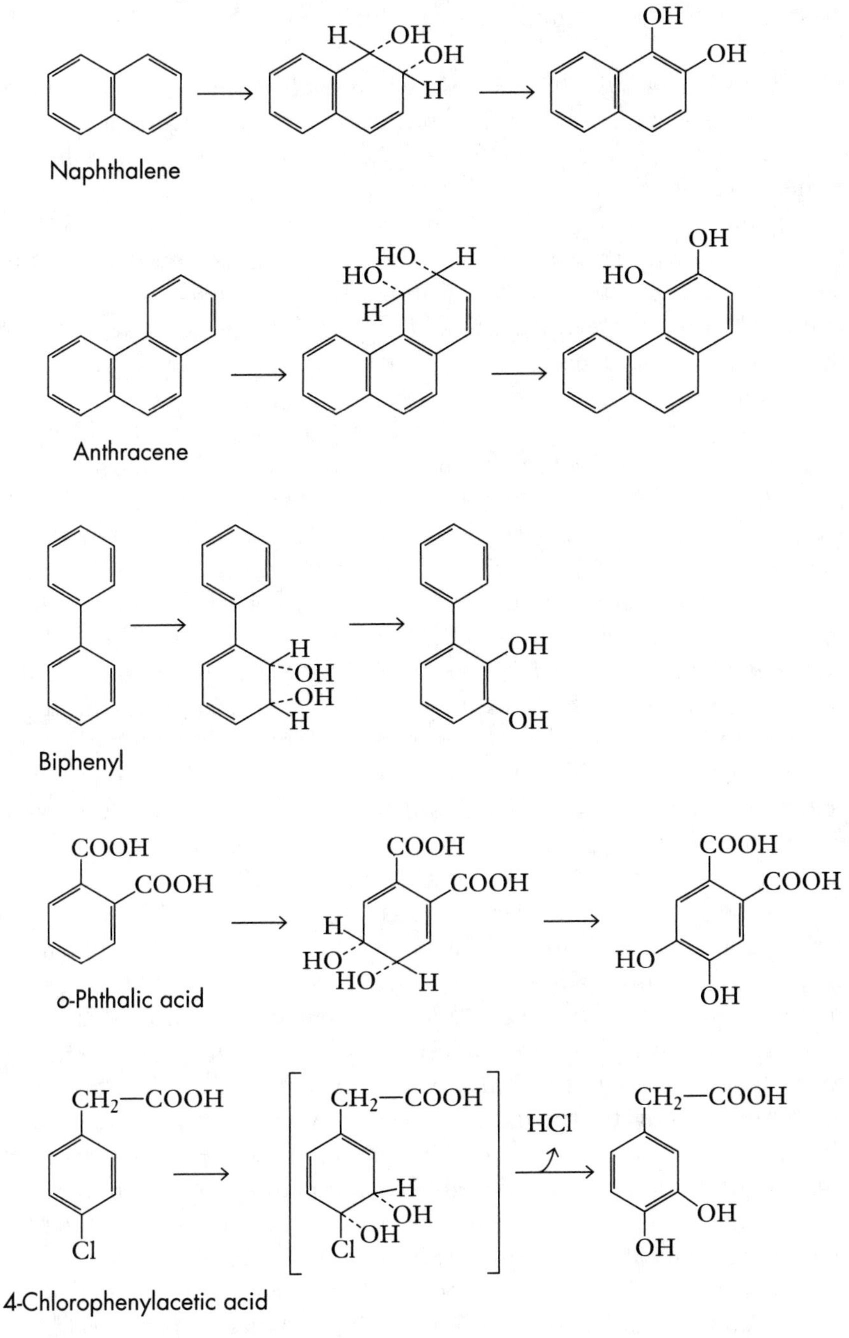

Figure 15.10
Initial steps in the metabolism of various aromatic hydrocarbons.

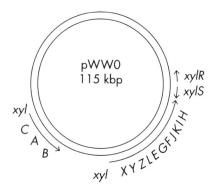

Figure 15.11
The TOL (pWW0) catabolic plasmid.

product of toluene degradation (see Figure 15.12), induces the *ortho* pathway.

The TOL plasmid has been shown to confer on the host the capacity to degrade not only toluene but also *m*- and *p*-xylene and other benzene derivatives (Figure 15.13). The genes encoding catabolic enzymes have been named the *xyl* genes. The *xyl* genes of TOL (pWW0) are organized into two operons referred to as the *upper* and *lower* (*meta*) pathways. The upper pathway, *xylCAB*, encodes the degradation of toluene and xylenes to benzoate and toluates (methylbenzoates), respectively. The lower pathway, *xylXYZLEGFJKIH*, encodes the degradation of benzoate and toluates to acetaldehyde and pyruvate (see Figure 15.12 and Table 15.6).

The lower pathway branches at 2-hydroxymuconic semialdehyde, and the branches rejoin at the common product 2-oxo-4-pentenoate (see Figure 15.12). In analogy to the role of the alternative *ortho* and *meta* pathways for the degradation of catechol, this branching broadens the range of substrates that *P. putida* mt-2 can utilize. For example, *m*-toluate is degraded by the *xylF* branch, whereas benzoate and *p*-toluate are degraded by the *xylGHI* branch. The relative affinity of the enzymes of these two branches of the lower pathway for a particular substrate determines the branch by which the substrate is catabolized.

The following model has been proposed for the regulation of the *xyl* genes (Figure 15.14). The TOL (pWW0) plasmid carries two unlinked regulatory genes, *xylR* and *xylS* (see Figure 15.11). XylR is expressed constitutively at a high level. When a substrate, such as toluene, enters the cell, it binds to the XylR protein to form a XylR–toluene complex. This complex binds to the promoter (P_{upper}) of the *xylCAB* operon and activates its transcription by influencing the binding of RNA polymerase there. The lower pathway is activated in an analogous manner. Benzoate, the product of the degradation of toluene by the upper pathway, binds to the XylS protein produced constitutively at a low level. The XylS–benzoate complex binds to the promoter (P_{meta}) of the lower pathway and activates the transcription of its genes. If benzoate, rather than toluene, is introduced as the starting substrate, only the lower (*meta*)-pathway genes are induced. This regulation appears appropriate because the upper-pathway genes specify enzymes involved in the production of benzoate (see Figure 15.12 and Table 15.6).

There are additional subtleties to the regulation. The XylR–toluene complex binds to the promoter (P_s) of the *xylS* operon as well as to the P_{upper} promoter, thus activating transcription of *xylS* and so achieving simultaneous activation of both the upper- and the lower-pathway genes. When the XylS protein is present in an elevated

Figure 15.12
Pathway of toluene degradation in *Pseudomonas putida* mt-2.

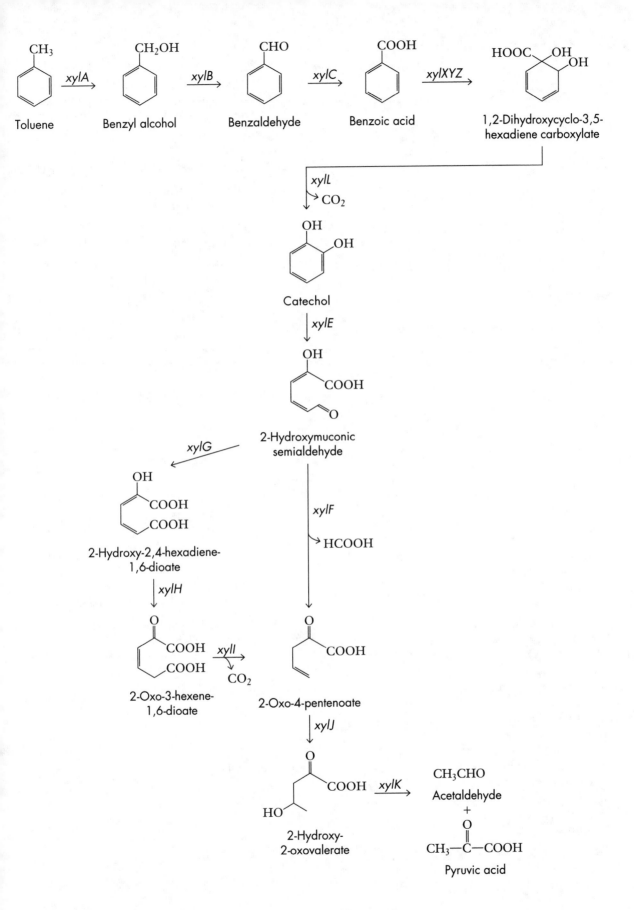

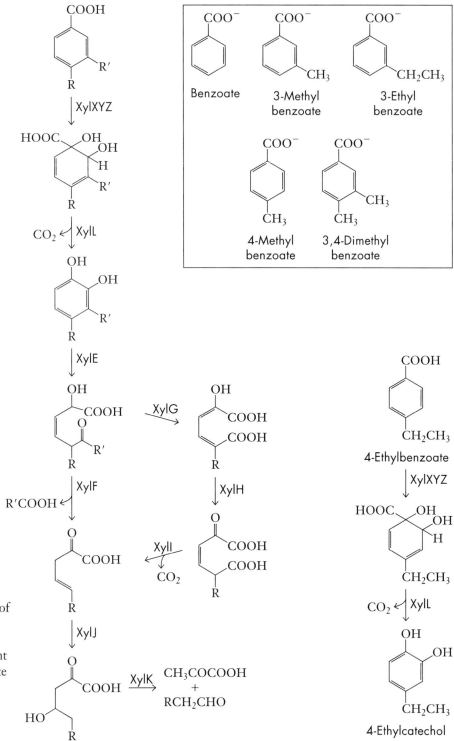

Figure 15.13
Pathway of degradation of various benzene derivatives in *Pseudomonas putida* mt-2. Structures of benzene derivatives that can be catabolized by *Pseudomonas putida* mt-2 are shown in the box. The catabolism of these derivatives follows the same pathway as that shown for toluene in Figure 15.12. Shown on the right is the conversion of 4-ethylbenzoate to 4-ethylcatechol in a *P. putida* strain with a mutant TOL plasmid in which the cleavage pathway is expressed constitutively.

Enzymes and Regulatory Proteins Encoded by Genes on the TOL Plasmid pWW0

TABLE 15.6

Gene	Enzyme or function
"Upper-pathway" operon	**Enzymes involved in the conversion of toluene and xylenes to benzoate and toluates**
xylA	Xylene oxygenase
xylB	Benzyl alcohol dehydrogenase
xylC	Benzaldehyde dehydrogenase
"Lower (meta)-pathway" operon	**Enzymes involved in the degradation of benzoate and toluates to acetaldehyde and pyruvate**
xylX,Y,Z	Toluate dioxygenase
xylE	Catechol 2,3-dioxygenase
xylF	2-Hydroxymuconic semialdehyde hydrolase
xylG	2-Hydroxymuconic semialdehyde dehydrogenase
xylH	4-Oxalocrotonate tautomerase
xylI	4-Oxalocrotonate decarboxylase
xylJ	2-Oxopent-4-enoate hydratase
xylK	2-Oxo-4-hydroxypentenoate aldolase
xylL	Dihydroxycyclohexadiene carboxylate dehydrogenase
	Proteins involved in controlling the transcription of the upper- and lower-pathway genes
xylR	Regulatory protein
xylS	Regulatory protein

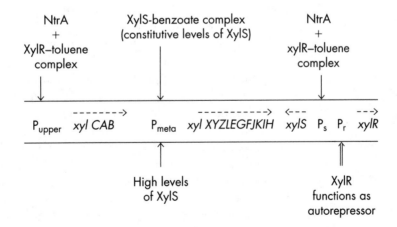

Figure 15.14

Model for *xyl* gene regulation in the TOL (pWW0) catabolic plasmid. Broken arrows indicate the direction of transcription. Solid arrows represent activation, and the open arrow represents repression of the operon indicated. NtrA is a sigma factor produced constitutively by the host. In the presence of toluene, the XylR–toluene complex activates transcription at the P_{upper} and P_s promoters. The consequent increased production of XylS leads to activation of transcription at P_{meta}. Alternatively, benzoate in the presence of constitutive amounts of XylS is able to activate transcription at P_{meta}. XylR functions as an autorepressor. Based on Burlage, R.S., Hopper, P.W., and Sayler, G.S. (1989), *Appl. Environ. Microbiol.* 55:1323–1328, Fig. 5.

amount, it activates the expression of the lower-pathway genes even in the absence of an inducer such as benzoate.

The proteins and promoter sequences that regulate the *xyl* genes in *P. putida* appear to share a common evolutionary origin with the ones used to regulate nitrogen metabolism in enteric bacteria. In enteric bacteria (*E. coli*, *Salmonella typhimurium*, *Klebsiella pneumoniae*, and so on), the transcription of certain genes, such as those for glutamine synthetase and nitrogenase, is regulated by the availability of combined nitrogen in the growth medium. Two gene products are required specifically for the synthesis of nitrogen-regulated proteins. These two proteins act at the transcriptional level. The first is the NtrA protein, an alternative subunit (sigma factor) for RNA polymerase. It allows RNA polymerase to recognize a particular class of promoters. The RNA polymerase–NtrA complex binds to these promoters but cannot initiate transcription in the absence of the second protein, NtrC. NtrC binds to enhancerlike sequences upstream of the nitrogen-regulated promoters, and in its presence the RNA polymerase–NtrA complex initiates transcription. In *P. putida*, the maximal expression of the *xylCAB* and *xylS* operons similarly requires NtrA sigma factor. In this system, XylR corresponds to the NtrC activator protein. These two proteins are not required for transcription of the lower-pathway genes. The two promoters at which NtrA function is required are indicated in Figure 15.14.

Before we close our discussion of the TOL (pWW0) plasmid, it will be instructive to summarize the key features that enable the TOL (pWW0)-carrying *P. putida* mt-2 to grow on many different benzene derivatives. The oxygenase (XylA) and dehydrogenases (XylB and XylC) that catalyze the conversion of various benzene derivatives to benzoic or toluic acids have a broad substrate specificity and act even on highly substituted compounds, such as 1,2,4-trimethylbenzene. At the level of catechol derivatives, where broad substrate specificity either is not achievable or is inappropriate, different enzymes catalyze the degradation of particular intermediates derived from benzenes with different substitution patterns. The various catechol derivatives are degraded either by the *ortho* pathway specified by chromosomal genes or by the *meta* pathway specified by pWW0 *xyl* genes (see Figure 15.12). The choice of pathway is determined by the ability of the enzymes of the pathway to act on a particular catechol derivative. Typically, catechols with alkyl substituents at the 3 or 4 position go through the *meta* cleavage pathway. Similarly, within the *meta* pathway, alternative branches handle different 2-hydroxymuconic semialdehyde derivatives. For example, *m*-toluate is degraded by the *xylF* branch, and benzoate and *p*-toluate are degraded by the *xylGHI* branch. *The versatile utilization of different benzene derivatives thus depends on the presence of multiple degradative pathways.* Wherever possible, a single enzyme with broad specificity is utilized.

GENETIC ENGINEERING IN CONTROL OF INDUSTRIAL POLLUTION

General Principles

Any individual industrial process generally produces a much simpler mixture of chemical waste products than the pools encountered at landfills, which contain the wastes of many different industries. Thus efforts at bioremediation are most effective when they are introduced at the end of *each* process that results in undesirable wastes. Pollution control by genetic engineering, for example, is likely to work best when pollutants are a known mixture of relatively concentrated organic compounds that are related to each other in structure, when conventional alternative organic nutrients are absent (or perhaps are added in controlled amounts only), and when there is no competition from indigenous microorganisms.

But considering the wide variety of existing organisms capable of degrading organic compounds, what can genetic engineering contribute to the control of pollution? Genetic engineering makes it possible to alter the properties of existing degradative enzymes, to modify regulatory mechanisms, and to assemble within a single organism degradative enzymes from phylogenetically distant organisms. The manipulations are precise, and the changes introduced in the genome can be checked in a rigorous manner. The task of engineering an organism for the degradation of a particular target compound may be very laborious. However, for that target, it is a task that need be done only once. The easiest way to create an appropriate genetically engineered strain is to begin with an organism that already possesses much of the necessary degradative enzymatic machinery. The successful example that follows highlights the extent to which the genetic engineering process depends on a thorough understanding of the genetics and biochemistry of the wild-type organism.

Horizontal Expansion of a Catabolic Pathway by Genetic Engineering

Through the genetic engineering of metabolic pathways, it is possible to extend the range of substrates that an organism can utilize. It is even possible to endow an organism with the ability to degrade a predetermined range of xenobiotics. As we have seen, *P. putida* carrying the TOL plasmid is able to grow on a variety of alkylbenzoates: benzoate, 3- and 4-methylbenzoate (3MB and 4MB), 3,4-dimethylbenzoate

(34DMB), and 3-ethylbenzoate (3EB). These alkylbenzoates are converted by the reactions shown in Figure 15.13 to intermediates of the tricarboxylic acid cycle. However, the organism cannot grow on the very closely related compound 4-ethylbenzoate (4EB). Because TOL-plasmid-bearing bacteria are able to metabolize a variety of alkylbenzoates, researchers assumed that the inability of these bacteria to degrade 4EB was not caused by a problem with its uptake into the cell or by any inherent toxicity to the bacteria. Instead, they reasoned, the inability to metabolize 4EB might have been due either to the failure of 4EB to serve as an effector for the induction of synthesis of the catabolic enzymes or to intermediates of 4EB breakdown not serving as substrates for the enzymes of the catabolic pathway, or to both of these causes. The possibilities were examined in turn.

Does 4-ethylbenzoate induce synthesis of the enzymes of the catabolic pathway for alkylbenzoates? In order for 4EB to induce synthesis of the enzymes of the catabolic pathway, it must bind to and activate XylS, and activated XylS must in turn stimulate transcription of the catabolic operon from its promoter P_{meta} (see Figure 15.14). A convenient means of monitoring transcriptional activity in response to experimental manipulations is the *transcriptional fusion procedure*, in which the promoter under investigation is fused to a test gene (such as *lacZ*) that codes for a readily assayable protein (such as β-galactosidase). The P_{meta} promoter was therefore linked *in vitro* to the *lacZ* gene on one hybrid plasmid, and the *xylS* gene coding for XylS, the positive activator of P_{meta}, was introduced into the same cell on a different, compatible hybrid plasmid. The levels of β-galactosidase synthesized by such bacteria, grown in the presence or absence of various benzoate derivatives, were then measured, and the relative increase, or induction ratio, in enzyme level induced by each derivative was determined. The data in Table 15.7 show that 4EB did not activate the *xylS* protein and was therefore unable to induce synthesis of the enzymes of the benzoate catabolic pathway.

Can the TOL-plasmid-encoded enzymes of the alkylbenzoate cleavage pathway catabolize 4EB? P. putida harboring a mutant TOL plasmid in which the cleavage pathway is expressed constitutively at high levels was used to determine which, if any, of the alkylbenzoate pathway enzymes are nonpermissive for the catabolism of 4EB. This plasmid enabled the bacterium to grow on 3MB, 4MB, and 34DMB but not on 4EB. However, when the organism was provided with glucose and 4EB, it grew at the expense of the sugar and accumulated 4-ethylcatechol (see Figure 15.13). *This shows that 4EB enters the cell and is metabolized by the initial enzymes of the catabolic pathway. However, catechol 2,3-dioxygenase is evidently unable to use 4-ethylcatechol as a substrate.*

The foregoing experiments revealed at least two of the biochemical obstacles to the degradation of 4EB through the TOL *meta*-degradation pathway. To redesign the pathway to enable it to degrade 4EB, it was necessary (1) to broaden the effector specificity of the *xylS* regulator

Activation of *xylS* Protein by Benzoate Derivatives

TABLE 15.7

Benzoate derivative	Induction ratio (relative increase in β-galactosidase)
None	1[a]
2-Methylbenzoate[b]	18
3-Methylbenzoate	17
4-Methylbenzoate	4
2,3-Dimethylbenzoate	10
3,4-Dimethylbenzoate	5
4-Ethylbenzoate	1

[a] *E. coli* K-12 bacteria containing two plasmids, one carrying $P_{meta}::lacZ$ fusion and the other carrying *xylS*, were cultured in medium containing or lacking the benzoate derivative indicated in the left column. A value of 1 indicates that the derivative did not induce the synthesis of β-galactosidase and thus does not serve as an effector of the *xylS* protein.

[b] For structures of the benozate derivatives, see Fig. 15.13.

SOURCE: Data from Ramos, J.L., and Timmis, K.N. (1987), Experimental evolution of catabolic pathways of bacteria, *Microbiol. Sci.* 4:228–237.

protein such that it would be activated by 4EB, and (2) to generate catechol 2,3-dioxygenase mutant enzymes capable of degrading 4-ethyl-catechol.

The strategy employed to isolate mutants in *xylS* that would produce an XylS protein that could be activated by 4EB is illustrated in Figure 15.15. Two compatible plasmids were introduced into an *E. coli* strain. A tetracycline resistance plasmid was modified so that the promoter responsible for transcription of the tetracycline resistance gene was replaced by the P_{meta} promoter of the TOL *meta*-cleavage operon. This plasmid was designated pJLR200. A second expression plasmid (pNM185) carried a kanamycin resistance gene and the *xylS* gene. *E. coli* cells containing these two plasmids are tetracycline-sensitive even though they are making the XylS protein. However, they become resistant if a *xylS* effector, such as benzoate, is added, because the XylS–effector complex binds to the P_{meta} promoter and allows expression of the tetracycline resistance gene.

E. coli cells containing plasmids pJLR200 and pNM185 were plated on nutrient agar containing ampicillin, kanamycin, tetracycline, 4EB, and the mutagen ethylmethane sulfonate. Cells that can grow under these conditions must be able to express the tetracycline resistance gene. Two classes of mutants were obtained. The first class grew on tetracycline-containing plates in the absence of 4EB. These mutants contained mutations in the P_{meta} promoter that allowed constitutive expression of the *tet* gene. A second class grew on tetracycline-containing plates only

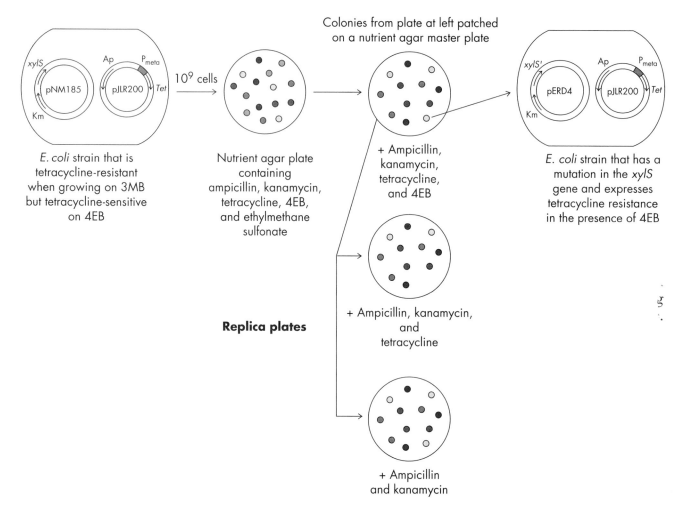

Figure 15.15
Strategy for the selection of mutants whose XylS protein would form a complex with 4EB that could bind at P_{meta} and allow the transcription of genes controlled by this promoter. Based on Ramos, J.L., and Timmis, K.N. (1987), *Microbiol. Sci.* 4:228–237, Fig. 5.

if they also contained 4EB. The mutation in these mutants lay within the *xylS* gene. The product of the mutant gene (*xylS'*), designated XylS' formed a complex with 4EB that was capable of binding at P_{meta} and inducing expression of the *tet* gene.

A plasmid carrying the mutant *xylS'* gene was introduced into *putida* carrying the TOL plasmid pWW0, but the bacteria were unable to grow on 4EB. However, when these organisms were grown a medium containing a utilizable carbon source (glucose), they syn sized the enzymes of the *meta* -cleavage pathway and accumulate ethylcatechol. Biochemical experiments revealed that 4-ethylcat

culture in the absence of methanogenesis. *Appl. Env. Microbiol.* 57:2287–2292.

Madsen, E.L., Sinclair, J.L., and Ghiorse, W.C., 1991. *In situ* biodegradation: Microbiological patterns in a contaminated aquifer. *Science* 252:830–833.

Pentachlorophenol Degradation in an Aquatic Environment

Pignatello, J.J., Martinson, M.E., Steiert, J.G., Carlson, R.E., and Crawford, R.L., 1983. Biodegradation and photolysis of pentachlorophenol in artificial freshwater streams. *Appl. Environ. Microbiol.* 46:1024–1031.

Pignatello, J.J., Johnson, L.K., Martinson, M.E., Carlson, R.E., and Crawford, R.L., 1985. Response of the microflora in outdoor experimental streams to pentachlorophenol: Compartmental contributions. *Appl. Environ. Microbiol.* 50:127–132.

Pignatello, J.J., Johnson, L.K., Martinson, M.E., Carlson, R.E., and Crawford, R.L., 1986. Response of the microflora in outdoor experimental streams to pentachlorophenol: Environmental factors. *Can. J. Microbiol.* 32:38–46.

Catabolic Plasmids

Burlage, R.S., Hooper, S.W., and Sayler, G.S., 1989. The TOL (pWW0) catabolic plasmid. *Appl. Environ. Microbiol.* 55:1323–1328.

Sayler G.S., Hooper, S.W., Layton, A.C., and King, J.M.H., 1990. Catabolic plasmids of environmental and ecological significance. *Microb. Ecol.* 19:1–20.

Degradation of Aromatic and Chloroaromatic Hydrocarbons by Bacteria

Rochkind-Dubinsky, L.M., Sayler, G.S., and Blackburn, J.W., 1987. *Microbiological Decomposition of Chlorinated Aromatic Compounds.* Marcel Dekker.

Reineke, W., and Knackmuss, H.J., 1988. Microbial degradation of haloaromatics. *Ann. Rev. Microbiol.* 42:263–287.

Harayama, S., and Timmis, K.N., 1989. Catabolism of aromatic hydrocarbons by *Pseudomonas*. In Hopwood, D.A., and Chater, K.F. (eds.), *Genetics of Bacterial Diversity.* pp. 151–174. Academic Press.

Smith, M.R., 1990. The biodegradation of aromatic hydrocarbons by bacteria. *Biodegradation* 1:191–206.

Commandeur, L.C.M., and Parsons, J.R., 1990. Degradation of halogenated aromatic compounds. *Biodegradation* 1:207–220.

Ambramowicz, D.A., 1990. Aerobic and anaerobic biodegradation of PCBs: A review. *Crit. Rev. Biotechnol.* 10:241–251.

Hardman, D.J., 1991. Biotransformation of halogenated compounds. *Crit. Rev. Biotechnol.* 1:1–40 (1991).

Genetic Engineering of Metabolic Pathways

Ramos, J.L., and Timmis, K.N., 1987. Experimental evolution of catabolic pathways of bacteria. *Microbiol. Sci.* 4:228–237.

Rojo, F., Pieper, D.H., Engesser, K.-H., Knackmuss, H.-J., and Timmis, K.N., 1988. Assemblage of ortho cleavage route for simultaneous degradation of chloro- and methylaromatics. *Science* 238:1395–1398.

Rojo, F., et al., 1988. Laboratory evolution of novel catabolic pathways. *Biotec* 2:65–74.

Wubbolts, M.G., and Timmis, K.N., 1989. Engineering *Pseudomonas* spp. for the production of *cis*-diols by biotransformation. In Hershberger, C.L., Queener, S.W., and Hegeman, G. (eds.), *Genetics and Molecular Biology of Industrial Microorganisms*, pp. 194–294. American Society for Microbiology.

Mechanisms of Dissolution (Leaching) of Metals from Ores

Brierley, C.L., 1982. Microbiological mining. *Sci. Amer.* 247(Aug.):44–53.

Brierley, C.L., Kelly, D.P., Seal, K.J. and Best, D.J., 1985. Materials and biotechnology. In Higgins, I.J., Best, D.J., and Jones, J. (eds.), *Biotechnology: Principles and Applications.* pp. 163–212. Blackwell Scientific.

Gentina, J.C., and Aceveda, F., 1985. Microbial ore leaching in developing countries. *Tr. Biotechnol.* 3:86–89.

Hutchins, S.R., Davidson, M.S., Brierley, J.A., and Brierley, C.L., 1986. Microorganisms in reclamation of metals. *Ann. Rev. Microbiol.* 40:311–336.

Torma, A.E., 1987. Impact of biotechnology on metal extractions. *Mineral Processing and Extractive Metallurgy Rev.* 2:289–330.

Rawlings, D.E., and Woods, D.R., 1988. Recombinant DNA techniques and the genetic manipulation of *Thiobacillus ferrooxidans*. In Thomson, J.A. (ed.), *Recombinant DNA and Bacterial Fermentation*, pp. 277–296. CRC Press.

Brierley, J.A., 1990. Biotechnology for the extractive metals industries. *JOM* 42:28–30.

Microorganisms in the Removal of Heavy Metals from Aqueous Effluents

Goldwater, L.J., 1971. Mercury in the environment. *Sci. Amer.* 224 (May):15–21.

Beveridge, T.J., and Doyle, R.J. (eds.), 1989. *Metal Ions and Bacteria*. Wiley.

Hammer, D.A., ed., 1989. *Constructed Wetlands for Wastewater Treatment: Municipal, Industrial, Agricultural*. Lewis.

Ehrlich, H.L., and Brierley, C.L. (eds.), 1990. *Microbial Mineral Recovery*. McGraw-Hill.

Volesky, B. (ed.), 1990. *Biosorption of Heavy Metals*. CRC Press.

Biological Detoxification of Wastewater from Precious Metal Processing

Whitlock, J.L., 1990. Biological detoxification of precious metal processing wastewater. *Geomicrobiol. J.* 8:241–249.

REVIEW QUESTIONS

The review questions have been formulated with two goals in mind. One, naturally, is to test the reader's understanding of the material presented in the book. The second is to provide enrichment in important areas that may have been covered only lightly in the text because of length limitations or in lectures because of time limitations. The questions are double-numbered. The numeral before the decimal point designates the chapter that deals most directly with the material the question addresses. Answers follow the questions in a separate section.

Many of the questions were taken from problem sets, midterms, and final examinations given to students in an upper-division course called "Applied Microbiology and Biochemistry," which we have taught for the past seven years at the University of California, Berkeley. These questions are of two general types. Conventional questions simply test recollection of material in particular chapters (an example is question 7.2). Questions that describe situations not treated directly in the book (or the course) require the student to combine this new information with the material presented in the book and to arrive at an answer through the application of analytic reasoning. Where a question raises an unresolved controversy, the answer presents the opposing points of view.

To emphasize the urgency and currency of the issues discussed in the book, we have based many of the questions on information and ideas presented in recent research papers and patent abstracts. Citations are provided for the source materials. Current issues of journals such as *Science, Nature, Applied and Environmental*

Microbiology, Applied Microbiology and Biotechnology, Bio/ Technology, Environmental Science and Technology, and *Journal of Bacteriology* are a rich source of questions on applied microbiology. Patent abstracts published in a number of journals (such as *Biotechnology Advances*) also provide interesting material for review questions.

Questions

1.1 In 1989, the National Research Council issued a report on risk assessment in the introduction of recombinant DNA (rDNA)-engineered organisms into the environment [*Field Testing Genetically Modified Organisms: Framework for Decisions* (National Academy Press)]. The consensus in this report is that "no conceptual distinction exists between genetic modification of plants and microorganisms by classical methods or by molecular techniques that modify DNA and transfer genes." Consequently, "[a]ssessment of the risk of introducing rDNA-engineered organisms into the environment should be based on the nature of the organism and the environment into which the organism is introduced, not on the method by which it was produced." A contrary view, termed "a horizontal approach," holds that there *is* something systematically similar and functionally important about the set of organisms whose only common characteristic is that they have been subject to manipulation by rDNA techniques. The latter view forms the basis of government regulations adopted in the United States in 1993 that apply to field trials of certain recombinant plants. Which view do you favor and why?

1.2 The coat protein of many viruses confers on the virus the specificity of interaction with the vector (for example, an insect, nematode, or fungus) that naturally transmits the virus. However, to protect plants against pathogenic viruses, plants are being engineered to express that virus's coat protein (capsid) gene. Such transgenic plants expressing a viral capsid protein exhibit resistance to that virus and to closely related strains but remain susceptible to other viruses. A.E. Greene and R.F. Allison [(1994), Recombination between viral RNA and transgenic plant transcripts, *Science* 263:1423–1425] describe the results of a study in which transgenic plants expressing the 3′ two-thirds of the cowpea chlorotic mottle virus (CCMV, an RNA virus) capsid gene were inoculated with a CCMV deletion mutant lacking the 3′ one-third of the capsid gene. The deletion mutant can multiply only in the inoculated cells, and systemic infections occur only if recombination restores a functional capsid gene. Such systemic infection took place in 4 of 125 inoculated transgenic plants.

Analysis of the viral RNA in these 4 plants confirmed that RNA recombination had united the incomplete transgenic messenger RNA for the capsid protein produced by the plant and the RNA of the challenging virus, making possible the synthesis of a functional capsid protein with restoration of the ability of the virus to cause systemic infection. In all instances, aberrant homologous recombination took place between the mRNA expressed by the plant and the RNA of the challenging deletion mutant CCMV. RNA sequencing showed that recombination took place within the overlapping region of the transgenic mRNA and the RNA of the virus but that this recombination resulted in an altered mRNA with several deletions in the capsid open reading frame. Nonetheless, the mutant capsid protein was fully functional.

The results of Greene and Allison's study have given rise to divergent views on the potential risks of using transgenic crops that express viral coat proteins. What might some of the concerns be?

2.1 "In their review on catabolic plasmids coding for functions of environmental and ecological significance, Sayler and collaborators [(1990), *Microbial Ecol.* 19:1–20] include a table of plasmids and their respective host species. Of the total number of strains in the list, three-quarters are *Pseudomonas*. Nearly half of them are strains of *P. putida*, and the next most abundant group is composed of anonymous '*Pseudomonas* species' for most of which there is not sufficient information to support their assignment to the genus as presently defined. This is indeed an unsatisfactory situation, only partly attenuated by the fact that the genus diagnosis has often been performed by workers for whom taxonomic research may at best represent a marginal activity" [N.J. Palleroni (1993), *Antonie van Leeuwenhoek* 64:231–251].

Suppose that some of the '*Pseudomonas* species' actually belong to other genera, does the misidentification matter?

3.1 A common way of making a "library" of chromosomal genes (a "genomic library") is to cut the chromosomal DNA partially with *Sau*3A, which cuts the DNA as indicated by the arrowheads in the upper sequence at right. The fragments are then inserted into a vector, which is cut open with *Bam*HI that cleaves the vector DNA at a single site as indicated in the lower sequence at right.

```
      ▼
   —GATC—
   —C TAG—
          ▲


      ▼
   —GGATCC—
   —CC TAGG—
          ▲
```

(a) Why is the vector cut with *Bam*HI instead of, say, *Eco*RI? (b) Why isn't the vector cut with *Sau*3A? (Consider how often a *Sau*3A site might occur in a random sequence of DNA.) (c) Why isn't the genomic DNA digested completely with *Sau*3A? (d) Why isn't the genomic DNA cut with *Bam*HI?

3.2 Steven Lindow and his associates cloned the ice-nucleation genes of *Pseudomonas syringae* by inserting fragments of the genomic

DNA of this bacterium into cosmid vectors, introducing recombinant constructs into an *E. coli* host, and then screening the *E. coli* progeny that contained recombinant DNA for the ability to induce ice nucleation. This involved floating, on a cold fluid at about −9°C, a sheet of aluminum foil with a few dozen water droplets each containing *E. coli* cells from an individual colony. (a) Why did they not use plasmid vectors such as pBR322? (b) Why did they not use expression vectors such as pUC18?

3.3   A paper entitled "Renaturation of a single-chain immunotoxin facilitated by chaperones and protein disulfide isomerase" was published in the journal *Bio/Technology* in 1992. This paper describes the production of a fusion protein in *E. coli*. The amino-terminal portion of the immunotoxin consists of a protein that binds specifically to certain human cancer cells, and the carboxyl-terminal portion consists of a *Pseudomonas* exotoxin, so that the immunotoxin is expected to bind to cancer cells and then to kill them through the action of the exotoxin. The properly folded protein contains three disulfide bonds in the amino-terminal domain and one such bond in the carboxyl-terminal domain.

The immunotoxin, however, is produced as inclusion bodies in *E. coli*. Renaturation of completely unfolded, reduced immunotoxin under optimal conditions yielded about 5% properly folded protein. When protein disulfide isomerase was present, the yield increased to about 20%, and when certain chaperonins were present, the yield was 15%. When both were present, the yield of the properly folded protein was about 30%. Explain these results.

4.1   Yeast has some advantages over *E. coli* if one is trying to express mammalian proteins that are to be used as therapeutic agents. What are the advantages?

4.2   A group of scientists were interested in producing a protein hormone of human origin in yeast *S. cerevisiae*. They started from a preparation of mRNA, isolated from the organ that secretes this hormone, and made a corresponding cDNA preparation. The cDNA was inserted into a YEp vector of the type shown in Figure 4.4, and the recombinant DNA was introduced into a yeast host strain.

(a) Why did they isolate mRNA from this specific organ rather than from other tissues? (b) Why was cDNA used as the source of DNA to be cloned, rather than the chromosomal DNA? (c) The cDNA (0.1 μg) was mixed with 0.3 μg of opened YEp vector, and the ligation reaction was carried out in a total volume of 0.02 ml. Why did they use this very small volume instead of a more conventional volume such as 1 ml or 5 ml? (d) What was the genotype of the host strain? (e) How were the cells containing plasmids selected? (f) How could the scientists have screened for cells containing the proper gene for this hormone?

4.3 The cDNA coding for the protein hormone mentioned in question 4.2 was inserted into a yeast expression plasmid. The human hormone is heavily glycosylated, but the product expressed in yeast contained little carbohydrate and was found mainly in the cytoplasm. (a) Why wasn't the protein glycosylated? Why wasn't the protein secreted? (b) What would you do to improve the efficiency of glycosylation and secretion of this protein in the yeast?

5.1 Suppose you work at a biotechnology company, and your superior has just told you to outline a way to produce a "better" diphtheria vaccine by the use of recombinant DNA methods. What would be your answer?

5.2 An envelope protein of polio virus is known to play a crucial role in the development of immunity against this pathogen. Some workers have cloned the gene for this protein, and inserted the DNA sequence into a recombinant plasmid already containing the gene for LamB, an *E. coli* outer membrane protein. The insertion was done so that the sequence coding for the polio protein was inserted, in the correct reading frame, into the middle of the gene for the LamB protein. This final construct was transformed into a strain of *Salmonella* that lacks enzymes for the synthesis of *p*-amino- and *p*-hydroxybenzoic acid. The LamB protein is an inducible protein and can become a major protein within the outer membrane when induced. What features would make this construct potentially more attractive as a polio vaccine than the same polio protein expressed in the *E. coli* cytoplasm?

5.3 In spite of their advantages, peptide vaccines have not shown protective activities comparable to those of traditional vaccines, especially under conditions close to "real-world" conditions. Explain why this is so.

6.1 The *Bacillus thuringiensis* toxins most widely used against mosquito larvae are from *Bacillus thuringiensis* serovar *israelensis*. A recent paper describes toxins from *Bacillus thuringiensis* serovar *medellin*, a strain isolated in Colombia. The toxins from the *medellin* strain are as effective against mosquito larvae as are those from the *israelensis* strain, but they show little immunological cross-reactivity when examined for reaction with antibodies raised against the *Bacillus thuringiensis* serovar *israelensis* toxins. Why are the serovar *medellin* toxins worth studying, even though they are no better against mosquitoes than the toxins from the serovar *israelensis?* Explain briefly.

6.2 A company has been inserting a *Bacillus thuringiensis* δ-endotoxin gene in the genomes of plants (in particular, cotton), in the hope of producing "supercrops" with built-in resistance to insects. Agricultural researchers have urged the company to sell its transgenic cotton seeds in mixtures with normal cotton seeds, when it introduces them in the mid-1990s, so that farmers won't plant entire fields of plants expressing *Bacillus thuringiensis* δ-endotoxin. What might be the reason for this recommendation?

6.3 The CryIIA protein is the predominant constituent of the cuboidal (P2) crystal of sporulating *Bacillus thuringiensis* subsp. *kurstaki* cells. This protein is toxic to both lepidopteran and dipteran insects. Recently, the 4.0-kb *Bam*HI-*Hin*dIII fragment encoding the *cryIIA* operon was inserted into the *B. thuringiensis-E. coli* shuttle vector pHT3101, giving rise to the plasmid pMAU1. pMAU1 was used to transform a strain of *Bacillus cereus* strain BT-8 that colonizes tomato leaves. This particular strain grows vegetatively on tomato leaves, reaches populations of 10^4 colony-forming units per leaf, and shows antifungal activity against several important tomato pathogens. Strain BT-8 transformed with plasmid pMAU1 expresses CryIIA at a low level [Moar, W.J., et al. (1994), *Appl. Environ. Microbiol.* 60:896–902]. What is the advantage of generating a recombinant version of *Bacillus cereus* strain BT-8 that is capable of expressing *B. thuringiensis* toxins?

7.1 The yeast *Kluyveromyces lactis* is used extensively in the food industry. [For a review, see Bonekamp, F.J., and Oosterom, J. (1994), *Appl. Microbiol. Biotechnol.* 41:1–3]. *K. lactis* itself has been used in the form of an inactivated powder as a health food and protein supplement in amounts up to 30 g per day for infants and for sick or underfed persons. *K. lactis* also produces an extracellular enzyme, lactase, that hydrolyzes milk lactose into its constituent monosaccharides glucose and galactose. *K. lactis* is the most widely used commercial preparation for the production of low-lactose milk for people incapable of tolerating lactose in their diet.

The same strain of *K. lactis* that is used to produce lactase has been used to engineer a recombinant strain that produces calf chymosin, a proteolytic enzyme with several disulfide bonds, for use in the manufacture of cheese. Was this a good choice of an organism for the production of chymosin? Why?

7.2 Microbial glucose isomerases are used in massive amounts to catalyze the conversion of glucose to fructose in the manufacture of high-fructose syrups. However, glucose isomerase is quite costly. How does the commercial process for the conversion of glucose to fructose minimize the expenditures on glucose isomerase?

7.3 The production by microorganisms of many catabolic enzymes of commercial importance depends on the presence of inducers; that is, the enzymes are not made in the absence of substrate. Examples are *Bacillus stearothermophilus* amylase, which degrades starch and which is induced by maltodextrins, products of starch degradation, and the *Klebsiella aerogenes* pullulanase, which degrades the polysaccharide pullulan and requires the product maltose for induction of its synthesis. (a) Why does the production of such enzymes depend on the addition of inducer? (b) How would one eliminate the dependence of enzyme synthesis on the presence of inducer? (c) Even when dependence on inducer has been eliminated, many enzymes of industrial importance are subject to regulation by *catabolite repression*. What is catabolite repression? (d) Explain how one might be able to avoid or eliminate catabolite repression.

8.1 B.L. Dasinger and colleagues [(1994), Composition and rheological properties of extracellular polysaccharide 105-4 produced by *Pseudomonas* sp. strain ATCC 53923, *Appl. Environ. Microbiol.* 60:1364–1366] describe "a novel extracellular polysaccharide (named EPS 105-4) with unusually potent thickening powers."

EPS105-4 contains the monosaccharides D-mannose, D-glucose, D-galactose, and D-glucuronic acid in the molar ratio of 1:4:1:2. Describe precisely what additional information is needed to determine what aspects of the structure of this polysaccharide are responsible for its "unusually potent thickening powers."

8.2 In the production of xanthan by *Xanthomonas juglandis* in continuous culture, the yield of polysaccharide depends on the nature of the limiting nutrient. When glucose is the limiting nutrient, the cell yield is 1.1 g/liter and the xanthan yield is 2.7 g/liter. When the nitrogen source is limiting, the cell yield is 1.6 g/liter and the xanthan yield is 7.0 g/liter. Explain these results.

8.3 The April 14, 1992, issue of *Science* included an article entitled "Polyhydroxybutyrate, a biodegradable thermoplastic, produced in transgenic plants" [Poirier, Y., et al. (1992), *Science* 256:520–523]. (a) What is polyhydroxybutyrate? (b) What kinds of organisms normally make polyhydroxybutyrate? Where in the cells of these organisms does the material accumulate? (c) What is the normal function of polyhydroxybutyrate? (e) Describe in detail the way the production of polyhydroxybutyrate in plants might have been achieved. (f) Why is the expression of polyhydroxybutyrate in plants an important achievement?

9.1 Plasmids are excellent vectors for the construction of recombinant DNA molecules and their introduction into host cells. The Ti plasmid of *Agrobacterium tumefaciens*, however, is difficult to use for such purposes, unless it is modified in several ways. List the reasons why the unmodified Ti plasmid is unsuitable as a vector of recombinant DNA.

9.2 Starch is a polymer that consists mainly of α-1,4-linked glucose residues, but there are also some α-1,6-linked branch points made by the "branching enzyme." Some companies are trying to produce potato containing starch with fewer branch points, because linear starch has some desirable properties compared with highly branched starch. If you were working in one of these companies, what would be your strategy?

9.3 During the last few years, at least half a dozen articles were published that dealt with the introduction of "foreign" DNA segments into chloroplast (plastid) DNA. (a) Why does anyone want to alter chloroplast DNA? (b) Would you predict that the standard *Agrobacterium*-based technique would result in the insertion of introduced DNA into the chloroplast genome? If your answer is "no," what could be the obstacles? What would you do to overcome these obstacles?

10.1 Let us say that a bacterium has been isolated from a piece of partially decomposed wood found at the mouth of a geothermal vent. This organism has been found to have an optimal growth temperature of 102°C. It produces an extracellular cellulase that is stable and active up to about 110°C. Do you think that this cellulase would have potential value in some industrial process? What would be the objective of such a process? Explain clearly why this cellulase would be particularly useful in place of some other, already available cellulase.

10.2 The recently issued U.S. Patent 5196069 is entitled "Apparatus and Method for Cellulose Processing Using Microwave Treatment." Why would it be interesting to investigate the effect of microwave radiation on cellulose?

10.3 A recent paper by V.V. Zverlov and colleagues [(1994), *Biotechnology Letters* 16:29–34] is entitled "Purification and cellulosomal localization of *Clostridium thermocellum* mixed linkage β-D-glucanase." What is the cellulosome? What is believed to be its primary biological role? The β-D-glucanase described in this paper cleaves both 1,3 and 1,4 linkages in glucans and is shown conclusively to be a component of the cellulosome. Glucans are important polysaccharide components of seeds. How does the finding of the β-D-glucanase in the cellulosome influence your view of the biological role of the cellulosome?

11.1 Recently, a cellulase gene from the Gram-positive bacterium *Cellulomonas uda* was introduced into *Zymomonas mobilis*. What is the practical purpose of the introduction of cellulase into *Zymomonas*?

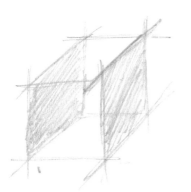

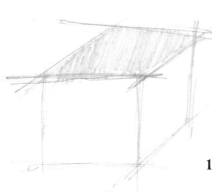

11.2 In a 1990 article entitled "Biofuels—real life implications" [*Biotechnology International*, pp. 325–329 (Century Press Ltd.)], J. Heaford wrote, "Forget calculations of biofuels versus oil prices, the answers differ according to the methodology. It is the long term trends which are significant." What long-term trends might make ethanol relevant someday as a fuel?

12.1 The amino acid arginine is made by an unbranched pathway that converts glutamic acid in four steps to ornithine, which is converted in a single step into citrulline, which in turn is finally converted in two enzymatic steps to arginine. It is known that *E. coli* can take up and utilize all the amino acids and intermediates just mentioned. You want to produce citrulline by fermentation using mutant strains of *E. coli*.

(a) How would you isolate *auxotrophic* mutant strains that will overproduce citrulline? (b) How would you carry out the actual fermentation process? Mention the types of media to be used and any nutrients that must be fed continuously to the culture. (c) It is known that arginine biosynthesis is regulated in *E. coli* both by repression and by feedback inhibition of the first enzyme. You note that the introduction of additional regulatory mutation(s) would reduce the production cost of citrulline. Why? (d) How would you isolate such regulatory mutants? (e) How would you find out which regulatory mechanism is altered in each of the mutants you have isolated? (f) How would you combine the two types of mutations into a single strain?

12.2 The amino acid tryptophan is synthesized by a branched pathway. In the common part of the pathway, phospho*enol*pyruvate and erythrose 4-phosphate are converted into 3-deoxy-D-*arabino*-heptulosonate 7-phosphate (DHAP) by DHAP synthase, and the 7-carbon sugar is converted in six steps to chorismate, the last common intermediate. In the tryptophan-specific branch, chorismate is first converted into anthranilate, which then reacts with phosphoribosylpyrophosphate to produce phosphoribosylanthranilate. The latter compound is in turn converted into tryptophan in four enzymatic steps (see Box 12.2).

Cloning and overexpression of some of the genes of tryptophan biosynthesis have been explored by several companies (as indicated by the eosinophilia–myalgia syndrome incident described on page 423), although such approaches are usually unnecessary for the production of most other amino acids.

In a report published in late 1993 in the journal *Bio/Technology*, Kyowa Hakko Co. scientists described their experience with the cloning and overexpression of tryptophan biosynthetic genes in *Corynebacterium glutamicum*.

(a) Why is the recombinant DNA approach thought to be necessary for the production of tryptophan? (b) Why do you think the scientists chose *C. glutamicum* rather than *E. coli* as the producing organism? (c) When the gene for DHAP synthase was cloned in a multicopy plasmid and the recombinant plasmid was introduced into the host strain, a large amount of chorismate was excreted into the medium. What was happening in these bacterial cells? (d) When the recombinant plasmid contained both the DHAP synthase gene and all the genes of the tryptophan-specific branch, less chorismate was excreted, but a large amount of anthranilate was excreted. How do you explain this observation? (e) What would you do to eliminate the excretion of intermediates such as anthranilate and increase the production of tryptophan?

12.3   5′-GMP is an important flavor enhancer that is thought to impart the flavor of shiitake mushrooms to cooked food. This compound is made by microorganisms by the pathway shown below. In a soil bacterium known to have rather simple regulatory mechanisms, the specific activity of IMP (inosine-5′-monophosphate) dehydrogenase was about 30 times higher than that of adenylosuccinate synthetase. Nevertheless, the first step taken by the scientists who created a mutant strain for the fermentative production of 5′-GMP was the mutational inactivation of the gene for adenylosuccinate synthetase. Why was this done?

In order to get a commercially profitable level of 5′-GMP production, a property of at least one enzyme had to be altered. Which enzyme is this? How were the strains with altered enzymes selected?

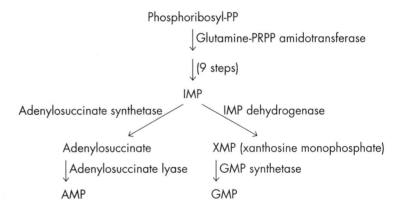

13.1   In going through the literature on the screening of new antibiotics, you find that scientists in a major pharmaceutical company have been screening for new β-lactam antibiotics that would inhibit the growth of an *Enterobacter cloacae* strain that is derepressed for the production of the chromosomally coded β-lactamase.

(a) Why did they not use organisms like *Staphylococcus aureus*, which are exquisitely susceptible to all kinds of β-lactams? (b) Why did they not use the wild-type *E. cloacae*, with its inducible β-lactamase? (c) When they finally found a new β-lactam compound with the screening procedure mentioned at the beginning, they also tried to make sure that the compound was active against *E. coli* harboring common R plasmids. Why? (d) Screening for β-lactams that are *active* against (that is, *inhibit* the growth of) organisms such as *E. cloacae* has its difficulties. What do you think is the most significant problem? (e) Instead of screening directly for compounds that would inhibit the *E. cloacae* strain, you want to devise a screening procedure that uses both the *E. cloacae* strain and a β-lactam-susceptible indicator strain such as *S. aureus*. This will allow you to discover new "lead" compounds, which may be made more active against the *E. cloacae* strain with further chemical modifications. How would you carry out this screening?

13.2 The following is a passage from a review by Professor Heinz G. Floss of Ohio State University [(1987), *Trends in Biotechnology* 5:111–115]. "Our own group is developing methods to interchange biosynthetic genes between *Streptomyces azurae*, the producer of thiostrepton, and *S. actuosus*, the producer of nosiheptide. These two highly modified peptide antibiotics share essential structural features, a macrocyclic ring containing a number of thiazole groups, but differ in several variable regions." (a) What would be the potential practical benefits of this research? (b) List two ways in which this genetic "interchange" might be achieved.

13.3 Many "natural" antibiotics produced by fungi or streptomycetes are often inactivated by enzymes coded by genes that are found on R factors. In contrast, enzymatic inactivation practically never occurs as an important mechanism of resistance against totally human-made chemotherapeutic agents, such as sulfonamides, trimethoprim, and fluoroquinolones. Explain why.

14.1 Cortisone is an important anti-inflammatory agent. The structure of this molecule is shown below. Cortisone has six asymmetric centers (indicated by asterisks). Consequently, 64 stereoisomers of this compound are possible. Only one of these (the one formed normally *in vivo*) has the full biological activity. The total chemical synthesis of cortisone requires many steps and is very costly. It can be significantly simplified by using microbial transformations. The first such transformation, introduced in 1952, was the hydroxylation of progesterone (an early intermediate in cortisone synthesis). What are three major advantages of such a microbial transformation over a purely chemical synthetic route?

Progesterone   11α-Hydroxyprogesterone   Cortisone

14.2 The recently issued U.S. Patent 5204251 is entitled "Process of Enzymatic Interesterification Maintaining a Water Content of 30–300 ppm (parts per million) using *Rhizopus*." *Rhizopus* is a fungus. The abstract reads, in part, "A process of an enzymatic interesterification in a low water content condition which comprises subjecting a reaction liquid containing (a) a fat or oil and (b) (i) another fat or oil; (ii) a fatty acid ester with a lower alcohol; or (iii) a fatty acid, to the action of a lipase in the form of dried *Rhizopus* cells while maintaining the water concentration between 5 and 1000 ppm in the reaction liquid (an organic solvent) during the reaction. An interesterification of fat or oil can be carried out efficiently in high yield under these conditions." (a) Write an equation illustrating one of the types of interesterification reactions described in this abstract. (b) Why does this reaction need to be carried out in a mixed aqueous-organic solvent with the water content being kept below 1000 ppm?

14.3 You have been asked to suggest a way of converting the triglyceride A (at left) very specifically to the compound B (at left). What *specific* method would you recommend?

15.1 What is a catabolic plasmid? Give an example of such a plasmid, and specify what benefits accrue to a cell that possesses the particular plasmid you have mentioned.

15.2 Extensive use of pentachlorophenol and polycyclic aromatic hydrocarbon-containing creosote in the wood-preserving industry has led to serious soil contamination by these compounds. A paper by M-P. Otte and colleagues entitled "Activation of an indigenous microbial consortium for bioaugmentation of pentachlorophenol/creosote contaminated soils" [(1994), *Appl. Microbiol. Biotechnol.* 40:926–932] describes a successful attempt to obtain a microbial population that rapidly degrades pentachlorophenol and polycyclic aromatic hydrocarbons in soil slurries in bioreactors. (Definitions: *Indigenous* means "occurring or living naturally in an area." *Bioaugmentation* in the context used here means "increasing the size of a microbial population.")

A

B

(a) Describe briefly what these investigators may have done to obtain a microbial population with such degradative capabilities. (b) Could one obtain a well-defined population of genetically engineered microorganisms with the same degradative capabilities as the population obtained by Otte et al.? If your answer is "yes," describe the precise steps you would need to take to obtain such a population. If your answer is "no," explain clearly why you do not think that the genetic engineering approach is feasible.

15.3 How do the microorganisms responsible for the solubilization of copper and uranium from low-grade ores obtain energy for growth? What do these organisms use as a source of cell carbon?

15.4 Early manufactured gas plant sites are heavily contaminated by polycyclic aromatic hydrocarbons, such as naphthalene, phenanthrene, benzo[*a*]pyrene, benzo[*a*]anthracene, and pyrene. These compounds pose a hazard because of their carcinogenicity and slow degradation in the environment. All of the compounds mentioned above are degradable by soil microorganisms. For example, many *Pseudomonas* species contain plasmids encoding genes for naphthalene degradation that also function in the degradation of anthracene and phenanthrene. Moreover, the naphthalene degradative genes are highly conserved.

Pseudomonas putida strain G7 87-kb plasmid NAH7 contains genes for eleven enzymes necessary for the degradation of naphthalene. These structural genes are encoded in two operons; *nahABCDEF* encodes the enzymes that degrade naphthalene to salicylate, and *nahGHINLJK* encodes the enzymes that oxidize salicylate to acetaldehyde and pyruvate. The *nah* operon is expressed at a low level without induction; in the presence of naphthalene, salicylate is produced and binds to the *nahR* gene product bound to the upstream promoter regions of both operons, activating transcription of the *nah* genes. A half-life of 12 minutes has been determined for the *nahA* mRNA.

Describe how you might estimate the expression of the *nah* catabolic genes *in situ* in soil contaminated with polyaromatic hydrocarbons. How would you establish that the level of expression of these genes, as determined by the procedure you suggest, correlates positively with naphthalene mineralization rates in the soil?

Answers

1.1 H.I. Miller and D. Gunary [(1993), Serious flaws in the horizontal approach to biotechnology risk, *Science* 262:1500–1501] examine the issue of a "vertical" approach versus the "horizontal" approach to risk assessment. They point out that "a tomato breeder or regulator of polio vaccines who wishes to assess the potential risks of a new rDNA-derived tomato or vaccine, respectively, would rely more on information about tomatoes and poliovirus manipulated by traditional techniques, than on rDNA-manipulated pigs or bacteria," and comment further that "given the kinds of organisms that have been modified and the traits introduced one might as well survey all experiments that were performed using plastic, as opposed to glass, pipettes; or ones that were begun on certain days of the week." They concur with the consensus view expressed in the National Research Council report and advocate the vertical approach. A rigorous demonstration that an rDNA-modified organism presents hazard(s) different from its non-rDNA counterpart under regular conditions of use requires specific experiments that provide reasonable opportunity to detect hazards associated with the particular introduced trait *under appropriate selective pressure.*

1.2 R. Hull [(1994), *Science* 264:1649–1650] comments that the transgenically produced coat protein might encapsidate the genome of an unrelated virus infecting the transgenic plant changing its vector specificity and, consequently, its host range. M. Gibbs [(1994), *Science* 264:1650–1651] points out that after infection of the transgenic plant by another virus, aberrant recombination events may give rise to a recombinant virus with a new combination of vector specificity and host range that may find a new niche. For a contrary view, see Falk, B.W., and Bruening, G. (1994), Will transgenic crops generate new viruses and new diseases? *Science* 263:1395–1396, and Falk, B.W., and Bruening, G. (1994), *Science* 264:1651–1652. This is an important issue because it affects the general question of the field release of transgenic plants engineered to express viral sequences of various types.

2.1 The misidentification of an organism is as offensive to a taxonomist as a dangling participle is to a grammarian. From a practical standpoint, misidentification is important because it conveys the erroneous impression that most catabolic plasmids are to be found in bacteria in the genus *Pseudomonas* and that they occur only infrequently in bacteria belonging to other genera. Such a misconception may influence the choice of bacterial strains to be screened for particular catabolic capabilities.

3.1 (a) Vector DNA should have the same "sticky" ends as the genomic DNA fragments so that annealing between them will occur and insertion of the fragments into the vector will be facilitated. *Bam*HI, but not *Eco*RI, produces the ends that are identical to those produced by *Sau*3A. (b) *Sau*3A recognizes a 4-base sequence. The probability of this particular sequence occurring in a random sequence is $1/4^4$, or once every 256 bases. Since most cloning vectors are several kilobases or larger in size, there will be many *Sau*3A sites in the vector. In contrast, *Bam*HI sites are much rarer (they occur once every 4096 bases on average), because this enzyme recognizes a 6-base sequence. (c) This would produce fragments of much too small a size. One would have to examine astronomical numbers of clones to find the gene of interest (see Box 3.1). In addition, since an "average" prokaryotic gene would be about 1 kb in size and an average eukaryotic gene is much larger, practically all fragments would contain only parts of a gene, and expression-based screening or selection of recombinant clones would become impossible. (d) If there is a *Bam*HI site within the gene of interest, this procedure will cut the gene in the middle, again making expression-based detection impossible. In contrast, with partial digestion with *Sau*3A, there will always be some fragments that contain the entire gene.

3.2 (a) Because the recombinant clones had to be screened by a rather inefficient procedure, it was essential to increase the amount of cloned DNA, in view of the considerations described in Box 3.1. Cosmids are far better than plasmids for this purpose. (b) When one is cloning large pieces of DNA, expression vectors such as pUC18 do not work because transcription termination signals are likely to exist between the strong promoter supplied by pUC18 and the gene of interest (here, the ice-nucleation gene).

3.3 Because there are eight cysteine residues in the protein, renaturation is likely to produce incorrectly folded proteins containing disulfide bonds between incorrect pairs of cysteine residues. Once these conformers are formed, they cannot be converted into the correct, active conformer, because they are stabilized by covalent, disulfide bonds.

Protein disulfide isomerase catalyzes the exchange of disulfide bonds and allows some of the incorrectly folded conformers to reach the correct conformation. Chaperonins retard the folding process, again increasing the chance that the correct conformation will be reached during the folding process (see Box 3.10).

4.1 See page 143. In addition, for therapeutic agents, it may be important that components of yeast cells are not toxic to mammals or humans. *E. coli* contains a powerful toxic component, lipopolysaccharide (endotoxin).

4.2 (a) To maximize the probability of isolating the correct genetic material, it is important to start from a tissue or organ in which the mRNA for this particular product is already enriched. (b) The gene is likely to contain introns, and yeast cells may not be able to splice RNA transcripts of such sequences correctly.

The gene might also be very large and difficult to handle. In addition, with cDNA the chances of getting the correct clone are much higher, as described in part (a). (c) The ligation of foreign DNA (here, cDNA) and the vector DNA is a bimolecular reaction, and its rate depends on the probability of collision between these two types of molecules. This probability becomes higher when their concentrations are made higher by the use of a very small reaction volume. (d) Because the plasmid in Figure 4.4 has LEU2 as the selective marker, the host must be leu2, to take advantage of this selection method. (e) The transformed yeast population was spread on minimal agar plates not containing leucine. (f) They could have used immunological detection of the hormone. Alternatively, they could have used DNA probes, if the nucleotide sequence of the gene (or even the amino acid sequence of the protein) is known.

4.3 (a) The cDNA should have contained the sequence coding for the signal sequence. The most likely scenario is that the signal sequence of this human (secreted) protein was poorly recognized by the secretion machinery of yeast cells, so most of the protein did not enter the endoplasmic reticulum-Golgi pathway and therefore was not glycosylated either. (Alternatively, the DNA sequence coding for the signal sequence could have become deleted accidentally during cloning, as with the prochymosin cDNA, page 137). (b) In either case, one would supply a signal sequence that is known to be recognized efficiently in yeast cells. DNA coding for such a segment should be inserted just ahead of the region coding for the mature hormone. One could also utilize a more elaborate scheme for secretion, such as the pre-pro-α-factor sequence (page 162).

5.1 The traditional diphtheria vaccine, an inactivated diphtheria toxin protein, is less than perfect because it has caused disasters when the inactivation was incomplete. Furthermore, the manufacture of this vaccine involves cultivating a pathogenic organism in large amounts. The most desirable way to make a recombinant vaccine would be to clone the gene for diphtheria toxin (say, in *E. coli*), to delete portions of this gene *in vitro* so that there would be no danger of reversion, to make sure that the protein folded correctly and presented correct immunological epitopes on the surface, and to produce this genetically inactive toxin in nonpathogenic organisms such as *E. coli* or yeast.

5.2 This construct has several major advantages. (1) Subunit vaccines are usually less active because the antigen does not occur, as it would on the surface of the real pathogen, as a dense array. In contrast, when the protein is expressed in the *Salmonella* strain, it will be a part of the LamB protein on the outer membrane. Thus very large number of copies of this protein will be expressed as a dense, two-dimensional array on the cell surface, presumably increasing its effectiveness as an immunogen. (2) *Salmonella* may contribute its adjuvant molecules (lipopolysaccharide, peptidoglycan fragments), further boosting the immune response. (3) It can be adminstered orally in small amounts, and *Salmonella* will grow to a limited extent in the host to produce sustained immune response. (4) There is no need to isolate the viral protein, in contrast to the protein produced in *E. coli* cytoplasm. This will drastically decrease the cost of vaccine and increase its stability during storage.

5.3 See Chapter 5, pages 196–197.

6.1 Different *Bacillus thuringiensis* δ-endotoxins recognize (and bind to) different receptors on the midgut epithelial cells of a particular insect or receptors present in different insects. This recognition event is central in determining the host range and toxicity of *B. thuringiensis* δ-endotoxins. Because antibodies raised against *B. thuringiensis* serovar *israelensis* toxins do not recognize the toxins from the *medellin* strain, it is likely that the *israelensis* and *medellin* toxins recognize different receptors in mosquito larvae and hence would be *complementary* weapons against mosquitoes.

6.2 This approach will leave some plants free of *Bacillus thuringiensis* δ-endotoxin so that they can serve as a refuge for susceptible insects. These susceptible insects will be much more numerous and will breed with any resistant ones that might arise, and because the resistance trait is believed to be recessive, it will take longer to show up in the population.

6.3 *Bacillus cereus* strain BT-8 grows vegetatively on tomato leaves for up to two weeks and achieves a high population density. Moreover, it has antifungal activity against several important tomato pathogens. If a recombinant strain of this organism were also to express a high level of insecticidal activity, it would provide longer-lived and broader protection against pests than would spraying tomato plants with *B. thuringiensis* crystal toxins and spores.

7.1 Because products from *K. lactis* are used for human consumption and are designated "Generally Recognized as Safe (GRAS)," *K. lactis* is an organism of choice for manufacturing recombinant enzymes, such as chymosin, to be used in food. Moreover, yeasts are particularly useful for the expression and secretion of proteins that contain disulfide bridges (see Chapter 4).

7.2 The commercial process utilizes enzyme immobilized on a solid support. A single batch of immobilized enzyme can be used multiple times. Moreover, procedures are available for desorbing the enzyme from the solid support, purifying it away from accumulated contaminants from the substrate feed that may interfere with the process, and reimmobilizing it.

7.3 See Chapter 7, pages 246–247.

8.1 From the information provided in the question, we do not know whether the four monosaccharides are distributed in a linear polymer or whether we are dealing with a branched polymer such as xanthan. No information is presented on the substituents on these monosaccharides (if any), their amounts, and where they are located. No structural information (from X-ray crystallography or nuclear magnetic resonance studies) is presented that might give insights into the structural basis of the "unusually potent thickening powers." Information is also needed on the dependence of the solution properties of EPS105-4 on salt concentration and temperature.

8.2 *Xanthomonas juglandis* utilizes glucose as a source of carbon for cell growth and energy and as the major building block for the synthesis of the extracellular polysaccharide xanthan. When glucose is the limiting nutrient, it is used preferentially to maintain cell growth and provide energy. The synthesis of xanthan is drastically limited under these conditions. When nitrogen is limiting and glucose is plentiful, the cells have the building blocks for xanthan synthesis in abundance and have energy to spare. The inadequate nitrogen supply limits cell growth. As a consequence, the amount of xanthan produced per cell increases dramatically.

8.3 See Chapter 8, pages 283–292.

9.1 (a) The Ti plasmid is too large for manipulation *in vitro*. Its large size also means that it contains multiple restriction sites for most restriction endonucleases, a distinct disadvantage (see Figure 3.11B). (b) Ti plasmid transfers T-DNA that contains information to produce diseases in plants. This is detrimental to our purpose, because we want to produce healthy transgenic plants, not crown galls.

9.2 The easiest strategy, which is used in many similar cases, is the antisense RNA approach. One would clone the gene for the branching enzyme and insert the gene in an inverse orientation behind a strong promoter such as the CaMV 35S promoter. The recombinant DNA construct would then be introduced into potato by using the *Agrobacterium* system. An alternative would be to introduce inactivating mutations into the gene *in vitro*, and try to get the mutated allele to replace the endogenous, wild-type allele in the potato plant. Methods for such gene replacement procedures are currently less well developed and are more difficult to perform.

9.3 (a) Chloroplast DNA codes for a number of proteins that play crucial roles in photosynthesis and associated processes. There is no other way to alter such processes than to alter the chloroplast

DNA. (b) It is somewhat unlikely, because the injected T-DNA appears to head for nucleus. Chloroplasts are surrounded by a double membrane system, which is far less porous than the nuclear membrane. Nuclear DNA is several orders of magnitude larger than chloroplast DNA, so if the T-DNA insertion occurs at random, this size difference alone would favor the insertion into nuclear DNA. One would first improve the chances of insertion of the injected DNA into chloroplast DNA by surrounding the foreign DNA segment with cloned chloroplast DNA; this would encourage incorporation through homologous recombination. (There are a few reports that the *Agrobacterium* system did work with this modification). Second, to overcome the membrane barrier, you could try bombardment with DNA-coated microprojectiles.

10.1 A cellulase that was stable and active up to about 110°C would be valuable in the conversion of lignocellulose to fermentation alcohol. It is possible that such an enzyme would eliminate the need for pretreatment of the lignocellulose (steam explosion, ball milling), because the structure of wood might be sufficiently disrupted by the continued exposure to high temperature during enzymatic cellulose degradation. Other available cellulases are inactivated at 110°C.

10.2 Microwave radiation might provide a rapid and straightforward method of disrupting the crystalline regions in cellulose and speeding up its enzymatic hydrolysis.

10.3 The cellulosome is a complex assemblage, present in clostridia, that contains a number of enzymes involved in the degradation of cellulose and hemicelluloses. The "textbook" role assigned to cellulosomes is to degrade the polysaccharides in wood. The finding that the cellulosome also contains an enzyme capable of degrading glucans, major polysaccharide components of seeds, suggests that the cellulosome should be regarded more broadly as an assemblage of enzymes that enables clostridia to utilize a wide range of carbohydrate polymers.

11.1 The ability to utilize only a few sugar substrates is a major disadvantage in the use of *Zymomonas mobilis* in large-scale production of fermentation ethanol. The introduction of a *Cellulomonas uda* cellulase gene into *Z. mobilis* is a step in the attempt to increase the range of carbohydrate substrates that *Z. mobilis* is able to ferment to ethanol.

11.2 Such trends might include a sharp increase in oil prices, a decrease in the price of agricultural land, a desire to minimize the increase in atmospheric CO_2 content, and a desire to decrease air pollution.

12.1 (a) Auxotrophic mutants will be selected, after mutagenesis, by adding penicillin (or ampicillin) to a culture growing in a minimal medium containing citrulline. Wild-type cells, as well as mutants blocked in steps earlier than citrulline, will grow and will thus be killed by penicillin. You must make sure that the mutant is an arginine auxotroph, because mutants auxotrophic for other amino acids, purines, pyrimidines, and vitamins will also survive this selection. You also should make sure that the correct enzyme step (not the last one just before arginine) is blocked, either by running an enzyme assay or by confirming the excretion of citrulline. (b) Fermentation has to be carried out in minimal medium, with a controlled, continuous addition of arginine, so that there will be no large excess of arginine at any given moment. (c) If regulatory mutations are present, you can use much less expensive media containing an excess of arginine. (d) Test many structural analogs of arginine. If some of them are shown to inhibit the growth of the wild-type *E. coli*, you should isolate mutants resistant to these analogs. These mutants are likely to be altered in the regulation of arginine biosynthesis. (e) If they are defective in feedback inhibition, the enzyme biosynthesis should still be repressed when the cells are grown in the presence of arginine. Thus, when the activity of one of the enzymes is measured in the extracts of these cells, it should be very low. In contrast, in repression-defective (constitutive) mutants, the level of such enzymes is always very high, even when the cells are grown in the presence of arginine. (f) In *E. coli*, several methods of classical genetics are available. Transduction is usually the method of choice, because it introduces little extraneous DNA.

12.2 (a) Classical approaches increased the amount of tryptophan excreted into the medium, but the amounts were not enough to produce desirable levels of commercial profit. This could have been because the classical approaches depend on the endogenous levels of enzymes present, which might not have been sufficient for high-level overproduction. (b) Probably because the regulatory mechanisms in *C. glutamicum* are much simpler than those in *E. coli* (see Box 12.2), although the cloning is certainly easier in *E. coli*. (c) One could increase the flow of material through the common part of the pathway, but because the branches (including the tryptophan branch) were tightly controlled, the common intermediate chorismate accumulated and was excreted. (d) One would normally assume that the first enzyme of the branch, the one that converts chorismate into anthranilate, is regulated most strongly within the branch. However, the excretion of anthranilate suggests that perhaps the *second* enzyme of the branch, which converts anthranilate to the next intermediate, is strongly regulated by tryptophan. Thus, instead of getting a large production of tryptophan, one gets the accumulation and excretion of the intermediate before this enzymatic step. (e) The problem must be that the genes came from the wild-type organism. One should first isolate regulation-defective mutants, altered in the feedback inhibition of the second enzyme (and possibly also the first enzyme) of the tryptophan branch, and clone the genes of the tryptophan branch from this mutant strain.

12.3 The preferential synthesis hypothesis suggests that AMP, the end product of the less favored branch, inhibits the first enzyme of the common pathway, glutamine-phosphoribosyl-PP amidotransferase. This branch was thus cut off to eliminate the feedback inhibition by AMP.

   GMP, as the product of the favored branch, should still inhibit the first enzyme of the branch, IMP dehydrogenase. Thus this enzyme had to be altered by selecting mutants that would grow in the presence of growth-inhibitory analogs of GMP.

13.1 (a) Using *S. aureus* as the indicator would result in the reisolation of common organisms producing classical penicillins and cephalosporins. (b) Using wild-type *E. cloacae* with its inducible β-lactamase could have resulted in the isolation of agents that could not induce the enzyme. Such compounds are not desirable, because if they are susceptible to hydrolysis by this enzyme, their clinical use will result in the selection of β-lactamase-constitutive strains (page 482). Thus there was a strong need to find compounds that are truly stable to the β-lactamase of this type. (c) Common R plasmids contain genes for class A β-lactamase, which is quite different in its enzymatic properties from the class C β-lactamase of *E. cloacae*. (d) *E. cloacae*, being a Gram-negative organism, is protected by the outer membrane barrier. Thus many antibiotics that are powerful inhibitors of intracellular targets may not show strong activities against intact *E. cloacae* cells. (e) One could use a mixture of *S. aureus* and *E. cloacae* as the indicator. If one uses cultural conditions so that only *S. aureus* would grow extensively, inhibition of *S. aureus* growth would mean that the antibiotic is not hydrolyzed by *E. cloacae* enzyme and has remained active.

13.2 (a) One would obtain hybrid antibiotics synthesized by combinations of genes from the two organisms. Such hybrid antibiotics cannot be obtained by mutagenesis of each of the parent species, nor are they likely to be found in nature. (b) Protoplast fusion and introduction of cloned DNA.

13.3 Production of the inactivating enzyme required many millions of years of evolution. In fact, genes for many of the inactivating enzymes appear to come from the producing organisms. There has been no time for microorganisms to develop inactivating enzymes against agents synthesized during the last half-century, although it has been much easier to alter the structure of the target protein (which already existed) in minor ways to make it resistant to these synthetic chemotherapeutic agents.

14.1 See Chapter 14, pages 519–521.

14.2 See Chapter 14, pages 532–533.

14.3 Hydrolyze the triglyceride with a 1,3-regioselective lipase.

15.1 See Chapter 15, pages 590–591.

15.2 (a) The investigators transferred the contaminated soil (as a slurry in water) to a reactor vessel. The consortium of microorganisms in the bioreactor degraded added pentachlorophenol (PCP) and polyaromatic hydrocarbons (PAH) at a rate that increased over a 35-day period by an order of magnitude for PCP and by 30- and 80-fold for phenanthrene and pyrene, respectively. This increase in the rate of degradation presumably reflects the increase in the population of the members of the microbial consortium that degraded PCP and PAH. (b) No. Contaminated soil from wood-preserving facilities contains a complex mixture of pollutants. In general, pure cultures of microorganisms are capable of mineralizing only a limited number of such compounds. Too little is known of the details of genetics and enzymology (within the consortium of indigenous microorganisms) that endow the consortium with the ability to degrade complex mixtures of pollutants, so that it is not possible to reproduce these capabilities by genetic engineering of well-characterized bacterial strains.

15.3 See Chapter 15, pages 609–611.

15.4 This problem was addressed in detail in a paper by T.J. Fleming, J. Sanseverino, and G.S. Sayler [(1993), Quantitative relationship between naphthalene catabolic gene frequency and expression in predicting PAH degradation in soils at town gas manufacturing sites, *Environ. Sci. Technol.* 27:1068–1074]. Fleming et al. developed a procedure for extraction of total RNA from soil samples. A solution containing the soil RNA is then hybridized under optimal conditions to radiolabeled *in vitro* transcribed antisense RNA probe complementary to *nahA*. The resulting double-stranded RNA hybrid is subjected to ribonuclease digestion, which degrades any nonhybridized single-stranded probe. The protected double-stranded hybrid is then quantified by comparison with protected *in vitro* transcribed sense standards via gel electrophoresis followed by autoradiography. (This method has a reported sensitivity of 0.1 picogram of specific RNA. It is better for absolute mRNA quantitation in soil extracts than methods based on the polymerase chain reaction, because contaminants from soil, such as humic acids, may interfere with the enzymatic steps in the PCR amplification.) The mRNA measurement provides an estimate of the instantaneous activity of soil microorganisms.

To measure the actual degradation of naphthalene, [1-^{14}C]naphthalene was added to soil samples, and its degradation with time to [^{14}C]CO$_2$ was followed as a function of time. *NahA* transcript levels correlated positively with [1-^{14}C]naphthalene mineralization rates. See the paper by Fleming et al. for details of this experiment.

INDEX

Page numbers in boldface indicate figures; t after a page number indicates a table.